D1703800

Solid State Electrochemistry I

Edited by
Vladislav V. Kharton

Further Reading

Kharton, V.V. (ed.)

Solid State Electrochemistry II

Electrodes, Interfaces and Ceramic Membranes

2011

ISBN: 978-3-527-32638-9

Endres, F., MacFarlane, D., Abbott, A. (eds.)

Electrodeposition from Ionic Liquids

2008

ISBN: 978-3-527-31565-9

Bard, A. J., Stratmann, M., Gileadi, E., Urbakh, M., Calvo, E. J., Unwin, P. R., Frankel, G. S., Macdonald, D., Licht, S., Schäfer, H. J., Wilson, G. S., Rubinstein, I., Fujihira, M., Schmuki, P., Scholz, F., Pickett, C. J., Rusling, J. F. (eds.)

Encyclopedia of Electrochemistry

11 Volume Set

2007

ISBN: 978-3-527-30250-5

Hamann, C. H., Hamnett, A., Vielstich, W.

Electrochemistry

2007

ISBN: 978-3-527-31069-2

Staikov, G. T. (ed.)

Electrocrystallization in Nanotechnology

2007

ISBN: 978-3-527-31515-4

Alkire, R. C., Kolb, D. M., Lipkowski, J., Ross, P. N. (eds.)

Diffraction and Spectroscopic Methods in Electrochemistry

2006

ISBN: 978-3-527-31317-4

Solid State Electrochemistry I

Fundamentals, Materials and their Applications

Edited by
Vladislav V. Kharton

WILEY-VCH Verlag GmbH & Co. KGaA

The Editor

Vladislav V. Kharton
University of Aveiro
Dept. of Ceramics and Glass Engineering CICECO
3810-193 Aveiro
Portugal

Cover Illustration
The picture in the background has been kindly provided by Ewald Bischoff, Barbara Panella and Michael Hirscher, Max-Planck-Institut für Metallforschung, Stuttgart, Germany.

Library of Congress Card No.: applied for

British Library Cataloguing-in-Publication Data
A catalogue record for this book is available from the British Library.

Bibliographic information published by the Deutsche Nationalbibliothek
The Deutsche Nationalbibliothek lists this publication in the Deutsche Nationalbibliografie; detailed bibliographic data are available on the Internet at http://dnb.d-nb.de.

Typesetting Thomson Digital, Noida, India
Printing betz-druck GmbH, Darmstadt
Binding Litges & Dopf GmbH, Heppenheim
Cover design Grafik-Design Schulz , Fußgönheim

Printed in the Federal Republic of Germany
Printed on acid-free paper

ISBN: 978-3-527-32318-0

Contents

Solid State Electrochemistry I: Fundamentals, Materials and their Applications. Edited by Vladislav V. Kharton

ISBN: 978-3-527-32318-0

Preface

Vladislav V. Kharton

Solid-state electrochemistry is an important and rapidly developing scientific field that integrates many aspects of classical electrochemical science and engineering, materials science, solid-state chemistry and physics, heterogeneous catalysis, and other areas of physical chemistry. This field comprises – but is not limited to – the electrochemistry of solid materials, the thermodynamics and kinetics of electrochemical reactions involving at least one solid phase, and also the transport of ions and electrons in solids and interactions between solid, liquid and/or gaseous phases, whenever these processes are essentially determined by the properties of solids and are relevant to the electrochemical reactions. The range of applications includes many types of batteries and fuel cells, a variety of sensors and analytical appliances, electrochemical pumps and compressors, ceramic membranes with ionic or mixed ionic-electronic conductivity, solid-state electrolyzers and electrocatalytic reactors, the synthesis of new materials with improved properties and corrosion protection, supercapacitors, and electrochromic and memory devices.

The first fundamental discoveries, which today are considered the foundation of solid-state electrochemistry, were made during the nineteenth century and the first half of the twentieth century by M. Faraday, E. Warburg, W. Nernst, C. Tubandt, W. Schottky, C. Wagner, and other famous scientists. Such investigations provided the background for the rapid progress achieved both in our understanding of solid-state electrochemical processes, and in the applied developments that were made during the latter part of the twentieth century. As the scope of this volume is limited, and so cannot provide an exhaustive analysis of the historical aspects and classical concepts, those readers interested in such information are referred to a range of well-known books [1–11]. It should be mentioned that, similar to any other scientific area, the continuous progress in solid-state electrochemistry leads both to new horizons and to new challenges. In particular, increasing demands for higher-performance electrochemical cells leads to a need to develop novel experimental and theoretical approaches for the nanoscale optimization of the cell materials and interfaces, for the analysis of highly nonideal systems, and for overcoming the numerous gaps in our knowledge, which became possible only due to recent achievements in the related areas of science and technology. Moreover, the increasing amount and diversity of available scientific information acquired during the past few decades

Solid State Electrochemistry I: Fundamentals, Materials and their Applications. Edited by Vladislav V. Kharton

ISBN: 978-3-527-32318-0

has raised the importance of systematization, the unification of terminology, and the standardization of experimental and simulation techniques.

The aim of this Handbook is to combine the fundamental information and to provide a brief overview of recent advances in solid-state electrochemistry, with a primary emphasis on methodological aspects, novel materials, factors governing the performance of electrochemical cells, and their practical applications. The main focus is, therefore, centered on specialists working in this scientific field and in closely related areas, except for a number of chapters which present also the basic formulae and relevant definitions for those readers who are less familiar with theory and research methods in solid-state electrochemistry. Since it has been impossible to cover in total the rich diversity of electrochemical phenomena, techniques and appliances, priority has been given to recent developments and research trends. Those readers seeking more detailed information on specific aspects and applications are addressed to the list of reference material below, which includes both interdisciplinary and specialized books [8–20].

The first volume of this Handbook contains brief reviews dealing with the general methodology of solid-state electrochemistry, with the major groups of solid electrolytes and mixed ionic-electronic conductors, and with selected applications for electrochemical cells. Attention is drawn in particular to the nanostructured solids, superionics, polymer and hybrid materials, insertion electrodes, electroanalysis and sensors. Further applications, and the variety of interfacial processes in solid-state electrochemical cells, will be examined in the second volume.

The chapters of the Handbook are written by leading experts in solid-state electrochemistry from Australia, China, France, Germany, Israel, Japan, Korea, Portugal, Russia, the United Kingdom, and the United States of America. Sadly, one of the authors of Chapter 3, Professor Alexander N. Petrov, died during the finalization stages of the Handbook. His professionalism and love of science will be well remembered by all of his colleagues, and the intellectual contributions made by Professor Petrov will continue to live on in the form of the inspiration that he has provided to his students, coworkers and, hopefully, the readers of this book.

References

1 Schottky, W., Uhlich, H. and Wagner, C. (1929) *Thermodynamik*, Springer, Berlin.
2 Kröger, F.A. (1964) *The Chemistry of Imperfect Crystals*, North-Holland, Amsterdam.
3 Delahay, P. and Tobias, C.W. (eds) (1966) *Advances in Electrochemistry and Electrochemical Engineering*, Wiley-Interscience, New York.
4 Kofstad, P. (1972) *Nonstoichiometry, Diffusion and Electrical Conductivity of Binary Metal Oxides*, Wiley-Interscience, New York.
5 Geller, S. (ed.) (1977) *Solid Electrolytes*, Springer, Berlin, Heidelberg, New York.
6 Rickert, H. (1982) *Electrochemistry of Solids. An Introduction*, Springer, Berlin, Heidelberg, New York.
7 Chebotin, V.N. (1989) *Chemical Diffusion in Solids*, Nauka, Moscow.
8 Bruce, P.G. (ed.) (1995) *Solid State Electrochemistry*, Cambridge University Press, Cambridge.
9 Gellings, P.J. and Bouwmeester, H.J.M. (eds) (1997) *Handbook of Solid State Electrochemistry*, CRC Press, Boca Raton.

10 Allnatt, A.R. and Lidiard, A.B. (2003) *Atomic Transport in Solids*, Cambridge University Press, Cambridge.

11 Bard, A.J., Inzelt, G. and Scholz, F. (eds) (2008) *Electrochemical Dictionary*, Springer, Heidelberg, Berlin.

12 West, A.R. (1984) *Solid State Chemistry and its Applications*, John Wiley & Sons, Chichester.

13 Goto, K.S. (1988) *Solid State Electrochemistry and Its Applications to Sensors and Electronic Devices*, Elsevier, Amsterdam.

14 Schmalzried, H. (1995) *Chemical Kinetics of Solids*, VCH, Weinheim.

15 Munshi, M.Z.A. (ed.) (1995) *Handbook of Solid State Batteries and Capacitors*, World Scientific, Singapore.

16 Vayenas, C.G., Bebelis, S., Pliangos, C., Brosda, S. and Tsiplakides, D. (2001) *Electrochemical Activation of Catalysis: Promotion, Electrochemical Promotion, and Metal-Support Interaction*, Kluwer/ Plenum, New York.

17 Hoogers, G. (ed.) (2003) *Fuel Cell Technology Handbook*, CRC Press, Boca Raton.

18 Wieckowski, A., Savinova, E.R. and Vayenas, C.G.,(eds) (2003) *Catalysis and Electrocatalysis at Nanoparticle Surfaces*, Marcel Dekker, New York.

19 Monk, P.M.S., Mortimer, R.J. and Rosseinsky, D.R. (2007) *Electrochromism and Electrochromic Devices*, 2nd edition, Cambridge University Press, Cambridge.

20 Zhuiykov, S. (2007) *Electrochemistry of Zirconia Gas Sensors*, CRC Press, Boca Raton.

List of Contributors

Alan Atkinson
Imperial College London
Department of Materials
London SW7 2AZ
United Kingdom

Doron Aurbach
Bar Ilan University
Department of Chemistry
52900
Ramat-Gan
Israel

Maxim Avdeev
Bragg Institute
Australian Nuclear Science and
Technology Organisation
1 PMB, Menai 2234
Australia

Alan V. Chadwick
University of Kent
School of Physical Sciences
Canterbury
Kent CT2 7NH
United Kingdom

Vladimir A. Cherepanov
Ural State University
Department of Chemistry
51 Lenin Av.
Ekaterinburg 620083
Russia

Jeffrey W. Fergus
Auburn University
Material Research and Education Center
275 Wilmore Laboratories
Auburn, AL
USA

Jacques Fouletier
LEPMI
ENSEEG-PHELMA
BP 75
38402 Saint Martin d'Hères Cedex
France

Véronique Ghetta
LPSC
53 Rue des Martyrs
38026 Grenoble Cedex
France

Solid State Electrochemistry I: Fundamentals, Materials and their Applications. Edited by Vladislav V. Khartan

ISBN: 978-3-527-32318-0

Michael Hermes
Ernst-Moritz-Arndt-Universität
Greifswald
Felix-Hausdorff-Straße 4
17489 Greifswald
Germany

Stephen Hull
The ISIS Facility
STFC Rutherford Appleton Laboratory
Didcot
Oxfordshire, OX11 0QX
United Kingdom

Nobuhito Imanaka
Osaka University
Department of Applied Chemistry
Faculty of Engineering
2-1 Yamadaoka, Suita
Osaka 565-0871
Japan

Vladislav V. Kharton
University of Aveiro
Department of Ceramics and Glass
Engineering
CICECO
3810-193 Aveiro
Portugal

John A. Kilner
Imperial College London
Department of Materials
London SW7 2AZ
United Kingdom

Sung-Woo Kim
Korea Atomic Energy Research Institute
1045 Daedeok-Daero
Yuseong-Gu
Daejeon 305-353
Republic of Korea

Seung-Bok Lee
Korea Institute of Energy Research
#71-2 Jang-Dong
Yuseong-Gu
Daejeon 305-600
Republic of Korea

Mikhael Levi
Bar Ilan University
Department of Chemistry
52900
Ramat-Gan
Israel

Joachim Maier
Max-Planck-Institut für
Festkörperforschung
Stuttgart
Germany

Fernando M.B. Marques
University of Aveiro
Department of Ceramics and Glass
Engineering
CICECO
3810-193 Aveiro
Portugal

Vladimir B. Nalbandyan
Southern Federal University
Department of Chemistry
Zorge 7
344090 Rostov-na-Donu
Russia

Alexander N. Petrov
Formerly Ural State University
Department of Chemistry
51 Lenin Av.
Ekaterinburg 620083
Russia

Su-Il Pyun
Korea Advanced Institute of Science and Technology
Department of Materials Science and Engineering
#373-1 Guseong-Dong
Yuseong-Gu
Daejeon 305-701
Republic of Korea

Shelley L.P. Savin
University of Kent
School of Physical Sciences
Canterbury
Kent CT2 7NH
United Kingdom

Fritz Scholz
Ernst-Moritz-Arndt-Universität Greifswald
Felix-Hausdorff-Straße 4
17489 Greifswald
Germany

Igor L. Shukaev
Southern Federal University
Department of Chemistry
Zorge 7
344090 Rostov-na-Donu
Russia

Shinji Tamura
Osaka University
Department of Applied Chemistry
Faculty of Engineering
2-1 Yamadaoka, Suita
Osaka 565-0871
Japan

Danmin Xing
Sunrise Power Co., LTD
Nation Engineering Research Center of Fuel Cells
No. 907, Huangpu Road
Hi-Tech Zone, Dalian 116025
P.R. China

Baolian Yi
Chinese Academy of Sciences
Dalian Institute of Chemical Physics
No. 457, Zhongshan Road
Dalian 116023
P.R. China

Andrey Yu. Zuev
Ural State University
Department of Chemistry
51 Lenin Av.
Ekaterinburg 620083
Russia

1
Fundamentals, Applications, and Perspectives of Solid-State Electrochemistry: A Synopsis

Joachim Maier

Abstract

The opening chapter of this Handbook highlights the characteristic features of solid-state electrochemistry, including basic phenomena, measurement techniques, and key applications. Materials research strategies that are based on electrochemical insight and the potential of nanostructuring are detailed in particular. Fundamental relationships between the decisive thermodynamic and kinetic parameters governing electrochemical processes are also briefly discussed.

1.1
Introduction

Electrochemistry refers to the conversion of electrical (chemical) information and energy into chemical (electrical) information and energy, the interconnection being anchored in the central thermodynamic quantity, the electrochemical potential (of a species k) $\tilde{\mu}_k = \mu_k + z_k F\phi$ where μ is the chemical potential, $z_k F$ the molar charge, and ϕ the electrical potential.

Solid-state electrochemistry, as a subsection of electrochemistry, emphasizes phenomena in which the properties of solids play a dominant role. This includes phenomena involving ionically and/or electronically conducting phases (e.g., in potentiometric or conductometric chemical sensors). As far as classical electrochemical cells are concerned, one refers not only to all-solid-state cells with solid electrolytes (e.g., ceramic fuel cells), but also to cells with liquid electrolytes, such as modern Li-based batteries in which the storage within the solid electrode is crucial [1–3].

Solid State Electrochemistry I: Fundamentals, Materials and their Applications. Edited by Vladislav V. Kharton

ISBN: 978-3-527-32318-0

1.2
Solid versus Liquid State

The key property of solids is their *rigidity*, which relies on the strong local bonds at a given temperature, and typically manifests itself in both long-range order and pronounced short-range order, the only exceptions being completely or partly amorphous materials, such as glasses or polymers. In many cases, no distinction can be made between intermolecular or intramolecular bonds, for example in NaCl or diamond crystals, which may even be termed three-dimensional (3-D) giant molecules. The strong bonds are typically associated with not only a great thermal stability but also a good mechanical stability in terms of shear resistance. The mechanical stability may not be pronounced in terms of fracture toughness; in fact, many crystals are brittle and can easily crack. Solids can often be used even at very high temperatures, they can be manufactured in highly reproducible fashion, and they may also be easily shaped and miniaturized.

In terms of ion conductivity, one huge advantage of solids is that of *transport selectivity*. The transference number of silver ions in α-AgI with its quasi-molten Ag^+ sublattice and its rigid I^- sublattice is unity [4] (see Chapter 7). This selectivity not only helps to avoid polarization effects, but is also of substantial advantage for chemical sensors. (Alternatively, such selectivity hampers the realization of supported electrolytes.) The greatest disadvantages of the strong bonds in this respect are the typically modest absolute values of the ionic conductivities. Exceptions here are superionic solids, such as the aforementioned α-AgI which has liquid-like silver ion conductivities. (The connection between bond-strength, thermal disorder and melting temperature was elucidated in Ref. [5]; see also Figure 1.1) [6]. To a substantial degree, this inherent problem makes solid-state electrochemistry a typical high-temperature science, with its specific advantages (fast reaction kinetics) and disadvantages (stability problems). A more modern strategy that allows solids to be electroactive even at room temperature – and which is outlined at the end of the chapter – is to improve overall transport in solids by down-sizing them. This emphasizes the significance of nanotechnology for solid-state electrochemistry (see also Chapter 4).

One major specificity of the ordered solid state is the appearance of quasi-free electronic conduction. Owing to the high number of overlapping orbitals, energy bands can form giving rise to excess electron conduction in the lowest not fully occupied band, and to hole conduction in the highest bond that is fully occupied at $T = 0$ K. (In reality, one finds the whole spectrum from delocalized motion to strongly localized polaron motion.) Hence, the "mixed conductor" is to the fore in solid-state electrochemistry from which – conceptually speaking – semiconductors and solid electrolytes emerge as limiting cases (see Chapter 3). There are various phenomena that are specific to mixed conductors, such as component permeation (transport of neutral component) or stoichiometry changes (storage of neutral component), which can be employed directly, in allowing for storage or separation, or indirectly by tuning transport properties.

Thermodynamically ionic carrier chemistry in normal (i.e., non-superionic) crystals is in fact similar to the electronic counterpart in semiconductors (the case of the superionic conductor may be compared rather with the metallic state; see

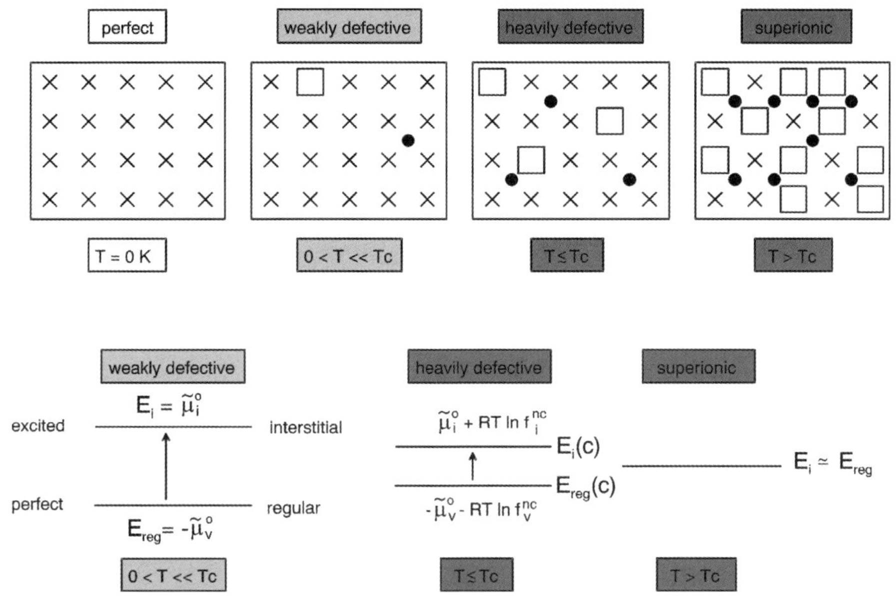

Figure 1.1 Depending on the bond strength, more or less point defects form on thermal excitation and will then alternatively interact and, owing to the formation avalanche, undergo sublattice melting or total melting [6]. The subscripts i and v refer to interstitial and vacancy defect; f denotes the activity coefficient owing to Coulomb interaction between these carriers. Reproduced with permission from Ref. [6]; © Wiley-VCH Verlag GmbH & Co. KGaA.

Figure 1.1). Accordingly – and, again, similar to the situation in semiconductor physics – these normal ion conductors do not show exceedingly high conductivity values, but in turn offer the advantage of a pronounced variability. In analogy to excess and holes being electronic excitations of the ground state, typical ionic carriers such as interstitials and vacancies represent ionic excitations [7]. Thermal energy may excite ions out of regular sites into interstitial positions, either leaving vacancies or producing separated pairs of vacancies or interstitials. If this inherent dissociation is not perceptible, then dopants can be added, either substitutionally or additively. In the case of polymer electrolytes, both anions and cations can be dissolved simultaneously, the ground state then being the undissociated ion pair. There is a strong correspondence of charge carriers in solids to the aqueous state: there, the point defects are H_3O^+ or OH^-, corresponding to excess proton and proton vacancy, respectively. In fact, these centers belong to larger clusters of perturbed structures. Similarly, interstitials and vacancies are the centers of a perturbation zone of a much larger perimeter. The "energy levels" in Figures 1.1 and 1.2 [3] correspond to the electrochemical potentials minus a configurational contribution, in dilute cases they contain the local chemical ("standard") potential μ° and $z_k F\phi$ (i.e., $\tilde{\mu}^\circ$), whereas in non-dilute cases they also contain the non-configurational portion of the activity coefficient. Solid-state ionics allows one to treat each solid as a solvent and, by applying defect chemistry to tune its properties as aqueous ionics, this can be achieved successfully with liquid water.

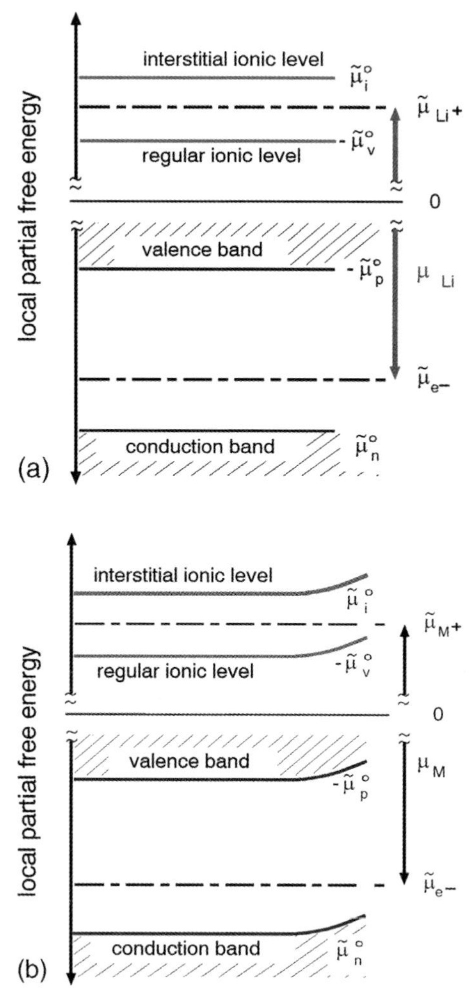

Figure 1.2 Representation of internal (Frenkel-) disorder in the (free) energy level diagram and its coupling with the fundamental electronic excitation in the bulk (a) and at boundaries (b) [3]. The illustrations correspond to particular cases when Li^+ (a) or generally monovalent M^+ cations (b) are excited in the lattice. ((a) Reproduced with permission from: J. Maier (2003) Defect chemistry and ion transport in nanostructured materials. Part II. Aspects of nano-ionics. *Solid State Ionics*, *157*, 327–334; © Elsevier Limited; (b) Reproduced with permission from Ref. [3]; © John Wiley & Sons, Limited.)

1.3
Thermodynamics and Kinetics of Charge Carriers

The statistics of the defects conceived as building elements (i.e., elements that can be added to the perfect structure to form the real structure) follow a Fermi–Dirac-type of statistics, whereas the statistics for the structure elements (i.e., elements that constitute the real structure) are of the Boltzmann type; hence, as a building element is a combination of two structure elements, consistency is preserved [8]. Figure 1.2a shows how ionic and electronic levels are connected through stoichiometry (μ of neutral component), and Figure 1.2b how they behave at a contact to a chemically compatible neighboring phase [3]. The utility of the energy level diagrams is particularly obvious at boundaries; for the description of bulk defect chemistry the usual approach of writing down explicitly defect chemical reactions that are dealt with by chemical thermodynamics is to be preferred on grounds of complexity. Note that at boundary zones the full account of Poisson's equation ($\rho = \Sigma_k z_k c_k = \nabla^2\phi/\varepsilon F$, where c is the molar concentration and ε the dielectric constant) must be made, while in the bulk it trivializes into the electroneutrality condition (charge density $= \rho = 0$).

In thermodynamic equilibrium, the equilibrium condition is $\nabla\tilde{\mu}_k = 0$ as regards the positional coordinate and $\Sigma_k \nu_{rk}\tilde{\mu}_k = 0$ (disappearance of the reaction sum of reaction r, where ν is the stoichiometric coefficient) as regards chemical displacement. Deviations lead to fluxes and generation/annihilation processes. Fluxes are – for not too-large driving forces – determined by the linear flux-driving force relationship $j_k \propto -\sigma_k \nabla\tilde{\mu}_k$. As a linear element of a Taylor expression, the partial conductivity σ_k refers to the equilibrium condition. Quite often, this equation is used even in cases where σ_k denotes a local non-equilibrium property, and this is allowed for local thermal equilibrium under small driving forces. At higher driving forces, however, higher orders in $\nabla\tilde{\mu}_k$ might be used or, preferably, the chemical kinetics approach via master equations might be applied [9]. In the latter case, local fields enter the rate coefficients and the driving force does not appear explicitly (only implicitly via concentrations and rate coefficients). We then obtain Butler–Volmer-type equations which can also be easily generalized for chemical reactions [10], viz.

$$j = \Re^\circ\left(g_1 \exp - \frac{\alpha\delta\Delta\varphi}{RT} + g_2 \exp - \frac{\beta\delta\Delta\varphi}{RT}\right) \tag{1.1}$$

where g_1 and g_2 refer to perturbations of the concentrations, $\Re^\circ$ is the exchange rate (the analogue to the conductivity for the reaction), and $\delta\Delta\varphi$ is the local drop of the applied bias. For the special case of transport, $g_1 = g_2 = 1$, $\alpha = \beta = 1/2$ and $\Re^\circ$ becomes the equilibrium conductivity.

As described in Ref. [11], such equations can be symmetrized for local thermal equilibrium to give, for example, for the transport case the relationship

$$j \propto -\sinh(\Delta\tilde{\mu}/2RT) \tag{1.2}$$

in the case of particle hopping (Δ refers to the variation over the elementary hopping distance). As the prefactor is now a generalized non-equilibrium conductivity, this is a useful local flux-driving force relation that does good service far from transport equilibrium [12]. A brief introduction into thermodynamics, together with several simple examples, can be found in Chapter 3.

A special problem consists of dealing with processes that do not lead to successful events, but rather to forward/backward hopping before the environment has acquired the opportunity to relax. Such phenomena give rise to frequency dependencies of the conductivity at very high frequencies, and form the transition to phonon dynamics [11].

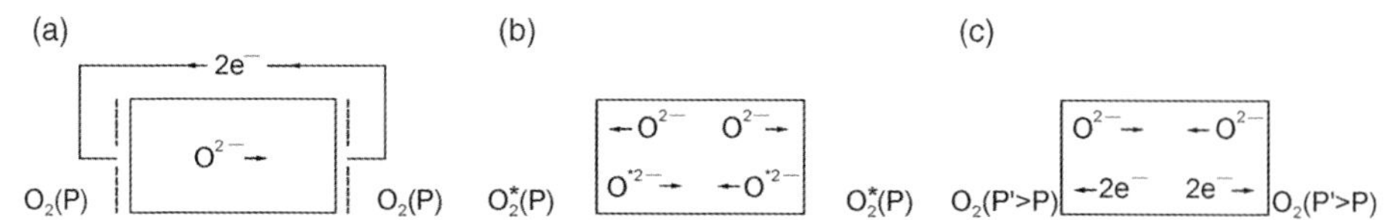

Figure 1.3 Three basic limiting experiments as discussed in the text. (a) Electrical; (b) Tracer; (c) Chemical [13]. The asterisk denotes an isotope. (Reproduced with permission from Ref. [13]; © Elsevier Limited.)

1.4 Usefulness of Electrochemical Cells

Figure 1.3 [13] shows the details of three key experiments in solid-state science representing limiting cases in practical experiments. The experiment shown in Figure 1.3c, refers to the pure chemical diffusion of an elemental compound such as oxygen in an oxide; this is a storage experiment for which a counter motion of ions and electrons is needed, as it changes the stoichiometry of the oxide (see Chapter 12). The experiment displayed in Figure 1.3b refers to a tracer exchange (Chapter 4); this requires the counter motion of two isotopes and not necessarily the explicit participation of electrons; the third experiment (Figure 1.3a) is the steady-state conductivity experiment in which the electrons are supplied by the outer circuit. These three processes can be connected with three diffusion coefficients (D^{δ}, D^{*}, D^{Q} for bulk transport) as far as the bulk is concerned. Accordingly, the three diffusion coefficients must be confronted with three relaxation rate constants (k^{δ}, k^{*}, k^{Q} for surface reaction), the relationships between which have been comprehensively discussed in Ref. [13].

Figure 1.4 [14] shows a typical battery cell in which conduction and storage occurs: conduction in the electrolyte, charge transfer through the boundary and storage in the electrodes. (The characteristic equations on the left-hand side refer to proximity to equilibrium.) Electrical and chemical resistors, electrostatic capacitors as well as chemical capacitors, are the basic ingredients of modeling electrochemical circuits even in complex cases. Owing to the normally huge value of the chemical capacitance, the time constant for the mass storage process dominates the overall process typically for Li-based batteries. Unlike the other processes addressed, this relaxation time depends sensitively on thickness (diffusion length), and can be efficiently varied by nanostructuring (see below).

As far as electrochemical cells relevant for applications or electrochemical measurements are concerned, we must distinguish between polarization cells, galvanic cells and open-circuit cells, depending on whether an outer current flows and, if so, in which direction this occurs. Table 1.1 provides examples of the purposes for which such cells may be used. In terms of application, we can distinguish between electrochemical sensors, electrochemical actors and galvanic elements such as batteries and fuel cells. These applications offer a major driving force for dealing with solid-state electrochemistry.

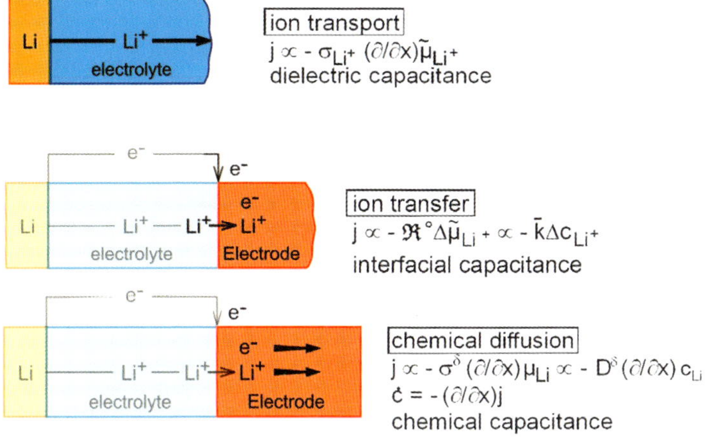

Figure 1.4 Resistive and capacitive processes in a Li-battery, and the targeted use of size dependence [14]. For simplicity, all relationships refer to proximity to equilibrium. Wherever the positional coordinate x appears, reduction of the transport path length is efficient. Moreover, the effective transport parameters can be sensitively varied by size effects, owing to local variations of defect chemistry. The superscript δ refers to the joint motion of Li^+ and e^-; D^{δ} is the chemical diffusion coefficient of Li, and σ^{δ} the ambipolar conductivity $(= \sigma_{e^-}\sigma_{Li^+}/(\sigma_{e^-} + \sigma_{Li^+}))$; $\bar{k}$ is the effective rate constant close to equilibrium. Reproduced with permission from Ref. [14]; © Elsevier Limited.

Figure 1.5 describes basic galvanic elements discriminated according to energy density (per mass) and temperature. All those galvanic cells that directly convert chemical into electrical energy, without thermal detours, are hence not bound by Carnot's efficiency, and offer high theoretical efficiencies. The application of solid

Table 1.1 An overview of electrochemical devices and measurement techniques based on various cell types [1].

Cell type	Measurement technique	Technological application
Polarization cell	Measurement of kinetic data by polarization	Electrochemical composition actors (electrolyzers, pumps, electrochromic windows), electrochemical composition sensors (amperometric, conductometric)
Current-generating cell	Measurement of kinetic data by depolarization	Electrochemical energy storage and conversion devices (batteries, fuel cells, supercapacitors)
Open-circuit cell	Measurement of thermodynamic formation data, transport number of electrons	Potentiometric composition sensors

Reproduced with permission from Ref. [1a]; © Springer.

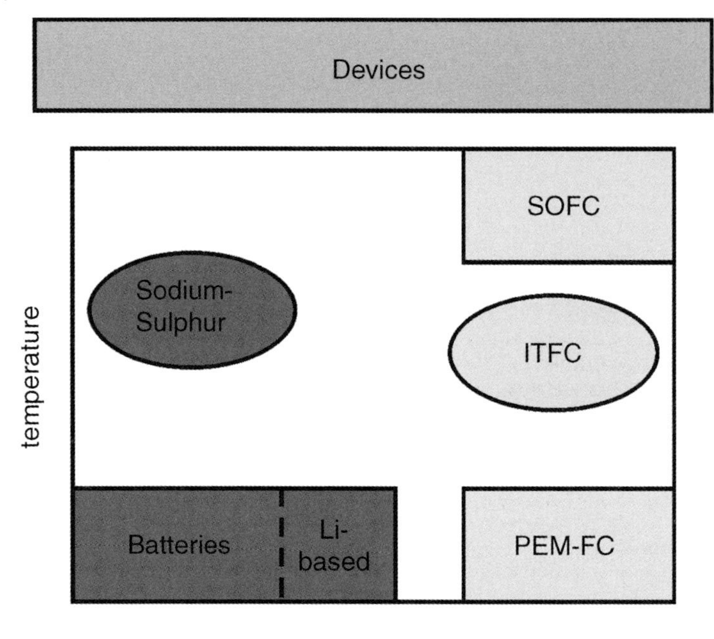

Figure 1.5 Selected galvanic cells (for details, see the text). SOFC = (high-temperature) solid oxide fuel cells; ITFC = intermediate-temperature fuel cells; PEM-FC = polymer electrolyte membrane fuel cells.

oxide fuel cells is driven by the fast electrode kinetics at high temperatures that enables the easy consumption of fuels such as hydrocarbons. The main advantage of near-room temperature fuel cells are the ease of performance and the lack of severe thermal stability problems. Here, however, the catalytic activity of the electrodes is much more demanding. Relevant batteries use solid storage materials, and are particularly relevant if Li is the element to be stored, owing to its extremely high reduction energy per mass.

Electrochemical cells can also be used for the precise determination of kinetic and thermodynamic parameters. Such cells can be classified according to the combination of reversible and blocking electrodes. Cell types and parameters to be determined are compiled in Table 1.2 [15–17]. The determination of kinetic parameters makes use of the condition that, in an experiment with a mixed conductor, the flux densities are composed of a drift and a stoichiometric term:

$$j_{\{\}} = -\frac{\sigma_{\{\}}}{\sigma}\frac{i}{z_{\{\}}F} - D^{\delta}_{\{\}}\nabla c_{\{\}}. \tag{1.3}$$

In Equation (1.3) only the total current i and the total conductivity σ carry no indices, the other quantities do ({ }). If associates do not play an important role, the indices simply refer respectively to ions or electrons or to the respective component

Table 1.2 Combination of reversible ((O^{2-}, e^-|, typically porous Pt) and blocking electrodes ((e^-|, i.e., only reversible for e^-, a typical example being graphite; or (O^{2-}|, i.e., only reversible for O^{2-}, a typical example being a Pt-contacted zirconia electrolyte) leads to a variety of measurement techniques applied to the oxide MO.

Cell	Quantities to be determined
$(O^{2-}, e^-|MO|e^-, O^{2-})$	Resistances and capacitances
$(O^{2-}, e^-|MO|e^-, O^{2-})'^{a}$	Transport numbers, chemical diffusion coefficients
$(O^{2-}, e^-|MO|e^-)$	Electronic conductivities as a function of activity, chemical diffusion coefficients
$(e^-|MO|e^-)$	Electronic conductivities, chemical diffusion coefficients
$(O^{2-}, e^-|MO|O^{2-})$	Ionic conductivities as a function of activity, chemical diffusion coefficients
$(O^{2-}|MO|O^{2-})$	Ionic conductivities, chemical diffusion coefficients
$(O^{2-}|MO|e^-)$	Stoichiometry, thermodynamic factor, chemical diffusion coefficients

[a] A different oxygen partial pressure was used on the right-hand side.

(in j, D^{δ}, c) [18]. If associates, however, play a substantial role, the respective "conservative ensemble" must be considered [19].

If, for example, oxygen vacancies ($V_O^{\cdot\cdot}$, denoting a missing O^{2-}) and electrons are present and then also associates such as $V_O^{\cdot}$ (one electron trapped by an oxygen vacancy) and $V_O^{\times}$ (two electrons trapped by an oxygen vacancy), the flux and conductivities in Equation (1.3) address the total oxygen ensemble, $\sigma_{\{O\}} \equiv \sigma_{V_O^{\cdot}} + 2\sigma_{V_O^{\cdot\cdot}}$ ($V_O^{\times}$ does not contribute being effectively neutral). The ensemble diffusion coefficient $D^{\delta}_{\{O\}}$ is essentially composed of $\sigma_{e'} - \sigma_{V_O^{\cdot}}$ and $\sigma_{V_O^{\cdot}} + 2\sigma_{V_O^{\cdot\cdot}}$, as well as of respective differential trapping factors [19]. (Of course, interaction can also formally be put into Onsager's cross-coefficient, but this does not provide any mechanistic insight [20].)

1.5 Materials Research Strategies: Bulk Defect Chemistry

Ultimately, we should be concerned with the strategies to optimize electrochemical parameters by materials research. The most momentous strategy is to seek new phases (new structures and compounds) and, indeed, many simple phases have already been explored. Yet, examining any higher compositional complexity increases the possibility that the new phases are chemically unstable when in contact with neighboring phases, and that is why the modification of given phases is of key significance.

One very decisive way to optimize a given phase (Figure 1.6) is the targeted variation of defect chemistry. The property window that can be addressed by defect chemical variation is often greater than the alteration of the mean property when going from one phase to the next. The key parameters are component activity (partial

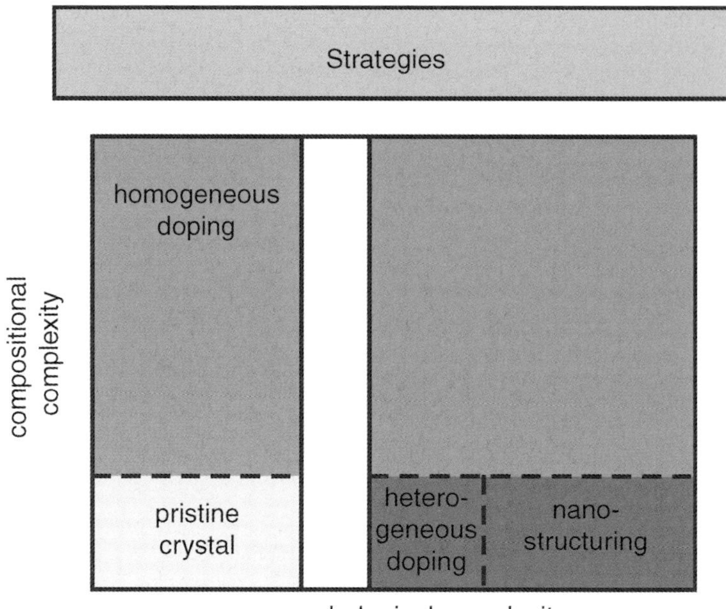

Figure 1.6 Strategies to optimize given phases.

pressures), the temperature, and the doping content. In reality, not all sublattices are in equilibrium with the neighboring phase, and these frozen concentrations then enter as the doping content C rather than being controlled by additional partial pressures ("component activities"). In the case of simple defect chemistry [21], which is characterized by the fact that all defects are randomly distributed (interactions can be taken into account by assuming randomly distributed associates), defect concentrations are typically given by

$$c_k(T, P, C) \propto P^{N_k} C^{M_k} \Pi_r K_r(T)^{\gamma_{rk}}. \tag{1.4}$$

where P is the component partial pressure and K_r the mass action constant of defect reaction r.

The exponents (N, M, γ) qualitatively follow simple rules and represent simple rational numbers [1, 3, 21]. Let us concentrate on the influence of the doping content a little more in detail, as it is – as far as application is concerned – the real relevant optimization parameter. The consequence of introducing a given dopant with known effective charge on any charge carrier k is simple; it is described by the "rule of homogeneous doping" [1, 3]

$$\frac{z_k \delta c_k}{z \delta C} < 0 \quad (\text{for any } k) \tag{1.5}$$

with z being the charge number. This means that an effectively positive (negative) dopant k increases the concentration of all negatively (positively) charged defects c_k, and decreases the concentration of all positively (negatively) charged ones individually.

If the power law of Equation 1.4 is valid, this can be rewritten as:

$$\frac{z_k}{z} M_k < 0 \quad (\text{for any } k). \tag{1.6}$$

Deviations from random distribution greatly modify and complicate the picture. Debye–Hückel corrections [9] do usually not lead very far; corrections by cube root terms do better service in many respects, but lose validity in heavily doped systems [5].

1.6 Materials Research Strategy: Boundary Defect Chemistry

So far, one important characteristic of solids that relies on the low mobility of at least one structure element, has not been addressed, namely the (meta)stability of higher-dimensional defects, and in particular of interfaces (see Figure 1.6). Although grain boundaries are often detrimental, there are cases in which they may be even deliberately generated with the purpose of improving transport properties. This includes the use of fine-grained ceramics, composites or heterolayers. Even in cases in which the interfaces do not give rise to significantly different mobilities, the conductivity effect at interfaces can be enormous. This is due to the greatly varied defect chemistry in boundary zones. Here, a similar rule – the rule of heterogeneous doping [1, 3] – is valid; that is:

$$\frac{z_k \delta c_k}{\delta \Sigma} < 0 \quad (\text{for any } k) \tag{1.7}$$

in which the charge of the dopant in Equation 1.5 is replaced by the surface Σ charge of the interface. By heterogeneous doping, the ion conductivities can be enormously modified, poor conductors may be changed into good conductors, and even the type of conduction mechanism can be changed, for example from vacancy to interstitial, from anion to cation conductivity, or from ionic to electronic [22].

1.7 Nanoionics

Even more intriguing are changes in which the local properties are varied throughout the sample; this is, for example, possible by curvature (capillary pressure), by the overlap of elastic effects or by space charge overlap [23]. In this way, not only synergistic transport phenomena but also synergistic storage phenomena can be verified, providing a bridge between multiphase systems to new artificial, "almost homogeneous", systems. Figure 1.7 [23] provides an overview of effects that can be dealt with in the context of what we refer to as "nanoionics". Most striking are the qualitatively novel conductors or storage materials which are arrived at in the limit of space charge overlap in two-phase systems. A systematic exploration of size effects in

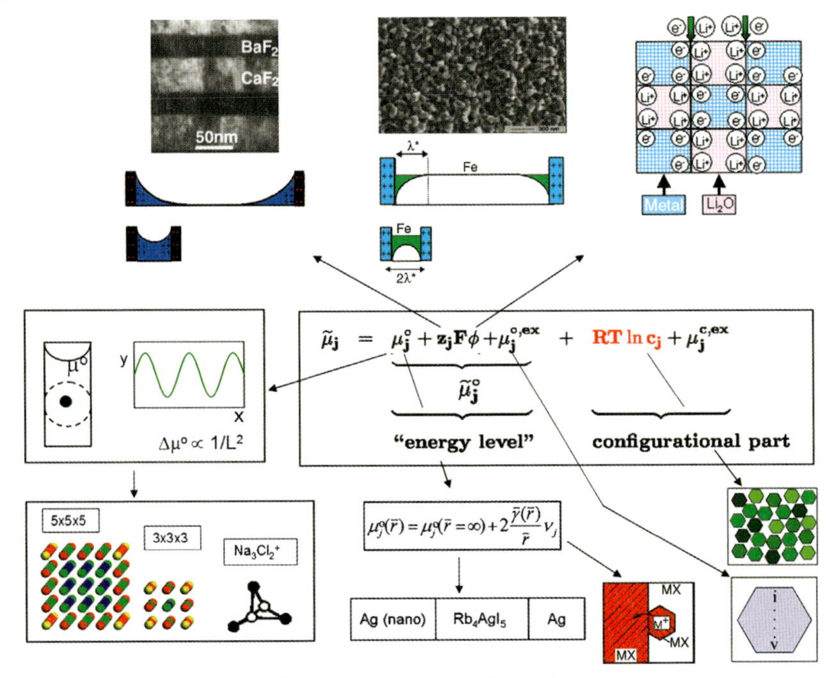

Figure 1.7 Typical true size effects (i.e., local property depends on size) on the ionic charge carrier j [23]. ($\bar{\gamma}$ = mean surface tension; $\bar{r}$ = mean radius of the particle with the composition M^+X^-.)

terms of Li-based batteries has recently been provided [14]; additional information can be found in Chapter 4 of this Handbook.

From a more conceptual point of view, it is the introduction of higher-dimensional defects that allows the transition to a "soft materials science", characterized by an enhanced information content even in systems in which the atomic bonds are not covalent. The future will be witness to increased research and applications in the field of metastable materials characterized by increased local complexity, with the possibility of further systematic collaboration with semiconductor physics and biology.

References

1 (a) Maier, J. (2005) Solid State Electrochemistry I: Thermodynamics and Kinetics of Charge Carriers in Solids, *Modern Aspects of Electrochemistry*, Vol. 38 (eds B.E. Conway, C.G. Vayenas and R.E. White), Kluwer Academic/Plenum Publishers, New York, pp. 1–173; (b) Maier, J. (2007) Solid State Electrochemistry II: Devices and Techniques, *Modern Aspects of Electrochemistry*, Vol. 41 (ed. C.G. Vayenas), Springer, New York, pp. 1–138.

2 Rickert, H. (1982) *Electrochemistry of Solids*, Springer-Verlag, Berlin.

3 Maier, J. (2004) *Physical Chemistry of Ionic Materials. Ions and Electrons in Solids*, John Wiley & Sons, Ltd, Chichester.

4 Geller, S. (1977) *Solid Electrolytes*, Springer-Verlag, Berlin.

5 Hainovsky, N. and Maier, J. (1995) *Phys. Rev. B*, **51**, 15789–15797.

6 Maier, J. and Münch, W. (2000) *Z. Anorg. Allg. Chem.*, **626**, 264–269.

7 Wagner, C. and Schottky, W. (1930) *Z. Phys. Chem. B*, **11**, 163–210.

8 Maier, J. (2005) *Z. Phys. Chem.*, **219**, 35–46.

9 Allnatt, A.R. and Lidiard, A.B. (1993) *Atomic Transport in Solids*, Cambridge University Press, Cambridge.

10 (a) Butler, J.A.V. (1924) *Trans. Faraday Soc.*, **19**, 734–739; (b) Butler, J.A.V. (1932), *Trans. Faraday Soc.*, **28**, 379–382.

11 Funke, K., Banhatti, R.D., Brückner, S., Cramer, C., Krieger, C., Mandanici, A., Martiny, C. and Ross, I. (2002) *Phys. Chem. Chem. Phys.*, **4**, 3155–3167.

12 Riess, I. and Maier, J. (2008) *Phys. Rev. Lett.*, **100**, 205901.

13 Maier, J. (1998) *Solid State Ionics*, **112**, 197–228.

14 Maier, J. (2007) *J. Power Sources*, **174**, 569–574.

15 Wagner, C. (1957) *Proceedings, 7th Meeting of the International Committee on Electrochemical Thermodynamics and Kinetics*, Butterworth, London.

16 Yokota, I. (1961) *J. Phys. Soc. Japan*, **16**, 2213–2223.

17 Maier, J. (1984) *Z. Physik. Chemie N. F.*, **140**, 191–215.

18 Wagner, C. (1975) *Prog. Solid State Chem.*, **10**, 3–16.

19 Maier, J. and Schwitzgebel, G. (1982) *Phys. Stat. Sol. (b)*, **113**, 535–547.

20 (a) Onsager, L. (1931) *Phys. Rev.*, **37**, 405–426; (b) Onsager, L. (1931), *Phys. Rev.*, **38**, 2265–2279.

21 Kröger, F.A. and Vink, H.J. (1956) in *Solid State Physics. Advances in Research and Applications*, Vol. 3 (eds F. Seitz and D. Turnbull), Academic Press, New York, pp. 307–435.

22 Maier, J. (1995) *Prog. Solid State Chem.*, **23**, 171–263.

23 Maier, J. (2005) *Nature Materials*, **4**, 805–815.

2
Superionic Materials: Structural Aspects

Stephen Hull

Abstract

Superionic conductors are solids which show high levels of ionic conductivity, which often approach values more typical of ionic liquids. This behavior is associated with the presence of extensive dynamic disorder within the crystalline lattice, and the nature of the defects has long proved a challenge to both experimental and computational approaches. This chapter provides a short introduction to the major techniques used to probe the structure–property relationships within superionic conductors, highlighting the roles of X-ray diffraction, neutron diffraction, EXAFS, NMR, and molecular dynamics methods. The relative advantages and limitations of each method are briefly described, followed by a discussion of the major families of highly conducting compounds and an assessment of the current state of knowledge concerning their structural properties.

2.1
Overview

The high levels of ionic conductivity associated with superionic conductors, which often approach values more typical of ionic liquids, are associated with the presence of extensive dynamic disorder within the crystalline lattice. As a consequence, a full understanding of the interplay between the structural and conducting properties of these important materials requires a detailed characterization of both the long-range arrangement of the ions within the crystal lattice and the short-range ion–ion correlations within the defects. The presence of significant levels of disorder, which can be both intrinsic (thermally induced) and extrinsic (due to chemical doping) in origin, presents a challenge to the conventional experimental methods used to provide structural information. In the case of superionic conductors, no one technique can provide a complete picture, and it is often necessary to employ several complementary approaches, of which X-ray diffraction (XRD), neutron diffraction,

Solid State Electrochemistry I: Fundamentals, Materials and their Applications. Edited by Vladislav V. Kharton

ISBN: 978-3-527-32318-0

extended X-ray absorption fine structure (EXAFS) and nuclear magnetic resonance (NMR) are arguably the most instructive. The relative advantages and limitations of each method are briefly described in the following section, followed by a summary of the role of computational methods in predicting the defect properties of compounds showing high ionic conductivities. The major families of highly conducting compounds will then be discussed, and the chapter will conclude with a brief description of the current status of the field of superionics.

2.2 Techniques

2.2.1 X-Ray and Neutron Diffraction

The X-ray diffraction technique is the most commonly used experimental method for investigating the crystal structures of crystalline solids. As the underlying theory and methods are detailed in several specific textbooks (e.g., Ref. [1]), only a brief description of the essential features will be provided at this point.

The diffraction of X-rays and neutrons of wavelength λ from a crystal is governed by Bragg's equation: $\lambda = 2d\sin\theta$, where d is the interplanar spacing and 2θ is the scattering angle. In a monochromatic experiment (i.e., constant λ), the scattering angle of the detector is scanned and the positions and intensities of the various Bragg peaks measured. The list of d-spacings of the Bragg peaks then determines the unit cell and the indexing of each Bragg peak in terms of its Miller indices h, k, and l (where $d = a/\sqrt{h^2 + k^2 + l^2}$ for a cubic crystal of lattice parameter a). The systematic absence of a certain class of Bragg peaks determines the space group symmetry (or, in some cases, a small number of possible alternatives), and the positions of the ions within the unit cell are then determined by various means (direct methods, Patterson techniques, trial-and-error model building) using the intensities of the Bragg peaks. The relative intensities of these peaks provides information concerning the positions, thermal vibrations and fractional occupancies of sites within the unit cell, and are generally obtained by fitting a trial structural model to the diffraction data (least-squares fitting to the extracted peak intensities for single crystals and to the whole diffraction pattern using Rietveld refinement for powder samples).

In general, single-crystal studies provide more accurate structural information than powder studies, because the diffraction information can be collected over three dimensions of reciprocal space, $\underline{Q}$, rather than being collapsed into a single dimension (i.e., $Q = |\underline{Q}|$). However, in the case of superionic conductors, the majority of published studies have been performed on powder samples, often because the presence of first-order transitions to highly disordered phases on increasing temperature would shatter single crystals. Diffraction studies of superionic conductors are often characterized by a rapid fall-off in intensity of the Bragg scattering with increasing Q, and the limited number of measurable peak intensities often restricts the complexity of structural models that can be applied in the data analysis (see

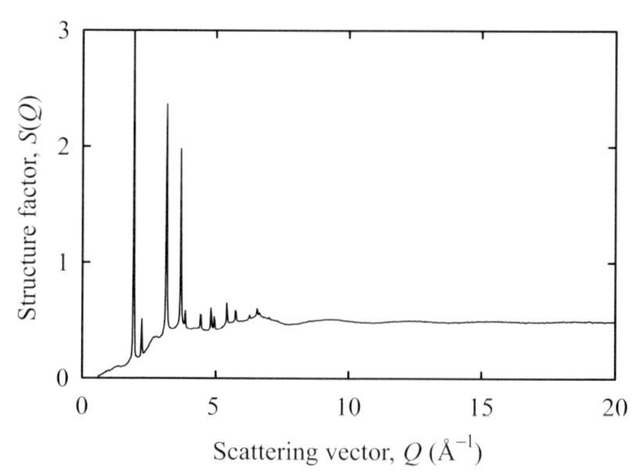

Figure 2.1 A neutron powder diffraction pattern collected from the superionic δ phase of Bi_2O_3 at 1033 K, illustrating the rapid decrease in the intensities of the Bragg peaks with increasing Q and the presence of diffuse scattering observed as broad undulations (S. Hull *et al.*, unpublished results).

Figure 2.1). An analysis of the Bragg diffraction provides a structural model of the unit cell contents, averaged over time and over all the unit cells in the sample. This information can be complemented by an analysis of the diffuse scattering, which is observed as broad humps of scattering in Q regions between the Bragg peaks (see Figure 2.1), and has its origin in the short-range instantaneous correlations between disordered ions.

The past few decades have witnessed the development of many new intense sources of X-rays and neutrons worldwide. The former are produced by the motion of electrons within a *synchrotron*, often via the use of sophisticated insertion devices (bending magnets, multipole wigglers) which produce highly collimated and extremely intense beams. In the case of neutrons, specialized reactor sources are gradually being superseded by intense spallation sources, in which pulses of neutrons are produced by bombarding a heavy metal target with high energy bursts of protons that have been accelerated in a synchrotron or linear accelerator. Whilst the neutron flux produced by these machines remains many orders of magnitude lower than the flux of X-ray photons produced at one of the latest synchrotron X-ray sources, neutron diffraction has played a major role in elucidating the structures of many compounds showing high levels of ionic conductivity. This is largely a consequence of the neutron's ability to use bulk samples, its increased sensitivity to the presence of lighter species such as Li^+, F^- and O^{2-}, and the absence of a form factor (as found in X-rays) which further hinders the ability to collect data to high Q values.

2.2.2
Extended X-Ray Absorption Fine Structure

The EXAFS technique exploits the absorption edges observed when the energy of an X-ray beam incident upon a sample coincides with the energy required to eject an electron from one of the sample's constituent atoms to a continuum state (i.e., ionization) [2]. A simple model for the absorption process predicts a monotonic decrease in the absorption coefficient with increasing energy at energies higher than the absorption edge. However, the absorption spectrum often shows oscillations in the immediate post-edge region, which can be analyzed by considering the ejected electron as a wave traveling outward from the central absorbing atom. In a condensed solid, the emitted electron wave is backscattered by the neighboring ions, such that the resultant interference gives rise to the oscillations in the post-edge absorption coefficient which extend towards higher energy by a several hundreds of eV. The analysis of the frequency and amplitude of the EXAFS pattern can provide information on the distance, type, and number of the nearest neighbors. In comparison with diffraction methods, EXAFS has the advantage of being atom-specific (by tuning the energy of the incident X-ray beam), although the structural information provided is restricted to the first few coordination shells. A brief overview on the use of EXAFS technique for the analysis of nanocrystalline materials is presented in Chapter 4.

2.2.3 Nuclear Magnetic Resonance

The NMR technique is used extensively within the field of chemistry, for example in the study of molecules in solution. In recent years, however, there has been a major development of solid-state NMR methods including, for example, magic angle spinning NMR (MAS-NMR) techniques, that have been applied to the study of ionically conducting systems [3] (see also Chapter 4). Unfortunately, a number of limitations have been demonstrated, including differences between the timescales probed by the NMR technique and those characteristic of ionic motion in solids, the limited temperature range accessible using NMR apparatus, and also difficulties in preparing samples containing nuclei with the required half-integer spin (such as ^{17}O). Nevertheless, several studies have recently been conducted with ionically conducting compounds, and which have resolved the different crystallographic sites occupied by mobile ions, thus determining those positions responsible for the ionic conduction and estimating the rates of exchange between sites; see, for example, the reports of Grey and coworkers on F^- conduction in $BaSnF_4$ [4] and O^{2-} conduction in $Bi_4V_2O_{11}$ and its chemical derivatives [5].

2.2.4 Computational Methods

Computer simulation methods have been extensively used to probe the structural behavior of ionically conducting solids, with particular emphasis on the preferred diffusion mechanisms and the nature of intrinsic (thermally induced) and extrinsic (generated by chemical doping) defects. As discussed elsewhere [6], these techniques can broadly be divided into three categories:

- Static lattice simulations, in which the energy of a configuration is minimized to compare, for example, the formation energies of different defect configurations.

- Molecular dynamics (MD) methods utilize the iterative solution of Newton's equations of motion for a large number of ions within a configuration. As energy is conserved (rather than minimized), it is possible to extract information concerning the dynamics of ions within the materials under investigation, including preferred diffusion pathways and any correlations between the motions of diffusing ions.

- Monte Carlo (MC) methods have been extensively employed to study systems containing high levels of ionic disorder. In this approach, a large number of configurations are generated by a succession of random moves, with the probability of including a given configuration determined by a Boltzmann factor.

Selected features of these computational methods, and examples of their application, are described in Chapters 5, 7 and 9.

In many cases, simulation methods are used in a complementary manner to experimental studies, with the validity of the calculations assessed by comparing simulated properties (e.g., crystal structure and activation energies) with those determined experimentally. The major factor in determining the reliability of all the simulation methods is the accuracy of the description of the interaction between the ions. The majority of studies of ionically conducting systems have utilized parameterized potentials containing explicit expressions for the various interactions (short-range repulsion, Coulomb, etc.), although recent advances in available computer power have enabled the application of *ab initio* methods (see Chapter 7).

2.3 Families of Superionic Conductors

This section provides an overview of the most important families of compounds which display high values of ionic conductivity. Inevitably, limitations of space mean that a number of interesting systems are not discussed, including several systems which are the subject of current research activity (e.g., the "LAMOX" compounds [7] and apatite-structured oxides [8]) which do not conveniently fit into one of the major groups discussed below.

2.3.1 Silver and Copper Ion Conductors

The family of Ag^+ and Cu^+ superionic conductors have been extensively studied for many decades, using a wide range of experimental and computational techniques; see also Chapter 7. They are principally of interest for fundamental reasons, as model systems in which to characterize the nature of the dynamic disorder and to probe the factors which promote high values of ionic conductivity within the solid state. Their commercial applications are generally limited by factors such as chemical stability, the high cost of silver, and their relatively high mass when compared, for example, to lithium-based compounds.

2.3.1.1 Silver Iodide (AgI)

Silver iodide (AgI) is arguably the most widely studied superionic conductor. The presence of a superionic transition at 420 K from the ambient temperature wurtzite-structured β phase (space group $P\bar{6}_3mc$) to the superionic α phase was studied by Tubandt [9], with the ionic conductivity increasing at the β → α transition by a factor of ~4×10^3 to a value of 1.3 S cm^{-1}. Transport measurements confirmed that the conduction within the α phase is predominantly ionic (rather than electronic) in nature, and due to the motion of the cations (rather than of the anions) [10]. X-ray powder diffraction studies showed that α-AgI possesses a body-centered cubic (*bcc*)-structured anion sublattice, with the two Ag^+ per unit cell proposed to be distributed over all the available interstices within this array [11, 12]. As illustrated in Figure 2.2, the *bcc* anion sublattice within space group $Im\bar{3}m$ comprises the 2(*a*) sites at 0, 0, 0 and $^1/_2$, $^1/_2$, $^1/_2$. The interstices can be described as octahedral (*oct*, 6(*b*) sites at 0, $^1/_2$, $^1/_2$, etc.), tetrahedral (*tet*, 12(*d*) sites at $^1/_4$, 0, $^1/_2$, etc.) and trigonal (*trig*, 24(*h*) sites at 0, *y*, *y*, etc., with $y = 3/8$).

A random distribution of Ag^+ over the six *oct*, 12 *tet* and 24 *trig* interstices, of the type originally proposed [11, 12], approximates to a "liquid-like" distribution of cations which, when coupled with the high values of ionic conductivity, led to the concept of a "molten sublattice" of Ag^+ within α-AgI. However, subsequent studies, using both powder and single-crystal samples and X-ray and neutron radiations, clearly showed that the Ag^+ preferentially occupy the *tet* positions [13–16]. The nature of the Ag^+ diffusion was the subject of some debate, with the time-averaged cation distribution extending in either the ⟨1 0 0⟩ or ⟨1 1 0⟩ directions (implying Ag^+ diffusion via the *oct* and *trig* positions, respectively). This issue was resolved by analysis of the total (Bragg plus diffuse) scattering components using the so-called reverse Monte Carlo (RMC) method [17, 18], which demonstrated that the Ag^+ spend around three-fourths of their time on the *tet* sites and predominantly hop between nearest-neighbor *tet* positions in ⟨1 1 0⟩ directions [19].

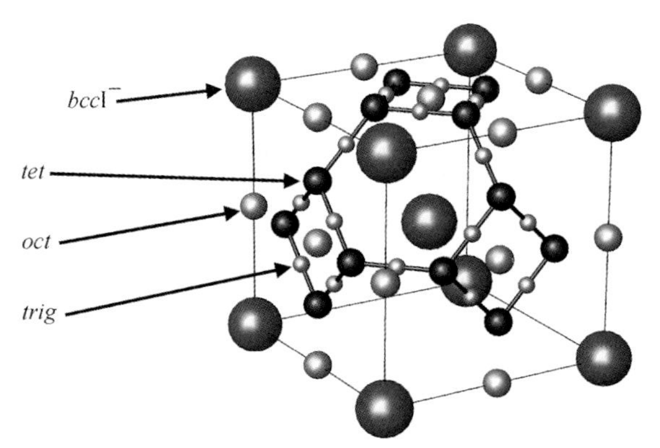

Figure 2.2 The *bcc* structure of α-AgI, showing the locations of the octahedral (*oct*), tetrahedral (*tet*), and trigonal (*trig*) interstices within the unit cell. The thick lines illustrate how Ag^+ diffusion occurs between the *tet* positions in ⟨1 1 0⟩ directions via the *trig* sites.

2.3.1.2 **Copper Iodide (CuI)**

At ambient temperature, the γ phase of CuI possesses the cubic zincblende structure, in which the Cu^+ are ordered over half the *tet* interstices within a face-centered cubic (*fcc*) sublattice of I^-. Formally, the anions are located in the 4(*a*) sites at 0, 0, 0, etc., of space group $F\bar{4}3m$, with two sets of *tet* interstices (the 4(*c*) sites at $^1/_4$, $^1/_4$, $^1/_4$, etc. and 4 (*d*) sites at $^3/_4$, $^3/_4$, $^3/_4$, etc.) and one set of *oct* cavities (the 4(*b*) sites at $^1/_2$, $^1/_2$, $^1/_2$, etc.). The Cu^+ occupy one set of *tet* interstices in γ-CuI and, on heating, CuI undergoes a phase transition to the β phase at 642 K. The structure of β-CuI can be described as an hexagonal close-packed (*hcp*) sublattice of anions, with the Cu^+ distributed in the approximate ratio 5 : 1 over two sets of *tet* interstices within space group $P\bar{3}m1$ [20]. On further heating, the anion sublattice reverts to an *fcc* arrangement at the transition to the superionic α phase at 680 K [20, 21].

The distribution of the mobile Cu^+ within α-CuI offers an interesting comparison with α-AgI, to probe the difference in superionic behavior between phases with *fcc* and *bcc* anion sublattices. α-CuI was initially proposed to be isostructural with the ambient temperature γ-CuI phase, but with large displacements of the Cu^+ in $\langle 1\,1\,1\rangle$ directions [22], whilst the results of EXAFS studies have suggested that the cations are disordered over all the *tet* and *oct* positions [23]. However, the findings of neutron diffraction studies have indicated that the Cu^+ are predominantly restricted to the *tet* positions within space group $Fm\bar{3}m$ (such that the *tet* sites form a single set in the 8(*c*) positions) [24, 25]. This structural model has been supported by computer simulations [26–28], with the Cu^+ shown principally to diffuse between the *tet* positions in $\langle 1\,0\,0\rangle$ directions. However, while the essential features of α-CuI are now well established, a number of details continue to attract attention, including differences between the partial pair distribution functions derived by computer simulations, neutron diffraction and EXAFS methods (for details, see Refs. [26, 29]).

2.3.1.3 **Other Ag^+ and Cu^+ Halides**

The two Ag^+ halides AgCl and AgBr adopt the rocksalt structure at ambient temperature (space group $Fm\bar{3}m$), in which all the *oct* 4(*b*) sites within an *fcc* array of anions are occupied by Ag^+. Neither compound undergoes a structural phase transition on increasing temperature, but both compounds show significant increases in ionic conductivity (up to $\sim$0.5 S cm^{-1}) over the temperature region $\sim$100–150 K below their melting points [9]. Powder neutron diffraction studies of AgBr performed at temperatures immediately below the melting point showed that the Ag^+ vibrate anisotropically about the *oct* positions, with an increasing occupancy of the *tet* positions [30]. This behavior was interpreted as a gradual transition to a superionic phase, which is interrupted by the melting transition before the fully highly disordered state is reached [31]. Similar behavior has been observed within the rocksalt structured phase of AgI, which is stable at pressures in excess of $\sim$0.4 GPa [32, 33].

The ambient temperature γ phases of CuCl, CuBr, and CuI all possess the cubic zincblende structure, but their structural behavior on increasing temperature is very different. CuCl transforms to its β phase at 681 K and melts at 703 K; β-CuCl adopts the wurtzite structure ($P6_3mc$), which is the hexagonal equivalent to the cubic zincblende structure (with the Cu^+ ordered over half the *tet* interstices within an *hcp* anion sublattice) [34]. β-CuI possesses a relatively high ionic conductivity ($\sigma_i \approx 0.1$ S cm^{-1} [35]), but neutron diffraction studies have indicated that the cations remain largely on the *tet* lattice sites [34]. By contrast, CuBr undergoes structural phase transitions to its β and α phases at 664 and 744 K, respectively. β-CuBr adopts a disordered wurtzite structure (with the Cu^+ distributed over the two sets of *tet* interstices in the approximate ratio of 3 : 1), and α-CuBr is isostructural with α-AgI [36]. Total neutron scattering studies indicated that the Cu^+ preferentially adopt a tetrahedral local environment, even though Br^- sublattice changes from *fcc* (γ) to *hcp* (β) to *bcc* (α) on heating [36].

2.3.1.4 Ag^+ Chalcogenides

The high-temperature superionic properties of the Ag^+ and Cu^+ chalcogenides (of general formula A_2X, with A = Ag, Cu and X = S, Se, Te) have been extensively studied. These compounds provide an interesting comparison with their halide counterparts, particularly when probing the effects of the doubled cation density on the ionic diffusion mechanisms and macroscopic ionic conductivity [37].

The three silver chalcogenides Ag_2S, Ag_2Se, and Ag_2Te all adopt low-symmetry, ordered phases under ambient conditions (space groups $P2_1/c$, $P2_12_12_1$, and $P2_1/c$, respectively), but all possess at least one high-temperature superionic phase. The ambient temperature γ-Ag_2S phase transforms to a *bcc*-structured (β) form at 452 K and an *fcc* structured (α) phase at 873 K, prior to melting at 1115 K. Ag_2Se possesses a single high-temperature superionic phase (α), which adopts a *bcc* anion sublattice and is stable between 406 K and the melting point of 1170 K. Ag_2Te transforms from its ambient temperature (phase to *fcc* (α) and *bcc* (γ) -structured superionic phases at 423 K and 1075 K, respectively, before melting at 1223 K (for further details, see Refs. [38, 39]).

Diffraction studies of the three *bcc*-structured phases (β-Ag_2S, α-Ag_2Se, and γ-Ag_2Te) showed that the Ag^+ are distributed predominantly over the *tet* interstices [40–43]. Whilst the number of *tet* sites per unit cell (12) still greatly exceeds the number of cations (four), long-range coulombic repulsions between cations might be expected to lead to short-lived, short-range correlations between the motions of individual Ag^+. This has been demonstrated by the observation of anisotropic diffuse scattering in both X-ray [40] and neutron [44] scattering experiments on the β phase of Ag_2S. Good agreement between the observed and calculated diffuse scattering was obtained using a structural model of local cation ordering comprising a microdomain of ordered Ag^+ sites, constructed using four connected cubo-octahedra of *tet* sites [40]. Neutron diffraction studies of the two *fcc*-structured superionic phases (α-Ag_2S and α-Ag_2Te) have shown that the cations are located predominantly on the *tet* interstices, with a fraction of Ag^+ found close to the *oct* positions [42, 43]. Both possess lower ionic conductivities than their *bcc*-structured counterparts, even though their order on increasing temperature is different in Ag_2S and Ag_2Te [37]. This is consistent with the lower number of available interstices within an *fcc*-structured sublattice (see Ref. [45]).

2.3.1.5 Cu^+ Chalcogenides

In comparison to the Ag^+ chalcogenides, the Cu^+ analogues have been less widely studied, being prone to significant non-stoichiometry (i.e., $Cu_{2-\delta}X$) and an increased electronic contribution to the conduction. The ambient temperature phases of $Cu_{2-\delta}S$ and $Cu_{2-\delta}Se$ both possess rather complex crystal structures, but undergo transitions on heating to higher symmetry phases showing extensive Cu^+ disorder. Neutron diffraction studies of β-$Cu_{2-\delta}S$, which is stable between 376 K and 703 K, showed that the anions adopt an *hcp* sublattice in space group $P6_3/mmc$, with the Cu^+ distributed over two sites displaced from the centers of the *tet* cavities [46]. Above 703 K, $Cu_{2-\delta}S$ adopts an *fcc*-structured α phase in which the cations occupy the 32(*f*) sites of space group $Fm\bar{3}m$, in *x*, *x*, *x*, positions with $x \approx 0.29$, corresponding to the *tet* cavities but displaced in ⟨1 1 1⟩ directions [47]. Cu_2Se transforms to a cubic α phase at 413 K, which appears isostructural with α-$Cu_{2-\delta}S$ [47]. However, single crystal X-ray diffraction studies and MD simulations have suggested that a small fraction of the cations in α-Cu_2Se occupy the *oct* positions [48], though this is inconsistent with reported EXAFS studies [49]. The structural properties of Cu_2Te are poorly understood and, whilst four or five phase transitions have been observed in the temperature range 430 K to 760 K (see Refs. [50, 51]), the crystal structures of most of the phases are poorly understood.

2.3.1.6 Silver Sulfur Iodide (Ag_3SI)

At temperatures above 519 K, Ag_3SI exists in a superionic (α) phase characterized by a *bcc*-structured sublattice formed by a random distribution of S^{2-} and I^- [52–54]. Thus, the averaged structure of α-Ag_3SI can be viewed as analogous to the superionic α phases of AgI and Ag_2S. Interest in this material stems from the fact that the high-temperature α phase can be quenched to room temperature, retaining the disordered anion array (α*-Ag_3SI), or annealed for prolonged periods at a temperatures of 450 K to form a phase with long-range ordering of the two cation species over the 0,0,0 and $^1/_2$, $^1/_2$, $^1/_2$ positions (β-Ag_3SI). The metastable α* phase possesses a high ionic conductivity under ambient conditions (~0.3 S cm^{-1}) which exceeds that of the annealed (anion ordered) β phase by almost two orders of magnitude [55].

The results of powder neutron diffraction studies have indicated that α-Ag_3SI possesses a highly disordered Ag^+ distribution, which resembles that of α-AgI and α-Ag_2S [56]. By contrast, the cations within β-Ag_3SI are localized in regions of the unit cell close to a subset of the *oct* cavities [56]. As discussed in greater detail elsewhere [38], ordering of the two anion species over the 0,0,0 and $^1/_2$, $^1/_2$, $^1/_2$ sites lowers the symmetry from $Fm\bar{3}m$ to $Pm\bar{3}m$, such that crystal structure is modified from one containing a network of numerous interconnected equivalent sites (6 × *oct* and 12 × *tet*) to one in which there are only three isolated, energetically favored cavities. This explanation for the dramatic difference between the Ag^+ ion conductivities of the α* and β phases of Ag_3SI has recently been supported by an MD study, in which configurations of 4 × 4 × 4 unit cells were constructed with the S^{2-} and I^- ordered and disordered over the 0, 0, 0 and $^1/_2$, $^1/_2$, $^1/_2$ sites (to represent the β and α* phases) [57].

2.3.1.7 Ternary AgI-MI_2 Compounds

The addition of aliovalent M^{2+} species to AgI forms a family of ternary compounds, which includes Ag_2CdI_4, Ag_2HgI_4, Ag_2ZnI_4, Ag_3SnI_5, and Ag_4PbI_6. Extensive ionic conductivity and diffraction studies at elevated temperatures have been performed. Seven superionic phases have been identified, formed with *bcc* (α-Ag_2CdI_4 [58] and ε-Ag_2HgI_4 [59]), *hcp* (δ-Ag_2HgI_4 [59] and α-Ag_2ZnI_4 [58]) and *fcc* (α-Ag_2HgI_4 [60], Ag_3SnI_5 [58] and Ag_4PbI_6 [61]) anion sublattices. These are analogous to the superionic phases found in the binary Ag^+ and Cu^+ halides, but with two cations (Ag^+/Cu^+) replaced by a single divalent one (M^{2+}) and a vacancy. In several cases, addition of the M^{2+} dopant ions lowers the superionic transition temperature compared to pure α-AgI, but at the cost of a significant reduction in the ionic conductivity [59].

2.3.1.8 Ternary AgI-MI Compounds

Over 30 ternary phases have been identified within the $(AgX)_x$-$(MX)_{1-x}$ and $(CuX)_x$-$(MX)_{1-x}$ (where $M = K^+$, Rb^+ and Cs^+; $X = Cl^-$, Br^-, and I^-) phase diagrams. Of these, $RbAg_4I_5$ has attracted extensive interest, because it possesses a high ionic conductivity at ambient temperature (0.21 S cm^{-1} [62]). The immobile Rb^+ within $RbAg_4I_5$ act as a "structural modifier", forcing the anion arrangement to adopt a structure which approximates to that of β-Mn (space group $P4_132$ [63, 64]). This sublattice contains non-intersecting, one-dimensional strings of tetrahedrally coordinated sites parallel to the three ⟨1 0 0⟩ directions, which are favorable conduction pathways for the Ag^+. Recent neutron scattering studies, using Maximum Entropy difference Fourier methods to investigate the time-averaged Ag^+ distribution, have also provided evidence for migration of Ag^+ between the channels [62].

2.3.1.9 Ternary Derivatives of Ag_2S

A number of ternary derivatives of Ag_2S have been investigated, with particular emphasis on the family of argyrodite-structured compounds within the series $Ag_{12-n}B^{n+}S_6$, where $B = Ta^{5+}$, Nb^{5+}, Ge^{4+}, Si^{4+}, Ti^{4+}, Ga^{3+} and Al^{3+}. The majority of the compounds adopt a cubic argyrodite structure at elevated temperatures (space group $F\bar{4}3m$), with the Ag^+ distributed over a large number of sites, but adopt lower symmetry variants at lower temperatures in which the cations are, at least partially, ordered over a subset of the sites [65]. As expected, the highest ionic conductivities are shown by the cubic phases, which also includes a number of Cu^+ analogues formed by the addition of halide anions (such as Cu_6PS_5Br [66]) in which the $F\bar{4}3m$ polytype is stabilized at ambient temperature.

2.3.2
Fluorite-Structured Compounds

The Ag^+ and Cu^+ superionic phases discussed above are generally characterized by an extensive diffusion of the cations within an immobile sublattice formed by the negatively charged counterions. In this section, we consider the family of compounds which crystallize in the fluorite crystal structure. These show similarly high levels of

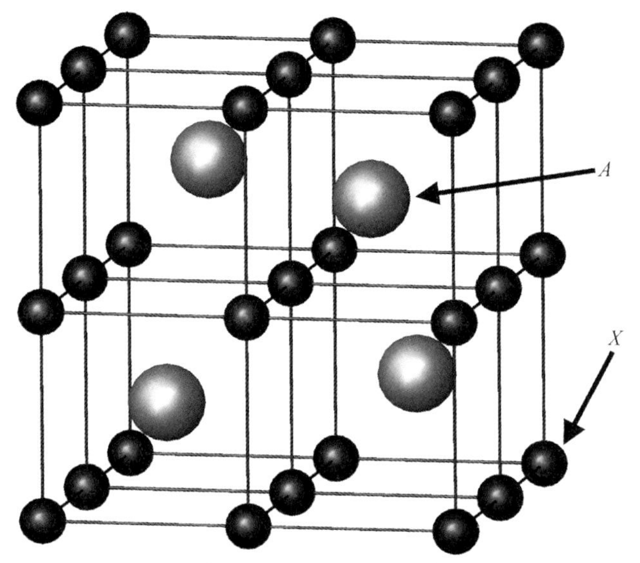

Figure 2.3 The cubic fluorite crystal structure of a compound of stoichiometry AX_2. The cations (A) occupy alternate cube centers within a simple cubic array of anions (X).

ionic conductivity at elevated temperatures, although charge transfer occurs via the motions of the anions; the level of ionic transport in particular systems and their applications are discussed in Chapters 9, 12 and 13.

2.3.2.1 The Fluorite Structure

The cubic fluorite crystal structure (space group $Fm\bar{3}m$) can be described as an *fcc* array of cations in which all the *tet* interstices are filled with anions and the *oct* sites are empty. Alternatively, as shown in Figure 2.3, the ionic arrangement can be viewed as a simple cubic array of anions with cations occupying alternate cube centers.

2.3.2.2 Halide Fluorites

The ionic conductivity of fluorite-structured halides such as SrF_2, CaF_2, BaF_2, β-PbF_2, and $SrCl_2$, shows a rapid but continuous increase on heating, reaching values very close to those shown by the liquid state [67–69]. The transition to the superionic state is also associated with peak in the specific heat C_p (the maximum value of which is generally taken to define the superionic transition temperature, T_c), anomalous behavior of the elastic constants and changes in the lattice expansivity (for details and references, see Refs. [38, 70]).

At temperatures well below T_c the defect concentration is very low and dominated by anion Frenkel pairs, with the interstitial anions located in the empty cube center positions [71]. However, neutron diffraction studies have shown that the empty cube center sites are not significantly occupied at temperatures in excess of T_c and, instead, the anion interstitials within the Frenkel pairs favor sites between the midpoint of the two nearest-neighbor anion sites and the center of an anion cube in a ⟨1 1 0⟩ direction [72]. This defect model was supported by analysis of the coherent diffuse neutron scattering, which also showed that the Frenkel defects are accompanied by relaxations of nearest-neighbor lattice anions in ⟨1 1 1⟩ directions towards the centers of the adjacent empty anion cubes [73]. Static energy calculations confirmed the stability of these clusters and suggested the favored location of the charge-compensating anion vacancies [74]. The concentration of Frenkel defects within the fluorite-structured halides at temperature close to T_c has been the subject of much debate, however, with values obtained by neutron diffraction studies around an order of magnitude higher than the few per cent predicted by MD simulations (for a further discussion of this issue, see Ref. [75]).

2.3.2.3 Lead Tin Fluoride ($PbSnF_4$)

Of all the fluorite-structured halides, β-PbF_2 has attracted the most attention, since it has the lowest superionic transition temperature ($T_c = 711$ K). The ionic conductivity and structural properties of a number of compounds based on PbF_2 have been studied, with the ternary compound $PbSnF_4$ possessing one of the highest values of ionic conductivity of any F^- ion conductor at ambient temperature ($\sigma_i = 10^{-3}$ S cm^{-1} [76]).

Under ambient conditions, α-$PbSnF_4$ possesses a crystal structure which is closely related to that of fluorite, but with cation ordering in the sequence PbPbSnSnPbPbSnSn ... along one of the pseudocubic ⟨1 0 0⟩ axes leading to tetragonal symmetry (space group *P4/nmm* [77]). While the anions between the two Pb^{2+} layers lie close to the *tet* sites occupied within the equivalent fluorite structure, the Sn^{2+}-Sn^{2+} layers are essentially empty [77, 78]. The displaced anions are located within the Pb^{2+} and Sn^{2+} sheets, leading to a highly disordered arrangement of F^-. The combined results of powder neutron diffraction studies and MD simulations have shown that anion diffusion occurs predominantly within these two-dimensional layers, but becomes more three-dimensional as the temperature is increased [78]. This behavior appears consistent with that of the isostructural compound $BaSnF_4$, which has been investigated using NMR methods [4].

2.3.2.4 Anion-Excess Fluorites

The cubic fluorite structure shows a remarkable ability to tolerate large deviations from its ideal AX_2 stoichiometry, with many grossly anion-deficient and anion-excess examples known to exist. In the latter case, cations of valence greater than 2^+ sit substitutionally on the host cation sites within a halide fluorite, with charge neutrality maintained by incorporating interstitials within the anion sublattice. At dopant levels of less than approximately 1%, the anion interstitials reside in the empty cube center positions and, depending on the relative sizes of the host and dopant cation species, are nearest neighbor or next nearest neighbor to the dopant cation [79]. These defects are associated with a large increase in the ionic conductivity until the dopant level reaches around 10%, beyond which the ionic conductivity falls due to trapping of the mobile anion interstitials within larger defect clusters (see, e.g., Ref. [80]). The nature of these defects has been studied using a wide variety of experimental and computational techniques, and there exists an extensive literature devoted to doped CaF_2, BaF_2, PbF_2, and $SrCl_2$ systems.

As an illustrative example, we consider here the material formed by doping CaF_2 with Y^{3+}. The results of the earliest neutron diffraction studies of $(Ca_{1-x}Y_x)F_{2+x}$ proposed that the excess anions were incorporated into the cubic fluorite lattice in the form of extrinsic defects which resembled the Frenkel clusters observed within the superionic phase of the pure fluorites (though without the need for the charge-compensating Frenkel vacancy) [81, 82]. However, subsequent single-crystal neutron diffraction studies, which included analysis of both the Bragg and diffuse scattering components, favored the presence of cubo-octahedral defects, in which six edge-sharing $(Ca/Y)F_8$ cubes are replaced by six corner-sharing square antiprisms. In this process, the central cavity changes from a cube of eight anions into a cubo-octahedron of 12 anions, so that four additional anions are incorporated into the lattice (or five if the central cavity is also filled) [83]. The presence of these clusters was supported by ^{19}F NMR [84] and EXAFS [85] studies, whilst static energy calculations showed that cubo-octahedral defects were favored over other defect types if the M^{3+} dopant cations are relatively small [86].

2.3.2.5 **Oxide Fluorites**

Fluorite-structured oxides, which include compounds such as cubic ZrO_2, PrO_2, CeO_2, ThO_2, UO_2, and PuO_2, are analogues of the fluorite-structured halides (see Section 2.3.2.2). Whether these compounds also undergo continuous superionic transitions at elevated temperatures is unclear, but their melting points are considerably higher than those of their halide counterparts. This implies that any superionic transition would occur at temperatures more difficult to access experimentally. Nevertheless, interest in a high-temperature superionic transition within fluorite-structured oxides has been motivated by the use of compounds such as UO_2 and PuO_2 within nuclear fission reactors. In particular, any excess specific heat contained within these materials due to the presence of high concentrations of Frenkel defects has safety implications if the operating temperature exceeds its normal range.

Data for the temperature dependence of the enthalpy of UO_2, obtained using drop calorimetry methods, showed a clear peak in the specific heat at $\approx$2610 K [87]. However, the structural origin of this feature was questioned by some groups, who proposed an alternative explanation in terms of electronic disorder (small polarons of the type $2U^{4+} \leftrightarrow U^{5+} + U^{3+}$), because the band gap of UO_2 ($\sim$2 eV) was much smaller than the formation energy for anion Frenkel defects ($\sim$5 eV) [88, 89]. However, the observation of similar behavior of the enthalpy of the isostructural compound ThO_2 at temperatures in the region of 2950 K favored the presence of Frenkel disorder, because the relative stability of the Th^{4+} valence state would lead to a much higher formation energy for small polarons [90]. The presence of an extensive lattice disorder within UO_2 and ThO_2 at elevated temperatures was confirmed by single-crystal neutron diffraction studies, with an increasing concentration of dynamic anion Frenkel defects observed at temperatures above $\sim$2000 K [91, 92]. Support has also been provided by recent MD studies, with the onset of superionic behavior within the simulations occurring at 2300 K [93].

2.3.2.6 Anion-Deficient Fluorites

Pure zirconia (ZrO_2) adopts a monoclinic baddelyite structure under ambient conditions (m-ZrO_2, space group $P2_1/c$), and transforms at ~1370 K to a tetragonal phase (t-ZrO_2, space group $P4_2/nmc$) which can be described as a fluorite lattice with alternate {1 1 0} planes of anions displaced in opposite directions along the [0 0 1] axis. At ~2643 K, ZrO_2 transforms to the cubic fluorite-structured form (c-ZrO_2), prior to melting at ~2988 K. The two high-temperature phases can be stabilized at ambient temperature by chemical doping with divalent (Ca^{2+}, Mg^{2+}, etc.) and trivalent (Y^{3+}, Sc^{3+}, Nd^{3+}, etc.) cations. This process generates the so-called "stabilized" tetragonal (t^*-ZrO_2) and cubic (c^*-ZrO_2) zirconias, in which the aliovalent dopant cations sit substitutionally on the host Zr^{4+} sites, with overall electrical neutrality being achieved by incorporating O^{2-} vacancies into the anion sublattice. The ionic conductivity of these anion-deficient fluorite compounds reaches significant levels at temperatures in excess of ~1200 K (as the anion vacancies become mobile), which makes them attractive candidates for many technological applications, including solid electrolytes within solid oxide fuel cells (SOFCs) [94]; see also Chapters 9 and 12.

The phase diagrams and ionic conductivities of anion-deficient ZrO_2 compounds have been widely investigated, to probe the influence of both dopant size and concentration (e.g., Ref. [95]). Whilst Sc^{3+} doped systems possess the highest values of ionic conductivity, factors such as cost and long-term stability have favored the use of $(Zr_{1-x}Y_x)O_{2-x/2}$ as the current "best" material, in which the $m \rightarrow t^*$ and $t^* \rightarrow c^*$ transitions are observed at $x \approx 0.05$ and $x \approx 0.16$, respectively [96]. On increasing dopant (anion vacancy) concentration, the ionic conductivity initially increases with x, but reaches a maximum close to the lower limit of stability of the c^* phase and then decreases rapidly [97]. Many theoretical studies have attempted to provide an explanation for this effect, and it is generally accepted that O^{2-} vacancies are trapped to form defect clusters (for further details, see Ref. [38]). However, a detailed structural description of the defects has proved controversial, with diffuse X-ray scattering indicating relaxations of the cations in ⟨1 1 0⟩ directions [98], whilst single-crystal neutron diffraction studies suggested that the anions are displaced from their lattice sites predominantly in ⟨1 0 0⟩ directions and, to a lesser extent, ⟨1 1 1⟩ directions [99]. A detailed single-crystal diffraction study, using both neutron and X-ray radiations, demonstrated that the defect structure of $(Zr_{1-x}Y_x)O_{2-x/2}$ is rather complex, with significant variations as a function of composition and temperature. With increasing x within the c^* phase, there is a tendency for anion vacancies to form pairs aligned along ⟨1 1 1⟩ directions, which aggregate to form larger defects of around 15–20 Å in diameter that act as highly effective traps for the mobile O^{2-} vacancies [100]. These defects show similarities to the ordered structure adopted by the compound $Zr_3Y_4O_{12}$ at higher Y^{3+} concentrations.

Numerous other anion-deficient fluorite-structured oxides have been investigated, with a view to identifying compounds that show ionic conductivities comparable to yttria-stabilized zirconia, but at a significantly lower temperature (see Chapter 9). Of these, CeO_2 doped with Gd_2O_3 is a promising candidate [101]. However, the reduction of Ce^{4+} to Ce^{3+} under low oxygen partial pressures leads to an additional electronic contribution to the conductivity, which may prove problematic for SOFC applications. Similar issues have also hindered the application of Bi_2O_3-based materials (these are discussed in the following subsection).

2.3.2.7 Bi_2O_3

Under ambient conditions, Bi_2O_3 adopts a rather complex monoclinic structure (space group $P2_1/c$), in which the Bi^{3+} are surrounded by an asymmetric polyhedron of O^{2-} characteristic of its $6s^2$ lone-pair electronic configuration. However, Bi_2O_3 undergoes a first-order phase transition at 1002 K to its superionic δ phase, which has the highest known oxide-ion conductivity ($\sim$1.5 S cm^{-1} [102]). δ-Bi_2O_3 adopts an anion-deficient cubic fluorite structure, with the extensive dynamic anion disorder attributed to the high concentration of intrinsic anion vacancies and the Bi^{3+} cation's ability to stabilize the relatively asymmetric local coordination environments which arise during diffusion of the surrounding anions.

The first structural model for the δ-Bi_2O_3 phase was proposed by Gattow [103], with the O^{2-} randomly distributed over the 8(*c*) fluorite anion lattice sites, such that each has a mean occupancy of $^3/_4$. Alternatively, the "Sillen" model [104] favored the inclusion of pairs of O^{2-} vacancies in $\langle 111 \rangle$ directions, whilst the "Willis" model was based on the partial (3/16) occupancy of 32(*f*) positions which are displaced from the lattice sites in $\langle 1\,1\,1 \rangle$ directions [105]. The results of powder neutron diffraction studies indicated that the time-averaged structure of δ-Bi_2O_3 has the anions distributed over both the 8(*c*) and 32(*f*) positions [106, 107]. The presence of short-range correlations between O^{2-} vacancies has been considered in a number of simulation studies comparing the relative stability of pairs of vacancies aligned in $\langle 1\,0\,0 \rangle$, $\langle 1\,1\,0 \rangle$ and $\langle 1\,1\,1 \rangle$ directions. Of these, the most recent studies favor the former arrangement [108, 109].

The ionic conductivity of the δ-Bi_2O_3 phase is almost two orders of magnitude higher than stabilized $(Zr_{1-x}Y_x)O_{2-x/2}$ at comparable temperatures, which has motivated many attempts to stabilize the fluorite-structured δ form of Bi_2O_3 at ambient temperature by the addition of isovalent (Y^{3+} and trivalent rare-earth) and aliovalent (Ba^{2+}, Nb^{5+}) cations. Whilst the former have proved particularly successful, the ionic conductivity is somewhat lower in doped fluorite-structured compounds such as Bi_3YO_6 (for a comprehensive review of these studies, see Refs [110–112]). In addition, the applications of Bi_2O_3 and its chemical derivatives have been limited due to the appearance of significant electronic conduction at low oxygen partial pressures [112].

2.3.2.8 Antifluorites

The antifluorite structure is generated from the fluorite arrangement by replacing the anions with cations, and *vice versa*. Although this structure is adopted by many chalcogenides of the alkali metals, lithium oxide (Li_2O) has attracted the most widespread attention. Experimental evidence for a superionic transition analogous to those observed in the halide and oxide fluorites was first provided by measurements of the ionic conductivity of Li_2O, which also confirmed that Li^+ are the mobile species [113, 114]. Single-crystal neutron diffraction studies indicated the presence of an increasing fraction of dynamic cation Frenkel defects at temperatures in excess of ~850 K [115]. The structure of these defects appeared similar to those found in the halide fluorites [73], but with somewhat smaller relaxations of the two nearest-neighbor cations in ⟨1 1 1⟩ directions (as might be expected in view of the smaller size of the mobile species in this antifluorite-structured system). Recent MD studies have also successfully reproduced the superionic transition within Li_2O [93].

2.3.2.9 The "Rotator" Phases

Lithium sulfate (Li_2SO_4) undergoes a superionic transition at 848 K to its α phase, which possesses an extremely high ionic conductivity (1–3 $S\,cm^{-1}$ [116]; see also Chapter 7). The crystal structure of α-Li_2SO_4 comprises an *fcc* array of isolated SO_4^{2-} tetrahedra, which rotate rapidly about the central S^{6+}. The Li^+ predominantly occupy the *tet* sites within this *fcc* array [117], so that the time-averaged ionic arrangement is equivalent to that of antifluorite-structured Li_2O if the O^{2-} ions are replaced by rotationally disordered SO_4^{2-} tetrahedra. However, the significantly higher ionic conductivity of α-Li_2SO_4 led to the suggestion that the rotating SO_4^{2-} tetrahedra enhance the Li^+ diffusion via a "paddle-wheel" mechanism, in which the SO_4^{2-} push the Li^+ through the lattice [118]. MD simulations supported the presence of correlations between the rotations of the SO_4^{2-} units and diffusion of the Li^+ [119], but the validity of the "paddle-wheel" mechanism has been challenged by doping studies. The addition of approximately 10% of Li_2WO_4 to Li_2SO_4 has been shown to increase the ionic conductivity by a factor of ~5, which contrasts with the decrease expected owing to the slower rotation of the heavier WO_4^{2-} tetrahedra [120].

2.3.3 Pyrochlore and Spinel-Structured Compounds

2.3.3.1 The Pyrochlore Structure

The cubic pyrochlore structure is adopted by many compounds of stoichiometry $A_2B_2X_7$, and is a superstructure of the cubic fluorite arrangement with a doubled unit cell dimension and space group $Fd\bar{3}m$. Each unit cell contains 16 *A* and 16 *B* cations, which are arranged in an ordered manner to form the *fcc* cation sublattice. The 56 *X* anions are located on two symmetry-independent sites, in 48(*f*) sites at x, $^1/_8$, $^1/_8$, etc., with $x \sim {}^3/_8$ (labeled *tet*1) and 8(*b*) sites at $^3/_8$, $^3/_8$, $^3/_8$, etc. (labeled *tet*2). A third set of sites, in 8(*a*) positions at $^1/_8$, $^1/_8$, $^1/_8$, etc. (labeled *tet*3), would be filled in the fluorite arrangement, but are empty in the ideal pyrochlore structure. The position of the anions on the *tet*1 sites tends to be displaced in ⟨1 0 0⟩ directions towards the vacant *tet*3 sites, to enable the smaller *A* cation species to adopt a distorted octahedral anions coordination, whilst the larger *B* cations possess a distorted cubic environment.

2.3.3.2 Oxide Pyrochlores

In the case of the rare-earth-doped zirconias, pyrochlore-structured phases are observed for relatively large rare-earth cations (larger than Tb^{3+}), with phases isostructural with $Zr_3Y_4O_{12}$ (see Section 2.3.2.6) observed for smaller dopants. Given the close relationship between the pyrochlore and fluorite structures, it might be expected that pyrochlore-structured compounds would adopt a fluorite form at elevated temperatures, with the *A* and *B* species randomly arranged over the cation sites to generate the required $Fm\bar{3}m$ symmetry. Whilst this behavior is observed experimentally [121], high values of oxygen-ion conductivity are also observed within the pyrochlore-structured phases at temperatures below the order → disorder transition. As an example, $Zr_2Gd_2O_7$ shows an ionic conductivity of approximately $8 \times 10^{-3}\,S\,cm^{-1}$ at 1000 K [122], even though the material does not transform to its fluorite-structured phase until ~1800 K [121]. The results of X-ray diffraction studies have indicated that the anions on the *tet*1 sites (see above) undergo anisotropic thermal vibrations preferentially towards the *tet*3 positions, suggesting that anion diffusion occurs via these empty sites [123]. This mechanism has been supported by both static energy calculations [124, 125] and MD simulations [126]. The pyrochlore to fluorite transition, and its influence on the anion conductivity, can also be investigated by chemical doping, an example being the system $(Zr_xTi_{1-x})_2Y_2O_7$. As x increases, the (mean) radii of the two cation sites become more similar, which favors a gradual transition from a fully ordered pyrochlore arrangement to a disordered fluorite-structured phase [127–129]. Selected data on the ionic conductivity of oxide pyrochlore phases are presented in Chapter 9.

2.3.3.3 The Spinel Structure

Spinel-structured compounds possess an ideal stoichiometry of A_2BX_4 and, in common with the pyrochlore arrangement, the immobile sublattice comprises an *fcc* array. However, spinel-structured compounds are generally cation (invariably Li^+) conductors, and it is the larger anions which adopt the *fcc* arrangement. Within the space group $Fd\bar{3}m$, the unit cell contains 32 *X* anions, so that there are a total of 32 octahedrally (*oct*) and 64 tetrahedrally (*tet*) coordinated interstices. The former comprises two sets of symmetry inequivalent sites (*oct*1 and *oct*2, in 16(*c*) sites at 0,0,0, etc. and 16(*d*) sites at $^1/_2$, $^1/_2$, $^1/_2$, etc., respectively), while the latter is formed by two eightfold sets (*tet*1 in 8(*a*) sites at $^1/_8$, $^1/_8$, $^1/_8$, etc. and *tet*2 in 8(*b*) sites at $^3/_8$, $^3/_8$, $^3/_8$, etc.), plus a single 48-fold set (*tet*3 in 48(*f*) sites at x, $^1/_8$, $^1/_8$, etc., with $x \sim {^3/_8}$). The so-called "normal" spinel structure is formed when the 16 *A* cations and eight *B* cations reside in the *oct*2 and *tet*1 positions, respectively, whilst the "inverse" spinels have half *A* cations occupying the *tet*1 positions and the remainder, plus the all the *B* cations, are randomly distributed over the *oct*2 sites. In the ideal case, the *oct*1, *tet*2, and *tet*3 positions are empty in both "normal" and "inverse" spinels.

2.3.3.4 Halide Spinels (LiM_2Cl_4, etc.)

The most widely studied ionically conducting compounds possessing the inverse spinel structure are the members of Li_2MCl_4 family of compounds, where *M* is one of the species Mg, Mn, Ti, Cd, Cr, Co, Fe, and V. The results of both neutron diffraction and Li NMR studies have indicated that the high ionic conductivity observed at elevated temperatures is due to diffusion of the Li^+ species, with those occupying the *tet*1 positions migrating to the empty *oct*1 sites, and implying that the conduction pathways are *tet*1 → *oct*1 → *tet*1 ... [130, 131]. At higher temperatures, many of the

Li_2MCl_4 compounds undergo a phase transition to a disordered cation-deficient rocksalt-structured form (space group $Fm\bar{3}m$), in which the two cation species are randomly distributed over all the *oct* sites, which then have an average occupancy of $^3/_4$ [132, 133].

2.3.3.5 Oxide Spinels: Li_2MnO_4

The compound $LiMn_2O_4$ adopts the "normal" spinel structure, with the Mn on the *oct*2 sites and the Li^+ located at the *tet*1 positions. In common with the "inverse" spinels discussed above, the favored diffusion route for Li^+ involves hops between the *tet*1 and "empty" *oct*1 positions, as shown by computer simulations [134] and ^7Li NMR studies [135]. $LiMn_2O_4$ has been the subject of extensive research interest for potential use within lightweight rechargeable batteries, with Li^+ removed from and inserted into the cathode material during the charge and discharge cycles of the battery, respectively. Charge balance is maintained through this process by changes in the charge of the manganese ions (which have an average valence of $+3.5$ in stoichiometric $LiMn_2O_4$). The Mn_2O_4 framework remains intact as the Li^+ content changes in $Li_xMn_2O_4$, with the cations gradually removed from the *tet*1 sites as the material tends towards the delithiated ($x=0$) compound λ-MnO_2 [136]. Conversely, Li^+ entering $LiMn_2O_4$ occupies the *oct*1 sites, with those Li^+ already present also migrating from the *tet*1 to the *oct*1 positions [136], until the compound reaches a composition $Li_2Mn_2O_4$ with an ordered rocksalt-like structure [137]. Selected data on the electrochemical behavior of these materials and their analogues are presented in Chapter 5.

2.3.4 Perovskite-Structured Compounds

2.3.4.1 The Perovskite Structure

The perovskite crystal structure is adopted by many oxides and fluorides of stoichiometry ABX_3. In many (though not all) cases, ionic conduction occurs via motion of the X anion species, which can be considered as diffusing through a *bcc* sublattice formed by an ordered arrangement of the A and B cations. As illustrated in Figure 2.4, an alternative description considers the smaller B cation species to be located at the center of an octahedron of X anions, with each BX_6 octahedron sharing each of its corners with a neighboring BX_6 octahedron. The A cations then reside in the cavity formed by eight corner-sharing BX_6 octahedra, such that it is surrounded by a cubo-octahedron of 12 anions. As first discussed by Goldschmidt, this arrangement places restrictions on the sizes of the A and B species that can form the cubic perovskite structure [138], which can be quantified using the so-called "tolerance factor", $t_G = (r_A + r_X)/\sqrt{2}(r_B + r_X)$ (where r_i is the ionic radius of the i-th species). The ideal cubic perovskite arrangement (space group $Pm\bar{3}m$) is usually observed for compounds with t_G values between $\sim$0.97 and $\sim$1.03. At lower values of t_G, cooperative rotations of the BX_6 octahedra brings some of the anions closer to the A cation, which lowers the symmetry and allows smaller species to be accommodated [139, 140].

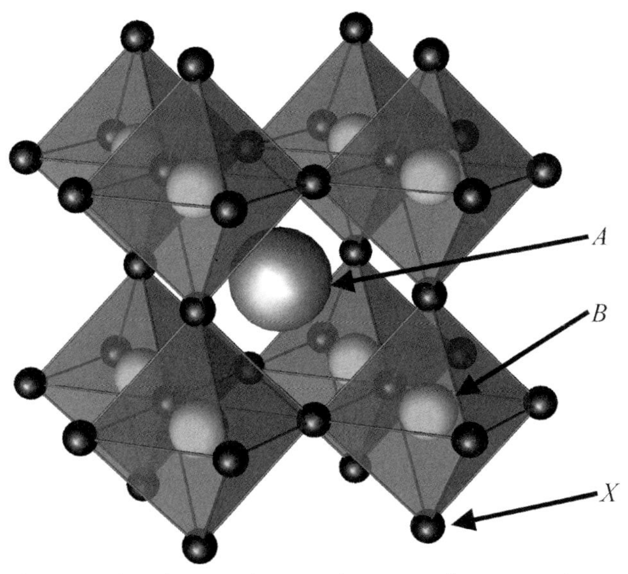

Figure 2.4 The cubic perovskite crystal structure of a compound of stoichiometry ABX_3. The larger A cations sit at the center of eight corner-linked BX_6 octahedra.

2.3.4.2 **Halide Perovskites**

In common with the fluorite-structured superionics discussed in Section 2.3.2, the onset of anion disorder within perovskite-structured compounds generally occurs at lower temperatures within halides than oxides, making the former preferable for experimental study. However, despite an extensive literature devoted to the high-temperature behavior of halide (and principally fluoride) perovskites using neutron diffraction, ionic conductivity measurements, NMR studies and MD simulations, there has been conflicting evidence both for and against high F^- ion conductivity in compounds such as $KMgF_3$, $NaMgF_3$, $KCaF_3$, and $KZnF_3$ (for further discussion and references, see Ref. [38]). As an example, the compound $NaMgF_3$ adopts an orthorhombic variant of the ideal cubic perovskite structure in space group *Pnma*. At around 1170 K it transforms to the ideal cubic $Pm\bar{3}m$ form and, on further heating, shows a rapid increase in ionic conductivity. Whilst the measured ionic conductivity has been reported to reach $\sim$1.3 S cm^{-1} [141] immediately below the melting point of 1303 K, this observation was not supported by subsequent ionic conductivity studies [142] or MD simulations [143].

2.3.4.3 **Cryolite (Na_3AlF_6)**

The cryolite structure is a superstructure of the perovskite arrangement with a doubled unit cell and space group $Fm\bar{3}m$. The ideal stoichiometry is A_3BX_6, such that the A cation species also occupy half of the B (octahedral) sites. The most widely studied example is the mineral cryolite itself, which has the chemical composition Na_3AlF_6 and possesses $P2_1/n$ symmetry at ambient temperature due to slight tilting and deformation of the AlF_6 octahedra [144]. On increasing temperature, Na_3AlF_6 shows abrupt increases in its ionic conductivity at 838 and 1153 K [145], with the former attributed to a structural transition to the cubic ($Fm\bar{3}m$) modification [146, 147] and the latter to a transition to a superionic phase characterized by extensive disorder of the anions [145]. However, single-crystal diffraction [144], NMR [148] and MD studies [149] have shown that this interpretation is incorrect, with the high ionic conductivity due to motion of the Na^+ rather than the F^-. Simulation studies indicated that the AlF_6 octahedra rotate about the central Al^{3+} at temperatures above 600 K, so that the structure can be viewed as an *fcc* array of spinning AlF_6 units. This model appears consistent with recent quasielastic neutron scattering studies of Na_3AlF_6 [150], although a full explanation of the temperature dependence of the ionic conductivity is still lacking.

2.3.4.4 **Oxide Perovskites**

Many perovskite-structured oxides exhibit high oxide-ion conductivities at elevated temperatures, and have attracted significant interest for use as solid electrolytes in, for example, SOFCs (see Chapters 9, 12 and 13). The compounds can be divided into camps with compositions $A^{3+}B^{3+}O_3$ or $A^{2+}B^{4+}O_3$, of which $LaGaO_3$ and $BaCeO_3$ are probably the most widely studied compounds in each group. Both adopt the *Pnma* distorted perovskite arrangement at ambient temperature, and undergo a series of phase transitions on heating, prior to transforming to the undistorted cubic form (see Ref. [151]). In addition, the perovskite-structured $MgSiO_3$ has attracted much attention as it is a major constituent of the Earth's lower mantle, and the presence of anion disorder would have implications for the geophysical behavior of our planet [152, 153].

In general, the ideal cubic perovskite structure (i.e., $t_G \approx 1$) is believed to favor high oxide ion mobility, because all the O^{2-} sites are energetically equivalent. However, other factors, such as the "free volume" available for diffusing anions, have also been shown to be important [154]. Anion diffusion within perovskites tends to occur in $\langle 1\,1\,0 \rangle$ directions along the edges of the *B*-centered octahedra [155]. The O^{2-} must then diffuse through a triangular face formed by two *A* and one *B* cations, although computer simulations have suggested that outward relaxations of both of the *A* and *B* cations would facilitate this motion [156, 157]. The oxide-ion conductivity within perovskites can be enhanced by doping the *A* and/or *B*-site with cation species of a lower valence [158], thus forming anion-deficient phases. The greatest improvements in ionic conductivity are generally achieved using a "radius-matched" strategy, with the dopant cation(s) of similar size to the host(s), in order to minimize any local distortions within the perovskite lattice. Typical examples are $Ba(Ce_{1-x}Y_x)O_{3-x/2}$ [159] and $(La_{1-x}Sr_x)(Ga_{1-y}Mg_y)O_{3-x/2-y/2}$ [160]. In common with the anion-deficient fluorite-structured oxides (see Section 2.3.2.6), the introduction of O^{2-} vacancies initially causes an increase in the ionic conductivity, although a maximum is observed at higher doping levels ($x \sim 0.2$), followed by a reduction at higher x values due to short-range interactions between anion vacancies [154].

2.3.4.5 Brownmillerites ($Ba_2In_2O_5$)

The brownmillerite structure is an orthorhombic superstructure of the cubic perovskite structure, with alternating layers of corner-sharing InO_6 octahedra and InO_4 tetrahedra in the [0 1 0] direction. The brownmillerite-structured compound $Ba_2In_2O_5$ shows high oxide-ion conductivities at elevated temperatures, reaching a value of around 0.1 S cm^{-1} at ~1200 K [161]. The onset of high oxide-ion conductivity is associated with a transition from the ambient temperature orthorhombic phase (space group *Icmm*) to an anion-deficient cubic perovskite structure [162–164], possibly via an intermediate tetragonal phase of symmetry *I4cm* [162, 164, 165]. Computer simulations indicate that an increasing fraction of InO_5 square pyramid entities is favored as the temperature increases [166].

2.3.4.6 BIMEVOXs

The compound Bi_2WO_6 can be considered to be an example of an $n = 1$ Aurivillius phase, formed by an intergrowth of n perovskite-like layers of composition $(WO_4)^{2-}$ and bismuth oxide $(Bi_2O_2)^{2+}$ sheets. The latter consist of edge-shared BiO_4 units, with the Bi^{3+} and four O^{2-} forming the apex and base, respectively, of a square pyramid. Replacing the W^{6+} in Bi_2WO_6 with lower-valence cations, such as V^{5+}, forms the compound $Bi_4V_2O_{11}$, in which the perovskite-like layer has a composition $VO_{3.5}$. At elevated temperatures, $Bi_4V_2O_{11}$ adopts an anion-disordered tetragonal phase (labeled γ) displaying relatively high ionic conductivities at temperatures in the region of 600–700 K [167]. On cooling, the γ phase transforms to poorly conducting phases, due to ordering of the vacancies over the anion sites. However, the high-temperature γ form can be stabilized at ambient temperature by replacing some of the V^{5+} with other metal cations (*Me*), such as Mg, Ni, Cu, and Zn. This generates the so-called BiMeVOx family of compounds [168]. A comprehensive review of the structural properties and ionic conductivities of the many BiMeVOx systems reported within the literature is available [169] (see also Chapter 9).

2.4
Current Status and Future Prospects

The crystal structures of many superionic compounds are now well established, including the nature of the dynamic disorder associated with the diffusion of ions through the crystalline lattice. However, as discussed above, there remain a number of materials in which the structural characterization is more controversial, with a degree of disagreement between the results provided by different experimental techniques. The inherent difficulties in studying systems containing high levels of dynamic disorder is undoubtedly a contributing factor, although significant advances in experimental techniques are currently being made. The latter point is illustrated by the development of new high count rate diffractometers at modern high-intensity X-ray and neutron sources. The ability to collect diffraction data of high statistical quality over a wide range of scattering vectors from such weakly scattering systems has, in turn, encouraged the development of novel data analysis methods tailored to the study of systems containing high concentrations of defects, including maximum entropy approaches (see, e.g., Refs [62, 107]) and Reverse Monte Carlo modeling of the "total" (Bragg plus diffuse) scattering [17, 18]. The availability of highly intense neutron and X-ray beams has also facilitated studies of the effects of hydrostatic pressure, which can provide an important insight into the nature of the ionic conduction mechanisms within different phases of the same material (e.g., Refs [32, 170]).

In parallel with the advances in experimental techniques, the increased availability of affordable computer power has facilitated a number of important developments in computer simulation methods. These include the use of bond valence formalisms to identify probable diffusion pathways [171], and also the so-called "configurational averaging" method, which randomly samples local minima on the potential energy surface and has proved valuable for the study of the disordered structure of several ionically conducting systems [26, 166]. The MD technique, with its ability to study the dynamics of the ions within the simulation, continues to be widely employed, with an increase in the use of *ab initio* methods. However, the computational efficiency provided by parameterized potentials (including those with the coefficients derived from *ab initio* calculations; e.g., Refs [75, 78]) still offer advantages for simulations requiring longer timescales and/or larger simulation boxes.

2.5 Conclusions

Detailed structural descriptions of ionically conducting compounds are essential, both to understand the fundamental nature of structure–property relationships within "model" superionic systems, and to inform strategies aimed at developing new materials with improved performance to meet the ever-increasing demands for applications within fuel cells, solid-state batteries, gas sensor properties, etc. In the former case, the original model of the superionic transition as "premelting", with one of the sublattices effectively molten within a crystalline lattice of immobile counterions, is no longer widely accepted. This is largely a consequence of diffraction studies which have demonstrated that, despite the high levels of dynamic disorder, the distribution of the mobile ions is concentrated at relatively well-defined crystallographic positions. Whilst the structural models provide key insights into the origins of the high ionic conductivity observed in many "model" compounds – including its variation as a function of temperature, pressure or chemical doping – there remains the challenge of repeating this success for more complex materials, many of which offer important technological applications. This includes a number of area in which research interest is currently expanding, including proton-conducting materials [172–174] and the so-called "nanoionics" [175].

References

1 Stout, G.H. and Jensen, L.H. (1989) *X-Ray Structure Determination. A Practical Guide*, John Wiley & Sons, Ltd, New York.
2 Koningsberger, D.C. and Prins, R. (1987) *X-Ray Absorption. Principles, Applications, Techniques of EXAFS, SEXAFS and XANES*, John Wiley & Sons. Ltd.
3 Brinkmann, D. (1992) *Prog. NMR Spectrosc.*, **24**, 527.
4 Chaudhuri, S., Wang, F. and Grey, C.P. (2002) *J. Am. Chem. Soc.*, **124**, 11746.
5 Kim, N. and Grey, C.P. (2002) *Science*, **297**, 1317.
6 Catlow, C.R.A. (1992) *Solid State Ionics*, **53–56**, 955.
7 Lacorre, P., Goutenoire, F., Bohnke, O., Retoux, R. and Laligant, Y. (2000) *Nature*, **404**, 856.
8 Islam, M.S., Tolchard, J.R. and Slater, P.R. (2003) *Chem. Commun.*, 1486.
9 Tubandt, C. and Lorenz, E. (1914) *Z. Phys. Chem.*, **87**, 513.
10 Jost, W. and Weiss, K. (1954) *Z. Phys. Chem. Neue Folge*, **2**, 112.
11 Strock, L.W. (1934) *Z. Phys. Chem. B*, **25**, 441.
12 Strock, L.W. (1936) *Z. Phys. Chem. B*, **31**, 132.
13 Bührer, W. and Hälg, W. (1974) *Helv. Phys. Acta*, **47**, 27.
14 Cava, R.J., Reidinger, F. and Wuensch, B.J. (1977) *Solid State Commun.*, **24**, 411.
15 Tsuchiya, Y., Tamaki, S. and Waseda, Y. (1979) *J. Phys. C: Solid State Phys.*, **12**, 5361.
16 Wright, A.F. and Fender, B.E.F. (1977) *J. Phys. C: Solid State Phys.*, **10**, 2261.
17 McGreevy, R.L. (2001) *J. Phys.: Condens. Matter*, **13**, R877.
18 Tucker, M.G., Keen, D.A., Dove, M.T., Goodwin, A.L. and Hui, Q. (2007) *J. Phys.: Condens. Matter*, **19**, 335218.
19 Nield, V.M., Keen, D.A., Hayes, W. and McGreevy, R.L. (1993) *Solid State Ionics*, **66**, 247.
20 Keen, D.A. and Hull, S. (1994) *J. Phys.: Condens. Matter*, **6**, 1637.
21 Bührer, W. and Hälg, W. (1977) *Electrochim. Acta*, **22**, 701.
22 Miyake, S., Hoshino, S. and Takenaka, T. (1952) *J. Phys. Soc. Japan*, **7**, 19.
23 Krug, J. and Sieg, L. (1952) *Z. Naturforsch.*, **7**, 369.
24 Chahid, A. and McGreevy, R.L. (1998) *J. Phys.: Condens. Matter*, **10**, 2597.
25 Keen, D.A. and Hull, S. (1995) *J. Phys.: Condens. Matter*, **7**, 5793.
26 Mohn, C.E. and Stølen, S. (2007) *J. Phys.: Condens. Matter*, **19**, article 466208.
27 Zheng-Johansson, J.X.M., Ebbsjö, I. and McGreevy, R.L. (1995) *Solid State Ionics*, **82**, 115.
28 Zheng-Johansson, J.X.M. and McGreevy, R.L. (1996) *Solid State Ionics*, **83**, 35.
29 Trapananti, A., di Cicco, A. and Minicucci, M. (2002) *Phys. Rev. B*, **66**, article 014202.
30 Nield, V.M., Keen, D.A., Hayes, W. and McGreevy, R.L. (1992) *J. Phys.: Condens. Matter*, **4**, 6703.
31 Andreoni, W. and Tosi, M.P. (1983) *Solid State Ionics*, **11**, 49.
32 Keen, D.A., Hull, S., Hayes, W. and Gardner, N.J.G. (1996) *Phys. Rev. Lett.*, **77**, 4914.
33 Mellander, B.-E. (1982) *Phys. Rev. B*, **26**, 5886.
34 Graneli, B., Dahlborg, U. and Fischer, P. (1988) *Solid State Ionics*, **28–30**, 284.
35 Boyce, J.B., Hayes, T.M. and Mikkelsen, J.C. Jr., (1981) *Phys. Rev. B*, **23**, 2876.
36 Nield, V.M., McGreevy, R.L., Keen, D.A. and Hayes, W. (1994) *Physica B*, **202**, 159.
37 Miyatani, S. (1981) *J. Phys. Soc. Japan*, **50**, 3415.
38 Hull, S. (2004) *Rep. Prog. Phys.*, **67**, 1233.
39 Wuensch, B.J. (1993) *Mater. Sci. Eng.*, **B18**, 186.
40 Cava, R.J. and McWhan, D.B. (1980) *Phys. Rev. Lett.*, **45**, 2046.

41 Cava, R.J., Reidinger, F. and Wuensch, B.J. (1980) *J. Solid State Chem.*, **31**, 69.
42 Hull, S., Keen, D.A., Sivia, D.S., Madden, P.A. and Wilson, M. (2002) *J. Phys.: Condens. Matter*, **14**, L9.
43 Keen, D.A. and Hull, S. (1998) *J. Phys.: Condens. Matter*, **10**, 8217.
44 Grier, B.H., Shapiro, S.M. and Cava, R.J. (1984) *Phys. Rev. B*, **29**, 3810.
45 Boyce, J.B. and Huberman, B.A. (1979) *Phys. Rep.*, **51**, 189.
46 Gray, J.N. and Clarke, R. (1986) *Phys. Rev. B*, **33**, 2056.
47 Oliveria, M., McMullen, R.K. and Wuensch, B.J. (1988) *Solid State Ionics*, **28–30**, 1332.
48 Kobayashi, M., Ishikawa, K., Tachibana, F. and Okazaki, H. (1988) *Phys. Rev. B*, **38**, 3050.
49 Boyce, J.B., Hayes, T.M. and Mikkelsen, J.C. (1981) *Solid State Ionics*, **5**, 497.
50 Asadov, Y.G., Rustamova, L.V., Gasimov, G.B., Jafarov, K.M. and Babajev, A.G. (1992) *Phase Transitions*, **38**, 247.
51 Vouroutzis, N. and Manolikas, C. (1989) *Phys. Status Solidi A*, **111**, 491.
52 Reuter, B. and Hardel, K. (1965) *Z. Anorg. Allg. Chem.*, **340**, 158.
53 Reuter, B. and Hardel, K. (1965) *Z. Anorg. Allg. Chem.*, **340**, 168.
54 Reuter, B. and Hardel, K. (1966) *Ber. Bunsen Ges.*, **70**, 82.
55 Chiodelli, G., Magistris, A. and Schiraldi, A. (1979) *Z. Phys. Chem. Neue Folge*, **118**, 177.
56 Hull, S., Keen, D.A., Gardner, N.J.G. and Hayes, W. (2001) *J. Phys.: Condens. Matter*, **13**, 2295.
57 Hull, S., Keen, D.A., Madden, P.A. and Wilson, M. (2007) *J. Phys.: Condens. Matter*, **19**, article 406214.
58 Hull, S., Keen, D.A. and Berastegui, P. (2002) *J. Phys.: Condens. Matter*, **14**, 13579.
59 Hull, S. and Keen, D.A. (2001) *J. Phys.: Condens. Matter*, **13**, 5597.
60 Hull, S. and Keen, D.A. (2000) *J. Phys.: Condens. Matter*, **12**, 3751.
61 Hull, S., Keen, D.A. and Berastegui, P. (2002) *Solid State Ionics*, **147**, 97.
62 Hull, S., Keen, D.A., Sivia, D.S. and Berastegui, P. (2002) *J. Solid State Chem.*, **165**, 363.
63 Bradley, J.N. and Greene, P.D. (1967) *Trans. Faraday Soc.*, **63**, 2516.
64 Geller, S. (1967) *Science*, **157**, 310.
65 Kuhs, W.F., Nitsche, R. and Scheunemann, K. (1979) *Mater. Res. Bull.*, **14**, 241.
66 Kuhs, W.F., Nitsche, R. and Scheunemann, K. (1978) *Acta Crystallogr.*, B34, 64.
67 Benz, R. (1975) *Z. Phys. Chem.*, **95**, 25.
68 Carr, V.M., Chadwick, A.V. and Saghafian, R. (1978) *J. Phys. C: Solid State Phys.*, **11**, L637.
69 Derrington, C.E. and O'Keeffe, M. (1973) *Nature Phys. Sci.*, **246**, 44.
70 Hayes, W. (1978) *Contemp. Phys.*, **19**, 469.
71 Lidiard, A.B. (1974) in *Crystals with the Fluorite Structure* (ed. W. Hayes), Clarendon Press, Oxford, p. 101.
72 Dickens, M.H., Hayes, W., Hutchings, M.T. and Smith, C. (1982) *J. Phys. C: Solid State Phys.*, **15**, 4043.
73 Hutchings, M.T., Clausen, K., Dickens, M.H., Hayes, W., Kjems, J.K., Schnabel, P.G. and Smith, C. (1984) *J. Phys. C: Solid State Phys.*, **17**, 3903.
74 Catlow, C.R.A. and Hayes, W. (1982) *J. Phys. C: Solid State Phys.*, **15**, L9.
75 Dent, A., Madden, P.A. and Wilson, M. (2004) *Solid State Ionics*, **167**, 73.
76 Réau, J.-M., Lucat, C., Portier, J., Hagenmuller, P., Cot, L. and Vilminot, S. (1978) *Mater. Res. Bull.*, **13**, 877.
77 Kanno, R., Ohno, K., Izumi, H., Kawamoto, Y., Kamiyama, T., Asano, H. and Izumi, F. (1994) *Solid State Ionics*, **70–71**, 253.
78 Castiglione, M., Madden, P.A., Berastegui, P. and Hull, S. (2005) *J. Phys.: Condens. Matter*, **17**, 845.
79 Corish, J., Catlow, C.R.A., Jacobs, P.W.M. and Ong, S.H. (1982) *Phys. Rev. B*, **25**, 6425.
80 Ivanov-Shits, A.K., Sorokin, N.I., Federov, P.P. and Sobolev, B.P. (1990) *Solid State Ionics*, **37**, 125.

81 Cheetham, A.K., Fender, B.E.F. and Cooper, M.J. (1971) *J. Phys. C: Solid State Phys.*, **4**, 3107.

82 Steele, D., Childs, P.E. and Fender, B.E.F. (1972) *J. Phys. C: Solid State Phys.*, **5**, 2677.

83 Hull, S. and Wilson, C.C. (1992) *J. Solid State Chem.*, **100**, 101.

84 Wang, F. and Grey, C.P. (1998) *Chem. Mater.*, **10**, 3081.

85 Catlow, C.R.A., Chadwick, A.V., Greaves, G.N. and Moroney, L.M. (1984) *Nature*, **312**, 601.

86 Bendall, P.J., Catlow, C.R.A., Corish, J. and Jacobs, P.W.M. (1984) *J. Solid State Chem.*, **51**, 159.

87 Ralph, J. and Hyland, G.J. (1985) *J. Nucl. Mater.*, **132**, 76.

88 MacInnes, D.A. (1978) *J. Nucl. Mater.*, **78**, 225.

89 Winter, P.W. and MacInnes, D.A. (1986) *J. Nucl. Mater.*, **137**, 161.

90 Fischer, D.F., Fink, J.L. and Liebowitz, L. (1981) *J. Nucl. Mater.*, **102**, 220.

91 Clausen, K., Hayes, W., Macdonald, J.E., Osborn, R. and Hutchings, M.T. (1984) *Phys. Rev. Lett.*, **52**, 1238.

92 Hutchings, M.T. (1987) *J. Chem. Soc.: Faraday Trans. II*, **83**, 1083.

93 Goel, P., Choudhury, N. and Chaplot, S.L. (2007) *J. Phys.: Condens. Matter*, **19**, article 386239.

94 Ishihara, T., Sammes, N.M. and Yamamoto, O. (2003) in *High Temperature Solid Oxide Fuel Cells. Fundamentals, Design and Applications* (eds S.C. Singhal and K. Kendall), Elsevier, Oxford.

95 Etsell, T.H. and Flengas, S.N. (1970) *Chem. Rev.*, **70**, 339.

96 Scott, H.G. (1975) *J. Mater. Sci.*, **10**, 1527.

97 Subbarao, E.C. and Ramakrishnan, T.V. (1979) in *Fast Ion Transport in Solids* (eds P. Vashishta, J.N. Mundy and G.K. Shenoy), Elsevier North Holland Inc., New York, p. 653.

98 Welberry, T.R., Butler, B.D., Thompson, J.G. and Withers, R.L. (1993) *J. Solid State Chem.*, **106**, 461.

99 Argyriou, D.M., Elcombe, M.M. and Larson, A.C. (1996) *J. Phys. Chem. Solids*, **57**, 183.

100 Goff, J.P., Hayes, W., Hull, S., Hutchings, M.T. and Clausen, K.N. (1999) *Phys. Rev. B*, **59**, 14202.

101 Steele, B.C.H. (2000) *Solid State Ionics*, **129**, 95.

102 Harwig, H.A. and Gerards, A.G. (1978) *J. Solid State Chem.*, **26**, 265.

103 Gattow, G. and Schroder, H. (1962) *Z. Anorg. Allg. Chem.*, **318**, 176.

104 Sillen, L.G. (1937) *Ark. Kemi. Mineral. Geol.*, **12**, 1.

105 Willis, B.T.M. (1965) *Acta Crystallogr.*, **18**, 75.

106 Battle, P.D., Catlow, C.R.A., Drennan, J. and Murray, A.D. (1983) *J. Phys. C: Solid State Phys.*, **16**, L561.

107 Yashima, M. and Ishimura, D. (2003) *Chem. Phys. Lett.*, **378**, 395.

108 Carlsson, J.M., Hellsing, B., Domingos, H.S. and Bristowe, P.D. (2002) *Phys. Rev. B*, **65**, 205122.

109 Walsh, A., Watson, G.W., Payne, D.J., Edgell, R.G., Guo, J.H., Glans, P.A., Learmonth, T. and Smith, K.E. (2006) *Phys. Rev. B*, **73**, article 235104.

110 Azad, A.M., Larose, S. and Akbar, S.A. (1994) *J. Mater. Sci.*, **29**, 4135.

111 Sammes, N.M., Tompsett, G.A., Näfe, H. and Aldinger, F. (1999) *J. Eur. Ceram. Soc.*, **19**, 1801.

112 Shuk, P., Wienhöfer, H.-D., Guth, U., Göpel, W. and Greenblatt, M. (1996) *Solid State Ionics*, **89**, 179.

113 Chadwick, A.V., Flack, F.W., Strange, J.H. and Harding, J. (1988) *Solid State Ionics*, **28–30**, 185.

114 Strange, J.H., Rageb, S.M., Chadwick, A.V., Flack, K.W. and Harding, J.H. (1990) *J. Chem. Soc.: Faraday Trans.*, **86**, 1239.

115 Farley, T.W.D., Hayes, W., Hull, S., Hutchings, M.T. and Vrtis, M. (1991) *J. Phys.: Condens. Matter*, **3**, 4761.

116 Kvist, A. and Lundén, A. (1965) *Z. Naturforsch.*, **20**, 235.

117 Kaber, R., Nilsson, L., Andersen, N.H., Lundén, A. and Thomas, J.O. (1992) *J. Phys.: Condens. Matter*, **4**, 1925.

118 Lundén, A. (1995) *Z. Naturforsch.*, **50**, 1067.

119 Ferrario, M., Klein, M.L. and McDonald, I.R. (1995) *Mol. Phys.*, **86**, 923.

120 Gundusharma, U.M., MacLean, C. and Secco, E.A. (1986) *Solid State Commun.*, **57**, 479.

121 Michel, D., Perez y Jorba, M. and Collongues, R. (1974) *Mater. Res. Bull.*, **9**, 1457.

122 van Dijk, M.P., de Vries, K.J. and Burggraaf, A.J. (1983) *Solid State Ionics*, **9–10**, 913.

123 Moriga, T., Yoshiasa, A., Kanamaru, F., Koto, K., Yoshimura, M. and Somiya, S. (1989) *Solid State Ionics*, **31**, 319.

124 van Dijk, M.P., Burggraaf, A.J., Cormack, A.N. and Catlow, C.R.A. (1985) *Solid State Ionics*, **17**, 159.

125 Wilde, P.J. and Catlow, C.R.A. (1998) *Solid State Ionics*, **112**, 173.

126 Wilde, P.J. and Catlow, C.R.A. (1998) *Solid State Ionics*, **112**, 185.

127 Glerup, M., Nielsen, O.F. and Poulsen, F.W. (2001) *J. Solid State Chem.*, **160**, 25.

128 Heremans, C., Wuensch, B.J., Stalick, J.K. and Prince, E. (1995) *J. Solid State Chem.*, **117**, 108.

129 Wuensch, B.J., Eberman, K.W., Heremans, C., Ku, E.M., Onnerud, P., Yeo, E.M.E., Haile, S.M., Stalick, J.K. and Jorgensen, J.D. (2000) *Solid State Ionics*, **129**, 111.

130 Soubeyroux, J.L., Cros, C., Gang, W., Kanno, R. and Pouchard, M. (1985) *Solid State Ionics*, **15**, 293.

131 Nagel, R., Groß, T.W., Günther, H. and Lutz, H.D. (2002) *J. Solid State Chem.*, **165**, 303.

132 Kanno, R., Takeda, Y. and Yamamoto, O. (1988) *Solid State Ionics*, **28–30**, 1276.

133 Steiner, H.J. and Lutz, H.D. (1992) *J. Solid State Chem.*, **99**, 1.

134 Ammundsen, B., Rozière, J. and Islam, M.S. (1997) *J. Phys. Chem. B*, **101**, 8156.

135 Verhoeven, V.W.J., de Schepper, I.M., Nachtegaal, G., Kentgens, A.P.M., Kelder, E.M., Shoonman, J. and Mulder, F.M. (2001) *Phys. Rev. Lett.*, **86**, 4314.

136 David, W.I.F., Thackeray, M.M., de Picciotto, L.A. and Goodenough, J.B. (1987) *J. Solid State Chem.*, **67**, 316.

137 Goodenough, J.B., Thackeray, M.M., David, W.I.F. and Bruce, P.G. (1984) *Rev. Chim. Minér.*, **21**, 435.

138 Goldschmidt, V.M. (1926) *Naturwissenschaften*, **14**, 477.

139 Glazer, A.M. (1972) *Acta Crystallogr.*, **B28**, 3384.

140 Howard, C.J. and Stokes, H.T. (1998) *Acta Crystallogr.*, B54, 782.

141 O'Keeffe, M. and Bovin, J.-O. (1979) *Science*, **206**, 599.

142 Andersen, N.H., Kjems, J.K. and Hayes, W. (1985) *Solid State Ionics*, **17**, 143.

143 Cheeseman, P.A. and Angell, C.A. (1981) *Solid State Ionics*, **5**, 597.

144 Yang, H., Ghose, S. and Hatch, D.M. (1993) *Phys. Chem. Minerals*, **19**, 528.

145 Landon, G.J. and Ubbelohde, A.R. (1957) *Proc. R. Soc. Lond. A*, **240**, 160.

146 Holm, J.L. (1965) *Acta Chim. Scand.*, **19**, 261.

147 Steward, E.G. and Rooksby, H.P. (1953) *Acta Crystallogr.*, **6**, 49.

148 Spearing, D.R., Stebbins, J.F. and Farnan, I. (1994) *Phys. Chem. Minerals*, **21**, 373.

149 Foy, L. and Madden, P.A. (2006) *J. Phys. Chem. B*, **110**, 15302.

150 Jahn, S., Olliver, J. and Demmel, F. (2008) *Solid State Ionics*, **179**, 1957.

151 Knight, K.S. (1994) *Solid State Ionics*, **74**, 109.

152 Cahn, R.W. (1984) *Nature*, **308**, 493.

153 Matsui, M. and Price, G.D. (1991) *Nature*, **351**, 735.

154 Hayashi, H., Inaba, H., Matsuyama, M., Lan, N.G., Dokiya, M. and Tagawa, H. (1999) *Solid State Ionics*, **122**, 1.

155 Kilner, J.A. (1981) *Philos. Mag. A*, **43**, 1473.

156 Cherry, M., Islam, M.S. and Catlow, C.R.A. (1995) *J. Solid State Chem.*, **118**, 125.

157 Islam, M.S. (2002) *Solid State Ionics*, **154–155**, 75.

158 Iwahara, H. (1992) *Solid State Ionics*, **52**, 99.

159 Bonanos, N., Knight, K.S. and Ellis, B. (1995) *Solid State Ionics*, **79**, 161.

160 Ishihara, T., Matsuda, H. and Takita, Y. (1994) *J. Am. Chem. Soc.*, **116**, 3801.

161 Goodenough, J.B., Ruiz-Diaz, J.E. and Zhen, Y.S. (1990) *Solid State Ionics*, **44**, 21.

162 Adler, S.B., Reimer, J.A., Baltisberger, J. and Werner, U. (1994) *J. Am. Ceram. Soc.*, **116**, 675.

163 Berastegui, P., Hull, S., García-García, F.J. and Eriksson, S.-G. (2002) *J. Solid State Chem.*, **164**, 119.

164 Speakman, S.A., Richardson, J.W., Mitchell, B.J. and Misture, S.T. (2002) *Solid State Ionics*, **149**, 247.

165 Hashimoto, T., Ueda, Y., Yoshinaga, M., Komazaki, K., Asaoka, K. and Wang, S.R. (2002) *J. Electrochem. Soc.*, **149**, A1381.

166 Mohn, C.E., Allan, N.L., Freeman, C.L., Ravindran, P. and Stølen, S. (2005) *J. Solid State Chem.*, **178**, 346.

167 Abraham, F., Debreuille-Gresse, M.F., Mairesse, G. and Nowogrocki, G. (1988) *Solid State Ionics*, **28–30**, 529.

168 Abraham, F., Boivin, J.C., Mairesse, G. and Nowogrocki, G. (1990) *Solid State Ionics*, **40–41**, 934.

169 Abrahams, I. and Krok, F. (2002) *J. Mater. Chem.*, **12**, 3351.

170 Parfitt, D.C., Keen, D.A., Hull, S., Crichton, W.A., Mezouar, M., Wilson, M. and Madden, P.A. (2005) *Phys. Rev. B*, **72**, 054121.

171 Adams, S. (2000) *Solid State Ionics*, **136–137**, 1351.

172 Iwahara, H. (1996) *Solid State Ionics*, **86–88**, 9.

173 Kreuer, K.D. (1996) *Chem. Mater.*, **8**, 610.

174 Nowick, A.S. and Du, Y. (1995) *Solid State Ionics*, **77**, 137.

175 Maier, J. (2005) *Nature Materials*, **4**, 805.

3
Defect Equilibria in Solids and Related Properties: An Introduction

Vladimir A. Cherepanov, Alexander N. Petrov[1]*, and Andrey Yu. Zuev*

Editorial Preface

Vladislav Kharton

Continuing the consideration of methodological aspects related to the analysis of electrochemical processes in solids, in this chapter is presented a brief overview of basic defect chemistry mechanisms and relationships between the defect thermodynamics and transport. The conceptual theoretical elements and fundamental relationships discussed here were established in classical reports made between the 1920s and 1960s (see, in particular, Refs. [1–14] and references cited therein). Progress during the following decades resulted in a deeper understanding of many particular systems and mechanisms involving defect chemistry, and in the development of a variety of advanced methods for their experimental study and simulations. It is commonly recognized that the simplest, limiting cases of defect reactions are rather exceptional; the description of defect formation- and diffusion-related processes in real systems often requires one to account for both short- and long-range defect interactions, the effects of the electronic subsystem, and microstructural and interfacial phenomena. These factors all lead to serious deviations from the hypothetic idealized situations. Nevertheless, approximate approaches still provide very useful tools, both for the researchers developing new materials and electrochemical systems, and for the newcomers to this scientific area. In this chapter are summarized some important definitions and formulae necessary to understand the solid-state electrochemical processes and research methodology; the discussion is based on examples relevant to numerous practical applications, such as fuel cells, batteries, and sensors.

1) Professor Alexander N. Petrov died on October 17th, 2008, shortly after the preparation of this chapter.

Solid State Electrochemistry I: Fundamentals, Materials and their Applications. Edited by Vladislav V. Kharton

ISBN: 978-3-527-32318-0

3.1 Introduction

One key feature of crystalline solids is the presence of "long-range order". This means that atoms (ions, molecules) are located at specific positions that can be reproduced by translation towards certain directions on a constant periodic distance. In such a way, the model of an ideal crystal can be reconstructed. However, this ideal state can be realized only at 0 K, when all atoms possess their lowest oscillation energy $\varepsilon^0 = h\nu_0/2$. Due to fluctuations at higher temperatures, any single atom in the lattice can obtain sufficient energy to leave its site, which will lead to defect formation. This process is energetically favorable even from the statistical thermodynamic point of view; it can be shown that the Gibbs energy for the formation of imperfect crystal at $T > 0$ K is negative [9, 15]. Another source of irregularities relates to the crystal growth conditions.

3.2 Defect Structure of Solids: Thermodynamic Approach

3.2.1 Selected Definitions, Classification, and Notation of Defects

The defects of crystal structure – that is, all types of irregularity of an ideal crystal lattice – can be classified based on their size and/or number of dimensions [9, 15–19]:

- Point defects (zero-dimensional defects) indicate a fault within one site of the crystal lattice. This can be either a vacant site (called a *vacancy*), or an atom (ion) in the *interstitial* site. The impurity species can be also treated as point defects. Often, electronic defects are also ascribed to this group, especially if the electronic charge carriers are localized.
- Line (one-dimensional) defects or dislocations correspond to deviations in atom lines packing. Two extreme cases of dislocations are well-known, namely "edge" and "screw". An edge dislocation can be represented as a half-plane of atoms that is lost at some line within the crystal, as shown schematically in Figure 3.1. The center of region, where the extra half-plane terminates, is the so-called "dislocation line". A screw dislocation (Figure 3.2) can be depicted if the block of an ideal crystal were to be partially cut along one of the atomic planes. Both parts would then be moved in opposite directions parallel to the cut plane, thus transforming successive planes into a helical surface [20, 21].
- Plane (two-dimensional) defects indicate a minor mismatch of neighboring planes that divide crystals into different domains. An appearance of such defects leads to the formation of a mosaic or domain structure.
- Pores or cavities (three-dimensional defects) represent microcracks, holes or "bubbles" of gaseous phase included within the crystal, and other impurity phases.

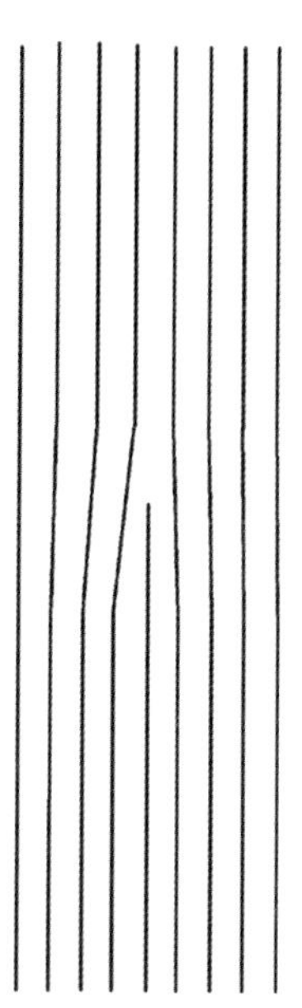

Figure 3.1 Projection of edge dislocation.

Only the point defects are thermodynamically equilibrium defects. All others, which sometimes are referred to as "biographical defects", depend on the prehistory of each sample – that is, on the method and conditions of preparation, heat treatments, and so on. Therefore, a thermodynamic approach can be applied basically for the point defects.

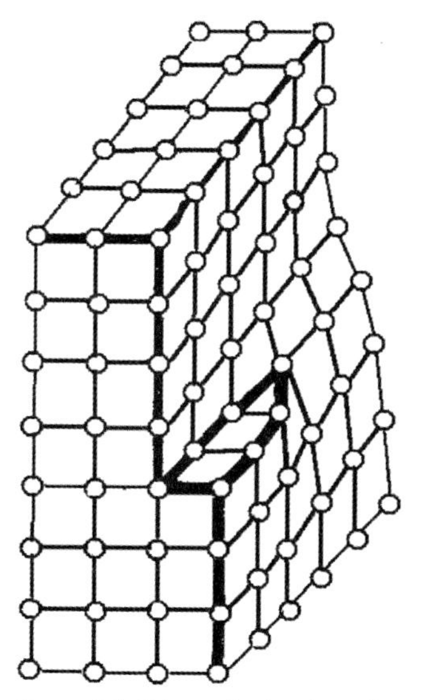

Figure 3.2 Schematic drawing of a screw dislocation.

In order to specify point defects, Kröger and Vink notation [9] is often used in the literature. In this notation, a vacant regular site of the crystal lattice (vacancy) is denoted as $V_A^{\bullet}$, $V_B^{\prime}$, or $V_M^{\times}$. Here, the subscript denotes the chemical identity of the sublattice where the vacancy is formed, while the superscript denotes an effective charge of the vacancy with respect to the ideal structure (point = unit positive charge; dash = negative charge; cross = neutral). Occasionally, the vacant sites may also be indicated by other symbols, such as square box (□). An *interstitial atom* (an atom located at an irregular position in the crystal lattice) is denoted by its chemical symbol, followed by the subscript "*i*", with its effective charge in the superscript. For example, a positively charged interstitial lithium cation is denoted as $Li_i^{\bullet}$.

3.2.2 Defect Formation and Equilibria

Normally, the point defects are expected to be in a local or global equilibrium state when the thermodynamic approach can be used. Within the framework of this approach, the defects and their simplest associates are treated as chemical species [9, 15–19]. Therefore, the chemical potential of each structural element (μ_i), which may correspond to atoms (ions) in their regular positions or defects, and the Gibbs energy change for any process involving the *i*-type species (ΔG), can be written as

$$\mu_i = \mu_i^o + kT\ln[i], \tag{3.1}$$

$$(\Delta G)_{p,T} = (\Sigma \nu_i \mu_i). \tag{3.2}$$

where ν_i are the stoichiometric coefficients, the term in square brackets corresponds to the species concentration, the activity coefficient is omitted for clarity, and μ_i^0 is the chemical potential of *i*-species at the standard conditions (in this case it assumes the concentration of defects equal to unity $[i] = 1$). The processes of the defect formation and disappearance can be presented in a form of *quasi-chemical* reactions and corresponding equilibrium constants (K):

$$(\Delta G^{\circ})_{p,T} = -RT\ln(K) \tag{3.3}$$

The rules for *quasi-chemical* reactions are the same as for the normal chemical reactions, namely mass balance and electroneutrality conditions; one extra requirement appears, however, for crystalline solids where the ratio of sites in the crystal structure should be constant and should satisfy to stoichiometric formula. This means that if, for instance, in the AB_2 crystal one site for A atom is formed, then automatically two B-sites appear as well, regardless of their occupancy. It should be noted that the point defects and/or the processes of their formation can also classified into two groups, namely stoichiometric and nonstoichiometric. The first type of process does not disturb the stoichiometric ratio of components constituting the crystal, which is a closed thermodynamic system; the second type leads to nonstoichiometric compounds by exchanging components between the crystal and its environment, while the crystal constitutes an open system. Let us now consider different possibilities for disordering processes in the hypothetical AB_2 crystal.

3.2.3
Formation of Stoichiometric (Inherent) Defects

3.2.3.1 Schottky Defects

The Shottky disordering mechanism presumes that atoms leave their sites in the crystal bulk and rebuild the crystal lattice on the surface. As a result, the vacancies form in both cation and anion sublattices. For example, in AB_2 crystal

$$A_A^{\times} + 2B_B^{\times} = V_A^{\times} + 2V_B^{\times} + A_A^{\times(\text{surf})} + 2B_B^{\times(\text{surf})}. \tag{3.4}$$

If it is assumed that the activities of species in their regular sites are close to unity, and that the ideal crystal without defects is initially denoted as *nil*, then Equation (3.4) can be simplified to the following form:

$$nil = V_A^{\times} + 2V_B^{\times}. \tag{3.5}$$

The equilibrium constant for this process can be written as follows:

$$K_{\text{Sh}} = [V_A][V_B]^2 = K_{\text{Sh}}^{\circ} \cdot \exp\left(-\frac{\Delta H_{\text{Sh}}^{o}}{kT}\right), \tag{3.6}$$

where $\Delta H_{\text{Sh}}^{\circ}$ is the enthalpy of Shottky defect formation. This type of defect is predominant in alkali halides.

3.2.3.2 Frenkel Defects

According to the Frenkel mechanism, an atom moves from its regular site into the nearest interstitial position; hence, two types of defect are formed in the crystal, namely the vacancy and the interstitial atom. If the Frenkel defects form in the A-sublattice of AB_2 crystal, this process and the corresponding equilibrium constant can be written as:

$$A_A^{\times} = V_A^{\times} + A_i^{\times}, \tag{3.7}$$

$$K_F = \left[V_A^{\times}\right]\left[A_i^{\times}\right] = K_F^{o} \cdot \exp\left(-\frac{\Delta H_F^{o}}{kT}\right). \tag{3.8}$$

As an example, Frenkel defects are formed in AgCl, with silver atoms displaced into the interstitial positions.

It should be noted that neither Schottky nor Frenkel disordering processes affect the stoichiometry of crystals.

3.2.3.3 Intrinsic Electronic Disordering

Within the framework of standard band approach, thermally activated electrons can jump from the valence band via the band-gap towards the conduction band.

Consequently, a free electron in the conduction band and an itinerant hole in the valence band appear simultaneously:

$$nil = e' + h^{\bullet}, \tag{3.9}$$

$$K_i = n \cdot p = K_i^o \exp\left(-\frac{E_i}{kT}\right). \tag{3.10}$$

Alternatively, the process of intrinsic electronic disordering can be represented within the framework of localized electrons model, for example:

$$2A_A^{\times} = A'_A + A_A^{\bullet} \quad \text{or} \quad A_A^{\times} + B_A^{\times} = A'_A + B_A^{\bullet} \tag{3.11}$$

3.2.3.4 Ionization of Defects

The effectively neutral atomic defects may trap or donate electrons, thus acquiring a charge, the sign of which will depend on the chemical nature of the defects and their surroundings. The following quasi-chemical reactions provide examples of the point-defect ionization processes:

$$M_i^{\times} = M_i^{\bullet\bullet} + 2e', \qquad K_{ion} = \left[M_i^{\bullet\bullet}\right] \cdot n^2 \left[M_i^{\times}\right]^{-1} = K_{ion}^o \exp\left(-\frac{E_a}{kT}\right) \tag{3.12}$$

$$V_M^{\times} = V_M^{//} + 2h^{\bullet}, \qquad K_{ion} = \left[V_M^{//}\right] \cdot p^2 \left[V_M^{\times}\right]^{-1} = K_{ion}^o \exp\left(-\frac{E_\beta}{kT}\right) \tag{3.13}$$

The possible ionization of the Schottky-type defects in NaCl, given by

$$V_{Na}^{\times} + V_{Cl}^{\times} = V'_{Na} + V_{Cl}^{\bullet} \tag{3.14}$$

can be explained in terms of the ionic structure of this crystal. As the NaCl crystal is built from the ions, the empty site does not compensate a cumulative effective charge of the surrounding ions, and hence this site or vacancy must possess an effective charge which is opposite to the charge of the absent cation or anion. In terms of the band theory, the examples presented above can be explained by the schemes shown in Figure 3.3a and b.

3.2.4 Influence of Temperature

In order to exclude the influence of gaseous phase at this stage, it is essential to take into consideration a simple example (e.g., silicon) as a semiconducting material with vacancies. Assuming that possible defects of the crystal structure are vacancies and electron defects, the processes of intrinsic electronic disordering, vacancy formation, and vacancy ionization can be written, respectively, as:

$$nil = e' + h^{\bullet}, \qquad K_1 = n \cdot p = K_1^o \exp\left(-\frac{\Delta H_1}{kT}\right), \tag{3.15}$$

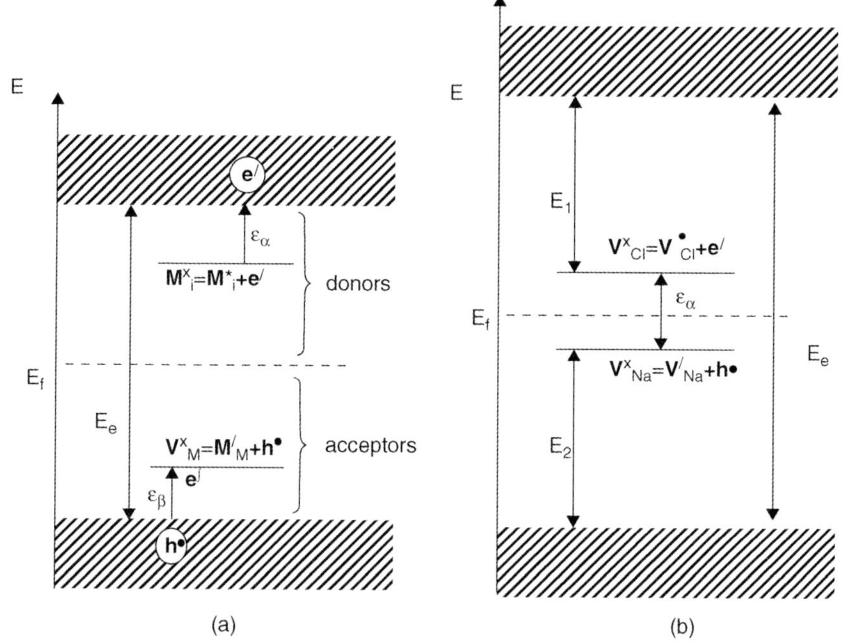

Figure 3.3 Examples of energy level diagrams for defect ionization. (a) Frenkel defects; (b) Schottky defects.

$$nil = V_{Si}^{\times}, \qquad K_2 = \left[V_{Si}^{\times}\right] = K_2^o \exp\left(-\frac{\Delta H_2}{kT}\right), \tag{3.16}$$

$$V_{Si}^{\times} = V_{Si}^{/} + h^{\bullet}, \qquad K_3 = \frac{[V_{Si}'] \cdot p}{[V_{Si}^{\times}]} = K_3^o \exp\left(-\frac{\Delta H_3}{kT}\right). \tag{3.17}$$

The concentrations of the charged defects are linked by the electroneutrality condition:

$$n + [V_{Si}^{/}] = p \tag{3.18}$$

This set of equations, including the expressions for equilibrium constants (Equations (3.15–3.17)) and the electroneutrality condition (Equation (3.18)), can be solved with respect to the concentrations of all defects considered:

$$p = \sqrt{K_1 + K_3 K_3}, \tag{3.19}$$

$$n = \frac{K_1}{\sqrt{K_1 + K_2 K_3}}, \tag{3.20}$$

$$\left[V_{Si}^{/}\right] = \frac{K_2 K_3}{\sqrt{K_1 + K_2 K_3}}. \tag{3.21}$$

It is worth noting here that the exact solution of a set of nonlinear equations for more complicated equilibria is often unachievable. In such cases, the approximation method implying a simplification of the overall electroneutrality condition using the only pair of predominant defects can be useful. This approach can be illustrated on the basis of the above example of a Si crystal. As the equilibrium constants (Equations (3.15–3.17)) are functions of temperature, the concentrations of different defects can alter in different ways, depending on the value of the pre-exponential factor K_i° and the enthalpy of the defects reaction, ΔH_i. As a result, it is possible to choose a temperature range where the overall electroneutrality condition (Equation (3.18)) can be approximated by pairing the predominant defects. In this case, two possible approximations can be suggested:

$$n = p \tag{3.22}$$

and

$$[V'_{\mathrm{Si}}] = p \tag{3.23}$$

Let us assume that $\Delta H_2 > \Delta H_1 > \Delta H_3$, as the energy of atomic disorder is normally larger than that needed for intrinsic electronic disordering; in turn, the latter is higher with respect to the energetic effects related to defect charging (see Figures 3.3 and 3.4). Under this assumption, it is possible to consider two alternative approximations.

- *Approximation 1:* $K_1 >> K_3$; consequently, the overall electroneutrality condition (Equation 3.18) can be written in the form of Equation 3.22. This approximation

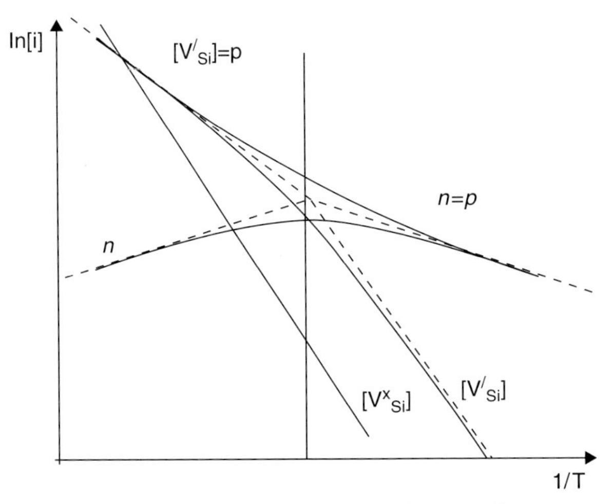

Figure 3.4 Defect concentrations in Si as a function of temperature. The solid lines correspond to the exact solution of Equations (3.19–3.21); the dotted lines are approximations according to Equations (3.24–3.27).

allows us to easily obtain the concentration of electronic defects according to Equation (3.15):

$$n = p = (K_1^o)^{1/2}\exp\left(-\frac{\Delta H_1}{2kT}\right). \qquad (3.24)$$

This result, in combination with Equation (3.17), leads to the following solution:

$$\left[V_{Si}^{/}\right] = K_2^o K_3^o (K_1^o)^{-\frac{1}{2}}\exp\left(-\frac{\Delta H_3 + \Delta H_2 - \Delta H_1/2}{kT}\right). \qquad (3.25)$$

- *Approximation 2:* $K_3 \gg K_1$; the overall electroneutrality condition can be simplified to Equation (3.23). According to this approach, the equilibrium concentration of different defects can be calculated as follows:

$$\left[V_{Si}^{/}\right] = p = (K_3^o)^{\frac{1}{2}} K_2^o \exp\left(-\frac{\Delta H_2 - \Delta H_3/2}{kT}\right), \qquad (3.26)$$

$$n = K_1^o (K_3^o)^{-\frac{1}{2}} (K_2^o)^{-1}\exp\left(-\frac{\Delta H_1 - \Delta H_2 + \Delta H_3/2}{kT}\right). \qquad (3.27)$$

The advantage of such an approximation method becomes clear when complex systems with a number of unknown parameters must be analyzed. It is very convenient to present the results graphically, using "log $[i] - 1/T$" plots, and one such example, comparing the exact solution and the approximation method mentioned above, is shown in Figure 3.4. As can be seen from these data, the electrons and holes represent the predominant defects in silicon at low temperatures, whereas silicon vacancies and holes dominate at high temperatures.

3.2.5
Nonstoichiometry: Equilibria with Gaseous Phase

Whilst the above-mentioned disordering processes do not disturb the crystal stoichiometry, the effects of interaction between the crystal and its environment must be taken into account, especially at elevated temperatures [9, 15–19]. In particular, this becomes necessary when one or several lattice constituents are volatile, for example, in the case of oxides, sulfides, and halides. In a heterogeneous system, the condition of phase equilibria is expressed by the equality of chemical potentials of a given component in both phases:

$$\mu_i^{Sol} = \mu_i^{Gas} = \mu_i^{\circ} + RT\ln P_i, \qquad (3.28)$$

where P_i is the partial pressure of i-component; the stoichiometric composition can be reached only at fixed T and P_i. Consequently, the stoichiometric composition of a compound with respect to a volatile constituent is rather the exception than the rule. In fact, the homogeneity ranges where the chemical composition of solids changes without alterations in the phase composition vary for different substances, from vanishingly small to significantly large values.

Let us now consider binary oxide $MO_{1\pm\delta}$ equilibrated with gaseous phase at constant temperature. The process of oxygen exchange between the solid and gaseous phases can be represented by the following reaction:

$$\mathrm{MO(sol)} = \mathrm{MO}_{1\mp\delta}(\mathrm{sol}) \pm \frac{\delta}{2}\mathrm{O}_2(\mathrm{gas}). \tag{3.29}$$

In order to simplify the task, we should accept several assumptions: (i) that the predominant atomic defects are vacancies in the metal and oxygen sublattices; (ii) that metal M is nonvolatile in the conditions under investigation; (iii) that the concentration of different defects in the crystal lattice is small enough, so that the atomic fraction of atoms in regular positions can be taken as $[M_M^\times] \approx 1$, $[O_O^\times] \approx 1$ and interaction of defects can also be neglected; and (iv) that the vacancies are completely ionized. Accordingly, the disordering processes can be presented as:

$$nil = V_M'' + V_O^{\bullet\bullet} \qquad K_S = [V_M''][V_O^{\bullet\bullet}], \tag{3.30}$$

$$nil = e' + h^\bullet \qquad K_i = np, \tag{3.31}$$

$$\frac{1}{2}\mathrm{O}_2 = \mathrm{O}_O^\times + V_M'' + 2h^\bullet \qquad K_1 = \left[V_M''\right]p^2 P_{\mathrm{O}_2}^{-\frac{1}{2}}, \tag{3.32}$$

$$nil = \frac{1}{2}\mathrm{O}_2 + V_O^{\bullet\bullet} + 2e' \qquad K_2 = \left[V_O^{\bullet\bullet}\right]n^2 P_{\mathrm{O}_2}^{\frac{1}{2}}. \tag{3.33}$$

The complete electroneutrality condition should now be written as:

$$n + 2[V_M''] = p + 2[V_O^{\bullet\bullet}]. \tag{3.34}$$

One important task here is to express each defect concentration as a function of the oxygen partial pressure at a given temperature. Hence, the set of nonlinear Equations (3.30)–(3.33) should be solved using the approximation method. The whole range of the oxygen chemical potentials can be tentatively divided into three regions:

- *Region 1.* At relatively high oxygen pressures, the process of Equation (3.32) should dominate, whereas the reaction Equation (3.33) is suppressed. As a result, the pair of predominant defects in this region includes vacancy in the metal sublattice and free hole, with the electroneutrality condition

$$2[V_M''] = p. \tag{3.35}$$

This leads to the following solution of Equations (3.30–3.33):

$$\left[V_M''\right] = \left(\frac{K_1}{4}\right)^{\frac{1}{3}} P_{\mathrm{O}_2}^{\frac{1}{6}}, \tag{3.36}$$

$$p = (2K_1)^{\frac{1}{3}} P_{\mathrm{O}_2}^{\frac{1}{6}}, \tag{3.37}$$

$$\left[V_O^{\bullet\bullet}\right] = K_S\left(\frac{K_1}{4}\right)^{-\frac{1}{3}} P_{\mathrm{O}_2}^{-\frac{1}{6}} = \frac{K_2}{K_i^2}(2K_1)^{\frac{2}{3}} P_{\mathrm{O}_2}^{-\frac{1}{6}}, \tag{3.38}$$

$$n = K_i(2K_1)^{-\frac{1}{3}} P_{\mathrm{O}_2}^{-\frac{1}{6}}. \tag{3.39}$$

- *Region 2*. In this intermediate range of the oxygen partial pressure, there are no dominating reactions and, therefore, the concentrations of metal and oxygen vacancies are comparable. As a consequence, the composition of the oxide phase is nearly stoichiometric. This allows the electroneutrality condition to be approximated by using two alternative expressions:

  (a) The concentration of ionic defects exceeds that of electronic defects – that is, $[V_M''] = [V_O^{\bullet\bullet}] \gg n, p$. Under these conditions, the defect concentrations can be expressed as:

$$\left[V_M''\right] = \left[V_O^{\bullet\bullet}\right] = K_S^{\frac{1}{2}}, \tag{3.40}$$

$$p = K_1^{\frac{1}{2}} K_i^{-\frac{1}{4}} P_{O_2}^{\frac{1}{4}}, \tag{3.41}$$

$$n = K_2^{\frac{1}{2}} K_i^{-\frac{1}{4}} P_{O_2}^{-\frac{1}{4}}. \tag{3.42}$$

  (b) The concentration of electronic defects exceeds that of ionic defects – that is, $n = p \gg [V_M''], [V_O^{\bullet\bullet}]$. This yields the following concentrations of the defects:

$$n = p = K_i^{\frac{1}{2}}, \tag{3.43}$$

$$\left[V_M''\right] = K_1 K_i^{-1} P_{O_2}^{\frac{1}{2}}, \tag{3.44}$$

$$\left[V_O^{\bullet\bullet}\right] = K_2 K_i^{-1} P_{O_2}^{-\frac{1}{2}}. \tag{3.45}$$

- *Region 3*. At relatively low oxygen chemical potential, the process of Equation (3.33) dominates, and the reaction in Equation (3.32) is suppressed. The crystal lattice releases oxygen atoms, and this oxygen loss is accompanied by the formation of oxygen vacancies and electrons as dominating defects. The electroneutrality condition can, therefore, be approximated by the following equation:

$$n = 2[V_O^{\bullet\bullet}] \tag{3.46}$$

  Under these conditions:

$$\left[V_O^{\bullet\bullet}\right] = \left(\frac{K_2}{4}\right)^{\frac{1}{3}} P_{O_2}^{-\frac{1}{6}}, \tag{3.47}$$

$$n = (2K_2)^{\frac{1}{3}} P_{O_2}^{-\frac{1}{6}}, \tag{3.48}$$

$$p = K_i (K_2)^{-\frac{1}{3}} P_{O_2}^{\frac{1}{6}}, \tag{3.49}$$

$$\left[V_M''\right] = K_S \left(\frac{K_2}{4}\right)^{-\frac{1}{3}} P_{O_2}^{\frac{1}{6}} = K_1 K_i^{-2} (2K_2)^{\frac{2}{3}} P_{O_2}^{\frac{1}{6}}. \tag{3.50}$$

The results obtained can be plotted as log $[i]$ versus log P_{O_2} dependencies (the so-called Brouwer diagram), as illustrated in Figure 3.5. The same approach can be used for any other volatile component, or even for several volatile species.

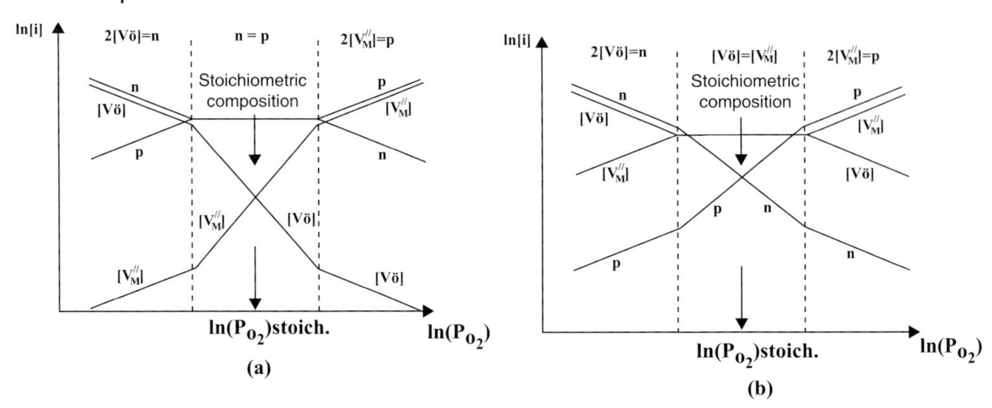

Figure 3.5 Defect concentrations as a function of the oxygen partial pressure (Brouwer diagrams) in binary $MO_{1\pm\delta}$ oxide. (a) Electronic defects are predominant in the intermediate range of oxygen partial pressures; (b) Atomic defects are predominant in the intermediate range of oxygen partial pressures.

3.2.6
Impurities and their Effects on Defect Equilibria

Impurities introduced into the crystal lattice can strongly affect the defect equilibria [9, 15–19]. This influence can be illustrated by the example of MO oxide matrix doped with Me_2O_3, assuming that predominant defects in pure MO are vacancies in the metal and oxygen sublattices (V_M'', $V_O^{\bullet\bullet}$) in accordance with Equation (3.30). The dissolution of Me_2O_3 is also assumed to occur via a substitution mechanism, that is, Me atoms incorporate into the M-sublattice:

$$\mathrm{Me_2O_3} \xrightarrow{\text{into MO}} 2\mathrm{Me}_M^{\bullet} + V_M'' + 3\mathrm{O}_O^{\times}, \qquad K_3 = [\mathrm{Me}_M^{\bullet}]^2[V_M'']. \tag{3.51}$$

Since the oxidation state of Me^{3+} is higher than that of Me^{2+}, the former acquires an effective positive charge. According to the structure conservation condition, an appearance of three new sites of oxygen must be accompanied by the formation of an equal number of the metal sites, with two sites occupied by Me^{3+} ions and one site vacant. The overall electroneutrality condition in this case should be written as

$$2[V_O^{\bullet\bullet}] + [\mathrm{Me}_M^{\bullet}] = 2[V_M'']. \tag{3.52}$$

By using the approximation method, the entire range of thermodynamic parameters can be divided into two regions (Figure 3.6):

1. The so-called intrinsic region, where the concentration of impurities is relatively small and the content of thermally induced vacancies is, in contrast, relatively high. In this case, the electroneutrality condition can be approximated by $[V_O^{\bullet\bullet}] = [V_M'']$.
2. The so-called impurity-control region, where the opposite situation takes place and the electroneutrality condition can be simplified to $[\mathrm{Me}_M^{\bullet}] = 2[V_M'']$.

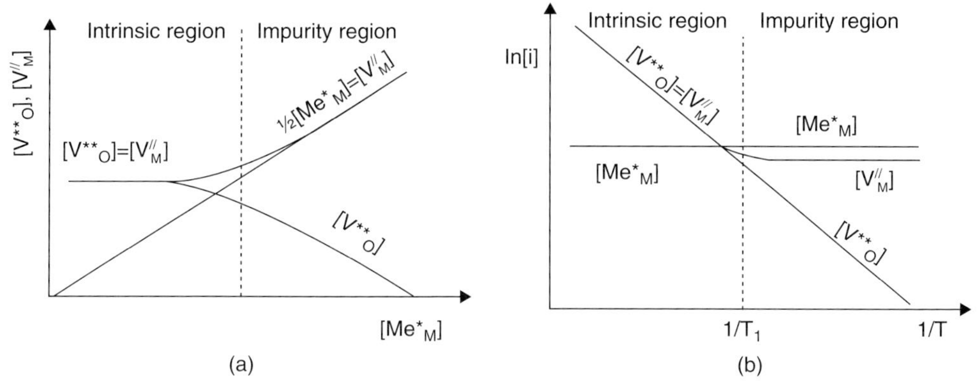

Figure 3.6 Defect equilibria in MO matrix doped by the Me_2O_3. (a) As a function of the impurity concentration (T = const); (b) As a function of temperature at constant Me_2O_3 content.

3.2.7
Crystallographic Aspects of Defect Interaction: Examples of Defect Ordering Phenomena

All of the results discussed in the previous sections in the framework of the simplified equilibrium thermodynamic approach were obtained by assuming that the point defects do not interact with each other. However, as they are effectively charged, the defects (especially when their concentration is high) will interact definitely, owing to coulombic or electron-exchange forces. As a result, various types of associate – or, in other words, *defect clusters* – are generated. When such associates can be described by using the model of interaction between simple point defects, the same thermodynamic approach can be applied to the defect structure analysis. The process of association can, therefore, be presented by a quasi-chemical reaction with the corresponding equilibrium constant. For example, $Y^{/}_{Zr}$ as a heterovalent impurity in yttria-stabilized zirconia ($Zr_{1-x}Y_xO_{2-x/2}$; YSZ) is expected to attract an oppositely charged oxygen vacancy $V^{\bullet\bullet}_{O}$ and so can serve as a trap for the ionic charge carriers, forming immobile clusters such as $(Y^{/}_{Zr}V^{\bullet\bullet}_{O})^{\bullet}$. In the case of CaF_2 doped with YF_3, an impurity defect $Y^{\bullet}_{Ca}$ and interstitial F'_i can also interact, forming neutral pairs:

$$Y^{\bullet}_{Ca} + F'_i \rightleftarrows (Y^{\bullet}_{Ca}F'_i)^{\times} \qquad (3.53)$$

An increase in the point defects concentration may also lead to their long-range ordering. In particular, when such ordering phenomena occur, the structure of highly deficient phases is usually described by a new unit cell, where the ordered vacant sites act not as the defects but rather as regular empty positions. One example of such transformation, demonstrating the relationships between perovskite ABO_3 and brownmillerite $A_2B_2O_5$ ($ABO_{2.5}$) structures (see Chapter 2), is presented in Figure 3.7. The orthorhombic unit cell of brownmillerite, where one in six of the anion sites are vacant with respect to perovskite, arises due to oxygen vacancy ordering, and is related to that of cubic perovskite as $a_{br} \approx \sqrt{2a_c}$, $b_{br} \approx \sqrt{2a_c}$, and $c_{br} \approx 4a_c$; the vacancies are ordered in alternate (001) BO_2 planes of the cubic

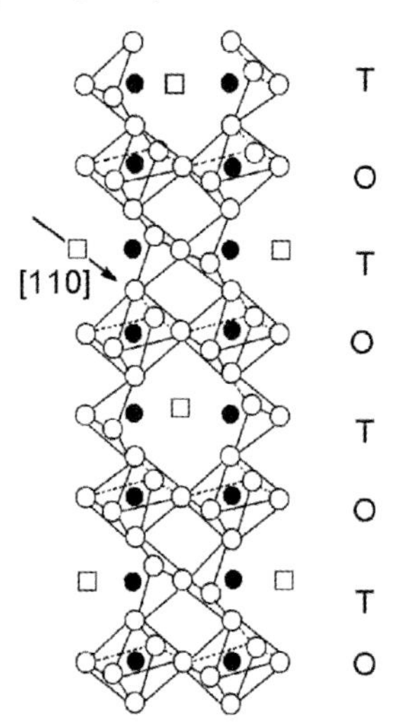

Figure 3.7 Brownmillerite structure as an example of long-range oxygen vacancy ordering. O = octahedral formed by oxygen (open circles) and B cations (closed circles). T = tetrahedra involving B-site cations. The squares denote oxygen vacancies. For clarity, the A-site cations are not shown.

perovskite structure in such a way that alternate [110] rows of oxygen atoms disappear. While the perovskite structure is built of BO_6 octahedra connected by corners (OOOO...), the brownmillerite network consists of alternating octahedra and tetrahedra (OTOT...). A number of other superstructures with the general formula $A_nB_nO_{3n-1}$, where $(n-1)$ is the number of corner-shared octahedra divided by one tetrahedron in the *c*-direction, have been found:

- $n = \infty$, ideal ABO_3 perovskite: OOOO ...;
- $n = 5$, $ABO_{2.8}$: OOOOT ... OOOOT ...;
- $n = 4$, $ABO_{2.75}$: OOOT ... OOOT ...;
- $n = 3$, $ABO_{2.67}$: OOT ...OOT ...;
- $n = 2$, brownmillerite $ABO_{2.5}$ OTOT ....

Depending on the oxygen partial pressure and temperature, numerous cobaltite-, ferrite-, and manganite-based systems exhibit formation of these structural types, although these are not the only possible superstructures, even among the perovskite derivatives. A variety of different ordering types and corresponding structures have been described elsewhere [22–24].

Since the studies of Magneli, Anderson, Wadsley, and others [16, 18, 25], an appearance of homologous series of phases by means of so-called crystallographic shear (CS) have become universally recognized. The most convenient structure to depict the CS formation is the ReO_3-type, which can be depicted as a three-dimensional (3-D) framework of corner-sharing octahedra. If oxygen vacancies are located along certain planes, then shifting two parts of the crystal on both sides of the plane in opposite directions along that plane becomes possible. This shift leads to the

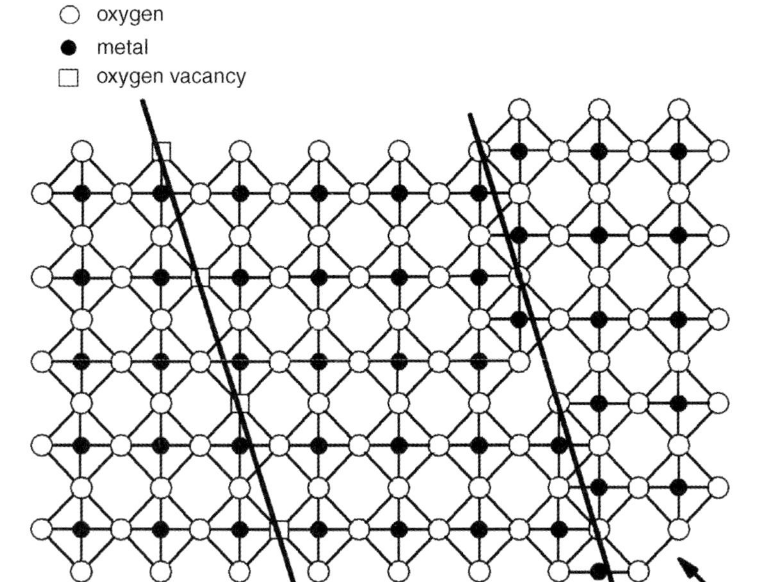

Figure 3.8 Formation of the crystallographic shear in the ReO_3 structure. The arrow indicates the direction of movement. The solid line is the crystallographic shear.

annihilation of oxygen vacancies by their being occupied with neighboring anions. As a result, the octahedra involved with CS planes are shared not only by corners but also by edges, as shown in Figure 3.8. Due to compression of the structure, the ratio M : O increases while there are no point defects such as vacancies in the crystal. However, depending on the direction, the CS may have different indices. The different values of CS repetition rate cause the appearance of different members of the homologous series M_nO_{3n-1} ($n = 8, 9, 10, 11, 12$, and 14) for the ReO_3-type structure.

3.2.8
Thermal and Defects-Induced (Chemical) Expansion of Solids

Thermal expansion is another fundamental property of materials that originates from the reinforcement of atomic vibrations, with an essential anharmonic component, leading to changing interatomic distances with temperature. This phenomenon can be quantified by using either a volumetric (α_V) or linear (α_L) coefficient of thermal expansion (CTE):

$$\alpha_V = \frac{1}{V_0}\left(\frac{\partial V}{\partial T}\right)_P, \tag{3.54}$$

$$\alpha_L = \frac{1}{L_0}\left(\frac{\partial L}{\partial T}\right)_P, \tag{3.55}$$

where V_0 and L_0 are the values of volume and length, respectively, in a selected initial state, and P is the total pressure. Note that, in many branches of materials science, the CTE is defined phenomenologically as:

$$\varepsilon = \frac{\Delta L}{L_0} = \alpha_L \Delta T, \tag{3.56}$$

where ε is the so-called *uniaxial strain*, and ΔL is the length increment corresponding to the temperature change ΔT. In general, the CTE is dependent on temperature, although for engineering materials used at elevated temperatures this effect should be minimized to as great an extent as possible. In the case of complicated devices such as solid oxide fuel cells (SOFCs) consisting of several different components, a close match in their thermal expansion coefficients is a mandatory requirement.

The thermal expansion of solids depends on their structure symmetry, and may be either isotropic or anisotropic. For example, graphite has a layered structure, and its expansion in the direction perpendicular to the layers is quite different from that in the layers. For isotropic materials, $\alpha_V \approx 3^* \alpha_L$. However, in anisotropic solid materials the total volume expansion is distributed unequally among the three crystallographic axes and, as a rule, cannot be correctly measured by most dilatometric techniques. The true thermal expansion in such case should be studied using *in situ* X-ray diffraction (XRD) to determine any temperature dependence of the lattice parameters.

The use of a CTE is often insufficient for an adequate description of solids when substantial amounts of defects are formed at elevated temperatures. As an example, this is the case of mixed-conducting oxides where the lattice volume is a function of both temperature and oxygen vacancies concentration. Such strain variations can often be quantified in terms of both the standard volumetric CTE (α_V) and volumetric chemical expansivity (α_C) induced by the vacancy formation [26]

$$\alpha_C = \frac{1}{V_0}\left(\frac{\partial V}{\partial X_{V_O}}\right)_{T,P} \tag{3.57}$$

where X_{V_O} is the oxygen vacancies mole fraction, and V_0 is the specific volume of an oxide with the stoichiometric composition regarding oxygen at a given temperature. The former quantity is defined, for example, as $\delta/3$ in $ABO_{3-\delta}$ perovskite phases, where δ is the oxygen nonstoichiometry. α_C defined by Equation (3.57) is valid under the assumption that the chemical expansion is uniaxial. By using the aforementioned definitions, the total derivative of the uniaxial strain ε in the absence of additional pressure or mechanical forces can be written as

$$d\varepsilon(T, X_{V_O}) = \frac{1}{3}\alpha_V dT + \frac{1}{3}\alpha_C dX_{V_O}. \tag{3.58}$$

This makes it possible to determine both thermal and chemical constituents of the uniaxial strain by dilatometric measurements, as shown for Sr-substituted cobaltites $La_{1-x}Sr_xCoO_{3-\delta}$ [27].

To date, partially substituted perovskites $ABO_{3-\delta}$, where A and B are rare-earth and 3d transition metal cations, respectively, are among the most extensively studied materials with respect to their chemical expansion. Nonetheless, information on the chemical expansion mechanisms remains limited. In particular, since the oxygen vacancy formation is accompanied by the reduction of 3d metal cations, there are at least two possible mechanisms. The first mechanism – which is often referred to as "dimensional" – is based on an increasing average size of the B-site cations due to the apparent substitution of "large" B^{n+} for smaller $B^{(n-1)+}$ owing to the reduction. The second mechanism refers to changing coulombic forces because of the oxygen vacancy formation. Many research groups [28–32] have ascribed this chemical expansion as mainly being due to the dimensional effect, although others [26, 27, 33, 34] have emphasized an insufficiency of the cation radius change to describe (quantitatively) the chemical expansion, and have suggested the involvement of other factors such as atomic packing, local structure, preferred coordination, association between dopants and vacancies, and a decrease in the binding energy of an oxide when the nonstoichiometry increases. As an example, Zuev *et al.* [35] showed that the isothermal chemical expansion of a perovskite-type $LaCoO_{3-\delta}$, parent composition for many promising mixed conductors, could be explicitly described by the mean ionic radius change within a relative narrow range of the nonstoichiometry variations [36]. Nevertheless, the question of the nature of chemical expansion in phases with a lower symmetry and wider nonstoichiometry domains remains, at present, unanswered.

3.3 Basic Relationships Between the Defect Equilibria and Charge Transfer in Solids

3.3.1 Phenomenological Equations

Placing a solid under an external field (chemical, electrical, thermal, etc.) results in the appearance of the corresponding flows of mass, electric charge, or heat. In linear non-equilibrium thermodynamics [37], these flows can be expressed as

$$\left.\begin{aligned} J_1 &= L_{11}F_1 + L_{12}F_2 + \ldots + L_{1m}F_m = \sum_{k=1}^{k=m} L_{1k}F_k \\ J_2 &= L_{21}F_1 + L_{22}F_2 + \ldots + L_{2m}F_m = \sum_{k=1}^{k=m} L_{2k}F_k \\ &\ldots \\ J_i &= L_{i1}F_1 + L_{i2}F_2 + \ldots + L_{im}F_m = \sum_{k=1}^{k=m} L_{ik}F_k \end{aligned}\right\} \tag{3.59}$$

where L_{jk} are the kinetic coefficients and F_k are the thermodynamic forces (gradients of chemical potential, electric potential, temperature, etc.). The diagonal kinetic coefficients correspond to so-called "pure" processes, such as diffusion or electric current which may occur under a single driving force, whereas other cross-coefficients reflect interference of the forces. One important property of the kinetic coefficients is their reciprocity. This means that, if the force F_k is able to evoke the flow J_j, then the force F_j can induce the flow J_k at an equivalent degree (Onsager reciprocal relations):

$$L_{jk} = L_{kj} \tag{3.60}$$

Note also that the diagonal coefficients ($L_{11}, L_{22} \ldots L_{ii}$) are always positive, in contrast to the cross-coefficients L_{jk}. For example, the heat conductivity or electrical conductivity coefficients have always a positive sign, whereas no sign may be compulsorily ascribed to the thermodiffusion or thermoelectrical coefficients without consideration of a particular system.

Another important relationship between the kinetic coefficients is the so-called "principle of symmetry", as formulated by P. Curie and introduced to nonlinear thermodynamics by Kondepudi and Prigogine [37]. As applied to thermodynamics, this postulates that a scalar quantity could not evoke a vector effect. For example, a scalar thermodynamic force – chemical affinity (driving the process of chemical reaction) that has very high isotropy symmetry – could not cause heat flow, which has a particular direction and is therefore anisotropic. Taking into account the reciprocal relationships, this can be formulated as

$$L_{jk} = L_{kj} = 0 \tag{3.61}$$

These relationships allow the description of key processes associated with the mass and charge transfer in electrochemical systems, particularly in ionic crystals.

3.3.2 Mass Transfer in Crystals

Let us consider an easiest case – the one-dimensional diffusion of uncharged particles M under the chemical potential gradient:

$$\nabla\mu_M \neq 0, \quad T = \text{Const}, \quad \nabla\varphi = 0 \tag{3.62}$$

$$J_M^{\text{dif}} = L_{11}F_1 = -L_{11}\nabla\mu_M = -L_{11}\frac{d\mu_M}{dx} \tag{3.63}$$

where φ is the electrical potential. Taking into account the standard expression for chemical potential similar to Equation (3.1), it is possible to transform Equation (3.63):

$$J_M^{\text{dif}} = -L_{11}\nabla\mu_M = -L_{11}\frac{kT}{c_M}\frac{dc_M}{dx}. \tag{3.64}$$

where c_M is the concentration of M particles. By comparing Equation (3.64) with the first Fick law

$$J_M^{\text{dif}} = -D_M\frac{dc_M}{dx}, \tag{3.65}$$

one can easily establish the relationship between the linear kinetic coefficient L_{11} and diffusion coefficient D_M:

$$D_M = L_{11}\frac{kT}{c_M}, \tag{3.66}$$

and finally represent the mass flow as:

$$J_M^{\text{dif}} = -\frac{D_M c_M}{kT}\frac{d\mu_M}{dx}. \tag{3.67}$$

Let us now analyze a more complicated situation with the transfer of charged species M_j^z (the charge z is not specified for generalization), when the mass transport may be driven by gradients of both chemical potential and electric field. If the chemical potential gradient is zero

$$\nabla\mu_M = 0, \quad T = \text{Const}, \quad \nabla\varphi \neq 0 \tag{3.68}$$

the situation is referred to as *migration* or *electromigration*. By definition, the density of mass flow J_M can be expressed via the number of particles crossing a unit surface per unit time. On the other hand, this flow may be represented as a product of the species concentration and their average velocity (v_M):

$$J_M^{\text{migr}} = c_M v_M \tag{3.69}$$

The velocity of charged particles is proportional to the electric field:

$$v_M = \pm u_M \frac{d\varphi}{dx} \tag{3.70}$$

where the proportionality constant u_M is called the *mobility*; the physical meaning of this relates to the velocity of M_j^z species under an unit field. Consequently, for the migration flow

$$J_M^{\text{migr}} = -\frac{z}{|z|}c_M u_M \frac{d\varphi}{dx}. \tag{3.71}$$

The term $z/|z|$ determines only the difference in the directions of cations and anions motion in the electric field.

The total mass flow under the chemical and electrical forces is expressed by the following equation, often referred to as Wagner's law:

$$J_M = J_M^{\text{dif}} + J_M^{\text{migr}} = L_{11}F_1 + L_{12}F_2 = -\frac{D_M c_M}{kT}\frac{d\mu_M}{dx} - \frac{z}{|z|}u_M c_M \frac{d\varphi}{dx}. \tag{3.72}$$

In the thermodynamic system where work is performed by the chemical and electrical forces, the total differential of Gibbs energy is written as:

$$dG = -SdT + VdP + \sum_j \mu_j dn_j + \sum_j \varphi dq_j \tag{3.73}$$

where $q_j = z_j e n_j$ is the charge of the given subsystem, and n_j is the number or concentration of the j-th charge carriers. At $T,P =$ Const, Equation (3.73) can be transformed:

$$dG = \sum_j \mu_j dn_j + \sum_j z_j e\varphi dn_j = \sum_j (\mu_j + z_j e\varphi) dn_j. \tag{3.74}$$

The right-hand term in parentheses expresses the joint effect of chemical and electrical forces, and is known as the *electrochemical potential*:

$$\tilde{\mu}_j = \mu_j + z_j e\varphi \tag{3.75}$$

where the gradient is the driving force of the j-th charge carriers:

$$\nabla\tilde{\mu}_j = \nabla\mu_j + z_j e\nabla\varphi. \tag{3.76}$$

Under thermodynamic equilibrium conditions, $dG_{p,T} = 0$ and, therefore:

$$\nabla\tilde{\mu}_j = \nabla\mu_j + z_j e\nabla\varphi = 0 \tag{3.77}$$

leading to zero mass flow:

$$J_j = J_j^{\text{dif}} + J_j^{\text{migr}} = 0 \tag{3.78}$$

By combining Equations (3.67), (3.71), (3.77) and (3.78), it is possible to obtain a very important expression which is often referred to as the Nernst–Einstein equation:

$$\frac{D_j}{kT} = \frac{u_j}{|z_j|e} \tag{3.79}$$

that shows the relationship between the diffusion coefficient and absolute mobility of the j-th species.

3.3.3 Electrical Conductivity. Transport under a Temperature Gradient

The charge flow under an electrical potential gradient, expressed in terms of the electrical current density (i) and total conductivity (σ), is described by the phenomenological Ohm law:

$$i = -\sigma\nabla\varphi \tag{3.80}$$

where

$$i = \sum_j i_j = \sum_j z_j e J_j. \tag{3.81}$$

The partial current densities (i_j) and partial conductivities (σ_j) also obey the Ohm law. The charge fraction transported by the j-th sort of carriers is referred to as the transference number, t_j:

$$t_j = \frac{i_j}{i} = \frac{\sigma_j}{\sum_j \sigma_j} = \frac{\sigma_j}{\sigma}. \tag{3.82}$$

The right-hand part of this equality is always valid in the differential form; in integral form, this equality is observed in electrochemical systems when the interfacial phenomena can be neglected. By combining Equations (3.71),(3.80) and (3.81), it is possible to obtain

$$i_j = -\frac{z_j}{|z_j|} z_j e u_j c_j \nabla \varphi. \tag{3.83}$$

Then, by comparing Equations (3.80) and (3.83)

$$\sigma_j = z_j e u_j c_j. \tag{3.84}$$

Analogously, Equations (3.71) and (3.72) for migrating M_j^z particles can also be transformed to the form:

$$J_M^{\mathrm{migr}} = -\frac{z_M}{|z_M|} \frac{\sigma_M}{z_M e} \frac{d\varphi}{dx}. \tag{3.85}$$

$$J_M = J_M^{\mathrm{dif}} + J_M^{\mathrm{migr}} = -\frac{\sigma_M}{(z_M e)^2} \left(\frac{d\mu_M}{dx} + z_M e \frac{d\varphi}{dx} \right). \tag{3.86}$$

The Nernst–Einstein equation (Equation (3.79)) may be rewritten in a commonly used form

$$\frac{D_j}{kT} = \frac{\sigma_j}{|z_j| e c_j}, \tag{3.87}$$

In addition, Equation (3.86) can be generalized for a more general case when chemical, electric, and temperature fields are combined:

$$J_M = -\frac{\sigma_M}{(z_M e)^2} \left(T \frac{d}{dx} \frac{\mu_M}{T} + z_M e \frac{d\varphi}{dx} + \frac{u_M^*}{T} \frac{dT}{dx} \right), \tag{3.88}$$

where u_M^* is the energy of M_j^z transfer. Finally, taking into account the Nernst–Einstein equation or Onsager reciprocal relations, Equation (3.88) is often written as

$$J_M = -\frac{D_M c_M}{kT} \left(T \frac{d}{dx} \frac{\mu_M}{T} + z_M e \frac{d\varphi}{dx} + \frac{u_M^*}{T} \frac{dT}{dx} \right). \tag{3.89}$$

3.3.4 Electrochemical Transport

3.3.4.1 Mass and Charge Transport under the Chemical Potential Gradient: Electrolytic Permeation

Let us consider one example of the mass and charge transfer in a simple oxygen-deficient oxide, $MO_{1-\delta}$. This example is relevant for the analysis of solid electrolyte applicability limits, electrolytic leakage phenomena in SOFCs and sensors, and for an understanding of the transport processes in dense ceramic membranes for oxygen and hydrogen separation (see Chapters 9, 12 and 13). Let us assume that the predominant defects are doubly ionized oxygen vacancies and electrons, formed according to the following reaction:

$$\frac{1}{2}O_{2(\text{gas})} + V_O^{\bullet\bullet} + 2e' \Leftrightarrow O_O^{\times} \qquad n = 2\left[V_O^{\bullet\bullet}\right] = (K)^{\frac{1}{3}} \cdot p_{O2}^{-\frac{1}{6}}. \tag{3.90}$$

If the oxygen partial pressures at the opposite membrane sides are different (in other words, the gradient of oxygen chemical potential is created), the mass and charge flows should arise as illustrated by Figure 3.9. When the membrane is dense enough to avoid transport in the pores, the following equation for the flow of j-species is valid:

$$J_j = -\frac{\sigma_j}{(z_j e)^2}\left(T\frac{d}{dx}\frac{\mu_j}{T} + z_j e\frac{d\varphi}{dx} + \frac{u_j^*}{T}\frac{dT}{dx}\right) \tag{3.91}$$

At constant temperature, the fluxes of oxygen vacancies and electrons are expressed as

$$J_{V_O^{\bullet\bullet}} = -\frac{\sigma_{V_O^{\bullet\bullet}}}{4e^2}\nabla\mu_{V_O^{\bullet\bullet}} - \frac{\sigma_{V_O^{\bullet\bullet}}}{2e}\nabla\varphi \tag{3.92}$$

$$J_{e'} = -\frac{\sigma_{e'}}{e^2}\nabla\mu_{e'} + \frac{\sigma_{e'}}{e}\nabla\varphi \tag{3.93}$$

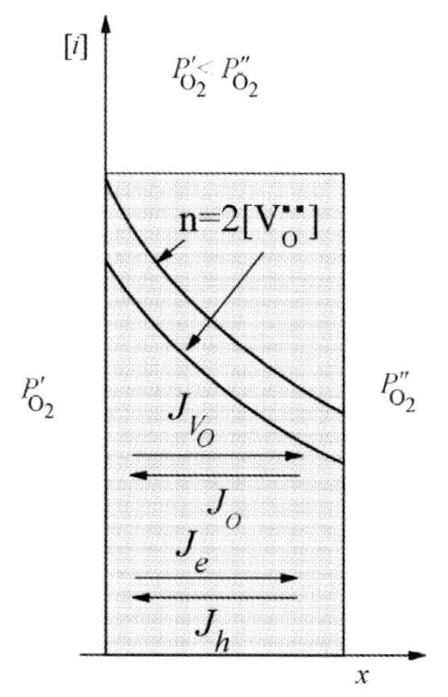

Figure 3.9 Defect concentration changes and directions of the flows in an oxide membrane placed under a gradient of the oxygen partial pressure.

Under steady-state conditions, the total electric current should be zero; then

$$2J_{V_O^{\bullet\bullet}} = J_{e'} \tag{3.94}$$

$$-\frac{\sigma_{V_O^{\bullet\bullet}}}{2e^2}\nabla\mu_{V_O^{\bullet\bullet}} - \frac{\sigma_{V_O^{\bullet\bullet}}}{e}\nabla\varphi = -\frac{\sigma_{e'}}{e^2}\nabla\mu_{e'} - \frac{\sigma_{e'}}{e}\nabla\varphi \tag{3.95}$$

Solving the latter equation relative to the electrical potential gradient yields

$$\nabla\varphi = \frac{1}{e}\cdot\frac{\sigma_{e'}}{\sigma_{V_O^{\bullet\bullet}}+\sigma_{e'}}\cdot\nabla\mu_{e'} - \frac{\sigma_{V_O^{\bullet\bullet}}}{\sigma_{V_O^{\bullet\bullet}}+\sigma_{e'}}\cdot\frac{1}{2e}\cdot\nabla\mu_{V_O^{\bullet\bullet}} \tag{3.96}$$

or

$$\nabla\varphi = \frac{t_{e'}}{e}\nabla\mu_{e'} - \frac{t_{V_O^{\bullet\bullet}}}{2e}\nabla\mu_{V_O^{\bullet\bullet}} \tag{3.97}$$

By introducing Equation (3.97) into Equation (3.93), and taking into account that $t_{e'} + t_{V_O^{\bullet\bullet}} = 1$:

$$J_{e'} = -\frac{\sigma_{e'}\cdot t_{V_O^{\bullet\bullet}}}{e^2}\cdot\nabla\mu_{e'} - \frac{\sigma_{e'}\cdot t_{V_O^{\bullet\bullet}}}{2e^2}\cdot\nabla\mu_{V_O^{\bullet\bullet}} \tag{3.98}$$

For the reaction of oxygen vacancies formation (Equation (3.90)), the condition of local or global equilibrium is given by

$$\frac{1}{2}\mu_{O_2} + \mu_{V_O^{\bullet\bullet}} + 2\mu_{e'} = \mu_{O_O^{\times}} \tag{3.99}$$

Assuming that the concentration of oxygen ions is essentially constant along the membrane (or expressing the vacancy activity as the ratio of vacancy and anion concentrations), it is possible to write:

$$\frac{1}{2}\nabla\mu_{O_2} = -\nabla\mu_{V_O^{\bullet\bullet}} - 2\nabla\mu_{e'} \tag{3.100}$$

$$J_{e'} = -\frac{\sigma_{e'}\cdot t_{V_O^{\bullet\bullet}}}{e^2}\cdot\nabla\mu_{e'} - \frac{\sigma_{e'}\cdot t_{V_O^{\bullet\bullet}}}{2e^2}\cdot\nabla\mu_{V_O^{\bullet\bullet}} = \frac{\sigma_{e'}\cdot t_{V_O^{\bullet\bullet}}}{2e^2}(-2\nabla\mu_{e'} - \nabla\mu_{V_O^{\bullet\bullet}}) = \frac{\sigma_{e'}\cdot t_{V_O^{\bullet\bullet}}}{4e^2}\nabla\mu_{O_2}. \tag{3.101}$$

Therefore, according to Equation (3.94), the flow of oxygen vacancies can be presented as:

$$J_{V_O^{\bullet\bullet}} = \frac{1}{2}J_{e'} = \frac{\sigma_{e'}\cdot t_{V_O^{\bullet\bullet}}}{8e^2}\nabla\mu_{O_2} \tag{3.102}$$

Consequently, the molar flow of molecular oxygen can be expressed via the partial conductivities:

$$J_{O_2} = -\frac{1}{2}J_{V_O^{\bullet\bullet}} = -\frac{\sigma_{e'}\cdot t_{V_O^{\bullet\bullet}}}{16F^2}\nabla\mu_{O_2} = -\frac{1}{16F^2}\frac{\sigma_{e'}\cdot\sigma_{V_O^{\bullet\bullet}}}{\sigma_{e'}+\sigma_{V_O^{\bullet\bullet}}}\nabla\mu_{O_2}. \tag{3.103}$$

Equation (3.103) (which is also known as Wagner's equation) may be transformed into the formula for steady-state oxygen permeation under conditions when the surface exchange limitations are negligible:

$$J_{O_2} = -\frac{RT}{16F^2L}\frac{\sigma_{e'}\cdot\sigma_{V_O^{\bullet\bullet}}}{\sigma_{e'}+\sigma_{V_O^{\bullet\bullet}}}\ln\frac{P''_{O_2}}{P'_{O_2}} \tag{3.104}$$

where L is the membrane thickness. If the electronic conductivity is predominant ($\sigma_{e'} \gg \sigma_{V_O^{\bullet\bullet}}$), as is typical for numerous perovskite and composite materials, then Equation (3.104) can be further simplified:

$$J_{O_2} = -\frac{RT}{16F^2L}\sigma_{V_O^{\bullet\bullet}}\ln\frac{P''_{O_2}}{P'_{O_2}} \tag{3.105}$$

In the case when the ionic conductivity prevails ($\sigma_{e'} \ll \sigma_{V_O^{\bullet\bullet}}$), the solution valid for solid electrolytes can be obtained:

$$J_{O_2} = -\frac{RT}{16F^2L}\sigma_{e'}\ln\frac{P''_{O_2}}{P'_{O_2}} \tag{3.106}$$

One important conclusion here is that electrolytic permeation is usually governed by the transport of the slowest component(s). More complicated cases of the ambipolar diffusion are considered in the following sections.

3.3.4.2 **Charge Transfer under Temperature Gradient and Seebeck Coefficient: Selected Definitions**

If a solid with mobile ionic and/or electronic charge carriers is placed in a temperature gradient, charge separation will occur inside the sample and a thermovoltage will appear (the Seebeck effect). For relatively small gradients, this voltage is proportional to the temperature difference; the proportionality constant ($\alpha = d\varphi/dT$) is referred to as the Seebeck coefficient. Most mobile species move towards the "cold" zone and, therefore, the sign of the cold end reflects the sign of dominant charge carriers. Let us consider a simple situation when electrons or holes dominate. Here, the partial current density can be expressed as:

$$i_j = z_j e J_j = -\frac{\sigma_j}{z_j e}\left(T\frac{d}{dx}\frac{\mu_j}{T} + z_j e\frac{d\varphi}{dx} + \frac{u_j^*}{T}\frac{dT}{dx}\right) \tag{3.107}$$

Taking into account that the chemical potential of *j*-type species depends on temperature and on the concentrations of all defects (c_k) that have influence on *j*-type defects:

$$\frac{d\mu_j}{dT} = \sum_k \frac{\partial\mu_j}{\partial c_k}\frac{\partial c_k}{\partial T} + \frac{\partial\mu_j}{\partial T} \tag{3.108}$$

Equation (3.107) can be transformed into

$$i_j = -\frac{\sigma_j}{z_j e}\left(\sum_k \frac{\partial\mu_j}{\partial c_k}\frac{dc_k}{dT} + T\frac{\partial}{\partial T}\left(\frac{\mu_j}{T}\right) + z_j e\frac{d\varphi}{dT} + \frac{u_j^*}{T}\right)\nabla T \tag{3.109}$$

The Gibbs–Helmholtz equation states that

$$\frac{\partial}{\partial T}\left(\frac{\mu_j}{T}\right) = -\frac{h_j}{T^2} \tag{3.110}$$

whilst the difference between the transport energy of charge carriers and their enthalpy in the quiescent state (h_j) can be expressed as

$$Q_j^* = u_j^* - h_j \tag{3.111}$$

where Q_i^* is the redundant heat transferred by the j-type species (transferred heat). As a result,

$$i_j = -\frac{\sigma_j}{z_j e}\left(\sum_k \frac{\partial \mu_j}{\partial c_k}\frac{dc_k}{dT} + z_j e\frac{d\varphi}{dT} + \frac{Q_j^*}{T}\right)\nabla T. \tag{3.112}$$

The measurements of thermovoltage correspond to the open-circuit conditions; that is, the current is equal to zero ($i_j = 0$), and it is therefore possible to express the derivative $d\varphi/dT$:

$$\alpha_{\text{hom}} = \frac{d\varphi}{dT} = \pm\sum_j \frac{t_j}{z_j e}\left(\sum_k \frac{\partial \mu_j}{\partial c_k}\frac{dc_k}{dT} + \frac{Q_j^*}{T}\right) \tag{3.113}$$

where t_j is the transference number and the sign corresponds to that of the charge carriers. By definition, α_{hom} relates to the gradient of electrical potential within the sample at the 1 K temperature gradient. However, any cell used for the measurements always comprises the sample and two electrodes, which are usually metallic and placed at different temperatures (T and $T + \Delta T$). Hence, the measured voltage between these electrodes is:

$$\Delta\varphi = V(T+\Delta T) - V(T) + \frac{d\varphi}{dT}\Delta T \tag{3.114}$$

where $V(T)$ is the contact potential drop between the electrode and sample at temperature T. These equations show that thermovoltage consists of two components related to the homogeneous coefficient $\alpha_{\text{hom}} = d\varphi/dT$ (Equation (3.113)) and the heterogeneous thermovoltage coefficient:

$$\alpha_{\text{heter}} = \frac{V(T+\Delta T) - V(T)}{\Delta T}. \tag{3.115}$$

If the temperature difference at the electrodes is not very large (a few degrees), then

$$\alpha_{\text{heter}} = \frac{dV}{dT} \tag{3.116}$$

Particular expressions for the Seebeck coefficient depend on the nature of the charge carriers.

For a non-degenerated semiconductor (n-type or p-type), the carriers are electrons with transport number t_- and holes with the transference number t_+. If the ionic conductivity is close to zero (i.e., $t_+ + t_- = 1$), the homogeneous thermovoltage coefficient can be written according to Equation (3.113) as:

$$\alpha_{\text{hom}} = -\frac{t_-}{e}\left(\frac{\partial \mu_-}{\partial n}\frac{dn}{dT} + \frac{Q_-^*}{T}\right) + \frac{t_+}{e}\left(\frac{\partial \mu_+}{\partial p}\frac{dp}{dT} + \frac{Q_+^*}{T}\right). \tag{3.117}$$

As the equilibrium between the sample and the metal electrodes is maintained by the electron exchange, the equality of the electrochemical potentials of electrons in these phases leads to the following expressions for the contact voltage drop and the heterogeneous thermovoltage coefficient:

$$V = \frac{1}{e}[\mu_-(M) - \mu_-(\text{sample})]. \tag{3.118}$$

$$\alpha_{\text{heter}} = \frac{1}{e}\frac{d\mu_-}{dT}. \tag{3.119}$$

Assuming that the intrinsic electron disorder proceeds as:

$$nil = e' + h^{\bullet}, \tag{3.120}$$

and, consequently,

$$\mu_- + \mu_+ = 0, \tag{3.121}$$

Equation (3.119) can be rewritten as:

$$\alpha_{\text{heter}} = \frac{t_-}{e}\frac{d\mu_-}{dT} + \frac{t_+}{e}\frac{d\mu_-}{dT} = \frac{t_-}{e}\frac{d\mu_-}{dT} - \frac{t_+}{e}\frac{d\mu_+}{dT}. \tag{3.122}$$

Introducing Equation (3.117) yields

$$\alpha = t_-\alpha_- + t_+\alpha_+, \tag{3.123}$$

where α_- and α_+ correspond to the n- and p-type contributions, respectively:

$$\alpha_- = -\frac{1}{e}\left(\frac{Q_-^*}{T} - \frac{\partial \mu_-}{\partial T}\right), \tag{3.124}$$

$$\alpha_+ = \frac{1}{e}\left(\frac{Q_+^*}{T} - \frac{\partial \mu_+}{\partial T}\right). \tag{3.125}$$

In the case of localized electron defects (so-called small polarons), the chemical potentials are linked with the concentrations as follows:

$$\mu_e = \mu_e^o + kT\ln\left(\frac{n}{N_-}\right) \quad \text{or} \quad \mu_h = \mu_h^o + kT\ln\left(\frac{p}{N_+}\right) \tag{3.126}$$

where N_- and N_+ are the numbers of sites available for the charge carrier location. Then

$$\alpha_- = -\frac{k}{e}\left[\frac{Q_-^*}{kT} + \ln\left(\frac{N_-}{n}\right)\right] \quad \text{and} \quad \alpha_+ = \frac{k}{e}\left[\frac{Q_+^*}{kT} + \ln\left(\frac{N_+}{p}\right)\right]. \tag{3.127}$$

When both contributions are significant

$$\alpha = \frac{t_+ k}{e}\left[\frac{Q_+^*}{kT} + \ln\left(\frac{N_+}{p}\right)\right] - \frac{t_- k}{e}\left[\frac{Q_-^*}{kT} + \ln\left(\frac{N_-}{n}\right)\right] \quad (3.128)$$

The concentrations of electrons and holes depend on the external thermodynamic parameters, such as T and Po2, and may often be widely varied within the phase stability limits. Equations (3.37) and (3.48) presented such dependencies at a constant temperature, but at relatively high and low oxygen pressures these concentrations can be expressed as

$$p = (2K_1)^{\frac{1}{3}} P_{O_2}^{\frac{1}{6}} = (2K_1^{\circ})^{\frac{1}{3}} \exp\left(-\frac{E_1}{3kT}\right) \cdot P_{O_2}^{\frac{1}{6}} = K_h \exp\left(-\frac{W_h}{kT}\right) \cdot P_{O_2}^{\frac{1}{6}}, \quad (3.129)$$

$$n = (2K_2)^{\frac{1}{3}} P_{O_2}^{-\frac{1}{6}} = (2K_2^{\circ})^{\frac{1}{3}} \exp\left(-\frac{E_2}{3kT}\right) \cdot P_{O_2}^{-\frac{1}{6}} = K_e \exp\left(-\frac{W_e}{kT}\right) \cdot P_{O_2}^{-\frac{1}{6}}. \quad (3.130)$$

By combining Equations (3.129) and (3.130) with Equation (3.127), it is possible to obtain the dependencies of the Seebeck coefficient on the oxygen partial pressure.

$$\alpha_e = -\frac{k}{e}\left(A_e - \frac{W_e}{kT} + \frac{1}{6}\ln P_{O_2}\right), \qquad \alpha_h = \frac{k}{e}\left(A_h - \frac{W_h}{kT} - \frac{1}{6}\ln P_{O_2}\right). \quad (3.131)$$

The isothermal dependencies of electronic conductivity and Seebeck coefficient on the oxygen pressure, along with the corresponding defect concentrations, are shown in Figure 3.10.

3.4 Examples of Functional Materials with Different Defect Structures

When considering electrical properties from a general point of view, all solids may be divided in two major groups, namely *conductors* and *insulators*.

In conductors, the charge can be transported by electrons (holes), by ions, or by both. The electronic conductors may exhibit metallic conductivity, semiconductivity, or even superconductivity. Dominant ionic conduction occurs in the solids known as solid electrolytes. Although many ionic crystals exhibit ionic transport (e.g., NaCl), only a few of them can be regarded as electrolytes. The solid electrolytes should possess a relatively high value of ionic conductivity, comparable to that of liquid electrolytes. They must also meet the requirement of unipolar ionic conduction – that is, to be insulators with respect to electronic transport. It is worth noting here that many solids with a really high ionic conductivity (known as *superionic conductors*) display also a small electronic conduction that cannot, however, be ignored. In a strict sense, such materials are not solid electrolytes; rather, there are also solids that exhibit simultaneously significant levels of both ionic and electronic transport, and are referred to as mixed conductors (MIECs). Each type of the above-mentioned conducting materials finds its own applications in a variety of solid-state electrochemical devices.

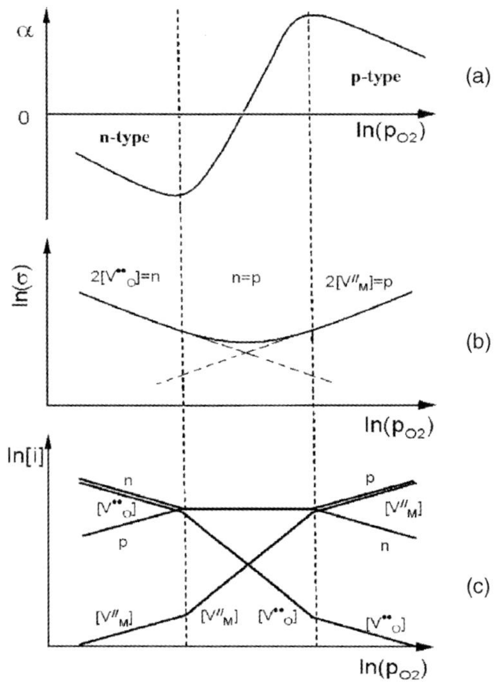

Figure 3.10 (a) Seebeck coefficient, (b) conductivity and (c) defect concentrations in an electrically conducting oxide.

Although the electrical properties of materials depend on numerous factors, it is generally recognized that the crystal and defect structures of solids are of key importance in our understanding of conductivity mechanisms.

3.4.1 Solid Electrolytes

As it is impossible to analyze all types of electrolyte within the limitations of this chapter, the reader is directed towards many comprehensive reviews where the known solid electrolytes are classified according to their technological functions [38], the nature of their transition to a highly conducting state [13], the constituent chemical species [39], or their crystal structures [40]. Other recent surveys have been devoted to systems with 3-D ionic migration [41] and to the electrolytes with a certain type of mobile species (e.g., oxygen anions [42]). Information on these groups of ionic conductors can be found in Chapters 7–9. Irrespective of classifications and microscopic mechanisms, the partial ionic conductivity (σ_{ion}) of a solid can be expressed as

$$\sigma_{\text{ion}} = |z_i| e c_i u_i \tag{3.132}$$

where z_i, c_i, and u_i are the charge number, concentration, and mobility of the ionic charge carriers i, respectively. The solid electrolyte phase α-AgI (see Chapter 2)

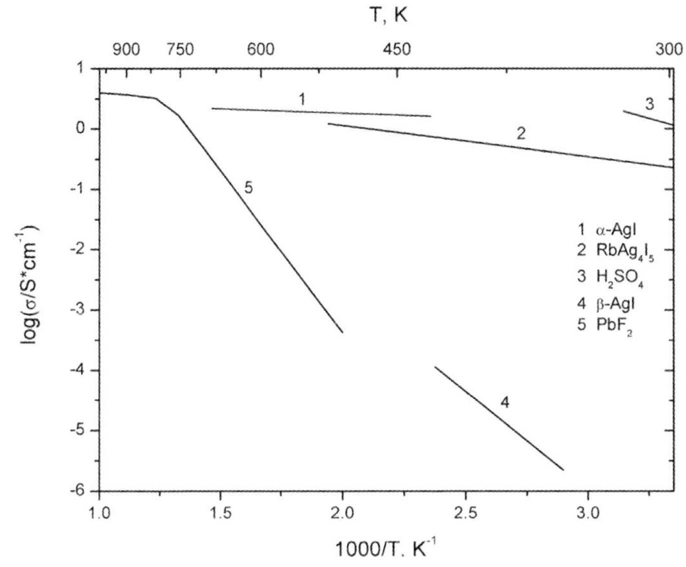

Figure 3.11 Conductivity in selected electrolytes.

has $\sigma_{\text{ion}} \approx 1\,\Omega^{-1}\,\text{cm}^{-1}$ and $c_i \approx 1/2$, compared to $\sigma_{ion} = 10^{-5}\,\Omega^{-1}\,\text{cm}^{-1}$ and $c_i = 10^{-4}$ for a typical "normal" ionic conductor NaCl at temperatures slightly below the melting point (see Figure 3.11). The commensurable ion mobility values for both of these solids means that high σ_{ion} values for a solid electrolyte correlate with the high degree of lattice disorder (i.e., the level of level of c_i), rather than with differences in ion mobilities. Nevertheless, when a new solid with a high ionic conductivity for a peculiar application is required, both factors should be maximized. In order to achieve these aims, any potential candidate must meet the following requirements:

1. Only one type of ionic species should be mobile, unlike liquid electrolytes.
2. There should be a sufficiently high concentration of mobile ions, and a sufficient number of unoccupied sites equivalent to those where the charge carriers are located (unlike most ionic crystals such as NaCl and MgO).
3. The energy of migration from one site to the equivalent unoccupied position should be small, preferably less than 1 eV. In other words, the presence of low-energy pathways or conduction channels is required.
4. Highly polarizable ions within the immobile sublattice are beneficial, as these can relatively easy deform, so as to allow diffusing ions to "squeeze" through smaller gaps.
5. A mixed ionic–covalent bonding character should be present in the solid, since such character allows the mobile ion to be stable in several different coordinations during the diffusion process.
6. There should be a low charge of the mobile ions, providing a minimum coulombic interaction energy during the diffusion process.

7. The radius of the charge carriers should be small, so as to facilitate diffusion through the gaps of the immobile sublattice.
8. There should be negligible electronic conductivity.

In particular, the small ionic radius of Ag^+ ($r_{Ag+} = 1.00$ Å) obviously favors the cation diffusion in α-AgI via migration in the sublattice formed by large I^- anions. However, Na^+ has a comparable size ($r_{Na+} = 0.99$ Å) while NaI is not a solid electrolyte, even at elevated temperature. On the other hand, if the aforementioned requirement of a small ionic radius were crucial, then solid electrolytes with mobile anions would be rather uncommon, as anions are usually larger than cations. As a matter of fact, F^- and O^{2-} conductors may outnumber cationic conductors; thus, the requirements mentioned above are somewhat interdependent, and there is no separate factor which can be used solely to explain fast ionic transport.

Two typical examples of solid electrolytes, which may be considered as prototypes of the important electrolyte families, are α-AgI and β-PbF_2. The ionic conductivity of β-PbF_2 increases rapidly with temperature, reaching a value close to that of the molten state (Figure 3.11). It is now generally recognized that the fast ionic transport in β-PbF_2 and other fluorite-structured compounds is observed due to anion diffusion solely, since both the energy required for cation defect formation and the electronic band gap are relatively large (see Ref. [41] and references therein). Repulsive interactions between the defects were shown to suppress the development of complete sublattice disorder such as observed in α-AgI. Below a characteristic temperature corresponding to the transition to superionic state (T_c), the principal defect clusters in β-PbF_2 are the anion Frenkel-type pairs with the interstitial anion in one of the empty anion cubic center positions (see Chapter 2), whilst the vacancy is, at least, more distant than the next nearest-neighbor anion site. When the defect concentration increases with temperature above approximately 1%, the presence of F^- vacancies at closer distances destabilizes the interstitials, leading to a fundamental change in the nature of defects [41]. Under ambient conditions, silver iodide (AgI) usually exists as a mixture of the β and γ phases, having the wurtzite and zincblende structures, respectively. As illustrated in Figure 3.11, AgI undergoes an abrupt superionic transition at ~420 K (Ref. [41] and references therein). The conductivity of the formed α phase is predominantly provided by Ag^+ cations, whilst the partial electronic conductivity is two orders of magnitude lower. The I^- sublattice in α-AgI contains a number of interstices achievable for the Ag^+ cations as sites. The highly disordered structure equally corresponds to a random distribution of cations available over the whole free volume, which is not occupied by anions. The properties of such solids are commonly described as being "molten sublattice" or "liquid-like". According to the different character of the superionic transition on heating (as illustrated by the examples of α-AgI and β-PbF_2), all solid electrolytes are often classified as two types: *type I* with an abrupt transition (e.g., AgI); and *type II*, with a continuous transition (likewise β-PbF_2). The particular features of these compounds, representing important families of solid electrolytes, are compared in Table 3.1.

One conventional method of stabilizing the high-temperature superionic phases under the application conditions is to dope them with isovalent or aliovalent ions,

Table 3.1 Comparison of the key features of superionic properties of AgI and β-PbF_2, as suggested in Ref. [41].

Compound	AgI	β-PbF_2
Family	Ag^+ and Cu^+ chalcogenides and halides	Fluorite-type halides and oxides
Other examples	CuI, CuBr, Ag_2S, etc.	CaF_2, $SrCl_2$, UO_2, CeO_2, etc.
Mobile species	Cations	Anions
Superionic transition	1st order structural (type-I)	Gradual transition (type-II)
Structure under ambient conditions	Zincblende, wurtzite, etc.	Cubic fluorite
Structure in the superionic state	*bcc*/*fcc* anion sublattice in combination with "liquid-like" cation sublattice	Fluorite structure and dynamic anion Frenkel-type defects

bcc = body-centered cubic; *fcc* = face-centered cubic.

which may be incorporated into either immobile or mobile sublattices. For example, the partial substitution of Rb^+ for Ag^+ in AgI yields $RbAg_4I_5$, the phase with highest cationic conductivity at room temperature as compared to known crystalline solids (0.25 S cm^{-1}; Figure 3.11), the activation energy for ionic transport of about 0.07 eV and a negligibly low level of the electronic conductivity ($\sim 10^{-9}$ S cm^{-1} at room temperature).

The structure of another promising family of ionic and mixed conductors, ABO_3 perovskites, is usually described using the Goldschmidt tolerance factor, $t_G = (r_A + r_O)/\sqrt{2}(r_B + r_O)$, where r_A, r_B, and r_O are the corresponding ionic radii. The stability of the cation-ordered perovskite arrangement does not coincide with the geometric requirements for fast ionic transport; as a rule, perovskite-type compounds display superionic behavior at significantly higher temperatures compared to the cation-disordered phases such as α-AgI. Nevertheless, electrolytes with high oxygen-ionic or protonic conductivity at elevated temperatures are necessary for many practical applications, including SOFCs, sensors, and steam electrolyzers. In order to identify promising perovskite-like oxide electrolytes, the following selection criteria can be used:

- A relatively low metal–oxygen binding energy, allowing O^{2-} anions to disengage and diffuse through the lattice.
- As a rule, B-site cations with variable valence states should not be introduced in the solid electrolyte compositions, as these species may contribute to the electronic conduction. Note that an opposite statement may be valid for MIECs.
- The tolerance factor should be relatively high, usually close to 0.96. This value often provides a compromise between the necessity of an ideal cubic lattice reached at $t_G = 1$, and that of a large "free volume" which grows with decreasing t_G. The requirement of possible maximum symmetry is associated with achieving energetic equivalence for all (or most) oxygen sites and an absence of geometric

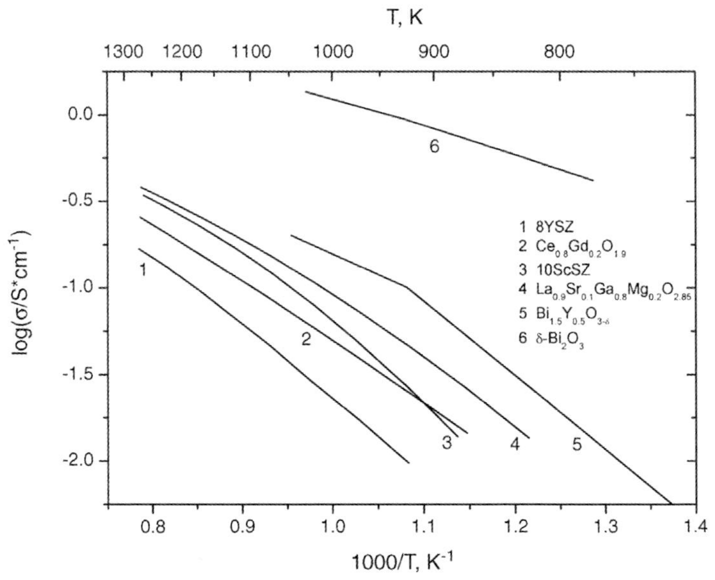

Figure 3.12 Conductivity of selected solid oxide electrolytes. 8YSZ and 10ScSZ correspond to 6 mol.% yttria- and 10 mol.% scandia-stabilized zirconias.

constraints; the maximum free volume is necessary to increase the space available for O^{2-} diffusion.

Examples of the oxide perovskites which essentially satisfy these criteria are acceptor-doped $BaCeO_3$ and $LaGaO_3$; the behavior of their derivatives is analyzed in Chapters 9, 12 and 13. As an example, Figure 3.12 compares the conductivities of several solid oxide electrolytes, including one representative of the lanthanum gallate-based family, $La_{0.9}Sr_{0.1}Ga_{0.8}Mg_{0.2}O_{2.85}$. One should note separately that the incorporation of dopants with a lower valence into both B- and A-sublattices leads to an increase of oxygen vacancy concentration due to the electroneutrality condition and, as a consequence, to the promotion of ionic conductivity at lower temperatures. The latter is the case when the dopant ionic radii are similar to those of the host cations. Moreover, the oxygen nonstoichiometry (δ) in an ion-conducting perovskite $ABO_{3-\delta}$ was shown to have optimum values close to 0.15–0.25, due to growing tendencies to vacancy interaction and clustering above this level.

The ionic transport properties of fluorite-type oxide phases (see Chapter 2), another important family of solid electrolytes, are also discussed in subsequent chapters. Briefly, for the well-known zirconia electrolytes, Zr^{4+} itself is too small to sustain the fluorite structure at moderate temperatures; doping with divalent (Ca^{2+}) or trivalent (e.g., Y^{3+}, Sc^{3+}, Yb^{3+}) cations stabilizes the high-temperature polymorph with the cubic fluorite-type structure. Due to the electroneutrality condition, anion vacancies are formed:

$$Y_2O_3 \xrightarrow{ZrO_2} 2Y_{Zr}^{/} + 3O_O^{\times} + V_O^{\bullet\bullet}, \quad (3.133)$$

which allow oxygen anions to migrate through the lattice and cause a substantial ionic conductivity at temperatures above 900–1000 K, whilst the electronic transport is sufficiently low. Owing to these properties being combined with a high thermomechanical stability, yttria-stabilized zirconia (YSZ) ceramics find important applications in SOFCs and oxygen sensors. As for the perovskites, all stabilized zirconias exhibit maxima in the conductivity versus dopant concentration dependencies, which is often explained by the interaction between the acceptor cations and charge-compensating oxygen vacancies. The origin of such an interaction is believed to be more complicated than simply a coulombic attraction, and is attributed to the elastic strain of the lattice due to a mismatch between the sizes of the host and dopant cations. Therefore, the best ion conductors in the fluorite-type systems are found among those in which the dopant and host cation radii fit each other to a maximum extent. Examples are $(Zr_{1-x}Sc_x)O_{2-x/2}$ (ScSZ) and $Ce_{1-x}Gd_xO_{2-x/2}$ (CGO).

3.4.2
Examples of Defect Chemistry in Electronic and Mixed Conductors

For any solid electrolyte, the contribution of electronic transport to the total conductivity should be negligible (normally <1%). Even a very low concentration of the electronic charge carriers may promote significant electronic conductivity, as the mobility of electronic carriers is much higher compared to that of ionic species. For this reason, most oxygen ion-conducting materials are in fact mixed conductors. However, the MIECs with a high level of both ionic and electronic transport find their own applications, particularly in the electrodes of fuel cells and batteries, in various sensors, and in dense gas-separation membranes. For high-temperature electrochemical cells, significant attention is focused on the perovskite and perovskite-related MIECs. For instance, the perovskite-type solid solutions derived from lanthanum cobaltite $LaCoO_{3-\delta}$ and manganite $LaMnO_{3-\delta}$ are widely considered for SOFC cathodes due to their electrocatalytic activity and high conductivity (see Chapters 9 and 12). Again, necessary properties are provided by extensive doping in both cation sublattices. The substitution of alkali earth cations (e.g., Sr or Ca) in the A sites leads to increasing hole concentration

$$\mathrm{Sr}_{\mathrm{La}}^{\times} + B_B^{\times} = B_B^{\bullet} + \mathrm{Sr}_{\mathrm{La}}^{/} \tag{3.134}$$

and/or oxygen vacancy formation

$$2\mathrm{Sr}_{\mathrm{La}}^{\times} + \mathrm{O}_{\mathrm{O}}^{\times} = 2\mathrm{Sr}_{\mathrm{La}}^{/} + V_{\mathrm{O}}^{\bullet\bullet} + \frac{1}{2}\mathrm{O}_2 \tag{3.135}$$

which is accompanied with an impressive growth of the electronic and ionic conductivities. A high reducibility of the 3d transition metal cations in the B-sublattice also favors vacancy formation at elevated temperatures

$$2B_B^{\bullet} + O_{\mathrm{O}}^{\times} = 2B_B^{\times} + V_{\mathrm{O}}^{\bullet\bullet} + \frac{1}{2}\mathrm{O}_2 \tag{3.136}$$

It is noteworthy that the ionic conductivity in heavily doped $La_{1-x}Sr_xCoO_{3-\delta}$ (LSC) and its derivatives is comparable to, or even higher than, that of most oxygen ionic conductors. In contrast, $La_{1-x}Sr_xMnO_{3-\delta}$ (LSM) has a relatively low ionic conductivity as compared to $La_{1-x}Sr_xCoO_{3-\delta}$; however, the thermal expansion of LSM is compatible with that of YSZ ($\alpha_L = (10–12) \times 10^{-6}\,K^{-1}$), whereas the CTEs of $La_{1-x}Sr_xCoO_{3-\delta}$ are excessively high ($>20 \times 10^{-6}\,K^{-1}$). Numerous efforts are therefore made to reduce thermal expansion of cobaltites, particularly via B-site doping.

In both perovskite- and fluorite-type phases, the ionic transport occurs by the anion-vacancy migration-based mechanism. Other mechanisms involving oxygen interstitials are known for perovskite-related oxides, such as K_2NiF_4-type $A_2BO_{4+\delta}$, where A corresponds to rare-earth or alkaline-earth cations and B = Ni, Cu, Co, and so on. In the latter structure, perovskite-like ABO_3 sheets are separated by rock-salt layers AO. Such lattices are able to incorporate an oxygen excess into the rock-salt layers; a network of interstitial positions can be partly occupied due to either: (i) oxygen uptake from environment; or (ii) via the intrinsic Frenkel mechanism when a vacancy is formed in the oxygen octahedron nearest to a given interstitial position. The first mechanism is often essentially relevant for the oxygen transfer along the rock salt planes; the second mechanism governs, in particular, anion migration in the perpendicular direction. The oxygen ionic conductivity in such phases has, therefore, an anisotropic character, unlike that of perovskites.

Perovskite-like oxides serve as the conventional example to illustrate how electrical properties can be varied from insulating via a pure oxygen-ion conductivity towards predominant electronic transport, depending on the nature of the B cations within the framework of the same crystal structure. An example of the electronically conducting phases refers to the materials derived from lanthanum chromite $LaCrO_{3-\delta}$, where the p-type electronic transport dominates over a very wide range of oxygen partial pressures, for example, $-16 \leq \log(Po_2, \text{atm}) \leq 0$ at 1270 K. This group of materials finds applications for SOFC current collectors and high-stability electrodes in sensors; another promising area relates to the developments of oxide SOFC anodes, but where the use of chromites is hampered due to low electrocatalytic activity. The holes localized on chromium cations (Cr^{4+}; or $Cr^{\bullet}_{Cr}$ in the Kröger–Vink notation) are formed due to acceptor-type doping, similar to Equation (3.134). However, the substitution of La for an alkaline-earth metal does not lead to the reaction analogous to Equation (3.136), as the enthalpy of B^{4+} formation from B^{3+} for B = Co or Mn cations is quite comparable to the ΔH values for the oxygen vacancy formation in LSC and LSM; in lanthanum chromite, the enthalpy of the latter process is significantly higher with respect to the oxidation $Cr^{3+} \rightarrow Cr^{4+}$. Thus, $La_{1-x}A_xCrO_3$ perovskites remain essentially oxygen-stoichiometric, even at relatively low oxygen chemical potentials and elevated temperatures. For instance, $La_{0.9}Ca_{0.1}$-CrO_3 retains its oxygen stoichiometry at 1273 K and a Po_2 of $\leq 10^{-12}$ atm. The defect structure associated with the $Cr^{3+} \rightarrow Cr^{4+}$ transition gives rise to an almost pure electronic conductivity provided by the hopping of small polarons $Cr^{\bullet}_{Cr}$ in the chromium sublattice, with a typical activation energy of 0.1 eV. As an example, the electronic conductivity of $La_{0.8}Ca_{0.2}CrO_{3-\delta}$ increases from ~10 S cm^{-1} at room temperature to approximately 100 S cm^{-1} at 1273 K in air. The ionic conductivity remains very low; for comparison, the oxygen vacancy diffusion coefficient D_{V_O} in $La_{0.7}Ca_{0.3}CrO_{3-\delta}$ is close to $10^{-10}\,cm^2\,s^{-1}$ at 1273 K [43], whereas in undoped $LaCoO_3$ [44] a value of $3 \times 10^{-6}\,cm^2\,s^{-1}$ is achieved at the same temperature. Due to the predominant electronic conduction and high stability under both oxidizing and reducing conditions, acceptor-doped lanthanum chromites currently represent state-of-the-art interconnect materials for SOFCs stacks, where a low oxygen permeability is necessary in order to avoid power losses.

References

1 Schottky, W., Uhlich, H. and Wagner, C. (1929) *Thermodynamik*, Springer, Berlin.
2 Wagner, C. (1930) *Z. Phys. Chem. B*, **11**, 139–51.
3 Wagner, C. (1935) *Phys. Z.*, **36**, 721–5.
4 Denbigh, K.G. (1951) *The Thermodynamics of the Steady State*, John Wiley & Sons, New York.
5 Hebb, M.H. (1952) *J. Chem. Phys.*, **20**, 185–90.
6 Kiujjola, K. and Wagner, C. (1957) *J. Electrochem. Soc.*, **104**, 379–87.
7 Yokota, I. (1961) *J. Phys. Soc. Jap.*, **16**, 2213–24.
8 Schmalzried, H. (1962) *Z. Elektrochem.*, **66**, 572–6.
9 Kröger, F.A. (1964) *The Chemistry of Imperfect Crystals*, North-Holland, Amsterdam.
10 Wagner, C. (1966) in *Advances in Electrochemistry and Electrochemical Engineering* (eds P. Delahay and C.W. Tobias), Wiley-Interscience, New York, pp. 1–46.
11 Wagner, J.B.(1973) in *Materials Science Research, Vol. 6: Sintering and Related Phenomena* (ed.G.C. Kuczynski), Plenum Press, New York, pp. 29–47.
12 Wagner, C. (1977) *Annu. Rev. Mater. Sci.*, **7**, 1–22.
13 Boyce, J.B. and Huberman, B.A. (1979) *Phys. Rep.*, **51**, 189–265.
14 Rickert, H. (1982) *Electrochemistry of Solids: An Introduction*, Springer, Berlin, Heidelberg, New York.
15 Majer, J. (2004) *Physical Chemistry of Ionic Materials*, John Wiley & Sons, Chichester.
16 West, A.R. (1999) *Basic Solid State Chemistry*, 2nd edn, John Wiley & Sons Ltd, Chichester.
17 Hannay, N.B. (1967) *Solid State Chemistry*, Prentice-Hall, New Jersey, USA.
18 Rao, C.N.R. and Gopalakrishnan, J. (1997) *New Directions in Solid State Chemistry*, Cambridge University Press,Cambridge.
19 Kofstad, P. (1972) *Non-Stoichiometry, Diffusion and Electrical Conductivity in Binary Oxides*,Wiley Interscience, New-York.
20 Kelly, A. and Groves, G.W. (1970) *Crystallography and Crystal Defects*, Longman, London.
21 Hirth, J.P. and Lothe, J. (1968) *Theory of Dislocations*, McGraw-Hill Book Co., New York.
22 Anderson, M.T., Vaughey, J.T. and Poeppelmeier, K.R. (1993) *Chem. Mater.*, **5**, 151–65.
23 Rao, C.N.R., Gopalakrishnan, J. and Vidyasagar, K. (1984) *Ind. J. Chem.*, **23**, 265–84.
24 van Doorn, R.H.E. and Burggraaf, A.J. (2000) *Solid State Ionics*, **128**, 65–78.
25 Andersson, S. and Wadsley, A.D. (1966) *Nature*, **211**, 581–3.
26 Adler, S.B. (2001) *J. Am. Ceram. Soc.*, **84**, 2117–19.
27 Chen, X.,Yu, J. and Adler, S.B. (2005) *Chem. Mater.*, **17**, 4537–46.
28 Armstrong, T.R., Stevenson, J.W., Pederson, L.R. andRaney, P.E. (1996) *J. Electrochem. Soc.*, **143**, 2919–25.

29 Larsen, P.H., Hendriksen, P.V. and Mogensen, M. (1997) *J. Therm. Anal.*, **49**, 1263–75.

30 Zuev, A., Singheiser, L. and Hilpert, K. (2002) *Solid State Ionics*, **147**, 1–11.

31 Miyoshi, S., Hong, J.-O., Yashiro, K et al. (2003) *Solid State Ionics*, **161**, 209–17.

32 Hilpert, K., Steinbrech, R.W., Boroomand, F. et al. (2003) *J. Eur. Ceram. Soc.*, **23**, 3009–20.

33 Atkinson, A. and Ramos, T.M.G.M. (2000) *Solid State Ionics*, **129**, 259–69.

34 Kharton, V.V., Yaremchenko, A.A., Patrakeev, M.V. et al. (2003) *J. Eur. Ceram. Soc.*, **23**, 1417–26.

35 Zuev, A.Yu., Vylkov, A.I., Petrov, A.N. and Tsvetkov, D.S. (2008) *Solid State Ionics*, **179**, 1876–79.

36 Zuev, A.Yu., Petrov, A.N., Vylkov, A.I. and Tsvetkov, D.S. (2007) *J. Mater Sci.*, **42**, 1901–8.

37 Kondepudi, D. and Prigogine, I. (1999) *Modern Thermodynamics From Heat Engines to Dissipative Structures*, John Wiley & Sons Ltd, Chichester.

38 Goodenough, J.B. (1984) *Proc. R. Soc. London, Ser. A*, **393**, 215–36.

39 Chandra, S. (1981) *Superionic Solids: Principles and Applications*, North-Holland, Amsterdam.

40 Hayes, W. (1978) *Contemp. Phys.*, **19**, 469–86.

41 Hull, S. (2004)*Rep. Prog. Phys.*, **67**, 1233–314.

42 Kharton, V.V., Marques, F.M.B. and Atkinson, A. (2004) *Solid State Ionics*, **174**, 135–49.

43 Yasuda, I., Ogasawara, K. and Hishinuma, M. (1997) *J. Am. Ceram. Soc.*, **80**, 3009–12.

44 Tsvetkov, D.S., Zuev, A.Yu., Vylkov, A.I. et al. (2007) *Solid State Ionics*, **178**, 1458–62.

4
Ion-Conducting Nanocrystals: Theory, Methods, and Applications

Alan V. Chadwick and Shelley L.P. Savin

Abstract

The aim of this chapter is to review the current state of knowledge in ionic materials with crystallite dimensions less than 100 nm, systems which sometimes are referred to as *nanoionics*. The chapter will detail the preparation, characterization and the important applications of these materials, especially in sensors, solid-state batteries, and fuel cells. Particular focus will be placed on ionic transport in these materials, as this is a topic of considerable contemporary interest, and where conflicting reports exist of enhanced diffusion in nanocrystals.

4.1
Introduction

Nanomaterials are systems that contain particles with one dimension in the nanometer regime. The past decade has witnessed a growing intense interest from biologists, chemists, physicists, and engineers in the application of these materials – the so-called "nanotechnology", which is sometimes referred to as "the next industrial revolution" [1]. The reasons for such interest are the unusual properties and potential technological applications that are exhibited by these materials when compared to their bulk counterparts [2–10]. In this chapter, attention will be focused on rather simple ionic solids, where the interatomic attractions are predominantly coulombic forces, and the dimensions are predominantly <100 nm. Such systems have been termed nanoionics [11, 12].

There are three dominant reasons for the study of these particular systems:

- The simplicity of the interatomic forces means that some of the generic features and problems of nanomaterials should be tractable to theoretical and/or computer modeling approaches for these particular systems.
- Many relatively simple ionic compounds, such as binary halides and oxides, in their bulk form have important technological applications as electrolytes, catalysts,

Solid State Electrochemistry I: Fundamentals, Materials and their Applications. Edited by Vladislav V. Kharton

ISBN: 978-3-527-32318-0

sensors, and ceramics, and there is the promise of improved performance by using them in nanocrystalline form.

- These materials were among the first nanomaterials studied over two decades ago as part of the pioneering investigations of Gleiter and coworkers [2, 3, 7], and there is sound background information on their on their preparation and characterization.

The key property of these systems on which attention will be focused is the transport of ions. This fundamental property underlies many of the important processes and applications associated with ionic solids, and is still not fully understood when the solids are in nanocrystalline form.

The origins of these unusual properties of nanomaterials are twofold:

- The fact that the dimensions of the particles approaches, or becomes smaller than, the critical length for certain phenomena (e.g., the de Broglie wavelength for the electron, the mean free path of excitons, the distance required to form a Frank–Reed dislocation loop, thickness of the space-charge layer, etc.).
- Surface effects dominate the thermodynamics and energetics of the particles (e.g., crystal structure, surface morphology, reactivity, etc.).

In nanostructured semiconductors it is the first of these factors which leads to special electrical, magnetic, and optical properties and to the possibility also of quantum dot (QD) devices. It is also an explanation of the unusual hardness, sometimes referred to as “super-hardness”, of nanocrystals [13].

The second factor can lead to nanocrystals adopting different morphologies to bulk crystals, with different exposed lattice planes leading to an extraordinary surface chemistry and catalytic activity [14]. The importance of surfaces and boundaries in nanocrystalline systems is demonstrated in Figure 4.1, which shows the fraction of atoms in these regions as a function of grain size.

The earliest experimental studies of self-diffusion in nanocrystalline metals yielded diffusion coefficients which were many orders of magnitude higher than the values found for bulk diffusion in single crystals [3, 15, 16], and even higher than the values found for grain boundary diffusion, which is usually regarded as the fastest diffusion process in a solid. The phenomenon has been regarded as generic to nanocrystals and independent of the interatomic bonding. An early explanation of the origin of this unusually fast atomic transport was that the interfaces between the grains in a nanocrystalline compact were highly disordered in comparison to the normal grain boundaries found in normal solids. The two types of interface are illustrated in Figure 4.2. The model assumed for a nanocrystalline sample is drawn schematically in Figure 4.2a, with extensive disorder in the interface that is several atoms in width. In this figure, the black circles represent atoms in the grains and the open circles are the atoms in the interfaces. In some of the early studies on nanocrystals this was intuitively assumed to be the case, and the interfaces were referred to as either “gas-like” or “liquid-like”. This structure would clearly account for rapid diffusion in nanocrystalline samples. More recently, however, an alternative view has emerged in which the nanocrystalline interface is similar to

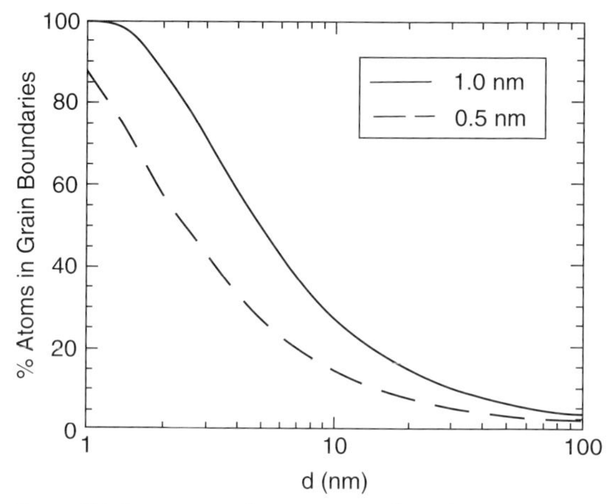

Figure 4.1 Percentage of atoms in grain boundaries as a function of grain size, assuming boundary widths of 0.5 and 1 nm. Reproduced with permission from Ref. [8].

a grain boundary in normal bulk materials, as shown in Figure 4.2b. In this case, the interfaces would exhibit usual behavior, although they would be present in unusually large numbers, and therefore the compacted nanocrystalline sample would show a higher diffusivity than would a coarse-grained counterpart. As many of the applications of ionic materials are based on their ability to transport charge, the use of nanocrystalline samples represents an obvious approach to improving performance.

The aim of this chapter is to present an overview of the current state of understanding of the structure and transport in ion-conducting nanocrystals. Interest here is at a fundamental level and, although a number of unusual architectures have recently

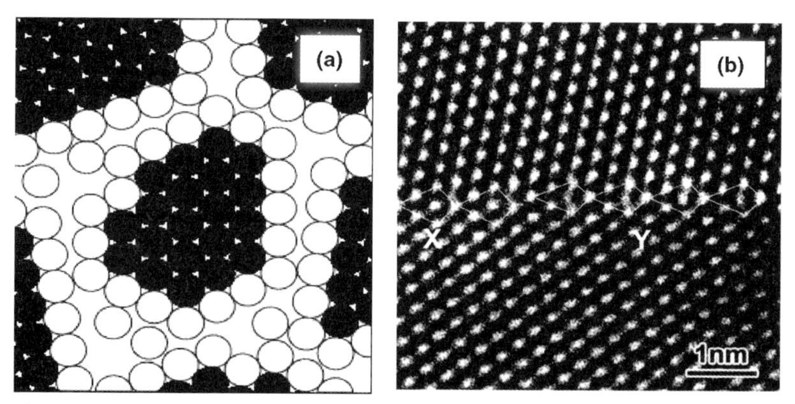

Figure 4.2 Two possible models for the interface between nanocrystalline grains.(a) A disordered interface [2]; (b) A "normal" grain boundary, a boundary in a ZnO bicrystal [17].

been developed for ionic and semi-ionic systems (e.g., nanowires [18], nanobelts [19], nanotubes [20], etc., and a range of nanocomposites), attention will be restricted to simple nanocrystals. This overview will first provide a survey of the models proposed to explain the reported unusual transport in nanoionic systems, after which the potential applications of the materials, notably in sensors, batteries, and fuel cells, are described, together with details of their preparation and characterization. A review is also provided of the experimental data acquired relating to transport in nanomaterials, in addition to an overview of the current state of knowledge and suggestions for future developments in this field.

4.2 Theoretical Aspects

Although the intuitive model of disordered interfaces between nanocrystals has provided an explanation of the reported high diffusion coefficients in these samples, it will be seen later that this is unrealistic. The nature of the interfaces depends on the method of sample preparation, and these can be highly ordered, akin to grain boundaries in bulk samples. Thus, other explanations must be sought and, in the case of ionic materials, these have been based on surface space-charge and surface texture approaches.

4.2.1 Space-Charge Layer

This is specifically a model that is applied to ionic crystals, and to outline the model it is necessary to begin with a brief description of basic defect thermodynamics and diffusion theory.

All real crystals above 0 K contain point defects which are thermodynamically inherent [21, 22]. In a monatomic crystal, the simplest defects are the *vacancy*, a lattice site that is empty, and the *interstitial atom*, an atom on an interstitial site in the lattice. The equilibrium concentration of these defects is thermally controlled and has an exponential dependence on temperature. For example, the site fraction of vacancies, c_v, in a pure monatomic crystal is given by:

$$c_v = \exp\left(\frac{-g_v}{kT}\right) \tag{4.1}$$

Here g_v is the Gibbs free energy to form a vacancy, k is the Boltzmann constant, and T is the temperature. Diffusion in a crystal lattice occurs by motion of atoms via jumps between these defects. For example, vacancy diffusion – the most common mechanism in close-packed lattices such as face-centered cubic (*fcc*) metals, occurs by the atom jumping into a neighboring vacancy. The diffusion coefficient, D, therefore will depend upon the probability that an atom is adjacent to a vacancy, and the probability that it has sufficient energy to make the jump over the energy barrier into the vacancy. The first of these probabilities is directly proportional to c_v, and the second is dependent on the energy of migration of the atom, Δg_m. Diffusion follows an Arrhenius law, that is

$$D = D_o \exp\left(\frac{-Q}{kT}\right), \tag{4.2}$$

where D_o is the pre-exponential factor and Q is the activation energy. For vacancy diffusion it follows that

$$Q = g_v + \Delta g_m. \tag{4.3}$$

The same defect thermodynamics and diffusion theory can be applied to ionic crystals with one important proviso, which is the need to account for the charges on the ions (and hence *effective charges* on the defects), and that the crystal must remain electrically neutral overall. This means that the defects will occur as multiplets to satisfy this later condition. For example, in a MX crystal they will occur as pairs: the *Schottky pair* – a cation vacancy and an anion vacancy; the *cation-Frenkel pair* – a cation vacancy and an interstitial cation; and the *anion-Frenkel pair* – an anion vacancy and an interstitial anion. The concentrations of the defects in the pair are related by a solubility product equation, which for Schottky pairs in an MX equation takes the form:

$$c_+ \cdot c_- = K_S = \exp\left(\frac{-g_S}{kT}\right) \tag{4.4}$$

Here, c_+ and c_- are the site fractions of the cation and anion vacancies, respectively, K_S is the equilibrium constant for the formation of the Schottky pair, and g_S ($= g_+ + g_-$, the sum of the individual defect formation energies) is the Gibbs free energy to form the pair. In the bulk of a pure crystal, the condition of electrical neutrality demands that the concentrations of each defect in the pair are equal; that is:

$$c_+ = c_- = K_S{}^{1/2} = \exp\left(\frac{-g_S}{2kT}\right) \tag{4.5}$$

The condition of electrical neutrality will not apply at the surface of a crystal, and since g_+ is not equal to g_- there will be an excess of one of the defects. This effect, which is referred to in the early literature as the Frenkel–Lehovec space charge layer – results in an electric potential at the surface of the crystal [23–25]. In this instance, the surface will not simply be the external surface but will also include internal surfaces such as grain boundaries and dislocations. The effect decays away in moving from the surface to the bulk, and can be treated by classical Debye–Hückel theory [26–29]. This leads to a Debye screening length, L_D, given by:

$$L_D = \left(\frac{\varepsilon_r \varepsilon_0 kT}{2q^2 c_b}\right)^{\frac{1}{2}} \tag{4.6}$$

Here, ε_o and ε_r are the permitivities of free-space and the sample, respectively, and c_b is the concentration of the bulk majority carrier with charge q. For a solid with $\varepsilon \sim 10$ and $c_b \sim 10^{22}\ \mathrm{m}^{-3}$ and $T \sim 600$ K, this leads to a Debye length of ~ 50 nm and a space charge width of approximately twice that value. The qualitative effect on the

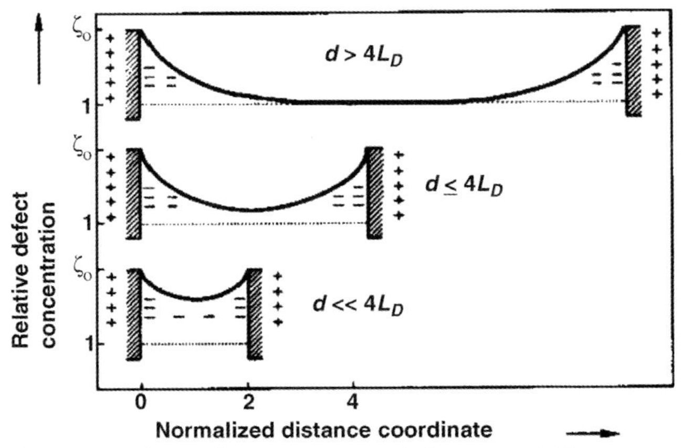

Figure 4.3 Defect profiles in structures with dimension, *d*. The bulk defect concentration is not reached when $d < < 4L_D$, where L_D is the Debye length. Reproduced with permission from Ref. [26].

relative concentrations of the two defects of the pair, ζ_o, is shown schematically in Figure 4.3 as a function of grain size. Clearly, this increased defect concentration will translate into enhanced ionic diffusion and conductivity in the space-charge region – that is, in the plane of the interface. If the grains are sufficiently small this can dominate the ionic transport in the sample and far outweigh the transport in the bulk of the grains.

The space-charge layer region described above applies to a sample made purely of small grains of one ionic compound. There will also be space-charge layers at the interfaces between ionic crystals and other materials, such as other ionic crystals, insulators, semiconductors, and metals. The theoretical treatments of the effects in these systems can be found in an extensive review [28]. There is a difference in the details for each situation, although essentially there will be a difference in the electrochemical potential of the ions across all the interfaces, leading to an imbalance in the relative defect concentrations in the surface of the ionic crystal and the generation of a space-charge layer. The approach was particularly successful in explaining the high ionic conductivity of materials known as "composite electrolytes", "heterogeneous electrolytes", or "dispersed second phase electrolyte" [30]. These materials consist of an ionic conductor mixed intimately with a fine-powdered insulator, such as alumina, silica, and titania. The ionic conductivity can be increased by up to two orders of magnitude compared to that found in the pure ionic material. This increased conductivity is due to an enhanced conductivity in the interfacial region, arising from the increase in charge-carrying defects in the ionic component. A combination of the space-charge layer approach and percolation theory has provided quantitative models of the dependence of the conductivity on composition, grain size of the insulator, and temperature [28]. Clearly, a similar approach can be applied to thin films of an ionic conductor laid down on an insulating surface, where an enhanced conductivity is possible due to the space-charge layer at the interface.

4.2.2
Surface Texture and Mismatch at Surfaces

The very earliest experimental investigations of diffusion in solids noted that diffusion along the surface and grain boundaries of solids differed from that in the bulk (a historical review can be found in Ref. [31]). In general, diffusion was found to occur faster in the grain boundaries [32], although the situation was complicated in some systems when impurities were segregated at the boundaries and blocked the diffusion. Although this phenomenon was identified in all types of solids, it was most widely studied in metals and hence the effect was shown not to be due to a space-charge layer at the interface. In simple terms, the origin of a faster diffusion in the grain boundaries was that the point defects responsible for diffusion (vacancies in the case of metals) had a lower formation and migration energy in the disordered interface region, where there were variations of lattice spacing and a high lattice strain. Thus, diffusion not only occurs faster in the grain boundaries but also has lower activation energy, Q. As an example, in the widely studied case of nickel oxide in comparison to lattice diffusion, both nickel and oxygen were seen to diffuse faster by several orders of magnitude in high-angle grain boundaries [33]. In some extreme cases, however, the high strain was seen to result in the formation of a different polymorph of the material. In recent years, studies of grain boundary diffusion have become more sophisticated [34]. Many of the early studies were performed on samples with an ensemble of grain boundaries and a range of orientations, whereas more recent studies of oriented metal bicrystals have been performed such that data now exist linking the dependence of diffusion on the mismatch tilt angle of the grain boundary. The results of these studies showed that, when a large mismatch occurred at the grain boundary (incoherent interface), diffusion was significantly faster than when there was a good match (coherent interface).

These effects can clearly be quite diverse and very sample-dependent. For example, in a compacted sample of nanocrystals of a single ionic compound there will be a wide variety of interfaces; moreover, even if these are like grain boundaries in the bulk samples, there will be a range of mismatch angles. Thus, the enhanced diffusion or conductivity that is measured will be an average over all the types of interface in the sample. In cases where the nanomaterial is prepared as a thin film on a substrate, the degree of mismatch at the interface will be a crucial factor in determining the atomic transport in the interfacial region. At a later date, these variations in interfacial structure led to problems when different research groups attempted to compare measurements of the same basic system (see below).

4.3
Applications and Perspectives

The obvious physical advantages of using nanomaterials to create devices relate to reductions in both size and mass. Consequently, increasing effort is being applied to build devices, at both the microscale and nanoscale, that can be applied to the new generation of microelectromechanical systems. Yet, surprisingly, these advantages arise basically from changes in architecture and scale rather than from any benefits accrued from the specific properties of nanomaterials, which form the focal point of this section.

4.3.1 Nanoionic Materials as Gas Sensors

Metal oxides have been investigated extensively as possible sensing materials for a wide range of gases [35, 36]. For example, tin oxide has been shown to respond to a range of flammable gases, including carbon monoxide [37–39], hydrocarbons [39–41], hydrogen [42, 43], as well as other pollutant gases such as hydrogen sulfide [44, 45] and oxides of nitrogen [38, 46]. In this role, these metal oxide solid-state gas sensors exploit the inherent properties of these materials, and notably that of semiconductivity.

Although the fact that significant changes occurred in the resistance of a semiconductor when a gas was adsorbed onto its surface was first reported over 55 years ago [47], the use of a metal oxide semiconductor was not suggested until 1962 [48]. Since that time, many commercial devices exploiting this phenomenon have been developed [49], and whilst several different oxides [47, 50–58] have been investigated in the role, tin oxide has been the most widely used.

When metal oxide semiconductors are exposed to oxidizing or reducing gases, a change is observed in their electrical resistance, and this forms the basis of their use as a sensor material. The mechanisms involved in the detection of flammable gases and vapors using metal oxide-based materials relate principally to the oxide's surface, where defects will not only modify the electronic properties but also act as catalytic sites for the combustion of flammable gases. Although, several models have been postulated to explain the observed response of tin oxide to flammable gases in particular [59–61], one important factor in the response of a material to a gaseous species is its morphology, as this will affect the distribution of the gas throughout the bulk of the solid. For example, if a material has regions of closed porosity, then its interior may be inaccessible to the gas in question, and only the conductivity of its exterior surface will respond to changes in the ambient atmosphere.

Ideally, sensor materials should exhibit a fast reversible response to a specific target gas; that is, they should be both sensitive and selective to the gaseous species in question. Many metal oxide materials respond not only to a wide range of gases and vapors (which means that they are essentially nonselective) but also to environmental factors such as oxygen concentration, pressure and humidity. Consequently, a range of techniques, such as doping or the addition of catalysts, have been investigated to improve both sensitivity and selectively. The most widely used catalysts in tin oxide-based materials are the noble metals Pd [39, 62–64], Pt [39, 64, 65], Au [39, 66], Rh [67], and Ru [65, 67]. The use of a catalyst results in an increase in the sensitivity and speed of the response to flammable gases at low temperatures, while the high-temperature response is diminished [68]. However, the selectivity is generally not

Table 4.1 The range of dopants added to tin oxide-based sensing materials, and their effects.

Target gas	Dopant	Effect of dopant
H_2S	CuO	Improves sensitivity [44, 69]
	ZrO_2	Improves sensitivity and selectivity over H_2 [45]
Ethanol	La_2O_3	Improves sensitivity [69, 70]
	Cd	Improves selectivity over CO, H_2, and i-C_4H_{10} [71]
CO	Si	Improves response [72]
	MoO_x	Reduces response [38]
	Bi	Improves selectivity relative to CH_4 [73]
	CuO	Reduces optimum operating temperature [74]
VOCs	Cr	Improves sensitivity [75]
NO_2	Si	Improves response [72]
	In	Improves sensitivity [76]
	MoO_x	Improves response of materials annealed at high temperature [38]
	Al	Improves sensitivity [76]
NO	CuO	Improves sensitivity [77]
CH_4	Os	Lowers optimum operating temperature [78]
	Ca	Improves sensitivity [41]
H_2	Cd	Improves selectivity over CO and i-C_4H_{10} in absence of ethanol [71]
	ZnO	Increases selectivity over CO [74]
	ThO_2	At temperatures $>250\,^\circ C$, resistance decreases; at temperatures $<80\,^\circ C$, resistance increases [43]

improved as the catalysts will oxidize most gases and vapors, but may also offer some resistance to poisoning by water vapor.

The use of other metal oxides or cations as dopants has been widely reported. The addition of a second metal oxide may provide unique adsorption sites on the surface, and thus allow the catalysis of specific reactions. The incorporation of dopants will modify the electronic properties of the material by creating discrete donor or acceptor levels within the band gap. The presence of a dopant may also introduce defects that affect the conductance, and perhaps also alter the morphology of the material. A range of dopants investigated in tin oxide-based sensing materials is provided in Table 4.1, along with the target gases and the effect of the dopant on both the sensitivity and/or selectivity of the material. It should be noted that many reports for improving both sensitivity and selectivity for different gases use the same dopant. However, these differences may result from the preparation and sample history, or simply because the response to only one gas was investigated.

It can be seen from the above discussion that the use of nanoionic materials for gas sensors is a natural extension to the findings already reported. Nanosized materials offer advantages in terms of improved sensor response due to the much higher surface areas available. However, the definition of a nanoionic sensor material can be very broad; the nanoionic component of the sensor material might refer to the "bulk" majority phase, but alternatively it could refer to a dispersed catalytic or dopant phase, or even a combination of both.

In the case of tin oxide-based sensors, the most widely recognized method for enhancing sensitivity is to use nanocrystalline tin oxide [42, 69, 79]. The sensitivity of tin oxide-based gas sensors has been shown to be enhanced for grain sizes of the order of $2L_D$ (i.e., ca. 5 nm), as the depletion layer becomes more dominant [42, 79]. This is in agreement with the results of Hall effect measurements which showed that, for a porous tin oxide thin film, the Debye length is approximately 3 nm at 300 °C [80]. While it is relatively easy to obtain both pure and doped nanocrystalline tin oxide with grain sizes of <5 nm, the high operating temperatures required for the combustion reactions with flammable gases tend to induce the growth of nanocrystals, and this will result in problems with both the sensitivity and reproducibility of the sensing materials. For example, grain growth in nanocrystalline tin oxide has been observed at temperatures as low as 400 °C [81], well within the typical operating temperatures, but this leads to a corresponding decrease in the gas sensitivity [82] due to a reduction in the surface area. While several methods have been investigated to overcome this limitation when using nanocrystalline tin oxide, two main approaches have been taken. The first method is to reduce the operating temperature of the sensor material by introducing secondary metal oxides or metal particles as surface catalysts [66, 78]. The second method is to physically inhibit the grain growth by introducing a second metal oxide such as SiO_2 [83, 84], Bi_2O_3 [85], or In_2O_3 [86] in the case of tin oxide. An added advantage of this approach is that the second metal oxide phase may not only inhibit grain growth but also improve the selectivity of the sensing material to a given species. For example, nanocrystalline TiO_2 has been investigated for its response to CO and NO_2 [87]. The introduction of either Ga_2O_3 or Er_2O_3 was also shown to inhibit the anatase-to-rutile transformation of the TiO_2 nanocrystals and to retard grain growth. The best sensitivity to CO was obtained when Ga_2O_3 was used, whereas the presence of Er_2O_3 resulted in the best sensitivity to NO_2 at 200 °C.

Despite many investigations having been carried out in the past on the use of nanocrystalline tin oxide in sensors, extensive research is continuing in this area. One recently developed method for obtaining a tin oxide-based sensor for hydrogen [88] which operates at 100–130 °C (rather than 350–400 °C [89]) involved combining a porous nanocrystalline thin film, prepared via a sol–gel route, with a pulse heating mode. This combination of low temperature and pulse heating also reduced the effects of humidity on sensor behavior, and improved the selectivity of the sensor to hydrogen over CO. Other approaches for improving the sensitivity and selectivity of nanocrystalline tin oxide-based sensors have included the use of surface modification to inhibit the response to moisture [90], thus improving the response to methane. The incorporation of other nanomaterials, such as single-walled carbon nanotubes (CNTs) has also been investigated [91]. The sensitivity of this nanocomposite sensor to hydrogen was greatly increased when compared to a pure nanocrystalline tin oxide sensor at 250 °C. This effect was attributed to the hollow CNTs allowing a much greater diffusion of the hydrogen to the bulk of the tin oxide film. Whilst the expected linear relationship between hydrogen concentration and sensor response was not achieved over the entire concentration range examined, this was considered due to issues with the desorption of hydrogen from the CNTs.

The resistance of the nanocomposite film was also greater than that of the pure tin oxide film, most likely because the presence of the CNTs had led to the introduction of pores. Nonetheless, this illustrates the potential of these nanocomposite-type materials – where both phases are nanocrystalline – in the field of gas sensors.

An alternative approach, in the case of nanocrystalline tungsten oxide and molybdenum oxide, is to improve the selectivity to a given analyte through careful control of the crystalline polymorph used [92]. For example, nanocrystalline thin films of hexagonal-WO_3 were prepared by spin coating with crystallite sizes of 30–50 nm [93]. These films were found to provide a linear response over a concentration range of 50–500 ppm ammonia at 300 °C. When the response of nanocrystalline hexagonal-tungsten oxide to NO_2 was investigated [94], the response to NO_2 was found to be optimal at 250 °C but to decrease to zero at room temperature. In contrast, the inclusion of a low concentration of metal-containing (Ag, Au) CNTs into the tungsten oxide matrix resulted in a sensor material that exhibited a response to NO_2 at room temperature. Again, this illustrates the potential of these materials in the field of gas sensors.

Other, simple binary oxides which exhibit semiconductivity have also been widely investigated. For example, when monitoring the effect of the thickness of nanocrystalline films of CdO on the response to liquefied petroleum gas (LPG), the response was seen to be governed by a combination of film thickness, particle size, film morphology, and operating temperature [95]. At the optimum operating temperature of 698 K, a maximum response was exhibited for a 1.5 μm-thick film, which also showed the most uniform morphology and offered the highest surface area for interaction with the LPG. While the response of thinner films was shown to depend on the crystallite size, the response decreased for thicker films.

Although the above-described examples have all been based on relatively simple nanocrystalline metal oxides, additional phases might have been introduced. The use of more complex metal oxides has also been investigated, with nanocrystalline thick films of both barium titanate [96] and cobalt titanate [97] having been considered as possible sensor materials. When the response of such barium titanate films doped with 10% CuO and 10% CdO was studied with respect to CO, LPG, H_2S, and H_2 [96], sensor selectivity was improved for LPG over the other gases at 250 °C. However, the addition of 0.3 wt% Pd resulted in an even greater selectivity to LPG at a lower temperature, of 225 °C.

Nanocrystalline $CoTiO_3$ thick films doped with La, Sm, Gd, Ho, Li, Na, K, Pb, and Sb at 2 atom% were shown to exhibit p-type semiconductivity and to respond to both ethanol and propylene [97]. Both, the particle size and response also depended on the dopant used, with the undoped material showing no pronounced selectivity to ethanol. The temperature of maximum selectivity was also shown to depend on the dopant used. La-doped $CoTiO_3$ was investigated further as it exhibited the greatest ethanol sensitivity, along with good response and recovery times. In contrast to many tin oxide-based sensors, this material showed a minimal response to changes in humidity, which suggests that it might be effective as a sensor material.

Finally, complex nanocrystalline oxide-based materials such as $Gd_{0.9}Sr_{0.1}CoO_3$ [98] and $La_{0.7}Pb_{0.3}Fe_{0.4}Ni_{0.6}O_3$ [99] have also been studied. In both cases, the response of the pure material to CO_2 and O_2 for the former, and to H_2S for the latter, were improved by the addition of Ag or Pd, respectively.

These examples illustrate the wide range of materials and target gases investigated using nanoionic sensors. However, in order to fully exploit this field, much remains to be done in terms of optimizing not only the sensitivity and selectivity of the material but also the fabrication of the devices, as well as acquiring a basic understanding of the sensing mechanisms.

4.3.2
Nanoionics as Battery Materials

In this section, attention will be focused on the rechargeable lithium ion batteries that are now commonplace, and power many types of portable electronic device. Nonetheless, research in this area remains extensive, as demonstrated by the numerous reports and reviews that have appeared during the past few years [100–106] (see also Chapters 5 and 7). The current research can be grouped into three broad categories: (i) to acquire an understanding of the fundamental mechanisms observed in lithium ion batteries; (ii) the exploration of new and improved materials for both the electrode and electrolyte components of the batteries; and (iii) the investigation of new and improved routes for the fabrication of battery materials. Nanoionic materials have roles to play in all three categories, and examples of each will be briefly discussed here.

Nanoionic binary oxides, such as CoO, NiO, and Co_3O_4 have been shown as interesting candidates for the negative electrode of lithium ion batteries when fully discharged to near-zero voltage versus Li [107–110]. They offer a new concept, in that they involve conversion reactions rather than intercalation or insertion at the positive electrode. The processes which use a nanoparticle CoO positive electrode are summarized in Figure 4.4. The electrochemical reduction of the transition metal elements to the metallic state occurs simultaneously with the formation of lithium oxide. The partial reversibility of the reduction process in cobalt oxides has been attributed to the nanosize Co particles which are produced during reduction and dispersed in the lithium oxide matrix. These nanosize Co particles possess an enhanced reactivity, while the complex Li_2O/Co system is surrounded by a solid electrolyte interface [110]. The electrochemical processes that occur during the charge/discharge cycles have been the focus of significant interest, in particular the use of *in situ* experiments carried out on prototype lithium batteries [111–114]. For example, X-ray absorption spectroscopy (XAS) has been used to investigate the insertion/extraction of lithium in $NiCo_2O_4$ spinel oxide [115]. Reversible capacities of approximately 884 $mA\,h\,g^{-1}$ versus Li [116] have been observed when $NiCo_2O_4$ spinel oxide has been used as the active electrode material in lithium cells. XAS spectra were collected on composite electrodes consisting of nanocrystalline $NiCo_2O_4$ mixed with polyvinylidene fluoride (PVDF) and carbon black supported on a copper foil and used as the positive electrode in Li- and Na-anode cells. The normalized Co K-edge spectra of the as-prepared and used electrodes show a very weak absorption pre-peak representing the transition of the 1 s electron to

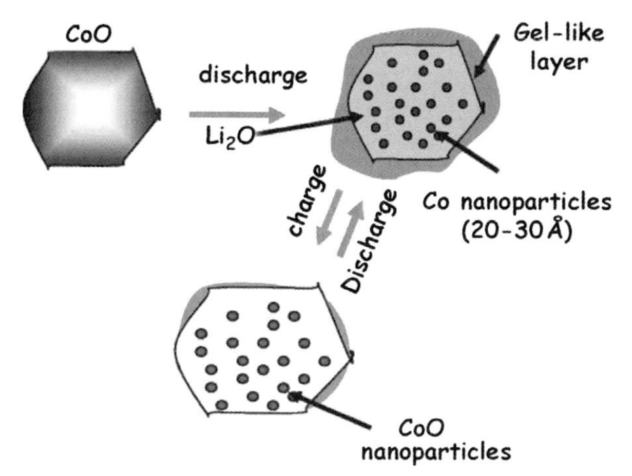

Figure 4.4 Schematic of the conversion reaction process showing both the formation of Co nanoparticles and of the gel-like polymeric layer, as well as their evolution upon subsequent charges/discharges [110].

an unoccupied e_g orbital of cobalt ions with a low-spin electronic configuration, and was ascribed to the presence of tetrahedral cobalt in the structure of $NiCo_2O_4$, which is commonly described as an inverse spinel [117]. The main absorption peak has been ascribed to the transition of a 1 s core electron to an unoccupied 4p bound state with t1g symmetry, and thus is electric dipole-allowed for both tetrahedrally and octahedrally coordinated cobalt atoms. A comparison of the X-ray absorption near edge structure (XANES) spectra of $NiCo_2O_4$ and Co_3O_4 shows that they are virtually identical, but with a very slight shift indicating that the Co oxidation states are the same in both (i.e., 8/3 or 2.667). Hence, the charge on the Ni is also 2.667, which suggests that some Ni^{3+} might be present.

The changes in the Co K-edge XANES spectra of the $NiCo_2O_4$ electrodes show the displacement of the edge to lower energies from the start of the discharge process; this indicates an immediate cobalt reduction process in $NiCo_2O_4$, resulting in metallic cobalt on approaching to zero volts. In addition, a progressive loss of signal intensity is observed, which can be related to the loss of p character in the 1 s $\rightarrow$ 4p transition, due to d-p mixing, as a consequence of the conversion to metal cobalt particles absorbing X-rays at the lower energy values. The reversibility of the reduction process after lithium extraction up to 1.5 V is shown by the fact that the spectra recovery partially their shape. However, the maximum oxidation state of the cobalt atoms is not fully recovered, which is in agreement with the non-recovered irreversible capacity. In addition, the changes in oxidation state of the nickel atoms were also followed using X-ray absorption spectroscopy. The extended X-ray absorption fine structure (EXAFS) results for the fully lithiated spinel, $NiCo_2O_4 + 9\,Li$ are particularly interesting. The Fourier transform shows that, while the electrode has clearly been converted to metallic nickel, the intensity of the plots is significantly attenuated compared to the nickel metal foil. This could be attributed to the presence of small or disordered particles. A detailed analysis showed that the particles would have to be extremely small (i.e., ~1 nm) for this attenuation to be due to solely to particle size. The presence of very disordered Co/Ni alloy particles would not be surprising as they are being formed in a disordered mixture of Li and lithium oxide and, as a result, the attenuation observed was attributed to a combination of these two possibilities.

Currently, new materials and adaptations to existing materials and their use in lithium ion batteries, are also undergoing extensive investigation. For example, the surface modification of cathode materials such as $LiCoO_2$, $LiNiO_2$ and $LiMn_2O_4$ with oxides such as MgO, Al_2O_3, ZnO, SnO_2 and ZrO_2 have been studied [118]. These coatings have been shown to act in a variety of ways, such as improving structural stability, reducing any phase transitions, and decreasing the disorder of cations in the crystal lattice, in addition to simply preventing direct contact with the electrolyte solution. These have the effect of improving the reversible capacity, the coulombic efficiency, the cycling behavior, and the high rate capability of these materials. However, our understanding of the different actions of different coatings and the mechanisms of these coatings remains incomplete.

Other types of positive electrode materials include $Li_xNi_2(MoO_4)_3$, where $0 \leq x \leq 4$, have also been investigated [119]. In this case, good discharge/charge profiles with a reversible capacity of 170 mA h g^{-1} over the potential window of 3.5–1.5 V after the first cycle were observed. However, it was found that the discharge capacity deteriorated slowly upon repeated cycling, this being tentatively attributed to a disproportion reaction. Nanocrystalline $Li_4Ti_5O_{12}$ has been investigated as a possible anode material [120], and showed an exceptional retention capacity at a discharge rate as high as 10 C in Li-test cells. In addition, carbon nanocoatings on $Li_4Ti_5O_{12}$ and TiO_2 investigated as cathode materials have been shown to result in a significant increase in the electron conductivity and a very large increase (>100%) in the reversible capacity at a C/3 rate [121]. It was also found that the gradual deterioration in the reversible capacity of TiO_2 could be prevented by using this type of carbon coating.

Finally, the use of various synthetic routes, including the sol–gel process [122], rheological phase [123], precipitation/decomposition [124], hydrothermal [125] sonochemical [126] and combustion reactions [121], are all currently being widely investigated for the production of nanoionic materials for use in lithium ion batteries.

4.3.3 Nanoionic Materials in Fuel Cells

Although fuel cells are not yet as widely used as lithium ion batteries, their potential commercial applications are finally being realized [127–130]. The basic fuel cell consists of an anode compartment (with the fuel), an electrolyte membrane, and a cathode compartment (with the oxidant). The systems of interest for this chapter are the solid oxide fuel cells (SOFCs), as these utilize a number of components that are ionic in nature (see Chapter 12). As with sensors and batteries, the application of nanoionic materials to SOFC research is of major interest, and can be grouped into two areas: (i) the design of new and improved materials for both the electrode and solid electrolyte components of the cells; and (ii) the investigation of new and improved routes for the fabrication of cell materials.

The most commonly used electrolyte materials in SOFCs are based on zirconia and ceria doped with a suitable cation, normally a rare earth (see Chapter 9). The properties that make these two materials attractive for use in fuel cells are discussed in Section 4.4.4, and it is sufficient to note that the most important feature is that they are good oxygen ion conductors. We will focus here on some recent investigations of these materials, with emphasis placed on their methods of preparation. A reduction in the particle sizes of zirconia and ceria has been reported as resulting in an increase in conductivity, which is considered important for improving the performance of SOFCs. Many methods have been developed to synthesize nanocrystalline ZrO_2 and CeO_2, and a brief review of the syntheses and the grain size dependant conductivity has recently been published [131]; hence, only a few examples will be provided here to illustrate the range of investigations being carried out.

- Homogeneous $Ce_{0.8}Sm_{0.2}O_{1.9}$ powders were prepared using a carbonate coprecipitation route to provide particles of approximately 12 nm [132]. The powders were found to possess a high sinterability, high conductivity, and a low activation energy, and were used in both the anode and electrolyte in single cells. The powders also showed promise for use in low-temperature (600 °C) SOFCs, with a maximum power density of 400 mW cm^{-2}.

- An alternative method for producing ceria-based nanocrystalline powders is that of urea-combustion [133]. In this case, the combustion synthesis of $(CeO_2)_{0.92}$ $(Ln_2O_3)_{0.04}(CaO)_{0.04}$ systems, where Ln = Y, Gd or Sm is described, but no investigation into their use in fuel cells was reported.

- A third technique used to prepare both Gd-doped CeO_2 and NiO/YSZ powders is that of aerosol flame deposition [134, 135]. In the case of NiO/YSZ powders used for the anodes (see Chapter 12), nanosized spherical particles were obtained and the particle size distribution could be limited by controlling the processing parameters [135]. Subsequently, the electrical conductivity of 70% NiO/YSZ was found to be 10^{-1} S cm^{-1} at 700 °C, with the material exhibiting typically metallic behavior. However, the use of such materials in fuel cells was not reported.

An example of the use of nanoionic materials as the cathode in a SOFC is that of nanotubes of $La_{0.6}Sr_{0.4}CoO_3$ and $La_{0.6}Sr_{0.4}Co_{0.2}Fe_{0.8}O_3$ [136]. These perovskite-type mixed oxides (see Chapters 9 and 12) are widely used as cathode materials, and the nanotubes were prepared by denitration, microwave irradiation and calcination at 800 °C. The shape and size of the nanotubes was controlled by the porous template used, although their application in a fuel cell was not reported.

4.4
Experimental Methods

In this section, attention will be focused on the preparation and characterization of nanocrystals, paying particular attention to methods that probe the microstructure and the atomic transport. It should be noted that an excellent review of the chemistry of nanocrystalline oxides, together with coverage of many of the characterization methods employed, is available [137].

4.4.1
Preparation of Nanoionic Materials

Numerous methods have been used to prepare nanoionic materials, and only the more generally employed procedures are considered here. Initially, the methods used to produce relatively large quantities of powdered sample will be considered.

Spray pyrolysis has been widely used to prepare nanocrystalline metal oxides such as Al_2O_3, ZnO, and ZrO_2 [138], and involves dispersing a solution of a chemical precursor as an aerosol. The droplets are transported to a hot zone where they decompose to form the required nanocrystalline oxide particles.

In contrast, *inert gas condensation* (IGC) has been used to prepare both nanocrystalline metallic and metal oxide materials [139]. In this case, the metal is evaporated inside an ultrahigh vacuum chamber, filled with an inert gas at low pressure. The vapors from the hot source migrate into the cooler gas, and eventually nucleate into a large number of clusters that grow, via coalescence and agglomeration. The clusters, entrained in the cooling gas, are then transported to a liquid nitrogen finger from which they can be scraped off and compacted. A schematic representation of the IGC set-up is shown in Figure 4.5. This technique enables the production of powders with a well-defined, narrow size distribution [2, 139]; however, as only small amounts of material are produced, several modifications have been developed aimed at increasing the yield [140–143].

Sol–gel methods, which offer control over both the composition at the molecular level and the structure [144, 145], have been used for many years. Typically, a metal alkoxide, $M(OR)_x$, undergoes replacement of the OR group by OH via controlled hydrolysis. This results in the formation of very small colloidal particles (the sol), which then form a gel via condensation reactions. The gel is then dried to produce an oxide (as in the case of silicon tetraethylorthosilicate), a hydroxide (zirconium iso-propoxide), or a mixed methoxy-hydroxide (magnesium methoxide). Finally, calcination at high temperatures is required to produce the oxide. This final step is difficult to control and presents two major problems, as illustrated by recent studies with ZrO_2 [146]. A sufficiently high calcination temperature is required to ensure that all of the residual OH are removed from the material; however, if the temperature is too high, the particle will grow and result in a loss of the nanocrystalline material. The surface energy of nanocrystals is such that this can occur at relatively modest temperatures (ca. 400 °C in the case of most nanocrystalline oxides, where measurable grain growth is observed after a few minutes [81]). This growth can be restricted

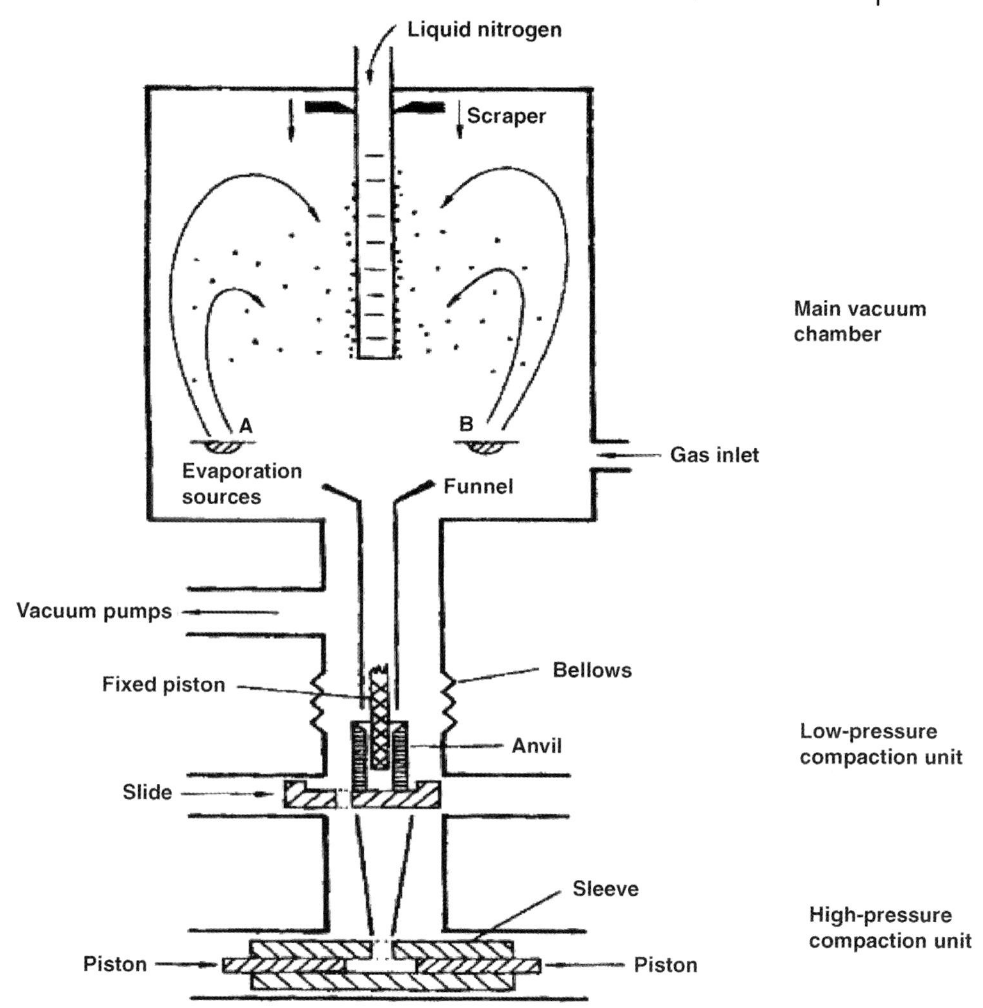

Figure 4.5 Inert-gas condensation facility for the synthesis of nanocrystalline particles. Reproduced with permission from Ref. [2].

by introducing a dispersion of very small particles of a second phase (in the case of oxides, an inert oxide such as silica or alumina) to stabilize the highly curved surface of a nanoparticles; this effect is known as *Zener pinning* [147]. The second phase can be introduced at the sol–gel stage by adding the corresponding alkoxide of the inert material. This method has proved very successful for a wide range of oxides [83, 84, 148, 149].

Finally, *mechanical attrition* has been widely used to produce all forms of nanocrystals. Typically, a high-energy ball mill is used to reduce the grain size of the bulk material [150–153], which means that almost every material is accessible. A further advantage is that control of the final grain size can be achieved by the careful choice of milling time. In addition, large amounts of material can readily be produced, while the use of double decomposition reactions [154, 155] makes it possible to produce materials *in situ*. However, a careful choice of the milling vial and balls is required as contamination can occur through abrasion of the milling media. One significant disadvantage of this system is that the milling can result in the presence of significant quantities of amorphous debris. This has been illustrated in recent studies on ball-milled Al_2O_3, where the sample was shown to consist of nanocrystalline grains embedded in an amorphous material [156].

In general, these methods are used for the production of nanocrystalline powders which may be further compacted via techniques such as hot-pressing [157, 158] or magnetic pulsed compaction [159, 160]. In addition, other types of nanoionic material maybe prepared, such as nanometer-thin films, using techniques including molecular beam epitaxy [161], pulsed laser deposition [162] or spin-coating methods [163]. Novel structures, such as *core-shell* [164–166] and *multi-layered* [167, 168] (so-called *onion structures*) materials, may also be produced in this way.

4.4.2 Determination of Particle Size and Dispersion

The most important measurement for any nanocrystalline compound – and one which is invariably the first to be undertaken after synthesis – is to determine the grain (crystallite) size, since this is fundamental to understanding the material's properties. Many techniques have been used to determine to characterize nanocrystallite size and shape, including optical absorption spectroscopy, dynamic light scattering (DLS), small angle X-ray scattering (SAXS), dark-field electron microscopy, low-frequency inelastic Raman scattering, proton-induced X-ray emission (PIXE), X-ray diffraction (XRD), transmission electron microscopy (TEM), and gas adsorption surface area measurements. The latter three methods will be discussed here, as they are: (i) the most commonly used; (ii) readily available in most laboratories; and (iii) relatively simple measurements to perform and analyze. Whilst each of these methods has its advantages and disadvantages, a number of well-known problems and artifacts are also encountered, and these will be highlighted along with the results of some studies conducted to compare and assess the different techniques.

4.4.2.1 Transmission Electron Microscopy

TEM images result from the direct observation of the particles, and provide information on both particle shape and size. In addition, high-resolution (HR) TEM images will reveal the degree of crystallinity of the particles and the presence of amorphous material in the sample. A good example of this is the study of nanocrystalline ball-milled lithium niobate, which shows an amorphous region on the surface of the crystallites [169]. The main disadvantage of the TEM techniques is that to obtain a statistical average and monitor the dispersion of the particle size can be very time-consuming, as typically the dimensions of several hundred particles must be determined. In some cases, a lack of contrast or particle overlap may also blur the boundaries between particles, such that an elaborate sample preparation procedure is often required.

4.4.2.2 **X-Ray Based Methods**

Both, SAXS and XRD, are indirect methods but offer the advantage of providing reliable statistical information on particle size. XRD is particularly attractive as it can be performed on a very basic laboratory-based powder diffractometer, and for this reason is the most commonly used method. The technique involves measuring the peak broadening of the diffraction lines which, for perfect crystals, would be sharp except for a very small inherent broadening due to the uncertainty principle (i.e., there is not an infinite number of diffracting planes). In practice, however, these are broadened due to the instrumental optics and crystallite size. The most common approach to determining the crystallite size is to use the Scherrer relationship [170–172]:

$$d_V = \frac{k\lambda}{\beta \cdot \cos\theta} \tag{4.7}$$

where, k is a constant that depends on the shape of the grain (k varies between 0.89 and 1.39, and for spherical crystal is near unity), λ is the wavelength of the X-ray source, β is the width of the peak (full width at half maximum; FWHM) after correcting for instrumental broadening and measured in radians, and θ is the Bragg angle of the diffraction peak. The crystallite size, d_V, is a volume-weighted average domain size; that is, it is the mean crystallite size in the direction perpendicular to the crystals planes that are causing the diffraction. Thus, some information on the shape of the grains is available from the measurement of d_V for different diffraction peaks. There can be an additional broadening of the diffraction peaks due to microstrain, ε_S, in the crystallites which can be expressed as [173]:

$$\beta_\varepsilon = 4\varepsilon_S \cdot \tan\theta \tag{4.8}$$

The two broadening effects can be separated by using a Williamson–Hall plot [174], which combines Equations (4.7) and (4.8) to give:

$$\beta\cos\theta = \frac{\lambda}{d_V} + 4\varepsilon_S \cdot \sin\theta \tag{4.9}$$

This assumes that the two broadening effects are simply additive, and a plot of $\beta \times \cos\theta$ versus $\sin\theta$ will yield ε_S and d_V from the slope and intercept, respectively. The assumption of linear additivity of the broadening effects assumes that both have a Lorentzian shape (Cauchy approximation). The three other cases, Gauss–Gauss, Gauss–Cauchy and Cauchy–Gauss, have each been considered (see for example Ref. [175]). It is usually found that ε_S increases with decreasing crystallite size. One common problem with XRD measurements of particle size is that they may provide a false impression of the sample if there is amorphous material present that will not diffract [176].

The surface areas of powders, S_{BET}, are commonly measured for catalysis by the adsorption of nitrogen gas at 77 K and applying the Brunauer–Emmett–Teller (BET) analysis (see, for example Ref. [177]). Fully automated systems are now available for the determination of S_{BET}. Classically, the area is related to the particle size, d_{BET}, by the relationship [178]:

$$S_{BET} = \frac{6}{\rho_d \cdot d_{BET}} \tag{4.10}$$

where, ρ_d is the density of the solid and the factor of 6 applies to spherical and cubic-shaped particles. If there is a large dispersion of particle size, then the BET method will overweigh the larger particles as they have a small surface area but contribute a major part of the mass.

When making a relative comparison of the sizes of particles within a given material prepared by different routes, it clearly does not matter which of these methods is used. However, a more positive note is that in cases where different methods have been applied to the same samples (albeit rather idealized systems), there has been a reasonable consistency in the particle sizes determined. For example, when measuring batches of anatase TiO_2 nanoparticles with sizes ranging from 12 to 120 nm, TEM, XRD and BET showed a satisfactory agreement [175]. Similarly, a good agreement between TEM and XRD measurements was found for anatase TiO_2 particles ranging from 3 to 35 nm [179]. Likewise, a good agreement between SAXS, TEM and XRD was found for highly monodisperse alloy nanoparticles of 5 and 8 nm. Clearly, if the sample has a high dispersion of sizes, and/or the shapes of the crystallites are not simple (e.g., platelets, needles, etc.), then the only reliable method would be TEM.

4.4.3 Characterization of Microstructure

Microstructure is the key to the important properties of nanocrystalline materials. It was noted earlier that simple geometric considerations lead to the conclusion that a large fraction of the atoms in a nanocrystal are in the surface (see Figure 4.1). However, the crucial questions involved here are first, the nature of the surface (in terms of the level of atomic order), and second the structure of the interface between grains. In the case of nanocrystalline films the interface between the grains and the substrate is equally crucial to the properties. Two extreme possibilities for the intergrain structure were shown in Figure 4.2, and discussed previously (see Section 4.1). In this section, attention will be focused on the techniques available to characterize the interfaces between grains.

High-resolution TEM can be used to provide the microstructural details in a simple pictorial manner, and example micrographs of nanocrystalline lithium niobate [169] are shown in Figure 4.6. The high-resolution (HR) TEM image in Figure 4.6b shows that the surfaces of the sol–gel prepared sample are highly ordered, whereas in Figure 4.6a the ball-milled material of similar grain size has a clearly visible amorphous region that is a few nanometers in thickness. This dependence of microstructure on the preparative route is a recurring theme, and in general, materials prepared using "soft" routes, such as IGC or chemical methods, such as

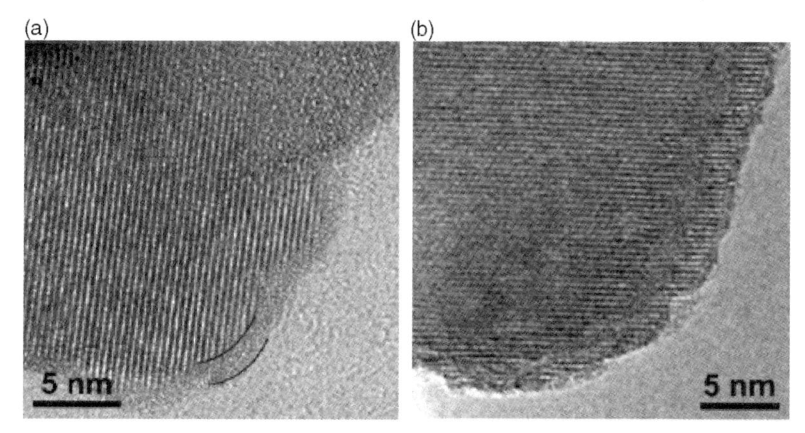

Figure 4.6 (a) HRTEM image of nanocrystalline $LiNbO_3$ prepared by ball milling for 32 h, with an average particle size of ca. 20 nm. A large amount of amorphous $LiNbO_3$ can be seen; (b) HRTEM image of nanocrystalline $LiNbO_3$ prepared chemically via complete hydrolysis of the lithium niobium double alkoxide. Although the grain boundary regions are somewhat disordered, very little amorphous material can be detected. Reproduced with permission from Ref. [169].

sol–gel synthesis, show no evidence of any highly disordered interfaces. For example, TEM measurements on both IGC- and chemically prepared nanocrystalline ceria showed the ~10 nm grains to have a high degree of perfection and to be separated by sharp boundaries [180]. Unfortunately, as very few data are available from HRTEM studies, other structural techniques have had to be used to explore the microstructure, such as electron diffraction [181], positron annihilation spectroscopy [182], EXAFS measurements [183–186] and nuclear magnetic resonance (NMR) spectroscopy [187, 188].

In an X-ray absorption experiment the absorption coefficient, μ_A, is measured as a function of incident photon energy as the energy is scanned across the value required for the emission of a core (K or L shell) electron; this known as the *absorption edge*. A typical spectrum is shown in Figure 4.7, where the data are for the Ti K-edge absorption in Ti_2O_3. The EXAFS represents oscillations beyond the absorption edge [189–191]; these oscillations arise when the emitted photoelectron wave is backscattered by neighboring atoms and interferes with the outgoing wave. The simplest case to consider here is backscattering by the shell of nearest-neighbor atoms. If the two waves are in phase, there will be constructive interference, a lower final state energy, and a higher probability for absorption. But, if the waves are out of phase, there will be destructive interference, a higher final state energy, and a lower probability for absorption. As the incident photon energy increases, so too does the energy of the emitted photoelectron, with consequential changes on its wavelength. As the distance between the target atom and its neighbors is fixed, there will be shifts in and out of phase, and hence the observation of the EXAFS oscillations would be a simple sine wave. There would also be backscattering from other shells of neighbors, and the EXAFS would show a combination of sine waves. The intensity of the oscillations would depend on the number and type of neighbors giving rise to the

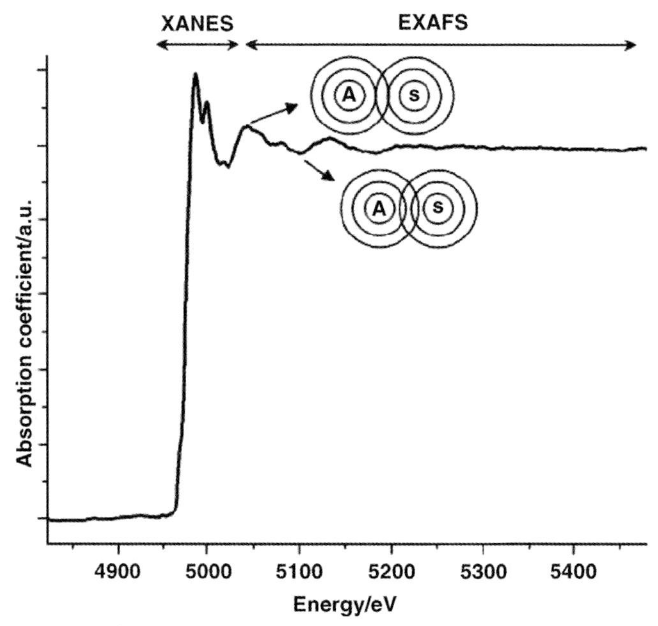

Figure 4.7 The raw Ti K-edge XAS spectrum of Ti_2O_3. The division of the spectrum into XANES and EXAFS regions is indicated. The figure also shows (schematically) the effect of the interference of the outgoing photoelectron wave from the target atom (A) and the backscattered wave from a neighboring atom (S).

backscattering and an EXAFS Debye–Waller factor (an uncertainty in the distance between target and scattering atoms arising from both static and dynamic disorder). EXAFS does not rely on long-range order and is sensitive to the local environment of the target atom out to 5 Å. The Fourier transform of the EXAFS yields a partial radial distribution function in real space, with peak areas proportional to the average coordination numbers and the Debye–Waller factors. One final point to note here is that *all* of the target atoms in the sample would contribute to the EXAFS analysis which would, therefore, measure the *average* environment of the target atom.

For a nanocrystalline sample, the EXAFS signal might be attenuated for two reasons: (i) the particle is so small that the average coordination numbers for the neighboring shells is reduced; or (ii) there is sufficient static disorder in the sample (e.g., at the interfaces) that the Debye–Waller factors are increased. At first sight it would appear that EXAFS has little to offer as a microstructural probe, although in order for condition (i) to be operative the particle size must be very small (typically <5 nm). Thus, at least in principle, EXAFS can be used to probe disorder in the interfaces of nanocrystals. However, the results obtained have been very confusing and the subject of much discussion. The EXAFS data for ZrO_2 represent a typical example, with such studies of this system having been conducted which claim evidence for disordered interfaces in nanocrystalline samples – that is, an attenuation of the EXAFS for the Zr–Zr correlation [192, 193]. However, similar measurements

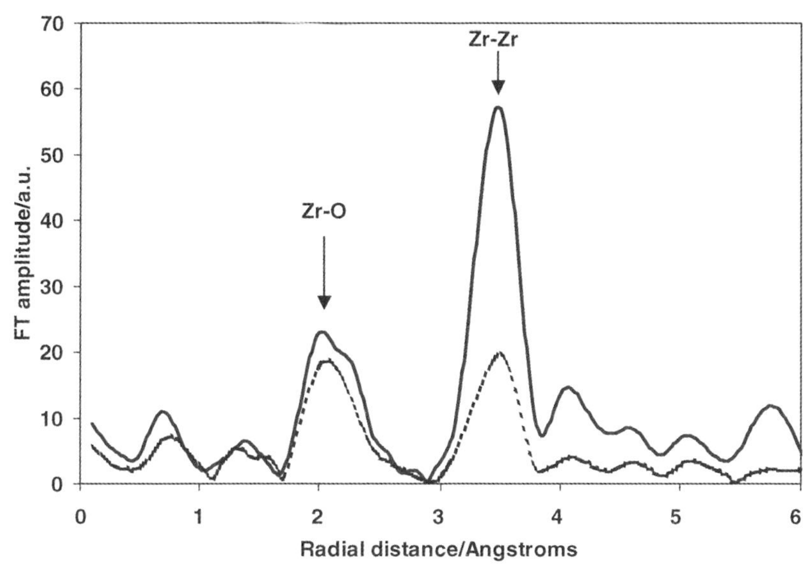

Figure 4.8 The Fourier transform of Zr K-edge EXAFS of bulk and ball-milled monoclinic ZrO_2 at 17 K. The Fourier transform has been corrected with the phase shift of the first shell. The solid line (––) is the bulk sample (110 nm); the dotted line (··· ···) is the ball-milled sample (13 nm).

on carefully prepared films, with particle sizes down to 6 nm, found the EXAFS to be indistinguishable from the bulk [194], and great care must be taken to ensure that all hydroxyl species are removed from the sample. In contrast, the EXAFS of ball-milled ZrO_2, with a grain size of 15 nm (too large to show any reduction of the average coordination number) showed a marked reduction in the Zr–Zr correlation [195] (see Figure 4.8). This was interpreted as the presence of amorphous material in the ball-milled sample, analogous to the study of ball-milled Al_2O_3 [195]. Similar effects were observed in the EXAFS analyses of other ball-milled oxides, for example $LiNbO_3$ [169, 196]. In general, the EXAFS analysis of sol–gel-prepared nanocrystalline oxides (ZrO_2, SnO_2, CeO_2, ZnO) showed no evidence of excessive disorder [185]. It is worth noting here that the EXAFS studies of nanocrystalline metals have also been controversial [197], although again sample preparation has been shown to be important. For example, the EXAFS studies of 13 nm grain size Cu, with a sample that had not been machined, showed a spectrum that was not attenuated and close to that for bulk Cu [197]. This provided evidence for interfaces that were similar to normal grain boundaries.

Today, magic-angle spinning (MAS) NMR is becoming an increasingly important spectroscopic tool for the characterization of solids. This is a spectroscopic technique that offers a local probe of the environment of a target nucleus based on the interactions between the nuclear spins. The versatility of the NMR approach is that it is possible to tune to various NMR interactions, such as the chemical shift (δ_C), the scalar coupling (J_c), the dipolar interaction (D_{int}), and the quadrupolar interaction (Q_{int}). MAS-NMR is now a well-established technique for characterizing zeolite catalysts and monitoring their synthesis using the ^{29}Si and ^{27}Al nuclei [198, 199]. However, a wider range of nuclei are becoming accessible, including ^{17}O [200], which is particularly important for nanoionic materials. These techniques are also becoming increasingly sophisticated [201, 202].

One relatively straightforward application of MAS-NMR is to monitor the signal of a target nucleus during synthesis of the nanocrystals. As the position of the resonance line depends on the local environment of the site, and the intensity of the line is proportional to the number of nuclei on that site, the evolution of the material can be followed. In this way it was possible to use ^{17}O NMR to show that the conventional method of producing nanocrystalline zirconia – by heating amorphous zirconium hydroxide (produced by the reaction of an aqueous solution of a zirconium salt with ammonia) and calcining at 500 °C – produces material that still contains hydroxyl entities and not pure zirconia [146]. A number of nanocrystalline oxides undergo phase changes during their production and subsequent growth, and these can be followed by using NMR. Good examples of this include the evolution of phases in nanocrystalline gallium oxide using the ^{71}Ga nucleus [188], and in nanocrystalline alumina using the ^{27}Al nucleus [203]. The ^{17}O NMR study of the synthesis of 2–3 nm nanocrystalline anatase phase titania from alkoxide precursors was particularly informative, as it was possible to differentiate between the bulk and surface O species. Today, these MAS-NMR techniques are becoming even more powerful and will undoubtedly in time play a major role in characterizing nanocrystalline materials, especially oxides.

4.4.4 Transport Measurements

A very wide range of techniques can be used to probe atomic transport in solids, and these have been detailed in various books [204–208] and reviews [21, 209–212] (see also Chapters 13, 8, 11 and 12). The most commonly used are tracer methods, ionic conductivity, and NMR measurements. Less commonly used (but more specialized) techniques include creep, quasi-elastic neutron scattering (QENS), and Mössbauer spectroscopy (MS). An elegant survey of the methods that have been used on nanoionic materials has been made by Heitjans and Indris [210]. The principles, procedures, and limitations of the more common techniques are outlined in the following sections.

Some general comments might be useful, however, before considering the individual methods. First, the techniques may be divided into: (i) *macroscopic* methods, which are used to measure the effect of long-range motion of atoms; and (ii) *microscopic* methods, which are used to measure the effect of jump frequencies of atoms [210, 212]. In principle, for a simple jump process via point defects in a solid, the two are interconnected by the classical Einstein–Smoluchowski equation [204]:

$$D = \frac{1}{6}\frac{a^2}{\tau_c} \tag{4.11}$$

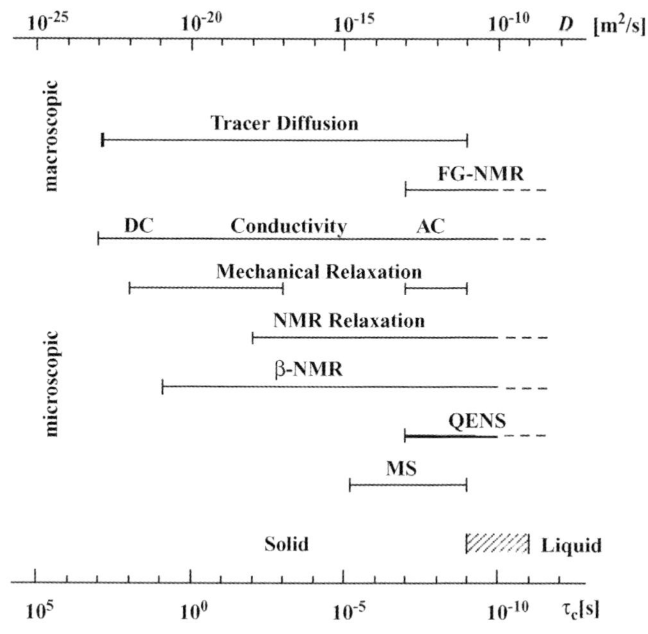

Figure 4.9 Typical ranges of the diffusivity D and motional correlation time τ_c of some macroscopic and microscopic methods, respectively, for studying diffusion in solids. FG-NMR = field gradient NMR; β-NMR = β-radiation-detected NMR; QENS = quasi-elastic neutron scattering; MS = Mössbauer spectroscopy. The hatched bar indicates the transition from the solid to the liquid, where the motional correlation time is reduced by about two orders of magnitude. Reproduced with permission from Ref. [210].

where D is the diffusion coefficient, a is the length of a jump, and τ_c is the motional correlation time (the time between diffusive jumps). The ranges of D and τ_c accessible to the various techniques are summarized schematically in Figure 4.9. First, although Equation (4.11) appears very simple, great care must be exercised when comparing transport data from different techniques, and how the raw data from any given technique are interpreted. Second, the diffusion coefficient in a crystalline solid by a single jump process is expected to show Arrhenius behavior (see Equation 4.2). In practice, however, most experimental methods will yield reasonably accurate values of the activation energy (Q) from the temperature dependence of a given property. This can be extremely useful in an ionic solid, as Q can be directly related to the energy of migration of the ions via point defects and/or the energy of formation of the point defect.

4.4.4.1 Tracer Diffusion

Tracer diffusion is the classical macroscopic technique [204–208], and is regarded as the most reliable method for obtaining diffusion coefficients. In these experiments, an isotopic tracer (usually a radioisotope) of the atom under study is diffused into the sample for a known time, and at a fixed temperature. Sections are then removed from the sample, the sections analyzed for the tracer concentration, the penetration profile determined, and D determined from the boundary conditions to Fick's second law of diffusion [213]. The simplest situation is when the tracer is deposited as a very thin layer on the surface and allowed to diffuse into the sample, without reaching the opposite face – this is termed the *infinitely thin layer on a semi-infinite solid* situation. In this case the appropriate solution is:

$$c(x,t) = \frac{M_t}{\sqrt{\pi D^* \cdot t}} \cdot \exp\left(\frac{-x^2}{4D^* \cdot t}\right) \tag{4.12}$$

where $c(x,t)$ is the concentration of tracer at distance x into the sample after an annealing time t, M_t is the initial surface concentration of the tracer, and D^* is the tracer diffusion coefficient. Hence, D can be obtained from the *penetration profile*, a plot of $\ln[c(x,t)]$ versus x^2. For penetration depths exceeding 1 μm, mechanical sectioning can be employed along with a radiochemical analysis of the sections. By contrast, secondary ion mass spectrometry (SIMS) profiling is applicable for penetration depths less than 1 μm. Here, the surface of the specimen is bombarded with a beam of primary ions, which results in a continuous atomization of the sample; the sputtered secondary ions can then be detected using mass spectrometry. For many elements there is a suitable radiotracer that is available, the notable exceptions being oxygen and lithium, given the current interest in oxides and lithium ion conductors. However, ^{18}O is readily available and has been widely used in SIMS experiments [214]. Clearly, all or part of the sample is destroyed by the sectioning, and this is a disadvantage of the method. A further disadvantage is that the experiment is very time-consuming, with annealing times of the order of months for samples with very low diffusivity; indeed, it is this factor that sets the lower limit of the measurable D^*.

For a single-crystal specimen, D^* is proportional to the diffusion coefficient of atoms through the bulk lattice, usually designated as D_B, and sometimes referred to as the volume diffusion coefficient; the proportionality coefficient is usually referred to as the Haven ratio (see below). One advantage of the radiotracer technique is that, since the profile is determined, it is often possible to separate out different diffusion processes (e.g., bulk, grain boundary, dislocation. surface diffusion, etc.), provided that they have sufficiently different diffusivities [215, 216]. The various pathways through which an atom may move in a compacted nanocrystalline sample are illustrated in Figure 4.10. In addition to bulk lattice diffusion, there will be contributions to the profile from diffusion along the grain boundary (characterized by diffusion coefficient, D_{gb}, and grain boundary width, δ) and from inter-agglomerate diffusion (characterized by diffusion coefficient, D_A, and the separation between agglomerates, δ_A). In principle, by varying the grain size, d, and the annealing time, t, it is possible to separate out the various diffusion coefficients, particularly if the specimen can be prepared without agglomerates and pores. However, problems may occur with grain growth and grain boundary migration, which can perturb the profiles.

4.4.4.2 NMR Spectroscopy Methods

NMR spectroscopy offers a range of methods for studying diffusion in the solid state which span an exceptionally large range of diffusivity [217, 218] (see Figure 4.9).

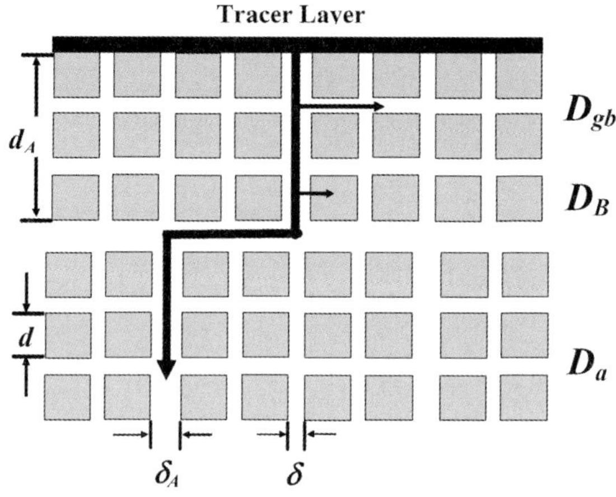

Figure 4.10 A schematic model for tracer diffusion in a compacted nanocrystalline sample. D_B is the bulk lattice diffusion coefficient, D_{gb} is the grain boundary diffusion coefficient, D_A is the inter-agglomerate diffusion coefficient, d is the grain size, δ is the width of the grain boundary, and δ_A is the separation between agglomerates.

These methods are nondestructive, and are particularly useful for atoms that have no – or only very short-lived – radioisotopes. There is a requirement for an isotope with a suitable nuclear spin, sensitivity and abundance, of which ^{19}F and ^{7}Li are important examples. These methods divide essentially into two classes: (i) *relaxation time experiments*, where the diffusive motion of spins results in the loss the coherence of the spins; and (ii) *field gradient experiments*, where the nuclear spin is used as a label on the atom, in a similar manner to a tracer experiment.

4.4.4.2.1 **Relaxation Time Experiments** In a relaxation time experiment, an external magnetic field (B_o) is applied to the sample and the spins distribute themselves among the available quantum levels set by the field (the *Zeeman levels*); thus, the sample has a net magnetization along the B_o direction. A pulse of radiofrequency (RF) radiation is then applied which perturbs the equilibrium of the spin system, causing the excitation of some spins from a lower to a higher quantum state. Following the RF pulse the spin system will return to equilibrium as the spins return from the higher to a lower quantum state and the magnetization is recovered. This relaxation is a resonance process, and requires interaction with an oscillating magnetic field at the frequency corresponding to the separation between the upper and lower quantum states for the spins, the Larmor frequency, ω_o. The diffusive motion of the atoms, with their nuclear spins, in the sample gives rise to an oscillating local magnetic field with a range of frequencies, the spectral density. There will be a component of the spectral density at the Larmor frequency, causing relaxation, and this component will depend on temperature. At a low temperature (slow motion), there will mostly low-frequency components, while at high temperatures (fast motion) there will be components over a wide range of frequencies.

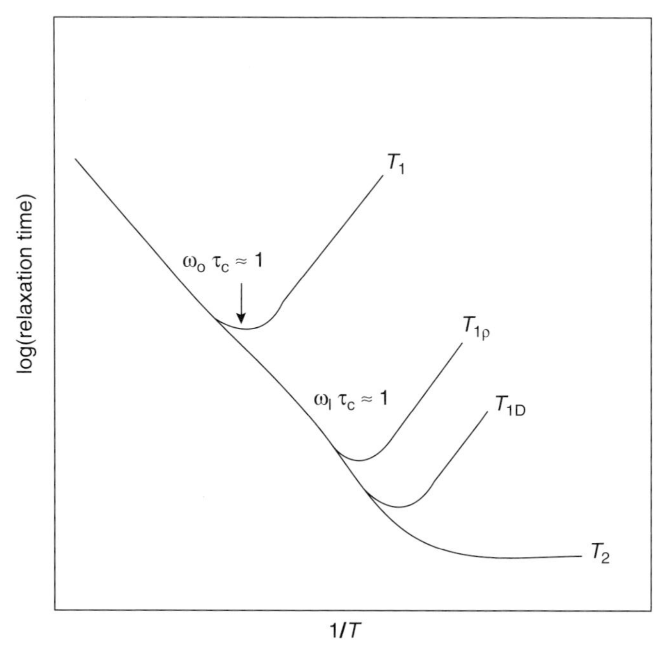

Figure 4.11 A schematic representation of the temperature dependence of NMR relaxation times.

The characteristic time for the relaxation process described here is referred to as T_1 (the spin-lattice relaxation time), as the spin system returns to equilibrium with the external surroundings. The idealized temperature dependence of T_1 is shown in Figure 4.11; this emphasizes the importance of a minimum in T_1, since at this temperature the diffusion correlation time τ_C is approximately equal to the Larmor frequency.

Other relaxation times that can be determined are: T_2 (the spin-spin relaxation time, the characteristic time for the spin system to come equilibrium following a perturbation); $T_{1\rho}$ (the spin-lattice relaxation time in the rotating frame); and T_{1D} (the spin-lattice relaxation time in the local dipolar field). The importance of the latter two times is that the magnetic fields are much smaller than the B_o field; consequently, the separation between the quantum levels is smaller and the measurements are sensitive to slower motions (i.e., lower diffusion coefficients). The relationship between these relaxation times and temperature is shown in Figure 4.11. A final point is that the line-width of the resonance line in a conventional NMR spectrum is inversely proportional to T_2, and hence is affected by diffusion. In a solid, the resonance line is very broad, but as the nuclei begin to diffuse with increasing temperature the line becomes narrowed; this is referred to as motional line narrowing. Thus, the line width is inversely proportional to D, while the measurements provide a simple and direct means of studying diffusion.

The exact relationship between the relaxation times and the diffusion coefficient is complex, due to temporal and spatial correlation effects, and has only been solved and tested thoroughly for a cubic system (i.e., BaF_2 single crystals [219]).

Thus, although the measurements provide accurate activation energies for diffusion they are difficult to correlate precisely with diffusion coefficients obtained with other techniques. However, the measurements are very rapid, the relaxation times being the order of seconds or less, and if there are two diffusion mechanisms operative with coefficients differing by an order of magnitude or more then they will have relaxation times that can be separated.

4.4.4.2.2 **Field Gradient Methods** Field gradient (FG) NMR experiments rely on a different principle from the relaxation time experiments. One method of measuring the T_2 relaxation time involves the *spin-echo* procedure [220]. This involves using an RF pulse at the Larmor frequency to align the spins along a particular direction at 90° to the B_o field, waiting a short time, τ, during which the spins will de-phase and the net magnetization decays, and then flipping the spins through 180° with another RF pulse; this means that they will then re-phase to produce an echo in the magnetization after time 2 τ. The magnitude of the echo compared to the original magnetization will be diminished by the natural T_2. However, if the spins are in a gradient of magnetic field and are diffusing within the sample during time τ, then the echo will be further attenuated. Essentially, the diffusion of a particular spin to another point in the sample with a different value of B_o means that the 180° pulse will not be on resonance for that spin, the spin will not be flipped by 180°, and it will not contribute to the echo.

An alternative method is the pulsed-field gradient (PFG), in which short pulses of a magnetic field gradient are applied after each RF pulse to de-phase and re-phase the spins. Again, translation diffusion of the spins will result in an attenuation of the echo and hence the nuclear spin is effectively being used as a tracer label for the atom. The extent of the attenuation can be directly related to the diffusion coefficient, and the measurements can be "calibrated" by measuring the echo attenuation in a system with an accurately known diffusion coefficient. This technique is used extensively in the study of liquids [221, 222], where diffusion is fast but can be extended to a slower diffusion ($D > 10^{13}\,\mathrm{m^2\,s^{-1}}$), as found in solids near their melting point, by using large field gradients [223, 224].

For ionic solids, measurement of the ionic conductivity, σ_i, has long provided a method for studying their atomic diffusion [25, 209, 225, 226] (see also Chapter 3). The measurements are usually made with an alternating current (AC) bridge operating at a fixed frequency, f (typically $>1\,\mathrm{kHz}$), to avoid polarization effects. The early studies were restricted to measurements on single crystals, and in this case σ_i and the tracer diffusion coefficient were seen to be related by the Nernst–Einstein equation [25]:

$$D^* = \frac{H_r \sigma_i k T}{N_i q_i^2} \tag{4.13}$$

where H_r is the Haven ratio, N_i is the particle density, and q_i is the charge of the mobile ion. Equation (4.13) assumes that only one of the ions in the crystal is mobile, and that there is only one mechanism of migration. In this case, the Haven ratio is related to the degree of correlation of the ionic jumps, the correlation factor, f_c [227].

For jumps involving single point defects, f_c is accurately known for the different crystal structures. Ionic conductivity measurements, coupled with other diffusion measurements, have proved to represent a very powerful method for identifying diffusion mechanisms. However, the requirement for single-crystal samples proved to be very restrictive in terms of the materials that could be investigated, and the approach has been used successfully only for very simple systems. Examples include the combination of conductivity and diffusion in the study of alkali and silver halides [226], and the combination of conductivity and NMR in the study of barium fluoride [219].

Electrical conductivity measurements using AC remain the dominant technique for studying ionic transport, and for single crystals the precision is 1 part in 10^4. However, as the majority of contemporary measurements are performed on polycrystalline samples, which frequently are sintered compacts, the simple approach for single-crystal samples with a fixed frequency bridge is not applicable. Currently, the most frequently used approach is that of complex-plane impedance analysis [212, 228–230]. For this, the application of a small, sinusoidally oscillating potential $\Delta E \sin \omega t$ with angular frequency $\omega \, (= 2\pi f)$ across the electrodes will give rise to a current $\Delta i \sin (\omega t + \theta)$ that is phase-shifted by an angle θ. The impedance Z can be characterized by a combination of the ratio $\Delta E/\Delta i$ and the phase angle θ. It is convenient to represent Z at a given ω as a complex function, and hence by a point on the complex impedance plane, where the y-axis is the imaginary component, Z'', and the x-axis is the real component, Z'. Thus, $|Z|$ is the length of the line from the origin to the point, while θ is the angle between this line and the x-axis. The locus of Z as ω is varied will depend on the components in the circuit between the electrodes. The advantage of the complex impedance representation is that a series combination of elements can be described by the vectorial addition of individual components. The situation is more complicated for a parallel combination of elements. For example, a simple combination of resistance R and capacitance C leads to a semicircle in the complex–impedance plot. The ideal plot for a ceramic pellet, as shown in Figure 4.12, consists of three semicircles, one from each component of the sample: the interior of the grains (the required bulk or lattice conductivity); the grain boundaries; and the sample electrode interface. For a parallel combination of elements it is often more useful to plot the admittance $Y \, (= 1/Z)$ in the complex plane – that is, the imaginary part B (the susceptance) versus the real part G (the conductance). All of these parameters are interrelated, and it is possible to present the basic measurement of $|Z|$, ω, and θ in a variety of formats; in fact, such interconversion is a built-in feature of some modern commercial impedance analyzers. The problems in many studies are, first, to identify the equivalent circuit that corresponds to the measurements, and second to identify the circuit elements with physical features or chemical processes in the electrode-sample arrangement.

4.4.4.2.3 **Creep Measurements** Creep measurements are basically very simple and involve measurement of the strain rate, $\dot{\varepsilon}$, at a fixed applied extensive or compressive stress, σ_S. Depending on the sample morphology, the applied stress, strain rate and temperature, one or more processes can be responsible for the creep, including

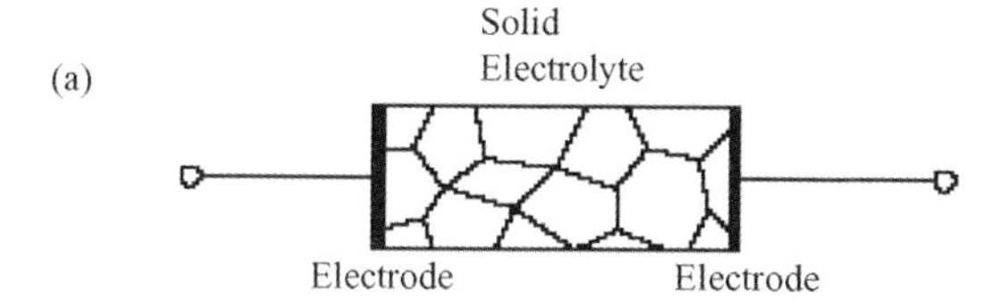

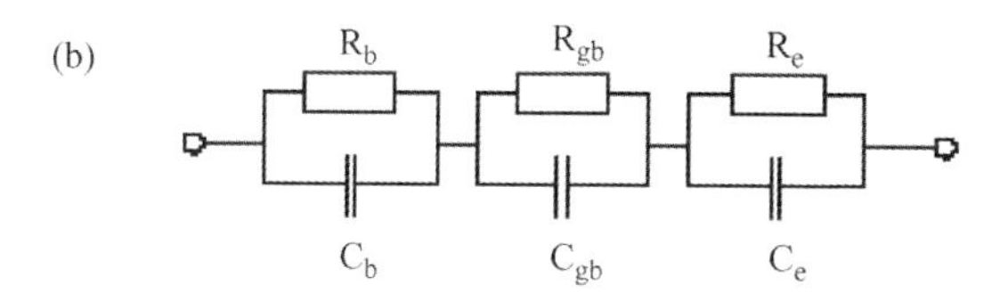

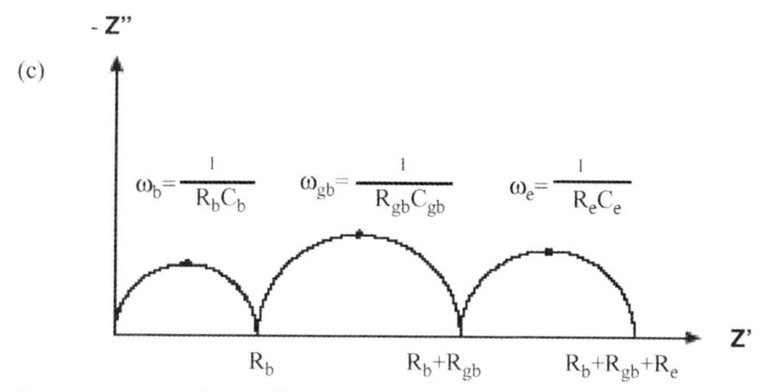

Figure 4.12 (a) Polycrystalline solid electrolyte with contacts; (b) Equivalent circuit with contributions from (the bulk of) the grains, the grain boundaries and the electrodes; (c) Impedance plot for the case $\omega_b \gg \omega_{gb} \gg \omega_e$. Reproduced with permission from Ref. [210].

grain boundary sliding, grain boundary diffusion, dislocation climb and bulk lattice diffusion via point defects. The main focus is normally the steady-state creep region where the strain rate is constant and very often follows a power law of the form:

$$\dot{\varepsilon} = C_S \cdot \sigma_S^n \tag{4.14}$$

where n is between 4 and 5. At high temperatures (above half of the melting temperature) and low stresses, steady-state creep occurs by diffusion of the vacancies in the sample. Here, the grains change shape by the migration of atoms via vacancies, and the term C_S is directly dependent on the diffusion coefficient. Such diffusional creep can arise from two possible mechanisms. First, the diffusion occurring through the bulk lattice, known as Nabarro–Herring creep [231, 232], when the strain rate depends on D_b. Second, the diffusion occurring along the grain boundaries, known as Coble creep [233], when the strain rate depends on D_{gb}. Whichever process is fastest will become dominant. However, in an ionic material the diffusion of cations and anions will be coupled as the system must maintain electrical neutrality. Hence, diffusional creep will be controlled by the slowest moving ion.

Exact relationships between creep rate and diffusion coefficients have been proposed, such as the Bird–Dorn–Mukherjee equation:

$$\dot{\varepsilon} = \frac{A_S D G_S b}{kT}\left(\frac{b}{D}\right)^p\left(\frac{\sigma_S}{G_S}\right)^n \quad (4.15)$$

where A_S is a dimensionless constant, G_S is the shear modulus, b is the magnitude of the Burgers vector, p is the inverse grain size exponent, and n is the stress exponent. For Nabarro–Herring creep $D = D_b$, $A = 28$, $p = 2$ and $n = 1$, whereas for Coble creep $D = D_{gb}$, $A = 33$, $p = 3$ and $n = 1$ [234]. Clearly, whilst it would be difficult to extract accurate diffusion coefficients from creep experiments, the temperature dependence of the strain rate readily provides activation energies for the diffusion. In practice, however, it can be very complicated to differentiate which of the two mechanisms is operative and dominant. In addition, pores and impurities in ceramic materials can mask intrinsic effects [13, 234].

The QENS and MS techniques of measuring diffusion have much in common in terms of the underlying principles [235, 236]. In both cases. the energy broadening of a resonance line is due to incoherent atomic motion. In QENS, the elastic line comes from the scattering of neutrons, whereas for MS the elastic line is the resonance line due to nuclear emission or resonant absorption of a gamma quantum. For QENS, a quantum of energy is transferred to the neutron, whereas for MS the quantum is the energy difference between the source and absorber. Both, QENS and MS measure the self-correlation function in time and space; hence, studies yield the jump frequency ($1/\tau_c$), and from investigations with single crystals the jump vector can be derived. Although these techniques can, in principle, both provide fundamental mechanistic information, they each have their limitations. While both are currently limited to solids where the diffusion is relatively fast (see Figure 4.9), MS is limited to the study of very few elements, with only Fe and Sn being readily accessible. The fact that more elements are accessible via QENS experiments, coupled with recent developments in both instrumentation and analysis, suggest that this technique will be used increasingly in the future.

4.5 Review of the Current Experimental Data and their Agreement with Theory

4.5.1 Microstructure

Based on the techniques described in Section 4.4 that are used to characterize these materials, it is clear that the microstructure of nanocrystalline ionic materials is heavily dependent on the method of synthesis. In the case of powdered samples, the "gentler" synthetic methods of spray pyrolysis, sol–gel and IGC tend to produce nanocrystals which are internally highly ordered, as well as surfaces that are ordered. The evidence for this is derived from a range of techniques, notably using HRTEM and EXAFS. In contrast, high-energy ball-milling produces materials with highly disordered surfaces, as well as a large fraction of amorphous material. A good example of the differences between the microstructures of sol–gel and ball-milled samples can be seen in the HRTEM images of lithium niobate in Figure 4.6 [169].

Although, the preparation of highly ordered nanocrystalline grains is relatively straightforward, for most transport experiments (e.g., tracer and conductivity measurements) they must be compressed into a dense pellet. Unfortunately, the densification process presents problems, most notably the removal of pores. Likewise, in procedures that involve annealing at elevated temperature there will also be grain growth. The question then arises as to the nature of the interfaces between the crystallites, as the grains will have virtually random orientations on compression. This will create grain boundaries with a range of angles of mismatch. An example of the microstructure of freshly prepared and annealed nanocrystalline pellets can be seen in calcium fluoride [237] and zirconia [238]. Given the wide range of mismatch between the grains, it is difficult to assess the detailed nature of the interfaces (dislocation content, lattice spacing, etc.) and the degree of strain. This is a very important point when interpreting transport data in these samples.

Recent transport measurements have been carried out on nanocrystalline thin films – either as single layers on an inert substrate or as multilayers – and in these cases the interfaces were more well-defined than in compacted samples. The examination of interfaces using TEM is also simpler to interpret, as the samples are generally more uniform. However, there is normally a lattice mismatch between the film and the substrate, or between alternating layers, such that the degree of mismatch may be large and lead to disorder and strain in the interface. The nature of the interface is therefore very dependent on the lattice parameters of the layers and the preparation conditions. These points must be borne in mind when discussing the transport results.

4.5.2 Transport

Although the emphasis here will, by necessity, be placed on more recent data, several key reviews of transport in nanocrystalline ionic materials have been presented, the details of which will be outlined first. An international workshop on interfacially controlled functional materials was conducted in 2000, the proceedings of which were published in the journal *Solid State Ionics* (Volume 131), focusing on the topic of atomic transport. In this issue, Maier [29] considered point defect thermodynamics and particle size, and Tuller [239] critically reviewed the available transport data for three oxides, namely cubic zirconia, ceria, and titania. Subsequently, in 2003, Heitjans and Indris [210] reviewed the diffusion and ionic conductivity data in nanoionics, and included some useful tabulations of data. A review of nanocrystalline ceria and zirconia electrolytes was recently published [240], as have extensive reviews of the mechanical behavior (hardness and plasticity) of both metals and ceramics [13, 234].

The approach taken here is to consider data according to the type of material and, within each type, to discuss the results in terms of experimental method – that is, tracer, conductivity, and NMR. Attention will be focused on relatively simple materials, although for details of composite electrolytes the reader is referred to more recent reviews and summaries [210, 241, 242].

4.5.2.1 Simple Halides

4.5.2.1.1 **Calcium Fluoride** This simple fluoride, in which the mobile species are the fluoride ion, has been extensively investigated by Heitjans and coworkers using samples prepared with IGC and employing ^{19}F NMR [243] and AC impedance spectroscopy [237, 244]. The NMR relaxation times revealed a highly mobile fluoride ion in the virgin sample at low temperatures, with an activation energy which was only 20% of that found in single crystals [243]. In addition, there was a narrow component in the line width due to the fast diffusion. On thermal cycling, the fast component diminished and the activation energy increased. These data were explained by fast fluoride ion diffusion along the interfaces which became less dominant as the grains grew on annealing. For the conductivity measurements, the powder was compressed into pellets of 96% theoretical density by the application of 2 GPa pressure, and Ag electrodes were applied [237]. The high-frequency component of the impedance spectrum, which represents conduction through the bulk and the grain boundaries, is shown in Figure 4.13, and is some four orders of magnitude

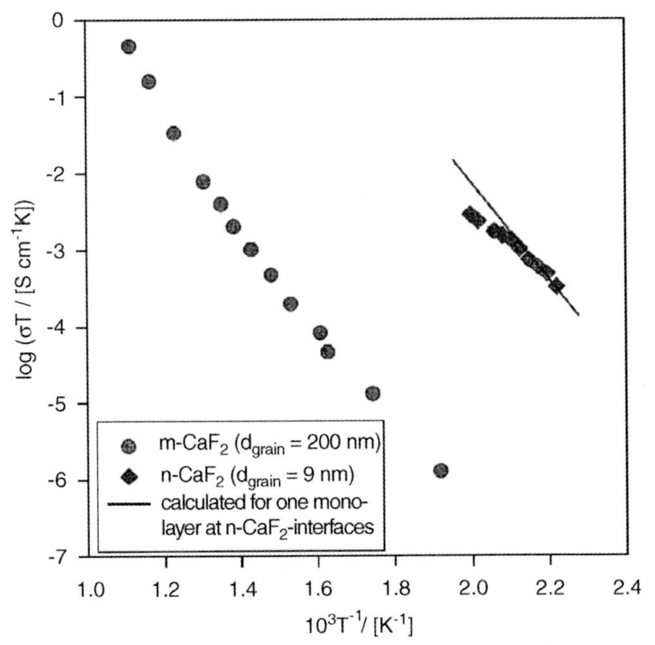

Figure 4.13 Temperature dependence of the conductivities of nanocrystalline and microcrystalline CaF_2. The line represents the estimated conductivity assuming a pronounced space-charge effect. Reproduced with permission from Ref. [237].

higher than for coarse-grained material. The increased conductivity was ascribed to a space-charge layer effect with an increased vacancy concentration in the interfaces. This would not be the extreme case where the space charge layers dominated (where $d \ll 4L_D$, as seen in Figure 4.3), as the level of sodium impurity (which creates fluoride ion vacancies) reduces L_D to ~1 nm, much less than the 9 nm grain size. On thermal cycling the high conductivity was reduced, while TEM revealed considerable grain growth and the formation of large pores in the sample.

4.5.2.1.2 **Calcium Fluoride-Barium Fluoride** Much of the current interest in nanoionic systems has resulted from the seminal work of Maier and coworkers on alternating thin layers of calcium and barium fluoride produced by molecular beam epitaxy (MBE) on silica and alumina substrates [245, 246]. The periodicity and thickness of the layers was varied systematically, and the conductivity measured parallel to the layers by using AC impedance spectroscopy. The essential features of the conductivity data are shown in Figure 4.14. As the period (twice the layer thickness) decreased down to ~100 nm, the conductivity increased in proportion to the number of heterojunctions. This was consistent with the space-charge layer model in which there is a redistribution of the fluoride ions across the interfaces inducing the generation of defects. When the period was in the range 15–100 nm, there was a more rapid increase in the conductivity ascribed to the layer thickness

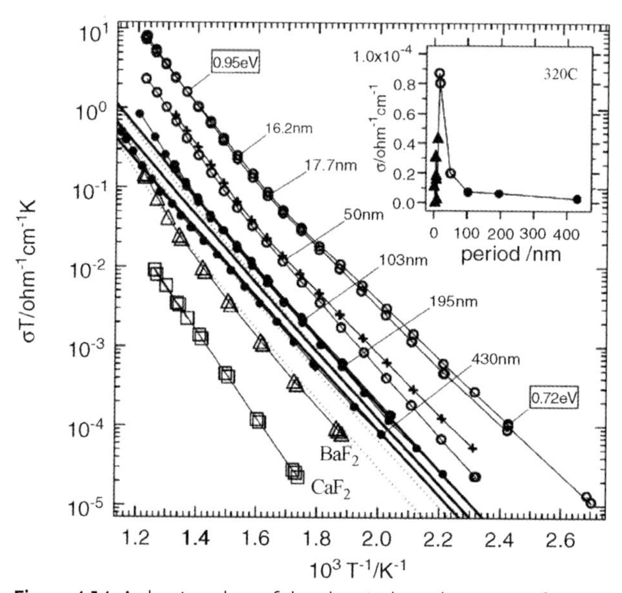

Figure 4.14 Arrhenius plots of the electrical conductivities of BaF_2–CaF_2 heterostructured thin films (•,○), and thin films of BaF_2 (□), CaF_2 (△) and their mixture $Ba_{0.5}Ca_{0.5}F_2$ (+). The numbers in the figure give the BaF_2/CaF_2 period. The inset shows the conductivity dependency on individual thickness of heterostructures. Reproduced with permission from Ref. [246].

being less than the Debye length, assessed as approximately 60 nm; that is, the space-charge layers were overlapping in the layer and penetrated the heterolayer package. For a spacing <5 nm, the conductivity was decreased due to a lack of continuity of the films. The detailed structure of the three heterolayers was studied later [247], with TEM revealing that samples with a layer spacing of 80 nm had interfaces that could be regarded as ideal, with regular arrays of misfit dislocations and their Burgers vectors on the interface. For a layer spacing of 10 nm, the interfaces were wavy with additional dislocations and Burgers vectors at $\sim 73^\circ$ to the interface; there were also drastic changes in the lattice parameters close to the interface. It was noted that these unusual interface structures could provide conduction processes in addition to the effect of the space-charge layer effect, and that further investigation was required to clarify this point. Heterolayers were also prepared on a range of substrates with different crystallographic orientations which produced layers with different orientations [248]. It was then possible to eliminate the effects of the interface at the substrate. In addition, layers with different orientations and a periodicity of 25 nm had the same conductivity, which indicated that the enhanced conductivity at this periodicity was not due to interfacial mismatch but could be explained by the space-charge layer model. Further support for the space-charge layer model was derived from conductivity measurements of the heterolayers perpendicular to the interfaces. For large interfacial spacings (>50 nm), the conductivity was limited by the higher-resistance layers of calcium fluoride. However, at very small spacings (<30 nm), the conductivity increased steeply and tended towards a saturation value, corresponding to the space-charge overlap situation and with an overall value that could be attributed to fluoride ion interstitials accumulated in the calcium fluoride.

Recently, a conductivity study of a nanocomposite of calcium fluoride (70 mol%) and barium fluoride (30 mol%) was reported which showed a similar feature to the heterolayers [249]. The sample was prepared by quenching the molten mixture, and consisted of randomly oriented thin lamellae (<100 nm) of $Ba_{0.96}Ca_{0.04}F_2$ and $Ca_{0.97}Ba_{0.03}F_2$. The ionic conductivity of the composite at 500 °C was 25- and 330-fold higher than those of the parent BaF_2 and CaF_2, respectively. As the cations were isovalent, this could not have been a doping effect. Rather, the authors suggested that the enhanced conductivity was due to the heterojunctions, and that the explanation might be similar to that proposed for the MBE heterolayers. These studies point the way towards simple preparation methods for bulk samples.

4.5.2.2 Oxides

4.5.2.2.1 Lithium Niobate Lithium niobate ($LiNbO_3$) is widely used as a photonic material, and has a wide range of applications in lasers, nonlinear optics, optical communications, optical memories, and diffractive optics [250]. It is not generally considered as an ionic conductor. However, since ^7Li NMR measurements first revealed a rapid Li ion motion in nanocrystalline powders prepared by ball-milling [251], the material has attracted considerable interest and has proved to be a good model system [169]. As stated earlier, it is now known that the ball-milled material has a considerable amorphous content, as shown by EXAFS [169, 252] and HRTEM [169],

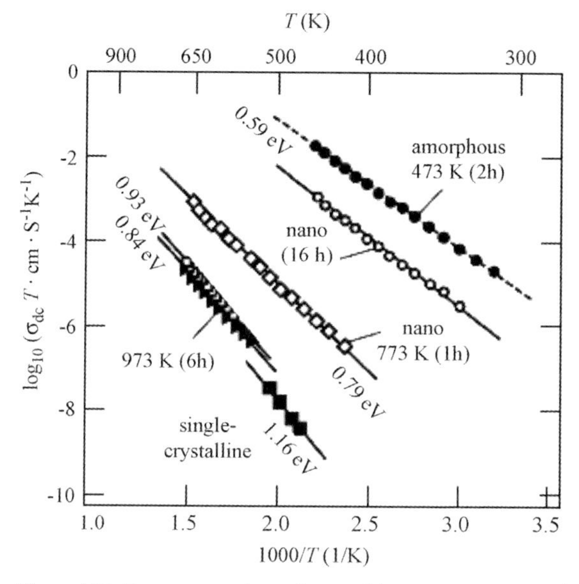

Figure 4.15 Temperature dependence of the DC conductivity of various nanocrystalline $LiNbO_3$ samples. The amorphous sample was prepared via the sol–gel route. The sample nano (16 h) was a ball-milled sample with crystallite 23 nm. The other nano samples were prepared via the sol–gel route and then annealed at the stated temperature for the time shown. Samples nano 773 K (1 h) and nano 973 K (6 h) had crystallite sizes of 27 and ~100 nm, respectively. The data on the line with activation energy of 0.93 eV are for a microcrystalline sample. Reproduced with permission from Ref. [169].

and the fast Li ion transport in the samples was due to ions moving in the amorphous phase. Nevertheless, ^{7}Li NMR line width measurements indicated that there was some enhanced Li ion motion in sol–gel-prepared samples [252]. Heitjans and coworkers [169] have reported a definitive study of the microstructure (EXAFS and HRTEM) and transport (^{7}Li NMR and AC complex impedance spectroscopy) of nanocrystalline $LiNbO_3$ prepared by ball-milling and sol–gel techniques. The conductivity data are summarized in Figure 4.15. Although the sol–gel and ball-milled samples have similar grain sizes of 27 and 24 nm, respectively, the latter has a conductivity which is similar in magnitude and activation energy to that of the amorphous sample. The sol–gel-prepared sample had a much lower conductivity, but this was still two orders of magnitude higher than that of either a large-grain powder or single crystals. The origin of this enhanced conductivity still originates from ions migrating in the interfacial regions; however, evidence obtained from the range of techniques was that these were more akin to grain boundaries in bulk samples rather than being heavily disordered or amorphous.

4.5.2.2.2 **Zirconia** Zirconia (see also Chapter 9) is one of the most important ceramic oxides due to its applications as both a structural and functional material. The structural applications of zirconia rely on it being both exceptionally hard and tough, especially when transformation-toughened [253], and it has a low thermal conductivity.

Thus, it used extensively as a coating in areas such as engines [254] and bio-implants [255], where the use of nanomaterial offers increased hardness and super-plasticity [13, 234]. The functional applications rely on the high oxygen ion conduction, with uses not only in oxygen sensors [256] but also in SOFCs [127, 257]. The major interest in this nanomaterial, however, has been to gain an improved conductivity. It is important to note that in both structural and functional applications, zirconia is doped with aliovalent cations. Pure, large-grained zirconia at atmospheric pressure has three polymorphs: an ambient temperature monoclinic phase which transforms at 1170 °C to a tetragonal, and then at 2370 °C to a cubic fluorite phase. Doping with lower-valency cations (e.g., Ca^{2+}, Mg^{2+}, Y^{3+} or trivalent rare earth ion) at low concentrations stabilizes the tetragonal phase and, at higher concentrations, also the cubic phase. For example, 3 mol% yttria forms tetragonal crystals, while 10 mol% yttria forms cubic crystals. Since the intrinsic point defects in zirconia are anion-Frenkel pairs, the doping increases the concentration of oxygen ion vacancies, which in turn enhances the oxygen transport. The high surface energy of nanocrystalline zirconia stabilizes the tetragonal phase in undoped material, with a transformation to the monoclinic phase when the particle size reaches ~30 nm [258, 259].

Recently, extensive studies have been conducted of atomic transport in bulk zirconia, including tracer diffusion studies of both cations [260] and anions [261], as well as conductivity measurements [262]. It is clear that the best oxygen ion conductors are those where the doping level is just sufficient to stabilize the cubic phase, and it has long been known that the grain boundaries are "blocking"; that is, the conductivity is lower in the boundaries due to segregation of the silica impurity [263–266]. However, by reducing the grain size in high-purity 15 mol % CaO-stabilized ZrO_2, the effect of silica was segregation was eliminated [266], despite the resultant grain boundary conductivity being two orders of magnitude lower than that of the bulk conductivity. A low grain boundary conductivity was also found in a study of microcrystalline 8 mol% Y_2O_3-stabilized ZrO_2, and accounted for by the oxygen vacancy depletion in the grain boundary space-charge layer [267]. In a conductivity study of nanocrystalline 1.7 and 2.9 mol% Y_2O_3 tetragonal stabilized ZrO_2 with average grain sizes of 25–50 nm, the grain boundary conductivity was also less than that for the bulk material [268]. An earlier conductivity study of yttria-doped tetragonal zirconia polycrystalline ceramics with an average crystallite size <200 nm also found no enhancement; rather, the authors concluded that smaller grain sizes might be required to produce such an effect [269]. Thus, the majority of the early conductivity studies of nanocrystalline materials showed no evidence of enhanced transport in zirconia.

The group of Kosacki and coworkers was the first to demonstrate a nanocrystalline effect on the conductivity of zirconia [270, 271]. The group measured the conductivity of 1 μm thin films of 16 mol% Y_2O_3-stabilized ZrO_2 nanocrystals on sapphire and alumina substrates prepared by a polymer spin-coating procedure [271]. The conductivity parallel to the surface of the film was found to increase with decreasing grain size of the film; ultimately, a grain size of 20 nm had a conductivity which was two to three orders of magnitude higher than that of coarse-grained samples and single crystals, and also a lower activation energy (0.93 eV compared to 1.23 eV). The impedance spectroscopy data suggested that the higher conductivity might have been due to an increased grain boundary conductivity. However, these data were clearly at variance with those of earlier studies, and it was suggested that the thin films might be affected by the preparation technique, or perhaps be more susceptible to contamination and humidity that might be affecting the results [239]. These criticisms proved to be less important to a subsequent conductivity study of thin epitaxial films of 9.5 mol%-Y_2O_3-stabilized ZrO_2 nanocrystals prepared by pulsed laser deposition on MgO single crystals [272]. Here, as the film thickness decreased, the conductivity parallel to the film surface increased, and it was concluded that the enhancement was due to the increasing importance of conduction along the YSZ-MgO interface. At a film thickness of 15 nm the conductivity was approximately two orders of magnitude higher than for the bulk material (see Figure 4.16). It was argued that the conductivity in the interface region, the thickness of which was estimated at about 1.6 nm, was some three orders of magnitude higher than the bulk conductivity. The results of these studies led to a plethora of conductivity investigations with nanocrystalline zirconia in relation to the interfacial effects.

Azad *et al.* [273] measured the conductivity of alternating MBE-deposited layers of Gd_2O_3-doped CeO_2 and ZrO_2 on single crystal alumina substrates; this was a similar experiment to that using CaF_2-BaF_2 multilayers [245]. Conductivity was seen to

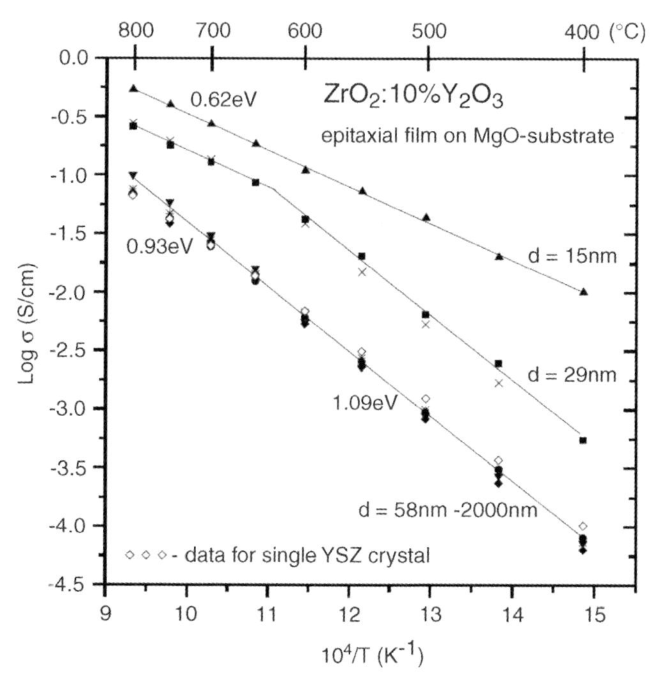

Figure 4.16 Temperature dependence of the electrical conductivity determined for epitaxial YSZ thin films with different thicknesses [272].

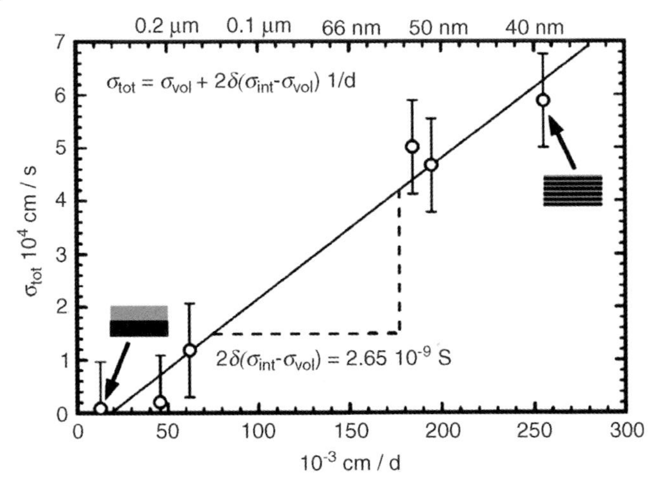

Figure 4.17 Total conductivity CSZ/Al_2O_3 multilayer systems at 575 °C as a function of the reciprocal thickness 1/d of the ionic conducting layers [274].

increase with decreasing layer thickness, down to an individual layer thickness of 15 nm, and the conductivity was higher than that in either of the two bulk components. Space-charge effects were ruled out as an explanation of the data, as the Debye length in these systems was on the order of 0.1 nm. The enhancement was assigned to a combination of lattice strain and extended defects due to lattice mismatch between the heterogeneous structures. Recently, a systematic conductivity study was made of pulsed laser-deposited multilayers of 8.7 mol% CaO-stabilized ZrO_2 and Al_2O_3 on single-crystal alumina substrates [274]. As the thickness of the layers was decreased from 0.78 μm to 40 nm, conductivity parallel to the layers was increased and shown to be proportional to the number of interfaces (see Figure 4.17). Again, the enhancement was not due to space-charge effects, as in this highly doped zirconia the Debye length was on the order of 0.1 nm. The most likely explanation for this strong change in conductivity and activation energy was found to be a high ionic mobility in the interface core regions due to a strong structural mismatch between the zirconia and Al_2O_3 layers.

Radiolabel tracer measurements using ^{18}O and SIMS analysis have been reported for a number of nanocrystalline zirconia samples [238, 275, 276]. The first measurements were for pure monoclinic zirconia samples, the nanocrystalline powders being prepared by direct current (DC) sputtering zirconium, oxidizing, and compressing into pellets with a relative mass density of 95–97% and a particle sizes of 80 or 300 nm. The results showed that the grain boundary diffusion was three to four orders of magnitude higher than for bulk diffusion, with activation energies of 1.95 and 2.29 eV, respectively. The same group reported analogous measurements for cubic 6.9 mol% Y_2O_3-stabilized ZrO_2 samples with grain sizes of 65–99 nm and a relative mass density of 96% [274, 275]. Again, the results showed the grain boundary diffusion to be about three orders of magnitude higher than for

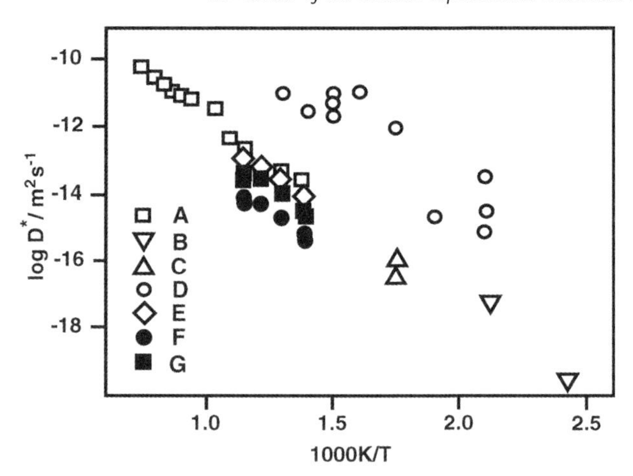

Figure 4.18 Tracer diffusion coefficients of oxygen, D^*, in YSZ as a function of inverse temperature. A, D^*_B single crystal 9.5 mol% Y_2O_3-doped [261]; B, D^*_B single crystal 9.5 mol% Y_2O_3-doped [275]; C, D^*_B single crystal 6.9 mol% Y_2O_3-doped [275]; D, D^*_B compacted nanocrystal 6.9 mol% Y_2O_3-doped [275]; E, D^*_B compacted nanocrystal 8 mol% Y_2O_3-doped [277]; F, D^*_{gb} [lower limit] compacted nanocrystal 8 mol% Y_2O_3-doped [277]; G, D^*_{gb} [upper limit] compacted nanocrystal 8 mol% Y_2O_3-doped [277]. The grain size and relative density in D were 65–100 nm and 96%. The grain size and relative density in E, F, and G were ≈50 nm and >99.9%, respectively.

bulk diffusion, with activation energies of 0.91 and 1.11 eV, respectively. This was clearly not consistent with the existing conductivity data [263]. It has been pointed out that, when comparing diffusion and conductivity data via the Nernst–Einstein equation, the same samples must be used for both experiments [276]. The most recent ^{18}O diffusion measurements have been for cubic 8 mol% Y_2O_3-stabilized ZrO_2 samples consolidated by spark plasma sintering [277]. The relative mass density was >99.9% and the grain size ~50 nm. The study results showed no enhancement of diffusion along the grain boundaries, and the conclusion was drawn that the earlier studies on less dense samples had been affected by porosity and microcracks. These diffusion data are shown in Figure 4.18. In fact, this study yielded a grain boundary diffusion which was less than that in the bulk and was consistent with the conductivity data and diffusion data measured for isolated grain boundaries [278].

4.5.2.2.3 **Ceria** Ceria (CeO_2) is similar to zirconia and, in pure form, has the fluorite structure (see Chapters 2 and 9). The dominant point defects are anion Frenkel pairs and, like zirconia, ceria can be doped with rare earth cations to increase the concentration of anion vacancies and increase the oxygen ion conductivity. There is considerable interest in the application of ceria in SOFCs, as it has a higher conductivity than zirconia and can therefore operate at lower temperatures [127]. Unlike zirconia, ceria is readily reduced at elevated temperatures, resulting in the loss of oxygen and the reduction of Ce^{4+} to Ce^{3+}. The overall effect is to increase n-type electronic conductivity [279].

Early complex impedance conductivity studies of pure ceria pellets with a grain size of ~10 nm identified a substantially increased electrical conductivity, but it was subsequently realized that this was due to an increase in electronic rather than in ionic conductivity [180]. It was proposed that the origin of this effect was based on a lowering of the reduction energy in the nanocrystals, as well as a lowering of the point defect formation energies at the surface. To date, several studies have been conducted with pure and lightly doped ceria that have confirmed the increased conductivity to be electronic in nature [280–283]; the phenomenon has also been successfully modeled in terms of a space-charge layer [281, 284, 285].

The effect of doping ceria with rare earth oxides is to reduce the susceptibility to reduction, and hence reduce the electronic conductivity to negligible levels at normal oxygen partial pressures. Several experimental conductivity studies of heavily doped (≥10 mol% rare earth oxide) nanocrystalline ceria (crystallite sizes ~10 nm) have been reported, including measurements on thin films of 20 mol% Gd_2O_3 samples [163, 286, 287], on 10 mol% Y_2O_3 and Sm_2O_3 pellets [288], on 20 mol% Gd_2O_3 pellets [289], and on 30 mol% Sm_2O_3-doped pellets [290]. The results were rather confusing, with some groups reporting (orders of magnitude) increases in conductivity as the crystallite size decreased, while others reported no significant difference from the microcrystalline counterparts. The situation has also been complicated by the variety of different preparation procedures employed, notably the pressing and sintering procedures for pellets which lead to grain growth during processing. The system is quite complex, with several techniques having shown there to be very significant segregation of the dopant at the surface of crystallites [286, 288, 291]; indeed, it is this effect which reduces the grain growth in ceria by Zener pinning [287]. Thus, one explanation of an increased conductivity in the nanocrystalline samples is that there is an enrichment of the dopant in the grain boundaries, which in turn substantially lowers the grain boundary resistance and hence increases the total conductivity [286]. At a high level of doping the space-charge layer should be very thin compared to the crystallite size. In the case of pressed pellets, the situation is unclear as witnessed by two recent reports. In one of these studies, 10 mol% rare earth oxide-doped pellets were sintered by fast-firing under high pressure to produce samples with crystallite sizes 25–65 nm and a relative density of 92–94% [288]. This resulted in an enhanced conductivity, as shown in Figure 4.19, which was higher than the bulk conductivity in large-grained material. This effect was attributed to an increased diffusion of oxygen ions along the grain boundaries [292]. In contrast, measurements on 30 mol% Sm_2O_3-doped ceria prepared by spark plasma sintering, with a crystallite size of 16.5 nm and a relative density >99%, showed the total conductivity to be close to the bulk conductivity [290]. The study results showed that decreasing the crystallite size reduced the grain boundary resistance to the extent that only one semi-circle was seen in the complex impedance spectra, and this was attributed to the bulk conductivity. However, there was no significant conductivity enhancement. The differences between the two studies indicated that processing plays a major role in the resultant transport behavior of pellets, with the most obvious effects being the role in grain growth and dopant segregation.

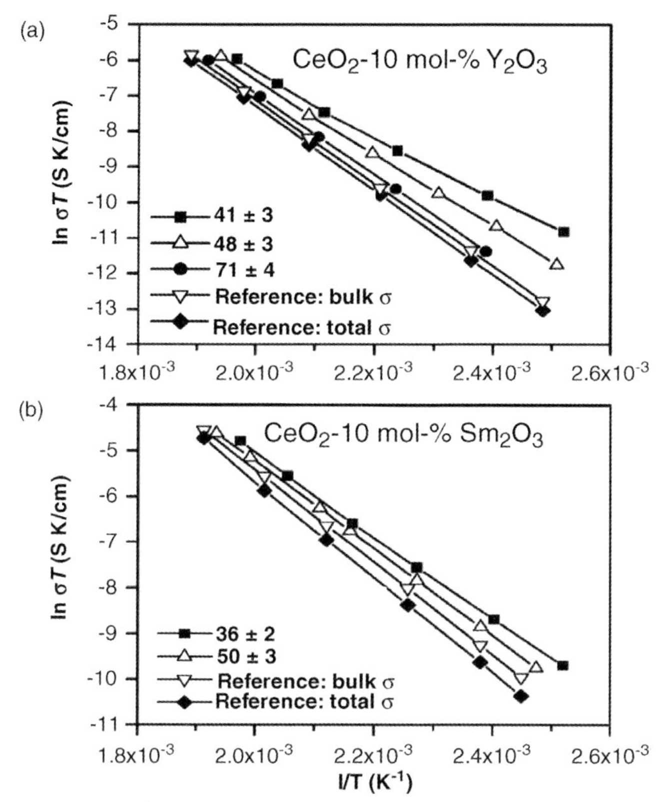

Figure 4.19 Arrhenius plots of CeO_2 10 mol% Y_2O_3 (a) and CeO_2 10 mol% Sm_2O_3 (b) ceramics, for different grain sizes. The numbers are the grain sizes (nm). Reproduced with permission from Ref. [288].

4.5.2.2.4 **Titania** Titania has three common polymorphs, *anatase, rutile,* and *brookite,* and is widely used as a white pigment. The technological applications are as a photocatalyst [293], the removal of organic pollutants from water, and in photovoltaic devices [294] which convert light to electricity. Bulk titania is a n-type semiconductor, the conductivity of which can be enhanced by either reduction or doping.

Most electrical measurements for titania concern the rutile phase, which is the stable phase for coarse-grained samples. Titania is readily prepared as nanocrystals in the anatase phase by a variety of techniques, one common procedure being the sulfate route [295]. An early conductivity study was reported for a nanocrystalline anatase ceramic with a grain size of 35 ± 10 nm [296]. An unusual result was the discovery of a domain of ionic conductivity at high oxygen partial pressures, but this was attributed to space-charge layer effects [28]. These created defects at the grain boundaries and at those boundaries providing fast diffusion paths, although proton conduction might represent another explanation [297]. An interesting point might be to explore the effect of grain size on the extent of the ionic domain in titania ceramics.

4.6
Overview and Areas for Future Development

In the foregoing sections we have presented the available information on the structure and transport in ion-conducting nanocrystals, and shall now attempt to summarize and assess the present state of knowledge.

In general, good progress has been made in our understanding of the microstructure of these materials. The dominant factor in determining the level of disorder in the systems is the method of preparation, and a range of techniques has shown that aggressive preparation methods (e.g., mechanical attrition) produce highly disordered samples with disarray at the interfaces together with a significant fraction of amorphous material. These samples are equivalent to the early pictures of nanocrystalline materials, in which the interfaces were considered as either "gas-like" or "liquid-like". In contrast, the "softer" preparation methods, such as IGC and sol–gel procedures, produce samples with highly crystalline grains, with the interfaces between grains being comparable to the grain boundaries in bulk samples. The extensive studies on lithium niobate represent good examples of the link between a preparation method and microstructure [169, 252].

Although samples of nanocrystalline materials can be prepared with interfaces comparable to normal grain boundaries, the results of recent studies have indicated that the interfaces must be considered in even more detail. The degree of mismatch between the crystallites and the resultant strain may be an important factor in determining properties, especially those involving atomic transport. This is particularly true for heterostructures, such as nanocrystalline films on substrates or alternating nanocrystalline film of different materials, with differences in the lattice parameters leading to disordered incoherent interfaces [274]. A second concern for doped materials, such as the rare earth-doped ceria and zirconia, is the degree of dopant segregation to the interfaces that occurs when the nanocrystals are pressed, with heating, to form pellets. Such segregation has been well-documented in earlier studies [291], and could well play a significant role in the transport properties.

The only way to rationalize the transport data for nanocrystalline ionic materials is to consider the systems separately, from the viewpoint of the level of defects in the crystals and the nature of the samples. In the case of lowly defective systems, such as the alkaline earth fluorides, there is good evidence for a conductivity enhancement, and the data can be explained in terms of models based on the space-charge layer. A key experiment here was the observation of enhanced conductivity in very thin alternating CaF_2/BaF_2 layers when measured perpendicular to the layers [298]. This was explained as being due to the space-charge layers overlapping and saturating the layers. However, this observation is difficult to explain in terms of a model based on surface mismatch.

Although extensive data are available on ceria and zirconia systems, the results have shown wide variations in behavior. In heavily rare earth-doped systems, the point defect concentrations are high and the Debye lengths are on the order of Ångstroms, which in turn makes the space-charge layer effects negligible. Thus, the enhanced conduction observed in thin films of these materials is best explained in terms of surface mismatch. This suggests that the magnitude of the enhancement will depend heavily on the preparation conditions – that is, the extent of lattice mismatch between layers or layer and substrate, the rate of deposition of the film, and so on. In the case of compacted pellets the data may be confused with reports of enhanced transport and no marked effect. The studies of heavily doped ceria and zirconia include results from diffusion and conductivity measurements. There is also an important difference between samples in terms of their relative densities. In studies where an enhanced diffusion in zirconia has been observed (e.g., Ref. [276]) and an enhanced conductivity in ceria (e.g., Ref. [288]), the relative densities were $\leq$98%. However, diffusion studies performed with pellets of nanocrystalline zirconia prepared by spark plasma sintering and with relative densities $\geq$99.9%, yielded no overall enhancement [277]. Similarly, the conductivity studies of spark plasma-sintered, heavily doped ceria with a relative density >99% showed no enhanced conductivity, but only a reduction in grain boundary resistance [290]. The difference in these diffusion studies can be rationalized as the less-dense samples providing pathways for the anomalous penetration of the tracer. The differences in conductivity studies may lie in different levels of surface segregation of the dopants in samples due to different levels of grain growth during sintering. Another possibility that has been suggested is a contribution from proton conduction [277], which is always an underlying concern in the study of oxides [299].

Ideally, the ambiguity in the data for sintered pellets of oxides as discussed above should be resolved, and at this point we make some suggestions, albeit tentative, to help with this dilemma. One problem here is that the studies have predominantly involved measurements of the conductivity and so are subject to difficulties in interpretation. Experiments which would be fruitful would be combined diffusion and conductivity studies of the same samples, as this would eliminate any differences in sample composition and microstructure. In addition, they would allow the realistic use of the Nernst–Einstein equation, and hence some insight into the transport mechanisms that are operative would be possible. It would also be useful to extend the measurements to smaller grain sizes, if possible to those <10 nm. To date, surprisingly few NMR diffusion studies have been conducted with nanocrystalline oxides, although ^{18}O studies of bulk oxides have proved fruitful in the case of coarse-grained samples [300–302]. The advantages of this technique, as outlined elsewhere [169], is that the measurements are less sample-dependent and if there are two separate mechanisms of transport – that is, along the boundaries and in the bulk crystallite – then these should be observable as different resonance lines or relaxation times.

The application of nanocrystalline metal oxides in sensor devices is now well-established, and should produce benefits in terms of improved sensitivity and speed of response. On a similar note, nanomaterials have become increasingly important in battery technology, particularly in the development of lithium solid-state batteries [106, 303, 304]. Nanocrystalline oxides offer many advantages in SOFCs, primarily by increasing the surface area of the materials and hence the catalytic activity [305, 306], and this is especially important for lowering the cell's operating temperature. Overall, however, it remains clear that further research into the applications of nanocrystalline ionic materials will identify a much wider usage for these devices.

References

1 Roco, M.C. (2002) *JOM-J. Minerals Metals and Materials Soc.*, **54**, 22.
2 Gleiter, H. (1989) *Prog. Mater. Sci.*, **33**, 223.
3 Gleiter, H. (1992) *Adv. Mater.*, **4**, 474.
4 Henglein, A. (1989) *Chem. Rev.*, **89**, 1061.
5 Weller, H. (1993) *Angew. Chem., Int. Ed. Engl.*, **32**, 41.
6 Siegel, R.W. and Fougere, G.E. (1995) *Nanostructured Mater.*, **6**, 205.
7 Gleiter, H. (2000) *Acta Mater.*, **48**, 1.
8 Siegel, R.W. (1991) *Annu. Rev. Mater. Sci.*, **21**, 559.
9 Würschum, R., Brossmann, U. and Schaefer, H.-E. (2002) in *Nanostructured Materials – Processing, Properties, and Applications* (ed. C.C. Koch), Noyes Publications, Norwich, p. 267.
10 Edelstein, A.S. and Cammarata, R.C. (eds) (2002) *Nanomaterials: Synthesis, Properties and Applications*, Institute of Physics, Bristol, UK.
11 Despotuli, A.L. and Nikolaichik, V.I. (1993) *Solid State Ionics*, **60**, 275.
12 Maier, J. (2005) *Nat. Mater.*, **4**, 805–15.
13 Tjong, S.C. and Chen, H. (2004) *Mater. Sci. Eng. R*, **45**, 1.
14 Klabunde, K.J., Stark, J., Koper, O., Mohs, C., Park, D.G., Decker, S., Jiang, Y., Lagadic, I. and Zhang, D. (1996) *J. Phys. Chem.*, **100**, 12142.
15 Horvath, J., Birringer, R. and Gleiter, H. (1987) *Solid State Commun.*, **62**, 319.
16 Schumacher, S., Birringer, R., Strauss, R. and Gleiter, H. (1989) *Acta Metall.*, **37**, 2485.
17 Sato, Y., Mizoguchi, T., Oba, F., Yodogawa, M., Yamamoto, T. and Ikuhara, Y. (2005) *J. Mater. Sci.*, **40**, 3059.
18 Liu, X.X., Jin, Z.G., Bu, S.J., Zhao, J. and Liu, Z.F. (2006) *J. Am. Ceram. Soc.*, **89**, 1226.
19 Wang, Z.L. (2004) *Annu. Rev. Phys. Chem.*, **55**, 159.
20 Gopal, K.S., Mor, K., Fitzgerald, A. and Grimes, C.A. (2007) *J. Phys. Chem. C*, **111**, 21.
21 Chadwick, A.V. (1994) *Encyclopedia of Applied Physics Volume 8*, VCH Publishers, New York, p. 193.
22 Agullo-Lopez, F., Catlow, C.R.A. and Townsend, P.D. (1984) *Point Defects in Materials*, Academic Press, London.
23 Lehovec, K. (1953) *J. Chem. Phys.*, **21**, 1123.
24 Frenkel, J. (1946) *Kinetic Theory of Liquids*, Oxford University Press, New York.
25 Lidiard, A.B. (1957) *Handbuch der Physik*, **XX**, 246.
26 Maier, J. (1987) *Solid State Ionics*, **23**, 59.
27 Maier, J. (1987) *J. Electrochem. Soc.*, **134**, 1524.
28 Maier, J. (1995) *Prog. Solid State Chem.*, **23**, 171.
29 Maier, J. (2000) *Solid State Ionics*, **131**, 13.
30 Wagner, J.B. (1989) in *High Conductivity Solid Ionic Conductors* (ed. T. Takahashi), World Scientific, Singapore, p. 149.
31 Atkinson, A. (1990) *J. Chem. Soc. Faraday Trans.*, **86**, 1307.
32 Kaur, I., Mishin, Y. and Gust, W. (1995) *Fundamentals of Grain and Interphase Boundary Diffusion*, John Wiley & Sons, Chichester, UK.
33 Atkinson, A. and Monty, C. (1989) in *Surfaces and Interfaces of Ceramic Materials* (eds L.C. Dufour *et al.*), Kluwer Academic, Dordrecht, The Netherlands, p. 273.
34 Mishin, Y. and Herzig, C. (1999) *Mater. Sci. Eng. A*, **260**, 55.
35 Eranna, G., Joshi, B.C., Runthala, D.P. and Gupta, R.P. (2004) *Crit. Rev. Solid State Mater. Sci.*, **29**, 111.

36 Moseley, P.T. (1997) *Meas. Sci. Technol.*, **8**, 223.

37 Williams, G. and Coles, G.S.V. (1999) *MRS Bull.*, **June**, 25.

38 Chiorino, A., Ghiotti, G., Prinetto, F., Carotta, M.C., Gnani, D. and Martinelli, G. (1999) *Sens. Actuators B*, **58**, 338.

39 Cabot, A., Arbiol, J., Morante, J.R., Weimar, U., Bârsan, N. and Göpel, W. (2000) *Sens. Actuators B*, **70**, 87.

40 Fukui, K. and Katsuki, A. (2000) *Sens. Actuators B*, **65**, 316.

41 Choi, S.-D. and Lee, D.-D. (2001) *Sens. Actuators B*, **77**, 335.

42 Xu, C., Tamaki, J., Miura, N. and Yamazoe, N. (1990) *Chem. Lett.*, **19**, 441.

43 Kanefusa, S., Nitta, M. and Haradome, M. (1979) *J. Appl. Phys.*, **50**, 1145.

44 Maekawa, T., Tamaki, J., Miura, N. and Yamazoe, N. (1991) *Chem. Lett.*, **20**, 575.

45 Kanefusa, S., Nitta, M. and Haradome, M. (1998) *IEEE Trans. Electron Devices*, **35**, 65.

46 Ghiotti, G., Chiorino, A. and Prinetto, F. (1995) *Sens. Actuators B*, **24–25**, 564.

47 Brattain, W.H. and Bardeen, J. (1953) *Bell Syst. Tech. J.*, **32**, 1.

48 Seiyama, T., Kato, A., Fujhshi, K. and Nagatani, M. (1962) *Anal. Chem.*, **34**, 1502.

49 Taguchi, N. (1972) G.B. Patent 1280809.

50 Martinelli, G., Carotta, M.C., Traversa, E. and Ghiotti, G. (1999) *MRS Bull.*, June, 30.

51 Meixner, H. and Lampe, U. (1996) *Sens. Actuators B*, **33**, 198.

52 Mabrook, M. and Hawkins, P. (2001) *Sens. Actuators B*, **75**, 197.

53 Chadwick, A.V., Russell, N.V., Whitham, A.R. and Wilson, A. (1994) *Sens. Actuators B*, **18–19**, 99.

54 Williams, D.E. (1999) *Sens. Actuators B*, **57**, 1.

55 Chadwick, A.V., Harsch, A., Russell, N.V., Tse, K.F., Whitham, A.R. and Wilson, A. (1995) *Radiat. Eff. Defects Solids*, **137**, 51.

56 Zakrzewska, K. (2001) *Thin Solid Films*, **391**, 229.

57 Pokhrel, S. and Nagaraja, K.S. (2003) *Sens. Actuators B*, **92**, 144.

58 Comini, E., Ferroni, M., Guidi, V., Faglia, G., Martinelli, G. and Sberveglieri, G. (2002) *Sens. Actuators B*, **84**, 26.

59 Windischmann, H. and Mark, P. (1979) *J. Electrochem. Soc.*, **126**, 627.

60 Orton, J.W. and Powell, M.J. (1980) *Rep. Prog. Phys.*, **43**, 1263.

61 McAleer, J.F., Moseley, P.T., Norris, J.O.W. and Williams, D.E. (1987) *J. Chem. Soc., Faraday Trans.*, **83**, 1323.

62 Ghiotti, G., Chiorino, A., Martinelli, G. and Carotta, M.C. (1995) *Sens. Actuators B*, **24–25**, 520.

63 Lu, F., Liu, Y., Dong, M. and Wang, X. (2000) *Sens. Actuators B*, **66**, 225.

64 Schweizer-Berberich, M., Zheng, J.G., Weimar, U., Göpel, W., Bârsan, N., Pentia, E. and Tomescu, A. (1996) *Sens. Actuators B*, **31**, 71.

65 Morimitsu, M., Ozaki, Y., Suzuki, S. and Matsunaga, M. (2000) *Sens. Actuators B*, **67**, 184.

66 Cabot, A., Diéguez, A., Romano-Rodríguez, A., Morante, J.R. and Bârsan, N. (2001) *Sens. Actuators B*, **79**, 98.

67 Pan, Q., Xu, J., Dong, X. and Zhang, J. (2000) *Sens. Actuators B*, **66**, 237.

68 McAleer, J.F., Moseley, P.T., Norris, J.O.W., Williams, D.E. and Tofield, B.C. (1988) *J. Chem. Soc., Faraday Trans.*, **84**, 441.

69 Yamazoe, N. (1991) *Sens. Actuators B*, **5**, 7.

70 Matsushima, S., Maekawa, T., Tamaki, J., Miura, N. and Yamazoe, N. (1989) *Chem. Lett.*, **18**, 845.

71 Tianshu, Z., Hing, P., Li, Y. and Jiancheng, Z. (1999) *Sens. Actuators B*, **60**, 208.

72 Comini, E., Faglia, G. and Sberveglieri, G. (2001) *Sens. Actuators B*, **76**, 270.

73 Rastomjee, C.S., Dale, R.S., Schaffer, R.J., Jones, F.H., Egdell, R.G., Georgiadis, G.C., Lee, M.J., Tate, T.J. and Cao, L.L. (1996) *Thin Solid Films*, **279**, 98.

74 Yu, J.H. and Choi, G.M. (2001) *Sens. Actuators B*, **75**, 56.

75 Sayago, I., Horrillo, M.C., Getino, J., Gutiérrez, J., Arés, L., Robla, J.I.,

Fernández, M.J. and Rodrigo, J. (1999) *Sens. Actuators B*, **57**, 249.
76 Sayago, I., Gutiérrez, J., Arés, L., Robla, J.I., Horrillo, M.C., Getino, J. and Agapito, J.A. (1995) *Sens. Actuators B*, **24–25**, 512.
77 Zhang, G. and Liu, M. (2000) *Sens. Actuators B*, **69**, 144.
78 Quaranta, F., Rella, R., Siciliano, P., Capone, S., Epifani, M., Vasanelli, L., Licciulli, A. and Zocco, A. (1999) *Sens. Actuators B*, **58**, 350.
79 Xu, C., Tamaki, J., Miura, N. and Yamazoe, N. (1991) *Sens. Actuators B*, **3**, 147.
80 Ogawa, H., Nishikawa, M. and Abe, A. (1982) *J. Appl. Phys.*, **53**, 4448.
81 Davis, S.R., Chadwick, A.V. and Wright, J.D. (1997) *J. Phys. Chem. B*, **101**, 9901.
82 Davis, S.R., Chadwick, A.V. and Wright, J.D. (1998) *J. Mater. Chem.*, **8**, 2065.
83 Chadwick, A.V., Savin, S.L.P., O'Dell, L.A. and Smith, M.E. (2006) *J. Phys. Condens. Matter*, **18**, L163.
84 Savin, S.L.P., Chadwick, A.V., O'Dell, L.A. and Smith, M.E. (2006) *Solid State Ionics*, **177**, 2519.
85 Sarala Devi, G., Manorama, S. and Rao, V.J. (1998) *J. Electrochem. Soc.*, **125**, 1039.
86 Shukla, S., Seal, S., Ludwig, L. and Parish, C. (1992) *Sens. Actuators B*, **9**, 71.
87 Mohammadi, M.R., Fray, D.J. and Ghorbani, M. (2008) *Solid State Sci.*, **10**, 884.
88 Adamyan, A.Z., Adamyam, Z.N., Aroutiounian, V.M., Arakelyan, A.H., Touryan, K.J. and Turner, J.A. (2007) *Int. J. Hydrogen Energy*, **32**, 4101.
89 Park, S.-S. and Mackenzie, J.D. (1996) *Thin Solid Films*, **274**, 154.
90 Chakraborty, S., Mandal, I., Ray, I., Majumdar, S., Sen, A. and Maiti, H.S. (2007) *Sens. Actuators B*, **127**, 554.
91 Gong, J., Sun, J. and Chen, Q. (2008) *Sens. Actuators B*, **130**, 829.
92 Gouma, P.I. (2003) *Rev. Adv. Mater. Sci.*, **5**, 123.
93 Balazsi, C., Wang, L., Zayim, E.O., Szilagyi, I.M., Sedlackova, K., Pfeifer, J., Toth, A.L. and Gouma, P.I. (2008) *J. Eur. Ceram. Soc.*, **28**, 913.
94 Balazsi, C., Sedlackova, K., Llobet, E. and Ionescu, R. (2008) *Sens. Actuators B*, **133**, 151.
95 Salunkhe, R.R. and Lokhande, C.D. (2008) *Sens. Actuators B*, **129**, 345.
96 Chaudhari, G.N., Bambole, D.R. and Bodade, A.B. (2006) *Vacuum*, **81**, 251.
97 Siemons, M. and Simon, U. (2007) *Sens. Actuators B*, **126**, 595.
98 Michel, C.R., Mena, E.L., Preciado, A.H.M. and de Leon, E. (2007) *Mater. Sci. Eng. B*, **141**, 1.
99 Jagtap, S.V., Kadu, A.V., Sangawar, V.S., Manorama, S.V. and Chaudhari, G.N. (2008) *Sens. Actuators B*, **131**, 290.
100 Patil, A., Patil, V., Shin, D.W., Choi, J.-W., Paik, D.-S. and Yoon, S.-J. (2008) *Mater. Res. Bull.*, **43**, 1913.
101 Aurbach, D., Levi, M.D. and Levi, E. (2008) *Solid State Ionics*, **179**, 742.
102 Ritchie, A. and Howard, W. (2006) *J. Power Sources*, **162**, 809.
103 Jiang, C., Hosono, E. and Zhou, H. (2006) *Nano Today*, **1**, 28.
104 Stura, E. and Nicolini, C. (2006) *Anal. Chim. Acta*, **568**, 57.
105 Aurbach, D. (2005) *J. Power Sources*, **146**, 71.
106 Tarascon, J.M. and Armand, M. (2001) *Nature*, **414**, 359.
107 Poizot, P., Laruelle, S., Grugeon, S., Dupont, L. and Tarascon, J.M. (2000) *Nature*, **407**, 496.
108 Dollé, M., Poizot, P., Dupont, L. and Tarascon, J.M. (2002) *Electrochem. Solid-State Lett.*, **5**, A18.
109 Larcher, D., Sudant, G., Leriche, J.B., Chabre, Y. and Tarascon, J.M. (2002) *J. Electrochem. Soc.*, **149**, A234.
110 Tarascon, J.M., Grugeon, S., Morcrette, M., Laruell, S., Rozier, P. and Poizot, P. (2005) *C. R. Chimie*, **8**, 9.
111 Deb, A., Bergmann, U., Cramer, S.P. and Cairns, E.J. (2007) *J. Electrochem. Soc.*, **154**, A534.
112 Holzapfel, M., Proux, O., Strobel, P., Darie, C., Borowski, M. and Morcrette, M. (2004) *J. Mater. Chem.*, **14**, 102.

113 Yoon, W.S., Grey, C.P., Balasubramanian, M., Yang, X.Q. and McBreen, J. (2003) *Chem. Mater.*, **15**, 3161.
114 Balasubramanian, M., Sun, X., Yang, X.Q. and McBreen, J. (2001) *J. Power Sources*, **92**, 1.
115 Chadwick, A.V., Savin, S.L.P., Fiddy, S.G., Alcantara, R., Fernandez Lisbona, D., Lavela, P., Ortiz, G.F. and Tirado, J.L. (2007) *J. Phys. Chem. C*, **111**, 4636.
116 Alcántara, R., Jaraba, M., Lavela, P. and Tirado, J.L. (2002) *Chem. Mater.*, **14**, 2847.
117 Knop, O., Reid, K.I.G., Sutarno, O. and Nakagawa, Y. (1968) *Can J. Chem.*, **46**, 3463.
118 Fu, L.J., Liu, H., Li, C., Wu, Y.P., Rahm, E., Holze, R. and Wu, H.Q. (2006) *Solid State Sciences*, **8**, 113.
119 Prabaharan, S.R.S., Michael, M.S., Ramesh, S. and Begam, K.M. (2004) *J. Electroanal. Chem.*, **570**, 107.
120 Singhal, A., Skandan, G., Amatucci, G., Badway, F., Ye, N., Manthiram, A., Ye, H. and Xu, J.J. (2004) *J. Power Sources*, **129**, 38.
121 Dominko, R., Gaberscek, M., Bele, M., Mihailovic, D. and Jamnik, J. (2007) *J. Eur. Ceram. Soc.*, **27**, 909.
122 Fu, L.J., Liu, H., Li, C., Wu, Y.P., Rahm, E., Holze F R. and Wu, H.Q. (2005) *Prog. Mater. Sci.*, **50**, 881.
123 Liu, H. and Tang, D. (2008) *Solid State Ionics*, **179**, 1897.
124 Wu, J.B., Tu, J.P., Wang, X.L. and Zhang, W.K. (2007) *Int. J. Hydrogen Energy*, **32**, 606.
125 Myung, S.-T., Komaba, S., Kurihara, K. and Kumagai, N. (2006) *Solid State Ionics*, **177**, 733.
126 Odani, A., Nimberger, A., Markovsky, B., Sominski, E., Levi, E., Kumar, V.G., Motiei, M., Gedanken, A., Dan, P. and Aurbach, D. (2003) *J. Power Sources*, **119–121**, 517.
127 Ormerod, R.M. (2003) *Chem. Soc. Rev.*, **32**, 17.
128 Haille, S.M. (2003) *Acta Mater.*, **51**, 5981.
129 Yano, M., Tomita, A., Sano, M. and Hibino, T. (2007) *Solid State Ionics*, **177**, 3351.
130 Acres, G.J.K. (2001) *J. Power Sources*, **100**, 60.
131 Omata, T., Goto, Y. and Otsuka-Yao-Matsuo, S. (2007) *Sci. Technol. Adv. Mater.*, **8**, 524.
132 Ding, D., Liu, B., Zhu, Z., Zhou, S. and Xia, C. (2008) *Solid State Ionics*, **179**, 896.
133 Chinarro, E., Jurado, J.R. and Colomer, M.T. (2007) *J. Eur. Ceram. Soc.*, **27**, 3619.
134 Im, J.M., You, H.J., Yoon, Y.S. and Shin, D.W. (2007) *J. Eur. Ceram. Soc.*, **27**, 3671.
135 Yoon, Y.S., Im, J.M., You, H.J. and Shin, D.W. (2007) *J. Eur. Ceram. Soc.*, **27**, 4257.
136 Sacanell, J., Bellino, M.G., Lamas, D.G. and Leyva, A.G. (2007) *Physica B: Condens. Matter*, **398**, 341.
137 Fernandez-Garcıa, M., Martınez-Arias, A., Hanson, J.C. and Rodriguez, J.A. (2004) *Chem. Rev.*, **104**, 4063.
138 Messing, G.L., Zhange, S.C. and Jaynthi, G.V. (1993) *J. Am. Ceram. Soc.*, **76**, 2707.
139 Gleiter, H. (1981) in *Deformation of Polycrystals: Mechanisms and Microstructures* (eds N. Hansen, A. Horsewell, T. Lefferes and H. Lilholt), Riso National Laboratory, Roskilde, Denmark, p. 15.
140 Ying, J. (1993) *J. Aerosol. Sci.*, **24**, 315.
141 Pearson, D.H. and Edelstein, A.S. (1999) *Nanostruct. Mater.*, **11**, 1111.
142 Gonzalez, G., Freites, J.A. and Rojas, C.E. (2001) *Scr. Mater.*, **44**, 1883.
143 Taneja, P., Chandra, R., Banerjee, R. and Ayyub, P. (2001) *Scr. Mater.*, **44**, 1915.
144 Brinker, C.J. and Scherer, J.W. (1990) *Sol–Gel Science: The Physics and Chemistry of Sol–Gel Processing*, Academic Press, Boston.
145 Pierre, A.C. (1998) *Introduction to Sol–Gel Processing*, Kluwer Academic Publishers, Boston.
146 Chadwick, A.V., Mountjoy, G., Nield, V.M., Poplett, I.J.F., Smith, M.E., Strange, J.H. and Tucker, M.G. (2001) *Chem. Mater.*, **13**, 1219.
147 Liu, Y. and Patterson, B.R. (1996) *Acta Mater.*, **44**, 4327.
148 Chadwick, A.V. and Savin, S.L.P. *J. Alloys Compd.* (in press).

149 Chadwick, A.V., Savin, S.L.P., O'Dell, L.A. and Smith, M.E. (2007) *ChemPhysChem.*, 8, 882.

150 Fecht, H.J. (1995) *Nanostruct. Mater.*, **6**, 33.

151 Koch, C.C. (1997) *Nanostruct. Mater.*, **9**, 13.

152 Cukrov, L.M., Tsuzuki, T. and McCormick, P.G. (2001) *Scr. Mater.*, **44**, 1787.

153 Indris, S., Bork, D. and Heitjans, P. (2000) *J. Mater. Synth. Process.*, **8**, 245.

154 Ding, J., Tsuzuki, T., McCormick, P.G. and Street, R. (1996) *J. Phys. D*, **29**, 2365.

155 Baburaj, E., Hubert, K. and Froes, F. (1997) *J. Alloys Compd.*, **257**, 146.

156 Scholz, G., Stosser, R., Klein, J., Silly, G., Buzaré, J.Y., Laligant, Y. and Ziemer, B. (2002) *J. Phys.: Condens. Matter*, **14**, 2101.

157 Weibel, A., Bouchet, R. and Knauth, P. (2006) *Solid State Ionics*, **177**, 229.

158 Bouchet, R., Weibel, A., Knauth, P., Mountjoy, G. and Chadwick, A.V. (2003) *Chem. Mater.*, **15**, 4996.

159 Huang, H., Kelder, E.M., Jak, M.J.G. and Schoonman, J. (2001) *Solid State Ionics*, **139**, 67.

160 Jak, M.J.G., Ooms, F.G.B., Kelder, E.M., Legerstee, W.J., Schoonman, J. and Weisenburger, A. (1999) *J. Power Sources*, **80**, 83.

161 Arthur, J.R. (2002) *Surf. Sci.*, **500**, 189.

162 Bouessay, I., Rougier, A., Poizot, P., Moscovici, J., Michalowicz, A. and Tarascon, J.-M. (2005) *Electrochim. Acta*, **50**, 3737.

163 Suzuki, T., Kosacki, I. and Anderson, H.U. (2002) *Solid State Ionics*, **151**, 111.

164 Adair, J.H., Li, T., Kido, T., Havey, K., Moon, J., Mecholsky, J., Morrone, A., Talham, D.R., Ludwig, M.H. and Wang, L. (1998) *Mater. Sci. Eng. R*, **23**, 139.

165 Sun, L., Wei, G., Song, Y., Liu, Z., Wang, L. and Li, Z. (2006) *Mater. Lett.*, **60**, 1291.

166 Tom, R.T., Sreekumaran Nair, A., Singh, N., Aslam, M., Nagendra, C.L., Philip, R., Vijayamohanan, K. and Pradeep, R. (2003) *Langmuir*, **19**, 3439.

167 Cushing, B.L., Kolesnichenko, V.L. and O'Connor, C.J. (2004) *Chem. Rev.*, **104**, 3893.

168 Ravel, B., Carpenter, E.E. and Harris, V.G. (2002) *J. Appl. Phys.*, **91**, 8195.

169 Heitjans, P., Masoud, M., Feldhoff, A. and Wilkening, M. (2006) *Faraday Discuss.*, **134**, 67.

170 Scherrer, P. (1918) *Nachr. Ges. Wiss. Göttingen*, **26**, 98.

171 Klug, H.P. and Alexander, L.E. (1974) *X-Ray Diffraction Procedures for Polycrystalline and Amorphous Materials*, 2nd edn, John Wiley & Sons, New York.

172 Warren, B.E. (1969) *X-Ray Diffraction*, Dover Publications, New York.

173 Balzar, B. (1999) *Defect and Microstructure Analysis from Diffraction*, Oxford University Press, New York.

174 Williamson, G.K. and Hall, W.H. (1953) *Acta Metall.*, **1**, 22.

175 Weibel, A., Bouchet, R., Boulc'h, F. and Knauth, P. (2005) *Chem. Mater.*, **17**, 2378.

176 Sing, K. (2001) *Colloids Surf., A*, **187**, 3–9.

177 Allen, T. (1999) *Particle Size Measurement*, 5th edn, vol. **1**, Kluwer Academic Publishers, Dordrecht, The Netherlands.

178 Uvarov, V. and Popov, I. (2007) *Mater. Charact.*, **59**, 883.

179 Borchert, H., Shevehenko, E.V., Robert, A., Mekis, I., Kornowski, A., Grubel, G. and Weller, H. (2005) *Langmuir*, **21**, 1931.

180 Chiang, Y.-M., Lavik, E.B., Kosacki, I., Tuller, H.L. and Ying, J.Y. (1997) *J. Electroceram.*, **1**, 7.

181 Weirich, Th.E., Winterer, M., Seifried, S., Hahn, H. and Fuess, F. (2000) *Ultramicroscopy*, **81**, 263.

182 Würschum, R., Soyez, G. and Schaefer, H.-E. (1993) *Nanostruct. Mater.*, **3**, 225.

183 Haubold, T., Birringer, R., Lengeler, B. and Gleiter, H. (1989) *Phys. Lett. A*, **135**, 461.

184 de Panfilis, S., d'Acapito, F., Haas, V., Konrad, H., Weissmüller, J. and Boscherini, F. (1995) *Phys. Lett. A*, **207**, 397.

185 Chadwick, A.V. and Rush, G.E. (2002) Characterisation of nanocrystalline metal oxides by XAS, in *Nanocrystalline Materials* (eds P. Knauth and J. Schoonman), Kluwer, New York, p. 133.

186 Chadwick, A.V. (2006) *Solid State Ionics*, **177**, 2481.

187 Scolan, E., Magnenet, C., Massiot, D. and Sanchez, C. (1999) *J. Mater. Chem.*, **9**, 2467.

188 O'Dell, L.A., Savin, S.L.P., Chadwick, A.V. and Smith, M.E. (2007) *Appl. Magn. Reson.*, **32**, 527.

189 Teo, B.K. and Joy, D.C. (eds) (1980) *EXAFS Spectroscopy; Techniques and Applications*, Plenum Press, New York.

190 Hayes, T.M. and Boyce, J.B. (1982) *Solid State Phys.*, **37**, 173.

191 Koningsberger, D.C. and Prins, R. (eds) (1988) *X-Ray Absorption*, John Wiley & Sons, New York.

192 Wang, Y.R., Lu, K.Q., Wang, D.H., Wu, Z.H. and Fang, Z.Z. (1994) *J. Phys.: Condens. Matter*, **6**, 633.

193 Qi, Z., Shi, C., Wei, Y., Wang, Z., Liu, T., Hu, T., Zhan, Z. and Li, F. (2001) *J. Phys.: Condens. Matter*, **13**, 11503.

194 Rush, G.E., Chadwick, A.V., Kosacki, I. and Anderson, H.U. (2000) *J. Phys. Chem. B*, **104**, 9597.

195 Scholz, G., Stösser, R., Klein, J., Silly, G., Buzaré, J.Y., Laligant, Y. and Ziemer, B. (2002) *J. Phys.: Condens. Matter*, **14**, 2101.

196 Chadwick, A.V., Pooley, M.J. and Savin, S.L.P. (2005) *Phys. Status Solidi C*, **2**, 302.

197 Stern, E.A., Siegel, R.W., Newville, M., Sanders, P.G. and Haskel, D. (1995) *Phys. Rev. Lett.*, **75**, 3874.

198 Klinowski, J. (1993) *Anal. Chim. Acta*, **283**, 929.

199 Bell, A.T. (1999) *Colloids Surf. A*, **158**, 221.

200 Ashbrook, S.E. and Smith, M.E. (2006) *Chem. Soc. Rev.*, **35**, 718.

201 Epping, J.D. and Chmelka, B.F. (2006) *Curr. Opin. Colloid Interface Sci.*, **11**, 81.

202 Bonhomme, C., Coelho, C., Baccile, N., Gervais, C., Azaïs, T. and Babonneau, F. (2007) *Acc. Chem. Res.*, **40**, 738.

203 O'Dell, L.A., Savin, S.L.P., Chadwick, A.V. and Smith, M.E. (2007) *Solid State Nucl. Magn. Reson.*, **31**, 169.

204 Philibert, J. (1991) *Atom Movement, Diffusion and Mass Transport in Solids*, Les Éditions de Physique, Paris.

205 Shewmon, P. (1989) *Diffusion in Solids*, 2nd edn, TMS, Warrendale, PA.

206 Borg, R.J. and Dienes, G.J. (1988) *An Introduction to Solid State Diffusion*, Academic Press, Boston.

207 Heitjans, P. and Kärger, J.(eds) (2006) *Condensed Matter - Methods, Materials, Models*, Springer, Berlin.

208 Mehrer, H. (2007) *Diffusion in Solids: Fundamentals, Methods, Materials, Diffusion-Controlled Processes. Springer Series in Solid-State Sciences*, Springer, Berlin.

209 Chadwick, A.V. (1991) *Philos. Mag. A*, **64**, 983.

210 Heitjans, P. and Indris, S. (2003) *J. Phys.: Condens. Matter*, **15**, R1257.

211 Funke, K. and Cramer, C. (1997) *Curr. Opin. Solid State Mater. Sci.*, **2**, 483.

212 Chadwick, A.V. (1990) *J. Chem. Soc., Faraday Trans.*, **86**, 11575.

213 Crank, J. (1995) *The Mathematics of Diffusion*, Clarendon Press, Oxford.

214 Kilner, J.A., Steele, B.C.H. and Ilkov, L., (1984) *Solid State Ionics*, **12**, 89.

215 Harrison, L.G. (1961) *Trans. Faraday Soc.*, **57**, 1191.

216 Mishin, Y. and Herzig, C. (1995) *Nanostruct. Mater.*, **6**, 859.

217 Chadwick, A.V. (1990) *J. Chem. Soc., Faraday Trans.*, **86**, 1157.

218 Heitjans, P., Schirmer, A. and Indris, S. (2005) in *Diffusion in Condensed Matter - Methods, Materials, Models* (eds P. Heitjans and J. Kärger), Springer, Berlin, chapter 9.

219 Figueroa, D.R., Strange, J.H. and Wolf, D. (1979) *Phys. Rev. B*, **19**, 148.

220 Hahn, E. (1950) *Phys. Rev.*, **80**, 580.

221 Price, W.S. (1997) *Concepts Magn. Reson.*, **8**, 299.

222 Price, W.S. (1998) *Concepts Magn. Reson.*, **10**, 197.

223 Stejskal, E.O. and Tanner, J.E. (1965) *J. Chem. Phys.*, **42**, 288.

224 Gordon, R.E., Strange, J.H. and Webber, J.B.W. (1978) *J. Phys. E*, **11**, 1051.

225 Bénière, F. (1972) *Physics of Electrolytes* (ed. J. Hladik), Academic Press, London, p. 203.

226 Corish, J. and Jacobs, P.W.M. (1973) in *Surface and Defect Properties of Solids*, Vol. 2 (eds M.W. Roberts and J.M. Thomas), The Chemical Society, London, p. 184.
227 Le Claire, A.D. (1970) in *Physical Chemistry–An Advanced Treatise*, Vol. 10 (eds H. Eyring, D. Henderson and W. Jost), Academic Press, New York, pp. 261–330.
228 Bauerle, J.E. (1969) *J. Phys. Chem. Solids*, **30**, 2657.
229 Archer, W.I. and Armstrong, R.D. (1981) in *Electrochemistry* (ed. H.R. Thirsk), The Chemical Society, London, p. 157.
230 MacDonald, J.R. (1987) *Impedance Spectroscopy*, John Wiley & Sons, New York.
231 Nabarro, F.R.N. (1948) *Report of a Conference on the Strength of the Solids*, The Physical Society of London, London, p. 75.
232 Herring, C.J. (1950) *J. Appl. Phys.*, **21**, 437.
233 Coble, R.L. (1963) *J. Appl. Phys.*, **34**, 1679.
234 Meyers, M.A., Mishra, A. and Benson, D.J. (2006) *Prog. Mater. Sci.*, **51**, 427.
235 Vogl, G. (1996) *Physica B*, **226**, 135.
236 Bée, M. (2003) *Chem. Phys.*, **292**, 121.
237 Puin, W., Rodewald, S., Ramlau, R., Heitjans, P. and Maier, J. (2000) *Solid State Ionics*, **131**, 159.
238 Brossmann, U., Wurschum, R., Sodervall, U. and Schaefer, H.E. (1999) *J. Appl. Phys.*, **85**, 7646.
239 Tuller, H.L. (2000) *Solid State Ionics*, **131**, 143.
240 Omata, T., Goto, Y. and Otsuka-Yao-Matsuo, S. (2007) *Sci. Technol. Adv. Mater.*, **8**, 524.
241 Knauth, P. (2000) *J. Electroceram.*, **5**, 111.
242 Agrawal, R.C. and Gupta, R.K. (1999) *J. Mater. Sci.*, **34**, 1131.
243 Puin, W., Heitjans, P., Dickenscheid, W. and Gleiter, H. (1993) in *Defects in Insulating Materials* (eds O. Kanert and J.-M. Spaeth), World Scientific, Singapore, p. 137.
244 Puin, W. and Heitjans, P. (1995) *Nanostruct. Mater.*, **6**, 885.
245 Sata, N., Ebermann, K., Eberl, K. and Maier, J. (2000) *Nature*, **408**, 946.
246 Sata, N., Jin-Phillipp, N.Y., Eberl, K. and Maier, J. (2002) *Solid State Ionics*, **154–155**, 497.
247 Jin-Phillipp, N.Y., Sata, N., Maier, J., Scheu, C., Hahn, K., Kelsch, M. and Ruhle, M. (2004) *J. Chem. Phys.*, **120**, 2375.
248 Guo, X.X., Sata, N. and Maier, J. (2004) *Electrochim. Acta*, **49**, 1091.
249 Sorokin, N.I., Buchinskaya, I.I., Fedorov, P.P. and Sobolev, B.P. (2008) *Inorg. Mater.*, **44**, 189.
250 Arizmendi, L. (2004) *Phys. Status Solidi A*, **201**, 253.
251 Bork, D. and Heitjans, P. (2001) *J. Phys. Chem. B*, **105**, 9162.
252 Pooley, M.J. and Chadwick, A.V. (2003) *Radiat. Eff. Defects Solids*, **158**, 197.
253 Hannink, R.H.J., Kelly, P.M. and Muddle, B.C. (2000) *J. Am. Ceram. Soc.*, **83**, 461.
254 Narendra, B., Dahotre, B. and Nayak, S. (2005) *Surf. Coat. Technol.*, **194**, 58.
255 Piconi, C. and Maccauro, G. (1999) *Biomaterials*, **20**, 1.
256 Maskell, W.C. (2000) *Solid State Ionics*, **134**, 43.
257 Singhal, S.C. (2002) *Solid State Ionics*, **152**, 405.
258 Garvie, R.C. (1965) *J. Phys. Chem.*, **69**, 1238.
259 Djurado, E., Bouvier, P. and Lucazeau, G. (2000) *J. Solid State Chem.*, **149**, 399.
260 Kilo, M. (2005) *Defects and Diffusion in Ceramics: An Annual Retrospective VII*, **242–244**, 185.
261 Manning, P.S., Sirman, J.D., De Souza, R.A. and Kilner, J.A. (1997) *Solid State Ionics*, **100**, 1.
262 Kharton, V.V., Marques, F.M.B. and Atkinson, A. (2004) *Solid State Ionics*, **174**, 135.
263 Beekmans, N.M. and Heyne, L. (1976) *Electrochim. Acta*, **21**, 303.
264 van Dijk, T. and Burggraaf, A.J. (1981) *Phys. Status Solidi A*, **63**, 229.
265 Badwal, S.P.S. (1995) *Solid State Ionics*, **76**, 67.
266 Aoki, M., Chiang, Y.M., Kosacki, I., Lee, I.J.R., Tuller, H. and Liu, Y.P. (1996) *J. Am. Ceram. Soc.*, **79**, 1169.

267 Guo, X., Sigle, W., Fleig, J. and Maier, J. (2002) *Solid State Ionics*, **154–155**, 555.

268 Mondal, P., Klein, A., Jaegermann, W. and Hahn, H. (1999) *Solid State Ionics*, **118**, 331.

269 Jiang, S.S., Schulze, W.A., Amarakoon, V.R.W. and Stangle, G.C. (1997) *J. Mater. Res.*, **12**, 2374.

270 Kosacki, I., Gorman, B. and Anderson, H.U. (1998) in *Ionic and Mixed Conductors, Vol. III* (eds T.A. Ramanarayanan, W.L. Worrell, H.L. Tuller, A.C. Kandkar, M. Mogensen and W. Gopel), Electrochemical Society, Pennington, New Jersey, p. 631.

271 Kosacki, I., Suzuki, T., Petrovsky, V. and Anderson, H.U. (2000) *Solid State Ionics*, **136–137**, 1225.

272 Kosacki, I., Rouleau, C.M., Becher, P.F., Bentley, J. and Lownde, D.H. (2005) *Solid State Ionics*, **176**, 1319.

273 Azad, S., Marina, O.A., Wang, C.M., Saraf, L., Shutthanandan, V., McCready, D.E., El-Azab, A., Jaffe, J.E., Engelhard, M.H., Peden, C.H.F. and Thevuthasan, S. (2005) *Appl. Phys. Lett.*, **86**, 131906.

274 Peters, A., Korte, C., Hesse, D., Zakharov, N. and Janek, J. (2007) *Solid State Ionics*, **178**, 67.

275 Knöner, G., Reimann, K., Röwer, R., Södervall, U. and Schaefer, H.-E. (2003) *Proc. Natl Acad. Sci. USA*, **100**, 3870–3873.

276 Brossmann, U., Knoner, G., Schaefer, H.E. and Wurschum, R. (2004) *Rev. Adv. Mater. Sci.*, **6**, 7.

277 De Souza, R.A., Pietrowski, M.J., Anselmi-Tamburini, U., Kim, S., Munir, Z.A. and Martin, M. (2008) *Phys. Chem. Chem. Phys.*, **10**, 2067.

278 Nakagawa, T., Sakaguchi, I., Shibata, N., Matsunaga, K., Yamamoto, T., Haneda, H. and Ikuhara, Y. (2005) *J. Mater. Sci.*, **40**, 3185.

279 Tuller, H.L. and Nowick, A.S. (1979) *J. Electrochem. Soc.*, **126**, 209.

280 Kim, S. and Maier, J. (2004) *J. Eur. Ceram. Soc.*, **24**, 1919.

281 Kim, S., Fleig, J. and Maier, J. (2003) *Phys. Chem. Chem. Phys.*, **5**, 2268.

282 Suzuki, T., Kosacki, I., Anderson, H.U. and Colomban, P. (2001) *J. Am. Ceram. Soc.*, **84**, 2007.

283 Tschöpe, A., Sommer, E. and Birringer, R. (2001) *Solid State Ionics*, **139**, 255.

284 Tschöpe, A. (2001) *Solid State Ionics*, **139**, 267.

285 Tschöpe, A. (2005) *J. Electroceram.*, **14**, 5.

286 Huang, H., Gür, T.M., Saito, Y. and Prinz, F. (2006) *Appl. Phys. Lett.*, **89**, 143107.

287 Rupp, J.L.M. and Gauckler, L.J. (2006) *Solid State Ionics*, **177**, 2513.

288 Bellino, M.G., Lamas, D.G. and Walsöe de Reca, N.E. (2006) *Adv. Funct. Mater.*, **16**, 107.

289 Chiang, Y.M., Lavik, E.B. and Blom, D.A. (1997) *Nanostruct. Mater.*, **9**, 633.

290 Anselmi-Tamburini, U., Maglia, F., Chiodelli, G., Tacca, A., Spinolo, G., Riello, P., Bucella, S. and Munir, Z.A. (2006) *Adv. Funct. Mater.*, **16**, 2363.

291 Lei, Y., Ito, Y. and Browning, N.D. (2002) *J. Am. Ceram. Soc.*, **85**, 2359.

292 Bellino, M.G., Lamas, D.G. and Walsöe de Reca, N.E. (2006) *Adv. Mater.*, **18**, 3005.

293 Herrmann, J.M. (1999) *Catal. Today*, **53**, 115.

294 Nazeeruddin, M.K., Kay, A., Rodicio, I., Humphrybaker, R., Muller, E., Liska, P., Vlachopoulos, N. and Gratzel, M. (1993) *J. Am. Chem. Soc.*, **115**, 6382.

295 Knauth, P., Bouchet, R., Schäf, O., Weibel, A. and Auer, G. (2002) in *Synthesis, Functionalization and Surface Treatments of Nanoparticles* (ed. M-.I. Baraton), American Science Publishers, Stevenson Ranch CA, chapter 8.

296 Knauth, P. and Tuller, H.L. (1999) *J. Appl. Phys.*, **85**, 897.

297 Knauth, P. (2006) *Solid State Ionics*, **177**, 2495.

298 Guo, X.X., Matei, I., Lee, J.-S. and Maiera, J. (2007) *Appl. Phys. Lett.*, **91**, 103102.

299 Norby, T., Widerøe, M., Glöckner, R. and Larring, Y. (2004) *Dalton Trans.*, **19**, 3012.

300 Fuda, K., Kishio, K., Yamauchi, S. and Fueki, K. (1985) *J. Phys. Chem. Solids*, **46**, 1141.

301 Fuda, K., Kishio, K., Yamauchi, S. and Fueki, K. (1985) *Solid State Commun.*, **53**, 83.

302 Fuda, K., Kishio, K., Yamauchi, S., Fueki, K. and Onoda, Y. (1984) *J. Phys. Chem.*, **45**, 1253.

303 Jamnik, J. and Maier, J. (2003) *Phys. Chem. Chem. Phys.*, **5**, 5215.

304 Arico, A.S., Bruce, P., Scrosati, B., Tarascon, J.M. and Van Schalkwijk, W. (2005) *Nature Mater.*, **4**, 366.

305 Laosiripojana, N. and Assabumrungrat, S. (2006) *Chem. Eng. Sci.*, **61**, 2540.

306 Serra, J.M., Uhlenbruck, S., Meulenberg, W.A., Buchkremer, H.P. and Stöver, D. (2006) *Top. Catal.*, **40**, 123.

5
The Fundamentals and Advances of Solid-State Electrochemistry: Intercalation (Insertion) and Deintercalation (Extraction) in Solid-State Electrodes

Sung-Woo Kim, Seung-Bok Lee, and Su-Il Pyun

Abstract

Over several decades, solid-state electrodes in which reversible intercalation (insertion) and deintercalation (extraction) of cationic guest atoms occur along with accompanying electron flow without any change of their crystal structure, have attracted great interest in fundamental and practical perspectives for improving the performance of rechargeable batteries. This chapter provides comprehensive reviews of principle and recent advances especially in thermodynamic and kinetic approaches to lithium intercalation into, and deintercalation from, transition metals oxides and carbonaceous materials. Thermodynamic properties such as chemical potential, entropy and enthalpy of lithium intercalation/deintercalation are first discussed, based on a lattice gas model with various approximations. Lithium intercalation/deintercalation involving an order–disorder transition or a two-phase coexistence caused by strong interaction of lithium ions in solid-state electrodes is explained, based on the lattice gas model and with the help of computational methods. Second, the kinetics of lithium intercalation/deintercalation is treated in detail on the basis of a "cell-impedance-controlled" model. Anomalous features of potentiostatic current transients obtained experimentally from transitional metal oxide and carbonaceous electrodes, which are hardly explained under a "diffusion control" model, are readily analyzed by the "cell-impedance-controlled" lithium transport concept, with the aid of computational methods.

5.1
Introduction

When cationic guest atoms such as lithium, hydrogen, and sodium reversibly enter or leave the host oxide crystal, along with an accompanying electron flow but without any change in crystal structure, the reaction is referred to as intercalation/deintercalation as follows [1, 2]:

$$\delta A^{+} + \delta e + \mathrm{MO_2} \leftrightarrow A_{\delta}\mathrm{MO_2} \tag{5.1}$$

Solid State Electrochemistry I: Fundamentals, Materials and their Applications. Edited by Vladislav V. Kharton

ISBN: 978-3-527-32318-0

where A^+ may represent a monovalent cation and M a transition metal. The metal oxide phase MO_2 can be generally replaced by carbon phase C, such as graphite. Here, the forward reaction is called intercalation (insertion, charging), while the backward reaction is termed deintercalation (desertion, extraction). Intercalation and deintercalation proceed reversibly in opposite directions. While insertion or extraction is a more general term to describe the reaction, whatever the crystal structure, intercalation or deintercalation is limited to expressing either insertion or extraction reactions, respectively, and is associated with layered and spinel structures.

The change of guest atom composition in the matrix atom is accompanied by an intercalation reaction within the same crystal structure of the matrix atom. The characteristic feature of intercalation/deintercalation is, first, that the reaction proceeds not only at the interface between electrolyte and host intercalation compound electrode but also even in the interior of the electrode.

Electrons participating in the intercalation/deintercalation reaction (Equation (5.1)) can be represented by a current-producing system. Second, it is characteristic that the current-producing system reversibly operated by a self-driven (galvanic) cell (discharging the battery) performs the electrical useful work $\Delta G = -zFE$ (where E is the EMF of the cell), because electrical potential difference is spontaneously developed between two electrodes. By contrast, when the cell is short-circuited – that is, when the two electrodes are not separated from each other but are directly in electrical contact – electrons do not appear explicitly but rather participate in corrosion (or permeation in the case of solid electrolyte cells). They perform no electrical useful work because the two electrodes have the same electrical potential.

It is possible to differentiate a current-consuming system from a current-producing system, which can be reversibly operated by an externally driven (electrolysis) cell (charging the battery). The current-producing system is composed of an anode (a negative electrode) and a cathode (a positive electrode) which are spatially separated by, for example, a lithium ion-containing nonaqueous electrolyte, whereby the forward intercalation reaction (Equation (5.1)) by lithium ions occurs spontaneously, not only into the cathode at the cathode/electrolyte interface, but also within the cathode. In contrast, the backward deintercalation reaction proceeds spontaneously, not only from the anode at the anode/electrolyte interface but also within the anode. The charging cell reverses the direction of the electron flow and the direction of the intercalation and deintercalation reaction (Equation (5.1)) occurring at the positive and negative electrodes [3].

In the current-flowing condition of the cell – which includes both the current-producing and current-consuming systems – the direction of electron flow across a planar (film) electrode is always just reverse to that of the lithium ion flow during the intercalation and deintercalation reaction (Equation (5.1)), regardless of the discharging and charging cell [4]. In contrast, a self-discharge may irreversibly occur in an open-circuit condition of the cell, which corresponds to corrosion.

Since the 1970s, the thermodynamics, kinetics and mechanism of lithium intercalation into, and deintercalation from, transition metal chalcogenides, transition metal oxides and carbonaceous electrodes have been of great importance in fundamental and practical perspectives. This has been the case in particular for improving the performance of lithium rechargeable batteries in terms of energy density, power density, and cycle life. Many reviews have described the various aspects of lithium intercalation compounds, such as the crystal structures, synthetic methods and their electrochemical performance, and many general overviews of design concept of lithium and lithium ion batteries are available [3, 5–9]. Whilst earlier reviews focused on the transition metal chalcogenides [5, 6], more recent reports have documented more the transition metal oxides and carbonaceous materials [3, 7–9]. Selected data on crystal structures and transport properties of these materials can be found in Chapters 2 and 7.

In this respect, this chapter details the fundamentals and most important advances in the research activities on lithium intercalation into and deintercalation from transition metals oxides and carbonaceous materials, especially from thermodynamic and kinetic points of view, including methodological overviews. The thermodynamics of lithium intercalation/deintercalation is first introduced with respect to a lattice gas model with various approximations, after which the kinetics of lithium intercalation/deintercalation are described in terms of a "cell-impedance-controlled" model. Finally, some experimental methods which have been widely used to explore the thermodynamics and kinetics of lithium intercalation/deintercalation are briefly overviewed.

5.2 Thermodynamics of Intercalation and Deintercalation

Consider an electrochemical cell with a solid host MO_2 as one electrode, an alkali metal A as the other electrode, and an electrolyte in which the monovalent cation A^+ is dissolved. The intercalation/deintercalation between the host MO_2 and the guest ion A^+ is given by Equation (5.1). On the other hand, the redox reaction between A and A^+ may be written as

$$A^+ + e^- \leftrightarrow A. \tag{5.2}$$

As described previously, the main aspect of intercalation/deintercalation from a thermodynamic view point is that the concentration of the guest ion can change, without any change in the space group and lattice parameter of the host structure. Under electrochemical equilibrium conditions, therefore, the galvanic potential difference between two electrodes – that is, the cell voltage – can be derived as:

$$E = (\phi^{A_\delta MO_2} - \phi^A) = -\frac{1}{F}\mu_A^{A_\delta MO_2}, \tag{5.3}$$

where $\phi^{A_\delta MO_2}$ and ϕ^A is the galvanic potential of $A_\delta MO_2$ and A, respectively, $\mu_A^{A_\delta MO_2}$ is the chemical potential of A in $A_\delta MO_2$, and F is Faraday's constant. As shown in Equation (5.3), one advantage in the investigation of the thermodynamics of the insertion compounds is that $\mu_A^{A_\delta MO_2}$ can be measured experimentally from the cell voltage when the alkali metal or a buffer mixture with known chemical activity of the potential-determining component is used as the reference electrode.

5.2.1
Simple Lattice Gas Model

In order to explore the thermodynamic properties, and especially the chemical potential of the intercalation compounds, a lattice gas model [10] has been adopted under the assumption that intercalated ions are localized at specific sites in the host lattice, with no more than one ion on any site, and that local and global electro-neutrality is observed and there is no strong interaction between the electrons and the intercalated ions. It should be noted that, in solid-state chemistry, this model is often referred to as "ideal solution approximation" when used to describe the thermodynamics of nonstoichiometric compounds. According to this model, the chemical potential of A in $A_\delta MO_2$ in Equation (5.3) can be divided into two terms as

$$E = -\frac{1}{F}\mu_A^{A_\delta MO_2} = \frac{1}{F}\mu_e^{A_\delta MO_2} - \frac{1}{F}\mu_{A^+}^{A_\delta MO_2}, \tag{5.4}$$

where $\mu_e^{A_\delta MO_2}$ and $\mu_{A^+}^{A_\delta MO_2}$ are the chemical potential of electrons and the intercalated ion A^+, respectively. In the case of metallic intercalation compounds, the chemical potential of electrons can be arranged as essentially constant because electrons added by intercalation should be located within kT of the Fermi energy.

The chemical potential – that is, the change in the Gibbs free energy G with the number of the intercalated ions (n) – can be divided into two parts related to the enthalpy (H) and entropy (S) variations:

$$\mu = \partial G/\partial n = \partial H/\partial n - T\partial S/\partial n. \tag{5.5}$$

The simplest way to approach the intercalation thermodynamics based on the lattice gas model is to assume that the intercalated ions do not interact with one another, that the available sites are equivalent and occupied by the ions at random, and that the chemical potential of electrons are constant. The entropy of distributing ions randomly on a fraction δ of the available sites (N) in the intercalation compounds is

$$S = -kN[\delta\log\delta + (1-\delta)\log(1-\delta)], \tag{5.6}$$

where k is Boltzmann's constant. Then, the partial entropy is

$$\partial S/\partial n = -k\log[\delta/(1-\delta)]. \tag{5.7}$$

In the simplest lattice gas model, where the sum of the energy increments associated with the ion site occupation and the chemical potential of electrons is denoted as E^0, Equation (5.4) is written as

$$E = E^0 - \frac{RT}{F}\ln\left(\frac{\delta}{1-\delta}\right), \tag{5.8}$$

where R is a gas constant and T represents the absolute temperature.

5.2.2
Consideration of Ionic Interaction Using the Lattice Gas Model

For most insertion compounds, the interaction of intercalated ions with each other in the host lattice is not negligible. In order to simply consider the contribution of ionic interaction in Equation (5.8), it is often assumed that each ion experiences a mean interaction or energy field from its neighboring ions, based on a mean-field theory [10]. According to this approximation, the contribution to the chemical potential is proportional to the fraction of sites occupied by the ions δ, and hence the interaction term is introduced into Equation (5.8) as

$$E = E^0 - \frac{RT}{F}\ln\left(\frac{\delta}{1-\delta}\right) + J(\delta-0.5), \quad (5.9)$$

where J is the interaction parameter between the intercalated ions.

The relationship between E and δ at various J values is shown in Figure 5.1. If there is no interaction ($J=0$), repulsive interaction ($J<0$), or small attractive interaction ($0<J<4RT/F$) between the intercalated ions, the intercalation/deintercalation occurs in a single phase. Interestingly, in the framework of this approximation the larger value of J than $4RT/F$ (at which the slope of E versus δ curve, $dE/d\delta$, is zero at $\delta=0.5$) might lead apparently to a minimum and a maximum in the E versus δ dependencies (the dashed line in Figure 5.1). When two phases with different compositions are in equilibrium, the chemical potential is constant according to Gibbs' phase rule. Therefore, the E versus δ curve should be constant in this two-phase region (the solid line in Figure 5.1).

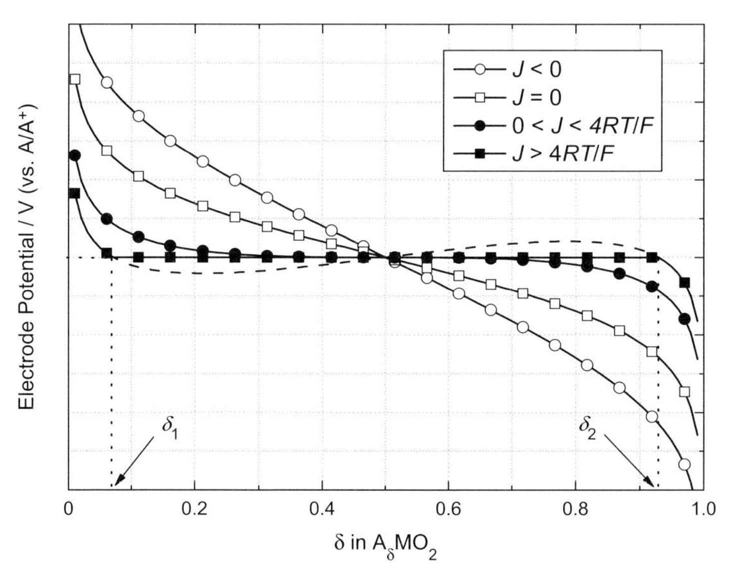

Figure 5.1 Relationships between the electrode potential E and content of guest ions A^+ (δ) intercalated into the host MO_2, calculated at various values of interaction parameter J.

5.2.3
Application to Lithium Intercalation/Deintercalation

As mentioned in Section 5.1, lithium intercalation compounds have for several decades attracted much interest for their application in advanced batteries with high power and energy density. Among these compounds, transition metal oxides such as $LiCoO_2$, $LiNiO_2$, $LiMn_2O_4$, and $LiNi_xCo_{1-x}O_2$, and carbonaceous materials such as graphite, hard carbon and soft carbon, are currently used commercially in the production of rechargeable lithium batteries.

Figure 5.2a–d shows the electrode potential versus lithium content curves for $Li_{1-\delta}NiO_2$ [11, 12], $Li_{1-\delta}CoO_2$ [13, 14], $Li_{1-\delta}Mn_2O_4$ [15–17], and graphite [18] electrodes, respectively, including the changes in the equilibrium phase composition as lithium intercalation/deintercalation occurs. For details concerning the chemical and physical properties of the equilibrium phases, the reader should consult the appropriate references. In Figure 5.2, it is clearly apparent that none of the electrodes shows a simple electrode potential versus lithium content curve. This is mainly attributed to a complexity in the arrangement of lithium ions interacting with each other in the lattice, which cannot be simply expressed by Equation (5.9). Consequently, many research groups have concentrated on analyzing the lithium intercalation/deintercalation involving phase transitions caused by interaction between lithium ions, with the aid of computational methods.

5.2.3.1 **Application of Lattice Gas Model with Mean Field Approximation**

In order to model the thermodynamics of lithium intercalation/deintercalation accounting for the ion interaction, a lattice gas model based on the mean field theory with several approximations for transition metal oxides [15, 19–23] and graphite [24] was applied. First, a Bragg–Williams approximation [19] was used for the structural analysis of $LiNiO_2$ including an order–disorder transition caused by a strong interaction between cations in the lattice. By using similar approximations, the electrode potential versus lithium content curves were also calculated for $Li_{1+x}Mn_{2-x}O_4$ [20] and $LiNi_xMn_{2-x}O_4$ [21] which had not been fully explained by Equation (5.9).

Recently, Pyun *et al.* simulated independently the thermodynamics of lithium intercalation into $Li_{1-\delta}Mn_2O_4$ [15]. In their model, the cubic-spinel structure of $LiMn_2O_4$ forming a diamond lattice was regarded as two interpenetrating face-centered cubic sublattices separated by 1/4, 1/4, 1/4 (as shown in Figure 5.3). The interaction between lithium ions was modeled by considering that each lithium ion had four nearest neighbors in the other sublattice and 12 neighbors in the second coordination sphere of the same sublattice. Based on the lattice gas model with Bragg–Williams approximation, the chemical potentials of lithium ions in the entire lattice (μ_{Li}) and in both these sublattices ($\mu_{Li,1}$ and $\mu_{Li,2}$, respectively) were derived as

$$\begin{aligned}\mu_{Li} &= \mu_{Li,1} = [U + 4J_1(1-\delta)_2 + 12J_2(1-\delta)_1] - T\{k\ln[\delta_1/(1-\delta)_1]\} \\ &= \mu_{Li,2} = [U + 4J_1(1-\delta)_1 + 12J_2(1-\delta)_2] - T\{k\ln[\delta_2/(1-\delta)_2]\}\end{aligned}, \quad (5.10)$$

where U is the site energy for a lithium ion in 8(a) site, $(1-\delta)_i$ lithium content in the i-th sublattice, and J_1 and J_2 represent the two-body interactions due to the nearest neighbors and the second-nearest neighbors, respectively.

By considering that $\mu_{Li,1} - \mu_{Li,2} = 0$ from Equation (5.10), one can iteratively calculate $(1-\delta)_1$ and $(1-\delta)_2$ by a bisection method [25]. Consequently, the electrode

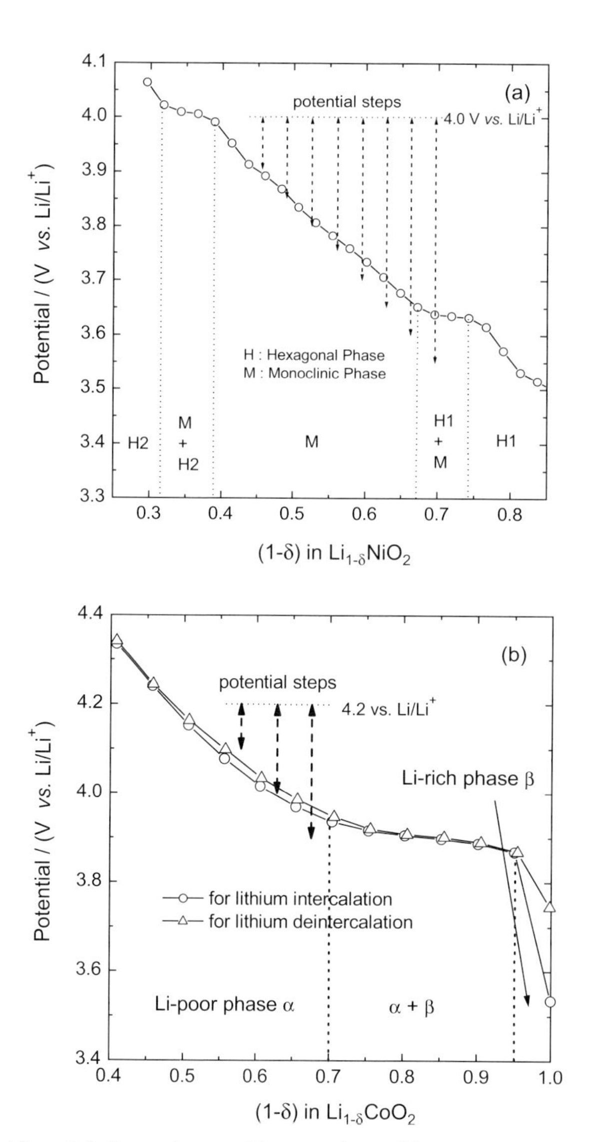

Figure 5.2 Electrode potential curves obtained for: (a) $Li_{1-\delta}NiO_2$; (b) $Li_{1-\delta}CoO_2$; (c) $Li_{1-\delta}Mn_2O_4$; and (d) graphite by using the galvanostatic intermittent titration technique. These include changes in the equilibrium phase as lithium intercalation/deintercalation proceeds. (Reproduced with permissions from (a) Ref. [11]; (b) Ref. [14]; (c) Ref. [17]; (d) Ref. [18].)

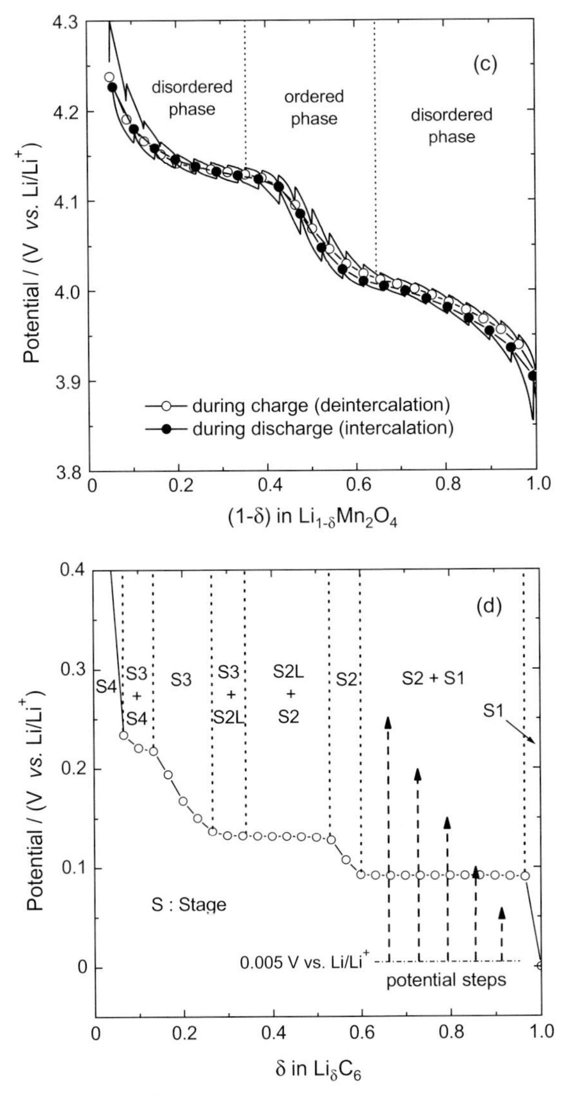

Figure 5.2 *(Continued)*

potential E and total lithium content $(1-\delta)$ in $Li_{1-\delta}Mn_2O_4$ are obtained from $E = -\mu_{Li}/e$ and $(1-\delta) = [(1-\delta)_1 + (1-\delta)_2]/2$, respectively. The partial molar enthalpy and entropy can also be derived from the first and second terms in Equation (5.10), respectively, given by

$$\Delta\tilde{H}_{Li} = \sum_i \Delta\tilde{H}_{Li,i}[d(1-\delta)_i/d(1-\delta)]/2, \quad (5.11)$$

$$\Delta\tilde{S}_{Li} = \sum_i \Delta\tilde{S}_{Li,i}[d(1-\delta)_i/d(1-\delta)]/2, \quad i = 1, 2. \quad (5.12)$$

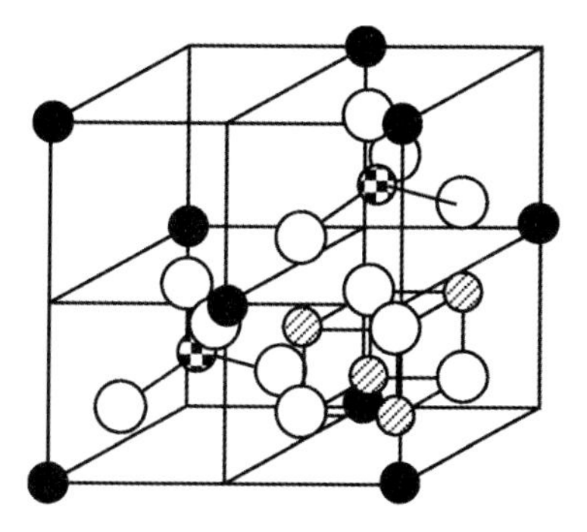

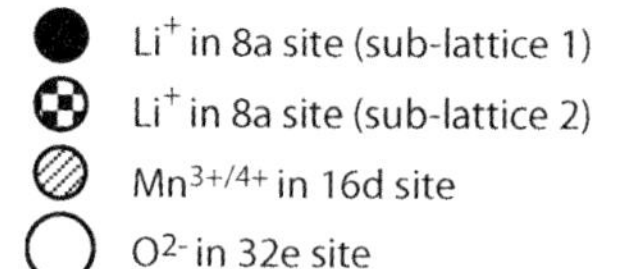

Figure 5.3 Schematic illustration of a cubic-spinel $LiMn_2O_4$ structure with a space group Fd3m.

The procedures of derivation and calculation of the above thermodynamic quantities are described in detail elsewhere [15].

Figure 5.4 represents the electrode potential versus lithium content curve and the plots of $(1-\delta)_1$ and $(1-\delta)_2$ with respect to $(1-\delta)$ calculated for the case of $U=-4.12$ eV, $J_1=37.5$ meV (repulsive interaction), $J_2=-4.0$ meV (attractive interaction), and $T=298$ K for $Li_{1-\delta}Mn_2O_4$. The theoretical electrode potential curve shows a steep potential drop at $(1-\delta)=0.5$, which is typical for ordering of lithium ions due to their strong interaction [15, 20, 22, 23]. The order–disorder phase transition occurs at the boundaries where $(1-\delta)_1$ and $(1-\delta)_2$ begin to deviate severely from $(1-\delta)$ values of approximately 0.15 and 0.85.

The theoretical plots of the partial molar enthalpy and entropy versus lithium content for $LiMn_2O_4$ are given in Figure 5.5a and b, respectively, with the calculated quantities being very consistent with experimental values. All partial molar quantities determined by the different methods showed a negative deviation below $(1-\delta)=0.5$ and a positive deviation above $(1-\delta)=0.5$ from the values calculated for the ideal solution. Based on the lattice gas model with the Bragg–Williams approximation, it is clearly envisaged that the abrupt rises in the partial molar quantities at $(1-\delta)=0.5$ in Figure 5.5a and b are indeed traced back to lithium ordering in the $LiMn_2O_4$ electrode.

Even though the mean field theory is known to be best suited for systems with long-range interaction of ions, the theory has been shown to successfully predict the thermodynamics of intercalation compounds with short-range interaction of guest ions, especially when the guest cations have homogeneous interactions with their neighbors throughout the lattice. Hence, the short-range interactions can be averaged.

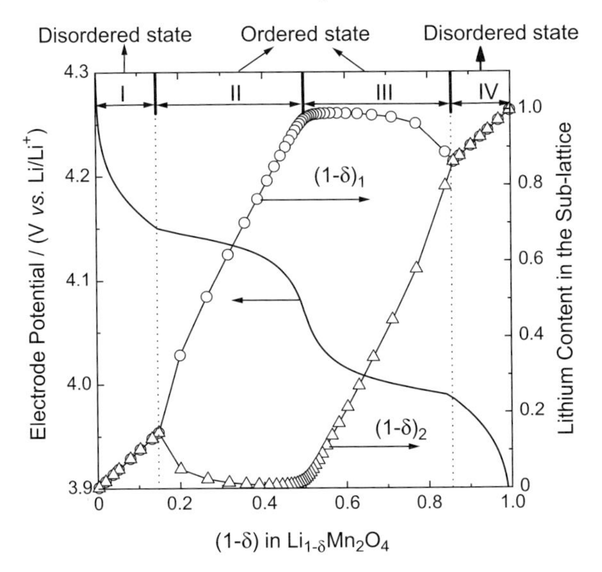

Figure 5.4 Electrode potential versus lithium content curve and plots of $(1-\delta)_1$ (O) and $(1-\delta)_2$ (△) with respect to $(1-\delta)$ calculated theoretically for the case of $U=-4.12$ eV, $J_1=37.5$ meV, $J_2=-4.0$ meV, and $T=298$ K. The vertical dashed lines, representing the boundaries between the ordered and disordered phases, were determined at the points where the sharp peaks appear in $-d((1-\delta))/dE$ versus E curve. (Reproduced with permission from Ref. [15].)

5.2.3.2 **Application of Lattice Gas Model with Monte Carlo Simulation**

Although the mean field theory discussed in Section 5.2.3.1 is well fitted for long-range interaction, the Monte Carlo method is best suited to short-range interactions (see Chapters 2 and 7). For transition metal oxides with a layered structure, Dahn *et al.* [19, 26, 27] first focused on an understanding of crystal structures by taking into account the short-range interaction between lithium ions using a Monte Carlo simulation. In the case of transition metal oxides with a cubic-spinel structure, Pyun *et al.* [16, 28] and Newman *et al.* [29] individually approached the analysis of intercalation-induced variations in thermodynamic properties (e.g., electrode potential, entropy, and enthalpy) by using the Monte Carlo method.

Considering the interaction of lithium ions as shown in Figure 5.3, the Hamiltonian H and the internal energy U of $Li_{1-\delta}Mn_2O_4$ are defined as [16]:

$$H = J_1\sum_{ij} c_i c_j + J_2\sum_{ik} c_i c_k - (\varepsilon + \mu_{Li})\sum_{i} c_i \tag{5.13}$$

and

$$U = J_1\sum_{ij} c_i c_j + J_2\sum_{ik} c_i c_k - \varepsilon\sum_{i} c_i, \tag{5.14}$$

respectively, where J_1 and J_2 are the effective pair-wise interaction parameters for the nearest and next-nearest neighbors, respectively; ε is the effective binding energy

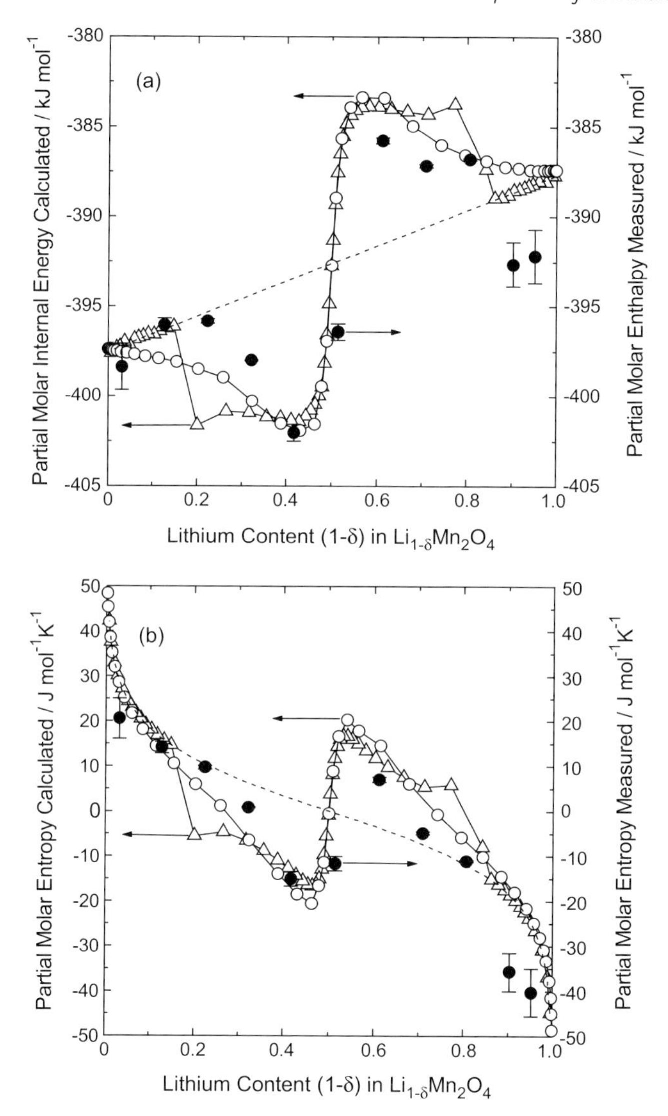

Figure 5.5 Plots of (a) the partial molar internal energy $\tilde{U}_{\mathrm{Li}}$ and (b) entropy $\tilde{S}_{\mathrm{Li}}$ of lithium ions at constant volume V and temperature T with respect to lithium content $(1-\delta)$, theoretically calculated from the mean-field approximation (△) and the Monte Carlo simulation (O) at $T = 298$ K. The partial molar enthalpy and entropy at constant pressure P and temperature T, determined from the experimentally measured temperature dependence of the electrode potential, are also plotted. The dashed lines represent the partial molar internal energy and entropy calculated for the ideal solution. (Reproduced with permission from Ref. [16].)

between lithium ion and manganese oxide matrix; c_i is the local occupation number of the site i; and c_j and c_k represent the local occupation numbers of the nearest and next-nearest neighbor sites, respectively. Here, c_i, c_j or $c_k = 1$ if the site is occupied by lithium ion, and c_i, c_j or $c_k = 0$ otherwise. As J_1, J_2, and ε cannot simply represent the direct interaction between the lithium ions and the oxide matrix, the effective values of these parameters should be determined by fitting the electrode potential versus lithium content theoretically calculated from the Monte Carlo simulation to the experimental dependence, under the assumption that these parameters are independent of the lithium content.

The simulation is performed in a grand canonical ensemble (GCE) where all microstates have the same volume (V), temperature and chemical potential under the periodic boundary condition to minimize a finite size effect [30, 31]. For thermal equilibrium at a fixed μ_{Li}, a standard Metropolis algorithm is repetitively employed with single spin-flip dynamics [30, 31]. When equilibrium has been achieved, the lithium content $(1-\delta)$ in the $Li_{1-\delta}Mn_2O_4$ electrode at a given μ_{Li} is determined from the fraction of occupied sites. The thermodynamic partial molar quantities of lithium ions are theoretically obtained by fluctuation method [32]. The partial molar internal energy $\tilde{U}_{Li}$ at constant V and T in the GCE is readily given by [32, 33]

$$\tilde{U}_{Li} = \frac{\mathrm{Cov}(U,N)}{\mathrm{Var}(N)}, \tag{5.15}$$

where Cov(U,N) is the covariance of U and N, and Var(N) represents the variance of N. $\tilde{U}_{Li}$ can be regarded as the partial molar enthalpy of lithium ions at constant pressure (P) and temperature, because only the useful work of the lattice can be done during the intercalation exclusive of the PV work. On the other hand, the partial molar entropy $\tilde{S}_{Li}$ at constant V and T is written as

$$\tilde{S}_{Li} = \frac{1}{T}\left[\frac{\mathrm{Cov}(U,N)}{\mathrm{Var}(N)} - \mu\right]. \tag{5.16}$$

The detailed simulation procedures and derivation of above thermodynamic quantities have been described elsewhere [16].

Pyun *et al.* [16] reported the electrode potential E versus lithium content $(1-\delta)$ curve and the plots of $(1-\delta)_1$ and $(1-\delta)_2$ with respect to $(1-\delta)$, theoretically obtained by the Monte Carlo simulations with $J_1 = 37.5$ meV (repulsive interaction), $J_2 = -4.0$ meV (attractive interaction), $\varepsilon = 4.12$ eV and $T = 298$ K for $Li_{1-\delta}Mn_2O_4$, as shown in Figure 5.6. As the lithium content increases, the theoretical electrode potential curve shows a steep potential drop at $(1-\delta) = 0.5$, typical for lithium ordering [15, 20, 22, 23]. This ordering can be envisaged by considering the difference between $(1-\delta)_1$ and $(1-\delta)_2$ at the same $(1-\delta)$ (Figure 5.6), with the aid of local cross-sectional snapshot of the equilibrium configuration of the cubic lattice (Figure 5.7).

For the thermodynamic second-order transition it might be expected that the susceptibility of the lattice gas, which is strongly related to Var(N), diverges at the transition points [34], and hence it is possible to establish the phase diagram for the order–disorder transition in $LiMn_2O_4$, as shown in Figure 5.8. As the intercalation proceeds along the isothermal line at $T = 298$ K (dotted line in Figure 5.8), the disorder to order transition occurs at $(1-\delta)_{tr(1)}$, after which disordering takes place at $(1-\delta)_{tr(2)}$. Similar phase diagrams for the order–disorder transition were also reported in other intercalation compounds, such as $LiTaS_2$ [35] and $LiCoO_2$ [36].

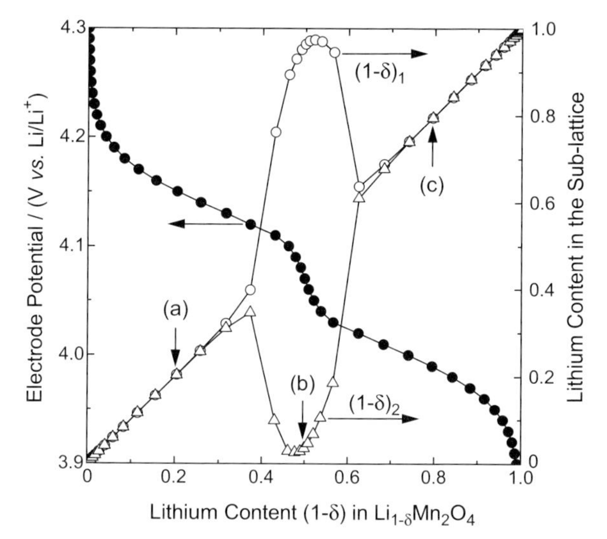

Figure 5.6 Electrode potential E versus lithium content $(1-\delta)$ curve (•) and the plots of $(1-\delta)_1$ (O) and $(1-\delta)_2$ (△) with respect to $(1-\delta)$, theoretically obtained from the Monte Carlo simulation with $J_1 = 37.5$ meV, $J_2 = -4.0$ meV, $\varepsilon = 4.12$ eV, and $T = 298$ K. The value of $(1-\delta)$ at the points (a), (b), and (c) are equal to 0.2, 0.5, and 0.8, respectively. (Reproduced with permission from Ref. [16].)

The simulated plots calculated from Equations (5.15) and (5.16) are shown in Figure 5.5a and b, respectively. The partial molar quantities determined from the Monte Carlo simulation are in a good agreement with experimental results. It is noteworthy that a similar conclusion was drawn from results obtained when using the mean-field approximation based upon the lattice gas model in Section 5.2.3.1.

Another interesting thermodynamic phenomenon caused by the strong interaction of lithium ions is a two-phase coexistence during lithium intercalation and deintercalation; that is, when intercalation/deintercalation proceeds in equilibrium between the Li-depleted and Li-rich phases (see Section 5.2.2). Pyun *et al.* also applied the Monte Carlo method to determine the mechanism of lithium intercalation into $Li_{1+\delta}[Ti_{5/3}Li_{1/3}]O_4$ in the two-phase domain [28]. For the cubic-spinel $Li_{1+\delta}[Ti_{5/3}Li_{1/3}]O_4$, both 8(a) and 16(c) sites are occupied by lithium ions; each 8(a) site is adjacent to four first-nearest 16(c) and four second-nearest 8(a) sites, and each 16(c) site is surrounded by two first-nearest 8(a) and six second-nearest 16(c) sites. According to the model [28], the lattice Hamiltonian is defined as

$$H = J_1 \sum_{ij} c_i c_j + J_2 \sum_{ik} c_i c_k + J_3 \sum_{jl} c_j c_l - (\varepsilon_{ij} + \mu_{Li}) \sum_{ij} c_{ij}, \tag{5.17}$$

where J_1 is the effective pair-wise interaction parameter between the first-nearest neighboring lithium ions in the 8(a) and 16(c) sites, J_2 is the interaction parameter between the 8(a) ions located in the second coordination sphere, J_3 is the same

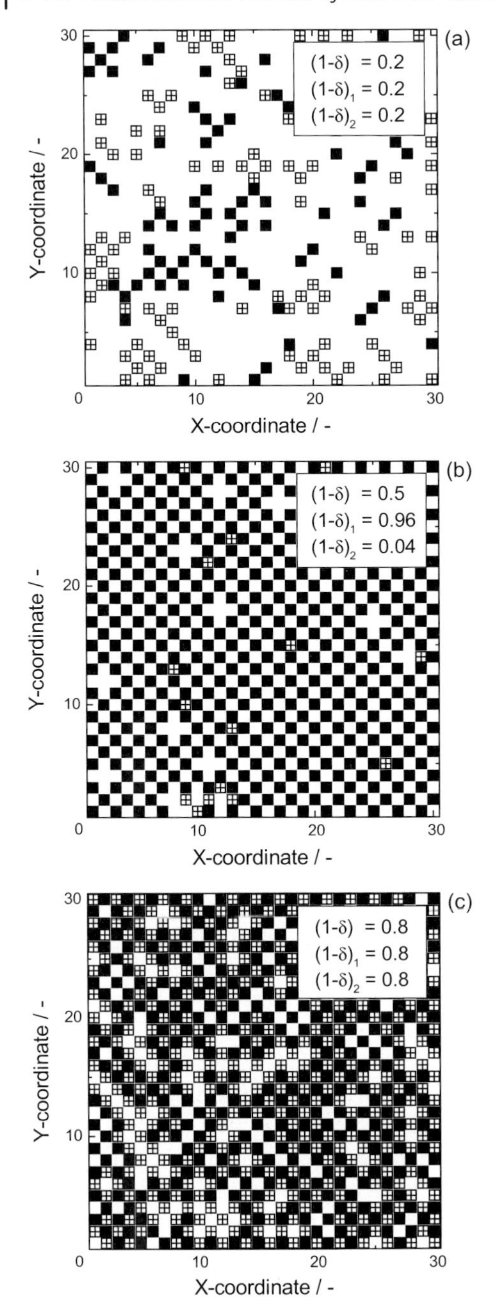

Figure 5.7 Local cross-sectional snapshots of the equilibrium configuration of the cubic lattice obtained from the Monte Carlo simulation at: (a) lithium content $(1-\delta) = 0.2$; (b) 0.5; and (c) 0.8. The closed square and the cross-centered square symbols represent lithium ions in the sublattices 1 and 2, respectively. (Reproduced with permission from Ref. [16].)

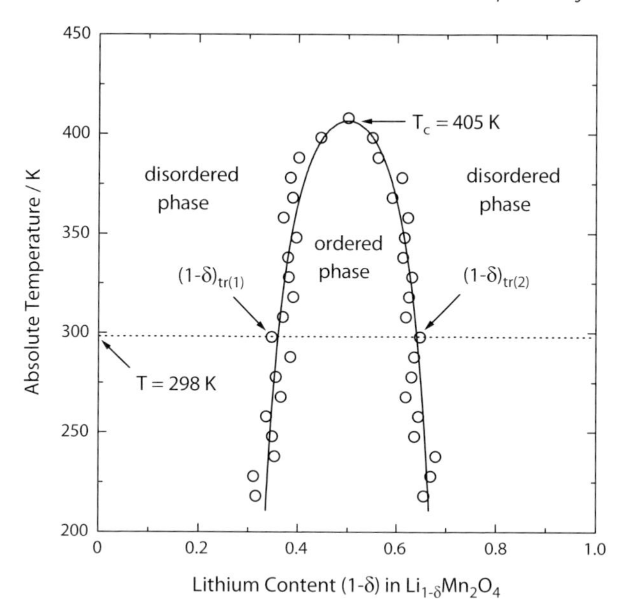

Figure 5.8 Theoretical phase diagram representing the order to disorder transition in the $LiMn_2O_4$ electrode. The dotted line represents the isotherm at $T = 298$ K. $(1-\delta)_{tr(1)}$ and $(1-\delta)_{tr(2)}$ indicate the disorder-to-order and the order-to-disorder transition points, respectively. The disordered phase is stable over the entire range of lithium content above the critical temperature T_c. (Reproduced with permission from Ref. [16].)

parameter for the 16(c) sites, ε_i and ε_j are the effective binding energies of lithium ions in the 8(a) site i and the 16(c) site j, respectively, with titanium oxide matrix; and c_i, c_j, c_k, and c_l represent the local occupation numbers of the corresponding sites.

Figure 5.9a compares the experimental E versus $(1 + \delta)$ dependencies (open and closed circles) and the curve calculated by the Monte Carlo method (solid line), taking $J_1 = 0.110$, $J_2 = 0.105$, $J_3 = 0.005$, $\varepsilon_i = 1.777$, and $\varepsilon_j = 1.677$ eV at $T = 298$ K for $Li_{1+\delta}[Ti_{5/3}Li_{1/3}]O_4$. The theoretical curve shows a wide potential plateau in the range of $(1 + \delta) = 1.06 \sim 1.94$, which coincides well with the experimental data. Figure 5.9a also shows that the theoretically calculated lithium content in the 8(a) sites, $(1 + \delta)_{8(a)}$, decreases with increasing $(1 + \delta)$ in the potential plateau region, whereas lithium content in the 16(c) sites $(1 + \delta)_{16(c)}$ increases.

In order to explain the variations of $(1 + \delta)_{8(a)}$ and $(1 + \delta)_{16(c)}$ in terms of the phase transformation between lithium-poor and lithium-rich phases, Figure 5.9b illustrates the local cross-sectional snapshot of the lithium ions configuration at $(1 + \delta) = 1.50$, as simulated by the Monte Carlo method. It is clear that the β phase existing in equilibrium with α phase, is sporadically embedded in the α matrix. Here, the α phase means the Li-poor region where lithium ions mainly reside in the 8(a) sites, while the β phase represents the Li-rich region where most lithium ions occupy the

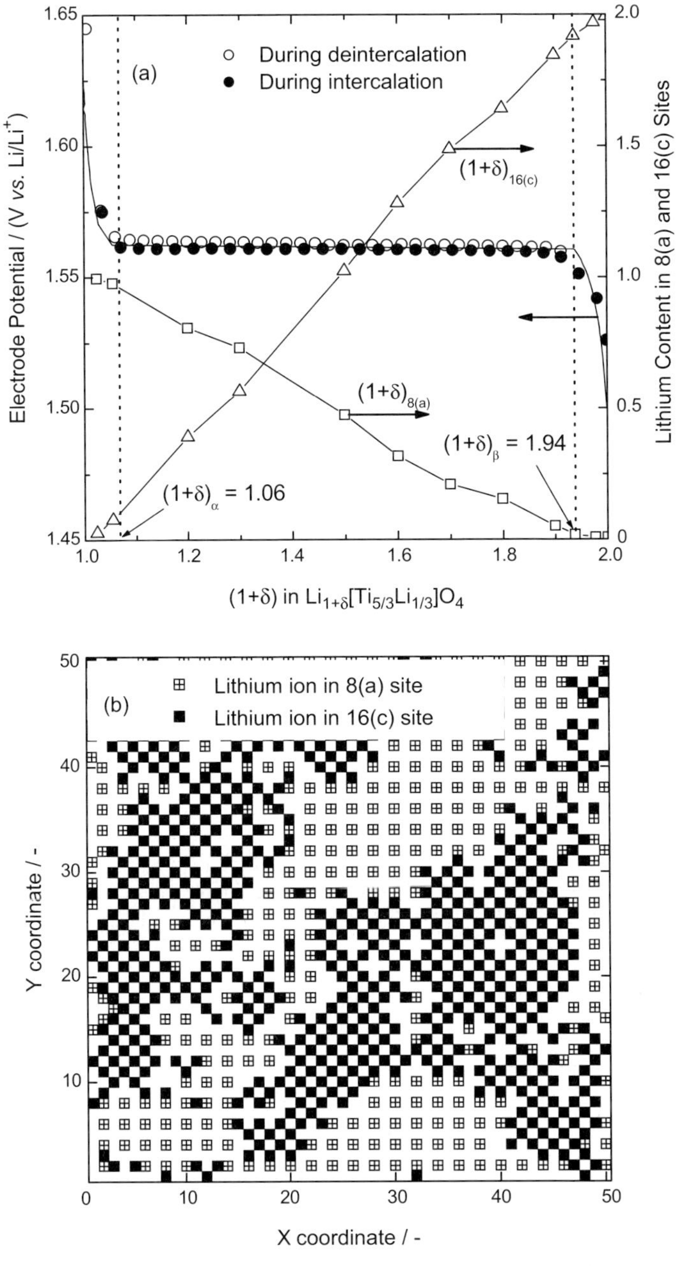

Figure 5.9 (a) Electrode potential E versus lithium content $(1 + \delta)$ curve (solid line) and the plots of $(1 + \delta)_{8(a)}$ (□) and $(1 + \delta)_{16(c)}$ (△) with respect to $(1 + \delta)$ theoretically calculated by the Monte Carlo method; (b) Local cross-sectional snapshot of the configuration of lithium ions at $(1 + \delta) = 1.50$. The electrode potential curves experimentally measured on the $Li_{1+\delta}[Ti_{5/3}Li_{1/3}]O_4$ film electrode during the lithium intercalation (O) and deintercalation (•) are also plotted. (Reproduced with permission from Ref. [28].)

16(c) sites, so as to avoid the repulsive interaction between lithium ions that increases the ensemble energy of the lattice.

5.2.3.3 Application of *Ab Initio* (First Principles) Method

As noted in Section 5.2.1, most thermodynamic approaches to lithium intercalation have been focused on the analysis of the chemical potential of lithium ions, under the assumption that the chemical potential of electrons is constant in the metallic intercalation compounds. When the electrons generated by the intercalation reaction are localized, however, the compound remains semi-conducting as intercalation proceeds. In this case, it is necessary also to take into account the coulombic interaction between electrons and ions.

In order to consider the contribution of electrons to the chemical potential of intercalation compounds, several groups have applied an *ab initio* or first principles calculation method to analyze the thermodynamics of lithium intercalation [37–44]. A typical *ab initio* (first principles) calculation for lithium intercalation consists of two steps: (i) energy calculation at 0 K to determine the ground states and relative energy difference between crystal structures; and (ii) the construction and calculation of a free energy model to determine the phase stability at non-zero temperature. Dahn *et al.* [37], Ceder *et al.* [38] and Benco *et al.* [41] all reported that the electrode potentials of transition metal oxides could be very well predicted by this method. In addition, Ceder *et al.* constructed (on a theoretical basis) the phase diagram of $LiCoO_2$ [40, 42, 43] and $LiMn_2O_4$ [44] during lithium intercalation involving the order–disorder phase transition, again in a good agreement with experimental findings. However, a two-phase coexistence in $LiCoO_2$ cannot be readily predicted by this model. Ceder *et al.* [40, 42] claimed that this originates from the metal–insulator transition around the two-phase region, and implied that the model should be improved further to consider varying interactions between the electrons and ions as lithium intercalation proceeded.

5.3 Kinetics of Intercalation and Deintercalation

The kinetics of lithium transport through transition metal oxides and carbonaceous materials has been extensively studied due to its great importance in the high-power applications of rechargeable batteries and to improve their charging and discharging rate. In most cases, the kinetic studies have been performed assuming a "diffusion control" concept [45–52], namely, that the diffusion of lithium in the electrode is very sluggish, whereas other reactions (including interfacial charge transfer) are too fast to affect the kinetics of lithium transport. Hence, lithium diffusion in the electrode would become a rate-controlled process of lithium intercalation.

However, various types of anomalous behavior of lithium transport have been reported in many current transients (CTs) and voltammetric curves obtained for a number of transition metal oxides and carbonaceous materials. Many research groups have attempted to disclose these atypical phenomena in various ways: (i) under the above "diffusion control" concept, combined with the effects of irregular geometry [53–55], electric field [48, 56], intercalation-induced stress [57–59], growth of a new phase in the electrode [56–63], and so on; and (ii) from other viewpoints, such as charge transfer-limited mechanisms [64–72] and trapping mechanisms [73–76].

Recently, it was reported by Pyun *et al.* that the CTs of transition metal oxides such as $Li_{1-\delta}CoO_2$ [14, 77–79], $Li_{1-\delta}NiO_2$ [11, 12], $Li_{1-\delta}Mn_2O_4$ [17, 80, 81], $Li_{1+\delta}[Ti_{5/3}Li_{1/3}]O_4$ [11, 28], V_2O_5 [11, 55] and carbonaceous materials [18, 82–84] hardly exhibit a typical trend of "diffusion-controlled" lithium transport – that is, Cottrell behavior. Rather, it was found that the current–potential relationship would hold Ohm's law during the CT experiments, and it was suggested that lithium transport at the interface of electrode and electrolyte was mainly limited by internal cell resistance, and not by lithium diffusion in the bulk electrode. This concept is referred to as "cell-impedance-controlled" lithium transport.

5.3.1
Diffusion-Controlled Transport

Diffusion-controlled lithium transport involves the following: the system is so kinetically facile that the equilibrium concentration of lithium is quickly reached at the interface between the electrode and electrolyte at a moment of potential stepping for CT experiments. The instantaneous depletion and accumulation in the lithium concentration at the interface caused by the chemical diffusion away from and to the interface (and to and away from the bulk electrode) is completely compensated by the supply and release away from and into the electrolyte, respectively. This condition is referred to as "real potentiostatic constraint" at the interface between the electrode and the electrolyte.

If the electrode material is assumed to be homogeneous, then the concentration gradient of lithium through the electrode is the only factor that drives lithium transport. Hence, lithium will enter/leave the planar electrode only at the electrode/electrolyte interface, and cannot penetrate into the back of the electrode. Under such an impermeable (impenetrable) constraint, the electric current (I) can be expressed by Equation (5.18) during the initial stage of diffusion, and by Equation (5.19) during the later stage [45]:

$$I(t) = FA_{ea}(c^1 - c^0)\left(\frac{\tilde{D}_{Li^+}}{\pi t}\right)^{1/2} \quad \text{(Cottrell relation)}, \quad t \ll \frac{l^2}{\tilde{D}_{Li^+}}, \tag{5.18}$$

$$I(t) = \frac{2FA_{ea}(c^1 - c^0)\tilde{D}_{Li^+}}{l} \exp\left(-\frac{\pi^2 \tilde{D}_{Li^+}}{4l^2} t\right), \quad t \gg \frac{l^2}{\tilde{D}_{Li^+}}, \tag{5.19}$$

where t is the injection/extraction time, A_{ea} is the electrochemically active area of the electrode, $\tilde{D}_{Li^+}$ is the chemical diffusivity of lithium ions, l is the thickness, and c^1 and c^0 represent the final and initial equilibrium concentrations of lithium in the electrode, respectively.

By using Equations (5.18) and (5.19), a number of CTs have been analyzed. The slopes of $I(t)$ versus $t^{-1/2}$ (or the values of $I(t) \cdot t^{1/2}$) and ln $I(t)$ versus t curves have been determined in the initial and later stages of lithium injection/extraction, respectively, to estimate the lithium chemical diffusion coefficients. However, it has been pointed out [48, 68, 85, 86] that there is a major discrepancy between the values of the chemical diffusivity as determined by CT technique using the "diffusion control" concept and the data obtained with other electrochemical methods, such as the galvanostatic intermittent titration technique (GITT) and electrochemical-impedance spectroscopy (EIS). In addition, anomalous behaviors in CTs, which are difficult to explain by using the conventional diffusion control concept, have been reported from various transition metal oxides and carbonaceous materials [55–63].

Serious efforts have been made to explain the atypical lithium transport behavior using modified diffusion control models. In these models the boundary conditions – that is, "real potentiostatic" constraint at the electrode/electrolyte interface and impermeable constraint at the back of the electrode – remain valid, while lithium transport is strongly influenced by, for example: (i) the geometry of the electrode surface [53–55]; (ii) growth of a new phase in the electrode [56–63]; and (iii) the electric field through the electrode [48, 56].

5.3.2 Cell-Impedance-Controlled Transport

5.3.2.1 Non-Cottrell Behavior

The primary abnormality of CTs is that of non-Cottrell behavior, and for this the CTs of $Li_{1-\delta}NiO_2$ can be taken as an example. Figure 5.10a presents the cathodic CTs obtained for the $Li_{1-\delta}NiO_2$ electrode, which show a simple decrease in logarithmic current with logarithmic time for the potential drops above plateau potential 3.63 V (versus Li/Li^+), whereas those curves are characterized by an inflexion point (or quasi-current plateau) below 3.63 V (versus Li/Li^+). In order to analyze the curves based upon the diffusion control concept, $I(t)\cdot t^{1/2}$ versus ln t plots of Figure 5.10b were reproduced from CTs of Figure 5.10a. Several groups [87–95] have proposed the use of this type of plot, the plateau of which has been said to be extremely narrow and to describe the semi-infinite planar diffusion that represents Cottrell behavior.

However, it is difficult to identify any region with Cottrell characteristics in Figure 5.10b. In fact, even the flattest region (see the inset!) in the figure shows only an upward convex shape with a local maximum. Under these circumstances, it is very doubtful that the pseudo-plateau in Figure 5.10b would be caused by the semi-infinite diffusion of lithium through the oxide. In other words, the physical meaning of the pseudo-plateau has been overestimated. Rather, it is more reasonable to suggest that the upward convex shape of the $I(t)\cdot t^{1/2}$ versus ln t plots simply reflects a continuous decrease in current with time. The simplicity of CTs implies that lithium transport runs without any critical transition in transport mechanism during the entire lithium intercalation.

The above point is equally valid for CTs of other transition metal oxides and carbonaceous material (graphite). For example, $I(t)\cdot t^{1/2}$ versus ln t plots from graphite electrode (Figure 5.11b) do not show any Cottrell region, as do the plots from $LiNiO_2$ in Figure 5.10b.

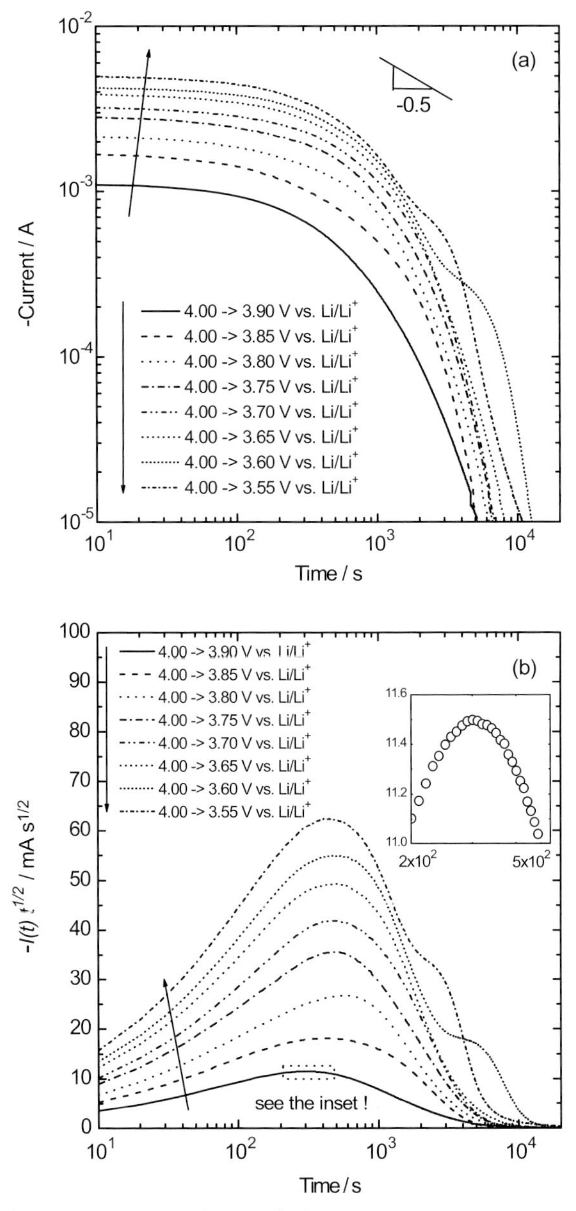

Figure 5.10 (a) Logarithmic cathodic current transients experimentally obtained from the $Li_{1-\delta}NiO_2$ electrode; (b) $I(t) \cdot t^{1/2}$ versus $\ln t$ plots reproduced from panel (a). (Reproduced with permission from (a) Ref. [12]; (b) Ref. [96].)

5.3.2.2 (Quasi-) Current Plateau

When one chooses the potential steps to encounter the plateau potential, the CTs deviate severely from the simple shape of those of single-phase electrodes. For

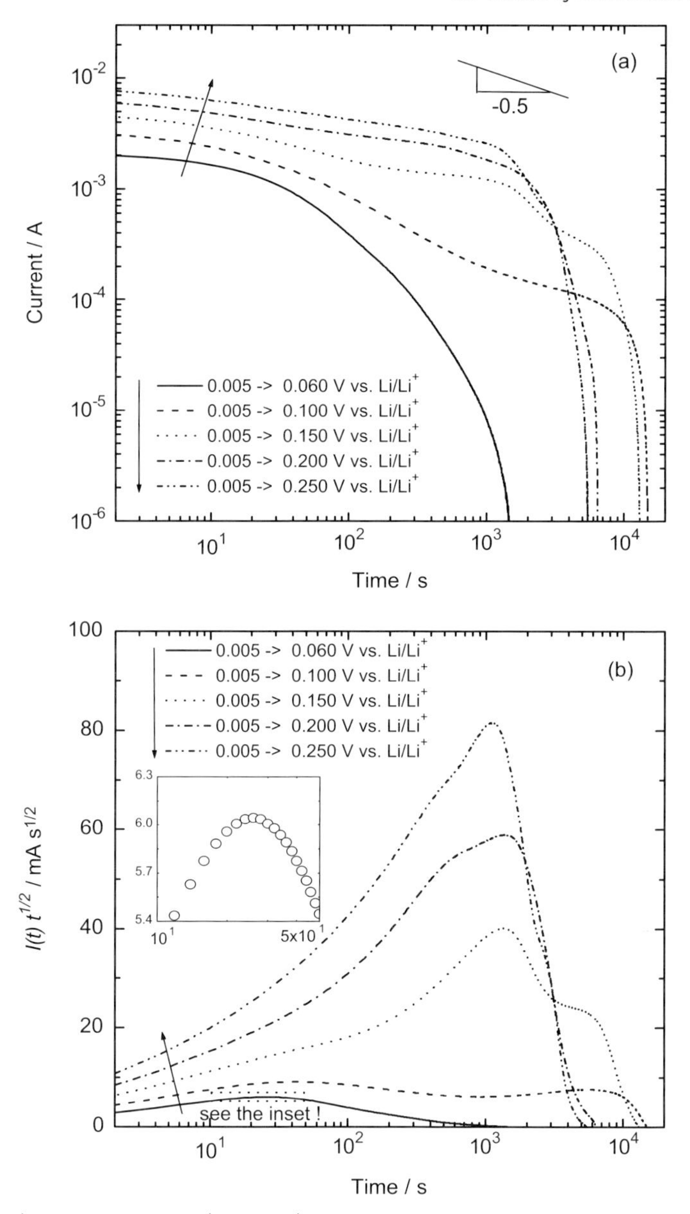

Figure 5.11 (a) Logarithmic anodic current transients experimentally obtained from the graphite electrode; (b) $I(t) \cdot t^{1/2}$ versus ln t plots reproduced from panel (a). (Reproduced with permission from (a) Ref. [18]; (b) Ref. [96].)

$Li_{1+\delta}[Ti_{5/3}Li_{1/3}]O_4$, two phases with different lithium content coexist in the potential plateau region, as shown in Figure 5.9a.

Figure 5.12a illustrates the cathodic CTs obtained from the $Li_{1+\delta}[Ti_{5/3}Li_{1/3}]O_4$ electrode at the potential drops from 1.700 V (versus Li/Li^+) down to various lithium

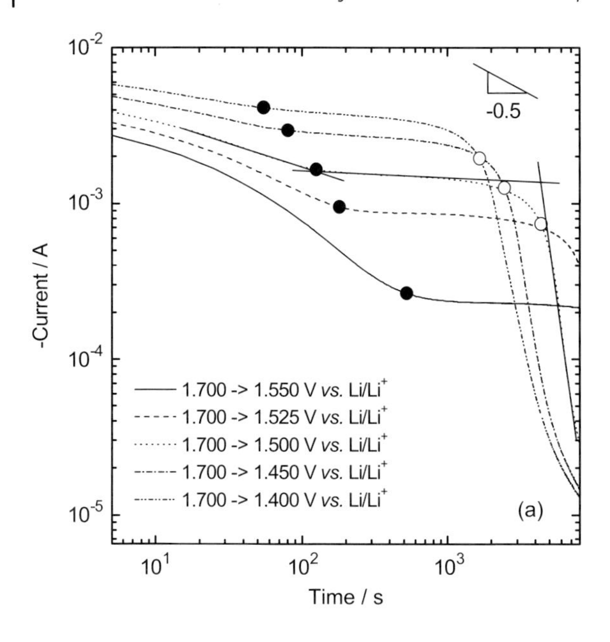

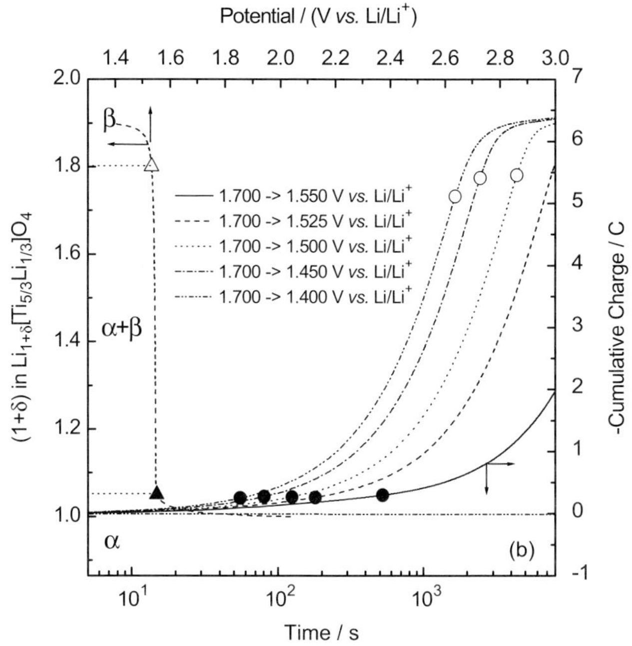

Figure 5.12 (a) Logarithmic cathodic current transients experimentally obtained from the $Li_{1+\delta}[Ti_{5/3}Li_{1/3}]O_4$ electrode at the potential drops from 1.700 V (versus Li/Li^+) to various lithium injection potentials below the plateau potential; (b) Cumulative charge versus time plots reproduced from panel (a), along with electrode potential curve of $Li_{1+\delta}[Ti_{5/3}Li_{1/3}]O_4$. (Reproduced with permission from (a) Ref. [11]; (b) Ref. [96].)

injection potentials below the plateau potential. The curves exhibit a three-stage behavior, with the logarithmic current first decreasing slowly, then remaining almost constant, and finally decaying steeply. To the present authors' knowledge, no reports have yet been made available that provide any clear explanation for this three-stage behavior, specifically the second stage – that is, the current plateau.

The onset time (solid circles)/end time (open circles) of the current plateau were determined graphically as the times at which the tangent line of the first-/second-stage curve intersects that of the second-/third-stage curve, as indicated in Figure 5.12a. It is worth noting here that the values of cumulative charge corresponding to the onset (solid circles in Figure 5.12b) and the end (open circles in Figure 5.12b) of the current plateau almost equal the maximum solubility limit of lithium in Li-poor α phase (solid triangle in Figure 5.12b) and the minimum solubility limit of lithium in Li-rich β phase (open triangle in Figure 5.12b), respectively, irrespective of the magnitude of the potential drop. This means that the phase transformation of α to β occurs during the "current plateau" interval in the CTs.

Bearing in mind that a "current plateau" implies a constant driving force for lithium intercalation, it is unlikely that the phase transformation from α to β is governed by diffusion-controlled lithium transport, where the rate of the phase boundary movement (i.e., the current) decreases significantly with time [60–63].

5.3.2.3 **Linear Relationship Between Current and Electrode Potential**

The most remarkable point worthy of mention in relation to the physical aspect of the CTs of transition metal oxides and graphite is the linear relationship between the initial current level I_{ini} and the potential step ΔE. In these studies, the term "initial current level" indicates the value of current at a time of 2–10 s during the CT experiments, rather than at a moment of application of potential step [96].

The solid circles in Figure 5.13a–e denote the initial current levels I_{ini} at various potential steps ΔE, calculated from the corresponding CTs. Invariably, all of the I_{ini} versus ΔE plots show a linear relationship. It should be mentioned that even diffusion-controlled CTs can exhibit this type of linear relationship, in the case that the electrode potential curves vary linearly with lithium stoichiometry, $\Delta E \propto (c^1 - c^0)$. However, the linear relationship between I_{ini} and ΔE is still valid for the electrodes (e.g., $Li_{1+\delta}[Ti_{5/3}Li_{1/3}]O_4$, $Li_{1-\delta}CoO_2$, $Li_\delta V_2O_5$ and graphite) where the electrode potential versus lithium stoichiometry curves deviate strongly from the linear relationship.

The above argument, along with the evidences presented in Sections 5.3.2.1–5.3.2.2, indicates that other transport mechanisms than diffusion-controlled lithium transport may dominate during the CT experiments. Furthermore, the Ohmic relationship between I_{ini} and ΔE indicates that internal cell resistance plays a critical role in lithium intercalation/deintercalation. If this is the case, it is reasonable to suggest that the interfacial flux of lithium ion is determined by the difference between the applied potential E_{app} and the actual instantaneous electrode potential $E(t)$, divided by the internal cell resistance R_{cell}. Consequently, lithium ions barely undergo any "real potentiostatic" constraint at the electrode/electrolyte interface. This condition is designated as cell-impedance-controlled lithium transport.

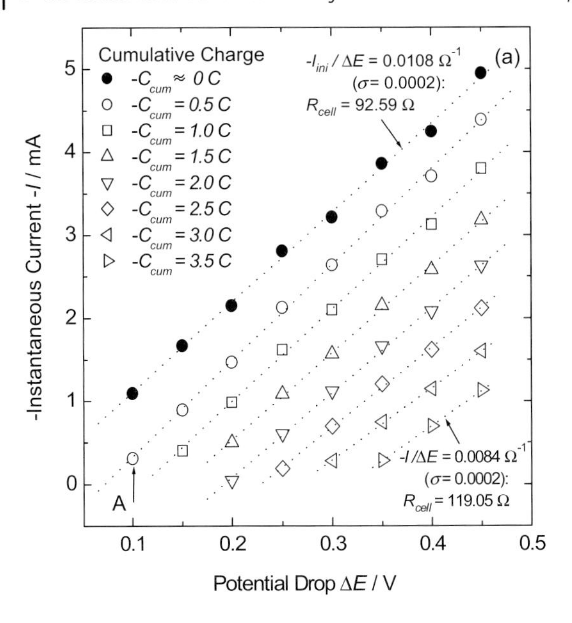

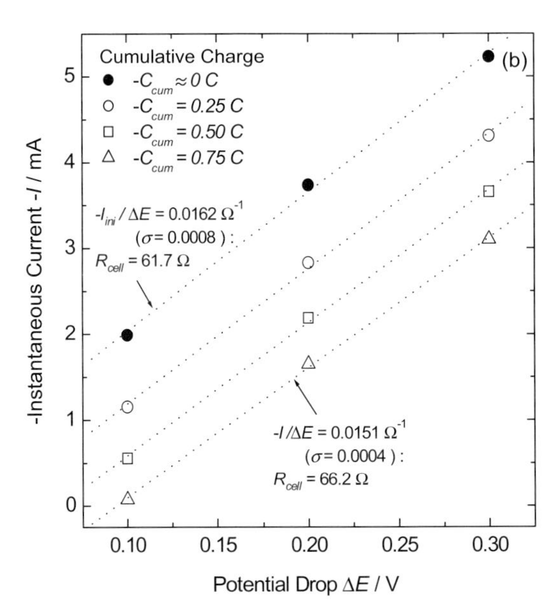

Figure 5.13 Dependence of instantaneous current level I on potential step ΔE at various values of cumulative charge, reproduced from the current transients of: (a) $Li_{1-\delta}NiO_2$; (b) $Li_{1-\delta}CoO_2$; (c) $Li_{\delta}V_2O_5$; (d) $Li_{1+\delta}[Ti_{5/3}Li_{1/3}]O_4$; and (e) graphite. (Reproduced with permission from Ref. [96].)

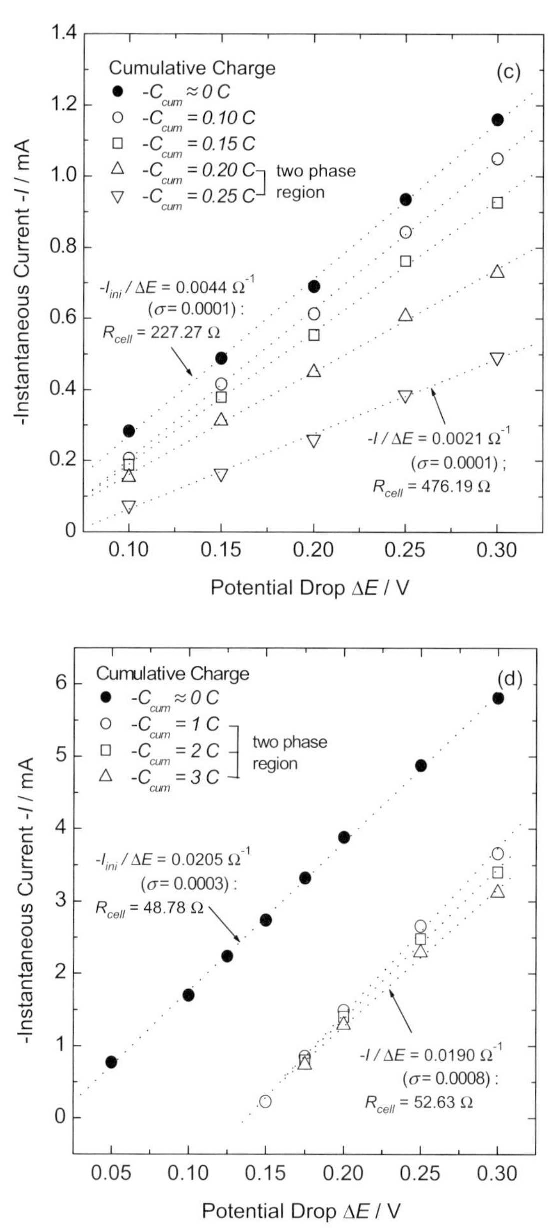

Figure 5.13 (*Continued*)

Now, let us examine the current–potential relationship with increasing lithium intercalation/deintercalation time. For this purpose we obtained (as a function of potential step) the values of current at the times when various amounts of cathodic or anodic charges had passed (open symbols in Figure 5.13a–e). For example, in order to

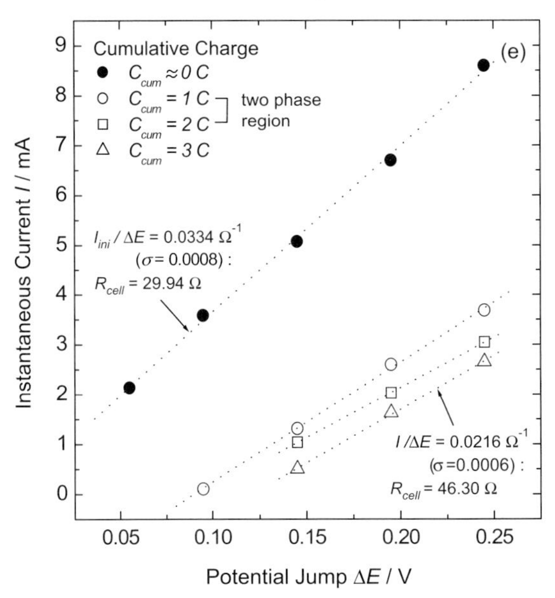

Figure 5.13 *(Continued)*

determine the open circle A in Figure 5.13a, the cumulative charge versus time plot is first obtained from the CT (–in Figure 5.10a) for $Li_{1-\delta}NiO_2$ electrode at the potential drop of 4.0 to 3.9 V (versus Li/Li^+) ($\Delta E = 0.1$ V). The time at which the cumulative charge reaches 0.5 C is then estimated. Finally, the value of current on the CT is determined just at time to 0.5 C.

Surprisingly, the linear relationship between current and potential (i.e., the Ohmic relationship) is clearly visible Figure 5.13a–e, regardless of the amount of cumulative charge or the intercalation/deintercalation time that has elapsed. This implies that the interfacial flux is determined by $[E_{app} - E(t)]/R_{cell}$, and that an infinite time is required to reach the equilibrium concentration of lithium at the electrode/electrolyte interface during the whole lithium intercalation/deintercalation; this is in contrast to the case of diffusion-controlled lithium transport. It should be stressed here that the validity of Ohm's law is extended to the cumulative charge range of the coexistence of two phases, indicating that lithium intercalation/deintercalation – even during the phase transition – is also limited by $[E_{app} - E(t)]/R_{cell}$.

For the sake of clarity of the above argument regarding cell-impedance-controlled lithium transport, it is very useful to determine experimentally the internal cell resistance as a function of the electrode potential, using EIS, and to compare this with the cell resistance as determined with the CT technique. Pyun *et al.* showed that internal cell resistances estimated via the I_{ini} versus ΔE plot at various lithium contents approximated satisfactorily values determined experimentally with EIS – the sum of the resistances from the electrolyte and conducting substrate, the resistance associated with the particle-to-particle contact among the oxide particles, and the resistance related to the absorption reaction of adsorbed lithium ion into the oxide [12]. This comparison tells us explicitly that the flux of lithium ion at the electrode/electrolyte interface is mainly limited by the internal cell resistance throughout the whole lithium intercalation/deintercalation.

5.3.3 Numerical Calculations

5.3.3.1 Governing Equation and Boundary Conditions

The governing equation for cell-impedance-controlled lithium transport is Fick's diffusion equation. The initial condition (I.C.) and the boundary conditions (B.C.) are given as

$$\text{I.C.}: c = c^0_{Li^+} \quad \text{for } 0 \leq r \leq R^* \text{ at } t = 0 \tag{5.20}$$

$$\text{B.C.}: I(t) = \frac{E_{app} - E(t)}{R_{cell}} \quad (\text{“cell-impedance controlled constraint”}) \quad \text{for } r = R^* \text{ at } t > 0 \tag{5.21}$$

$$-zFA_{ea}\tilde{D}_{Li^+}\frac{\partial c}{\partial r} = 0 \quad (\text{impermeable constraint}) \quad \text{for } r = 0 \text{ at } t \geq 0, \tag{5.22}$$

where r is the distance from center of the oxide or graphite particle, and R^* represents the average radius of the particles.

5.3.3.2 Calculation Procedure of Cell-Impedance-Controlled Current Transients

The model parameters are determined in the following manner: the functional relations $E = f(1-\delta)$ and $R_{cell} = f(E)$, which are incorporated into B.C. of Equation (5.21), are obtained by the polynomial regression analysis of the electrode potential curves and the R_{cell} versus E curves determined from the I_{ini} versus ΔE plots, respectively. It should be borne in mind that $(1-\delta)$ does not represent the average lithium content in the electrode, but the lithium content at the surface of the electrode. In other words, the electrode potential $E(t)$ in Equation (5.21) is the potential at the electrode surface. As the relationship $E = f(1-\delta)$ includes information about the phase transition, we can consider the effect of phase transition on the theoretical CT with the functional relationship $E = f(1-\delta)$, without taking any of the intercalation isotherm.

The chemical diffusivity of lithium ions $\tilde{D}_{Li^+}$ in the transition metal oxides and graphite is taken as $10^{-8} \sim 10^{-9}\,cm^2\,s^{-1}$ [13, 97–100]; on the basis of scanning electron microscopy (SEM) inspections, the average radius R^* is estimated to be 1–10 μm. The electrochemically active area A_{ea} is calculated from the radius R^*, and the theoretical density of the particles considered assuming that A_{ea} is identical to the total surface area of the electrode comprised of the spherical particles.

The calculation procedure of CTs is as follows: at $t = 0$, the values of lithium content $(1-\delta)$ over the electrode (or the electrode potential E) and of internal cell resistance R_{cell} are first initialized to be those values at initial electrode potential E_{ini}. When the infinitesimal time Δt has elapsed – that is, just after the potential step from E_{ini} to the final electrode potential $E_{fin} = E_{app}$ is applied – the flux at the electrode/electrolyte interface $r = R^*$ (or current I) is calculated by $I = (E - E_{app})/R_{cell}$ (see Equation (5.21)). Subsequently, the lithium concentration at the electrode surface is evaluated. Next, E, R_{cell} and $(1 - \delta)$ inside the electrode are calculated and re-evaluated after Δt. The above process is iterated until the desired time elapses. Ultimately, the theoretical CTs are obtained in the form of a plot of I versus t.

5.3.3.3 Theoretical Current Transients and their Comparison with Experimental Values

Figures 5.14a and 5.15a depict, on a logarithmic scale, the theoretical CTs of $Li_{1-\delta}NiO_2$ and graphite electrodes, respectively, determined from the numerical solution to diffusion equations for the conditions of Equations (5.20)–(5.22), by taking the values described in Section (5.3.3.2). The theoretical CTs of Figures 5.14a and 5.15a agree fairly with experimentally acquired data (see Figures 5.10a and 5.11a, respectively).

First, there is no Cottrell region during the whole lithium intercalation/deintercalation, which is substantiated by there being no region with constant value of $I(t)\cdot t^{1/2}$ in the $I(t)\cdot t^{1/2}$ versus ln t plots of Figures 5.14b and 5.15b. Moreover, Figures 5.14b and 5.15b show the shoulders and local maximum of more than one at the potential drops where the electrode undergoes the phase transition, in the same manner as the experimental $I(t)\cdot t^{1/2}$ versus ln t plots (see Figures 5.10b and 5.11b, respectively).

The theoretical CTs from $Li_{1+\delta}[Ti_{5/3}Li_{1/3}]O_4$ (Figure 5.16) feature the clear current plateau during the phase transition when the potential steps encounter the plateau potential, as did the experimental CTs in Figure 5.12. The same current plateaux during the phase transition were observed in the cell-impedance-controlled CTs from $Li_{1-\delta}NiO_2$ (Figure 5.14a) and graphite (Figure 5.15a). The occurrence of a current plateau in the cell-impedance-controlled CTs is due to the almost invariable electrode potential E or potential difference $(E_{app} - E)$ during the phase transition; that is, the constant driving force for lithium intercalation/deintercalation during the course of phase transformation. Nevertheless, it is noted that the current plateaux in the theoretical CTs appear more clearly when compared to those in the experimental CTs. The most likely reason for this slight discrepancy is the size distribution of the oxide and graphite particles.

5.3.3.4 Extension of Cell-Impedance-Controlled Lithium Transport Concept to the Disordered Carbon Electrode

Now, we can consider the kinetics of lithium intercalation/deintercalation of amorphous carbon electrodes having different lithium intercalation sites, as compared to graphite electrode in terms of cell-impedance-controlled lithium transport.

Figure 5.17 presents the electrode potentials obtained from galvanostatic intermittent discharge curves measured on disordered carbons electrodes MCMB800, MCMB1000, and MCMB1200 which have been heat-treated at 800, 1000, and 1200 °C, respectively. None of the electrode potential curves show any "potential plateau"; rather they run continuously throughout the whole lithium deintercalation.

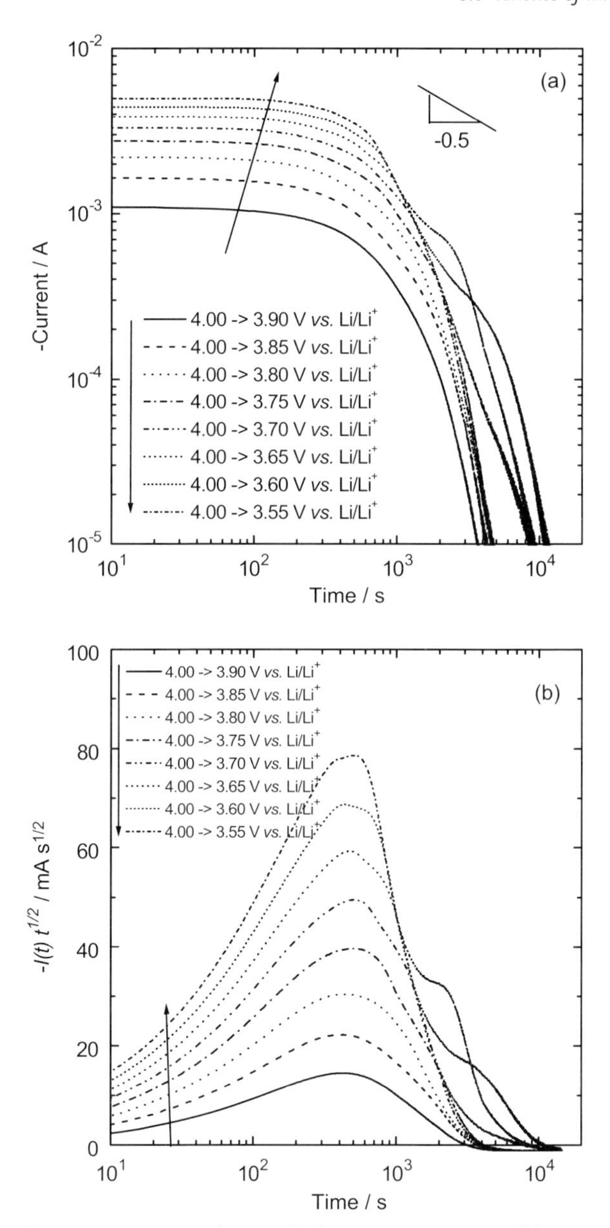

Figure 5.14 (a) Logarithmic cathodic current transients of the $Li_{1-\delta}NiO_2$ electrode, theoretically determined by means of numerical analysis based upon the "cell-impedance-controlled" lithium intercalation; (b) $I(t) \cdot t^{1/2}$ versus ln t plots reproduced from panel (a). (Reproduced with permission from (a) Ref. [12]; (b) Ref. [96].)

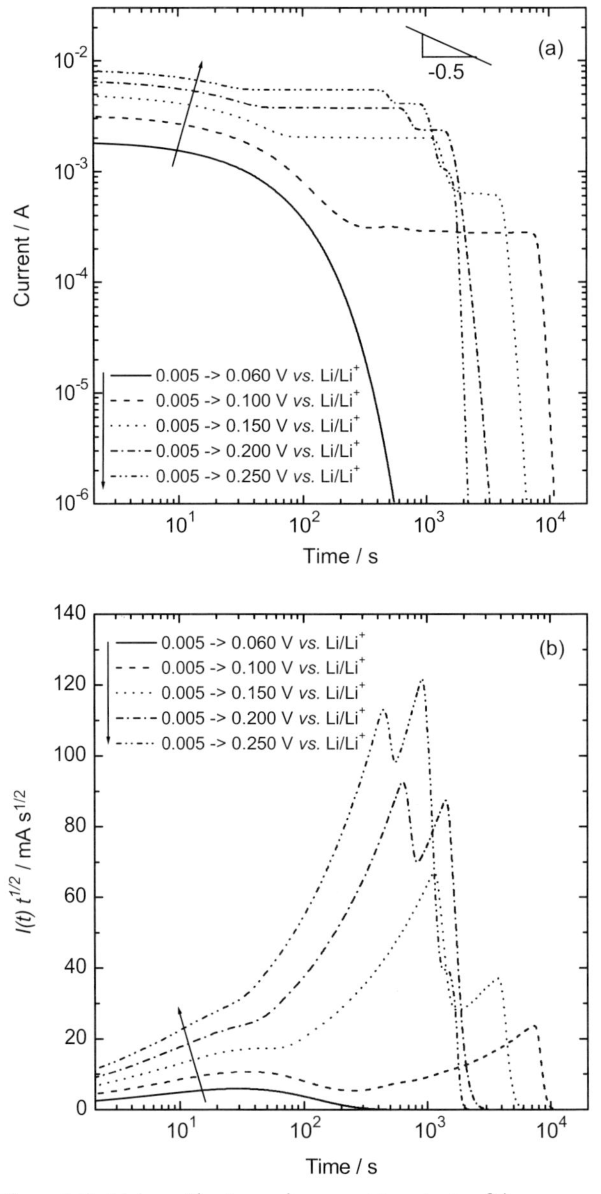

Figure 5.15 (a) Logarithmic anodic current transients of the graphite electrode, theoretically determined by means of numerical analysis based upon the "cell-impedance-controlled" lithium deintercalation; (b) $I(t) \cdot t^{1/2}$ versus ln t plots reproduced from panel (a). (Reproduced with permission from (a) Ref. [18]; (b) Ref. [96].)

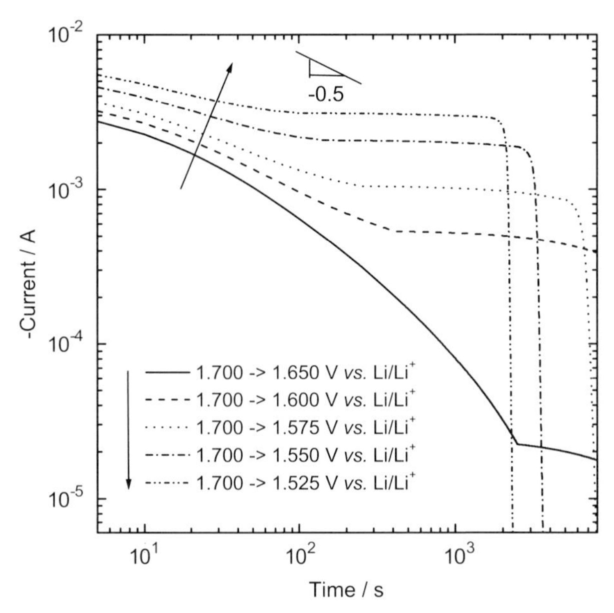

Figure 5.16 The logarithmic cathodic current transients of the $Li_{1+\delta}[Ti_{5/3}Li_{1/3}]O_4$ electrode at the potential drops from 1.70 V (versus Li/Li^+) to various lithium injection potentials below the plateau potential, theoretically determined by means of numerical analysis based upon the "cell-impedance-controlled" lithium intercalation. (Reproduced with permission from Ref. [11].)

This means that the MCMB which has been heat-treated below 1200 °C has a very low degree of crystallinity; consequently, lithium is deintercalated from the MCMB particles, without the formation of any thermodynamically stable phase.

Figure 5.18a–c shows, on a logarithmic scale, the anodic CTs obtained experimentally from the MCMB800 electrode at potential jumps of 0.3, 0.2, and 0.05 V (versus Li/Li^+) to various lithium extraction potentials. Under the cell-impedance-controlled constraint, it is generally accepted that an inflexion point – that is, the "quasi-current plateau" – should be necessarily observed in the CT, only when a "potential plateau" indicating the coexistence of two phases appears in the electrode potential curve (as discussed in Section 5.3.3.3). Interestingly, the CTs of Figure 5.18a–c exhibit inflexion points, even though the electrode potential curve measured on the electrode does not show any potential plateau throughout the whole lithium intercalation/deintercalation.

Pyun *et al.* [18, 82–84] have suggested that this abnormal behavior in CTs involving the inflexion point could be reasonably explained in terms of the difference in the kinetics of lithium deintercalation from two different lithium deintercalation sites having different activation energies for lithium deintercalation. The McNabb–Foster equation [101, 102] was modified to satisfy spherical symmetry and to represent the coexistence of two different types of trap site, and was also used as a governing

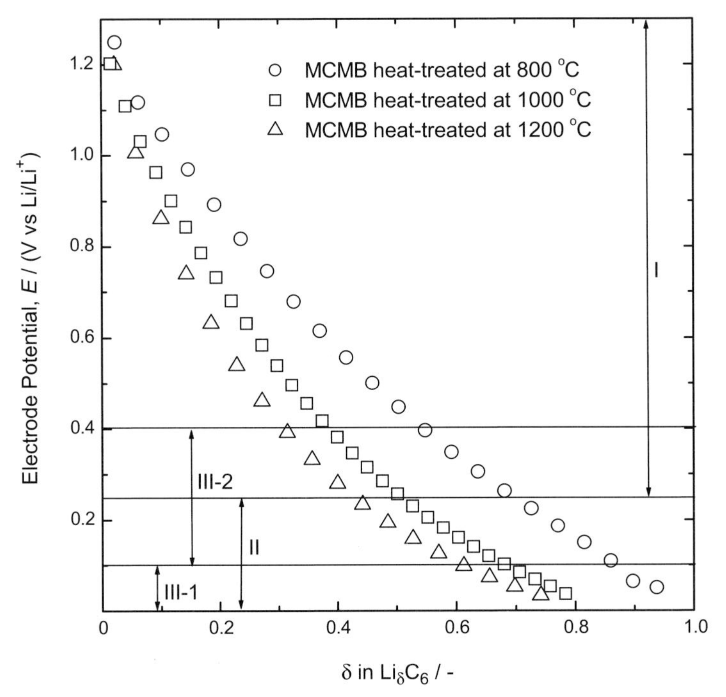

Figure 5.17 The galvanostatic intermittent discharge (electrode potential) curve measured on the PVDF-bonded MCMB800 (O), MCMB1000 (□) and MCMB1200 (△) composite electrodes in 1 *M* $LiPF_6$-EC/DEC solution. Regions I, II, III-1, and III-2 represent the potential ranges necessary for lithium deintercalation from the sites for Type I, II, III-1, and III-2, respectively. (Reproduced with permission from Ref. [82].)

equation for lithium transport as follows:

$$\tilde{D}_{\mathrm{Li}}\left[\frac{2}{R^{*}}\left(\frac{\partial c}{\partial r}\right)+\frac{\partial^{2} c}{\partial r^{2}}\right]=\frac{\partial c}{\partial t}+N_{\mathrm{I}}\frac{\partial \theta_{\mathrm{I}}}{\partial t}+N_{\mathrm{II}}\frac{\partial \theta_{\mathrm{II}}}{\partial t} \tag{5.23}$$

with the rate of reversible trapping into and escaping from the trap sites described by

$$\frac{\partial \theta_{\mathrm{I}}}{\partial t}=k_{\mathrm{I}}c(1-\theta_{\mathrm{I}})-p_{\mathrm{I}}\theta_{\mathrm{I}} \tag{5.24}$$

$$\frac{\partial \theta_{\mathrm{II}}}{\partial t}=k_{\mathrm{II}}c(1-\theta_{\mathrm{II}})-p_{\mathrm{II}}\theta_{\mathrm{II}}, \tag{5.25}$$

where t is the deintercalation time, N_{I} and N_{II} are the concentrations of the trap sites of I and II, respectively, θ_{I} and θ_{II} are the occupancy fractions of the trap sites of I and II, respectively, k_{I} and k_{II} are the capture rates, and p_{I} and p_{II} represent the release rates from the trap sites of I and II, respectively. I.C. and B.C. were formulated as shown in Equations 5.20–5.22.

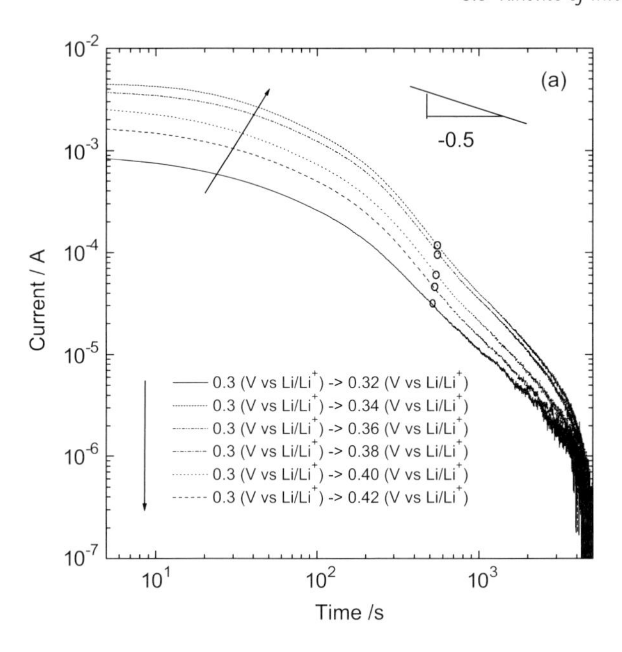

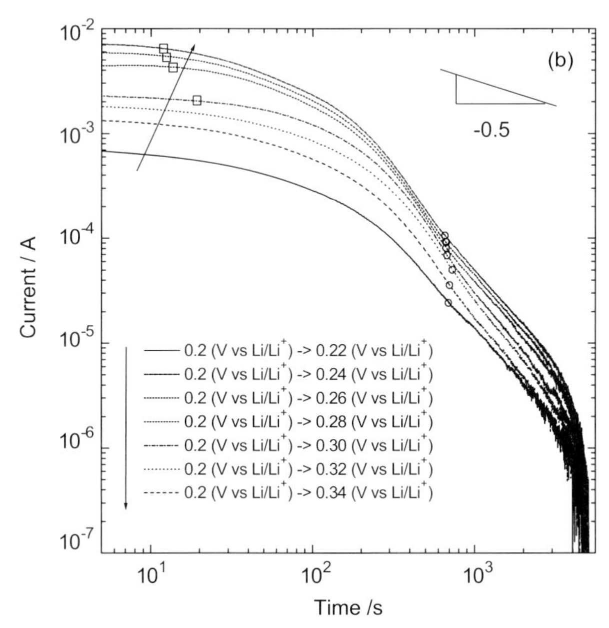

Figure 5.18 The anodic current transient experimentally obtained from the PVDF-bonded MCMB800 composite electrode at the potential jumps of $E =$ (a) 0.30, (b) 0.20, and (c) 0.05 V (versus Li/Li^+) to various lithium extraction potentials E_{ext}, as indicated in figure. The symbols □ and O denote inflexion points. (Reproduced with permission from Ref. [82].)

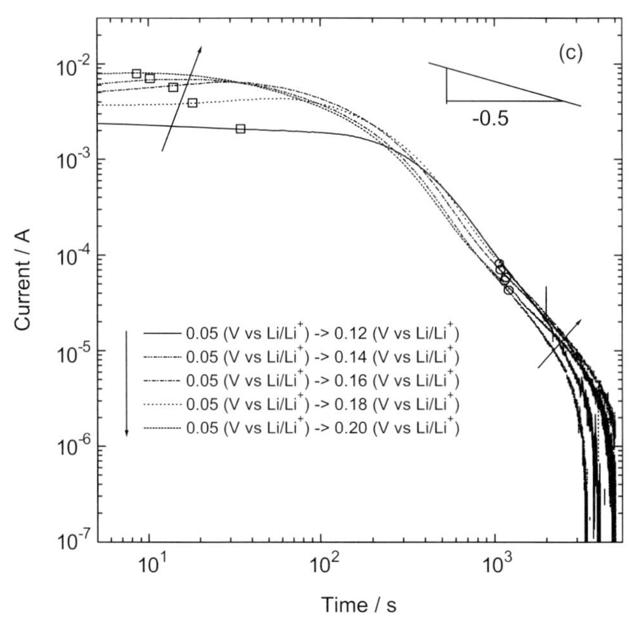

Figure 5.18 (*Continued*)

From the numerical solution to the modified McNabb–Foster equation, the anodic CTs determined at the potential jumps of 0.3, 0.2, and 0.05 V (versus Li/Li^+) to various lithium extraction potentials E_{ext} are illustrated in Figure 5.19a–c. The anodic CTs theoretically calculated based upon the modified McNabb–Foster equation, along with the cell-impedance-controlled constraint in Figure 5.19a–c, almost coincide with experimental CTs (see Figure 5.18a–c). This strongly indicates that the appearance of an inflexion point in the CT is due to the lithium deintercalation from two different kinds of sites with clearly distinguishable activation energies.

5.3.4
Statistical Approach with Kinetic Monte Carlo Simulation

5.3.4.1 Calculation Procedure of Cell-Impedance-Controlled Current Transients with Kinetic Monte Carlo Method

Recently, Pyun *et al.* applied a kinetic Monte Carlo (KMC) method to explore the effect of phase transition due to strong interaction between lithium ions in transition metal oxides with the cubic-spinel structure on lithium transport [17, 28, 103]. The group used the same model for the cubic-spinel structure as described in Section 5.2.3, based on the lattice gas theory. For KMC simulation in a canonical ensemble (CE) where all the microstates have equal V, T, and N, the transition state theory is employed in conjunction with spin-exchange dynamics [104, 105].

The flux of lithium ions at the electrolyte/electrode interface, which corresponds to the number of lithium ions extracted from the electrode as well as within the

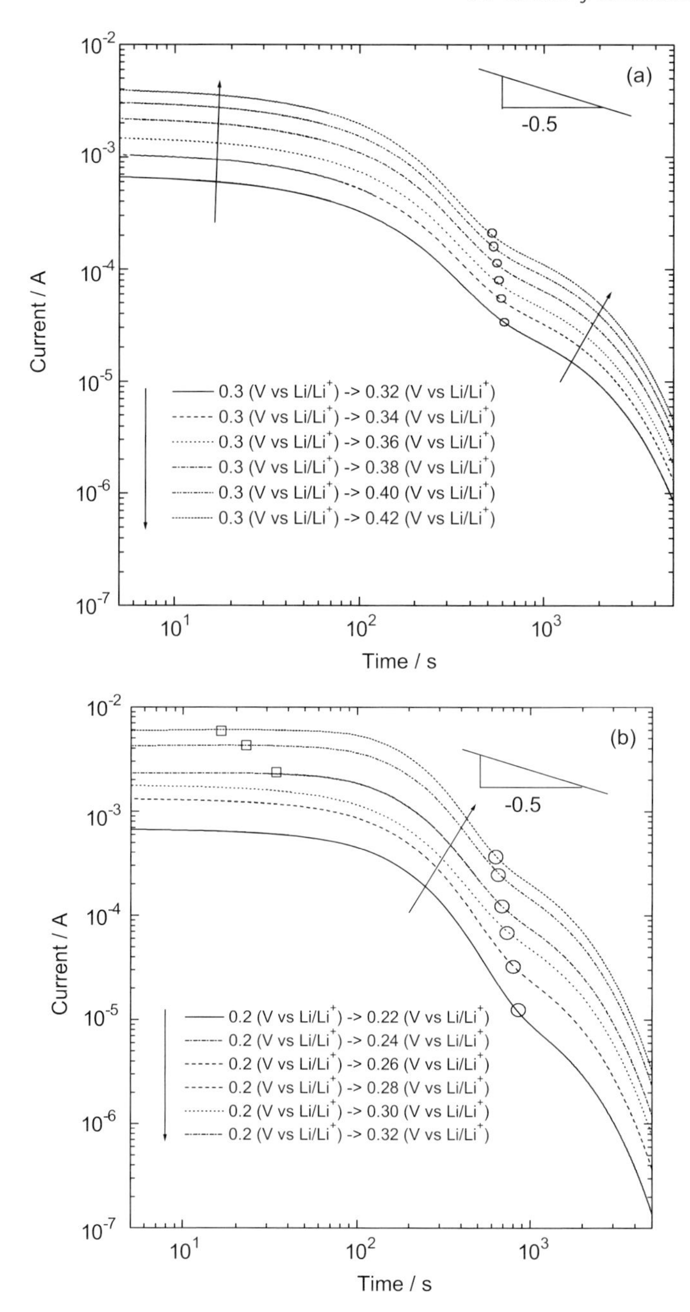

Figure 5.19 The anodic current transient theoretically calculated based upon the modified McNabb–Foster equation at the potential jumps of $E = $ (a) 0.30, (b) 0.20, and (c) 0.05 V (versus Li/Li^+) to various lithium extraction potentials E_{ext} under the "cell-impedance-controlled" constraint, by taking $A_{ea} = 1.56\,cm^2$, $\tilde{D}_{Li} = 2 \times 10^{-10}\,cm^2\,s^{-1}$, $R_{cell} = 27.8\,\Omega$, and $R^* = 5\,\mu m$. Symbols □ and O denote inflexion points. (Reproduced with permission from Ref. [82].)

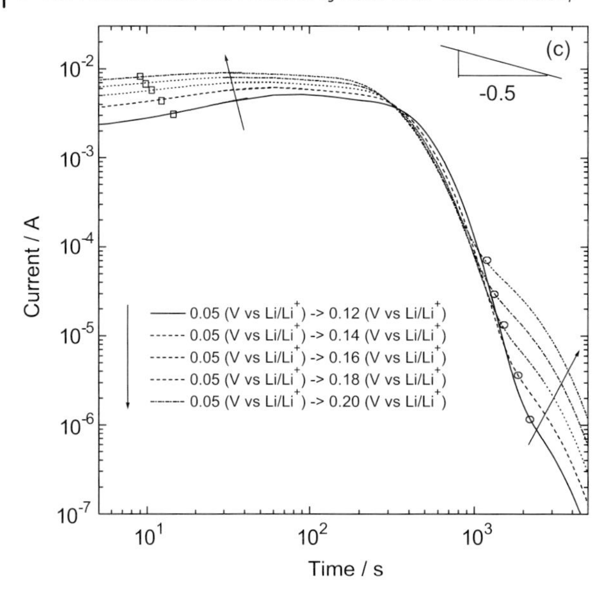

Figure 5.19 (*Continued*)

electrode per unit Monte Carlo step (MCS) time, is theoretically calculated by solving the master equation as a discrete expression of the Fick's diffusion equation given by:

$$\begin{aligned}\frac{\partial P(\{c\},t)}{\partial t} &= \sum[W_{tr}(\{c'\}\rightarrow\{c\})P(\{c'\},t)-W_{tr}(\{c\}\rightarrow\{c'\})P(\{c\},t)]\\ &= -\sum j(\{c\},t) = -J(\{c\},t)\end{aligned} \tag{5.26}$$

under the cell-impedance-controlled constraint of transition probability at the electrode/electrolyte interface

$$W_{tr} = f|E-E_{\text{app}}| = \frac{|E-E_{\text{app}}|}{R_{\text{cell}}}, \tag{5.27}$$

where $\{c\}$ and $\{c'\}$ are a sequence of the lattice configurations generated one after another, corresponding to a set of the local site-occupation numbers; that is, $\{c\} = \{c_1, c_2, \ldots c_i, c_j, c_k \ldots\}$, $W_{tr}(\{c\} \rightarrow \{c'\})$ is the transition probability from the configuration of lattice $\{c\}$ to $\{c'\}$, $P(\{c\}, t)$ is the probability that the configuration of lattice $\{c\}$ occurs at t MCS time, $J(\{c\}, t)$ is the flux of lithium ions per unit jump length in the lattice with the configuration $\{c\}$ at t MCS time, and f represents the conversion factor which is inversely proportional to R_{cell}.

5.3.4.2 Theoretical Current Transients and their Comparison with Experimental Data

Figure 5.20 shows the experimental anodic CTs of the $Li_{1-\delta}Mn_2O_4$ electrode at the potential jumps of 3.90 V (versus Li/Li^+) to various lithium extraction potentials. None of the CTs ever follow the Cottrell behavior – that is, there is no linear relationship between logarithmic current and logarithmic time with a slope of -0.5 in the

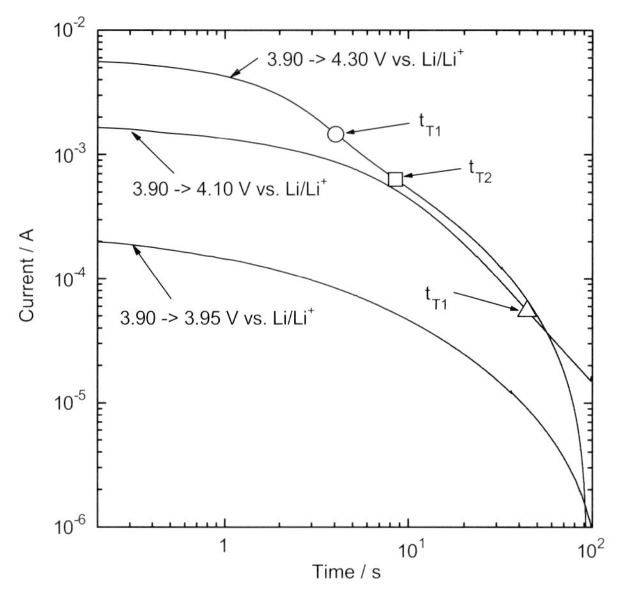

Figure 5.20 Anodic current transients on a logarithmic scale experimentally measured on the $Li_{1-\delta}Mn_2O_4$ electrode at the potential jumps of 3.90 V (versus Li/Li^+) to various lithium extraction potentials. (Reproduced with permission from Ref. [17].)

early stage. Moreover, the CT obtained at the potential step that encounters the disorder to order transition point of 4.01 V (versus Li/Li^+); that is, at the potential jump of 3.90–4.10 V (versus Li/Li^+), shows an inflection point followed by an upward, concave-shaped current drop. The CT obtained at the potential step across both the disorder-to-order and the order-to-disorder transition points of 4.01 and 4.13 V (versus Li/Li^+), respectively – that is, at the potential jump of 3.90–4.30 V (versus Li/Li^+) – exhibits two inflection points divided by an upward, concave-shaped current drop.

Figure 5.21a presents, on a logarithmic scale, the anodic CTs calculated on a theoretical basis, with and without considering the interaction between lithium ions in the $Li_{1-\delta}Mn_2O_4$ electrode under the cell-impedance-controlled constraint with the conversion factor $f = 0.2$ at the potential step across the disorder-order and backward transition points. In the case when no interaction is assumed, the theoretical CT does not display any transition time, but rather shows a monotonic increase of its slope from an almost flat value to one of infinity.

The theoretical CT calculated considering strong ion interaction shows two inflection points (t_{T1}, marked with an open circle and t_{T2}, marked with an open square), and exhibits an upward, concave shape in the time interval between $t = t_{T1}$ and $t = t_{T2}$. In Figure 5.21b, the CT calculated at the potential jump of 3.90 to 4.30 V (versus Li/Li^+) shows both inflection points t_{T1} and t_{T2}, while only one inflection point appears in the CT calculated at the potential jump of 3.90 to 4.10 V (versus

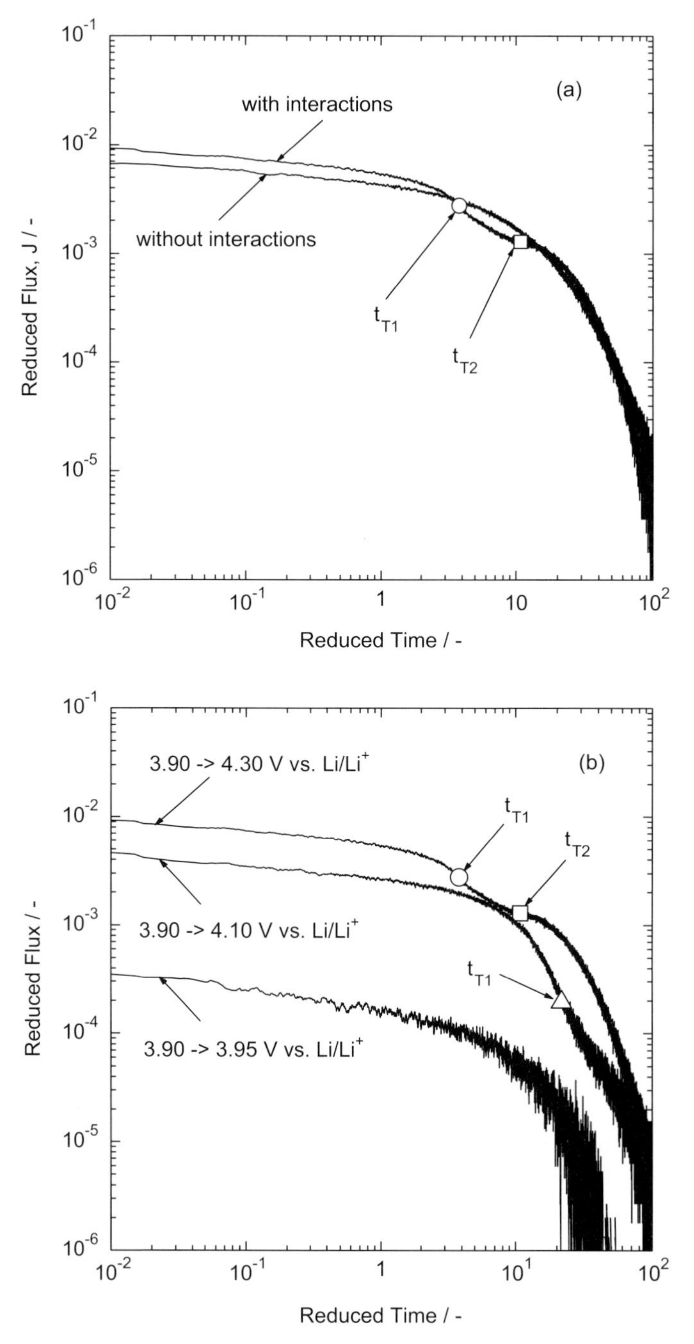

Figure 5.21 Anodic current transients calculated: (a) At the potential step across the disorder-to-order and the order-to-disorder transition points, with and without considering the interaction between lithium ions; (b) By jumping the electrode potential 3.90 V (versus Li/Li^+) to various lithium extraction potentials, considering the interaction between lithium ions under the "cell-impedance-controlled" constraint with $f = 0.2$. (Reproduced with permission from Ref. [17].)

Li/Li^+). All CTs calculated theoretically and accounting for the interaction under the cell-impedance-controlled constraint coincided well in shape with the experimentally measured CTs.

5.4 Methodological Overview

5.4.1 Galvanostatic Intermittent Titration Technique (GITT) in Combination with EMF-Temperature Measurement

The GITT is one of frequent methods to investigate steady-state or equilibrium electrode potentials and diffusion coefficients as function of lithium content in a lithium intercalation electrode. The detailed experimental procedures to determine these thermodynamic and kinetic parameters have been well documented in previous studies [45]. From repeated coulombic titration processes of lithium in the electrode, by application of a constant current with a low value and sufficient time interval to reach equilibrium (i.e., to obtain uniform distribution of the lithium ions throughout the electrode), it is possible to obtain the electrode potentials at various lithium contents, as depicted in Figure 5.2.

In addition, when applying a sufficiently small current, the chemical diffusion coefficient $\tilde{D}$ can be also derived for an instantaneous planar source of diffusion species with I.C.

$$c(x,t) = c^0 \quad (0 \le x \le t) \quad \text{at}\, t = 0, \tag{5.28}$$

with impermeable B.C.

$$\left(\frac{\partial c}{\partial x}\right)_{x=l} = 0 \quad (t \ge 0) \quad \text{at}\, x = l, \tag{5.29}$$

and with galvanostatic B.C.

$$-\tilde{D}\left(\frac{\partial c}{\partial x}\right)_{x=0} = \frac{I_0}{zFA_{ea}} \quad (t \ge 0) \quad \text{at}\, x = 0, \tag{5.30}$$

where c^0 is the initial lithium content at the electrode/electrolyte interface and I_0 is the applied constant current (the other quantities are explained in Section 5.3.1), By solving Fick's second equation, the chemical diffusion coefficient is written as [45]:

$$\tilde{D} = \frac{4}{\pi}\left(\frac{V_m}{FA_{ea}}\right)^2 \left[I_0\left(\frac{dE}{d\delta}\right) / \left(\frac{dE}{d\sqrt{t}}\right)\right]^2 \quad \text{for} \quad t \ll l^2/\tilde{D}, \tag{5.31}$$

where V_m is molar volume of the intercalation compound, $dE/d\delta$ the slope of the coulometric titration curve, and $dE/d\sqrt{t}$ represents the slope of the straight line obtained from the evolution of the relaxation potential versus $\sqrt{t}$. This method has been widely used to measure chemical and component diffusivities in various transition metal oxides [106–108]

In combination with GITT, the enthalpy and entropy of the intercalation can be calculated when the temperature dependence of the electrode potential is measured at a fixed composition after the electrode has reached equilibrium [15, 16, 109–111] by using the known relationships between thermodynamic parameters, such as $\Delta\tilde{G}_{\text{Li}} = -FE$, $\Delta\tilde{H}_{\text{Li}} = -F[E - T(\partial E/\partial T)]$ and $\Delta\tilde{S}_{\text{Li}} = F(\partial E/\partial T)$. On altering the temperature of the electrode, the shift of the electrode potential is so rapid that it takes generally about 1 h to stabilize the electrode potential at constant temperature. The results of EMF-temperature measurements are known to be satisfactorily reproducible on heating and cooling for the lithium intercalation compounds.

5.4.2
Electrochemical AC-Impedance Spectroscopy

The diffusion coefficient of lithium ions in the intercalation electrodes also can be measured, using EIS, by the analysis of Warburg impedance representing diffusion through the electrode with I.C. in Equation (5.28), impermeable B.C. in Equation (5.29), and sinusoidal oscillation B.C.

$$c(0, t) = \Delta c^0 \exp(j\omega t) \quad (t \geq 0) \quad \text{at } x = 0, \tag{5.32}$$

where ω is the angular frequency. By solving Fick's second law under these conditions, the Warburg coefficient σ_ω can be expressed as [112]:

$$\sigma_\omega = \left[\frac{V_m}{FA_{ea}}\left(\frac{\partial E}{\partial \delta}\right)\frac{1}{\sqrt{2\tilde{D}}}\right]. \tag{5.33}$$

The Warburg coefficient can be experimentally determined from the average value for the slope of the Z' versus $\omega^{-1/2}$ plot, and for the slope of the $-Z''$ versus $\omega^{-1/2}$ plot. In many reports, the diffusion coefficients of lithium ions have been determined for various transition metal oxides [13, 97, 113, 114] and graphite [59] electrodes by using EIS.

5.4.3
Potentiostatic Current Transient Technique

The potentiostatic current transient (PCT) technique has been known as the most popular method to understand lithium transport through an intercalation electrode, based on the assumption that lithium diffusion in the electrode is the rate-determining process of lithium intercalation/deintercalation [45]. By solving Fick's second equation for planar geometry with I.C. in Equation (5.28), impermeable B.C. in Equation (5.29), and potentiostatic B.C.

$$c(0, t) = c_s \quad (t \geq 0) \quad \text{at } x = 0, \tag{5.34}$$

where c_s is lithium content at the electrode/electrolyte interface, the current can be expressed as Equations (5.18) and (5.19) in Section 5.3.1. Using Equations (5.18) and (5.19), a number of PCTs have been analyzed. The slopes of the $I(t)$ versus $t^{-1/2}$ and $\ln I(t)$ versus t curves have been determined in the initial and later stages of lithium intercalation/deintercalation, respectively, to estimate the chemical diffusivity in various lithium intercalation electrodes [45–52].

As discussed in Section 5.3, however, various types of anomalous behavior of lithium transport have been reported in PCTs measured for many transition metal oxides and carbonaceous materials [53–63]. Recently, Pyun *et al.* [11, 12, 17, 18, 28, 78–84, 96, 103] reported that those anomalies in PCTs are barely explained using the conventional diffusion-controlled constraint, and attempted to interpret such behaviors under the cell-impedance-controlled lithium transport (see Section 5.3). More recently, Pyun *et al.* applied PCT techniques to effectively elucidate the reaction mechanisms of hydride-forming metal electrodes for rechargeable metal-hydride (MH) batteries [115] and solid oxide electrodes for fuel cells [116, 117]. One of the major advantages of the PCT technique is that the theoretical treatment involved is simpler and easier than that for the cathodic potentiodynamic polarization method which is conventionally used in this research area. For instance, it is possible to acquire information about the relative contributions of the migration and charge transfer processes to total oxygen reduction kinetics under the constraint of the oxygen migration coupled with charge transfer [115]. In fact, the oxygen reduction mechanism on the Pt/C disk electrode [117] and on the dense $(La_{0.85}Sr_{0.15})_{0.9}MnO_3$/yttria-stabilized zirconia (YSZ) composite [118] and $La_{1-x}Sr_xCo_{0.2}Fe_{0.8}O_3$ [119] electrodes were successfully investigated using the cathodic PCT technique.

In addition, the cathodic PCT technique offers an exceptionally powerful tool for understanding the kinetics of the cathode reaction, in the case where electrochemical reactions are self-enhanced over long periods of time ($\approx$4 h) under the cathodic polarization. In contrast to the cathodic potentiodynamic polarization curves with a short measuring time ($\approx$10 min), the cathodic PCTs allow observation of variations in the steady-state current with polarization time, which may provide valuable information when analyzing the reaction rate under cathodic polarization [120].

5.5 Concluding Remarks

In this chapter we have summarized the fundamentals and recent advances in thermodynamic and kinetic approaches to lithium intercalation into, and deintercalation from, transition metals oxides and carbonaceous materials, and have also provided an overview of the major experimental techniques. First, the thermodynamics of lithium intercalation/deintercalation based on the lattice gas model with various approximations was analyzed. Lithium intercalation/deintercalation involving phase transformations, such as order–disorder transition or two-phase coexistence caused by strong interaction of lithium ions in the solid electrode, was clearly explained based on the lattice gas model, with the aid of computational methods.

The chapter then detailed the kinetics of lithium intercalation/deintercalation based on a cell-impedance-controlled model. Anomalous features of the experimental CTs from various transitional metal oxides and carbonaceous materials were introduced, and the physical aspects of experimental CTs were discussed in terms of internal cell resistance. Based on the cell-impedance-controlled lithium transport concept, it is possible to analyze, quantitatively, the CTs and to understand the anomalous features that barely emerge under the diffusion-controlled concept.

Finally, a brief overview was presented of important experimental approaches, including GITT, EMF-temperature measurement, EIS and PCT, for investigating lithium intercalation/deintercalation. In this way, it is possible to determine – on an experimental basis – thermodynamic properties such as electrode potential, chemical potential, enthalpy and entropy, as well as kinetic parameters such as the diffusion coefficients of lithium ion in the solid electrode. The PCT technique, when aided by computational methods, represents the most powerful tool for determining the kinetics of lithium intercalation/deintercalation when lithium transport cannot be simply explained based on a conventional, diffusion-controlled model.

References

1 Budnikov, P.P. and Ginstling, A.M. (1968) *Principles of Solid State Chemistry-Reactions in Solids*, Gordon and Breach Science Publishers, New York.

2 Shoellhorn, R. (1982) *Intercalation Chemistry* (eds M.S. Whittingham and A.J. Jacobson), Academic Press, New York.

3 Manthiram, A. (2004) *Materials Aspects: An Overview in Lithium Batteries – Science and Technology* (eds G.-A. Nazri and G. Pistoia), Kluwer Academic Publishers, New York.

4 Lee, S.-J., Pyun, S.-I. and Shin, H.-C. (2009) *J. Solid State Electrochem.* **13**, 829.

5 Whittingham, M.S. (1978) *Prog. Solid State Chem.*, **12**, 1.

6 Abraham, K.M. (1981) *J. Power Sources*, **7**, 1.

7 Whittingham, M.S. (1994) *Solid State Ionics*, **69**, 173.

8 Wakihara, M. and Yamamoto, O. (1997) *Lithium Ion Batteries – Fundamentals and Performance*, Wiley-VCH, Weinheim.

9 Thackeray, M.M. (1997) *Prog. Solid State Chem.*, **25**, 1.

10 Mckinnon, W.R. and Haering, R.R. (1983) Physical Mechanisms of Intercalation, in *Modern Aspects of Electrochemistry* (eds R.E. White and J.O'M. Bockris), Plenum Press, New York.

11 Shin, H.-C., Pyun, S.-I., Kim, S.-W. and Lee, M.-H. (2001) *Electrochim. Acta*, **46**, 897.

12 Lee, M.-H., Pyun, S.-I. and Shin, H.-C. (2001) *Solid State Ionics*, **140**, 35.

13 Choi, Y.-M., Pyun, S.-I. and Moon, S.-I. (1996) *Solid State Ionics*, **89**, 43.

14 Shin, H.-C. and Pyun, S.-I. (1999) *Electrochim. Acta*, **45**, 489.

15 Pyun, S.-I. and Kim, S.-W. (2000) *Mol. Cryst. Liq. Cryst.*, **341**, 155.

16 Kim, S.-W. and Pyun, S.-I. (2001) *Electrochim. Acta*, **46**, 987.

17 Kim, S.-W. and Pyun, S.-I. (2002) *Electrochim. Acta*, **47**, 2843.

18 Pyun, S.-I., Lee, S.-B. and Chang, W.-Y. (2002) *J. New Mater. Electrochem. Systems*, **5**, 281.

19 Li, W., Reimers, J.N. and Dahn, J.R. (1992) *Phys. Rev. B.*, **46**, 3236.

20 Gao, Y., Reimers, J.N. and Dahn, J.R. (1996) *Phys. Rev. B*, **54**, 3878.

21 Zheng, T. and Dahn, J.R. (1997) *Phys. Rev. B*, **56**, 3800.

22 Kudo, T. and Hibino, M. (1998) *Electrochim. Acta*, **43**, 781.

23 Abiko, H., Hibino, M. and Kudo, T. (1998) *Electrochem. Solid-State Lett.*, **1**, 114.

24 Yamaki, J., Egashira, M. and Okada, S. (2000) *J. Electrochem. Soc.*, **147**, 460.

25 Atkinson, K.E. (1985) *Elementary Numerical Analysis*, John Wiley & Sons, New York.

26 Reimers, J.N., Li, W. and Dahn, J.R. (1993) *Phys. Rev. B*, **47**, 8486.

27 Li, W., Reimers, J.N. and Dahn, J.R. (1994) *Phys. Rev. B*, **49**, 826.

28 Jung, K.-N., Pyun, S.-I. and Kim, S.-W. (2003) *J. Power Sources*, **119–121**, 637.

29 Wong, W.C. and Newman, J. (2002) *J. Electrochem. Soc.*, **149**, A493.

30 Binder, K. and Heerman, D.W. (1997) *Monte Carlo Simulation in Statistical Physics: An Introduction*, Springer-Verlag, Berlin.

31 Gould, H. and Tobochnik, J. (1988) *An Introduction to Computer Simulation Methods: Applications to Physical Systems*, Addison-Wesley.

32 Lim, S.H., Hasebe, M., Murch, G.E. and Oates, W.A. (1990) *Philos. Mag. B*, **62**, 173.

33 Landau, L.D. and Lifshitz, E.M. (1980) *Statistical Physics*, Pergamon Press, Oxford.

34 Binder, K. and Landau, D.P. (1980) *Phys. Rev. B*, **21**, 1941.

35 Dahn, J.R. and McKinnon, W.R. (1984) *J. Phys. C*, **17**, 4231.

36 Reimers, J.N. and Dahn, J.R. (1992) *J. Electrochem. Soc.*, **139**, 2091.

37 Reimer, J.N. and Dahn, J.R. (1993) *Phys. Rev. B*, **47**, 2995.

38 Aydinol, M.K., Kohan, A.F. and Ceder, G. (1997) *J. Power Sources*, **68**, 664.

39 Aydinol, M.K. and Kohan, A.F. (1997) *Phys. Rev. B*, **56**, 1354.

40 Van der Ven, A., Aydinol, M.K. and Ceder, G. (1998) *Phys. Rev. B*, **58**, 2975.

41 Benco, L., Barras, J.L., Atanasov, M., Daul, C.A. and Deiss, E. (1998) *Solid State Ionics*, **112**, 255.

42 Ceder, G. and Van der Ven, A. (1999) *Electrochim. Acta*, **45**, 131.

43 Van der Ven, A. and Ceder, G. (1999) *Phys. Rev. B*, **59**, 742.

44 Van der Ven, A., Marianetti, C., Morgan, D. and Ceder, G. (2000) *Solid State Ionics*, **135**, 21.

45 Wen, C.J., Boukamp, B.A., Huggins, R.A. and Weppner, W. (1979) *J. Electrochem. Soc.*, **126**, 2258.

46 Bard, A.J. and Faulkner, L.R. (1980) *Electrochemical Methods: Fundamentals and Applications*, John Wiley & Sons, New York.

47 Lindström, H., Södergren, S., Solbrand, A., Rensmo, H., Hjelm, J., Hagfeldt, A. and Lindquist, S.-E. (1997) *J. Phys. Chem. B*, **101**, 7710.

48 Sato, H., Takahashi, D., Nishina, T. and Uchida, I. (1997) *J. Power Sources*, **68**, 540.

49 de Albuquerque Maranhão, S.L. and Toressi, R.M. (1998) *Electrochim. Acta*, **43**, 257.

50 McGraw, J.M., Bahn, C.S., Pariaal, P.A., Perkins, J.D., Reasey, D.W. and Ginley, D.S. (1999) *Electrochim. Acta*, **45**, 187.

51 Nishizawa, M., Koshika, H., Hashitani, R., Itoh, T., Abe, T. and Uchida, I. (1999) *J. Phys. Chem. B*, **103**, 4933.

52 Waki, S., Dokko, K., Itoh, T., Nishizawa, M., Abe, T. and Uchida, I. (2000) *J. Solid State Electrochem.*, **4**, 205.

53 Isidorsson, J., Strømme, M., Gahlin, R., Niklasson, G.A. and Granqvist, C.G. (1996) *Solid State Commun.*, **99**, 109.

54 Strømme Mattsson, M., Niklasson, G.A. and Granqvist, C.G. (1996) *Phys. Rev. B*, **54**, 2968.

55 Pyun, S.-I., Lee, M.-H. and Shin, H.-C. (2001) *J. Power Sources*, **97–98**, 473.

56 Choi, Y.-M., Pyun, S.-I. and Paulsen, J.M. (1998) *Electrochim. Acta*, **44**, 623.

57 Bae, J.-S. and Pyun, S.-I. (1996) *Solid State Ionics*, **90**, 251.

58 Pyun, S.-I. and Choi, Y.-M. (1997) *J. Power Sources*, **68**, 524.

59 Pyun, S.-I. and Ryu, Y.-G. (1998) *J. Power Sources*, **70**, 34.

60 Shin, H.-C. and Pyun, S.-I. (1999) *Electrochim. Acta*, **44**, 2235.

61 Funabiki, A., Inaba, M., Abe, T. and Ogumi, Z. (1999) *Carbon*, **37**, 1591.

62 Funabiki, A., Inaba, M., Abe, T. and Ogumi, Z. (1999) *Electrochim. Acta*, **45**, 865.

63 Funabiki, A., Inaba, M., Abe, T. and Ogumi, Z. (1999) *J. Electrochem. Soc.*, **146**, 2443.

64 Levi, M.D. and Aurbach, D. (1997) *J. Electroanal. Chem.*, **421**, 79.

65 Levi, M.D. and Aurbach, D. (1997) *J. Phys. Chem. B*, **101**, 4630.

66 Levi, M.D. and Aurbach, D. (1999) *Electrochim. Acta*, **45**, 167.

67 Levi, M.D., Salitra, G., Markovsky, B., Teller, H., Aurbach, D., Heider, U. and Heider, L. (1999) *J. Electrochem. Soc.*, **146**, 1279.

68 Zhang, D., Popov, B.N. and White, R.E. (2000) *J. Electrochem. Soc.*, **147**, 831.

69 Noel, M. and Rajendran, V. (2000) *J. Power Sources*, **88**, 243.

70 Hjelm, A.-K., Lindbergh, G. and Lundqvist, A. (2001) *J. Electroanal. Chem.*, **506**, 82.

71 Hjelm, A.-K., Lindbergh, G. and Lundqvist, A. (2001) *J. Electroanal. Chem.*, **509**, 139.

72 Krtil, P. and Fattakhova, D. (2001) *J. Electrochem. Soc.*, **148**, A1045.

73 Mattsson, M.S. and Niklasson, G.A. (1999) *J. Appl. Phys.*, **85**, 8199.

74 Mattsson, M.S., Isidorsson, J. and Lindström, T. (1999) *J. Electrochem. Soc.*, **146**, 2613.

75 Mattsson, M.S. (2000) *Solid State Ionics*, **131**, 261.

76 Fu, Z.-W. and Qin, Q.-Z. (2000) *J. Phys. Chem. B*, **104**, 5505.

77 Pyun, S.-I. and Shin, H.-C. (2000) *Mol. Cryst. Liq. Cryst.*, **341**, 147.

78 Pyun, S.-I. and Shin, H.-C. (2001) *J. Power Sources*, **97–98**, 277.

79 Go, J.-Y., Pyun, S.-I. and Shin, H.-C. (2002) *J. Electroanal. Chem.*, **527**, 93.

80 Pyun, S.-I. and Kim, S.-W. (2001) *J. Power Sources*, **97–98**, 371.

81 Pyun, S.-I., Kim, S.-W. and Ko, J.-M. (2002) *J. New Mater. Electrochem. Systems*, **5**, 135.

82 Lee, S.-B. and Pyun, S.-I. (2002) *Electrochim. Acta*, **48**, 419.

83 Lee, S.-B. and Pyun, S.-I. (2003) *J. Solid State Electrochem.*, **7**, 374.

84 Chang, W.-Y., Pyun, S.-I. and Lee, S.-B. (2003) *J. Solid State Electrochem.*, **7**, 368.

85 Striebel, K.A., Deng, C.Z., Wen, S.J. and Cairns, E.J. (1996) *J. Electrochem. Soc.*, **143**, 1821.

86 Uchida, T., Morikawa, Y., Ikuta, H., Wakihara, M. and Suzuki, K. (1996) *J. Electrochem. Soc.*, **143**, 2606.

87 Levi, M.D., Levi, E.A. and Aurbach, D. (1997) *J. Electroanal. Chem.*, **421**, 89.

88 Levi, M.D. and Aurbach, D. (1997) *J. Phys. Chem. B*, **101**, 4641.

89 Markovsky, B., Levi, M.D. and Aurbach, D. (1998) *Electrochim. Acta*, **43**, 2287.

90 Aurbach, D., Levi, M.D., Lev, O., Gun, J. and Rabinovich, L. (1998) *J. Appl. Electrochem.*, **28**, 1051.

91 Aurbach, D., Levi, M.D., Levi, E., Teller, H., Markovsky, B. and Salitra, G. (1998) *J. Electrochem. Soc.*, **145**, 3024.

92 Yang, W., Zhang, G., Lu, S., Xie, J. and Liu, Q. (1999) *Solid State Ionics*, **121**, 85.

93 Lu, Z., Levi, M.D., Salitra, G., Gofer, Y., Levi, E. and Aurbach, D. (2000) *J. Electroanal. Chem.*, **491**, 211.

94 Varsano, F., Decker, F., Masetti, E. and Croce, F. (2001) *Electrochim. Acta*, **46**, 2069.

95 Levi, M.D., Lu, Z. and Aurbach, D. (2001) *Solid State Ionics*, **143**, 309.

96 Shin, H.-C. and Pyun, S.-I. (2003) Mechanisms of Lithium Transport through Transition Metal Oxides and Carbonaceous Materials, in *Modern Aspects of Electrochemistry* (eds C.G. Nayenas, B.E. Conway and R.E. White), Kluwer Academic Plenum Publishers, New York.

97 Choi, Y.-M. and Pyun, S.-I. (1997) *Solid State Ionics*, **99**, 173.

98 Dickens, P.G. and Reynolds, G.J. (1981) *Solid State Ionics*, **5**, 331.

99 Zaghib, K., Simoneau, M., Armand, M. and Gauthier, M. (1999) *J. Power Sources*, **81–82**, 300.

100 Takami, N., Satoh, A., Hara, M. and Ohsaka, T. (1995) *J. Electrochem. Soc.*, **142**, 371.

101 McNabb, A. and Foster, P.K. (1963) *Trans. Met. Soc. AIME*, **227**, 618.

102 Pyun, S.-I. and Yang, T.-H. (1998) *J. Electroanal. Chem.*, **441**, 183.

103 Kim, S.W. and Pyun, S.I. (2002) *J. Electroanal. Chem.*, **528**, 114.

104 Binder, K. (1984) *Applications of the Monte Carlo Method in Statistical Physics*, Springer-Verlag, Berlin, Germany.

105 Nassif, R., Boughaleb, Y., Hekkouri, A., Gouyet, J.F. and Kolb, M. (1998) *Eur. Phys. J. B*, **1**, 453.

106 Bae, J.-S. and Pyun, S.-I. (1995) *J. Alloys Compd.*, **217**, 52.

107 Choi, Y.-M., Pyun, S.-I., Bae, J.-S. and Moon, S.-I. (1995) *J. Power Sources*, **56**, 25.

108 Choi, Y.-M. and Pyun, S.-I. (1998) *Solid State Ionics*, **109**, 15.

109 Honders, A., der Kinderen, J.M., van Heeren, A.H., de Wit, J.H.W. and Broers, G.H.J. (1984) *Solid State Ionics*, **14**, 205.

110 Pereira-Ramos, J.P., Messina, R., Piolet, C. and Devynck, J. (1988) *Electrochim. Acta*, **33**, 1003.

111 Kumagai, N., Fujiwara, T., Tanno, K. and Horiba, T. (1993) *J. Electrochem. Soc.*, **140**, 3194.

112 Ho, C., Raistrick, I.D. and Huggins, R.A. (1980) *J. Electrochem. Soc.*, **127**, 343.

113 Pyun, S.-I. and Bae, J.-S. (1996) *Electrochim. Acta*, **41**, 919.

114 Pyun, S.-I. and Bae, J.-S. (1996) *J. Alloys Compd.*, **245**, L1.

115 Lee, J.-W. and Pyun, S.-I. (2005) *Electrochim. Acta*, **50**, 1777.

116 Lee, S.-J. and Pyun, S.-I. (2007) *Electrochim. Acta*, **52**, 6525.

117 Lee, S.-K., Pyun, S.-I., Lee, S.-J. and Jung, K.-N. (2007) *Electrochim. Acta*, **53**, 740.

118 Kim, J.-S. and Pyun, S.-I. (2008) *J. Electrochem. Soc.*, **155**, B8.

119 Kim, Y.-M., Pyun, S.-I., Kim, J.-S. and Lee, G.-J. (2007) *J. Electrochem. Soc.*, **154**, B802.

120 Kim, J.-S., Pyun, S.-I., Shin, H.-C. and Kang, S.-J.L. (2008) *J. Electrochem. Soc.*, **155**, B762.

6
Solid-State Electrochemical Reactions of Electroactive Microparticles and Nanoparticles in a Liquid Electrolyte Environment

Michael Hermes and Fritz Scholz

Abstract

Many electrochemical conversions of solid compounds and materials, including for example the corrosion of metals and alloys or the electrochemical conversions of most battery materials, take place within a liquid electrolyte environment, with the classic approach to investigation comprising macro-sized electrodes. However, in order to obtain a comprehensive understanding of the mechanism of these solid-state electrochemical reactions, the simple technique of immobilizing small amounts of a solid compound/material on an inert electrode surface provides an easy, yet sometimes exclusive, access to their study. In this chapter is presented a survey of the recent developments of this approach, which is referred to as the voltammetry of immobilized microparticles (VIM). Attention is also focused on progress in the field of theoretical descriptions of solid-state electrochemical reactions.

6.1
Introduction

The electrochemical conversions of solid compounds and materials that are in direct contact with electrolyte solutions or liquid electrolytes (ionic liquids), belong to the most widespread reactions in electrochemistry. Such conversions take place in a wide variety of circumstances, including the majority of primary and secondary batteries, in corrosion, in electrochemical machining, in electrochemical mineral leaching, in electrochemical refining (e.g., copper refining), and in electrochemical surface treatments (e.g., the anodization of aluminum).

The study of these reactions forms part of the main body of electrochemical research conducted during the past 200 years. Naturally, the majority of the studies were performed with macro-sized electrodes, mainly because the technical systems were of such size, although an additional reason was that the experimental handling

Solid State Electrochemistry I: Fundamentals, Materials and their Applications. Edited by Vladislav V. Kharton

ISBN: 978-3-527-32318-0

is normally much simpler than in experiments using microparticles and/or nanoparticles. However, a very serious issue here is that the kinetics may differ for macroparticles and micro/nanoparticles, and in order to obtain a full understanding of such systems it would be necessary to perform these experiments in both size domains. As a consequence of this, very remarkable differences have been observed, one example being the case of silver metal oxidation and silver halide reduction [1, 2]. Experiments with single micro/nanoparticles, or assemblies of such particles on inert electrodes, provide information that is not accessible with macroelectrodes, mainly because the latter possess a much larger number of nuclei in nucleation-growth processes. Another desirable feature of studies with micro/nanoparticles is the small currents employed and the high conversion efficiency, which allows not only a much better signal resolution but also the application of faster measurements. Consequently, experimental time spans required will be much shorter.

Only within the past two decades has the great potential of electrochemical studies of single micro/nanoparticles (or of their assemblies) been truly realized, and important advances have been made subsequently, both with respect to the experiments as well as to the theoretical basis of this topic. Among the different experimental approaches developed to assess the electrochemistry of micro/nanoparticles, that of mechanically immobilizing assemblies of micro/nanoparticles onto inert electrodes has proved to be especially advantageous, for the following reasons:

- The particles are freely accessible to the electrolyte solution, and are simultaneously in good electronic contact with the inert electrode material.
- The particles are not covered by any liquid binder (e.g., mineral oil), as in the case of paste electrodes; neither are they in contact with any solid binder (as in case of solid composite electrodes).
- The particles are very easy to prepare and to renew.
- The experiments can easily be combined with *in situ* spectroscopy.
- The experiments lend themselves to being applied in the analysis of very small amounts of sample; a particular benefit is in the analysis of precious objects in the areas of art and archeology.

Although the first reports of this approach involved studies with metal alloys [3] and minerals [4], within a few years the technique has been extended to a wide variety of research areas. As these findings have been summarized in several reviews [5–8] and also in a monograph [9], attention will be focused here on more recent developments, notably on the mechanical immobilization of particles on electrodes. Today, a huge amount of information is available for electrochemical systems comprising particles enclosed in polymer films or other matrices (see Refs [10–16]). Originally, the main aim of such particle enclosure was to achieve specific electrode properties (e.g., functionalized carbon/polymer materials as electrocatalysts [17, 18]; solid-state, dye-sensitized solar cells [19]), rather than to study the electrochemistry of the particles. This situation arose mainly because the preparation of these composites was too cumbersome for assessing the particles' properties. The techniques also suffered from interference caused by the other phases that constituted the electrode.

6.2
Methodological Aspects

The basic technique of mechanical immobilizing particles has been described in detail elsewhere [7, 9], and needs not be repeated at this stage.

Here, we shall discuss some new developments relating to the quantitative evaluation of data and, for the sake of completeness, also briefly mention previously developed approaches. As the technique of mechanical immobilizing particles on electrode surfaces does not allow the amount of deposited particles to be controlled,[1)] the following approaches for quantitative evaluation have been developed:

i. When the peak potential (or any other characteristic potential) of an electrochemical reaction of particles depends on the amount of particles, it is possible to determine the peak potentials and to integrate the currents so as to obtain a set of data "peak potential–charge", where the charge can be converted to amounts of substances via Faraday's law. If this is done for a number of deliberately different samples (electrodes with immobilized particles), it is possible to compare chemically different samples by comparing the peak potentials for the *same* charges. This provides the necessary standardization to avoid any dependence on amounts. An example of this approach is the analysis of a series of solid solutions of copper sulfide/selenides [20], and the determination of transformation enthalpies of mineral phases [21].

ii. For the quantification of two-component alloys, the peak ratios can be used, provided that standards for calibration are available [3]. Further examples are described in Ref. [9].

iii. In the case of powder mixtures of two electroactive compounds, the peak ratios can also be used. provided that standard mixtures have been measured for calibration [22].

iv. A very skillful evaluation technique was recently devised by Doménech-Carbó and coworkers [23], who applied the so-called H-point standard addition method developed earlier by Bosch and Campins [24]. In short, the problem of determining the amounts of two electroactive compounds A and B in a mixture is solved by the admixture of a known amount of a third electroactive compound, R, and by the standard addition of A. When all these probes are measured, the concentrations of A and B in the primary sample can be determined with very high precision. Bosch-Reig *et al.* [23] have applied the H-point standard addition technique to the analysis of madder samples made from alizarin and purpurin, and have used morin as the inner standard (R). For

1) If particles are immobilized by the evaporation of a small volume (e.g., several microliters) of a suspension, the amount of particles can be controlled. However, it cannot be taken as granted that all immobilized particles will remain on the electrode surface when the electrode is introduced into the electrolyte solution. Neither is it guaranteed that all particles, even if they remain on the electrode surface, will have such intimate contact with the latter that they can react electrochemically.

artificial mixtures of alizarin and purpurin, the deviations of analysis from the given compositions did not exceed 0.7%; this is a low value indeed, bearing in mind that the analysis was performed with trace amounts of the dye mixtures (most likely some tens or hundreds of nanograms).

v. The quantification of one electrochemical active component in an electrochemically inactive matrix can also be accomplished with the help of admixture of a standard. In solid solutions (mixed crystals) of copper hexacyanoferrate (*hcf*) and copper hexacyanocobaltate (*hcc*) (i.e., $Cu(hcf)_x(hcc)_{1-x}$), only the *hcf* system is electrochemically active [with the exception of very high *hcc* contents, where the copper (II)/(I) system can also be observed]. The dependence of the *hcf* peak currents on the composition of the mixed phases has been studied by adding nickel hexacyanoferrate to all samples in a ratio of 1 : 1; that is, by preparing powder mixtures which contain two phases, namely the solid solution phase and Ni(*hcf*) [25]. The latter compound can be used as an inner standard because the *hcf* system of Ni*hcf* has a formal potential that is distinctly different from the formal potential of the *hcf* system in the copper phases. Similarly, the dependence of peak currents of the *hcf* system of mixed nickel *hcf-hcc* phases could be studied by adding a constant amount of copper *hcf*.

6.3 Theory

The voltammetric behavior of surface-immobilized microparticles of redox active solid materials has been extensively studied by the groups of Scholz (Greifswald, Germany), Bond (Melbourne, Australia), Grygar (Rez, Czech Republic), Komorsky-Lovrić and Lovrić (Zagreb, Croatia), Doménech-Carbó (Valencia, Spain), Marken (Bath, UK), and others. Theoretical aspects, however, have been addressed only in some reports. Recently, the Compton group (Oxford, UK) made several reports on the theory of microparticle-modified electrodes, and these will mainly be discussed at this point.

6.3.1 General Theoretical Treatment

Independent of the nature of the respective solid electroactive crystal, it is a main feature of these systems that the electron transfer takes place at the three-phase boundary microcrystal | electrode | electrolyte, and the (frequently even electrochemically reversible) electron transfer is accompanied by the transfer of cations or anions between the microcrystal and the electrolyte phase, in order to maintain electroneutrality. This situation of a three-phase electrode, where the immobilized electroactive solid phase is simultaneously in contact with an electron-conducting phase (e.g., a metal or graphite) and also with an electrolyte solution phase, is illustrated in Figure 6.1.

Assuming an insertion electrochemical reaction, the electron transfer between Phases I and II must be accompanied by an ion transfer between the Phases II and III, as the respective net charge of the microparticle must be balanced to maintain the

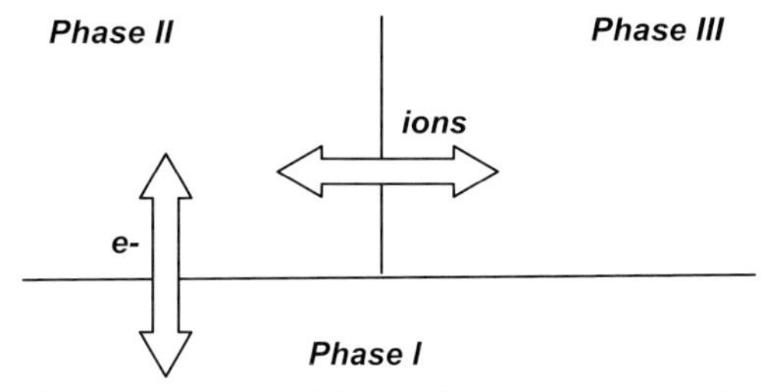

Figure 6.1 Scheme of the simultaneous electron and ion transfer at a three-phase electrode.

electroneutrality of Phase II. The formal thermodynamic treatment comprises an overall equilibrium equation as follows:

$$\mathrm{Ox}^{x+}_{\text{phase II}} + ne^{-}_{\text{phase I}} + nC^{+}_{\text{phase III}} \leftrightarrows \mathrm{Red}^{(x-n)+}_{\text{phase II}} + nC^{+}_{\text{phase II}}, \tag{6.1}$$

with the Nernst equation being:

$$E = E^{\ominus}_{\mathrm{Ox/Red}/C} + \frac{RT}{nF}\ln\frac{a_{\mathrm{Ox}^{x+}_{\text{phase II}}}\, a^{n}_{C^{+}_{\text{phase III}}}}{a_{\mathrm{Red}^{(x-n)+}_{\text{phase II}}}\, a^{n}_{C^{+}_{\text{phase II}}}}. \tag{6.2}$$

Formally, two separate equilibria can be written, one for the transfer of electrons, and one for that of ions:

$$\mathrm{Ox}^{x+}_{\text{phase II}} + ne^{-}_{\text{phase I}} \leftrightarrows \mathrm{Red}^{(x-n)+}_{\text{phase II}} \tag{6.3}$$

and

$$C^{+}_{\text{phase III}} \leftrightarrow C^{+}_{\text{phase II}}. \tag{6.4}$$

The respective Nernst equations are:

$$E_{\mathrm{I/II}} = E^{\ominus}_{\mathrm{Ox/Red}} + \frac{RT}{nF}\ln\frac{a_{\mathrm{Ox}^{x+}_{\text{phase II}}}}{a_{\mathrm{Red}^{(x-n)+}_{\text{phase II}}}}, \tag{6.5}$$

and

$$E_{\mathrm{II/III}} = E^{\ominus}_{C} + \frac{RT}{F}\ln\frac{a_{C^{+}_{\text{phase III}}}}{a_{C^{+}_{\text{phase II}}}}. \tag{6.6}$$

The standard potentials are interrelated by:

$$E^{\ominus}_{\mathrm{Ox/Red}/C} = E^{\ominus}_{\mathrm{Ox/Red}} + E^{\ominus}_{C}. \tag{6.7}$$

A more general explanation of the charge transfer reactions and charge transport pathways at three-phase electrodes is given in Figure 6.2. This scheme comprises the cases of insertion electrochemistry, as well as that of a dissolution of the electroactive phase.

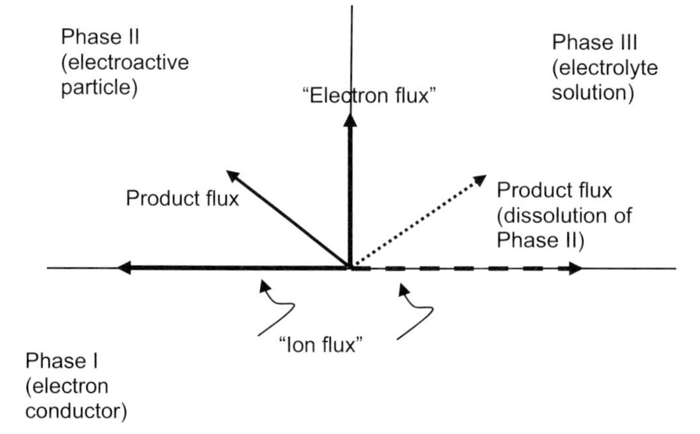

Figure 6.2 Possible pathways of the charge transfer reactions and the charge transport processes proceeding at a three-phase electrode consisting of an electrochemically active Phase II, and electrolyte solution (Phase III), and an electron conductor (Phase I). The electron flux shows the direction in which electrons can be transferred across the interface I/II and transported within Phase II. The ion flux shows the direction in which ions can be transferred across interface II/III and within Phase II (full arrow) or within Phase III (dashed arrow). (Depending on the reaction, electrons and ions can also flow in the opposite directions.) The product fluxes indicate in which direction the products may move.

The electrochemical reduction of a solid compound characterized by mixed ionic/electronic conductivity, immobilized on an electrode surface and in contact with an electrolyte solution, has been further studied on a theoretical basis [26]. Here, the coupled uptake or expulsion of electrons from the electrode and of cations from the electrolyte solution according to

$$\mathrm{Ox_{solid}} + \mathrm{e^-_{electrode}} + \mathrm{C^+_{solution}} \leftrightarrows \mathrm{RedC_{solid}} \tag{6.8}$$

has been considered. The simultaneous coupled diffusion following Fick's law will, at least for the beginning of the reaction, be possible only at the three-phase junction. The chronoamperometric curves, resulting from an applied potential step, have been analyzed. A transition of the initial three-phase reaction to a two-phase reaction occurs, if the reaction zone expands and the overall electric conductivity is sufficient. During the initial three-phase reaction, the reaction zone will preferably expand along the direction of the fastest transport process – that is, either parallel to the interface I-II when the ion transport is the fastest process, or perpendicular to that interface when the electron transport is fastest. Any further procedure of the solid-state reaction into the microcrystal will be determined by the electron and ion transport rates, because the advancement of the reaction layer is only possible by simultaneously maintaining the charge neutrality. That situation leads, in the case of a potential step experiment, to the concentration profiles shown in Figure 6.3.

Assuming a three-dimensional (3-D) crystal, two representative diffusion profiles are given in Figure 6.4.

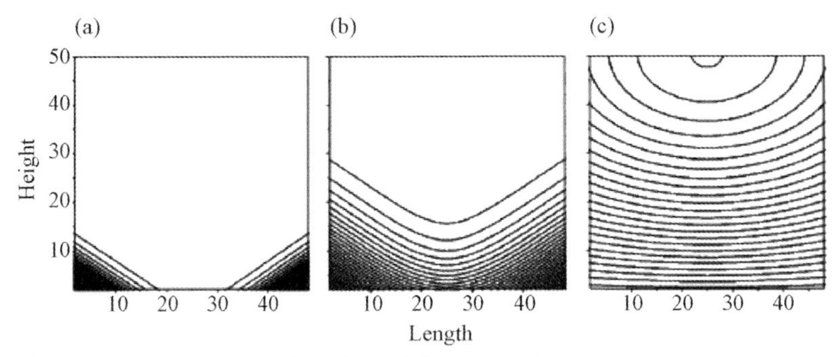

Figure 6.3 Concentration profile of a partially converted two-dimensional model of a microcrystal at different times after a potential step of 3 s (a), 10 s (b), and 40 s (c); $D_{e-} = 5 \times 10^{-9}\,cm^2\,s^{-1}$, $D_{C+} = 10^{-8}\,cm^2\,s^{-1}$ [26].

Simulated cyclic voltammograms (cf. Figure 6.5) for differently shaped cuboids of a uniform volume show that the respective ratios of microcrystal height, length and breadth determine the time of the total conversion of the crystals, as different diffusion regimes dominate.

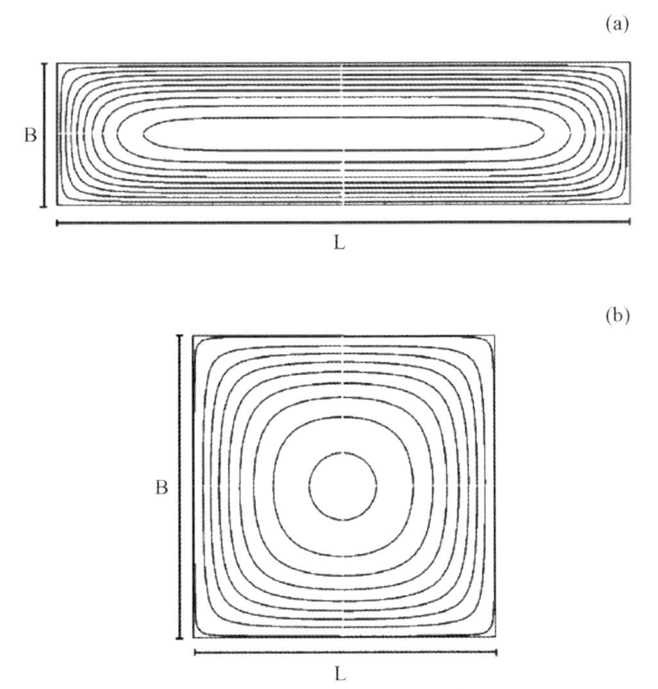

Figure 6.4 (a) Concentration profile inside partially (10%) converted crystals; (b) Cross-section through the x-y plane at half-height. Crystal of $L = H = 10\,\mu m$ and $B = 40\,\mu m$ (a), and $L = B = H = 10\,\mu m$ size (b). See Ref. [26].

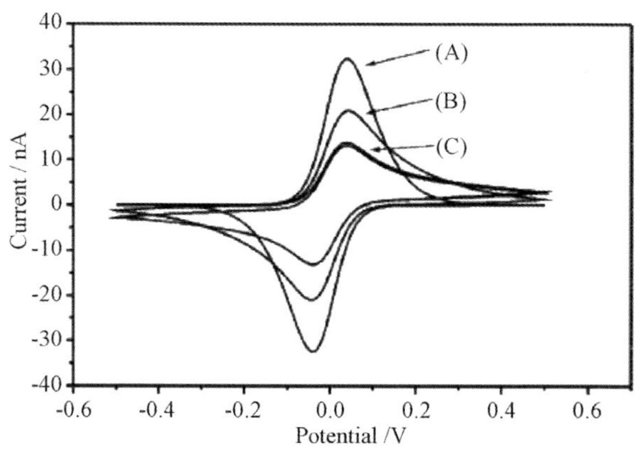

Figure 6.5 Cyclic voltammograms of differently shaped cuboid particles of constant volume with a square base of different size: $D_{e-} = D_{C+} = 10^{-8}\,cm^2\,s^{-1}$, $V_m = 153.8\,cm^3\,mol^{-1}$; (A) $L = B = 28\,\mu m$, $H = 10\,\mu m$; (B) $L = B = H = 20\,\mu m$; (C) $L = B = 16\,\mu m$, $H = 31\,\mu m$ [26].

In an experimentally focused publication, the behavior of a three-phase electrode with a rather well-defined three-phase boundary has been studied [27]. For this, white elemental phosphorus was oxidized at a graphite electrode that was partly embedded in the solid phosphorus, and partly immersed in an aqueous electrolyte solution (see Figure 6.6c).

Here, it could be shown that the currents are in fact proportional to the length of the three-phase-boundary, and the reaction can be explained by the ingress of protons into the phosphorous phase from which electrons are transferred to the graphite

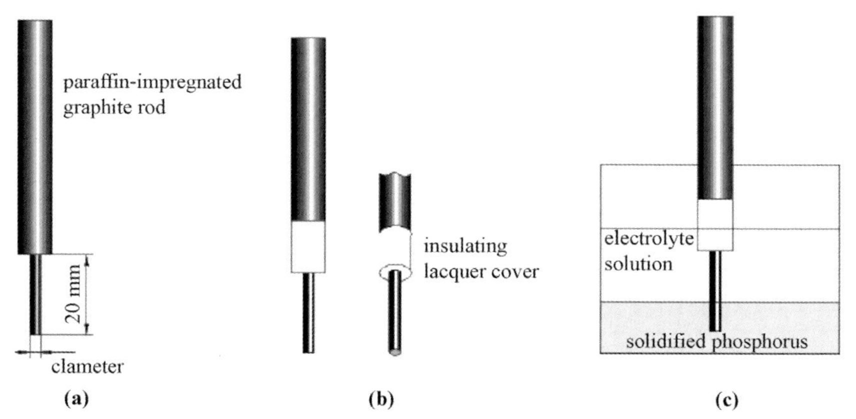

Figure 6.6 (a) Paraffin-impregnated graphite rod with cylindrically reduced tip; (b) Same electrodes, partially covered by an insulating lacquer; (c) Scheme of the electrochemical cell with a paraffin-impregnated graphite rod as working electrode, partially embedded in solidified white phosphorus [27].

contact. Thus, a new phase consisting of more or less protonated phosphorous oxides, most probably with varying oxidation states of phosphorus, is produced.

6.3.2 Voltammetry of Microparticle-Modified Electrodes

6.3.2.1 Adsorbed (Surface-)Electroactive Microparticles on Solid Electrodes

Oldham has produced a mathematical treatment for the quasi-reversible electrochemical redox reaction, being a surface-confined process – that is, the case of an "adsorbed" redox species [28]. Most notably, in that case, the peak current is linearly dependent on the scan rate. For the case of immobilized hemispherical dendrimers containing electroactive groups at their surface, Amatore *et al.* observed a linear dependence of peak current with the square root of scan rate in cyclic voltammetry [29, 30], and this was explained by "charge hopping" between the redox sites. For many other cases of functionalized carbon nanotubes (or modified graphite electrodes), Compton *et al.* also identified a square root of scan rate dependency of peak currents [31, 32]. These findings have been explained on a theoretical basis by the (non-Cottrellian) propagation of charge on the carbon particle surface [33, 34], whereby electrons are assumed to hop from one chemical moiety to another, all being attached to the particle surface, as depicted in Figure 6.7.

The charge movement can be treated diffusion-like as electron hopping with coupled solution phase ion motion, contributing to the effective diffusion coefficient [34] (as in the studies cited for the general treatment of a three-phase electrode in Section 6.3.1).

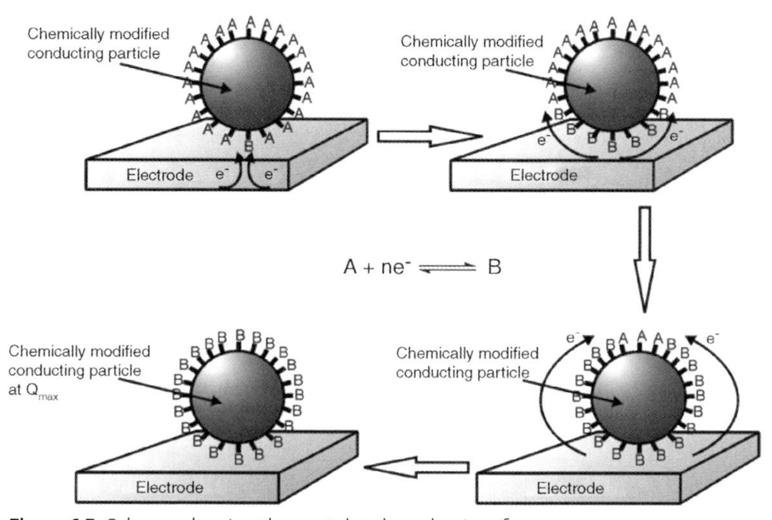

Figure 6.7 Scheme showing the postulated mechanism for charge diffusion on the surface of a sphere. The corresponding diffusion of ions in solution is assumed to be facile, and is not shown [34].

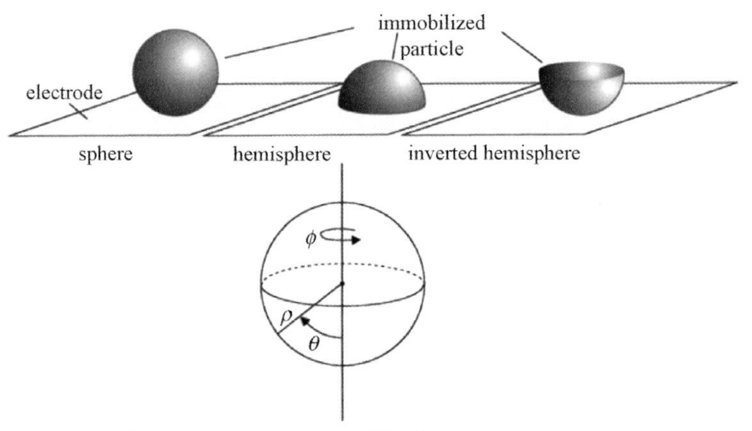

Figure 6.8 Different geometries of immobilized particles on the electrode surface, and the spherical coordinate system [33].

Thompson and Compton investigated, from a theoretical standpoint, the case of a spherical microparticle with an electroactive compound on its surface and attached to a solid electrode surface [33, 34]. The movement of charge was assumed to start exclusively from the contact point (or line) between the microparticle and electrode (i.e., at the three-phase boundary, if an electrolyte phase is considered) and to proceed over the particle surface only (see also Section 6.3.1). In Ref. [33], the idealized microparticle geometries of a full sphere, a hemisphere, and an inverted hemisphere have been considered (cf. Figure 6.8).

Butler–Volmer kinetics has been assumed for the electron transfer, and the electron hopping was modeled as a diffusional process (given in spherical coordinates) confined to the microparticle surface. The hopping of charge over the surface layer provides an apparent flux of charge equivalent to that given by Fick's law of diffusion in a spherical shell [29, 30]. From a full implicit finite difference approach, the surface coverage profiles and current–voltage responses (concentration profiles over the course of a cyclic voltammogram) were obtained, which are dependent mainly on the normalized sweep rate, ν_{norm}, and the normalized heterogeneous rate constant K_0, or:

$$\nu_{\mathrm{norm}} = \frac{F}{RT}\frac{\rho^2}{D}\nu, \tag{6.9}$$

and

$$K_0 = \frac{k_0\rho}{D}, \tag{6.10}$$

where ρ denotes the microparticle sphere radius, ν the voltage scan rate, D is the diffusion coefficient, and k_0 the heterogeneous rate constant, respectively. Figure 6.9 shows simulated cyclic voltammograms (CVs) for the case of the full sphere.

For high values of K_0, electrochemical reversibility can be reached at the point of three-phase contact, while for decreasing K_0, the slower electrode kinetics causes a

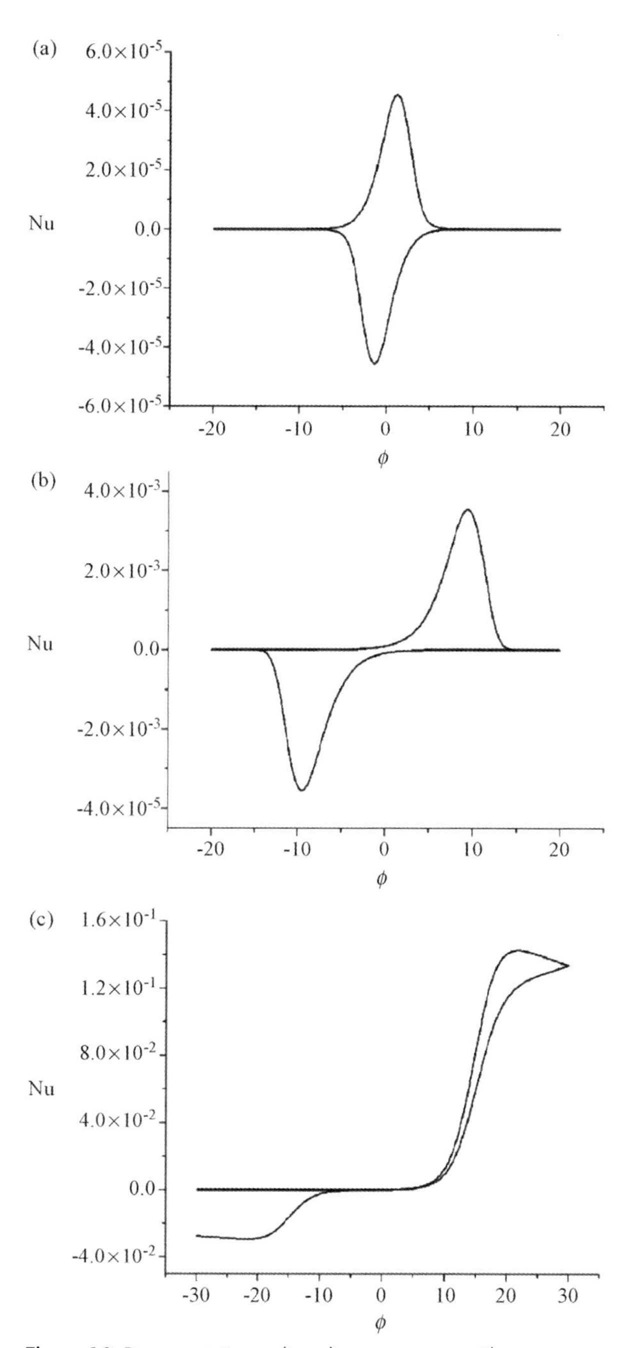

Figure 6.9 Representative cyclic voltammograms, with dimensionless current and potential, for the full sphere for $K_0 = 0.05$. (a) $\nu_{norm} = 1 \times 10^{-4}$; (b) $\nu_{norm} = 1 \times 10^{-2}$; (c) $\nu_{norm} = 10$ [33].

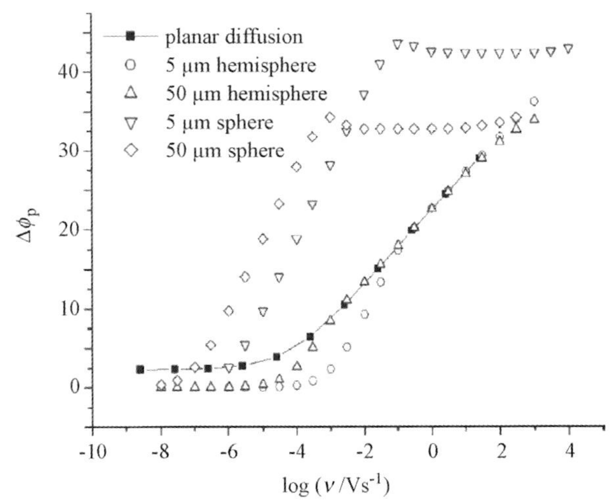

Figure 6.10 Plots of peak-to-peak separation against log ν_{norm} for $k_0 = 1 \times 10^{-5}$ and different sphere radii. Results for spheres and hemispheres are compared with the case for planar diffusion only [33].

higher peak-to-peak separation, $\Delta\phi_p$, for a given ν_{norm}. The hemispherical microparticle exhibits the most Cottrell-like response for the voltammetric (and chronoamperometric; see Ref. [34]) behavior. The peak-to-peak separation tends towards zero for the slowest scan rates, as the immobilized microparticle acts like an adsorbed redox species in a thin layer-like diffusion regime. Decreasing K_0 below a certain value leads to a non-symmetric CV.

The plot of $\Delta\phi_p$ versus log ν_{norm} for microparticles shows that, for higher scan rates, a planar diffusion-like behavior is observed. From the respective ratios with real dimensional variables, the effect of the microparticles size is visible (Figure 6.10).

The peak-to-peak separation of 50 μm and 5 μm microparticle spheres and hemispheres are compared with the case of the planar diffusion. The hemispherical particle exhibits a peak separation of zero at low scan rates, as it behaves as an adsorbed molecule. With increasing scan rates, $\Delta\phi_p$ increases and finally approaches the value(s) for the planar diffusion model. The larger the diameter of the hemispheres and spheres, the earlier the planar diffusion behavior is reached with increasing scan rates, as the curvature of larger particles has a lesser effect on the diffusion.

From the simulated surface coverage profiles and cyclic voltammograms of the three considered microparticle geometries and different ν_{norm} values, the following trends can be seen. There is thin layer-like behavior at low scan rates, symmetrical waves with increasing peak-to-peak separations at increasing scan rates, and asymmetry of the peak current appears, when the coverage of the product species on the sphere (and inverted hemisphere, see below) is no longer roughly uniform to equatorial point, the three-phase-contact point, leading to a reduced back scan current (cf. Figure 6.9). The inverted hemisphere geometry shows near-identical responses as for the sphere,

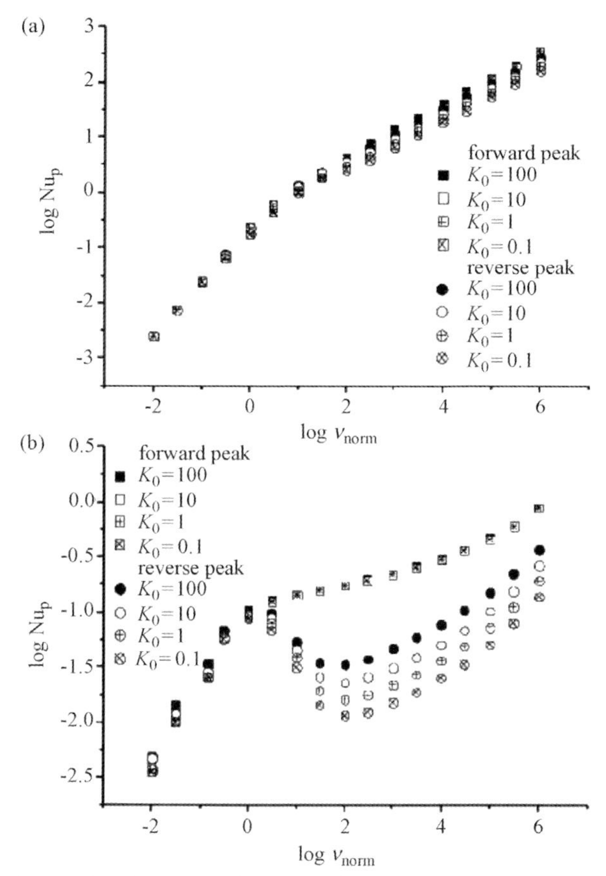

Figure 6.11 Plots of the logarithm of the dimensionless peak current log Nu_p against log ν_{norm} at different K_0 values for (a) the hemisphere case and (b) the full-sphere model [33].

while for the hemisphere there are much smaller planar-diffusion-like peak-to-peak separations, with much higher symmetrical responses.

Plots of the ratio of the logarithm of normalized peak currents (log Nu_p) against the scan rate parameter ν_{norm} for the sphere and hemisphere geometries (Figure 6.11) show that the gradient at low scan rates tends towards unity for both cases, according to the behavior of the (semi-)sphere as a single "adsorbed molecule". For the hemisphere case (Figure 6.11a) the slope changes to 0.5 at higher scan rates, corresponding to a simple planar diffusion.

One important finding of this study is that, for the case of full-sphere geometry (Figure 6.11b), the planar diffusion model is not an appropriate approximation. Rather, the microparticle sphere exhibits a region, where the reverse scan current decreases with scan rate, as the geometry favors a convergent diffusion of charge – that is, towards the equatorial point of the three-phase contact. This causes a "dilution" of the product species near this point at scan rates above the thin layer cell behavior, and the back-scan current is relatively diminished. Higher scan rates will finally outrun this effect, at which point the peak current will again begin to increase.

6.3.2.2 Voltammetry at Random Microparticle Arrays

Solid electrodes, modified with microparticles attached onto the surface, can be assumed as the result of the usual abrasive transfer by rubbing, for example, a graphite rod over nanogram amounts of the respective sample. These adhered microparticles are randomly distributed over the surface, and the whole set-up can be understood and theoretically treated as random array of microdisk electrodes (unlike regular arrays, see: Section 6.3.2.3) (Figure 6.12).

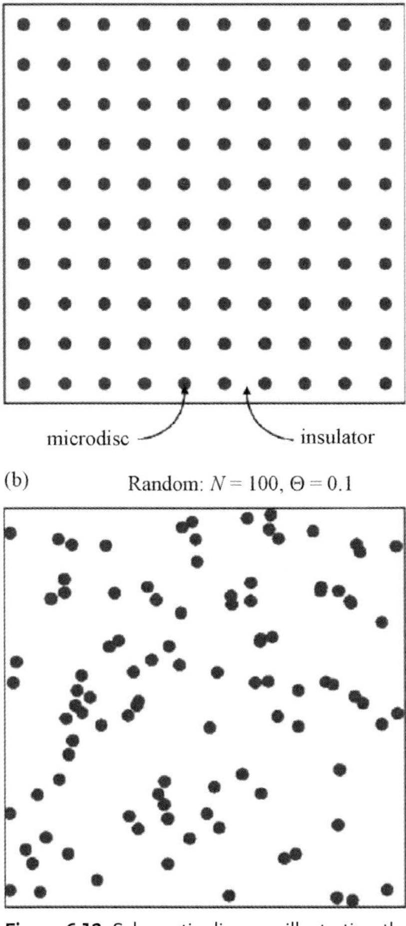

Figure 6.12 Schematic diagram illustrating the difference in distributions between a regular and random array of microdisc electrodes of equal size, R_b, and macroscopic coverage, Θ [35].

All microparticles on the array surface will contribute to the observed current–voltage (I–E) response. To describe the random distribution, the following parameters are defined [35]. The microscopic coverage θ refers to the fractional coverage of an individual diffusion domain (see next section):

$$\theta = \frac{\pi R_b^2}{\pi R_0^2} = \frac{R_b^2}{R_0^2}, \tag{6.11}$$

where R_b is the microparticle (disk) radius and R_0 the domain radius. The active area of a single domain is thus $\theta\pi R_0^2$, or πR_b^2. The macroscopic coverage Θ comprises the fractional coverage of the whole array surface; that is:

$$\Theta = \frac{N\pi R_b^2}{A_{\text{elec}}}, \tag{6.12}$$

where N is the number of microparticles in the array and A_{elec} is the total surface area including the insulating parts (see figure above). Hence, $\Theta\, A_{\text{elec}}$ equals the active array area (i.e., for a regular array of microparticles, $\Theta = \theta$). For a given value of Θ, a random array may consist of a wide range of θ values. When comparing the voltammetric characteristics of random versus regular arrays of microparticles, it can be said that both arrays (e.g., with same Θ) possess differences in voltammetry. A random array will never comprise an I–E response greater in magnitude than that of a regular array of the same macroscopic coverage Θ, but will be, at the utmost, the same magnitude.

6.3.2.2.1 **The Diffusion Domain Approach** The so-called diffusion domain approach was first proposed by Amatore *et al.* [36], and has proved highly useful in several theoretically based reports on this subject to model the diffusion current at those randomly distributed spherical micro- (or nano-) particle arrays [35, 37–39].

The electrode surface can be understood as an ensemble of independent cylindrical diffusion domains of radius r_0 with the respective solid microparticle at the center (Figures 6.13 and 6.14).

These zones are approximated as being cylindrical, with the particle situated at the symmetry axis. If a random spatial distribution of microparticles is assumed, the respective diffusion domains (cylinders) are of different sizes, with a probability distribution function as follows [41]:

$$f(s) = \frac{343}{15}\sqrt{\frac{7}{2\pi}}\left(\frac{s}{\langle s\rangle}\right)^{5/2}\exp\left(-\frac{7}{2}\frac{s}{\langle s\rangle}\right), \tag{6.13}$$

where s is the diffusion domain area (of a specific particle), and $\langle s\rangle$ the mean area (as the average of all particles). The mass transport can be simulated at a range of differently sized diffusion domains, and the current response of each domain is then weighted according to Equation 6.2. The total voltammetric response for the whole array results from a summation of the individual currents. The diffusion domain approach can help to simplify the 3-D diffusion process at/towards the microparticles since, by considering the highlighted plane in Figure 6.14, the experiment can be simulated two-dimensionally.

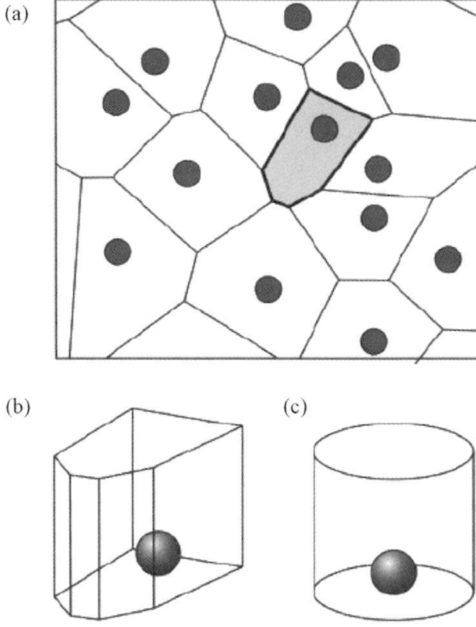

Figure 6.13 (a) Schematic diagram of diffusion domains for a random distribution of spherical nanoparticles (gray circles); (b) The highlighted diffusion domain; (c) A cylindrically approximated diffusion domain [37].

6.3.2.2.2 **The Diffusion Categories** Following a classification in a recent review [42], the situation of the microparticle-modified electrode can be described and treated as that of a spatially heterogeneous, or partially blocked, electrode. The solid macro-electrode is partially covered with microparticles, which – for the sake of theoretically

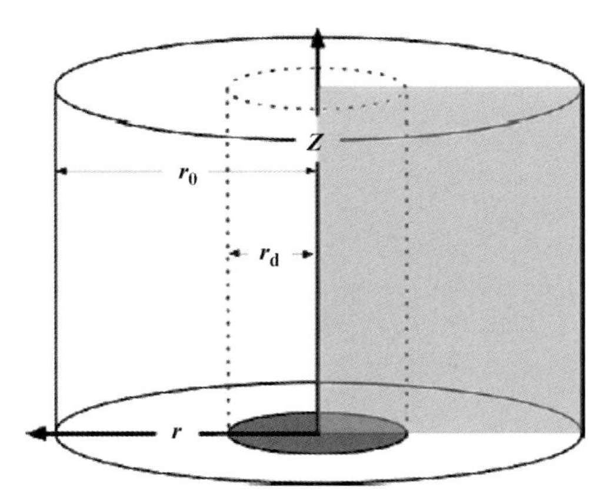

Figure 6.14 Coordinate system used to model the diffusion domain for a cylindrically approximated diffusion domain. The plane to be simulated is shaded [40].

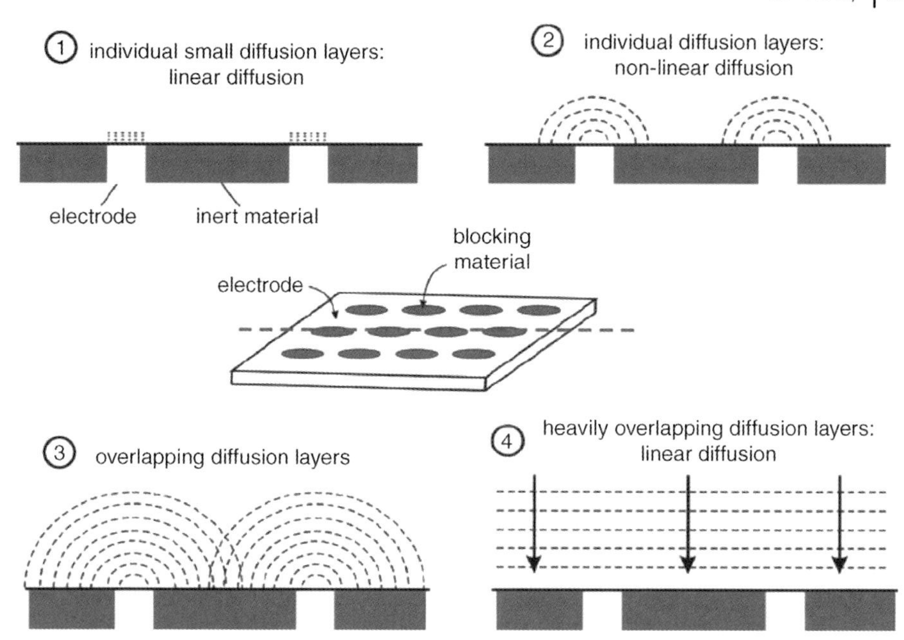

Figure 6.15 Voltammetric behavior of a heterogeneous electrode with active and inert parts, illustrating cases 1, 2, 3, and 4 (see text) [42].

treatability – are assumed to be inert, so that they partially block the electrode towards electrolytic reactions, but subsequently (as in reality) they are assumed to be electroactive. The electrode reaction is supposed to follow Butler–Volmer kinetics. Four main categories can be discussed, if the same total coverage θ of the electrode surface is considered to be realized by microparticle(s) of different numbers and sizes – that is, one "macro-block" particle, or increasingly smaller particles, but a larger number of them.

- Category 1: here, both the blocked and unblocked surface are macroscopically large (cf. Figure 6.15). The unblocked electrode experiences linear diffusion, and the observed voltammograms will be the same as for the unblocked case, but with currents reduced by the factor $(1 - \theta)$.
- Category 2: this comprises electroactive zones of "micro" size, which are separated by sufficiently large inert blocking material. This electrode as a whole behaves as a collection of isolated microelectrodes, each of which exhibits convergent diffusion (radial and axial).
- Category 3: here, the electroactive parts of the macroelectrode behave like microelectrodes. A convergent diffusion regime results from the size, but the scale of the insulating parts is sufficiently small, so that the adjacent diffusion layers begin to overlap with each other.

- Category 4: this comprises heavily overlapping diffusion layers of the electroactive "microelectrode"-sized zones; thus, the heterogeneous electrode as a whole behaves almost like an unblocked electrode. This limiting case was discussed earlier by Amatore who concluded, in accordance with these results, that linear diffusion characteristics – albeit with an apparently reduced electrochemical rate constant $k_0(1-\theta)$ – appears at a partially blocked electrode of coverage θ [36].

Categories 1 and 4 exhibit both the characteristics of linear diffusion, and consequently the simulation or fitting of the voltammetric responses can be performed with available simulation/fitting tools (e.g., Ref. [43]), according to and based upon the theory outlined in Section 6.3.1. Electrodes following category 2 respond as an array of diffusionally independent microelectrodes, while category 3 must include the beginning overlap of adjacent microelectrode diffusional fields. Both features must be considered for simulation of the voltammetric signals. The diffusional distance, δ, as derived from the Einstein equation:

$$\delta = \sqrt{2Dt} \tag{6.14}$$

or, considering the potential width of the respective cyclic voltammograms, that is,

$$\delta = \sqrt{2D\frac{\Delta E}{\nu}}, \tag{6.15}$$

can be compared with the size of the microparticles on the electrode surface, to index (assign) the electrochemical experiment under consideration into the given model categories 1 to 4. In another theoretical study conducted by the same group [37], simulated concentration profiles of the four model categories were given (Figure 6.16; here, σ_{SR} refers to the dimensionless scan rate, and R_0 to the diffusion domain size).

This approach has been tested on regular microdisk arrays – that is, arrays of microdisks separated from each other by a distance of 10 μm or more [42]. However, this holds true for both random assemblies of microelectrodes (RAM electrodes) as well as electrochemically investigated microelectrodes (e.g., by Fletcher and Horne [44]), as a wide enough spacing between microdisks guarantees resistive and diffusive independence. Figure 6.17 shows experimental cyclic voltammograms from the oxidation of aqueous potassium hexacyanoferrate (II) at a gold microdisk array comprising a cubic distribution of 10 μm-diameter disks with center-to-center nearest-neighbor separations of 100 μm, as well as the respective simulated signals.

From the block radius of 45 μm, the δ values calculated from Equation 6.15 are, for example, 125 μm and 80 μm for scan rates of 10 and 25 mV s^{-1}, respectively, and therefore the electrode can be assigned to category 3. For 100 mV s^{-1} (and above) however, with $\delta = 40$ μm, this electrode is showing category 2 behavior. The scan rate behavior is changing accordingly.

Streeter *et al.* used the model system of glassy carbon microspheres each of 15 ± 5 μm radius, covered with Pd, and abrasively transferred and attached onto the surface of a basal plane pyrolytic graphite (BPPG) electrode. In this way, it was possible to compare these experiments with numerical simulation results of the discussed main diffusional regimes of microparticle modified electrodes [37]. The

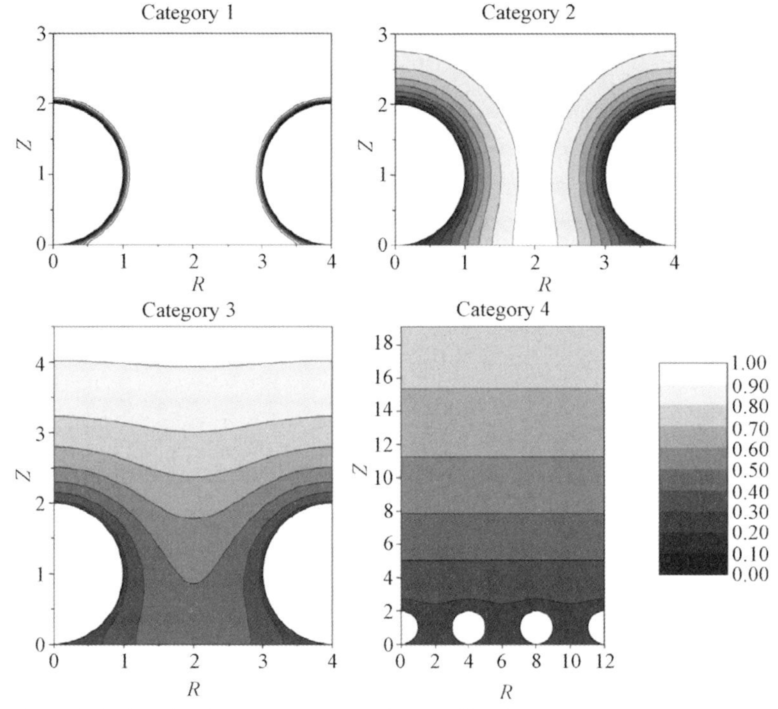

Figure 6.16 Simulated concentration profiles at a diffusion domain containing a spherical particle. Category 1: $\sigma_{SR} = 1000$. Category 2: $\sigma_{SR} = 10$. Category 3: $\sigma_{SR} = 1$. Category 4: $\sigma_{SR} = 0.01$. For all categories $R_0 = 2$. Concentration profiles were taken at the linear sweep's peak potential [37].

experimental cyclic voltammograms for electrodes with three different fractional surface coverages

$$\Theta = \frac{N\pi r_S}{A} \tag{6.16}$$

(where r_S is the spherical particle radius, A is the area of the supporting electrode, and N the number of spherical particles present on its surface) were compared in Figure 6.18. The respective reaction is the electrocatalytic reduction of protons at the Pd surfaces. Note that the electron transfer at these applied potentials takes place at the interface between the microsphere and the solution, and *not* at the electrochemically inert BPPG.

Different surface densities and different scan rates were studied. The cyclic voltammograms in Figure 6.18a, with a surface coverage of 0.445, refer to the diffusion behavior expected for category 4. That is, the high particle density causes a considerable overlapping of neighboring diffusion zones to form a near-linear planar net diffusion layer. The respective dependence of peak current, i_p, versus square root of scan rate, ν, is linear, supporting this assumption. The respective plot for Figure 6.18c of $\Theta = 7.64 \times 10^{-3}$ deviates from linearity, especially at higher scan

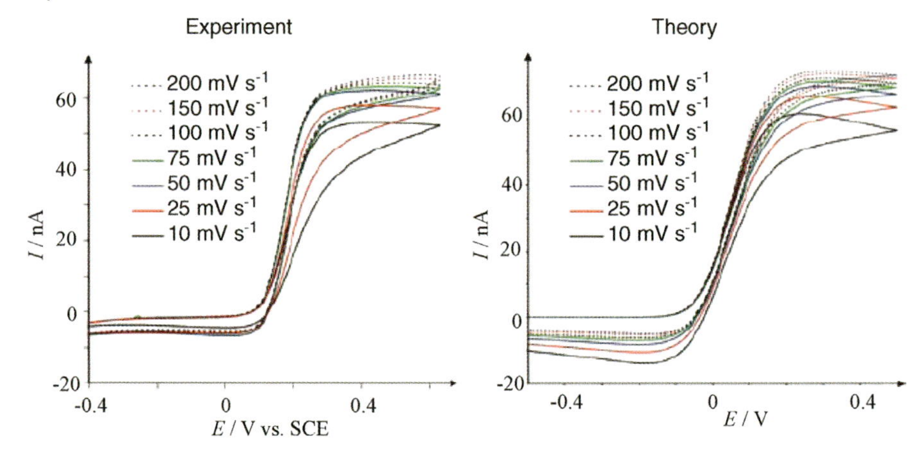

Figure 6.17 The simulation of microdisk arrays addressing the issue of diffusional independence. Shown are experimental voltammetric results obtained from the oxidation of 1 mmol l^{-1} ferrocyanide in 0.1 mol l^{-1} KCl at a gold microdisc array comprising a cubic distribution of 10 μm-diameter discs with center-to center nearest-neighbor separations of 100 μm. Scan rates used (from bottom to top of the curves) were 10, 25, 50, 75, 100, 150, and 200 mV s^{-1}. [42].

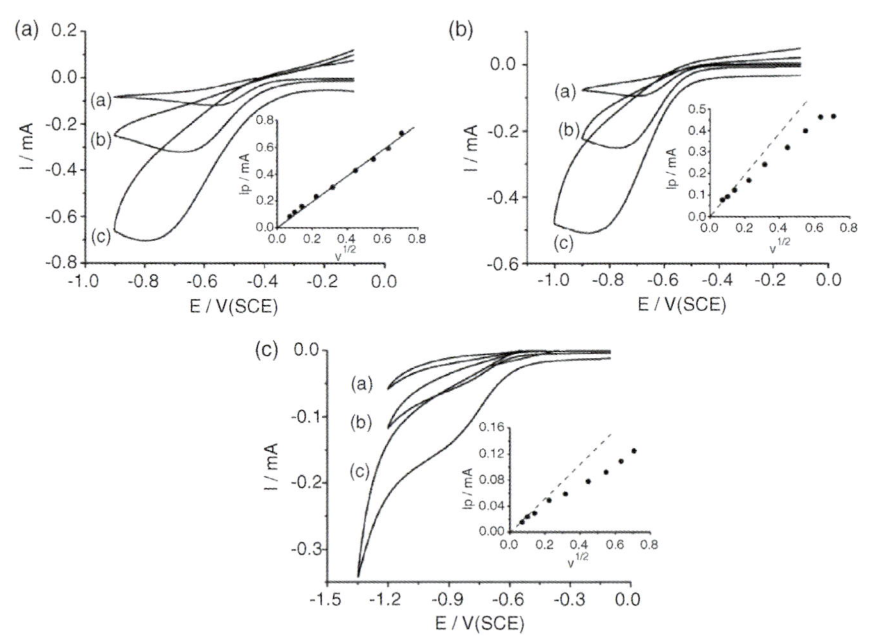

Figure 6.18 Cyclic voltammetry in 3 m*M* HCl, 0.1 *M* KCl at a 5 mm-diameter BPPG electrode modified by the abrasive attachment of different amounts of Pd-CMs. The fractional surface coverage of particles was: (a) 0.445; (b) 0.118; and (c) 7.64×10^{-3} cm^2. For each graph, the curves (a), (b), and (c) correspond respectively to voltammograms recorded at 10, 100, and 500 mV s^{-1} [37].

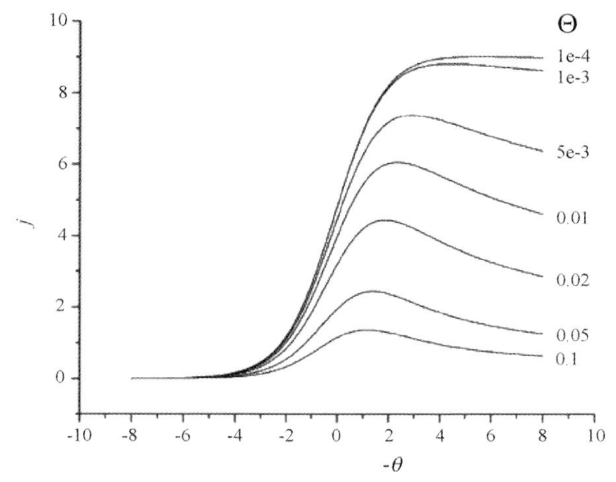

Figure 6.19 Simulated linear sweep voltammograms of a reversible electron transfer at a spherical particle modified electrode. Scan rate $\sigma_{SR} = 0.01$; Θ varies from 10^{-4} to 0.1 [37].

rates; moreover, the peak shape suggests near-steady-state diffusion. The diffusional behavior can thus be assigned to category 2 – that is, diffusionally independent microparticles, and with a much less significant overlapping of diffusion layers. Figure 6.18b, where $\Theta = 0.118$, finally comprises category 3 diffusion, when the plot shows a deviation of linearity, thus from overall planar diffusion, but the transient voltammograms suggest considerable overlap of diffusion zones.

For the numerical simulation,[2)] a single step (fast and reversible) electron transfer at the spherical particle surfaces has been assumed, with surface concentrations following the Nernst equation [37]. The microparticles were modeled as being randomly distributed on the supporting planar surface, such that the diffusion zones of the individual microparticles are allowed to overlap. The respective flux–potential plots for different values of Θ show at $\Theta = 10^{-4}$ and below, almost the same response as for an isolated spherical particle (see category 2). The wave shape approximates that of steady-state diffusion, while the diffusion limiting current can be given as:

$$i_{\lim} = -8.71 nFDr_S c_b, \tag{6.17}$$

where c_b is the bulk concentration and r_s is radius of the microparticle.

With increasing particle density, the simulated voltammograms decrease in magnitude and become more peak-shaped (Figure 6.19). The diffusional behavior changes into category 3. For very high surface coverage, the diffusional regime of category 4 is active. For the here-assumed reversible electron transfer, the current can be given by the Randles–Ševčik equation.

In conclusion, the highest total current output – from both theoretical and experimental standpoints – is observed with a high surface coverage of

2) The numerical simulation for this study is based on the alternating direction implicit finite difference method, using an expanding simulation grid with highest mesh density in the region of the particle [37].

Table 6.1 Linear sweep and cyclic voltammetry characteristics associated with the four categories (see text), where δ is the size of the diffusion zone, R_b is the microdisc radius, d is the center-to-center separation, I_p is the peak current, I_{lim} is the limiting current, and ν is the scan rate [35].

	Category			
Property	**1**	**2**	**3**	**4**
δ vs. R_b	$\delta < R_b$	$\delta > R_b$	$\delta > R_b$	$\delta > R_b$
δ vs. d	$\delta < d$	$\delta < d$	$\delta > d$	$\delta >> d$
Type of response	Clear peak $\rightarrow I_p$	Steady state $\rightarrow I_{lim}$	Slight peak to clear peak $\rightarrow I_p$	Clear peak $\rightarrow I_p$
Scan rate dependence?	Yes	No	Yes	Yes
Current dependence	$I_p \propto \nu^{0.5}$	$I_{lim} \propto R_b$	—	$I_p \propto \nu^{0.5}$

microparticles, corresponding to a net planar diffusion regime (category 4), whereas the highest rates of mass transport to a single microparticle sphere were found with a low surface coverage (category 2). The cyclic and linear sweep voltammetric characteristics associated with diffusion categories are summarized in Table 6.1 [35].

6.3.2.2.3 **Voltammetric Sizing** The average size of inert microparticles deposited on an electrode surface has been determined using cyclic voltammetric measurements [45]. This method was proposed to be used for smaller particles (i.e., if optical microscopy is not possible) as an inexpensive alternative solution to scanning electron microscopy (SEM).

The microparticles are assumed to cover a solid electrode surface, such as edge-plane graphite. A simple one-electron redox system:

$$A + e^- \xrightarrow{k_{0,\alpha}} B \tag{6.18}$$

is considered, where k_0 is the electron transfer rate constant and α is the symmetry coefficient. Experimental cyclic voltammograms are recorded on a redox couple present in solution at different scan rates, for example, between 0.2 and 2 V s^{-1}, and are compared with cyclic voltammograms from digital simulations. The diffusion coefficients of the A and B components (D_A and D_B, respectively) and kinetic parameters k_0 and α can thus be determined. When an inert material is distributed over the electrode surface, nonlinear diffusion will act during the experiment, resulting in a depletion above the blocking material; that is, a (partially) blocked electrode is present (see Section 6.3.2.2.2). The size of the blocks compared to the size of diffusion layer thickness δ defines the extent of this depletion. Therefore, the size of microparticles and their total area affect the resulting cyclic voltammogram, and their values – provided the total mass and density are known – are accessible.

6.3.2.3 **Voltammetry at Regularly Distributed Microelectrode Arrays (Microarrays, Microbands)**

Girault *et al.* earlier proposed cyclic voltammetry simulation for a regular microdisk array, based upon a 3-D cubic packing geometry [46]. Later, studies performed by Davies and Compton led to the proposal of a theoretical treatment for cyclic and linear sweep voltammetry of regular (as well as random) arrays of microdisk electrodes, following the diffusion domain approach (see Section 6.3.2.2.1) [35]. Of special attention here were the discussion of the diffusion zone(s) and its overlap (see Section 6.3.2.2.2) at microdisk arrays, in order to provide conditions for the fabrication of microdisk arrays. The center-to-center separation d between individual microparticles (i.e., microelectrodes) is highly relevant to the voltammetric characteristics of the array electrode. Preferably, d must be sufficiently large so as to avoid the diffusion zone overlap with a following "shielding effect" (see Section 6.3.2.2.2), but not too large as to result in an unnecessary waste of surface area. It can be concluded, that category 2 is the most common and most favored category associated with microelectrode arrays, as this results in the highest faradaic current/background current ratio. Category 2 type arrays exhibit scan rate-independent voltammograms with steady-state characteristics.

Davies and Compton simulated[3] a regular array of microelectrodes, varying the diffusion domain approach, as illustrated in Figure 6.20.

In view of the above-mentioned points, the offset of category 2/onset of category 3 behavior (beginning overlap of diffusion zones) is a crucial criterion in the voltammetry of regular arrays of microparticles/microelectrodes. Usually, a linearly dependent expression on the microparticle disk radius R_b for the size of the diffusion zones is given and used, for example, $d > 20\ R_b$ [47]. Davies *et al.*, in contrast, proposed a condition to ensure diffusion of category 2 behavior as follows [35]:

$$d > 2\sqrt{2D\frac{\Delta E}{\nu}}, \tag{6.19}$$

(see case C3 in Ref. [35]). The limiting current, I_{lim}, is calculated as follows:

$$I_{\text{lim}} = 2\pi R_b F D_A [A]_{\text{bulk}}, \tag{6.20}$$

with all symbols having their usual or previously defined meanings. Upon reaching the limiting current plateau, the Nernst diffusion layer thickness δ_N is only dependent on R_b (not on D or ν). Independent of the size of individual microdisks, a center-to-center separation of at least 560 μm would be required for example parameters: $D = 10^{-5}\ \text{cm}^2\,\text{s}^{-1}$, and $\nu = 5\ \text{mV}\,\text{s}^{-1}$.

The dependence of the microparticle (microdisk) size on respective linear sweep voltammograms is illustrated in Figure 6.21.

With decreasing R_0, or equivalently decreasing d, the voltammograms for each set-up changes: the magnitudes decrease, and limiting currents gradually are replaced by peak-shaped curves. That is, the diffusion category changes from type 2 to 3.

3) For simulation, a finite difference-based approach with expanding grid has been employed.

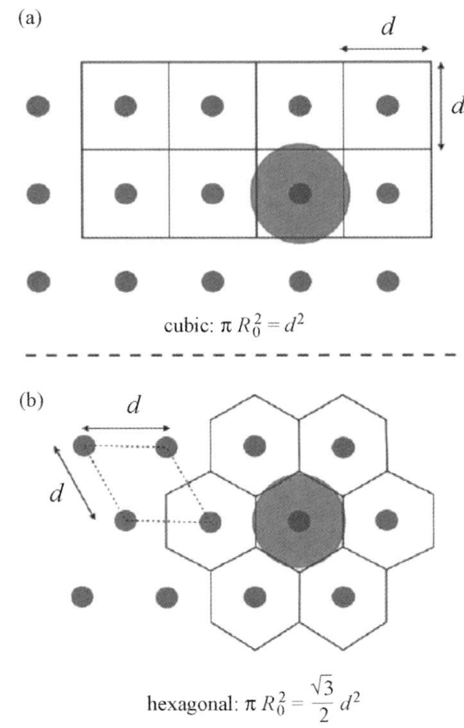

Figure 6.20 (a) Cubic and (b) hexagonal arrays of microdisk electrodes, with their corresponding diffusion domains [35].

Moreover, a category 3 to 2 transition can be found in Figure 6.21b at $R_0 = 1.5\,\mu m$. The "critical" value of R_0, below which transition to category 3 diffusion occurs, is highly dependent on the microparticle disk size.

The scan rate dependence is shown for two regular arrays of 1 μm microdisks, respectively, but with different values of R_0 (or center-to-center distance d).

Whereas, Figure 6.22a shows clearly category 2 behavior for all scan rates, the set-up in Figure 6.22b allows a transition into category 3 (or probably category 4) diffusion by decreasing the scan rate. The timescale of the experiments increases, which results in larger diffusion zones, that begin to show "shielding"/overlapping effects. In other words, a sufficiently high scan rate will ensure category 2 type voltammetry with a limiting current, as given above.

Finally, the diffusion coefficient D also affects the voltammetric characteristics, as its decrease will significantly reduce the area of the respective diffusion zone, and *vice versa*.

6.3.2.4 **The Role of Dissolution in Voltammetry of Microparticles**

During a solid-state electrochemical reaction of microparticles (but also of adhered films, solid paste electrodes, etc.), the process may not be totally confined to the solid state, but be accompanied by dissolution processes. This may lead to typical voltammetric features of both, solid-state and solution-phase reactions. Schröder

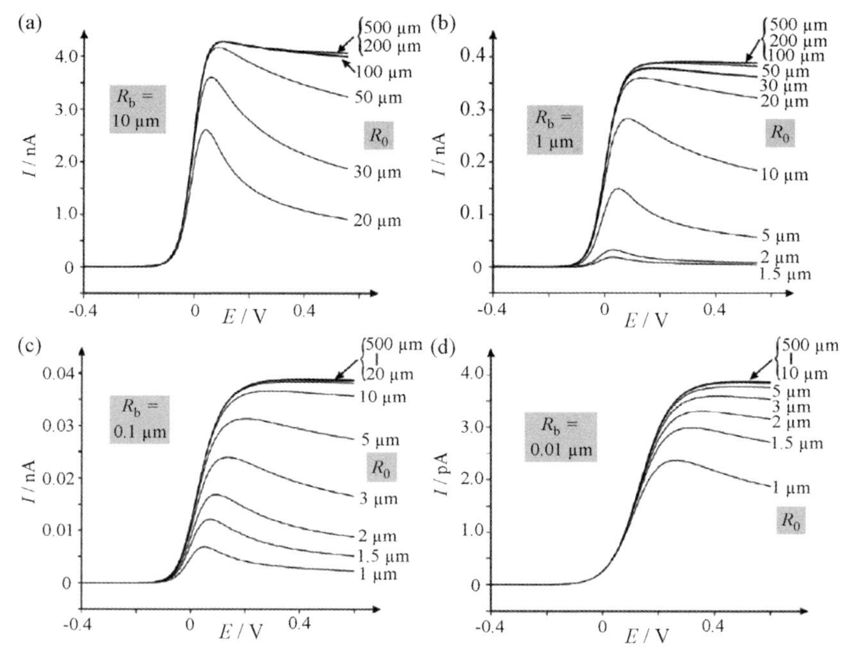

Figure 6.21 Simulated linear sweep voltammograms for diffusion domains (see text). The system parameters are $\nu = 0.1\,V\,s^{-1}$, $D = 10^{-5}\,cm^2\,s^{-1}$, $k^0 = 1\,cm\,s^{-1}$ and $[A]_{bulk} = 1\,mM$. Parts (a)–(d) correspond to domains containing discs of radius R_b = (a) 10 μm, (b) 1 μm, (c) 0.1 μm, and (d) 0.01 μm. In all four parts the domain radius, R_0, is varied from 500 to 1 μm (where appropriate) [35].

and Scholz referred to a dissolution, reprecipitation and ion-exchange mechanism in the context of studies on hexacyanoferrate-, octacyanotungstate-, and octacyanomolybdate- adhered microparticles [48]. For the voltammetry of C_{60} microparticles adhered to an electrode and in contact with dichloromethane, Bond *et al.* proposed a mechanism involving the slow dissolution of adhered C_{60}, but a rapid dissolution of the reduced form C_{60}^{-} [49]. In a recent report, the same group described systematically the influence of dissolution using microparticles adhered to electrode surface and in contact with several aqueous electrolyte solutions, in which variable levels of solubility were given [50]. Microparticles of TCNQ and $[PD][PF_6]_3$[4)] have been studied in contact with aqueous electrolyte solutions, but also with ionic liquid in addition to (or instead of) the aqueous solution, thus giving rise to different reaction pathways due to different solubility conditions [50].

It can be stated that the nature of the voltammetric response is very sensitively dependent on the solubility of the solid phase, in either (or both) oxidized and reduced forms, and even to least extent. The dissolution process may be associated

4) 7,7,8,8-Tetracyanoquinodimethane and 1,3,5-tris(3-pyridiniumferrocenylmethylamine)-2,4,6-triethyl benzene hexafluorophosphate, respectively.

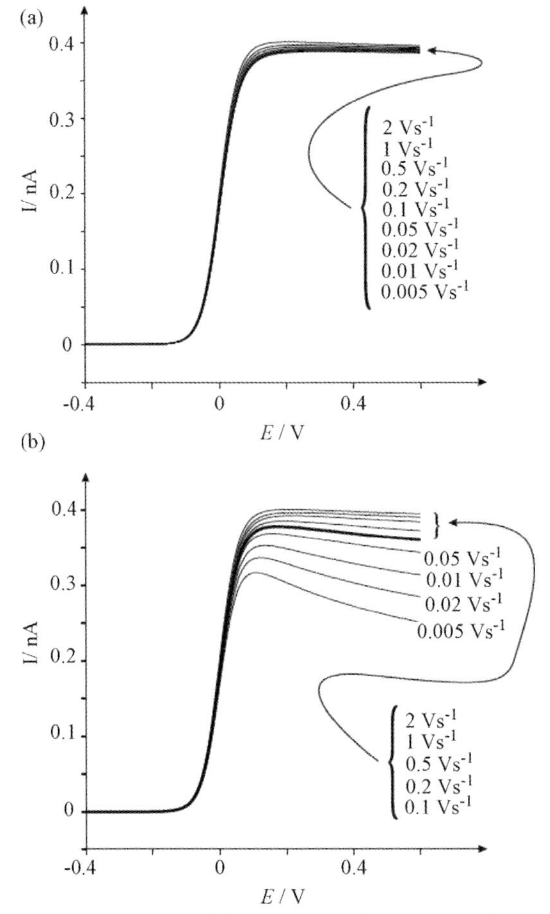

Figure 6.22 Simulated linear sweep voltammograms for a microdisk of radius $R_b = 1\,\mu m$ in a diffusion domain of radius: (a) $R_0 = 200\,\mu m$ (effectively a single unshielded microdisk); and (b) $R_0 = 30\,\mu m$. $D = 10^{-5}\,cm^2\,s^{-1}$, $k_0 = 1\,cm\,s^{-1}$ and $[A]_{bulk} = 1\,mM$. In both parts the scan rate is varied from 0.005 to $2\,V\,s^{-1}$, as indicated [35].

with enhanced currents, or (other) changes of voltammetric characteristics. Depending on the respective solubility conditions, the diffusion regime may change, although multiple reaction pathways may also occur.

6.3.3 Voltammetric Stripping of Electroactive Microparticles from a Solid Electrode

6.3.3.1 Microparticles Within a Carbon Paste Electrode

An early attempt to explain, on a theoretical basis, the stripping voltammetric (anodic stripping voltammetry; ASV) response of, for example, electroactive metal microparticles within a carbon paste electrode was made by Brainina *et al.* [51, 52]. Here, the metal microparticles (powder) are considered to be uniformly distributed within the carbon paste, which is assumed as a continuous, conducting medium. For both reversible and irreversible conditions, Butler–Volmer kinetics and uniform planar diffusion over the whole electrode were assumed. The peak potential of the stripping process under electrochemically reversible conditions is given as

$$E_p = E^0 + \frac{RT}{F} \ln c_i^0 d_p v^{1/2} + \text{const.}, \quad (6.21)$$

where c_i^0 is the initial concentration of metal particles to be stripped, within the upper surface layer of the electrode (in mol cm^{-3}), and d_p is the particle diameter (in cm).

The case of electrochemically irreversible stripping using the planar diffusion model is consistent with the results of a recent study [40] (see also Section 6.3.3.2).

6.3.3.2 Microparticles on a Solid Electrode Surface

A recent report outlined the theoretical treatment of a solid three-phase electrode approach (as detailed above), describing the anodic or cathodic voltammetric stripping process associated with the respective solid-state reaction of microparticles in the micrometer range [40]. The stripping process – that is, dissolution of the solid deposit by a cathodic or anodic sweep – was generally treated to form a consistent theory on three-phase electrodes such as abrasive-treated solid electrodes, and therefore also including many frequently used solid materials/electrodes such as edge plane pyrolitic graphite or boron-doped diamond electrodes.

The stripping phase has been modeled using a two-dimensional (2-D) simulation approach.[5] For this model, each solid particle on the electrode surface is assumed to have a cylindrical form with minimal height, thus comprising "flat disks" (see Figure 6.23).

The electrode surface A_{elec} was assumed to contain N electroactive metal or metal oxide centers, respectively, which can be not only uniformly but also (mimicking more realistic experimental conditions) randomly distributed; an example is the results of atomic force microscopy (AFM) studies on microparticle electrodes [53]. Here, the diffusion domain approach (as described in Section 6.3.2.2.1) has been employed; that is, the electrode surface is assumed to be an arrangement of independent diffusion domains of radius r_0. If all particles are of the same radius, r_d, but are distributed in a random manner, then a distribution of diffusion domains with different domain radii, r_0, follows. The local position-dependent coverage is given by Γ. The electroactive microparticle "flat disks" of the radius r_d are located in the center of the respective diffusion domain cylinder. The simulated (linear sweep voltammetric) reaction follows a one-electron transfer, and species B is stripped from the electrode surface into the solution, forming A, or:

$$\mathrm{B(s)} \pm \mathrm{e}^- \rightarrow \mathrm{A(aq)} \quad (6.22)$$

5) Fully implicit finite difference, combined with Newton's method, accompanied by Thomas' algorithm.

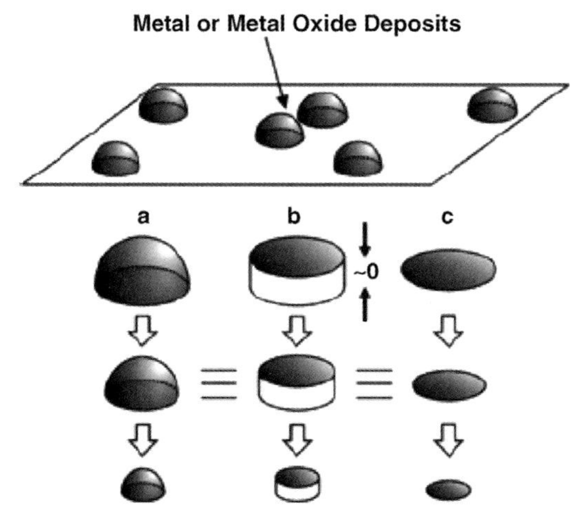

Figure 6.23 Schematic diagram showing the model for the stripping phase of anodic and cathodic stripping voltammetry [40].

Three mathematical models were considered. For model A, a "monolayer" system is considered; that is, the coverage of the electrode by the solid particle centers is to be uniform and, at a specific point, cannot exceed a maximum value Γ_{max}. Model B assumes a "thin-layer" system, but with no constrictions on the maximum electrode surface coverage. Model C finally considers a "thick-layer" case, such that the stripping or dissolution of one of multiple layers will mostly leave over a layer underneath.

The case of irreversible stripping conditions for all three models conforms to Brainina's planar diffusion model approach [51, 52]. The peak potential can be given for a one-electron transfer by:

$$E_p = E^0 + \frac{RT}{(1-\alpha)F}\ln\frac{(1-\alpha)F\Gamma_{max}}{RTk_d} + \frac{RT}{(1-\alpha)F}\ln\nu. \tag{6.23}$$

This is not dependent on the size of domains – that is, the size distribution of the diffusion domains. However, by assuming an electrochemically reversible stripping, E_p is given by (adapted from Refs [51, 52] to this simulation model):

$$E_p = E^0 + \frac{RT}{F}\ln\frac{\Gamma_{max}r_d^2}{r_0^2}\nu^{1/2} + \text{const.} \tag{6.24}$$

Here, in the micrometer size range, the hemispherical (convergent) diffusion becomes predominant, and different peak potentials are received (see Figure 6.24).

Microscopically, the peak potential increases with metal (microparticle) center radius r_d, but not (as found by Brainina [51]) linearly with $\ln(r_d)$. As can be explained on the basis of the simulated concentration profiles, this is due to the nature of diffusion; that is, with increasing r_d the spherical nature of diffusion decreases in favor of the planar one. The resultant slower transport provides a better chance for a re-deposition, thus shifting E_p to more positive values (Figure 6.25).

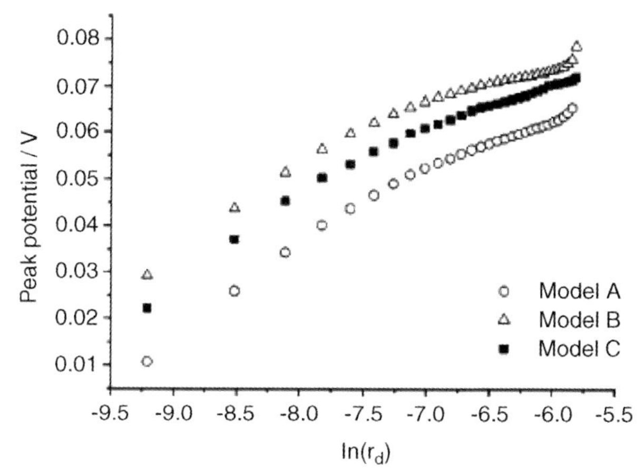

Figure 6.24 Plot showing the variation in peak potential with increasing metal center radius, r_d, within a fixed domain with $r_0 = 3 \times 10^{-3}$ cm, $\nu = 0.1\,V\,s^{-1}$, $D = 1 \times 10^{-6}\,cm^2\,s^{-1}$, $k_l = 1\,cm\,s^{-1}$, and $\Gamma_{max} = 1 \times 10^{-5}\,mol\,cm^{-2}$ [40].

The scan rate dependence on peak potential makes it clear that, at high microscopic coverage, a near-linear dependence of E_p on $\ln(\nu)$ applies, with the slope approaching $RT/2F$. At low θ, however, a hemispherical diffusion becomes predominant. With higher scan rates, the diffusion layer will become thinner and the diffusion regime more planar. Adjacent diffusion domains will then overlap with each other.

The result is that the present authors' model and Brainina's earlier assumption become increasingly identical and will lead to the same results: (i) at high microscopic coverage; (ii) with large metal (microparticle) center radii; and/or (iii) at slow scan rates. In other words, the role of spherical diffusion becomes more important if the metal (microparticle) center radii and the microscopic coverage are reduced. Spherical diffusion leads to non-uniform local coverage (σ_{LC}), while on the metal center

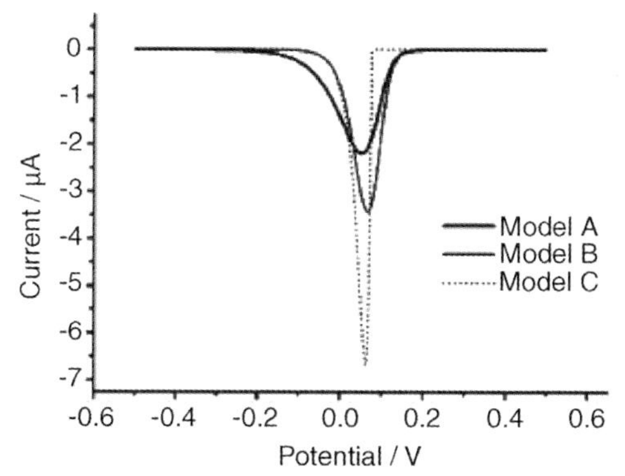

Figure 6.25 Example stripping voltammogram for each model. In all cases, $r_d = 1 \times 10^{-3}$ cm, $r_0 = 3 \times 10^{-3}$ cm, $\nu = 0.1\,V\,s^{-1}$, $D = 1 \times 10^{-6}\,cm^2\,s^{-1}$, $k_l = 1\,cm\,s^{-1}$, and $\Gamma_{max} = 1 \times 10^{-5}\,mol\,cm^{-2}$ [40].

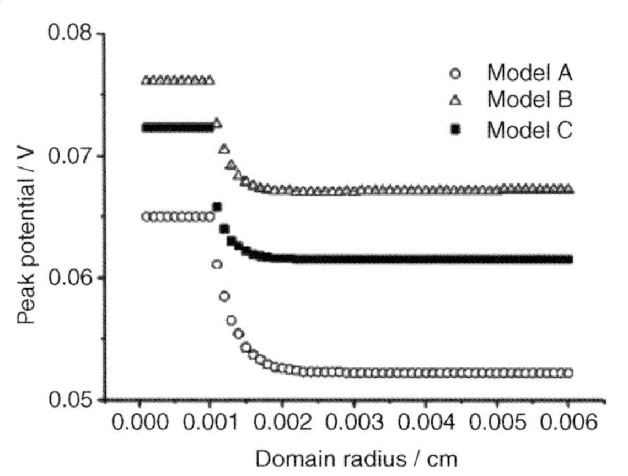

Figure 6.26 Plot showing the variation in peak potential with increasing domain radius r_0. In all cases, $r_d = 1 \times 10^{-3}$ cm, $v = 0.1\,\mathrm{V\,s^{-1}}$, $D = 1 \times 10^{-6}\,\mathrm{cm^2\,s^{-1}}$, $k_l = 1\,\mathrm{cm\,s^{-1}}$, and $\Gamma_{max} = 1 \times 10^{-5}\,\mathrm{mol\,cm^{-2}}$ [40].

surface a "shrinking disk" effect takes place – that is, the material is stripped faster from the edge of the metal center than from the middle.

In Figure 6.26, the increasingly predominant planar (and perpendicular to the electrode surface) diffusion behavior is visible, as θ tends towards 1, such that the peak potential E_p for all microparticle radii tends towards the same value, as holds true for the macroelectrode.

The general trend for the case of a random distribution of particles on the electrode surface (i.e., planar electrode surface A_{elec}, randomly covered with microparticles) is as for microscopic domains (the following highlights the dependence of global coverage Θ on peak potential E_p; see Figure 6.27). With increasing global coverage, the peak potential increases, and tends finally to that of a macroelectrode. The current

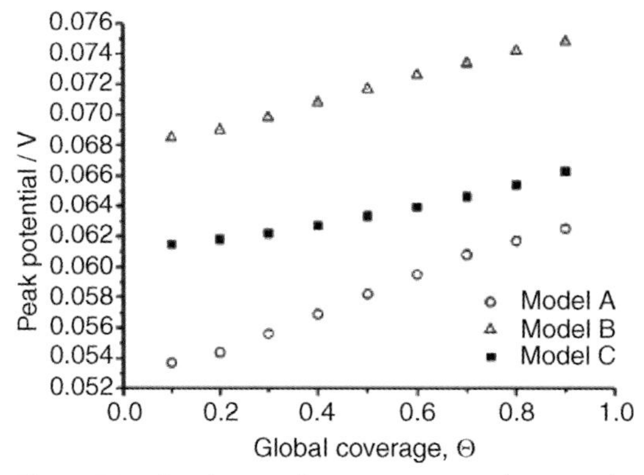

Figure 6.27 Plot showing the variation in peak potential with increasing global coverage, Θ, for models A, B, and C. In all cases, $r_d = 1 \times 10^{-3}$ cm, $v = 0.1\,\mathrm{V\,s^{-1}}$, $D = 1 \times 10^{-6}\,\mathrm{cm^2\,s^{-1}}$, $k_l = 1\,\mathrm{cm\,s^{-1}}$, $\Gamma_{max} = 1 \times 10^{-5}\,\mathrm{mol\,cm^{-2}}$, and $A_{elec} = 1\,\mathrm{cm^2}$ [40].

is also changing linearly with Θ, as the total metal (microparticle) area on the electrode surface will be changed.

A study to compare these simulation results [40] with experimentally acquired data has been reported, where the "monolayer" model A discussed above was applied to the (under potential deposition and) stripping of a monolayer of lithium at platinum polycrystalline and monocrystalline [(1 0 0), (1 1 0), and (1 1 1)] electrodes in tetrahydrofuran [54].

The specific case of stripping from a random array of microdisks electrode – that is, microdisks which are randomly arranged as part of the electrode surface – was treated on a theoretical basis by Fletcher and Horne [44] (see also Section 6.3.2.2).

6.3.4 Voltammetry of Single Microparticles (Microcrystals, Nanocrystals) on Solid Electrodes

The following results are applicable to any electrode modified with a sparse distribution of microparticles. The mass transport to a single, diffusionally independent microparticle can (in theory) be treated on an equal basis as a microparticle within an independent diffusional zone in the experimental time scale with respect to its neighbors. Therefore, many theoretical results produced for microparticle arrays of diffusional categories 1 and 2 (see Section 6.3.2.2.2) are also valid for single particles.

Based on the results of numerical simulations, Streeter and Compton [55] discussed the current response from different microparticle (or nanoparticles) of various shapes, including hemispheres, distorted spheres, and distorted hemispheres (see Figure 6.28). Here, the planar electrode beneath the particle is assumed to be conducting, and there is a point (or line) of electrical three-phase contact. However, the electrode is also electrochemically inactive, and electron transfer is feasible on the particle surface only. The current is simulated numerically[6] by solving the mass transport equations in cylindrical polar coordinates,[7] which is applicable (in principle) to any curved particle with an axis of symmetry, thus including spheroid and hemispheroid shapes. Its shape is described by a dimensionless parameter B_L, which is the ratio of the lengths b and a (see Figure 6.28). Mass transport obeys Fick's second law of diffusion, assuming an n-electron transfer reaction: $A + ne^- \rightleftarrows B$, where the A and B species are soluble.

Hence, the analytical expression for diffusion-limiting current at a hemispherical electrode is given by [55]:

$$i_{lim} = -2\pi nFDr_s[A]_{bulk}, \qquad (6.25)$$

(with all symbols having their usual meanings). The results of a numerical simulation in this study have been compared to (and confirmed by) that of Equation 6.25. For a

6) Alternating direction implicit finite difference method in conjunction with the Thomas' algorithm has been applied.

7) Alternatively, this system could be modeled by solving the diffusion equations in spherical radial coordinates [56].

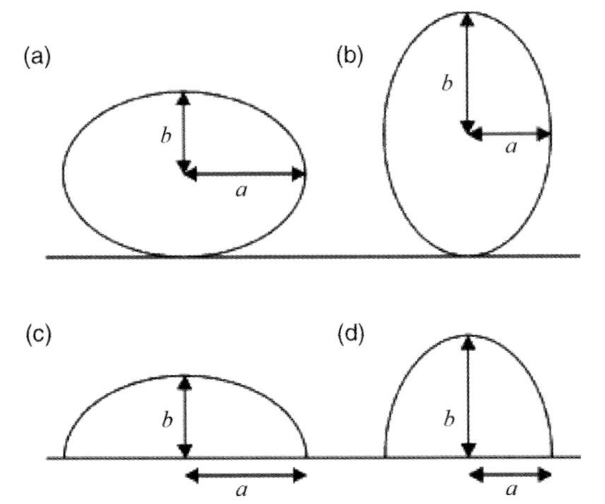

Figure 6.28 Various shaped particles on a supporting planar surface. (a) Oblate spheroid; (b) Prolate spheroid; (c) Oblate hemispheroid; (d) Prolate hemispheroid [55].

spherical particle, the diffusion-limiting flux was found to be:

$$i_{\lim} = -8.71nFDr_S[\mathrm{A}]_{\mathrm{bulk}}, \tag{6.26}$$

and for spheroids and hemispheroids:

$$i_{\lim} = -jnFD[\mathrm{A}]_{\mathrm{bulk}}\sqrt{\frac{a^2+b^2}{2}}. \tag{6.27}$$

Figure 6.29 shows some example linear sweep voltammograms assuming different scan rates (σ_{SR} refers to the dimensionless scan rate; $\sigma_{SR} = (F/RT)(\nu r_S^2/D)$). As the experimental time scale decreases, the diffusional behavior changes from near-steady-state to near-planar diffusion. With respect to the different shapes of microparticles, the mass transport-limiting current was found to be fairly consistent; that is, a difference of less than 2% for sphere and hemispheres of equal surface area.

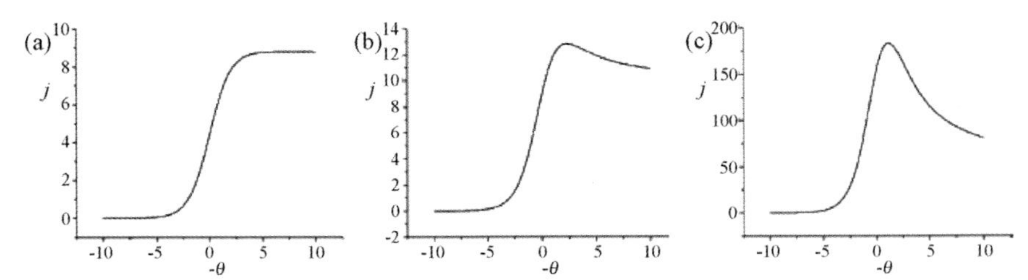

Figure 6.29 Simulated voltammetry for a reversible electrode transfer at the spherical particle. The following scan rates are used: (a) $\sigma_{SR} = 10^{-3}$; (b) $\sigma_{SR} = 1$; (c) $\sigma_{SR} = 1000$ [55].

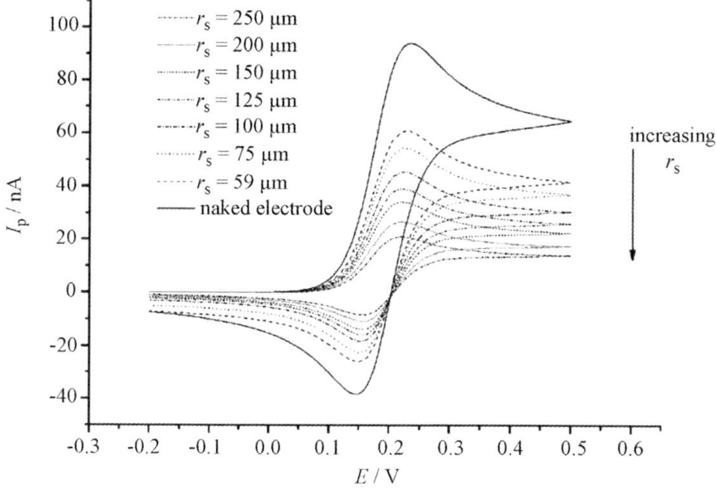

Figure 6.30 Simulated voltammograms for different radii of the sphere positioned in the center of the electrode. For the simulation the following parameters are used: $[A]_{bulk} = 3 \times 10^{-6}$ mol cm^{-3}, $r_e = 59$ μm, $\nu = 0.1$ V s^{-1}, $\alpha = 0.5$, $D = 0.63 \times 10^{-5}$ cm^2 s^{-1}, $k_0 = 0.05$ cm s^{-1}, $E_{start} = -0.2$ V, $E_{stop} = 0.5$ V, and $E_0 = 0.19$ V [57].

6.3.4.1 **Voltammetric Sizing of a Microparticle Sphere**

Voltammetric measurements could enable the size determination (see also Section 6.3.2.2.3) of single particles of known shape, such as a sphere on a microelectrode [57]. The respective microparticle is positioned centrally on a microelectrode, after which cyclic voltammetric measurements are recorded over a wider range of scan rates (e.g., from 2 mV s^{-1} to 1.5 V s^{-1}). With slower scan rates, the microparticle sphere increasingly perturbs the voltammetric signal, while the diffusion layer changes simultaneously (in both size and shape) from an almost flat layer to an approximately hemispherical shape. The sphere affects the voltammetric response, in that there is a relationship between the microparticle size and the peak current (see Figure 6.30).

The peak current increases (tends towards the value obtained for a bare electrode) with decreasing scan rates, as the mass transport will not be hindered by small sphere sizes. With increasing scan rates, the diffusion layer thickness decreases, and minimizes the effects of mass transport on the voltammetric response. Eventually, numerical simulations[8] of cyclic voltammograms as a function of scan rate will provide information on the microparticle sphere radius.

8) The alternating direction implicit (ADI) method combined with a modified Thomas algorithm was used.

6.4 Examples and Applications

Since its introduction in 1989 [3, 4], the voltammetry of (immobilized) microparticles (earlier termed abrasive stripping voltammetry) has attracted considerable attention and initiated a wide range of experimental and theory-based studies. As outlined in Section 6.1, the first decade of investigations has been well reviewed [5–8], while a recent monograph [9] has included the details of any diverse applications up until late 2004. Consequently, whilst the following sections will focus on the period between 2005 and the present, any earlier contributions deemed relevant to an understanding of the current investigations will also be cited or discussed.

6.4.1 Analytical Studies of Objects of Art

The value of voltammetric studies of immobilized microparticles for investigating objects of art was demonstrated by the group of Doménech-Carbó in Valencia. These authors described the analysis of Maya Blue, which had been used in wall paintings found at several archeological sites at Campeche and Yucatán in Mexico [58–62]. Figure 6.31 shows a wall painting from the Early Classical Maya period. The Maya

Figure 6.31 Wall painting from the archaeological site of Mayapán (Campeche region, Mexico) showing the typical Maya Blue. Courtesy of Institute of Restoration of Cultural Heritage, Polytechniqual University of Valencia, Spain.

Blue pigment was produced by the ancient Mayas from indigo, which they extracted from *Indigofera suffruticosa*, and a local clay called palygorskite, a fibrous phyllosilicate [ideally: $(Mg\text{-}Al)_4Si_8(O,OH,H_2O)_{24}\cdot nH_2O$]. Maya Blue is famous for its stability and a hue that ranges from bright turquoise to greenish blue. Unfortunately, neither the procedure used by the ancient Mayas is known, nor how they produced the different hues. Although many questions were raised regarding the composition of Maya Blue, Doménech-Carbó and colleagues [58–62] were able to prove that Maya Blue consisted of indigo and dehydroindigo attached to palygorskite. They were also able to reconstruct the preparation technique by investigating the solid-state electrochemistry of Maya Blue prepared in their laboratory, according to different recipes. Subsequently, the application of hierarchical cluster analysis and other chemometric methods to the voltammetric data allowed distinctions to be made among samples from different archeological sites and periods. In combination with other, non-electrochemical techniques, these authors were even able to deduce information on the temporal developments of the preparation techniques.

In another report from the same group, different anthraquinone-based dyes were studied using square-wave voltammetry following the mechanical immobilization of microparticles on the surface of graphite/polyester composite electrodes [63].

Inorganic pigments and corrosion products of metals have also been studied intensively by using voltammetry of immobilized microparticles. In this way, it was possible to identify hematite as the main corrosion product of a modern steel sculpture in Valencia [64]. Iron (III) oxides and oxide-hydrates are characterized by reductive dissolution signals with very distinct peak potentials and signal shapes.

Interestingly, electrochemistry also allows information to be extracted regarding the nature of *binders* used in paint formulations. The binders, which include vegetable oils, casein, egg, and bovine gelatin, react with the surface of mineral particles in such a way that metal ions are released to the adjacent binder layers. This in turn leads to additional reduction signals that can be detected and reflect the nature of the metal-binder species [65]. This serves as a form of speciation analysis with respect to metal ions in solid phases.

6.4.2 Metal Oxide and Hydroxide Systems with Poorly Crystalline Phases

Metal oxide and hydroxide systems serve many functions, including roles as pigments, in mineralogy, and also in catalysis. The classic techniques used in such investigations have included diffraction (especially X-ray diffraction; XRD), thermal analysis, transmission electron microscopy, Fourier transform infrared (FTIR) spectroscopy and Raman spectroscopy (see also Chapters 2 and 4). Until the introduction of voltammetry in the analysis of immobilized microparticles, electrochemical studies had been confined to solid electrolyte cells (Chapter 12), normally functioning at elevated temperatures. Unfortunately, these studies proved to be inapplicable for analytical characterization, and consequently a series of systematic studies was undertaken using immobilized microparticles of iron oxides and oxide-hydrates (for reviews, see

Refs. [6–9, 66]). Following the application of aqueous electrolyte solutions, the major potential of electrochemical studies became more clear, in that:

- Electrochemistry could also be applied to X-ray amorphous phases, for which it may yield unambiguous information.
- Electrochemical signals may be used for the quantitative analysis of solid solution systems.
- Electrochemistry could also provide interesting information on the reactivity of these phases (i.e., electrochemical kinetics).
- Electrochemistry could provide granulometric information, as the reactivity is often size-dependent.

Kovanda *et al.* [67] have employed not only the voltammetry of immobilized microparticles but also several solid-state analytical techniques when studying the thermal behavior of a layered Ni–Mn double hydroxide. The important advantage of these electrochemical experiments was that they proved the presence of substantial amounts of amorphous compounds. Moreover, distinct signals could be identified for the reductive dissolution of the different phases.

In another study, Grygar *et al.* [68] investigated the formation of mixed Cu–Mn, Cu–Mn–Al, Cu–Mg–Mn, and Cu–Mg–Mn–Al oxides by the calcination of amorphous basic carbonates, or by the calcination of hydrotalcite-like precursors. A combination of XRD and electrochemical measurements (again reductive dissolutions in aqueous electrolytes) led to the crystalline and amorphous phases being identified, and their formation during the thermal treatment elucidated.

Of particular interest was the way in which detailed information could be derived from voltammetric studies of clay minerals treated aerobically with iron (II) solutions. This caused the precipitation of a thin, active layer of iron (III) oxi(hydroxides) which could later be used for the sorption of arsenate(V) ions [69]. By employing the voltammetry of immobilized microparticles, it was possible to distinguish different iron species, namely: (i) ion-exchangeable, labile, or sorbed iron (III) ions; (ii) ferrihydrite or lepidocrocite; and (iii) crystalline hematite or goethite. Cepriá *et al.* subsequently employed the voltammetry of immobilized microparticles in the phase analysis of iron (III) oxides and oxi(hydroxides) in binary mixtures, as well as in cosmetic formulations [70].

The voltammetry of immobilized microparticles was also used for monitoring stream sediment samples in the vicinity of an old, unmonitored municipal landfill in Prague, in the Czech Republic, the aim being to detect the presence of Fe and Mn oxides [71]. The technique proved useful for detecting low concentrations of both goethite and hematite in paleosol and loess samples, where the total content of Fe oxides was <2 wt%, well below the detection limit of powder XRD. When using this technique, goethite could also be distinguished from hematite due to a different electrochemical reactivity of its product of thermal dehydroxylation [72].

Others [73] have also used the voltammetry of microparticles of common sulfide minerals to examine the influence of pyrrhotite content on the electrochemical behavior of pyrite-pyrrhotite mixtures. For this, artificial electrodes created from these two minerals were used, in addition to ore sediment samples [73].

6.4.3 Electrochemical Reactions of Organometallic Microparticles

The organic and organometallic complexes of transition metals are especially important in catalysis and photovoltaics, on the basis of their redox and electron-mediating properties. Whilst most complex compounds can be studied in (organic) solution-phase experiments, their solid-state electrochemistry (often in an aqueous electrolyte solution environment) is in general also easily accessible by attaching microcrystalline samples to the surface of electrodes. Quite often, the voltammetric characteristics of a complex in the solid state will differ remarkably from its characteristics monitored in solution. Consequently, chemical, physical or mechanistic data are each accessible via the voltammetry of immobilized microparticles.

A selection of organic and organometallic compounds studied in recent years using the voltammetry of immobilized microparticles is listed in Tables 6.2 and 6.3; however, only selected contributions will be described briefly in the following sections.

Whilst conducting a series of experimental and theoretically based studies, the group of Bond highlighted the value of voltammetric measurements with microparticles attached to an electrode in contact not only with aqueous or organic electrolyte solutions but also with ionic liquids such as *N,N*-dimethylammonium *N′,N′*-dimethylcarbamate (DIMCARB or BMIM[9]). In this way, both thermodynamic and kinetic information could be obtained [81, 92–94, 107, 108]. Within this context, two benefits should perhaps be mentioned:

- Neutral organometallic compounds are mostly sparingly soluble only (or after extensive period of sonication) in ionic liquids.
- In contrast, the respective charged compounds formed in the voltammetric experiment are often shown to be far more soluble and to dissolve rapidly in viscous ionic liquids.

An additional benefit is that only sub-microgram quantities of the solid are needed, whereas in a conventional voltammetric set-up volumes of at least 1–10 ml of the ionic liquid, containing millimolar concentrations of dissolved compound, would be required. The principal processes involved in the normal oxidation step of microparticles of redox active species at the electrode surface, and in contact with an ionic liquid, are shown in Figure 6.32.

Likewise, Komorsky-Lovrić *et al.* investigated the behavior of lutetium bisphthalocyanine with the voltammetry of microparticles [108]. This solid-state reaction (which may be studied with either square-wave or cyclic voltammetry) was shown to proceed via the simultaneous insertion/expulsion of anion ions. The oxidation was found to have quasi-reversible characteristics in electrolyte solutions containing perchlorate, nitrate, and chloride, whereas bromide and thiocyanate

9) 1-*n*-Butyl-3-methylimidazolium hexafluorophosphate [BMIM][PF_6].

Table 6.2 A selective overview of solid-state electrochemical studies on organometallics.

Metal	Compound	Reference(s)
Ru	Tris(2,2′-bpy) $Ru^{(II)}$ PF_6	[74]
Ru	*cis*-$Ru^{(II)}(dcbpy)_2(NCS)_2$[a]	[75]
Ru, Os	$[Ru(bpy)_2\ Os(bpy)_2\ (\mu\text{-}L)]\ [PF_6]$[b]	[76]
Ru	Ru(III) dithiocarbamate	[77]
Ru	Ru(III) diphenyl dithiocarbamate	[78]
Ru	$[Ru(bpy)_3]\ [PF_6]_2$	[74]
Ru	$[Ru(bpy)_3]_3\ [P_2W_{18}O_{62}]$	[79]
Fe	$Fe(\eta^5\text{-}C_5Ph_5)\ (\eta^6\text{-}C_6H_5)\ C_5Ph_4$	[80]
Fe	1,3,5-tris(3-(ferrocenylmethyl)amino-pyridiniumyl)-2,4,6-triethylbenzene $[PF_6]_3$	[81]
Fe	Bis(η^5-Ph_5-cyclopentadienyl)-$Fe^{(II)}$ $Fe^{(III)}$	[82]
Fe	Decamethylferrocene	[83]
Fe, Co	Ferrocene, Cobaltocene	[84, 85]
Fe	Decaphenylferrocene	[86]
Cu	Azurin	[87]
Co, Mn	Co and Mn phthalocyanines	[88, 89]
Co	$[Co(mtas)_2]\ X_n$[c,d]	[90]
Mn	*fac*-$Mn(CO)_3\ (\eta^2\text{-}Ph_2PCH_2\ PPh_2)\ Cl$	[91]
Mn	*cis*-$[Mn(CN)(CO)_2\ \{P(OPh)_3\}\ (dpm)]$[e]	[92]
Mo	$[Bu_4N]_4[S_2Mo_{18}O_{62}]$ (α isomer)	[93]
W	$[Bu_4N]_4[SiW_{12}O_{40}]$ (α, β, γ^* isomers)	[93]
W	$[Bu_4N]_4[S_2W_{18}O_{62}]$ (α, γ^* isomers)	[93]
Mo, W	$[Bu_4N]_2[M_6O_{19}]$, (M = Mo or W)[f]	[94]
Mo, W	$[Bu_4N]_4[\alpha\text{-}SiM_{12}O_{40}]$[f]	[94]
Mo, W	$[Bu_4N]_4[\alpha\text{-}S_2M_{18}O_{62}]$[f]	[94]
W	*cis*-$[W(CO)_2\ (dpe)_2]$[g]	[92]
Re	*cis-trans*-$[Re(CO)_2\ (P\text{-}P)_2]^+$ and *trans*-$[Re(CO)\ (P\text{-}P)_2]X$[h,i]	[95]
Re	*trans*-Re(Br) (CO)X[j,k]	[96]
Cr	*cis-trans*-$Cr(CO)_2\ (dppe)_2$	[96–99]
Cr	$[(C_4H_9)_4N]\ [Cr(CO)_5I]$	[100]
Cu	$[Cu_2(H_3L)(OH)]\ [BF_4]_2$, $[Cu_4L(OH)]\ [NO_3]_3$[l]	[101]
Cu	$Cu^{(II)}$ 5,10,15,20-tetraphenyl-21H,23H-porphyrin	[102, 103]
Os	$[Os(bpy)_2\text{-}4\text{-}tet\text{-}Cl]\ (ClO_4)$[m]	[104]
Os	$[Os(Ome\text{-}bpy)_3]\ [PF_6]_2$[n]	[105]
Os	$[Os(bpy)_2Cl\ 4\text{-}bpt\ Os(bpy)_2Cl]\ [PF_6]$[o]	[106]

[a] dcbpy = 2,2′-bipyridine-4,4′-dicarboxylic acid.
[b] L = 1,4-Dihydroxy-2,5-bis(pyrazol-1-yl)benzene.
[c] mtas = bis(2-(dimethylarsino)phenyl)-methylarsine.
[d] $X = BF_4^-$, $n = 3$; $X = ClO_4^-$, $n = 2, 3$; $X = BPh_4^-$, $n = 2$.
[e] dpm = $Ph_2PCH_2PPh_2$.
[f] M = Mo or W.
[g] dpe = $Ph_2P(CH_2)_2PPh_2$.
[h] P-P = diphosphine.
[i] X = Cl, Br.
[j] X = $(dppe)_2$, (dppz)/(dppm), (dppe)/(dppm).
[k] dppz = $(Ph_2P)_2\ C_6H_4$.
[l] L = O_4N_4.
[m] tet = 3,6-bis(4-pyridyl)-1,2,4,5-tetrazine.
[n] OMe = 4,4′-dimethoxy.
[o] bpt = 3,5-bis(pyridin-4-yl)-1,2,4-triazole.

Table 6.3 A selective overview of solid-state electrochemical studies on organic compounds.

Trivial name or compound class	Compound	Reference(s)
	Azobenzene	[110, 111]
	Carbazole/polycarbazole	[112]
Cocaine		[113]
DPPH	2,2-diphenyl-1-picrylhydrazyl	[114]
	Diphenylamine	[115, 116]
Fullerene	C_{60}	[117–120]
	$C_{60} \subset L_2$ [a]	[121]
HQBpt [b,c]		[122]
	5-Aminosalicylic acid, Ciprofloxacin, Azithromycin	[123]
	Quinhydrone, Acridine, Famotidine, Probucol, Propylthiouracil, Thionicotinoylanilide	[124]
	Lutetium bisphthalocyanine	[108]
Organic dyes and pigments		[125]
Simvastatin		[109]
TCNQ	7,7,8,8-Tetracyanoquinodimethane	[126–131]
Indigo		[124, 132]
	Pb dithiocarbamate, Hg dithiocarbamate	[133, 134]
Perylene	*N,N'*-bis(4-cyanophenyl)-3,4,9,10-perylene-bis(dicarboximide)	[135]
Naphthalene	*N,N'*-bis(4-cyanophenyl)-1,4,5,8-naphthalene-diimide)	[135]
	Tetraphenylviologen	[96]
	1,3,5-Tris[4-[(3-methylphenyl)-phenylamino]phenyl]benzene	[136]
TTF	Tetrathiafulvalene	[137–139]

[a] L = *p*-benzyl-calix[5]arene)$_2$]·8 toluene.
[b] HBpt = 3,5-bis(pyridine-2-yl)-1,2,4-triazole.
[c] H_2Q = 1,4-hydroquinone.

electrolytes did not support the intercalation reaction. When the same group studied the voltammetric characteristics of the lipid-lowering drug simvastatin [109], the oxidation reaction of microparticles attached to a graphite electrode and contacting an aqueous electrolyte solution, featured an irreversible behavior, the potential of which was measured at approximately 1.1 V (versus Ag|AgCl; 3 *M* KCl).

Doménech-Carbó *et al.* also showed the voltammetry of immobilized microparticles to be valuable in the unambiguous identification of dyes such as Curcuma and Safflower in microsamples of works of art and archaeological artifacts (see also Section 6.4.1) [140]. Here, the use of square-wave voltammetry in aqueous acetate or phosphate buffers led to the appearance of well-defined oxidation peaks of the dyes in the potential region of +0.65 to +0.25 V (versus Ag|AgCl).

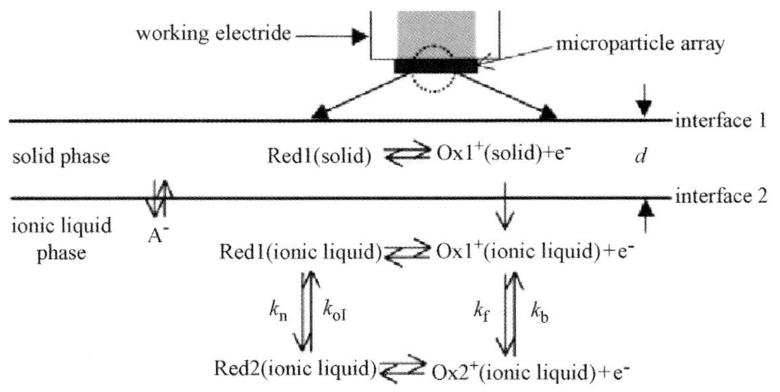

Figure 6.32 Schematic representation of the principal processes assumed to influence the voltammetric response when a microparticle modified electrode is placed in contact with an ionic liquid and when dissolved electrogenerated species $Ox1^+$ (ionic liquid) undergoes a square reaction scheme [92].

Cepria and coworkers used the voltammetry of immobilized microparticles to detect and quantify the cadmium pigments (e.g., cadmium sulfide and cadmium sulfoselenide) used in artists' paints, as well as in glasses, plastics, ceramics, and enamels [141]. For this, a simple, fast and reliable technique was developed that proved to be especially applicable for valuable art objects, as it was minimally invasive and required only nanogram quantities of material (see also Section 6.4.1). For quantification purposes, an abrasive stripping scan was used from +0.3 V to −1.0 V, following a 10 s pre-treatment step at −1.5 V. The Cd oxidation peak was evaluated with respect to an internal AgCl calibration standard.

Solid microparticles of three common *drugs* have been studied voltammetrically, including 5-aminosalicylic acid (an active component in the therapy of inflammatory bowel disease) and two broad-spectrum antibiotics, ciprofloxacin and azithromycin [123]. Minute amounts of these electroactive compounds using for example, square-wave voltammetry, makes them easily detectable either alone, or within their normal matrices (e.g., pills, capsules). Depending on the specific amounts present in dosage formulations, the voltammetry of microparticles represents an excellent approach for the rapid detection of an active compound in a formulation, prior to a quantitative analysis itself. Among other drugs or physiologically relevant compounds studied and/or screened in this way are cocaine [113] and pesticides [142]. The voltammetry of immobilized microparticles has also been used successfully for the direct identification of famotidine, probucol, propylthiouracil, nicotinoylanilide, thionicotinoylanilide [124], or benzocaine, cinchocaine, lidocaine, procaine, and codeine [110]. Initially, the methodology was proposed for rapidly determining the possible composition of suspected illegal powder samples encountered on the street. Recently, Oliveira *et al.* used square-wave voltammetry (but using boron-doped diamond electrodes) for the highly sensitive electroanalytical determination of lidocaine in pharmaceutical preparations (gels) [143].

6.4.4
Selected Other Applications

Extensive studies have been performed on boron-doped diamond films (see e.g., Ref. [144]); these are usually polycrystalline thin films that are grown on conductive Si substrates and doped with boron at a nominal atomic concentration of between 10^{19} and $10^{21}\,cm^{-3}$. These materials possess a high degree of corrosion resistance, thermal stability, hardness, light weight, and electrical conductivity. A recent study was conducted to determine the spatial surface activity of the boron-doped diamond (SiBDD) electrodes [145]. Whereas, the normal pretreatment of pyrolytic graphite or glassy carbon would increase the surface defect density, and in turn also the heterogeneous electron transfer rate [146], a combination of cyclic voltammetry and AFM showed the electrodeposited Pt microparticle growth to be independent (or at least less dependent) on the surface microstructure, and to be heavily influenced by the spatial distribution of the boron dopant [145].

The growth of electrodeposited Pt clusters was investigated with regards to the number and position of active sites on the underlying diamond; a typical AFM image and the respective cyclic voltammogram is shown in Figure 6.33. Those Pt microparticles with a round shape could be distinguished from BDD facets and steps. Moreover, the deposition of Pt microparticles seemed not to be related to the geometric characteristics of the BDD crystals. Other interesting results related to the microparticle shape; when measuring the diameters and heights of particles, using AFM, most had a raft-like rather than a hemispherical shape (Figure 6.34) (see also Section 6.3.4). The electrodeposition of Pt was also seen to occur preferentially at sites with the highest conduction properties due to boron doping. In other words, only a limited number of BDD sites was present at which the electrochemical processes would occur preferentially.

Many electrochemical studies relate to mesoscopic (i.e., nanosized) metal particles [147]. In one study, the nucleation and growth processes of Ag particles

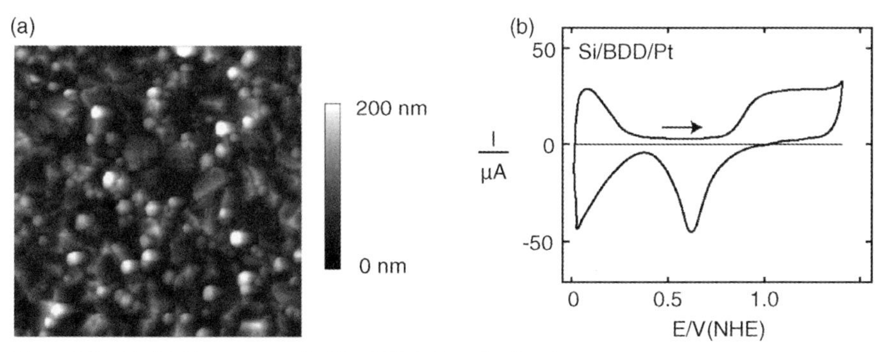

Figure 6.33 (a) AFM image recorded for SiBDD/Pt with a low Pt coverage, image size 5.2 μm × 5.2 μm; (b) Cyclic voltammogram recorded for SiBDD/Pt in 0.05 *M* H_2SO_4 at $50\,mV\,s^{-1}$ scan rate (low Pt coverage) [145].

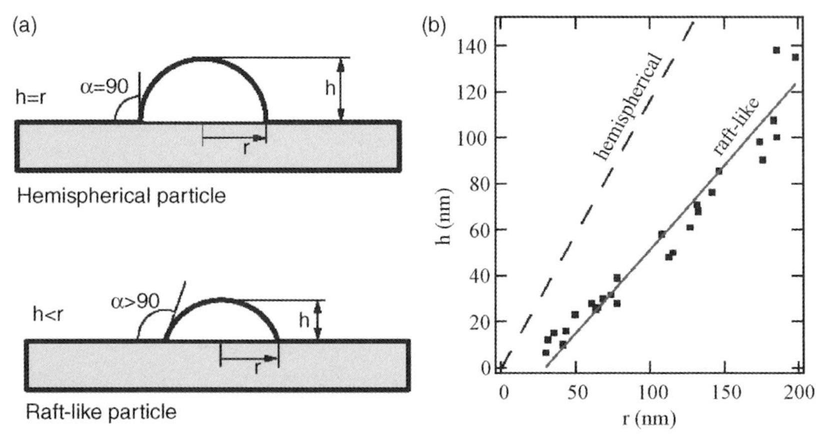

Figure 6.34 Pt particles on a surface can form hemispherical or raft-like particles. (a) The two geometrical models, with *r* representing the radius of contact of the particle and *h* its height; (b) Height versus radius graph of the Pt particles obtained from the analysis of sections across 25 particles. All sections were performed on the same AFM image [145].

were studied not on a solid substrate but rather at a small liquid | liquid interface [148]. Such interfaces (which were polarized externally with 1,2-dichlorethane/water) were supported at the tip of tapered circular pipettes with radii of 0.4–6 μm. The electrochemical process involved the reduction of silver ions in the aqueous phase, with butylferrocene as the organic phase electron donor. The silver nucleation and growth process was studied using cyclic voltammetry and potential step chronoamperometry.

The voltammetry of immobilized microparticles was, in several cases, successfully used as part of a series of standard analytical tasks, including the determination of "plant-available" metals in soils and sediments [149]. Highly polluted (metal contents up to a few grams per kilogram) forest/tilled soils and stream sediments from a mining and smelting area were subjected to single-extraction procedures to determine the extractable contents of Cd, Cu, Pb, and Zn. As with earlier studies [72], the content and crystallinity of the Fe and Mn oxides/hydroxides in the solid samples were determined using voltammetry of microparticles and diffuse-reflectance spectrometry [149].

Although not related directly to this chapter, current research into electrode materials for *lithium batteries* today involves hundreds of research groups worldwide, the aim being to develop improved cathode materials such as $LiMn_2O_4$, $LiFePO_4$, $LiMn_{1-x-y}Ni_xCo_yO_2$, $LiMn_{0.5}Ni_{0.5}O_2$, $LiMn_{1.5}Ni_{0.5}O_4$, $LiNi_{1-x}MO_2$ (M = Co, Al), Li_xVO_y, or $Li_xM_yVO_z$ (M = Ca, Cu) (for a recent review, see Ref. [150]; see also Chapters 5 and 7). Today, the voltammetry of microsized and nanosized particles of cathode materials represents a standard experimental approach (in addition to X-ray photoelectron and FTIR spectroscopy, electron microscopy and Raman spectroscopy), as it allows assessment of the electrode kinetics relating to film formation on the

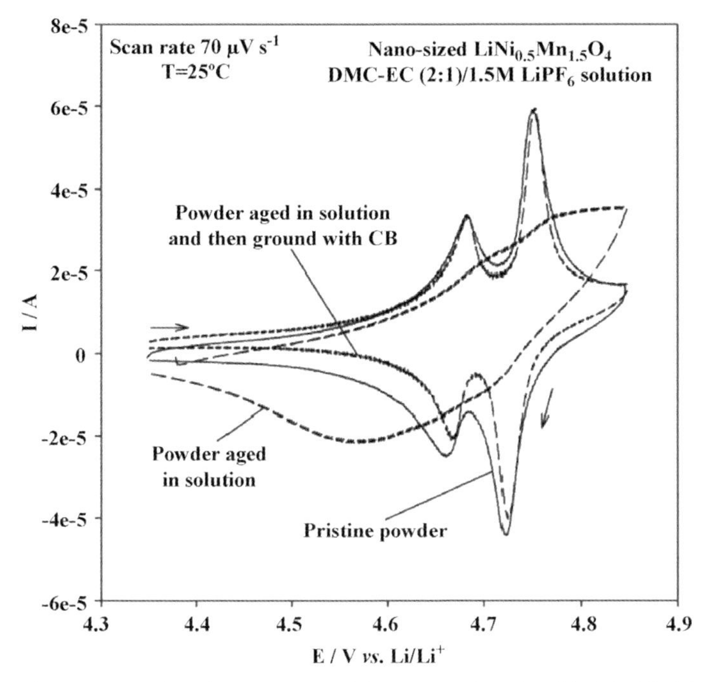

Figure 6.35 Cyclic voltammograms of thin electrodes (on Al foil current collectors) comprising nanosized particles of the pristine $LiNi_{0.5}Mn_{1.5}O_4$ powder, of a powder aged in a dimethyl carbonate (DMC)- ethylene carbonate (EC) (2 : 1)/1.5 *M* $LiPF_6$ solution at $T = 30\,°C$ for 35 days, and of the same aged powder ground with carbon black (15%). All the electrodes were free of poly(vinylidene fluoride) binder. The voltammograms were recorded in DMC-EC (2 : 1)/1.5 *M* $LiPF_6$ solutions at $T = 25\,°C$ using three-electrode flooded cells, at the potential scan rate of $70\,\mu V\,s^{-1}$ [150].

electrode surfaces, or the possible dissolution of transition metal ions. The cyclic voltammograms of electrodes comprising $LiMn_{1.5}Ni_{0.5}O_4$ microparticles, and resulting from different backgrounds of preparation or handling, are shown in Figure 6.35. Although two reversible redox couples (Ni^{2+}/Ni^{3+} and Ni^{3+}/Ni^{4+}) are visible here, those electrodes composed of an aging material would lose this voltammetric feature. This arbitrary example illustrates the extreme sensitivity and high value of these investigations.

References

1 Hasse, U., Wagner, K. and Scholz, F. (2004) *J. Solid State Electrochem.*, **8**, 842–53.

2 Hasse, U., Fletcher, S. and Scholz, F. (2006) *J. Solid State Electrochem.*, **10**, 833–40.

3 Scholz, F., Nitschke, L. and Henrion, G. (1989) *Naturwissenschaften*, **76**, 71.

4 Scholz, F., Nitschke, L., Henrion, G. and Damaschun, F. (1989) *Naturwissenschaften*, **76**, 167–168.

5 Scholz, F. and Lange, B. (1992) *Trends Anal. Chem.*, **11**, 359–67.
6 Scholz, F. and Meyer, B. (1994) *Chem. Soc. Rev.*, **23**, 341–7.
7 Scholz, F. and Meyer, B. (1998) Voltammetry of Solid Microparticles Immobilized on Electrode Surfaces, in *Electroanalytical Chemistry, Series of Advances*, Vol. 20 (eds A.J. Bard and I. Rubinstein), Marcel Dekker, Inc., New York, p. 1.
8 Grygar, T., Marken, F., Schröder, U. and Scholz, F. (2002) *Coll. Czech. Chem. Commun.*, **67**, 163–208.
9 Scholz, F., Schröder, U. and Gulaboski, R. (2005) *Electrochemistry of Immobilized Particles and Droplets*, Springer, Berlin, Heidelberg.
10 Pournaghi-Azar, M.H. and Habibi, B. (2007) *J. Electroanal. Chem.*, **601**, 53–62.
11 Arrieta Almario, A.A. and Vieira, R.L. (2006) *J. Chil. Chem. Soc.*, **51**, 971–7.
12 Alfonso, G., del Valle, M.A., Soto, G.M., Cotarelo, M.A., Quijada, C. and Vazquez, J.L. (2006) *Polym. Bull.*, **56**, 201–10.
13 Mourato, A., Wong, S.M., Siegenthaler, H. and Abrantes, L.M. (2006) *J. Solid State Electrochem.*, **10**, 140–7.
14 Golikand, A.N., Golabi, S.M., Maragheh, M.G. and Irannejad, L. (2005) *J. Power Sources*, **145**, 116–23.
15 Luque, G.L., Ferreyra, N.F. and Rivas, G.A. (2007) *Talanta*, **71**, 1282–7.
16 Bian, C., Yu, Y. and Xue, G. (2007) *J. Appl. Polym. Sci.*, **104**, 21–6.
17 Tarasevich, M.R., Kuzov, A.V., Klyuev, A.L. and Titova, V.N. (2007) *Elektrokhimicheskaya Energetika*, **7**, 156–60.
18 Beard, K.D., Monnier, J.M. and Van Zee, J.W. (2007) *Nanotechnology*, **2**, 87–99.
19 Zhang, C., Wang, K., Hu, L., Kong, F. and Guo, L. (2007) *J. Photochem. Photobiol. A*, **189**, 329–33.
20 Meyer, B., Zhang, S. and Scholz, F. (1996) *Fresenius' J. Anal. Chem.*, **356**, 267–70.
21 Meyer, B. and Scholz, F. (1997) *Phys. Chem. Miner.*, **24**, 50–2.
22 Scholz, F., Lange, B., Jaworski, A. and Pelzer, J. (1991) *Fresenius' J. Anal. Chem.*, **340**, 140–4.
23 Bosch-Reig, F., Doménech-Carbó, A., Doménech-Carbó, M.T. and Gimeno-Adelantado, J.V. (2007) *Electroanalysis*, **19**, 1575–84.
24 Bosch, F. and Campins, P. (1988) *Analyst*, **113**, 1011.
25 Widmann, A., Kahlert, H., Petrovic-Prelevic, I., Wulff, H., Yakhmi, J.V., Bagkar, N. and Scholz, F. (2002) *Inorg. Chem.*, **42**, 5706–15.
26 Schröder, U., Oldham, K.B., Myland, J.C., Mahon, P.J. and Scholz, F. (2000) *J. Solid State Electrochem.*, **4**, 314–24.
27 Hermes, M. and Scholz, F. (2000) *Electrochem. Commun*, **2**, 845–50.
28 Myland, J.C. and Oldham, K.B. (2005) *Electrochem. Commun.*, **7**, 282.
29 Amatore, C., Bouret, Y., Maisonhaute, E., Goldsmith, J.I. and Abruna, H.D. (2001) *Chem. Eur. J.*, **7**, 2206.
30 Amatore, C., Bouret, Y., Maisonhaute, E., Goldsmith, J.I. and Abruna, H.D. (2001) *ChemPhysChem*, **2**, 130.
31 Leventis, H.C., Wildgoose, G.G., Davies, I.G., Jiang, L., Jones, T.G.J. and Compton, R.G. (2005) *ChemPhysChem*, **6**, 590.
32 Leventis, H.C., Streeter, I., Wildgoose, G.G., Lawrence, N.S., Jiang, L., Jones, T.G.J. and Compton, R.G. (2004) *Talanta*, **63**, 1039.
33 Thompson, M. and Compton, R.G. (2006) *ChemPhysChem*, **7**, 1964–70.
34 Thompson, M., Wildgoose, G.G. and Compton, R.G. (2006) *ChemPhysChem*, **7**, 1328–36.
35 Davies, T.J. and Compton, R.G. (2005) *J. Electroanal. Chem.*, **585**, 63–82.
36 Amatore, C., Saveant, J.-M. and Tessier, D. (1983) *J. Electroanal. Chem.*, **147**, 39.
37 Streeter, I., Baron, R. and Compton, R.G. (2007) *J. Phys. Chem. C*, **111**, 17008–14.
38 Streeter, I., Fietkau, N., del Campo, J., Mas, R., Mũnoz, F.X. and Compton, R.G. (2007) *J. Phys. Chem. C*, **111**, 12058–66.

39 Ordeig, O., del Campo, J., Munoz, F.X., Banks, C.E. and Compton, R.G. (2007) *Electroanalysis*, **19**, 1973–86.

40 Ward Jones, S.E., Chevallier, F.G., Paddon, C.A. and Compton, R.G. (2007) *Anal. Chem.*, **79**, 4110–19.

41 Járai-Szabó, F. and Néda, Z. (2004) *arXiv Condens. Matter*, e-prints, **06**, 116.

42 Davies, T.J., Banks, C.E. and Compton, R.G. (2005) *J. Solid State Electrochem.*, **9**, 797–808.

43 DigiSim™, Bioanalytical Systems, Inc., www.bioanalytical.com.

44 Fletcher, S. and Horne, M.D. (1999) *Electrochem. Commun.*, **1**, 502–12.

45 Davies, T.J., Lowe, E.R., Wilkins, S.J. and Compton, R.G. (2005) *ChemPhysChem*, **6**, 1340–7.

46 Lee, H.J., Beriet, C., Ferrigino, R. and Girault, H.H. (2001) *J. Electroanal. Chem.*, **502**, 138.

47 Alfred, L.C.R. and Oldham, K.B. (1995) *J. Electroanal. Chem.*, **396**, 257.

48 Schröder, U. and Scholz, F. (2000) *Inorg. Chem.*, **39**, 1006.

49 Bond, A.M., Feldberg, S.W., Miao, W., Oldham, K.B. and Raston, C.L. (2001) *J. Electroanal. Chem.*, **201**, 22.

50 Zhang, J. and Bond, A.M. (2005) *J. Electroanal. Chem.*, **574**, 299–309.

51 Brainina, K.Z. and Vydrevich, M.B. (1981) *J. Electroanal. Chem.*, **121**, 1–28.

52 Brainina, K.Z. and Lesunova, R.P. (1974) *Zh. Anal. Khim.*, **29**, 1302.

53 Hyde, M.E., Banks, C.E. and Compton, R.G. (2004) *Electroanalysis*, **16**, 345–54.

54 Paddon, C.A. and Compton, R.G. (2007) *J. Phys. Chem. C*, **111**, 9016–18.

55 Streeter, E. and Compton, R.G. (2007) *J. Phys. Chem. C*, **111**, 18049–54.

56 Svir, I.B. (2001) *Analyst*, **126**, 1888–91.

57 Fietkau, N., Chevallier, F.G., Jiang, L., Jones, T.G.J. and Compton, R.G. (2006) *ChemPhysChem*, **7**, 2162–7.

58 Doménech-Carbó, A., Doménech-Carbó, M.T. and Vázquez, M.L. (2006) *J. Phys. Chem. B*, **110**, 6027–39.

59 Doménech-Carbó, A., Doménech-Carbó, M.T. and Vázquez de Agredos Pascual, M.L. (2007) *J. Phys. Chem. C*, **111**, 4585–95.

60 Doménech-Carbó, A., Doménech-Carbó, M.T. and Vázquez de Agredos Pascual, M.L. (2007) *Anal. Chem.*, **79**, 2812–21.

61 Vázquez de Ágredos Pascual, M.L. (2007) Caracterización químico-analítica de azul maya en la pinturas mural de las tierras bajas mayas. Ph.D. Thesis, Universidad Politécnica de Valencia.

62 Domenech, A., Domenech-Carbo, M.T. and Agredos Pascual, M.L.V. (2007) *J. Solid State Electrochem.*, **11**, 1335–46.

63 Doménech-Carbó, A., Doménech-Carbó, M.T., Saurí-Peris, M.C., Gimeno-Adelantado, J.V. and Bosch-Reig, F. (2003) *Anal. Bioanal. Chem.*, **375**, 1169–75.

64 Doménech-Carbó, A., Roig Salom, J.L. and Doménech-Carbó, M.T. (2006) *Arché*, **1**, 167–70.

65 Doménech-Carbó, A., Doménech-Carbó, M.T., Mas-Barberá, X. and Ciarrocci, J. (2007) *Arché*, **2**, 121–4.

66 Grygar, T. (1995) *Collect. Czech. Chem. Commun.*, **60**, 1261–73.

67 Kovanda, F., Grygar, T. and Dorničák, V. (2003) *Solid State Sci.*, **5**, 1019–26.

68 Grygar, T., Rojka, T., Bezdička, P., Večerníková, E. and Kovanda, F. (2004) *J. Solid State Electrochem.*, **8**, 252–9.

69 Doušová, B., Grygar, T., Martaus, A., Fuitová, L., Koloušek, D. and Machovič, V. (2006) *J. Colloid Interface Sci.*, **302**, 424–31.

70 Cepriá, G., Usón, A., Pérez-Arantegui, J. and Castillo, J.R. (2003) *Anal. Chim. Acta*, **477**, 157–68.

71 Ettler, V., Matura, M., Mihaljevic, M. and Bezdicka, P. (2006) *Environ. Geol.*, **49**, 610–619.

72 Grygar, T. and van Oorschot, I.H.M. (2002) *Electroanalysis*, **14**, 339–44.

73 Almeida, C.M.V.B. and Giannetti, B.F. (2003) *J. Electroanal. Chem.*, **553**, 27–34.

74 Ramaraj, R., Kabbe, Ch. and Scholz, F. (2000) *Electrochem. Commun.*, **2**, 190–4.

75 Bond, M., Deacon, G.B., Howitt, J., MacFarlane, D.R., Spiccia, L. and

Wolfbauer, G. (1998) *J. Electrochem. Soc.*, **146**, 648–56.
76 Bond, A.M., Marken, F., Williams, Ch.T., Beattie, D.A., Keyes, T.E., Foster, R.J. and Vos, J.G. (2000) *J. Phys. Chem.*, **104**, 1977–1983.
77 Nalini, B. and Sriman Narayanan, S. (1998) *Bull. Electrochem.*, **14**, 267–70.
78 Nalini, B. and Sriman Narayanan, S. (1998) *Bull. Electrochem.*, **14**, 241–5.
79 Fay, N., Dempsey, E., Kennedy, A. and McCormac, T. (2003) *J. Electroanal. Chem.*, **556**, 63–74.
80 Bond, A.M., Fiedler, D.A., Lamprecht, A. and Tedesco, V. (1999) *Organometallics*, **18**, 642–9.
81 Zhang, J., Bond, A.M., Belcher, J., Wallace, K.J. and Steed, J.W. (2003) *J. Phys. Chem. B*, **107**, 5777–86.
82 Bond, A.M., Lamprecht, A., Tedesco, V. and Marken, F. (1999) *Inorg. Chim. Acta*, **291**, 21–31.
83 Bond, A.M. and Marken, F. (1994) *J. Electroanal. Chem.*, **372**, 125–35.
84 Bond, A.M. and Scholz, F. (1991) *Langmuir*, **7**, 3197–204.
85 Zhang, J. and Bond, A.M. (2003) *Anal. Chem.*, **75**, 2694–702.
86 Zhang, J., Bond, A.M., Schumann, H. and Sühring, K. (2005) *Organometallics*, **24**, 2188–96.
87 Guo, S., Zhang, J., Elton, D.M. and Bond, A.M. (2004) *Anal. Chem.*, **76**, 166–77.
88 Komorsky-Lovric, Š. (1995) *J. Electroanal. Chem.*, **397**, 211–5.
89 Komorsky-Lovric, Š., Lovric, M. and Scholz, F. (1997) *Mikrochim. Acta*, **127**, 95–9.
90 Downard, A.J., Bond, A.M., Hanton, L.R. and Heath, G.A. (1995) *Inorg. Chem.*, **34**, 6387–95.
91 Eklund, J.C. and Bond, A.M. (1999) *J. Am. Chem. Soc.*, **121**, 8306–12.
92 Zhang, J. and Bond, A.M. (2004) *J. Phys. Chem. B*, **108**, 7363–72.
93 Zhang, J., Bhatt, A.I., Bond, A.M., Wedd, A.G., Scott, J.L. and Strauss, C.R. (2005) *Electrochem. Commun.*, **7**, 1283–90.
94 Zhang, J., Bond, A.M., MacFarlane, D.R., Forsyth, S.A., Pringle, J.M., Mariotti, A.W.A., Glowinski, A.F. and Wedd, A.G. (2005) *Inorg. Chem.*, **44**, 5123–32.
95 Bond, A.M., Colton, R., Humphrey, D.G., Mahon, P.J., Snook, G.A., Tedesco, V. and Walter, J.N. (1998) *Organometallics*, **17**, 2977–85.
96 Snook, G., Bond, A.M. and Fletcher, S. (2003) *J. Electroanal. Chem.*, **554–555**, 157–65.
97 Bond, A.M., Colton, R., Daniels, F., Fernando, D.R., Marken, F., Nagaosa, Y., Van Steveninck, R.F.M. and Walter, J.N. (1993) *J. Am. Chem. Soc.*, **115**, 9556–62.
98 Bond, A.M., Colton, R., Marken, F. and Walter, J.N. (1994) *Organometallics*, **13**, 5122–31.
99 Shaw, S.J., Marken, F. and Bond, A.M. (1996) *J. Electroanal. Chem.*, **404**, 227–35.
100 Bond, A.M., Colton, R., Mahon, P.J. and Tan, W.T. (1997) *J. Solid State Electrochem.*, **1**, 53–61.
101 Marken, F., Cromie, S. and McKee, V. (2003) *J. Solid State Electrochem.*, **7**, 141–6.
102 Zhuang, Q.K. and Scholz, F. (2000) *J. Porphyrins Phthalocyanins*, **4**, 202–8.
103 Zhuang, Q.K. (2000) *Mikrochem. J.*, **65**, 333–40.
104 Foster, R.J., Keyes, T.E. and Bond, A.M. (2000) *J. Phys. Chem. B*, **104**, 6389–96.
105 Keane, L., Hogan, C. and Forster, R.J. (2002) *Langmuir*, **18**, 4826–33.
106 Walsh, D.A., Keyes, T.E. and Forster, R.J. (2002) *J. Electroanal. Chem.*, **538–539**, 75–85.
107 Hultgren, V.M., Mariotti, A.W.A., Bond, A.M. and Wedd, A.G. (2002) *Anal. Chem.*, **74**, 3151.
108 Komorsky-Lovric, Š., Quentel, F., L'Her, M. and Scholz, F. (2008) *J. Solid State Electrochem.*, **12**, 165–9.
109 Komorsky-Lovric, Š. and Nigovic, B. (2006) *J. Electroanal. Chem.*, **593**, 125–30.
110 Komorsky-Lovric, Š., Vukašinovic, N. and Penovski, R. (2003) *Electroanalysis*, **15**, 544–47.

111 Komorsky-Lovric, Š. (1997) *J. Solid State Electrochem.*, **1**, 94–9.

112 Inzelt, G. (2003) *J. Solid State Electrochem.*, **7**, 503–10.

113 Komorsky-Lovric, Š., Galic, I. and Penovski, R. (1999) *Electroanalysis*, **11**, 120–3.

114 Zhuang, Q., Scholz, F. and Pragst, F. (1999) *Electrochem. Commun.*, **1**, 406–10.

115 Inzelt, G. (2002) *J. Solid State Electrochem.*, **6**, 265–71.

116 Fehér, K. and Inzelt, G. (2002) *Electrochim. Acta*, **47**, 3551–9.

117 Suarez, M.F., Marken, F., Compton, R.G., Bond, A.M., Miao, W.J. and Raston, C.L. (1999) *J. Phys. Chem.*, **103**, 5637–44.

118 Bond, A.M., Miao, W. and Raston, C.L. (2000) *J. Phys. Chem. B*, **104**, 2320–9.

119 Bond, A.M., Feldberg, S.W., Miao, W., Oldham, K.B. and Raston, C.L. (2001) *J. Electroanal. Chem.*, **501**, 22–32.

120 Tan, W.T., Lim, E.B. and Bond, A.M. (2003) *J. Solid State Electrochem.*, **7**, 134–40.

121 Bond, A.M., Miao, W., Raston, C.L. and Sandoval, C.A. (2000) *J. Phys. Chem. B*, **104**, 8129–37.

122 Keyes, T.E., Foster, R.J., Bond, A.M. and Miao, W. (2001) *J. Am. Chem. Soc.*, **123**, 2877–84.

123 Komorsky-Lovric, Š. and Nigovic, B. (2004) *J. Pharm. Biomed. Anal.*, **36**, 81–9.

124 Komorsky-Lovric, Š., Mirceski, V. and Scholz, F. (1999) *Mikrochim. Acta*, **132**, 67–77.

125 Grygar, T., Kuckova, S., Hradil, D. and Hradilova, J. (2003) *J. Solid State Electrochem.*, **7**, 706–13.

126 Bond, A.M., Fletcher, S., Marken, F., Shaw, S.J. and Symons, P.G. (1996) *J. Chem. Soc., Faraday Trans.*, **92** (20), 3925–33.

127 Bond, A.M., Fletcher, S. and Symons, P.G. (1998) *Analyst*, **123**, 1891–904.

128 Suárez, M.F., Bond, A.M. and Compton, R.G. (1999) *J. Solid State Electrochem.*, **4**, 24–33.

129 Wooster, T.J., Bond, A.M. and Honeychurch, M.J. (2003) *Anal. Chem.*, **75**, 586–92.

130 Neufeld, A.K., Madsen, I., Bond, A.M. and Hogan, C.F. (2003) *Chem. Mater.*, **15**, 3573–85.

131 Wooster, T.J. and Bond, A.M. (2003) *Analyst*, **128**, 1386–90.

132 Bond, A.M., Marken, F., Hill, E., Compton, R.G. and Hügel, H. (1997) *J. Chem. Soc., Perkin Trans.*, **2**, 1735–42.

133 Bond, A.M. and Scholz, F. (1991) *J. Phys. Chem.*, **95**, 7460–5.

134 Bond, A.M. and Scholz, F. (1991) *Langmuir*, **7**, 3197–204.

135 Uzun, D., Ozser, M.E., Yuney, K., Icil, H. and Demuth, M. (2003) *J. Photochem. Photobiol. A Chem.*, **156**, 45–54.

136 Rees, N.V., Wadhawan, J.D., Klymenko, O.V., Coles, B.A. and Compton, R.G. (2004) *J. Electroanal. Chem.*, **563**, 191–202.

137 Shaw, S.J., Marken, F. and Bond, A.M. (1996) *Electroanalysis*, **8**, 732–41.

138 Wooster, T.J., Bond, A.M. and Honeychurch, M.J. (2001) *Electrochem. Commun.*, **3**, 746–52.

139 Tan, W.T., Ng, G.K. and Bond, A.M. (2000) *Malaysian J. Chem.*, **2**, 34–42.

140 Domenech-Carbo, A., Domenech-Carbo, T., Sauri-Peris, C., Gimeno-Adelantado, J.V. and Bosch-Reig, F. (2005) *Mikrochim. Acta*, **152**, 75–84.

141 Cepria, G., Garcia-Gareta, E. and Perez-Arantegui, J. (2005) *Electroanalysis*, **17**, 1078–84.

142 Reddy, J., Hermes, M. and Scholz, F. (1996) *Electroanalysis*, **8**, 955.

143 Oliveira, R.T.S., Salazar-Banda, G.R., Ferreira, V.S., Oliveira, S.C. and Avaca, L.A. (2007) *Electroanalysis*, **19**, 1189–4.

144 Angus, J.C. and Hayman, C.C. (1988) *Science*, **241**, 913–21.

145 Riedo, B., Dietler, G. and Enea, O. (2005) *Thin Solid Films*, **488**, 82–6.

146 Bodalbhai, L. and Brajter-Toth, A. (1990) *Anal. Chim. Acta*, **231**, 191–201.

147 (a) Song, Y. and Murray, R.W. (2002) *J. Am. Chem. Soc.*, **124**, 7096; (b) Cheng, W.L., Dong, S.J. and Wang, E.K. (2002) *Electrochem. Commun.*, **4**, 412.

148 Guo, J., Tokimoto, T., Toman, R. and Unwin, P.R. (2003) *Electrochem. Commun.*, **5**, 1005–10.

149 Ettler, V., Mihaljevic, M., Sebek, O. and Grygar, T. (2007) *Anal. Chim. Acta*, **602**, 131–40.

150 Aurbach, D., Markovsky, B., Salitra, G., Markevich, E., Talyossef, Y., Koltypin, M., Nazar, L., Ellis, B. and Kovacheva, D. (2007) *J. Power Sources*, **165**, 491–9.

7
Alkali Metal Cation and Proton Conductors: Relationships between Composition, Crystal Structure, and Properties

Maxim Avdeev, Vladimir B. Nalbandyan, and Igor L. Shukaev

Abstract

The aim of this chapter is to provide an overview of materials where fast transport of alkali metal cations and protons is observed. A general discussion of factors affecting conductivity and techniques used to study ion migration paths is followed by a review of the large number of cation conductors. Materials with large alkali ions (Na-Cs) are often isostructural and therefore examined as a group. The lithium conductors with unique crystal structure types and proton conductors with unique conduction mechanisms are also discussed.

7.1
Principles of Classification, and General Comments

The aim of this chapter is to provide a brief introduction into the vast field of solid alkali cation and proton conductors, considering also their Ag^+- and Tl^+-conducting analogues. More detailed information and an extensive bibliography can be found elsewhere [1–21]. The analysis of the relationships between transport properties, crystal structure and composition of these materials requires, initially, a formulation of the major classification principles and criteria.

7.1.1
Physical State

Single crystals are the most valuable subjects for investigating the bulk properties of materials. In particular, they clearly demonstrate the conduction anisotropy (if any), and are free from grain-boundary effects. For technical utilization, however, they are usually impractical due to their unsuitable shape and size, and high cost of production.

Ceramics – that is, sintered polycrystalline bodies (often containing some glass phase) – are much more suitable for technical purposes because they may be

Solid State Electrochemistry I: Fundamentals, Materials and their Applications. Edited by Vladislav V. Kharton

ISBN: 978-3-527-32318-0

fabricated at a reasonable cost in a variety of desired shapes, from mere discs to closed-end tubes and thick-film electrode/electrolyte assemblies. At the same time, the properties of ceramics depend heavily on processing conditions determining porosity, grain-boundary phases, grain orientation (see Section 7.1.5), and size. Impedance spectroscopy analysis enables the elimination of grain-boundary contributions to the total resistance of the polycrystalline materials, and many groups have reported only bulk (i.e., grain interior) conductivity. However, in this chapter total conductivities, which are more important for practical applications, are analyzed.

Cast bodies may be prepared rarely, only when the material melts congruently at a relatively low temperature. These are polycrystalline, similar to ceramics, but may exhibit a higher degree of grain orientation and usually have larger grains and less mechanical strength. *Cold-pressed powder compacts* may have sufficient density, mechanical strength and conductivity only with very plastic materials. Most compounds considered in this chapter, especially those with high connectivity (see Section 7.1.4), have rigid grains; their powder compacts are porous and fragile, and possess much lower conductivities compared to ceramics. *Thin films, composites, polymers* and *glasses* are beyond the scope of this chapter; hence, only bulk single-phase crystalline materials are considered here.

7.1.2 Type of Disorder

It is common practice to divide all high-conductivity electrolytes into two classes, namely those with intrinsic and extrinsic disorder. However, it is suggested that a more detailed classification might include four classes

(i) *Stoichiometric compounds with intrinsic disorder.* These have a fixed rational composition (dictated by the oxidation states of the components) which, however, does not correspond to the multiplicities of the crystallographic sites. Mobile ions are distributed randomly over a larger number of energetically equivalent (or almost equivalent) positions, and may hop from one to another. Examples are $Na_5RSi_4O_{12}$ and $Na_3Sc_2(PO_4)_3$ (see Section 7.4), and high-temperature forms of AgI and Li_2SO_4 (see Figure 7.1).

(ii) *Stoichiometric compounds with heterovalent (aliovalent) dopants.* In these materials, the positional disorder and ionic conductivity (as in the case discussed above) are provided by doping a stoichiometric phase with an additional component. Without the dopant, the phase is ordered and has poor ion diffusivity. Within the framework of the point-defect approach, the ionic charge carriers are either vacancies or interstitial ions. This model, however, is not applicable to the high-conductivity materials with a great degree of disorder and strong interactions between the defects. For example, in the $Na_{1+x}Zr_2Si_xP_{3-x}O_{12}$ system ($0 \leq x \leq 3$), compositions with $x \approx 0$ may be considered as Si-doped $NaZr_2P_3O_{12}$ with interstitial Na^+, and compositions with $x \approx 3$, as P-doped $Na_4Zr_2Si_3O_{12}$ with sodium vacancies, but the both descriptions fail for the most-

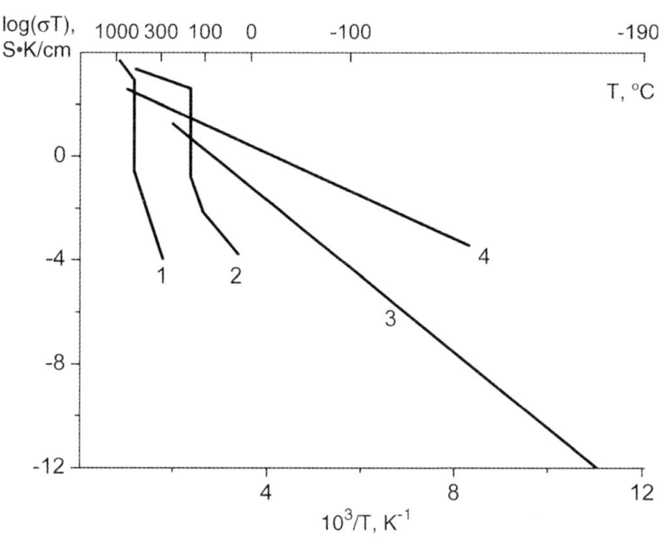

Figure 7.1 Arrhenius plots for ionic conductivity. (1) Li_2SO_4 [2]; (2) AgI [2]; (3) Li_3N [22]; (4) Na-β-alumina [23]. Plots (1) and (2) produced with polycrystals; plots (3) and (4) produced with single crystals; conductivity perpendicular to the hexagonal axis.

conducting material, $x \approx 2$, which is structurally related to $Na_3Sc_2(PO_4)_3$ (see Section 7.4.2).

(iii) *Nonstoichiometric phases.* In these materials, deviation from the ideal stoichiometry and resulting lattice disorder, similar to those in class (ii), are intrinsic features of the phase [as in class (i)], rather than result of doping. In other words, there is a "self-doping": an excess of the intrinsic component acts as a dopant. Classical examples are β- and β″-aluminas (see Section 7.4.1) that do not exist in stoichiometric ordered forms.

(iv) *Stoichiometric ordered ionic conductors* here. No considerable disorder can be detected (all sites are completely filled), but a significant ionic conductivity is still observed due to cooperative phenomena. Rare examples are hollandite-type phases (see Section 7.4.6). In some instances, the ionic conductivity may even decrease with heterovalent doping creating cation vacancies.

Some of the class (i) compounds are formed on a polymorph transformation from a low-temperature ordered phase. This has led to a popular view ("superionic transition fallacy" [22]) that every solid electrolyte is a result of "superionic phase transition", analogous to superconductive and ferroelectric transitions. In fact, the last two types of phase transition are always within the same structure type, with only subtle changes in bond lengths and small (if any) latent heats, but with drastic changes in their electrical properties. On the other hand, the drastic changes of ionic conductivity (Figure 7.1, graphs 1 and 2) are only observed at reconstructive phase transitions (e.g., from *hcp* to *bcc* in AgI, or from monoclinic to cubic in Li_2SO_4), with

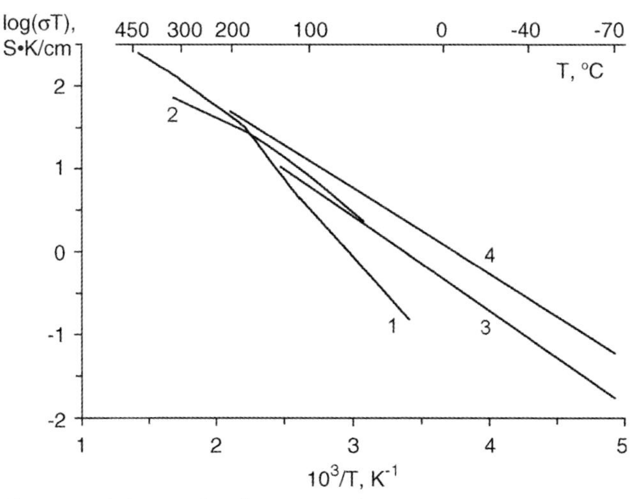

Figure 7.2 Arrhenius plots for ionic conductivity. (1) $Na_3Zr_2Si_2PO_{12}$ ceramics [24, 25]; (2) $Na_5YSi_4O_{12}$ ceramics [26]; (3) and (4) $Na_5YSi_4O_{12}$ single crystal, parallel and perpendicular to the main axis, respectively [27].

substantial latent heats. Most solid electrolytes of all the classes (i)–(iv) do not show any reconstructive transitions. Their conductivity usually varies smoothly with temperature from very low to very high values (Figure 7.1, graphs 3 and 4; Figure 7.2, graphs 3 and 4). Thus, there is no principal difference between "superionic" and "normal ionic" states.

In some materials, however, second-order phase transitions (or first-order transitions with very small latent heats) may occur within the same structure type. These are manifested in changing the slope of the Arrhenius plot (or, sometimes, in minor drops of the conductivity). One typical example is a distortion-type phase transition in $Na_3Zr_2Si_2PO_{12}$ from a high-temperature rhombohedral phase to low-temperature monoclinic polymorph at 180–200 °C (Figure 7.2, graph 1) established by numerous structural, calorimetric, and dilatometric studies [2, 24].

For ceramic samples, similar deflections of the Arrhenius plots may be due not only to phase transitions but also to grain-boundary effects (see also Chapter 4). These become significant at low temperatures, increasing the total resistance and apparent activation energy (E_a). On heating, the grain boundary resistance decreases more rapidly with respect to the grain interior, and can be often neglected above a certain temperature. This is illustrated by graphs 2–4 in Figure 7.2, where changing the slope at approximately 175 °C is observed only for ceramics, but not for single crystals, indicating that it is not associated with a phase transition. Nevertheless, a second-order transition (R32 $\leftrightarrow$ R$\bar{3}$c) does occur at about 150 °C, although this is only detected by the disappearance of certain X-ray reflections and is not accompanied by any thermal, electrical, or lattice parameter anomalies [28, 29].

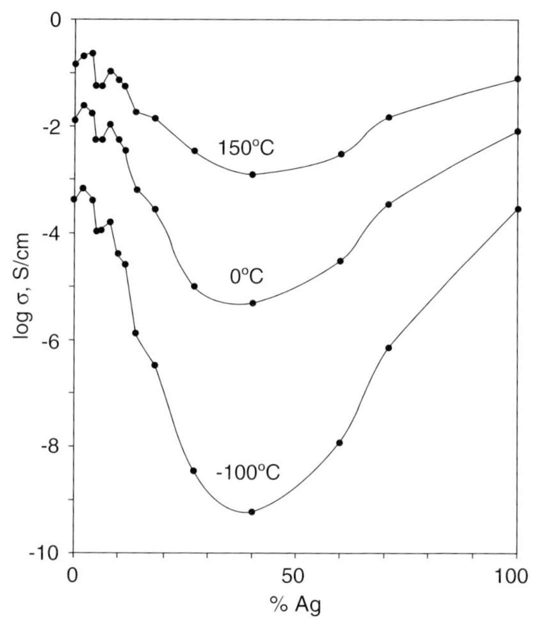

Figure 7.3 Ionic conductivity isotherms for (Na,Ag)-β-alumina solid solutions [33].

7.1.3 Type of Charge Carrier

Most solid electrolytes are *unipolar conductors* with only one type of mobile ion. Therefore, they are classified as anionic and cationic conductors, while the latter are further classified as proton conductors, Li^+ conductors, Na^+ conductors, and so on. When various types of potentially mobile cation are mixed in a solid solution, the activation energies are usually higher and ionic conductivities lower than expected from interpolation between the properties of the end members (see Figure 7.3 for an example). This *mixed alkali effect* (although the ions are not necessary alkali) has long been known in glasses [30]; it is also characteristic of the mixed cation β- and β″-aluminas, analogous polygallates and polyferrites [1, 2, 31–37], doped mixed alkali monoferrites, monoaluminates, monogallates [38, 39], and pyrophosphates [40, 41]. It is less pronounced in the higher-conductivity phases (β″-alumina type) and at higher temperatures. There are numerous theoretical approaches to explain this effect [42–46].

7.1.4 Connectivity of the Rigid Lattice

The *connectivity* (D) is a number of spatial dimensions (0, 1, 2, or 3) in which the atoms are connected infinitely by strong (mostly covalent) bonds. Discrimination between "strong" and "weak" bonds may be based on bond valences estimated as the formal valence of the ion divided by its coordination number (assuming equal bonds within the coordination group). For example, in $Na_5YSi_4O_{12}$ [28], Na–O, Y–O and Si–O bond valences are, respectively, $1/6 = 0.17$, $3/6 = 0.5$, and $4/4 = 1$. One may consider both Na–O and Y–O bonds as "weak"; then, it is a $D = 0$ structure, sodium yttrium cyclosilicate. However, by taking physical properties into account, another approach is more realistic, with only Na–O bonds weak; then, it is a $D = 3$ structure, sodium yttrosilicate.

The structures with $D = 3$ (frameworks) are usually most stable mechanically, thermally, and chemically. The layered ($D = 2$) and chain ($D = 1$) structures, and especially structures with finite anions ($D = 0$) with similar bond strength and polarity, are more labile: many of these may absorb water into the interlayer or interchain space, or hydrolyze to produce amorphous products or even be water-soluble. In addition, structures with $D = 1$ or 2 usually display strong anisotropy of the conductivity and/or thermal expansion (although some exceptions are possible). Multiple examples with $D = 0$, 2, and 3 are listed in Sections 7.4–7.6.

7.1.5
Connectivity of the Migration Paths

As for Section 7.1.4, the *connectivity* (*C*) of migration paths is a number of spatial dimensions (0, 1, 2, or 3) in which the positions of mobile ions are connected infinitely via suitable pathways (see Sections 7.2 and 7.3). Obviously, the materials with $C = 0$ are poor ionic conductors, but for $C = 3$ the conductivity of high-quality ceramics should be almost equal to that of a single crystal (Figure 7.2, graphs 2–4). With $C = 2$, conductivity of the single crystal in the "easy" direction and that of the ceramics with randomly oriented grains, differ only by a factor of 2–4 [2, 8, 47], but with $C = 1$ this difference is much greater, by up to two to three orders of magnitude [48, 49]. Thus, the electrolytes with $C = 1$ may be used only as single crystals or grain-oriented ceramics. It should be noted that there are no interdependences between *D* and *C*. Examples of sodium-deficient but low-conducting ($C = 0$) phases with $D = 0, 2,$ and 3 are $Na_{1-2x}Mn_xCl$ [50], $Na_{2-x}Ti_{3-x}Nb_xO_7$ [51], and $Na_{1-2x}Pb_xNbO_3$ [52], respectively; examples of layered ($D = 2$) phases with $C = 1$, 2, and 3 are $Na_{4+x}(Ti,M)_5O_{12}$ [53], $Na_xCr_xTi_{1-x}O_2$ [54], and $Li_3Zn_{0.5}Nb_2O_7$ [55], correspondingly. For details and further examples, see Sections 7.4–7.6.

7.1.6
Stability to Oxidation and Reduction

This property is important from a practical standpoint as solid electrolytes must provide a stable operation of the electrochemical cells, often in contact with strong oxidizing and reducing agents (oxygen, halogens, hydrogen, alkali metals, etc.), and often at elevated temperatures. Of the electronegative components, only fluoride, oxide and, occasionally, nitride and chloride are appropriate. However, essentially rigid frameworks are rare for halogens due to their low valence. Thus, most attention is usually paid to oxides. Among the electropositive components of the rigid lattice, Si

(4 +), B(3 +), Al(3 +), Be(2 +), Mg(2 +), Zr(4 +), Hf(4 +), Ta(5 +) and some rare earths (3 +) seem to be most reliable because they provide strong bonding to oxygen and high stability to oxidation and reduction. Other electropositive components are not sufficiently stable to reduction [e.g., S(6 +), P(5 +), and Sn(4 +)], or to oxidation [e.g., Cr(3 +) and Mn(2 +)], or both [e.g., Mn(3 +) or Co(3 +)]. The majority of transition elements change their oxidation state very easily and, even when the structure is not completely destroyed, the resultant mixed-valence states usually give rise to electronic conduction.

7.1.7
A Comment on the Activation Energy

Microscopically, the activation energy (E_a) is an energetic barrier for the elementary act of ion transport and, obviously, depends on the crystal structure, in particular, on the bottleneck size (see Section 7.2.2). Macroscopically, it is determined from the slope of the Arrhenius plot, assuming that the E_a is independent of temperature. This assumption is not necessary true, especially in the vicinity of a phase transition when the structure changes rapidly. For example, dilatometric studies of $Na_3Zr_2Si_2PO_{12}$ [24] show anomalously strong expansion between 150 and 200 °C (i.e., just below the phase transition point), and this corresponds to the region of elevated slope of the Arrhenius plot (Figure 7.2, graph 1). Here, the high *apparent* E_a does not mean a narrow bottleneck and poor conductivity; rather, it means a drastic *increase in conductivity* due to rapidly widening bottleneck and decreasing true E_a with decreasing the monoclinic distortion of the parent rhombohedral structure. At the transition point, the apparent E_a changes from 39–41 to 19–23 kJ mol^{-1} [24, 25], although the structure remains essentially the same. A similar behavior has been observed in $Na_3Sc_2(PO_4)_3$ [56].

7.2
Crystal-Chemistry Factors Affecting Cationic Conductivity

7.2.1
Structure Type

The main factor determining the possibility of high cationic conductivity in a crystalline phase is its structure type. This determines possible sites for potentially mobile cations, their coordination, type of bottlenecks (see Section 7.2.2), connectivity, and so on. Other factors (which are discussed in Sections 7.2.3–7.2.4) are also very important, but only valid within a given structure type. For example, the perovskite type, with its cubo-octahedral cation sites connected via square bottlenecks (Figure 7.4), is absolutely inappropriate for transport of large cations such as Na^+ or K^+, and any substitutions may hardly change this situation to a radical degree.

Most ion-conducting structure types may only be discovered by way of serendipity. For example, the structure of one of the best solid electrolytes, β-alumina,

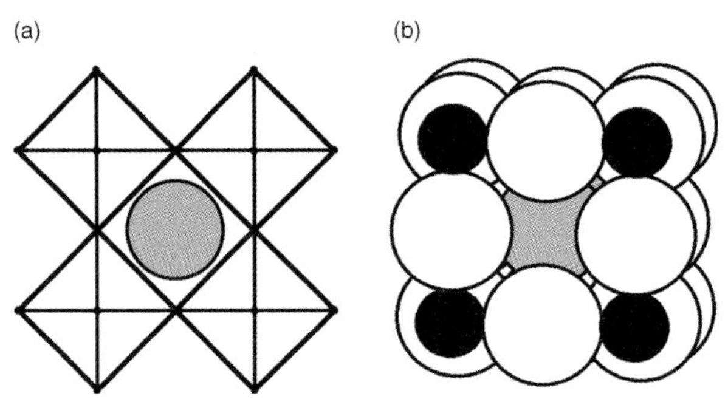

Figure 7.4 Perovskite structure (e.g., $NaNbO_3$). (a) Polyhedral view, making an illusion of a channel for Na^+; (b) Space-filling model, showing Na^+ in a closed cage. Na = gray, Nb = black, O = white.

$Na_2O{\cdot}nAl_2O_3$, $n \approx 9$ (see Section 7.4.1), be it not found experimentally, could hardly be predicted from a theoretical standpoint. With its very low sodium content (e.g., compared to $NaAlO_2$), it might seem less probable that sodium sites are connected into an infinite network. The low O : Al ratio (~1.56) indicates that the average coordination number of O with respect to Al should be at least 2.6 (if all Al^{3+} ions were tetrahedrally coordinated), or even 3.9 (if all Al^{3+} were octahedral). This might predict a great extent of condensation of AlO_n polyhedra, and the formation of diffusion paths appears doubtful. The real structure (Figure 7.5), indeed, contains highly condensed spinel-type blocks, but these are interleaved with loosely packed sodium-conducting layers providing a high conductivity. It would be even more difficult to predict, theoretically,

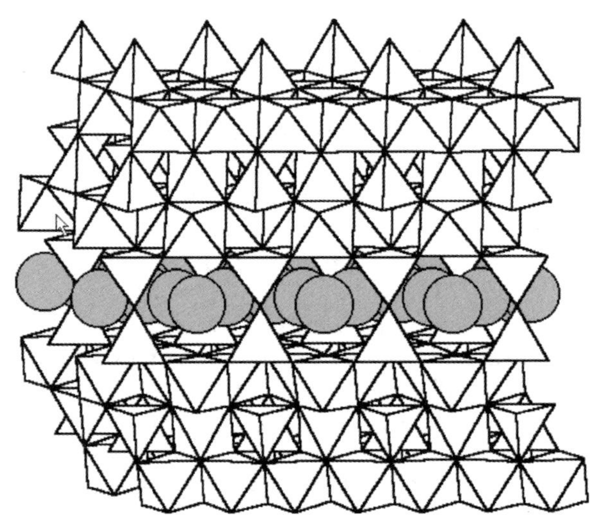

Figure 7.5 Crystal structure of sodium β-alumina [57]; one of the two Na^+-conducting planes, together with the two adjacent spinel blocks. Na ions = gray; AlO_4 tetrahedra and AlO_6 octahedra = white.

the thermodynamic stability with respect to various hypothetical combinations of phases having the same gross composition, such as $NaAlO_2 + 4Al_2O_3$ or $Na_2O{\cdot}nAl_2O_3$ with $n = 8$ and 11, or 7 and 11. Consequently, it is possible to define the main ways of finding new, high-conductivity crystalline electrolytes:

- Selection of an appropriate structure type among those found experimentally, by using the criteria discussed in Section 7.3, and further modifications (see Sections 7.2.2–7.2.4).
- Construction of a new structure type from the structurally related blocks found experimentally, that is, making a heteropolytype. A rare successful example of such an approach is synthesis of the $A_4M_{11}O_{30}$ series [58], built from the previously known $A_2M_7O_{18}$ and $A_2M_4O_{12}$ structures (A = Rb, Cs, Tl; M = Nb or Ta combined with W, Mo or Ti). Note, however, that the same structure type was simultaneously found empirically [59, 60].
- Free experimental search in systems selected according to certain criteria, like those discussed in Section 7.1.6.

7.2.2 Bottleneck Concept and Size Effects

One of the main obstacles in the migration of large cations such as Na^+, K^+, or NH_4^+ (not Li^+ or H^+) in rigid lattices is a too-small size of bottlenecks, or diffusion channels, between the cation sites, as shown above for the perovskite type (Figure 7.4). This factor is most critical for high-connectivity structures having $D = 3$ or 2. For $D = 1$ no data are available, while for $D = 0$ this factor may sometimes be of minor importance as small isolated ions (e.g., SO_4^{2-}) may rotate in the crystal structures, providing the so-called "paddle-wheel" mechanism [1, 2, 5, 61].

For oxides, many dozens of structure types have short unit cell axes of ~3 Å (an edge of an oxygen octahedron or tetrahedron) or ~4 Å (a diagonal of an octahedron). In these specific instances, the bottleneck sizes may be estimated without the complete determination of crystal structures. It has been shown [62] that all "4 Å structures" (especially those with $D = 3$) inhibit the motion of large cations (Na^+, K^+, but not Li^+; for example, see Figure 7.4). All "3 Å structures" containing large cations provide their high mobility along this short axis and, if $D = 3$, this mobility can only be one-dimensional ($C = 1$). This rule is illustrated by multiple examples in Sections 7.4.4 and 7.4.6.

For optimal ionic conductivity, the ion size should fit the size of the migration channel and the channel should be smooth, without very wide and very narrow sections. For β-alumina, the optimum sizes of mobile ions are those of Na^+ and Ag^+ (Figure 7.6). The small Li^+ ions are shifted off the conduction plane shown in Figure 7.5, and are strongly attached to the "wall" of the migration channel [65]. As might be expected from Figure 7.6, a high pressure will increase Li^+-ion conductivity due to adjusting the channel size to the Li^+ size, and decrease K^+-ion conductivity in β-alumina crystals [66]. For the $A_4Nb_6O_{17}$ structural family [67, 68],

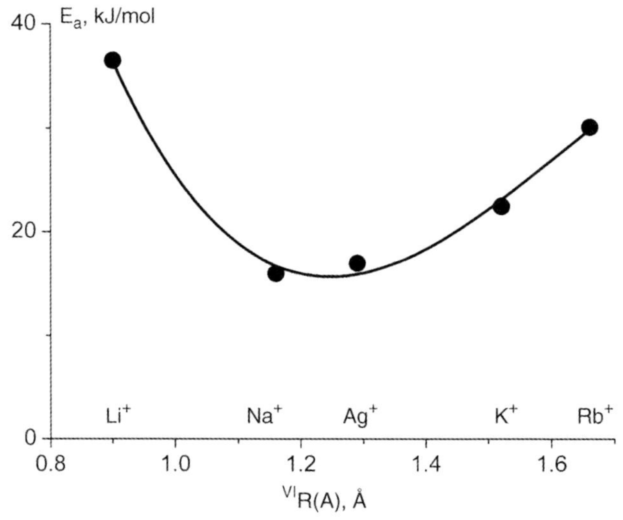

Figure 7.6 Activation energies of ionic diffusion in β-alumina crystals [63] versus octahedral radii of the mobile ions [64].

the optimum ion radius is that of A = K^+, as conductivities with A = Na^+ and Tl^+ are essentially lower.

For a given mobile ion, the channel size may be adjusted by substitutions in the rigid lattice. When $D = 3$ or 0, it seems obvious that increasing the size of a rigid lattice ion should cause a more or less isotropic expansion and, thus, a widening of the bottlenecks. This is illustrated by the increase in lattice parameters and ionic conductivities if comparing $K_{1-x}R_{1-x}T_xO_2$ with R = Al,Ga,Fe and T = Si,Ge,Ti [1], and also by many other examples (Figures 7.7 and 7.8).

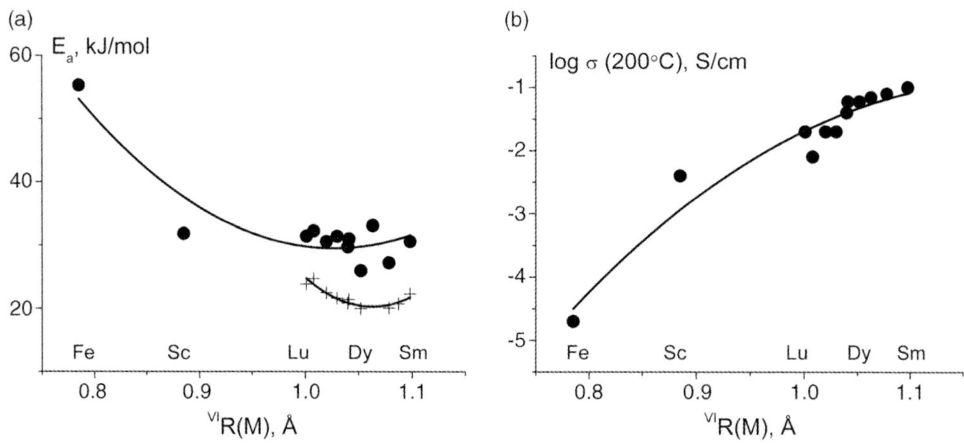

Figure 7.7 (a) Activation energies and (b) conductivities at 200 °C of $Na_5MSi_4O_{12}$ ceramics versus octahedral radii of M [64]. The crosses indicate data from Ref. [29]; filled circles indicate data from Ref. [69].

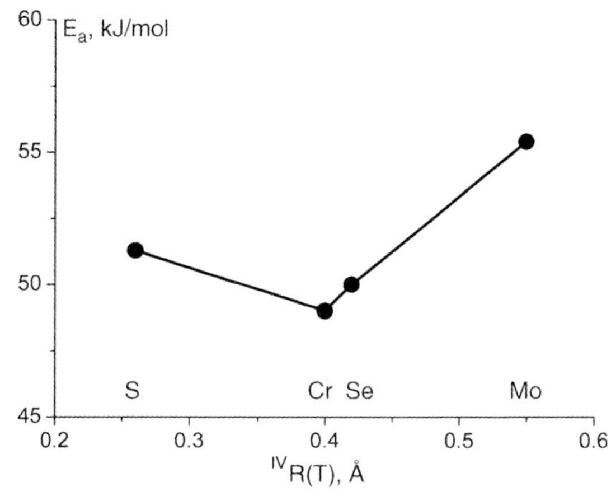

Figure 7.8 Activation energy of ionic conductivity of orthorhombic $Li_{3.5}Si_{0.75}T_{0.25}O_4$ [1] versus tetrahedral radius of the dopant T(6+) [64]. The plot differs from the original one [1] due to different values of radii.

For layered structures, however, the situation is less straightforward. In $A_x(M,M')O_2$, based on the octahedral brucite-like layers (Figure 7.9), increasing the hexagonal lattice parameter a [due to an increase in the size of $(M,M')O_6$ octahedra] leads to a *decrease* in the interlayer distance; the mobile cation A resides in a trigonal prism or antiprism and, when its basal edge (lattice parameter a) increases, its height should

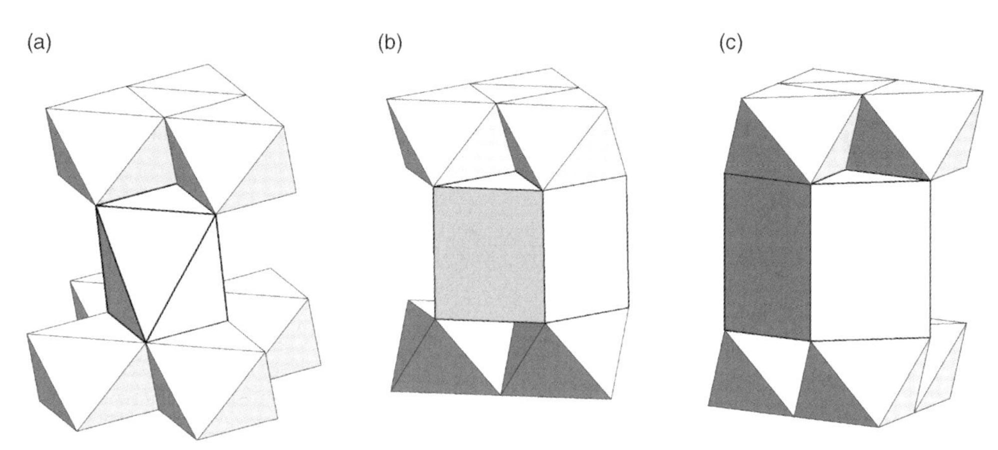

Figure 7.9 Polyhedral presentations of the three cation-conducting polytypes $A_x(M,M')O_2$ based on brucite-like octahedral layers. (a) O3; (b) P2; (c) P3. Small M and M′ cations are in octahedra, larger A cations are in antiprisms or prisms. The symbols denote coordination of the alkali ion (*P*rism or expanded *O*ctahedron) and number of layers in the hexagonal unit cell (2 or 3) [72].

decrease to maintain the normal A–O distance. This, in turn, makes the bottlenecks narrower. As an illustration, in the O3-type $Na_{0.8}(M_yM'_{1-y})O_2$, where MM′ = NiTi, FeTi, FeSb, NiSb [70, 71], a increases, c remains almost constant, and the ionic conductivity decreases, despite an increasing electronegativity of the MM′ pair (see Section 7.2.4).

7.2.3 Site Occupation Factors

In the classical theory of point defects (see Chapter 3), increasing ionic conductivity due to heterovalent doping in a disordered lattice should be associated mainly with rising pre-exponential factors, depending on the concentration of charge carriers, whereas the transport mechanism and E_a should remain essentially unchanged [3]. However, this model becomes invalid for fast ion-conducting materials where the charge carrier concentration is very high. At high doping levels (or in intrinsically nonstoichiometric phases), collective effects become important; the structures change significantly with changing composition, and maximum conductivities are usually attained due to minimum activation energies. In many cases, increasing the nonstoichiometry parameter x leads not only to a modification of the parent structure, but also to phase transition(s) into a structurally related phase with a higher conductivity, such as $\beta \rightarrow \beta''$ in $A_{1+x}(R,M)_{11}O_{17}$ (A = Na, K; R = Al, Ga), O3 → P2 in $Na_{1-x}(Ti,M)O_2$, $Li_4SiO_4 \rightarrow \gamma$-$Li_3PO_4$ in $Li_{4-x}(Si,T)O_4$ (see Sections 7.4 and 7.5).

7.2.4 Electronegativity, Bond Ionicity, and Polarizability

In a mixed oxide A_xMO_n, where A is a mobile cation and MO_n denotes a rigid lattice, any increase in the electronegativity of M is expected to shift electron density from O towards M and, thus, to increase the O–A bond ionicity, which should be favorable for the A mobility [72–74]. Such effects are not very strong and can be separated from other factors (structure type, bottleneck size, site occupancy) by comparing isostructural phases containing cations of similar size and oxidation state: Nb and Ta (5+), Zn or Ni and Mg (2+), Sn and Zr or Hf (4+). Indeed, introducing the first ion from each pair provides somewhat higher alkali ion or proton conduction in $K_{1-x}(M,M')O_2$ (M,M′ = In, Sn, Zr, Hf, Zn, Mg) [73, 74], $AMWO_6$ and AM_2O_5F (A = Rb, Tl, Cs; M = Nb, Ta) [75], $H_2M_2O_6{\cdot}xH_2O$ (M = Nb, Ta) [76], and $Na_{0.9}M_{0.45}Ti_{1.55}O_4$ (M = Ni, Mg) [49]. Unfortunately, the most electronegative cations are usually just least stable to reduction (see Section 7.1.6). For two ions with similar radii (Tl^+ and Rb^+), the former is more polarizable and has a considerably higher mobility in the same rigid lattice of the pyrochlore type [75]. The high electronegativities of Ag^+ and Cu^+ are extremely important from another point of view, namely for the formation of fast ion-conducting iodide and sulfide phases which are structurally different from the alkali compounds.

7.3
Crystal Structural Screening and Studies of Conduction Paths

As discussed in Section 7.2, the topology and geometry of the rigid sublattice, crystal chemical characteristics of its components (primarily electronegativity and polarizability), and availability of vacant sites in the mobile sublattice determine whether the material can demonstrate a high cationic conductivity. Among these factors, the question of whether the topology and geometry of the framework are suitable for diffusion is the least obvious, as the electronegativities and polarizabilities are tabulated and the presence of vacancies in the cation sublattice is usually known from structural studies, or even from the gross formula. Hence, it becomes necessary to consider major methods that may be used in order to identify and characterize diffusion pathways.

7.3.1
Topological Analysis with Voronoi Tessellation

In computational geometry, the Voronoi tessellation (VT) is defined as subdividing space containing n sites into n regions, so that any point of each region is closer to its respective site than to any other site. One example of a Voronoi diagram on a plane with randomly distributed sites is shown in Figure 7.10. The line segments and vertices of a Voronoi diagram consist of the points at the farthest distance from any two or three adjacent sites, respectively. This feature of the Voronoi network results in a wide range of applications ([77] and references therein). For three-dimensional (3D) space, VT is performed in a similar fashion, producing the

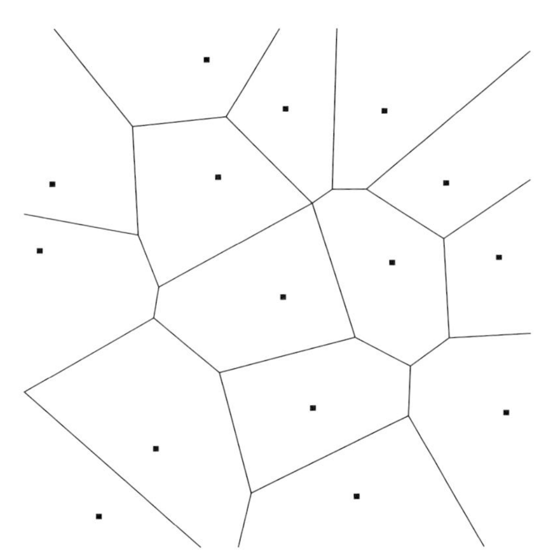

Figure 7.10 Two-dimensional Voronoi tessellation.

so-called Voronoi polyhedra which are bound by faces equidistant from any given pair of sites; the vertices and edges of the polyhedra are furthest from the surrounding sites. The radii of circumspheres centered at Voronoi polyhedra vertices characterize the size of inter-site voids, while the radii of circumcircles centered at the polyhedra edges characterize the bottleneck size for the channels connecting voids. Originally, VT was used to study atomic arrangements in amorphous systems [78, 79], but was then further developed for locating and characterizing voids (cavities) and channels in amorphous [80, 81] and crystalline structures [82–85] and in isolated macromolecules [86, 87]. However, it should be pointed out that, for atomic structures where the components are significantly different in size, a simple bisecting partitioning based only on the position of atomic centers results in an unphysical picture, with Voronoi polyhedra faces lying within larger atoms. For such systems, the problem is solved either by using a "radical planes" approach, when the distance from atoms to the faces of Voronoi polyhedra is proportional to relative atomic radii [88, 89], or by more mathematically rigorous approach using hyperbolic surfaces [90].

In practice, at the first stage of the VT analysis, an infinite 3D graph containing all the Voronoi polyhedra edges and vertices is constructed. The vertices and edges corresponding to cavities and bottlenecks of unsuitable size are then filtered out. Finally, the resultant conductivity graph is analyzed in terms of dimensionality, connectivity of fragments, radius of cavities circumsphere, bottleneck radius, pathway length, and so on. Software developed specifically for the analysis of crystal or molecular space includes Voro3D [91], PROVAT [92], and MOLE [93], all of which focus on the analysis of voids and channels in biomolecules, and also TOPOS [82, 94], which was designed for the analysis of periodic crystal structures.

Application of the VT is illustrated using $Na_5YSi_4O_{12}$ as an example. The rhombohedral structure contains $Si_{12}O_{36}{}^{24-}$ rings packed into columns and connected by octahedral Y^{3+} into a 3-D framework (Figure 7.11a). There are six nonequivalent sodium sites. Sites 4–6 (two of them being only partially occupied), with very short separations of 0.81–2.85 Å and high thermal parameters, are connected into non-intersecting oblique diffusion pathways between the columns (Figure 7.11b), parallel to $\langle 2\,1\,1\rangle$ (in the hexagonal notation) – that is, the edges of a rhombohedron. Sites 1–3 are within the columns and have greater Na–Na distances; the possibility of their transport is not clear. Moreover, if the paths shown in Figure 7.11 are the only existing paths, the structure cannot be a true 3-D ($C = 3$) conductor: long single-crystal bars parallel or perpendicular to the principal axis will be nonconductive, and only those parallel to one of the three channel axes will be conductive.

The analysis of the VT has shown that the bottleneck radius for Na(1)-Na(3) transport along the principal axis is 2.08 Å, and that for the oblique $\langle 2\,1\,1\rangle$ path is 2.22 Å. Our analysis of a great number of sodium-containing oxide compounds [84] has shown that these values are characteristic of poor and good Na^+ ion conductors, respectively. Thus, Na(1)-Na(3) sites cannot provide a high mobility, and only Na(4)-Na (6) sites are responsible for ionic conduction. The most interesting result, however, is that there are no bottlenecks at all between the two Na(5) sites of the different $\langle 2\,1\,1\rangle$

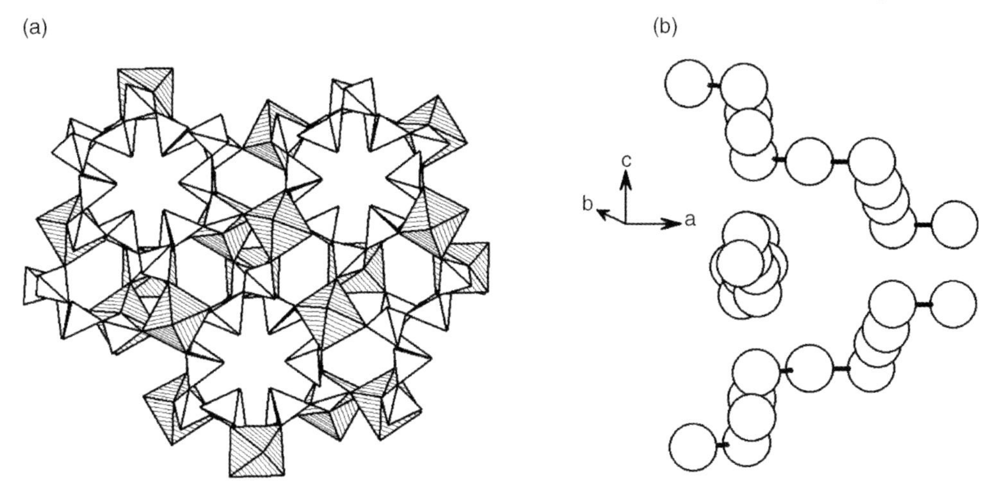

Figure 7.11 Crystal structure of $Na_5YSi_4O_{12}$ [28]. (a) The framework viewed along the threefold axis (sodium ions omitted); (b) Sodium sites 4, 5 and 6, connected into three infinite nonintersecting chains.

chains, although the distance between these sites is fairly long, at 3.39 Å. In other words, Na(5) sits in the narrowest section of the channel along the c-axis. Thus, the paths shown in Figure 7.11 are really interconnected and ion transport along [0 0 1] is possible. Yet, the c-axis conductivity is lower than that in the perpendicular direction (see Figure 7.2), and this is one more illustration of the fact that too wide channels are not favorable for ion transport.

The algorithms of Voronoi polyhedra construction may easily be used for the high-throughput analysis of new materials to identify the presence of an infinite network of channels with suitable size for ionic conductivity or molecular diffusion. At the same time, the approach has a limited applicability for those systems where the topological and geometric factors do not play a significant role, such as proton conductors, or where the framework is not truly immobile and participates in the conduction process, for example, via a paddle-wheel mechanism.

7.3.2
Topological Analysis with Bond-Valence Maps

Among the first theories describing behavior of atoms in ionic crystals were Pauling's principles, which determined the structure of complex ionic crystals [95]. The second of these rules postulates that ionic charges are compensated locally by the nearest counterions; that is, the bond valence sum of all bonds of a given ion should be equal to its formal charge or valence. With growing numbers of analyzed crystal structures, it became clear that the definition of bond valence given in Section 7.1.4 was adequate only for the simplest cases with uniform coordination polyhedra. For the general case of distorted polyhedra, bond valence must be weighed against the cation–anion distance. A number of "bond valence–bond distance" functions have been proposed over the past few decades, the most successful being that of Brown [96], in the form that combines sufficient accuracy and simplicity of use:

$$\nu_{ij} = \exp\left(\frac{R_0 - R_{ij}}{b}\right) \tag{7.1}$$

where ν_{ij} is the individual bond valence, R_0 is the tabulated parameter, R_{ij} is the observed bond length, and b is a constant with a typical value 0.37 Å. Now, the bond valence analysis is extensively used for testing structure solutions for correctness, locating light atoms, identifying atoms with different oxidation states in charge ordered systems, and so on (see Ref. [96] and references therein).

As follows from Equation 7.1, a bond valence may be calculated at any point at the distance R_{ij} from a given atom. A 3D grid of points for which a sum of bond valences to the nearest anions is calculated within a given distance threshold may be used to visualize the spatial distribution of bond valence in crystal structure, and is called a bond valence map. These maps may be constructed for both crystalline and amorphous materials, given the structural model obtained either using diffraction methods or reverse Monte Carlo (RMC) modeling, respectively (see Refs [97, 98] and references therein). Iso-surfaces interpolated through the points with bond valence sums equal to the formal charge of a mobile cation can be used to identify the most energetically favorable sites. Qualitatively, these iso-surfaces (or equivalently, iso-surfaces of low bond valence mismatch $\Delta\nu = |\nu_i - \nu_{\text{ideal}}|$) which extend through the whole crystal structure, indicate pathways suitable for ion migration. Several computer programs have been developed for calculating and visualizing bond valence maps, examples being VALE [99] and VALMAP [100]. An example of pathways modeling with bond valence maps is presented in Figure 7.12. In addition to such qualitative topological analysis that produces results very similar to those of VT, it was proposed recently that ionic conductivity and activation energy might be predicted by analyzing the structure volume fraction accessible by infinite pathways [101].

As with VT, identifying conduction pathways with bond valence maps provides accurate predictions for ionic conductors (both crystalline and amorphous) with a percolation mechanism of conductivity, but is less successful when modeling proton conductors.

7.3.3 Static First-Principles Calculations and Molecular Dynamics Modeling

The behavior of atoms in condensed matter is ultimately defined by the cooperative interaction of nuclei and electrons. If it was possible to accurately describe this interaction for a material on length and time scales that are studied experimentally in a laboratory or are used in practice, any physical property could be predicted *ab initio* given just its chemical composition (see also Chapter 5). Unfortunately, a many-body Schrödinger equation is not analytically solvable for any system of acceptable size. The most successful approach to treating many-body problems is the density functional theory (DFT) developed by Kohn *et al.* [103, 104]. The DFT reduces many-particle electron–electron interactions to a single electron potential expressed

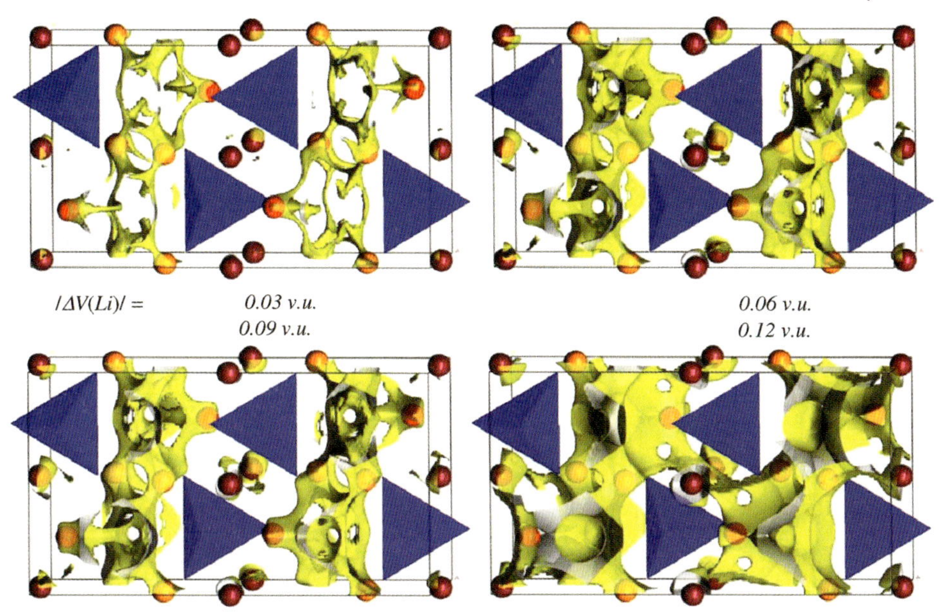

Figure 7.12 Bond valence model of Li pathways in Li_4GeS_4 depicted as iso-surfaces for different values of the lithium BV mismatch in the range 0.03–0.12 valence units. Reproduced with permission from Ref. [101].

as the exchange-correlation functional, depending only on the electron density distribution. The exact form of the functional that would reproduce the solution of the Schrödinger equation for the same system is not known, but a number of approximate functionals that can be treated numerically have been proposed, such as local density approximation (LDA), generalized gradient approximation (GGA), and their extensions. For more information, the reader is referred to the extensive literature on the subject (e.g., Ref. [105]). At this point, the application of DFT will be limited to ionic conductors.

The procedure of modeling diffusion processes with static DFT calculations is somewhat similar to the bond valence mapping, in which bond valence sums are calculated on a mobile ion placed at arbitrary positions in crystal structure. Let us consider an ionic conductor having a number of vacancies in the connected network of mobile ion sites. Crystallographic cells of such materials are often described with fractional atomic sites occupancies. Therefore, at first, a supercell containing whole atoms must be constructed by expanding the basic unit cell. Next, for all nonequivalent distributions of mobile ions over available sites, the total energy is calculated and the most stable configuration determined. Those configurations with the energies closest to the ground state are the most probable intermediate states during diffusion process, and indicate the most energetically favorable diffusion paths. The following (usually even more tedious and time-consuming) stage is the analysis of total energy as a function of the mobile ion position during a single act of hopping from the occupied site to the adjacent vacant site. In return, however, as the result of such energy profiling as a function of the mobile ion position along the migration path between two sites, the value of the activation barrier energy that is linked to the hopping rate (Γ) is obtained, according to transition-state theory [106]:

$$\Gamma = \nu_0 \exp\left(-\frac{\Delta E}{k_B T}\right) \tag{7.2}$$

where ν_0 is the attempt frequency ($\sim 10^{12}$–$10^{13}\,s^{-1}$), ΔE is the activation barrier energy, k_B is the Boltzmann constant, and T is the temperature. The hopping rate, in turn, may be translated into the jump diffusion coefficient that directly affects ionic conductivity properties:

$$D_J = \Gamma l^2, \tag{7.3}$$

where l is the hopping distance. Recent examples of modeling diffusion processes in ionic and mixed conductors, for example, Li_xCoO_2 [107], $LiFePO_4$ [108], Li_{1+x}-Ti_2O_4 [109], and Li_xWO_3 [110], have demonstrated good agreement with the experimental values obtained using impedance spectroscopy or nuclear magnetic resonance (NMR).

In order to model dynamic behavior on the atomic level, special techniques were developed that now constitute the molecular dynamics (MD) branch of computer simulations. The concept of MD is very simple, being based on Newton's equations of motion. For a set of atoms with given positions and velocities, the forces on each atom are calculated, after which the atoms are allowed to move to new positions during a time period on the scale of femtoseconds. In this way the acceleration remains approximately constant, and such a cycle of calculations is repeated as many times as required. In the simplest approach of classical MD, the forces are based on empirical interatomic potentials; this makes MD calculations relatively fast, such that systems have been studied with up to $\sim 10^{12}$ particles [111] or up to the microsecond range on the time scale. However, such an approach with fixed potentials becomes inadequate when the character of atomic interactions changes during the course of the simulated process, or extreme conditions (temperature, pressure) are to be simulated when the potentials are poorly known. A significantly more computationally expensive (but more accurate) approach is the so-called *ab initio* MD (AIMD), as developed by Car and Parrinello [112]. AIMD combines the classical description of nuclei with a DFT treatment of electronic density. Knowledge of the effective potentials is no longer required as the forces exerted on nuclei are calculated on-the-fly using DFT. Electronic degrees of freedom also provide AIMD with an ability to simulate chemical reactions. The only disadvantage of AIMD is that it is two to three orders of magnitude slower than classical MD, and this affects the accessible size of the system and the time scale.

Despite the rapid development of both computing hardware and numerical algorithms, the modeling of real experiments (10^{-6}–10^{-1} m length scale and 10^{-1}–10^{3} s time scale) with (AI)MD is not yet feasible. Nevertheless, a variety of material characteristics may be extracted using MD simulations, such as thermal expansion coefficients, vibrational density of states, and time dependence of mobile ion mean square displacement that may be translated into a diffusion coefficient. The pair correlation function, in combination with the analysis of ion trajectories, provides a direct means of visualizing migration paths and determining the positions of intermediate saddle-points. The "adiabatic trajectory method" [113], which is very similar to point-by-point barrier energy profiling, may be used to obtain microscopic barrier energies; from a series simulations of the system at different temperatures, an ion diffusion activation energy may be determined. Several recent examples of MD modeling for materials with different mobile ions have included $La_{0.6}Li_{0.2}TiO_3$ [114], yttria-stabilized zirconia [115], and proton-conducting perovskites H^+-ABO_3 [116].

7.3.4
Analysis of Diffraction Data with Maximum Entropy Method

The classical crystal structural analysis using X-ray or neutron diffraction data is based on the reconstruction of scattering density (electron or nuclear, respectively) in the unit cell via an inverse Fourier transform of experimentally measured structure factors. In order to reduce a number of parameters, for a long time the method remained "atom-centric" – that is, the electron density distribution was approximated by a set of atoms each described by a tabulated value of an atomic X-ray scattering factor or a coherent neutron scattering length. The dynamics was simulated by a displacement from an equilibrium position either by isotropic (sphere) or anisotropic (ellipsoid) thermal displacement parameters. Obviously, such a model is often inadequate for description of ionic conductors with inherent static and/or dynamic disorder of mobile ions, especially at high temperatures. To some extent, the distribution of scattering density related to ion mobility may be revealed via the careful analysis of difference Fourier maps (e.g., Ref. [117]), or by using the probability density function [118, 119]. However, these methods are affected by the structural model used, and also suffer from truncation effects as a Fourier summation must be carried out over an infinite set of terms; this is not available in practice and is especially reduced for powder diffraction experiments due to peak overlap.

An alternative approach would be to model scattering density distribution over a 3-D grid, mapping the unit cell. As the number of values to be determined will, in general, significantly exceed the number of observed structure factors, there will be an infinite number of solutions. The most probable solution (or the least committed with respect to missing experimental data) would be that having the maximum entropy (hence maximum entropy method; MEM) for the information on electron density in all grid points. Therefore, the task of scattering density reconstruction is solved by maximizing the entropy functional and simultaneously reproducing the observed structure factors. For technical details and a short historical review of MEM application in crystal structural analysis, the reader is referred to Ref. [120]. As MEM analysis results in a model of continuous distribution of scattering density in crystal structure, it becomes the perfect tool for studies of ionic conductors. In fact, the method is currently implemented in a number of computer programs that can be used for analysis of either single-crystal or powder diffraction data, both in 3-D and superspace; examples include ENIGMA [121], PRIMA + RIETAN-2000 [122], and BayMEM [123]. Given high-quality diffraction data and a sufficient contribution of mobile ions to the total X-ray or neutron scattering, MEM analysis can be used for the direct visualization of migration paths in crystal structures. The analysis of data collected at variable temperatures is especially interesting, as it allows one to follow the development of disorder on heating. Relevant examples include $Li_{1-y}Ni_{0.5}Mn_{0.5}O_2$ [124], $Ce_{0.93}Y_{0.07}O_{1.96}$ [125], and $Na_{0.74}CoO_2$ [126].

The capabilities of the methods described in Sections 7.3.1–7.3.4 are summarized in Table 7.1.

Although MD modeling is capable of very accurate predictions of ion diffusion topology, and may be used for estimating absolute conductivity values, its very high computational costs mean that it is not the ideal tool for screening purposes. Until now, the most efficient workflow for identifying crystal structures with potentially high ionic conductivity has consisted of a VT analysis, followed by ionic conductivity measurements. However, insights into the microscopic mechanisms of ionic conduction may be obtained via further MD modeling and MEM analyses of experimentally acquired diffraction data.

Table 7.1 Capabilities of methods for the conduction paths analysis.

	Voronoi tessellation	**Bond valence mapping**	***Ab initio* static calculations and MD**	**MEM**
Conductors with percolation mechanism	Yes	Yes	Yes	Yes
Conductors with non-percolation mechanisms	No	No	Yes	Yes
Non-crystalline conductors	Yes[a]	Yes[a]	Yes	No
Non-ambient conditions	Yes	No[b]	Yes	Yes
Computational cost	Very low	Very low	High	Low
Quantitative information obtained	Void size, bottleneck radii	Activation energy and conductivity (via empirical models)	Diffusion coefficient, thermal expansion coefficient[c], activation energy[c]	Scattering density distribution

[a] Given a model obtained by reverse Monte-Carlo modeling.
[b] Only very limited data on temperature and pressure dependence of bond valence parameters are available.
[c] From a series of MD simulations for different temperatures.

7.4 Conductors with Large Alkali Ions

7.4.1 β/β″-Alumina, β/β″-Gallates and β/β″-Ferrites [1–3, 6, 7, 23, 31–37, 47, 57, 63, 65, 66, 127–135]

Four structural families of materials with an Na^+-ion conductivity in excess of 0.1 S cm^{-1} at 300 °C are known presently (Sections [7.4.1–7.4.4]; see also Table 7.2). Of these, the β- and β″-aluminas have been studied most extensively. The isostructural sodium gallates, potassium, rubidium and cesium gallates, aluminates, and ferrites are also known. The ideal formulas for the β and β″ structures are $AM_{11}O_{17}$ and $A_2M_{11}O_{17}$, respectively, but all of the compounds are intrinsically nonstoichiometric and the actual alkali content is always greater than unity but less than 2. The most obvious charge compensation mechanism is a lower-valence cation substitution for trivalent M^{3+}, namely $A_{1+x}(M^{3+}_{11-x}M^{2+}_{x})O_{17}$ (M^{2+} = Mg, Co, Ni, Zn) and $A_{1+x}(M^{3+}_{11-x/2}Li^{+}_{x/2})O_{17}$. Without these dopants, the ferrites are charge-compensated by Fe^{2+} (giving rise to electronic conductivity), and sodium gallates, by sodium substitution for gallium. For aluminates, these variants are impossible. As a result,

Table 7.2 Properties of best Na^+-ion conducting ceramics of various structure types.

Formula	*D*	*C*	σ (S cm^{-1}) 25 °C	σ (S cm^{-1}) 300 °C	E_a (kJ mol^{-1})
β″-Alumina [37, 47]	3	2	$(2–3)\times10^{-3}$	0.15–0.4	14–17 (>250 °C) 27–28 (<200 °C)
β-Alumina [47]	3	2	$(1–8)\times10^{-3}$	0.08	14–25
$Na_3Zr_2Si_2PO_{12}$ [24, 25, 137, 140]	3	3	$(3–10)\times10^{-4}$	0.14–0.22	18–20 (>200 °C)
$Na_{3.2}Hf_2Si_{2.2}P_{0.8}O_{12}$ [141]	3	3	0.0023	0.22	18 (>200 °C)
$Na_3Sc_2P_3O_{12}$ [2, 142–145]	3	3	10^{-6}–10^{-4}	0.04–0.17	14–18 (>170 °C)
$Na_3Fe_2P_3O_{12}$ [2]	3	3	10^{-8}	0.01	
$Na_5GdSi_4O_{12}$ [69, 148, 149]	3	3	0.0016[b]	0.18–0.30[b]	18–27
$Na_5YSi_4O_{12}$ [26, 69, 149]	3	3	$(0.8–1.4)\times10^{-3}$	0.13–0.15	14–24
P2 $Na_{0.5}Ni_{0.25}Mn_{0.75}O_2$ [155]	2	2	0.002	0.20[a]	26
P2 $Na_{0.6}Cr_{0.6}Ti_{0.4}O_2$ [54]	2	2	0.003	0.12	23
P2 $Na_{0.64}Ni_{0.32}Ti_{068}O_2$ [152]	2	2	5×10^{-4}[b]	0.08	30
P3 $Na_{0.5}Cr_{0.5}Ti_{0.5}S_2$ [150]	2	2		0.008 (500 K)	34
Cubic $NaSbO_3$ and $Na_7Sb_6O_{18}F$ [67, 137]	3	3	4×10^{-6}–10^{-4}	0.03–0.08	34–49
O3 $Na_{0.8}Fe_{0.8}Ti_{0.2}O_2$ [70, 151]	2	2	2×10^{-5}	0.006–0.010	33
O3 $Na_{0.72}Fe_{0.72}Mn_{0.28}O_2$ [155]	2	2	1.2×10^{-5}	0.018[a]	41
$Na_{2.4}P_{0.4}S_{0.6}O_4$ [165]	0	3	10^{-7}[b]	0.012	42–82
$Na_4TiP_2O_9$ [166]	1	3		0.008	30 (>290 °C)
$Na_{1.85}Zn_{0.925}Si_{1.075}O_4$ [159]	3	3	10^{-6}	0.007	47

[a] With 3–4% electronic contribution.
[b] Extrapolated value.

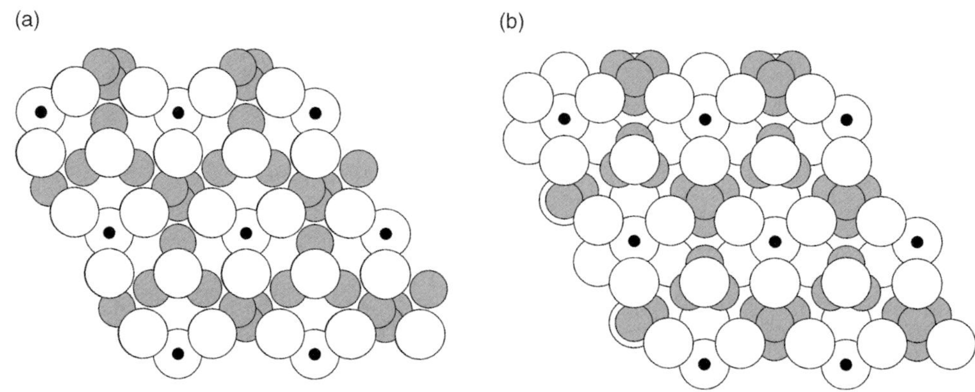

Figure 7.13 Conducting layers in the structures of Na-β-alumina [57] (a) and Mg- or Zn-stabilized Na-β″-alumina [130, 131] (b) , viewed along the hexagonal axis. White balls = oxygen; small black balls = aluminum; gray balls = sodium sites (only partially filled).

non-doped Na-β″-alumina is thermodynamically unstable and only appears in an intergrowth with the β-alumina; β-alumina itself is charge-compensated by an interstitial oxygen anion in the conduction plane bonded to two Frenkel-type Al defects; that is, Al^{3+} cations displaced from the normal octahedral sites into interstitial tetrahedra: $Na_{1+x}Al_{11}O_{17+x/2}$.

The crystal structure of β-alumina is shown in Figure 7.5. The β″ structure consists of the same bidimensional spinel blocks connected by bridging oxygens in the third dimension; thus, both structures are 3D but quasi-layered, and differ in the stacking mode of the spinel blocks and structure of the conducting layers (Figure 7.13). In the β structure, the layers are built from trigonal prisms, O_6, having also bridging oxygens as additional neighbors. In the β″ structure, the layers consist of tetrahedra (O_4) and trigonal antiprisms (O_6), with two additional neighbors. As these polyhedra are only partially occupied, each cation has some adjacent polyhedra occupied and some empty. This results in an unbalanced A^+–A^+ repulsion which shifts the cations from the centers of their polyhedra: an ideal site splits into several adjacent sites (gray balls in Figure 7.13) with low occupancies. Short distances between these sites exclude their simultaneous occupation and show easy pathways for cation transport. Some information on transport properties of this family is represented in Figures 7.1, 7.3, 7.6, 7.14 and Table 7.2.

7.4.2
Nasicon Family [1, 2, 4, 6, 7, 24, 25, 136–147]

At high temperatures, the $Na_{1+x}Zr_2Si_xP_{3-x}O_{12}$ system represents a continuous series of rhombohedral solid solutions in the whole range $0 \leq x \leq 3$. The framework is built of ZrO_6 octahedra sharing all vertices with SiO_4/PO_4 tetrahedra (Figure 7.15a). Compositions in the vicinity of $x = 2$ exhibit very high ionic

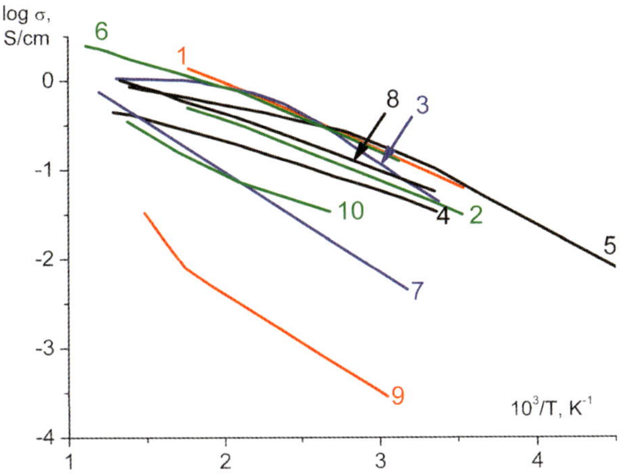

Figure 7.14 Single-crystal ionic conductivities. Plot (1) Na-β″-gallate; (2) Na-β-gallate [129]; (3) Na-β″-alumina; (4) Na-β-alumina [47]; (5) K-β″-alumina, Mg-stabilized [132]; (6) same [133]; (7) K-β-alumina, Mg-doped [133]; (8) K-β″-gallate, Na-stabilized [31]; (9) K-β-gallate, Zn-doped [134] (10) K-β″-ferrite, Cd-stabilized [135].

conductivity (Figures 7.2 and 7.16), and are called "NASICON", an acronym for *Na SuperIonic CONductor* [136, 137].

At $x=0$, all octahedral Na(1) sites shown as gray balls in Figure 7.15 are filled, while the other sites are empty. At $x=3$, both Na(1) and eight-coordinate Na(2) sites are completely occupied. As a result, both of these end members are poor ionic conductors. In the intermediate region, a partial occupation of the Na(2) sites creates

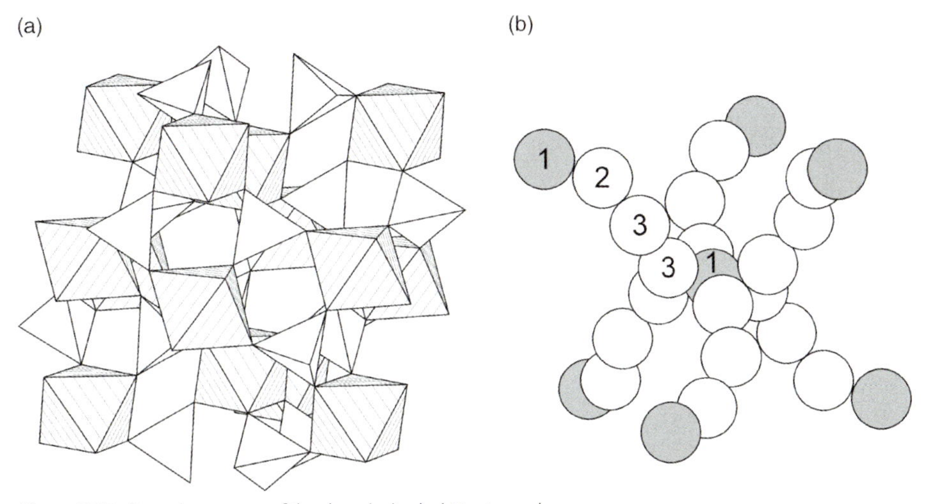

Figure 7.15 Crystal structure of the rhombohedral Nasicon phase $Na_3Zr_2Si_2PO_{12}$ at 350 °C [138]. (a) The framework (sodium cations omitted); (b) Sodium sites only (framework omitted).

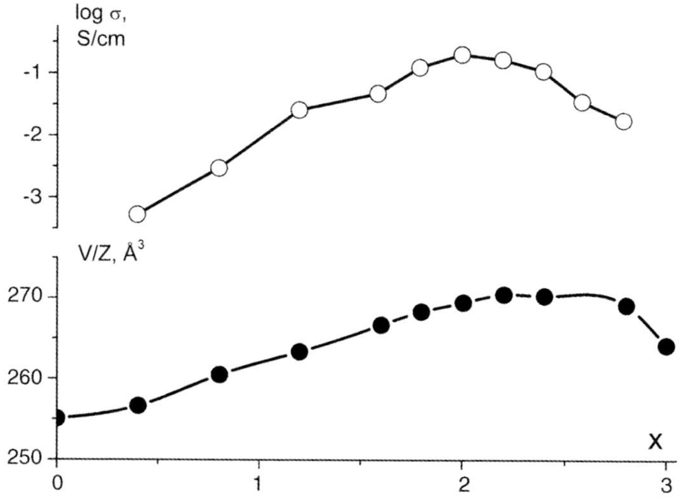

Figure 7.16 Compositional dependence of room-temperature formula volume [136] and conductivity at 300 °C [137] in the $Na_{1+x}Zr_2Si_xP_{3-x}O_{12}$ system.

an unbalanced Na^+–Na^+ repulsion similar to that discussed in Section 7.4.1. Some Na(1) ions are then displaced from their ideal positions towards the bottlenecks; this causes the bottlenecks to become wider and produces maximum volume, maximum conductivity (Figure 7.16), site splitting, and the low-temperature monoclinic distortion discussed in Sections 7.1.2 and 7.1.7. Note that the same intermediate compositions exhibit maximum thermal expansion [24], so that the volume anomaly should enhance with temperature. At high temperatures, the monoclinic distortion vanishes due to a fast Na^+ motion; in the diffraction data, this motion is described either by strongly anisotropic thermal parameters of Na(1) and Na(2), up to connection of their densities into a continuous "basin" [139], or by the appearance of an intermediate site, Na(3) (see Figure 7.15). At 350 °C, the occupancies of Na(1), Na(2), and Na(3) are 0.4, 0.68, and 0.106, respectively [138].

Very similarly, the $Na_{1+x}Hf_2Si_xP_{3-x}O_{12}$ system [140, 141] shows maximum volume, minimum E_a and maximum conductivity at $x = 2.2$. An additional NASICON analogue is *Phoscan*, $Na_3Sc_2P_3O_{12}$; however, the polymorphism of this is more complex and dependent on the processing conditions [2, 142–145].

Although a great number of other NASICON-type Na^+-ion conductors have been studied [1, 2, 4, 146, 147], all have exhibited lower conductivities (see Table 7.2), mostly due to geometric constraints, as their formula volumes (203–261 Å³) are less than the optimum value (see Figure 7.16), or due to a non-optimum sodium content. However, neither these factors nor the electronegativity can explain the decrease in conductivity when Ge is substituted for Si [140, 146], when As is substituted for P [140], or when In is substituted for Sc and Ga is substituted for Cr in $Na_{1+x}M^{3+}_xZr_{2-x}(PO_4)_3$ [147]. In all of these pairs, lower conductivities were observed with framework "cations" having filled the d-shell.

7.4.3 Sodium Rare-Earth Silicates [1, 2, 7, 9, 26–29, 69, 148, 149]

The main characteristics of the $Na_5MSi_4O_{12}$ family are presented in Sections 7.1.2, 7.1.4 and 7.3.1; see also Figures 7.2, 7.7, 7.11 and Table 7.2). This structure type exists with the majority of rare earths M^{3+} cations, except for the five largest [69], but could not be prepared when Li or K was completely substituted for Na.

7.4.4 Structures Based on Brucite-Like Octahedral Layers [1, 2, 7, 8, 54, 70–74, 150–157]

As for β- and β″-alumina, the structures shown in Figure 7.9 have the same space groups ($P6_3/mmc$ and $R\bar{3}m$, respectively), and contain conductive layers consisting of trigonal prisms (in P2 and P3) or octahedra and tetrahedra (in O3 or α-$NaFeO_2$ type). However, in contrast to the aluminas, these layers contain no bridging oxygen, the rigid layers are thinner, and the concentration of the conducting layers is, therefore, greater. The absence of the bridging oxygen anions has two opposing effects on the ionic transport. First, the interlayer distance becomes shorter, which is unfavorable; second, bottlenecks in the prismatic structures become rectangular (whereas they are triangular in β-, β″-alumina and O3), and this is favorable. As a result, the P2- and P3-type phases are always better ion conductors than O3-type phases built of the same components, and compare well with β-aluminas. The general formulas for these families are $A_x(M^{3+}_xM^{4+}_{1-x})O_2$, $A_x(M^{2+}_{x/2}M^{4+}_{1-x/2})O_2$, $A_x(M^{+}_{x/3}M^{4+}_{1-x/3})O_2$, and $A_x(M^{2+}_{(1+x)/3}M^{5+}_{(2-x)/3})O_2$, where $A^+ = Na$, K; $M^+ = Li$; $M^{2+} = Mg$, Co, Ni, Zn, Ca; $M^{3+} = Cr$, Fe, In; $M^{4+} = Mn$, Ti, Sn, Zr, Hf; $M^{5+} = Sb$, although not all combinations thereof may be realized. Prismatic coordination is usually less stable for ionic compounds than octahedral coordination, due to shorter O–O distances and a greater repulsion, and P2 or P3 phases do not exist in the vicinity of $x = 1$. Only O3 type (and, sometimes, tetrahedral structures, see Section 7.4.5) appear there.

However, the A^+–A^+ repulsion is also significant as the hexagonal parameter a is rather short (ca. 3 Å). When x diminishes, this parameter usually becomes even shorter due to the smaller radii of the higher-valence cations listed above. The P2 and P3 structures may then be stabilized because the number of prisms is twice the number of octahedra in O3; distributing A cations over these prisms makes some A–A distances greater than parameter a, a feature impossible at $x = 1$ and in the O3 structure [154]. The A–A repulsion is illustrated by the neutron diffraction data for P2-type titanates at 10–300 K indicating off-center location of Na^+ ions within their prisms of two types [153]. For Na compounds, the formation of P2 or P3 oxide phases requires values of the crystallographic a parameter not exceeding ~2.99 Å, whereas it may be even 3.23 Å for K-containing compounds [72]. High temperatures are also favorable for the prismatic structures due to fast ion motion and entropic stabilization [157].

Thus, the main origin for rising ionic conductivity with decreasing x is not merely a growing concentration of vacancies but, mainly, an increase in the bottleneck radius owing to the rearrangement of O3 to P2 or P3 and, within a given structure type,

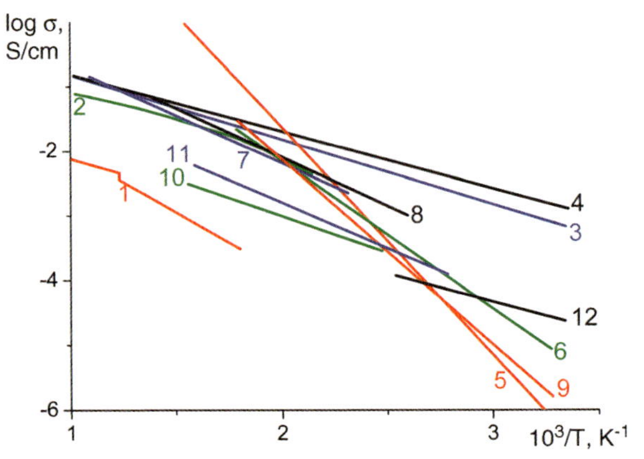

Figure 7.17 Potassium-ion conductivity of ceramic materials of various structure types. Cristobalite-related: $KAlO_2$ (1) and $K_{0.9}Al_{0.95}P_{0.05}O_2$ (2) [162]; $K_xAl_xTi_{1-x}O_2$ ($x = 0.1$–0.25) (3) [1, 161]; $K_{0.86}Fe_{0.86}Ti_{0.14}O_2$ (4) [1]; P2-type: $K_{0.72}In_{0.72}Sn_{0.28}O_2$ (5) [73]; $K_{0.70}Zn_{0.35}Sn_{0.65}O_2$ (6) [74] and $K_{0.56}Ni_{0.52}Sb_{0.48}O_2$ (7) [156]; P3-type: $K_{0.59}Mg_{0.53}Sb_{0.47}O_2$ (8) [156] and $K_{0.58}Zn_{0.29}Sn_{0.71}O_2$ (9) [74]; framework pyrochlore type: $KTi_{0.5}W_{1.5}O_6$ (10) and $KAl_{0.33}W_{1.67}O_6$ (11) [167]; layered $K_4Nb_6O_{17}$ (12) [67].

owing to shorter a (see Section 7.2.2), and off-center displacements (as discussed in Sections 7.4.1 and 7.4.2). The best known sodium and potassium ion conductors of the above-discussed types are compared in Table 7.2 and Figure 7.17.

7.4.5
Cristobalite-Related Tetrahedral Frameworks [1, 2, 8, 158–164]

Until now, three structural families with high potassium ion conductivity have been identified; two of these were discussed in Sections 7.4.1 and 7.4.4, while the third family is detailed in this section. This family is also characterized by maximum Rb^+ and Cs^+ transport [1, 2].

Cristobalite is the high-temperature form of SiO_2. Compounds AMO_2 (A = Na, K, Rb, Cs; M = Al, Ga, Fe) and some quaternary oxides (e.g., Na_2RXO_4 (R = Be, Zn; X = Si, Ge [159]) have tetrahedral $MO_{4/2}$ frameworks of the same topology, but are "stuffed" with A cations. $AAlO_2$ (A = Rb, Cs) [158] have ideal cubic structures with 12-coordinate A in the form of Laves' tetrahedron with wide hexagonal bottlenecks (Figure 7.18). With smaller A cations, the cubic phases are only stable at high temperatures [1, 160, 161], whereas room-temperature structures are puckered orthorhombic to provide normal A–O bond distances and lower coordination numbers. Here, again, drastic increases in alkali ion conductivity on heterovalent doping ($A_{1-x}M_{1-x}T_xO_2$, T = Si, Ge, Ti [1, 2], $A_{1-x}M_{1-x/2}E_{x/2}O_2$, E = P, Nb, Ta [162, 164] or $A_{1-2x}R_xMO_2$, R = Ba, Pb [163]) is associated with a decreasing E_a, owing to wider bottlenecks (especially with the largest T and R), and with the stabilization of high-temperature phases. Maximum conductivities observed on excess A_2O against the above formulas were explained by a combination of M and O vacancies formed

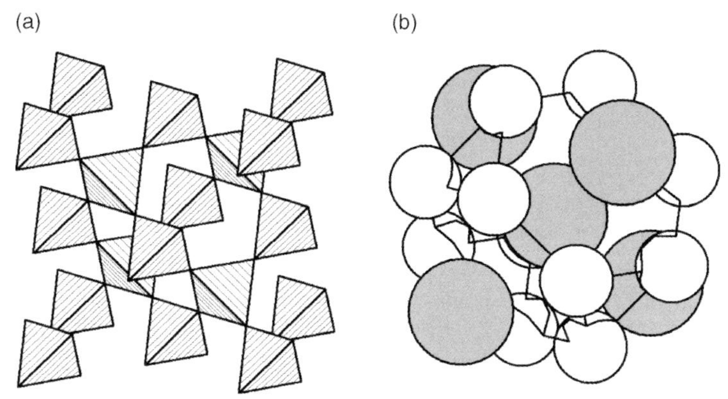

Figure 7.18 Idealized crystal structure of "stuffed cristobalite" AMO_2 typified by $CsAlO_2$ [158]. (a) The tetrahedral MO_2 framework (A cations omitted); (b) Environment of an A^+ cation (gray): 12 O^{2-} and four A^+ (M cations omitted). The room-temperature structure of $K_{1-x}Al_{1-x}Ti_xO_2$ ($x = 0.1$–0.2) only differs by splitting of O and K sites [161].

on A^+ insertion [1], although this seems very questionable from the crystal chemistry point of view. A very high volatility of A_2O at high sintering temperatures (1100–1300 °C) is a more plausible explanation. In the caption to Figure 7.17, this A_2O excess has been subtracted.

7.4.6
Other Materials

The best Na^+, K^+, Rb^+, and Cs^+ ion conductors all belong to the five families described above. Some other high-conductivity materials are represented in Table 7.2 and Figure 7.17, but descriptions of their structures are omitted here. Only a very specific class of oxide structures with a 3 Å translation (types I–XIV in Figure 7.19 [48, 49, 53, 168–187]) will be reviewed briefly. This short distance precludes the location of any atom or ion between two large alkali ions, guaranteeing their free motion in that direction [62]. The freedom may, however, be restricted by impurities, crystal defects (e.g., anions on cation sites [188]) and grain-boundaries in ceramics.

For single crystals of the types II [174] and III [48] with Na^+, types I [171–173] and IV [178] with Na^+ and K^+, type V with K^+ and Cs^+ [179], AC conductivities of the order of 10^{-2} S cm^{-1} at 200–300 °C (or even at 25 °C [171, 178]) have been reported. The ceramics' conductivity is significantly lower: (2–6) $\times 10^{-4}$ S cm^{-1} in types VII–IX [177], 10^{-5}–10^{-4} S cm^{-1} in types III and VI [49, 177, 180], and 10^{-6}–10^{-4} S cm^{-1} in type I [168–170], all at 300 °C. In contrast to all phases discussed in Sections 7.4.1–7.4.5, the type I $K_xAl_xTi_{4-x}O_8$ [168, 170] (and, probably, $K_xZn_{x/2}Ti_{4-x/2}O_8$ [170]) and type III $Na_xMg_{x/2}Ti_{2-x/2}O_4$ [49] showed the highest conductivity, with maximum filling of the alkali channels ($x = 1$ and $x = 0.95$, respectively). Obviously, ion conduction in the 1-D channels is a cooperative process (possibly,

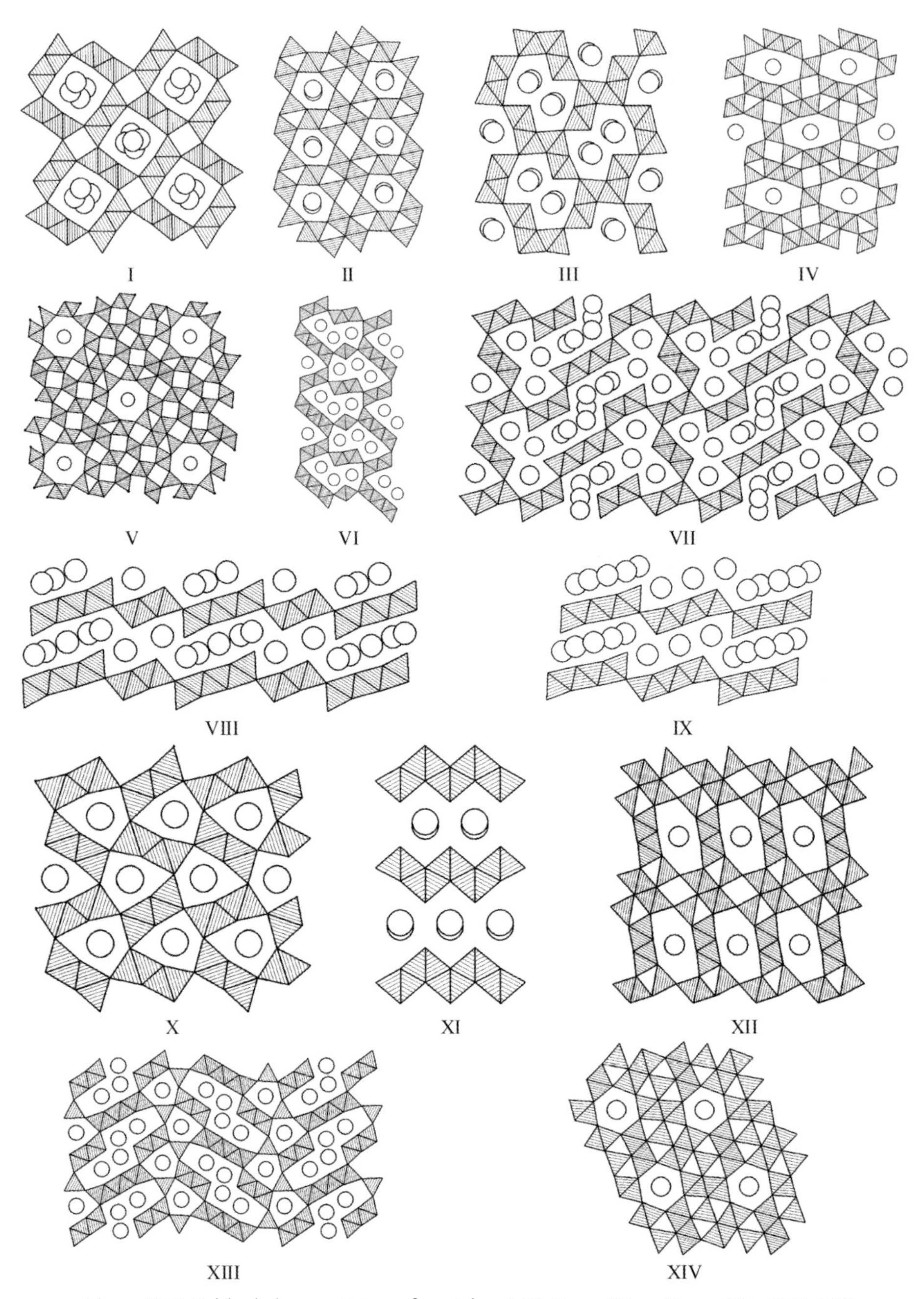

Figure 7.19 Polyhedral presentation of crystal structures viewed along the short axes of 2.88–3.01 Å. (I) hollandite [168–173]; (II) $Na_{0.7}Ga_{4.7}Ti_{0.3}O_8$ [174]; (III) $Na_x(M,Ti)_2O_4$ [48, 49, 175–177]; (IV) $A_{1-x}M_{5-x}Ti_{2+x}O_{12}$ (A = Na, K; M = Al, Ga) [178]; (V) $A_xGa_{16+x}Ti_{16-x}O_{56}$ (A = K, Rb, Cs) [179]; (VI) $Na_{2+x}(M,Ti)_4O_9$ [180–182]; (VII) $Na_{8+x}(Mg_{x/2}Ti_{11-x/2})O_{26}$ [177, 181]; (VIII) $Na_{4+x}(M,Ti)_5O_{12}$ (M = Mg, Zn, Fe. . .) [53, 177]; (IX) $Na_{4+x}(Mg,Ti)_6(O,F)_{14}$ [53, 177]; (X) $NaFeTiO_4$ [183]; (XI) $A_x(M,Ti)_2O_4$ (A = K, Rb, Cs; M = Mg, Zn, Fe, Li . . .) [184]; (XII) $Na_xGa_{4+x}Ti_{2-x}O_{10}$ [185]; (XIII) $Na_4Mn_4Ti_5O_{18}$ [186]; (XIV) $K_{0.8}Mg_{1.7}(Cr,Fe,Ti,Zr)_6O_{12}$ [187].

soliton-like) and not due to vacancy hopping. Macroscopic transport in the 1D channels has been demonstrated by fast ion exchange in type III and VI single crystals of 2–3 mm length along the channel axis [175, 182]. For types XI–XIV, ionic conductivity has not yet been reported, but may be anticipated. The narrowest bottleneck is characteristic of type X and it, indeed, is a poor ion exchanger [175]. Types VIII and IX (and probably also XI), although layered, should be considered as quasi-1D conductors [53].

7.5 Lithium Ion Conductors

7.5.1 General Comments

Li^+ ion conduction is known in a great variety of structure types [1–7, 9–13]. Here, attention is focused on the structures where the conductivity of at least 0.01 S cm^{-1} at 300 °C has been attained. Some relevant examples are listed in Table 7.3. Li^+ is a very poor X-ray scatterer and, thus, its positions and occupancies may be accurately determined only by neutron diffraction; X-ray data should be regarded with caution, especially in the presence of heavy elements, such as tantalum, rare earths, niobium, zirconium, indium, and tin.

7.5.2 Garnet-Related Mixed Frameworks of Oxygen Octahedra and Twisted Cubes [189–202]

Garnet is a cubic mineral (space group $Ia\bar{3}d$) with the typical formula $Ca_3Al_2Si_3O_{12}$, where the coordination numbers of Ca, Al and Si are 8, 6, and 4, respectively. For the cation-excess $Li_5La_3Ta_2O_{12}$ (Figure 7.20), the $La_3Ta_2Li(1)_3O_{12}$ array is of inverse garnet type with Li(1) on the Si sites, whereas additional Li(2) and Li(3) sites are split within an interstitial octahedral void sharing its opposite faces with two Li(1)

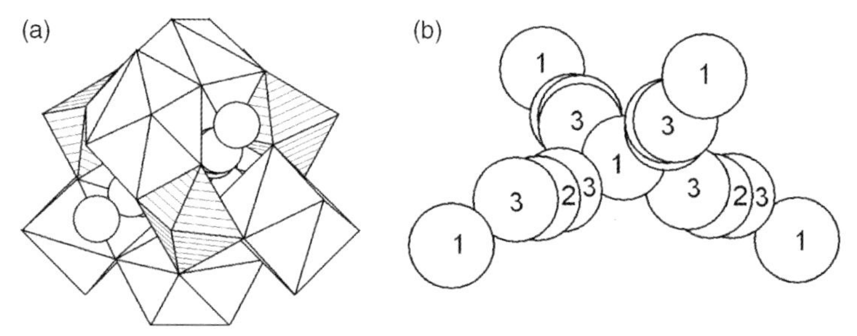

Figure 7.20 Crystal structure of $Li_5La_3Ta_2O_{12}$ [192]. (a) Fragment of the framework: The TaO_6 octahedra are hatched; the LaO_8 polyhedra are white; the balls indicate Li; (b) Li sites only.

tetrahedra. These sites form a continuous net of diffusion paths with intersite distances of 1.57 and 0.44 Å and Li(1), Li(2), and Li(3) occupancies of 0.80, 0.14 and 0.15, respectively. The sum occupancy of the interstitial octahedron is, therefore, $0.14 + 2\times0.15 = 0.44$. Of course, the real Li–Li distances are greater (~2.4 Å). Several isostructural phases are known, with M = Nb, Zr, Sb, Bi, W, and Te substituting for Ta, R = Ca, Sr, Ba, Pr, Nd, Sm, Eu, In, and K substituting for La, and the Li content varying between three and seven atoms per formula unit. An increasing Li content leads to depopulation of the tetrahedral sites and a higher occupancy of the octahedra. The highest ionic conductivities are found for largest R = Ba, La, largest M = Zr, Bi, and largest Li content (Table 7.3). An important feature of the Zr- and Ta-containing compounds is their redox stability.

Table 7.3 Properties of best polycrystalline Li^+ ion conductors of various structure types.

Formula	*D*	*C*	σ (S cm^{-1}) 25 °C	σ (S cm^{-1}) 300 °C	E_a (kJ mol^{-1})
$Li_7La_3Zr_2O_{12}$ [197]	3	3	$(2–5)\times10^{-4}$	0.07	30
$Li_6BaLa_2Ta_2O_{12}$ [191]	3	3	$(4–5)\times10^{-5}$	0.04	39
$Li_5La_3Bi_2O_{12}$ [194]	3	3	2×10^{-5}	0.04	45
$Li_5La_3Nb_2O_{12}$ [189, 194]	3	3	8×10^{-6}	0.02	41
$Li_5La_3Ta_2O_{12}$ [189, 194]	3	3	$(1–2)\times10^{-6}$	0.01	54
R-$LiZr_2P_3O_{12}$ [206]	3	3	$<10^{-5}$	0.014	40
R-$Li_{1.3}(Al_{0.3}Ti_{1.7})P_3O_{12}$ [207]	3	3	7×10^{-4}	0.09	
P-$Li_3Fe_2P_3O_{12}$ [2]	3	?	$\sim10^{-7}$	0.013	
P-$Li_3Sc_2P_3O_{12}$ [208]	3	?	$<10^{-10a}$	0.010	
P-$Li_{2.8}Sc_{2.8}Zr_{0.2}P_3O_{12}$ [208]	3	?	$\sim10^{-4a}$	0.016	
$Li_{0.18}La_{0.61}TiO_3$ [220]	3	3	5×10^{-5}	0.010	25 (100–250 °C)
$Li_{0.25}La_{0.58}TiO_3$ [213, 214]	3	3		0.04	
$Li_{0.33}La_{0.56}TiO_3$ [215, 224]	3	3		0.05–0.25	
$Li_{0.33}Sr_{0.55}Zr_{0.44}Ta_{0.56}O_3$ [12]	3	3	10^{-5}	0.03	35
$Li_{0.375}Sr_{0.438}Zr_{0.25}Ta_{0.75}O_3$ [227]	3	3	10^{-4}		
$Li_3Zn_{0.5}Nb_2O_7$ [55]	2	3	10^{-4}	0.024	31 (>160 °C)
$Li_{3.5}Si_{0.5}V_{0.5}O_4$ [230, 231]	0	3	5×10^{-6a}	0.01–0.05	35–44
$Li_{3.5}Zn_{0.25}GeO_4$ [229, 232]	0	3	10^{-6}	0.02–0.13	
$Li_{3.75}Ge_{0.75}V_{0.25}O_4$ [234]	0	3	6×10^{-6}	0.10	52
$Li_{3.70}Ge_{0.85}W_{0.15}O_4$ [234]	0	3	4×10^{-5}	0.07	40
$Li_{3.25}Ge_{0.25}P_{0.75}S_4$ [238]	0	3	2.2×10^{-3}		20
$Li_{4.275}Ge_{0.61}Ga_{0.25}S_4$ [239]	0	3	7×10^{-5}	9×10^{-3}	
$Li_{3.4}Si_{0.4}P_{0.6}S_4$ [240]	0	3	6×10^{-4}	0.13	28
Li_3N (with impurities) [2, 8]	0	2	$10^{-6}–6\times10^{-4}$		24–54
Li_6FeCl_8 [245, 246]	0	3		0.011	
Li_2CdCl_4 [247]	0	3		0.08	49 (>270 °C)
$LiInBr_4$ supercooled [248]	0	3	10^{-3}		
$Li_3InBr_{6-x}Cl_x$ ($0 \leq x \leq 3$) [248, 249]	0	?	$5\times10^{-5}–2\times10^{-3}$		20
$Li_4B_7O_{12}Cl$ [252]	3	3	2×10^{-6}	0.012	47
$Li_{2.25}C_{0.75}B_{0.25}O_3$ [253]	0	?	5×10^{-7}	0.01	56

[a]Extrapolated value.

7.5.3 Mixed Frameworks of Oxygen Octahedra and Tetrahedra

Lithium β-alumina (see Sections 7.1.3, 7.2.2 and Figure 7.6) is a metastable phase which can only be prepared by ion exchange. For sodium-free single crystals, very high conductivities (2.7×10^{-3} and $0.18\ S\,cm^{-1}$ at 25 and 300 °C, respectively) and $E_a = 23\ kJ\,mol^{-1}$ were reported [203], but no data are available for ceramics.

Two topologically different framework types with the general formula $Li_xM_2T_3O_{12}$ [1, 2, 6, 10–13, 204–212] are based on "lanterns" of two MO_6 octahedra corner-linked with three TO_4 tetrahedra (Figure 7.21, I). One of these frameworks (R), typified by $LiTi_2(PO_4)_3$, is of the NASICON type (see Figure 7.15), with the "lanterns" in parallel orientation; in another framework (P), typified by $Li_3M_2(PO_4)$ (M = Sc, Cr, Fe), the "lanterns" are inclined alternatively in opposite directions (Figure 7.21, II). In their most symmetrical, high-temperature forms, the frameworks R and P are, respectively, rhombohedral ($R\bar{3}c$ or R3) and orthorhombic Pbcn (or Pcan, or Pbna with interchanged axes). On cooling, they usually undergo triclinic or monoclinic distortions. In addition, an intergrowth of the structures R and P has been found in $Li_2FeTi(PO_4)_3$ [213].

For $Li_3In_2(PO_4)_3$ and $LiZr_2(PO_4)_3$, both type R and type P structures are known under ambient conditions, depending on the preparation temperature, with the type P phases being 2–4% denser. For the Zr compound, type P can only be prepared below 900 °C and the quenched high-temperature form is of the type R; however, for

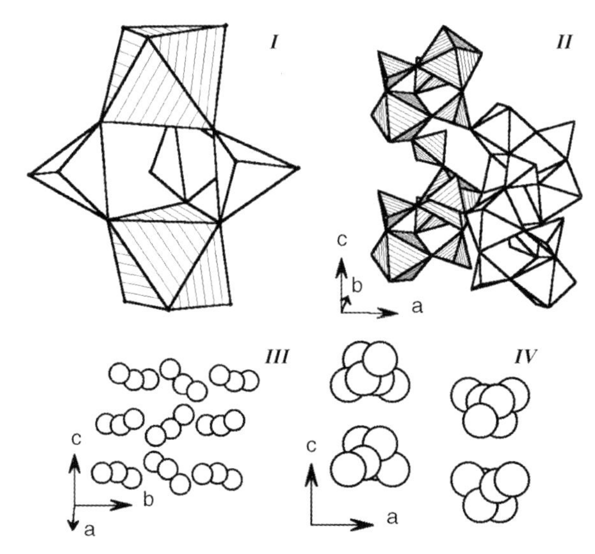

Figure 7.21 Orthorhombic crystal structure of $Li_3Sc_2(PO_4)_3$ (Pbcn) at 300 °C [209]. (I) A $M_2T_3O_6O_{12/2}$ "lantern", a common building block of NASICON and $Li_3M_2(PO_4)_3$ family; (II) A fragment of $Sc_2(PO_4)_3$ framework; the shaded and unshaded areas are identical "lanterns" in different orientations; (III, IV) Partially occupied Li sites viewed in different directions (framework omitted); the intersite distances along *b* are 1.77, 1.52, and 2.44 Å, whereas the shortest Li–Li distances along *c* are 3.52 and 3.72 Å, and along *a*, 3.98 Å.

the In-containing compound the situation is reversed. The high-temperature forms are more conductive in both cases.

For the stoichiometric $LiM_2(PO_4)_3$ and $LiM'M''(PO_4)_3$ (M = Ti, Ge, Sn, Zr; M′ = Al, Cr, Fe; M″ = Nb, Ta, Sb) of type R, the phase with M = Zr exhibits a maximum Li^+ conduction; transport in others may be drastically enhanced by lower-valence substitutions for M accompanied by adding Li^+ (Table 7.3).

In the high-temperature forms of type P $Li_3M_2(PO_4)_3$ (M = Sc,Cr,Fe), the Li^+ ions are disordered and provide a substantial conductivity. The transport is anisotropic, as might be anticipated from Figure 7.21 (III and IV). On cooling, a series of phase transitions leads to Li ordering, lowering the conductivity. However, these transitions are eliminated and the orthorhombic phases stabilized down to room temperature by heterovalent substitutions, drastically enhancing the room-temperature conductivity (Table 7.3).

7.5.4
Octahedral Framework and Layered Structures

Fast Li^+ transport is characteristic of the cation-deficient perovskite-type titanates, niobates, and tantalates. Li^+ is too small for the cubo-octahedral void of the perovskite structure (see Figure 7.4); the stabilization can be achieved introducing a large trivalent or bivalent cation, such as a rare-earth, alkaline-earth or, probably, Bi^{3+} or Pb^{2+}. The typical formulas are $R_{2/3-x/3}Li_x\square_{1/3-2x/3}TiO_3$ and $R_{1/3-x/3}Li_x\square_{2(1-x)/3}MO_3$ (R = La, Pr, Nd; M = Nb, Ta; $\square$ is a cation vacancy) [1, 2, 5, 7, 9–13, 214–229]. The highest room-temperature conductivities [$(1–3)\times10^{-3}$ S cm^{-1}] have been reported for ceramics based on lithium lanthanum titanate [216, 223, 224]. However, these values correspond to the grain interior resistance; the total (grain + boundary) resistance is usually one to two orders of magnitude greater [11, 216].

Although the ideal perovskite structure is simple cubic, $La_{2/3-x/3}Li_xTiO_3$ usually have multiple cells with small orthorhombic, tetragonal, or rhombohedral distortions due to tilting octahedra and partial Li/La/vacancy ordering, and also show a complex phase transitions pattern [216–218]. Nevertheless, the conductivity anisotropy is negligible [220]. The highest conductivity was found in the quenched, high-temperature disordered form. The Arrhenius plots are usually nonlinear, which is ascribed to either a continuous opening of the bottlenecks [220] or to the transition from cooperative to noncorrelated ion hopping [221] on heating. As the conductivity decreases at high pressures [222], the migration is controlled by the bottleneck size, being optimum when the R cations are slightly greater than La^{3+} (minor substitutions of Sr for La [224]) and the octahedral cation is slightly smaller than Ti (minor substitution of Al for Ti [222, 223]). The optimum vacancy fraction in the $LaTiO_3$-based systems is reported to be 0.08 [224]. Although tantalates and niobates [226–229] have poorer transport properties, one important advantage of tantalates is their greater stability to reduction.

Nonstoichiometric $Li_{4-2x}M_xNb_2O_7$ (M = Zn,Mg; $x\approx0.5$) [55] comprise Nb_2O_7 layers built of edge- and corner-shared NbO_6 octahedra (Figure 7.22). Li^+ ions, partially substituted by M^{2+}, are distributed over face-shared octahedral sites between the layers and *inside* them, thus providing migration paths not only in the interlayer distance but also *through* the layers. As a result, hot-pressed grain-oriented

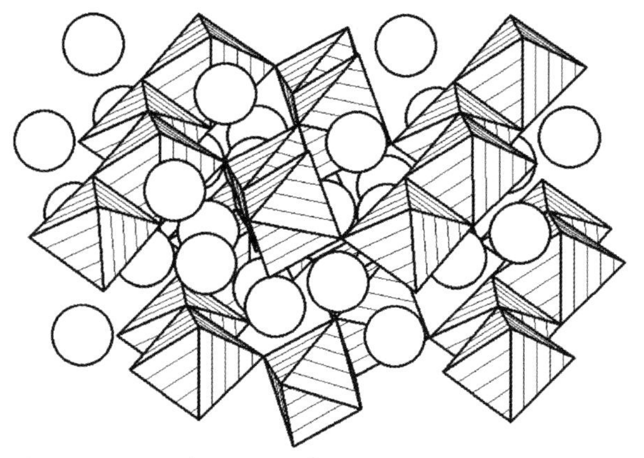

Figure 7.22 Crystal structure of $Li_3Zn_{0.5}Nb_2O_7$ [55]. The circles represent partially occupied mixed Li/Zn sites.

ceramics exhibit isotropic ionic conduction (Table 7.3). Only a small portion of Nb could be substituted by Ta in this structure.

7.5.5 Structures Based on Isolated Tetrahedral Anions [1, 2, 5, 10–12, 230–241]

Li_2SO_4 (see Figure 7.1) and mixed sulfates exhibit high cation conductivity only at elevated temperatures. The more important Li^+ conductors of this class are based on lithium orthogermanate, orthosilicate and orthophosphate and their thio-analogues. Although pure Li_4TX_4 and Li_3PS_4 (T = Si,Ge; X = O,S) are poor cation conductors, their conductivity is markedly enhanced by heterovalent doping with both lithium excess and lithium deficiency; examples include $Li_{4+x}M^{3+}_xSi_{1-x}O_4$, $Li_{4-3x}M^{3+}_xSiO_4$ (M = Al, Ga), $Li_{4-2x}M^{2+}_xTO_4$ (M = Zn, Mg), $Li_{4-x}R^{5+}_xT_{1-x}O_4$ (R = P, V), $Li_{4-2x}R^{6+}_xT_{1-x}O_4$ (R = S, Se, Cr, Mo, W) and $Li_{3-2x}Zn_xPO_4$ (see Table 7.3). The increase in conductivity is contributed by the created point defects, but originates primarily from structural rearrangements. The best Li^+ conductors of this family have γ-Li_3PO_4-type structure stabilized in wide concentration ranges between three and four Li^+ ions per formula unit. The oxogermanates are generally more conductive than oxosilicates, owing to the size effect and an easier stabilization of the γ structure. The Zn-modified germanate has been called LISICON (*Li* Super*Ionic* *CON*ductor). Its crystal structure (Figure 7.23) comprises a hexagonal close packing of oxygen with Ge in tetrahedral voids, and Li and Zn partially occupying almost all other tetrahedral and octahedral voids (except for those having common faces with GeO_4 tetrahedra). First principles modeling of the Li^+ diffusion in γ-Li_3PO_4 [236] predicts a considerable anisotropy, but the single-crystal conductivity of the γ-type $Li_{3.34}P_{0.66}Ge_{0.34}O_4$ has been found almost isotropic [237]. Thio-LISICONS, $Li_{4-x}Ge_{1-x}P_xS_4$ [238], $Li_{4-2x}Zn_xGeS_4$, $Li_{4+x+y}Ge_{1-x-z}Ga_xS_4$ [239], $Li_{4+x}Si_{1-x}Al_xS_4$, $Li_{4-x}Si_{1-x}P_xS_4$ [240], and $Li_{4+5x}Si_{1-x}S_4$ [241], are similar to their oxo-counterparts, but often show complicated superstructures and/or slight monoclinic distortions. The larger and more polarizable S^{2-} ions provide a higher Li^+ ion conductivity, especially at low temperatures.

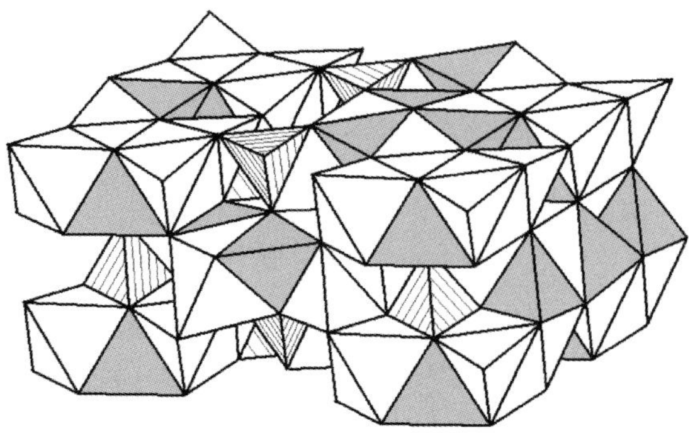

Figure 7.23 Crystal structure of LISICON $Li_{3.5}Zn_{0.25}GeO_4$ [234]. Striped tetrahedra = GeO_4; gray tetrahedra = $(Li,Zn)O_4$; the white polyhedra are partially occupied by lithium. The structures of $Li_{3.75}Ge_{0.75}V_{0.25}O_4$ and $Li_{3.70}Ge_{0.85}W_{0.15}O_4$ [235] differ only in minor details.

7.5.6
Structures with Isolated Monatomic Anions

The most conductive material of this class is lithium nitride [1–3, 7, 10, 13, 22, 242]. Its hexagonal structure (Figure 7.24) contains two types of lithium ion, with coordination numbers 2 and 3; only the latter contributes to the transport process and, thus, conductivity perpendicular to the principal axis is two orders of magnitude higher than that parallel to *c*. Stoichiometric Li_3N does not contain lithium defects,

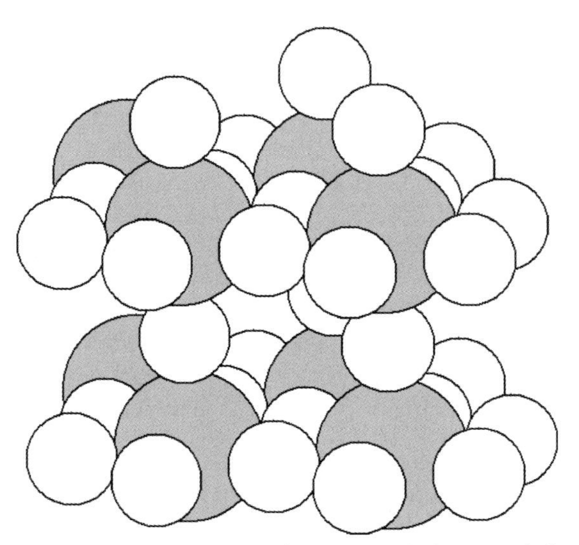

Figure 7.24 Crystal structure of Li_3N [242]. The large gray balls represent N; the small white balls represent lithium. Some 3% lithium vacancies have been found in the Li_2N layer.

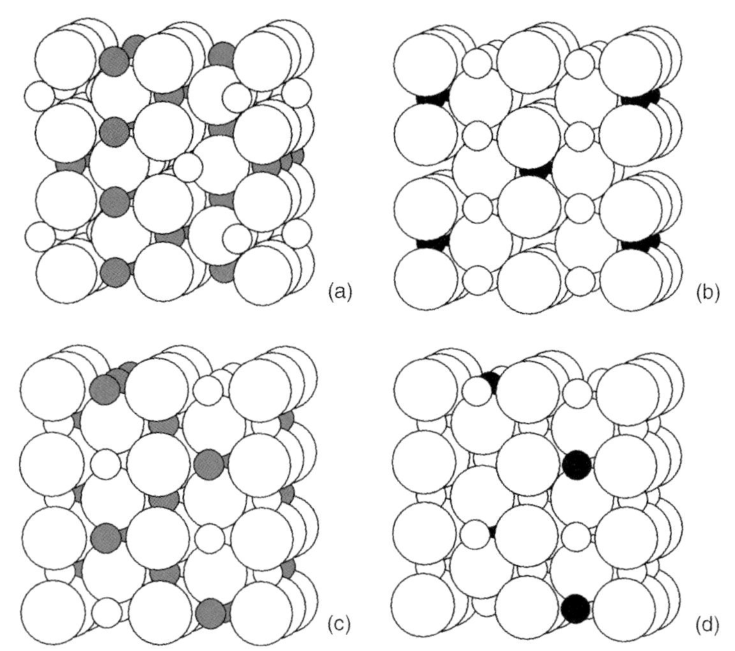

Figure 7.25 The main structure types of lithium M^{2+} halides [243, 244]. (a) Inverse spinel Li_2CdCl_4, Fd3m; (b) Li_2MgBr_4 at room temperature, Cmmm; (c) Li_2MgBr_4 at 400 °C, Fd3m; (d) Suzuki phase Li_6MgBr_8, Fm3m. Large balls represent Cl or Br; small white balls represent Li sites [50% filled in panel (c]; black balls represent Mg or Cd; gray balls represent mixed Li/M^{2+} sites.

and should have low conductivity; however, the conductivity is strongly enhanced by unavoidable impurities of H (as NH) and O producing lithium vacancies.

High Li^+ conductivity has also been found in mixed chlorides and bromides Li_2MX_4 and Li_6MX_8, where M = Mg, Mn, Fe, Cd; X = Cl, Br [1, 2, 5, 243–247]. All of these are based on cubic close packing of the X^- anions. Li_2MX_4 are usually referred to as inverse spinels, with one Li on tetrahedral sites and another Li mixed with M on octahedral sites (Figure 7.25a). However, this is true only for part of the compounds; others have an orthorhombic superlattice of the rock salt type, with all cations in octahedra (Figure 7.25b). On heating, both types usually transform to a disordered structure (Figure 7.25c), having the same space group as spinels but with all cations in octahedra, similar to rock salt but with one of four octahedra vacant. At higher temperatures, a further transformation into disordered LiX-based solid solutions is observed for some compounds. The ionic conduction is slightly increased with cation deficiency: $Li_{2-2x}M_{1+x}X_4$.

Similarly, Suzuki phases Li_6MX_8 are rock salt derivatives with ordering of Li, M and vacancies on the octahedral sites and doubled cubic cell edge (Figure 7.25d). On heating, these undergo a gradual disordering and finally become LiX-based solid solutions [245], sometimes with an intermediate exsolution of Li_2MX_4-based phase [246].

Lithium transport in all of these phases occurs via the shared faces of octahedra and tetrahedra. In normal spinels (e.g., Li_2ZnCl_4), where the tetrahedra are occupied by M^{2+}, the conductivity is much lower. Diffuse order–disorder transitions in both Li_2MX_4 and Li_6MX_8 result in strong deviations from linearity on their Arrhenius plots.

In the cation-deficient, spinel-type phase, $LiInBr_4$ [248], In^{3+} occupies half of octahedral sites whereas Li^+ has not been localized and provides a very high conductivity. There is also one low-temperature polymorph of which the conductivity is at least three orders lower. On heating, a phase transition occurs at 43 °C but, due to a considerable hysteresis, the spinel phase is preserved down to −13 °C on cooling. Li_3InBr_6 might be expected to contain discrete $[InBr_6]^{3-}$ units, but in its high-temperature monoclinic superionic phase, In^{3+} cations are disordered over two octahedral sites. As for $LiInBr_4$, this phase appears at 41 °C on heating but is preserved down to −13 °C on cooling [248]. The substitution of Cl for 50% Br shifts the phase transition point below room temperature [249].

7.5.7 Other Structures

Room-temperature conductivities of 10^{-3} S cm^{-1} or more have been announced for the cubic argyrodite-type phases $Li_6(PS_4)SX$ (X = Cl, Br, I) [250]. Detailed information for this is not available, although Ag^+ and Cu^+ cation conductors of this type are well known.

Lithium boracites, $Li_4B_7O_{12}X$ (X = Cl, Br, I), are cubic or pseudo-cubic framework structures built of BO_4 tetrahedra and planar BO_3 groups sharing all corners to form an open 3D framework. The large voids are filled with X anions tetrahedrally coordinated by four Li(1) sites and, in longer distances, octahedrally surrounded by six Li(2) sites. Li(1) is tetrahedrally coordinated by one Cl and three O. These Li(1) sites are only 25% filled at room temperature. A major amount of lithium, Li(2), occupies the distorted octahedra of four O and two *cis*-Cl. On heating, a progressive redistribution from Li(1) to Li(2) site is observed [251].

Monoclinic Li_2CO_3 is a poor ionic conductor, although the conductivity is increased by three to four orders of magnitude in the $Li_{2+x}C_{1-x}B_xO_3$ solid solutions [253]. In the pure carbonate, the LiO_4 tetrahedra share an edge to form isolated pairs [254]; the interstitial lithium ions most likely provide a continuous linkage between the edge-shared tetrahedra.

7.6 Proton Conductors

7.6.1 General Remarks [14–21, 255]

Both X-ray and neutron diffraction analyses are ineffective for locating hydrogen in crystals, because of its low electron density and strong inelastic scattering, respectively (not to mention its high mobility). Rather, the most reliable method is the

Figure 7.26 Cartoons representing the two mechanisms of proton transport. Upper: Grotthuss; lower: vehicle [255]. See Section 7.6.1 for further details.

neutron diffraction of *deuterated* crystals. In many cases, the hydrogen positions are presumed rather than experimentally established.

Two entirely different mechanisms have been established for proton transport (Figure 7.26):

- The Grotthuss mechanism, which is the best known, includes proton transfer along the hydrogen bond between a proton donor and a proton acceptor species (e.g., H_3O^+ and H_2O or H_2O and OH^-), followed by reorientation of the resultant species in preparation for the next step. Stronger hydrogen bonds (shorter O–O distances) provide easier proton transfer (either thermally activated hopping or quantum tunneling), but a more difficult reorientation. Thus, there appears to be a somewhat optimum O–O distance which is neither very short nor very long. This simple representation may be seriously modified by many-body interactions and thermal motion within the "rigid lattice" [14, 15, 21].

- The so-called vehicle mechanism relates to the transfer of voluminous proton-containing ions, such as H_3O^+ or NH_4^+. The size of both cations is similar to that of K^+ and Rb^+ and, thus, conditions for their optimized transport are similar to those discussed in Sections 7.2.2, 7.2.3 and 7.4, although it may be complicated by directional hydrogen bonds. The vehicle mechanism is generally favored by higher temperatures, lower pressures and higher proton concentration [21], although information on dominating mechanisms is often contradictory.

Regardless of the microscopic phenomena, protonic conductivity is critically sensitive to the water content inside crystals and on their *surface*. Intrinsically nonconductive materials may apparently exhibit proton transport in wet environments due to adsorbed and/or condensed water. Consequently, numerous reports on the conductivity of compacted powders at 90–100% relative humidity, when vapor condensation in pores cannot be avoided, are excluded from consideration. Heating or cooling may cause H_2O loss or uptake from the atmosphere, thus altering the conditions for proton transport in crystals. In such situations, the apparent E_a found

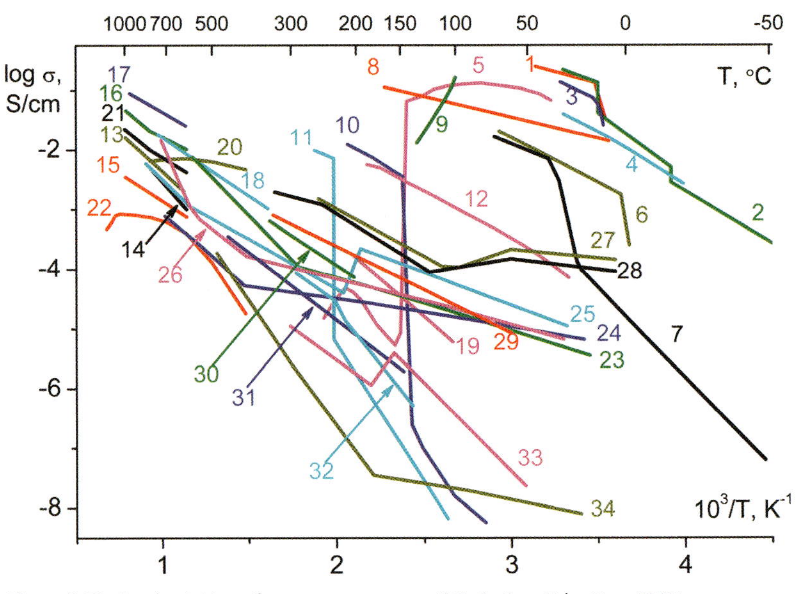

Figure 7.27 Conductivities of some representative crystalline proton conductors. Data for Nafion polymer membrane are included for comparison. (1) $H_3PW_{12}O_{40}\cdot29H_2O$, powder compact [256]; (2) the same [257]; (3) $H_3PW_{12}O_{40}\cdot28H_2O$, single crystal [258]; (4) $H_4SiW_{12}O_{40}\cdot28H_2O$, single crystal [258]; (5) $H_4SiW_{12}O_{40}\cdot18H_2O$, powder compact [261]; (6) $HUO_2PO_4\cdot4H_2O$, grain-oriented powder compact [262]; (7) $HUO_2AsO_4\cdot4H_2O$, powder compact [263]; (8) Nafion [264] (obviously, in saturated water vapor); (9) Nafion, p $(H_2O) = 10^5$ Pa [21]; (10) $CsHSO_4$, single crystal along both *a* and *c* [265]; (11) CsH_2PO_4, single crystal [268]; (12) $Cs(HSO_4)_{0.6}((H_2PO_4)_{0.4}$, powder compact [269]. Plots (13)–(22) represent ceramics: (13) $SrCe_{0.95}Yb_{0.05}O_{3-\delta}$ [272]; (14) $SrCe_{0.9}Y_{0.1}O_{3-\delta}$ [274]; (15) $SrZr_{0.95}Y_{0.05}O_{3-\delta}$ [272]; (16) $BaCe_{0.9}Nd_{0.1}O_{3-\delta}$ [272]; (17) $BaCe_{0.8}Y_{0.2}O_{3-\delta}$ [272]; (18) the same [275]; (19) $BaCe_{0.9}Y_{0.1}O_{3-\delta}$ [15]; (20) $BaZr_{0.9}Y_{0.1}O_{3-\delta}$ [276]; (21) $La_{0.9}Sr_{0.1}ScO_{3-\delta}$ [277]; (22) $La_{0.99}Ca_{0.01}NbO_{4-\delta}$, [289]. Plots (23)–(26) represent single crystals: (23) $H(H_2O)n$-β''-alumina, Mg-stabilized [296]; (24) $H(H_2O)_n$-β-alumina, Mg-doped [296]; (25) NH_4-β''-alumina, Mg-stabilized [297]; (26) NH_4-β-alumina, Mg-doped [297]; (27)–(31) ceramics: (27) $H_2Nb_2O_6\cdot H_2O$ [76]; (28) $H_2Ta_2O_6\cdot H_2O$ [76]; (29) NH_4NbWO_6 [301]; (30) NH_4TaWO_6 [302]; (31) $(NH_4)_4Ta_{10}WO_{30}$ [304]; (32) α-$Zr(HPO_4)_2$, grain-oriented powder compact [307]; (33) $H_3OZr_2(PO_4)_3$, powder compact [305]; (34) $H_{0.8}Zr_{1.8}Nb_{0.2}(PO_4)_3\cdot0.4H_2O$, powder compact [306].

from the slope of the Arrhenius plot is totally unconnected with the true microscopic value (not to mention the problems discussed in Sections 7.1.2 and 7.1.7). Finally, although a huge number of proton conductors have been discovered, only a few representative crystalline materials of various classes are reviewed below. A comparison of their conductivities (Figure 7.27) shows that, although a level of 0.01–0.1 S cm^{-1} may be achieved at 0–260 °C, or at 600–1000 °C, the challenge remains to prepare materials having such conductivities in combination with a sufficient stability in the intermediate-temperature region. This forms the basis for classifying all of these materials into the following three groups.

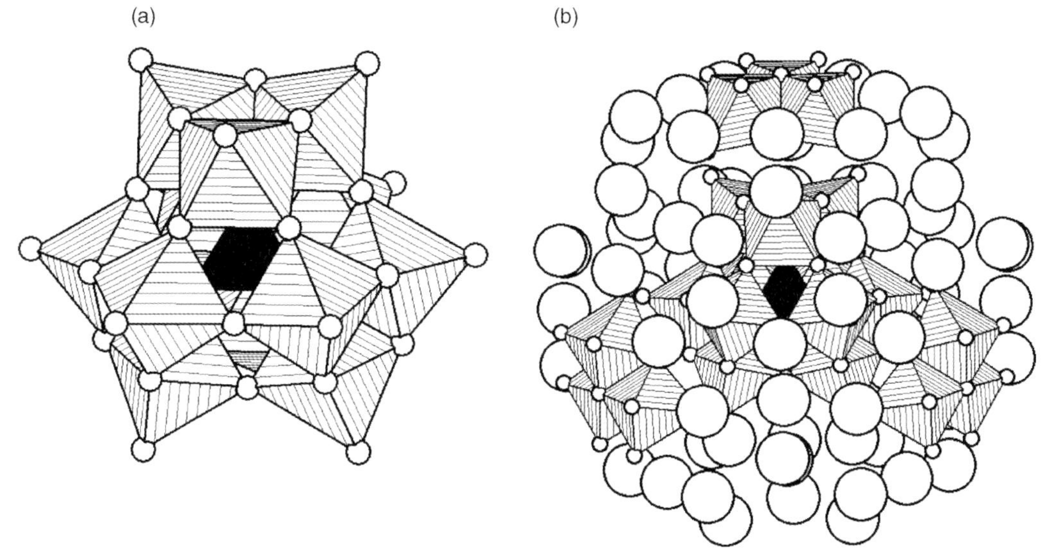

Figure 7.28 Crystal structure of $H_3PW_{12}O_{40}{\cdot}29H_2O$ [259]. (a) Polyhedral view of the Keggin anion; the PO_4 tetrahedron is shown black; the WO_6 octahedra are hatched; (b) The same anion with adjacent H_2O/H_3O^+ species (hydrogen not localized) and parts of the four surrounding anions.

7.6.2
Low-Temperature Proton Conductors: Acids and Acid Salts

Quite naturally, the best proton conductors are strong acids and acid salts. However, these compounds are usually thermally unstable, as they easily lose water and decompose at temperatures below 100–270 °C. The highest protonic conductivity at room temperature is known for fully hydrated 12-tungsto- and 12-molybdophosphoric acids, $H_3[PM_{12}O_{40}]{\cdot}(29 \pm 1)H_2O$ (M = Mo,W) [256–260]. Their Keggin-type anions (Figure 7.28a) are embedded into a quasi-liquid array of water molecules and oxonium ions (Figure 7.28b). The anionic charge of −3 is distributed over 12 terminal oxygen atoms; thus, each oxygen atom bears only −1/4, as in the perchlorate ion, and these acids are as strong as perchloric acid. At a relative humidity of ≥80%, both acids are stable at least to 80 °C, but at lower humidities and/or higher temperatures they rapidly convert to lower hydrates [260] with a decrease in conductivity. By comparison, 12-tungstosilicic acid is less conductive, but more thermally stable [258, 261].

In the tetragonal $HUO_2XO_4{\cdot}4H_2O$ (X = P,As) [262, 263], each oxygen atom of an XO_4^{3-} anion is bound to a UO_2^{2+} cation, forming an infinite layer of tetrahedra and octahedra. The remnant charge of the oxygen is again only −1/4, which is too small for a covalent binding proton, in contrast to other acid salts discussed below. The protons are delocalized within quasi-liquid layers of water molecules and oxonium ions and provide a high bidimensional conductivity.

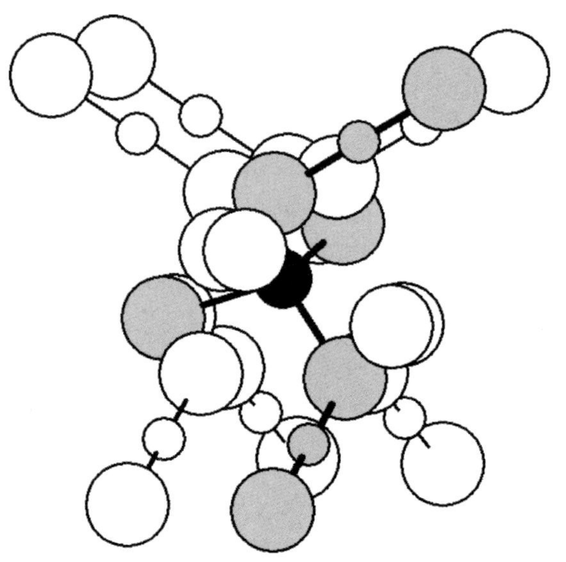

Figure 7.29 Crystal structure of $CsDSO_4$ at 175 °C [266]. The black ball represents sulfur; the large and small balls represent oxygen and hydrogen sites, respectively (all 25% occupied). One of the possible momentary arrangements is shown by gray balls and bold lines. Cesium and minority deuterium sites are not shown.

The high-temperature tetragonal phase of $CsHSO_4$ [265, 266] is, apparently, the best-studied and most conductive of the numerous anhydrous acid sulfates, selenates, phosphates and arsenates of alkali metals and thallium. It exists in a narrow range between the transition to the low-conducting monoclinic phase at 141 °C and melting at 217 °C, and exhibits essentially isotropic protonic conductivity owing to rapid reorientations of the anion. According to the neutron diffraction data (Figure 7.29), the hydrogen ion is delocalized over four equivalent O–O contacts of ~2.6 Å and, with much lower probability, four more O–O contacts of ~2.8 Å. In CsH_2PO_4, the disordered proton-conductive cubic phase exists above 230 °C [267, 268], but in the mixed-anion salts, $Cs(HSO_4)_{1-x}((H_2PO_4)_x$ ($0.25 \leq x \leq 0.75$), this phase may be supercooled down to room temperature [269]. Recently, comparable conductivity values were also found in the family of acid phosphites $AH(PO_3H)$ (A = K, Rb, Cs, NH_4) [270].

7.6.3
High-Temperature Proton Conductors: Ceramic Oxides

Since the pioneering studies conducted by Iwahara's group [271], hundreds of reports have been devoted to proton-conducting, acceptor-doped mixed oxides with perovskite structures (see Figure 7.4). Unfortunately, only a small amount of the available data can be reviewed here [14, 15, 19–21, 272–287]; additional information is available in Chapters 12 and 13. These materials are oxygen-deficient solid solutions, primarily of the families such as $A^{2+}M^{4+}_{1-x}R^{3+}_{x}O_{3-x/2}$ and $A_3Ca_{1+x}Nb_{2-x}O_{9-3x/2}$ (A = Ba, Sr, Ca; M = Ce, Zr, Hf, Sn, Th; R = Sc, Y, Ln, In). When sintered in a dry, hydrogen-free atmosphere, they do not contain hydrogen but easily react with moisture or hydrogen at moderately high temperatures and acquire protonic conductivity. Cerates are more conductive but less stable chemically than zirconates, and are decomposed by CO_2 or high concentrations of water vapor to form alkaline-earth carbonates or hydroxides [19, 272, 275, 278, 279]. Some solid solutions (e.g., $BaCe_{0.3}Zr_{0.5}Y_{0.2}O_{3-\delta}$ [275] and $BaZr_{0.4}Ce_{0.4}In_{0.2}O_3$ [280]) may maintain the relatively good chemical stability of barium zirconate with improved electrical conductivity.

Proton conduction in perovskites is assumed to be of the Grotthuss type [15, 21], although average O–O separations in zirconates and cerates are greater than typical hydrogen bond lengths (2.4–2.9 Å [288]), namely 2.9–3.1 Å along the octahedron edge and 4.1–4.4 Å between the octahedra. In addition, the hydrogen bond along the edge should be strongly bent due to repulsion between proton and octahedral cation. In noncubic perovskites, whilst some of the O–O distances become shorter, any continuous path must also include longer O–O distances. The cubic perovskites should, therefore, be better proton conductors, and a shortening of the O–O distances should be *dynamic*, due to thermal motion. The H^+-M^{n+} repulsion contributes to the activation energy of the proton transport. Indeed, $A^+M^{5+}O_3$-based materials are much poorer proton conductors than $A^{2+}M^{4+}O_3$-based phases. Consequently, the $A^{3+}M^{3+}O_3$ perovskites appear to show promise [15, 21], although they are usually distorted as all of the A^{3+} cations are too small for the ideal O_{12} cubo-octahedron.

Due to low dopant concentrations and high temperatures, the reactions of the doped perovskites with gases may be adequately described by the point defect model and quasi-chemical equilibria:

$$1/2\,O_2(g) + V_O^{\bullet\bullet} = O_O^x + 2\,h^\bullet; \quad H_2(g) + 2\,O_O^x + 2\,h^\bullet = 2\,OH_O^\bullet;$$

$$H_2O\,(g) + O_O^x + V_O^{\bullet\bullet} = 2\,OH_O^\bullet.$$

Thus, a trivalent dopant substituting for Ce(4+) or Zr(4+) [or a divalent dopant substituting for Nb(5+) or La(3+)] may be compensated by charge carriers of the three types: oxygen vacancies, electron holes, and interstitial protons in the form of hydroxyls. Thus, the materials may exhibit mixed (anion, hole, and proton) conductivity, depending on external conditions (see Chapter 13). The first contribution is dominant at low $p(O_2)$ and low $p(H_2O)$; the second at high $p(O_2)$; and the third at low $p(O_2)$ and high $p(H_2O)$. In addition, a Ce(4+) to Ce(3+) reduction occurs at very low $p(O_2)$, giving rise to *n*-type conductivity [281, 282], although more recent studies have tended to interpret this contribution, proportional to $p(O_2)^{-1/4}$, as protonic conductivity [273, 274]. At temperatures below 310 °C in wet air, hydroxyl ion conduction has been proposed in Yb-doped $SrCeO_3$ [283].

The borderlines between "low" and "high" partial pressures may depend on the material and temperature. Usually, hole conduction becomes significant or even dominant at p(O_2) ≥1 Pa [273, 274, 276, 277, 282, 284], but Pr- and Tb-doped cerates display several orders lower hole conductivities even in air [285, 286], as the two Ln (3+) are readily oxidized to Ln(4+) and therefore act as "traps" for holes. In Figure 7.27 (graphs 13–21), only the conductivities in essentially ionic regimes are depicted, although these may still contain some oxide-ion contribution. This is more probable at higher temperatures, as the sorption of water or oxygen is usually exothermic and the E_a of oxygen transport is usually greater than that of protonic conductivity. For example, under reducing conditions, when the conductivity is purely ionic, $BaCe_{0.85}R_{0.15}O_{3-\delta}$ (R = Sc,Y,Pr ... Lu) exhibit transport number for protons t_H = 0.46–0.66 at 700 °C and only 0.02–0.09 at 1000 °C [285]. Very similar results have been reported by different authors for $BaCeO_3$ doped with 10% Yb [19] or Nd [287], where t_H is close to unity only below 600 °C. The oxide ion contribution becomes lower for smaller cations [19]; the corresponding transference numbers were reported as low as 0.02 in $BaZr_{0.9}Y_{0.1}O_{3-\delta}$ at 700 °C [276], close to zero in $SrCe_{0.95}Tb_{0.05}O_{3-\delta}$ at 900 °C [286], and less than 0.05 in $SrCe_{0.95}Sc_{0.05}O_{3-\delta}$ up to 800 °C [271] and in $La_{0.8}Sr_{0.2}ScO_{3-\delta}$ up to ~650 °C [277]. It should be noted that the equilibration of a dense ceramics with the atmosphere may take a significant time – from 1 h at 500 °C [19] to tens of hours for similar compositions, even at 800–1000 °C [283]. Hence, at reduced temperatures the conductivity of ceramics may appear independent of the atmosphere for periods of measurements due to a "frozen-in" defect concentration created at high temperatures.

Similar principles hold for non-perovskite, acceptor-doped oxides, in particular, fergusonite-type $Ln_{1-x}Ca_xNbO_{4-\delta}$ [289], fluorite-type $La_{5.8}WO_{11.7}$ [290], β-K_2SO_4-type $La_{1-x}Ba_{1+x}GaO_{4-\delta}$ [291], monazite-type $Ln_{1-x}Sr_xPO_{4-\delta}$ [292], aragonite-type $La_{1-x}Sr_xBO_{3-\delta}$ [290, 293], pyrochlore type $La_2Zr_{2-x}Y_xO_{7-\delta}$ [294], and $La_{2-x}Ca_xM_2O_{7-\delta}$ (M = Zr,Ce) [295].

7.6.4
Intermediate-Temperature Proton Conductors

This class of material comprises structures with $D = 3$ or 2 (thus, more thermally stable than the $D = 0$ structures discussed in Section 7.6.2) containing intrinsic protonic species (H, OH, H_3O, or NH_4), in contrast to extrinsic proton conductors (as discussed in Section 7.6.3). Therefore, their protonic conductivities are independent of the surrounding atmosphere, although their decomposition temperatures (as listed below in this section) should be increased at high partial pressures of water and/or ammonia. In addition, they do not contain basic components such as SrO or BaO that cause the instability of perovskites in CO_2 and steam.

Rigid frameworks usually require a high synthesis temperature, in contradiction to the desirable high concentration of intrinsic protons. Only a few materials (mostly phosphates, as listed at the end of this section) may be prepared by direct low-temperature "wet" synthesis. A more general method includes the high-temperature preparation of ceramics or single crystals of hydrogen-free alkali or thallium(1+) compounds, followed by the low-temperature ion-exchange treatment of these precursor bodies in acids or molten ammonium nitrate.

The first (and subsequently most extensively studied) family of proton conductors prepared in this way were the hydronium and ammonium β- and β″-aluminas [296–298]. On heating, these gradually lose NH_3 and/or H_2O. The first decomposition stage (below 300–400 °C) is reversible, but thereafter stacking faults appear and a metastable nonstoichiometric spinel crystallizes at approximately 600 °C. Nevertheless, considerable protonic conductivity is preserved up to 600–800 °C (Figure 7.27, graphs 23–26). Whilst these results were obtained for single crystals, for ceramics the highest reported working temperatures were only 200–300 °C [299, 300]. The most serious obstacle here was cracking of the ceramics due to anisotropic changes of the hexagonal lattice parameters during ion exchange, thermal cycling and dehydration–deammoniation.

This problem has been surmounted by the use of cubic pyrochlore-type ceramics, $H_2M_2O_6{\cdot}H_2O$ [76] and NH_4MWO_6 [301–303], prepared by ion exchange in $Tl_2M_2O_6$ and $RbMWO_6$, respectively (M = Nb, Ta; graphs 27–30 in Figure 7.27). The pyrochlore structure comprises a framework of corner-shared octahedra (Figure 7.30a) with wide intersecting channels where mobile ions and/or molecules reside. A wider working range (up to 350–400 °C, and even to 550 °C for short periods) but somewhat lower conductivity has been achieved with the NH_4-exchanged pyrochlore-related but more condensed rhombohedral framework structure (Figure 7.30b). The $(NH_4)_4Ta_{10}WO_{30}$ ceramics [304] lose part of the NH_3 on first heating; the dominant protonic conductivity then becomes well reproducible on thermal cycling between 150 and 450 °C (graph 31 in Figure 7.27), and is independent of atmosphere (dry or wet air, N_2 or H_2).

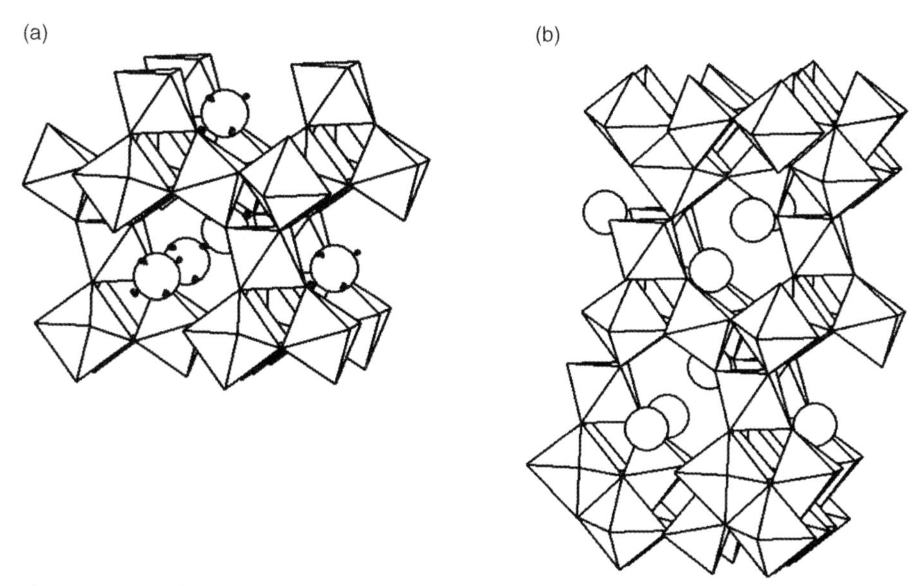

Figure 7.30 Crystal structures of (a) ND_4NbWO_6 [302] and (b) $Rb_4Nb_{10}WO_{30}$ [58]. Large balls represent N and Rb, respectively; small black balls represent deuterium sites (1/3 occupied).

NASICON-type $H_3OZr_2(PO_4)_3$ [305] and its solid solutions [306] were prepared by ion exchange from Li precursors. In addition, $NH_4Zr_2(PO_4)_3$ could be synthesized hydrothermally and transformed to $H_3OZr_2(PO_4)_3$ by deammoniation at 440–600 °C and hydration [305]. Layered α-$Zr(HPO_4)_2$ could also be prepared from solution [307]. All of these phosphates demonstrated a good thermal stability but low proton conductivity (graphs 32–34 in Figure 7.27). In addition, they could only be prepared in powdered form, and not as ceramics.

Unprecedented high protonic conductivities, of 0.1–0.2 S cm^{-1} at 200–350 °C in unhumidified air, were recently reported [308, 309] for SnP_2O_7 with trivalent dopants (Al or In). The published formulas, $Sn_{1-x}M^{3+}{}_xP_2O_7$, are, however, not charge-balanced, and the presence of x protons is implied. It is likely that the observed high conductivities might be due to residual liquid phosphoric acid used in excess for the synthesis; these materials require careful re-examination.

References

1 Burmakin, E.I. (1992) *Alkali Ion Conducting Solid Electrolytes*, Nauka, Moscow (in Russian).
2 Ivanov-Schitz, A.K. and Murin, I.V. (2000) *Ionika Tverdogo Tela (Solid State Ionics)*, Vol. 1, St. Petersburg University Press, Russia (in Russian).
3 Mehrer, H. (2007) *Diffusion in Solids*, Springer-Verlag, Berlin-Heidelberg.
4 Knauth, P. and Tuller, H.L. (2002) *J. Am. Ceram. Soc.*, **85**, 1654–1680.
5 Hull, S. (2004) *Rep. Prog. Phys.*, **67**, 1233–1314.
6 Kumar, P.P. and Yashonath, S. (2006) *J. Chem. Sci.*, **118**, 135–154.
7 Nalbandyan, V.B. and Subba Rao, G.V. (2003) Relational Database on Ionic Conductors (First 17 entries), ICDD, http://www.icdd.com/membership/minutes/pdf/Ceramics-2003-5.pdf.
8 Nalbandyan, V.B. (2004) Relational Database on Ionic Conductors (11 new entries), ICDD, http://www.icdd.com/membership/minutes/pdf/Ceramic-ION.pdf.
9 Adachi, G., Imanaka, N. and Tamura, S. (2002) *Chem. Rev.*, **102**, 2405–2429.
10 Thangadurai, V. and Weppner, W. (2002) *Ionics*, **8**, 281–292.
11 Robertson, A.D., West, A.R. and Ritchie, A.G. (1997) *Solid State Ionics*, **104**, 1–11.
12 Sebastian, L. and Gopalakrishnan, J. (2003) *J. Mater. Chem.*, **13**, 433–441.
13 Thangadurai, V. and Weppner, W. (2006) *Ionics*, **12**, 81–92.
14 Kreuer, K.-D. (1996) *Chem. Mater.*, **8**, 610–641.
15 Kreuer, K.-D. (2000) *Solid State Ionics*, **136–137**, 149–160.
16 Alberti, G. and Casciola, M. (2001) *Solid State Ionics*, **145**, 3–16.
17 Yaroslavtsev, A.B. and Kotov, V.Yu. (2002) *Russ. Chem. Bull.*, **51**, 555–568.
18 Baranov, A.I. (2003) *Cryst. Reports*, **48**, 1012–1037.
19 Oesten, R. and Huggins, R.A. (1995) *Ionics*, **1**, 427–437.
20 Norby, T., Widerøe, M., Glöckner, R. and Larring, Y. (2004) *Dalton Trans.*, 3012–3018.
21 Kreuer, K.-D., Paddison, S.J., Spohr, E. and Schuster, M. (2004) *Chem. Rev.*, **104**, 4637–4678.
22 Huggins, R.A. and Rabenau, A. (1978) *Mater. Res. Bull.*, **13**, 1315–1325.
23 Kummer, J.T. (1972) *Prog. Solid State Chem.*, **7**, 141–175.

24 Boilot, J.P., Salanie, J.P., Desplanches, G. and Le Potier, D. (1979) *Mater. Res. Bull.*, **14**, 1469–1477.

25 Baur, W.H., Dygas, J.R., Whitmore, D.H. and Faber, J. (1986) *Solid State Ionics*, **18/19**, 935–943.

26 Hong, H.Y.-P., Kafalas, J.A. and Bayard, M. (1978) *Mater. Res. Bull.*, **13**, 757–761.

27 Beyeler, H.U., Shannon, R.D. and Chen, H.Y. (1980) *Appl. Phys. Lett.*, **37**, 934–935.

28 Maximov, B.A., Petrov, I.V., Rabenau, A. and Schulz, H. (1982) *Solid State Ionics*, **6**, 195–200.

29 Atovmyan, L.O., Filipenko, O.S., Ponomarev, V.I., Leonova, L.S. and Ukshe, E.A. (1984) *Solid State Ionics*, **14**, 137–142.

30 Hughes, K. and Isard, J.O. (1972) *Physics of Electrolytes*, Vol. 1 (ed. J. Hladik), Academic Press, London, New York.

31 Chandrashekhar, G.V. and Foster, L.M. (1978) *Solid State Commun.*, **27**, 269–273.

32 Foster, L.M., Anderson, M.P., Chandrashekhar, G.V., Burns, G. and Bradford, R.B. (1981) *J. Chem. Phys.*, **75**, 2412–2422.

33 Bruce, J.A., Hunter, C.C. and Ingram, M.D. (1983) *Solid State Ionics*, **9/10**, 739–741.

34 Tsurumi, N., Singh, G. and Nicholson, P.S. (1987) *J. Solid State Chem.*, **66**, 372–375.

35 Nariki, S., Ito, S., Uchinokura, K. and Yoneda, N. (1989) *J. Electrochem. Soc.*, **136**, 2093–2099.

36 Barkovskii, A.I., Volkova, N.F., Kaul, A.R. and Tret'yuakov, Yu.D. (1986) *Neorgan. Mater.*, **22**, 1854–1856. (in Russian).

37 Wasiucionek, M., Garbarczyk, J. and Jakubowski, W. (1984) *Solid State Ionics*, **14**, 113–116.

38 Smirnov, N.B., Burmakin, E.I., Esina, N.O. and Shekhtman, G.Sh. (1996) *Russ. J. Electrochem.*, **32**, 502–504.

39 Smirnov, N.B., Burmakin, E.I. and Shekhtman, G.Sh. (1996) *Russ. J. Electrochem.*, **32**, 514–517.

40 Smirnov, N.B., Burmakin, E.I., Antonov, B.D. and Shekhtman, G.Sh. (2000) *Russ. J. Electrochem.*, **36**, 736–740.

41 Smirnov, N.B., Burmakin, E.I., Antonov, B.D. and Shekhtman, G.Sh. (2001) *Russ. J. Electrochem.*, **37**, 322–325.

42 Bondarev, V.N. (1996) *Solid State Ionics*, **89**, 93–98.

43 Meyer, M., Maass, P. and Bunde, A. (1998) *J. Chem. Phys.*, **109**, 2316–2324.

44 Wang, Y. and Cormack, A.N. (1999) *Radiat. Eff. Defects Solids*, **151**, 209–214.

45 Habasaki, J. and Hiwatari, Y. (2000) *Phys. Rev. E.*, **62**, 8790–8793.

46 Swenson, J. and Adams, S. (2003) *Phys. Rev. Lett.*, **90**, 155507.

47 Sudworth, J.L. and Tilley, A.R. (1985) *The Sodium Sulfur Battery*, Chapman & Hall, London, New York.

48 Belyaev, I.N., Dugin, V.E. and Nalbandyan, V.B. (1983) *Neorgan. Mater.*, **19**, 313–316. (in Russian).

49 Nalbandyan, V.B., Medvedev, B.S., Rykalova, S.I. and Isakova, S.Yu. (1988) *Neorgan. Mater.*, **24**, 1030–1034. (in Russian).

50 Friauf, R.J. (1972) *Physics of Electrolytes*, vol. **1** (ed. J. Hladik), Academic Press, London, New York.

51 Kikkawa, S., Yasuda, F. and Koizumi, M. (1985) *Mater. Res. Bull.*, **20**, 1221–1227.

52 Francombe, M.H. and Lewis, B. (1957) *J. Electronics*, **2**, 387–403.

53 Avdeev, M.Yu., Nalbandyan, V.B., Beskrovnyi, A.I. and Balagurov, A.M. (2001) *Russ. J. Inorg. Chem.*, **46**, 489–495.

54 Avdeev, M.Yu., Nalbandyan, V.B. and Medvedev, B.S. (1997) *Inorg. Mater.*, **33**, 500–503.

55 Shukaev, I.L., Shilov, G.V., Avdeev, M.Yu., Nalbandyan, V.B., Medvedev, B.S., Atovmyan, L.O. and Balagurov, A.M., (2000) in *Solid State Chemistry, 2000, Book of Abstracts* (eds P. Bezdicka and T. Grygar), Institute of Inorganic Chemistry, Academy of Sciences of the Czech Republic, Rez, p. 19.

56 Tkachev, V.V., Ponomarev, V.I. and Atovmyan, L.O. (1984) *Zh. Strukt. Khimii*, **25**, 128–132, (in Russian).

57 Peters, C.R., Bettman, M., Moore, J.W. and Glick, M.D. (1971) *Acta Crystallogr. B*, **27**, 1826–1834.

58 Michel, C., Guyomarc'h, A. and Raveau, B. (1977) *J. Solid State Chem.*, **22**, 393–403.

59 Gasperin, M. (1977) *Acta Crystallogr. B*, **33**, 398–402.

60 Fallon, G.D. and Gatehouse, B.M. (1977) *J. Solid State Chem.*, **22**, 405–409.

61 Kaber, R., Nilsson, L., Andersen, N.H., Lunden, A. and Thomas, J.O. (1992) *J. Phys.: Condens. Matter*, **4**, 1925–1933.

62 Nalbandyan, V.B. and Belyaev, I.N. (1985) *Neorgan. Mater.*, **21**, 1006–1010. (in Russian).

63 Yao, Y.F.Y. and Kummer, J.T. (1967) *J. Inorg. Nucl. Chem.*, **29**, 2453–2475.

64 Shannon, R.D. (1976) *Acta Crystallogr. A*, **32**, 751–767.

65 Edström, K., Gustafsson, T., Thomas, J.O. and Farrington, G.C. (1997) *Acta Crystallogr. B*, **53**, 631–638.

66 Radzilowski, R.H. and Kummer, J.T. (1971) *J. Electrochem. Soc.*, **118**, 714–716.

67 Singer, J., Fielder, W.L., Kautz, H.E. and Fordyce, J.S. (1976) *J. Electrochem. Soc.*, **123**, 614–617.

68 Belyaev, I.N., Nalbandyan, V.B., Trubnikov, I.L. and Medvedev, B.S. (1984) *Russ. J. Inorg. Chem.*, **29**, 144–145.

69 Shannon, R.D., Taylor, B.E., Gier, T.E., Chen, H.-Y. and Berzins, T. (1978) *Inorg. Chem.*, **17**, 958–964.

70 Smirnova, O.A., Fuentes, R.O., Figueiredo, F., Kharton, V.V. and Marques, F.M.B. (2003) *J. Electroceram.*, **11**, 179–189.

71 Politaev, V.V. and Nalbandyan, V.B. (2009) *Solid State Sciences*, **11**, 144–150.

72 Hagenmuller, P. (1983) *Solid State Chemistry 1982* (ed. R. Metselaar), Elsevier, Amsterdam, pp. 49–73.

73 Delmas, C., Fouassier, C., Reau, J.-M. and Hagenmuller, P. (1976) *Mater. Res. Bull.*, **11**, 1081–1086.

74 Maazaz, A., Delmas, C., Fouassier, C., Reau, J.-M. and Hagenmuller, P. (1979) *Mater. Res. Bull.*, **14**, 193–199.

75 Sleight, A.W., Gulley, J.E. and Berzins, T. (1977) *Adv. Chem. Series*, **163**, 195–204.

76 Nalbandyan, V.B., Trubnikov, I.L., Bukun, N.G. and Medvedev, B.S. (1986) *Inorg. Mater.*, **22**, 736–740.

77 Hwang, Y.K. and Ahuja, N. (1992) *Computing Surveys*, **24**, 219.

78 Bernal, J.D. (1959) *Nature*, **183**, 141.

79 Bernal, J.D. and Mason, J. (1960) *Nature*, **188**, 910.

80 Whittaker, E.J.W. (1978) *J. Non-Cryst. Solids*, **28**, 293.

81 Elam, W.T., Kerstein, A.R. and Rehr, J.J. (1984) *Phys. Rev. Lett.*, **52**, 1516.

82 Blatov, V.A., Ilyushin, G.D., Blatova, O.A., Anurova, N.A., Ivanov-Schits, A.K. and Dem'yanets, L.N. (2006) *Acta Crystallogr. B*, **62**, 1010.

83 Kuppers, H., Lieba, F. and Spek, A.L. (2006) *J. Appl. Crystallogr.*, **39**, 338.

84 Shukaev, I.L. and Nalbandyan, V.B. (2002) *Fundamental'nye problemy ioniki tverdogo tela (Fundamental Problems of Solid State Ionics)*, Chernogolovka, Russia, pp. 20–21. (in Russian).

85 Fischer, W. (1986) *Crystal Res. Technol.*, **21**, 499.

86 Richards, F.M. (1974) *J. Molec. Biol.*, **82**, 1.

87 Poupon, A. (2004) *Curr. Opin. Struct. Biol.*, **14**, 233.

88 Fischer, W. and Koch, E. (1979) *Z. Kristallogr.*, **150**, 245.

89 Gellatly, B.J. and Finney, J.L. (1982) *J. Non-Cryst. Solids*, **50**, 313.

90 Medvedev, N.N., Voloshin, V.P., Luchnikov, V.A. and Gavrilova, M.L. (2006) *J. Comput. Chem.*, **27**, 1676.

91 Dupuis, F., Sadoc, J.F., Jullien, R., Angelov, B. and Mornon, J.P. (2005) *Bioinformatics*, **21**, 1715.

92 Gore, S.P., Burke, D.F. and Blundell, T.L. (2005) *Bioinformatics*, **21**, 3316.

93 Petrek, M., Kosinova, P., Koca, J. and Otyepka, M. (2007) *Structure*, **15**, 1357.

94 Blatov, V.A., Shevchenko, A.P. and Serezhkin, V.N. (1999) *Russ. J. Coord. Chem.*, **25**, 453.

95 Pauling, L. (1929) *J. Am. Chem. Soc.*, **51**, 1010.

96 Brown, I.D. (2002) *The Chemical Bond in Inorganic Chemistry: The Bond Valence Model*, Oxford University Press, New York.

97 Adams, S. and Swenson, J. (2004) *Ionics*, **10**, 317.

98 Adams, S. (2000) *Solid State Ionics*, **136**, 1351.

99 Nayal, M. and Dicera, E. (1994) *Proc. Natl Acad. Sci. USA*, **91**, 817.

100 Gonzalez-Platas, J., Gonzalez-Silgo, C. and Ruiz-Perez, C. (1999) *J. Appl. Crystallogr.*, **32**, 341.

101 Adams, S. and Swenson, J. (2000) *Phys. Rev. Lett.*, **84**, 4144.

102 Adams, S. (2006) *J. Power Sources*, **159**, 200.

103 Hohenberg, P. and Kohn, W. (1964) *Phys. Rev.*, **136**, B864.

104 Kohn, W. and Sham, L.J. (1965) *Phys. Rev.*, **140**, A1133.

105 Koch, W. and Holthausen, M.C. (2001) *A Chemist's Guide to Density Functional Theory*, 2nd edn, Wiley-VCH, Weinheim.

106 Vineyard, G.H. (1957) *J. Phys. Chem. Solids*, **3**, 121.

107 Van der Ven, A. and Ceder, G. (2001) *J. Power Sources*, **97–98**, 529.

108 Ouyang, C.Y., Shi, S.Q., Wang, Z.X., Huang, X.J. and Chen, L.Q. (2004) *Phys. Rev. B*, **69**, 104303.

109 Anicete-Santos, M., Gracia, L., Beltran, A., Andres, J., Varela, J.A. and Longo, E. (2008) *Phys. Rev. B*, **77**, 085112.

110 Gracia, L., Garcia-Canadas, J., Garcia-Belmonte, G., Beltran, A., Andres, J. and Bisquert, J. (2005) *Electrochem. Solid State Lett.*, **8**, J21.

111 Kadau, K., Germann, T.C. and Lomdahl, P.S. (2006) *Int. J. Mod. Phys. C*, **17**, 1755.

112 Car, R. and Parrinello, M. (1985) *Phys. Rev. Lett.*, **55**, 2471.

113 Wang, C., Zhang, Q.M. and Bernholc, J. (1992) *Phys. Rev. Lett.*, **69**, 3789.

114 Katsumata, T., Inaguma, Y., Itoh, M. and Kawamura, K. (2002) *Chem. Mater.*, **14**, 3930.

115 Devanathan, R., Weber, W.J., Singhal, S.C. and Gale, J.D. (2006) *Solid State Ionics*, **177**, 1251.

116 Munch, W., Kreuer, K.D., Seifert, G. and Maier, J. (2000) *Solid State Ionics*, **136**, 183.

117 Sommariva, M. and Catti, M. (2006) *Chem. Mater.*, **18**, 2411.

118 Bachmann, R. and Schulz, H. (1984) *Acta Crystallogr. A*, **40**, 668.

119 Boysen, H. (2003) *Z. Kristallogr.*, **218**, 123.

120 Merli, M. and Pavese, A. (2006) *Z. Kristallogr.*, **221**, 613.

121 Tanaka, H., Takata, M., Nishibori, E., Kato, K., Iishi, T. and Sakata, M. (2002) *J. Appl. Crystallogr.*, **35**, 282.

122 Izumi, F. and Dilanian, R. (2002) *Recent Res. Dev. Phys.*, **3**, 699.

123 van Smaalen, S., Palatinus, L. and Schneider, M. (2003) *Acta Crystallogr. A*, **59**, 459.

124 Kobayashi, H., Arachi, Y., Kageyama, H. and Tatsumi, K. (2004) *J. Mater. Chem.*, **14**, 40.

125 Yashima, M., Kobayashi, S. and Yasui, T. (2007) *Faraday Discuss.*, **134**, 369.

126 Takahashi, Y., Akimoto, J., Kijima, N. and Gotoh, Y. (2004) *Solid State Ionics*, **172**, 505.

127 Collongues, R., Gourier, D., Kahn, A., Boilot, J.P., Colomban, Ph. and Wicker, A. (1984) *J. Phys. Chem. Solids*, **45**, 981–1013.

128 Foster, L.M., Campbell, D.R. and Chandrashekhar, G.V. (1978) *J. Electrochem. Soc.*, **125**, 1689–1695.

129 Foster, L.M. (1979) in *Fast Ion Transport in Solids* (eds P. Vashishta, J.N. Mundy and G.K. Shenoy), Elsevier, Amsterdam, pp. 249–254.

130 Collin, G., Boilot, J.P., Colomban, Ph. and Comes, R. (1980) *Phys. Rev. B*, **22**, 5912–5923.

131 Brown, G.M., Schwinn, D.A., Bates, J.B. and Brundage, W.E. (1981) *Solid State Ionics*, **5**, 147–150.

132 Briant, J.L. and Farrington, G.C. (1980) *J. Solid State Chem.*, **33**, 385–390.

133 Baffier, N., Badot, J.C. and Colomban, Ph. (1981) *Mater. Res. Bull.*, **16**, 259–265.

134 Chandrashekhar, G.V., Foster, L.M. and Burns, G. (1981) *Solid State Ionics*, **5**, 175–178.

135 Nariki, S., Ito, S., Uchinokura, K. and Yoneda, N. (1987) *J. Crystal Growth*, **85**, 483–487.

136 Hong, H.Y.-P. (1976) *Mater. Res. Bull.*, **11**, 173–182.

137 Goodenough, J.B., Hong, H.Y.-P. and Kafalas, J.A. (1976) *Mater. Res. Bull.*, **11**, 303–320.

138 Boilot, J.P., Collin, G. and Colomban, Ph. (1987) *Mater. Res. Bull.*, **22**, 669–676.

139 Kohler, H., Schulz, H. and Melnikov, O. (1983) *Mater. Res. Bull.*, **18**, 1143–1152.

140 Cava, R.J., Vogel, E.M. and Johnson, D.W. (1982) *J. Am. Ceram. Soc.*, **65**, C157–C159.

141 Vogel, E.M., Cava, R.J. and Rietman, E. (1984) *Solid State Ionics*, **14**, 1–6.

142 Hong, H.Y.-P. (1979) *Fast Ion Transport in Solids* (eds P. Vashishta, J.N. Mundy and G.K. Shenoy), Elsevier, Amsterdam, pp. 431–433.

143 Kalinin, V.B., Lazoryak, B.I. and Stefanovich, S.Yu. (1983) *Kristallografiya*, **28**, 264–270. (in Russian).

144 Atovmyan, L.O., Bukun, N.G., Kovalenko, V.I., Korosteleva, A.I., Tkachev, V.V. and Ukshe, E.A. (1983) *Elektrokhimiya*, **19**, 933–937 (in Russian).

145 Collin, G., Comes, R., Boilot, J.P. and Colomban, Ph. (1986) *J. Phys. Chem. Solids*, **47**, 843–854.

146 Hirata, Y., Kitasako, H. and Shimada, K. (1988) *J. Ceram. Soc. Japan*, **96**, 620–623.

147 Winand, J.M., Rulmont, A. and Tarte, P. (1990) *J. Mater. Sci.*, **25**, 4008–4013.

148 Bentzen, J.J. and Nicholson, P.S. (1980) *Mater. Res. Bull.*, **15**, 1737–1745.

149 Yamashita, K. and Nicholson, P.S. (1985) *Solid State Ionics*, **17**, 115–120.

150 Chaloun, O.A., Chevalier, P., Trichet, L. and Rouxel, J. (1981) *Solid State Ionics*, **2**, 231–235.

151 Burmakin, E.I. and Shekhtman, G.Sh. (1985) *Elektrokhimiya*, **21**, 752–757. (in Russian).

152 Nalbandyan, V.B. and Shukaev, I.L. (1992) *Russ. J. Inorg. Chem.*, **37**, 1231–1235.

153 Avdeev, M.Yu., Nalbandyan, V.B. and Beskrovnyi, A.I. (1998) 6th European Powder Diffraction Conference (EPDIC-6), Scientific Program and Abstracts, Budapest, Hungary, August 22–25, p. 235.

154 Shilov, G.V., Nalbandyan, V.B., Volochaev, V.A. and Atovmyan, L.O. (2000) *Int. J. Inorg. Mater.*, **2**, 443–449.

155 Medvedeva, L.I., Nalbandyan, V.B. and Medvedev, B.S. (2002) *Fundamental'nye problemy ioniki tverdogo tela (Basic Problems of the Solid State Ionics)*, Chernogolovka, Russia, p. 74 (in Russian).

156 Smirnova, O.A., Nalbandyan, V.B., Avdeev, M., Medvedeva, L.I., Medvedev, B.S., Kharton, V.V. and Marques, F.M.B. (2005) *J. Solid State Chem.*, **178**, 172–179.

157 Smirnova, O.A., Avdeev, M., Nalbandyan, V.B., Kharton, V.V. and Marques, F.M.B. (2006) *Mater. Res. Bull.*, **41**, 1056–1062.

158 Langlet, G. (1964) *Compt. Rend. Acad. Sci.*, **259**, 3769–3770.

159 Frostäng, S., Grins, J. and Nygren, M. (1988) *Chem. Scripta*, **28**, 107–110.

160 Husheer, S.L.G., Thompson, J.G. and Melnitchenko, A. (1999) *J. Solid State Chem.*, **147**, 624–630.

161 Burmakin, E.I., Voronin, V.I., Akhtyamova, L.Z., Berger, I.F. and Shekhtman, G.Sh. (2005) *Russ. J. Electrochem.*, **41**, 783–788.

162 Burmakin, E.I. and Shekhtman, G.Sh. (2005) *Russ. J. Electrochem.*, **41**, 1341–1344.

163 Burmakin, E.I., Nechaev, G.V. and Shekhtman, G.Sh. (2007) *Russ. J. Electrochem.*, **43**, 121–124.

164 Burmakin, E.I. and Shekhtman, G.Sh. (2007) *Russ. J. Electrochem.*, **43**, 983–986.

165 Irvine, J.T.S. and West, A.R. (1987) *J. Solid State Chem.*, **69**, 126–134.

166 Ivanov-Shitz, A.K., Sigarev, S.E. and Belov, O.I. (1986) *Fizika Tverdogo Tela*, **28**, 2898–2900. (in Russian).

167 Subramanian, M.A., Subramanian, R. and Clearfield, A. (1985) *Solid State Ionics*, **15**, 15–19.
168 Takahashi, T., Kuwabara, K. and Aoyama, H. (1974) *Nippon Kagaku Kaishi*, **12**, 2291–2296.
169 Takahashi, T. and Kuwabara, K. (1978) *Electrochim. Acta*, **23**, 375–379.
170 Reau, J.-M., Moali, J. and Hagenmuller, P. (1977) *J. Phys. Chem. Solids*, **38**, 1395–1398.
171 Yoshikado, S., Ohachi, T., Taniguchi, I., Onoda, Y., Watanabe, M. and Fujiki, Y. (1983) *Solid State Ionics*, **9/10**, 1305–1309.
172 Watelet, H., Besse, J.-P., Baud, G. and Chevalier, R. (1982) *Mater. Res. Bull.*, **17**, 863–871.
173 Yoshikado, S., Michiue, Y., Onoda, Y. and Watanabe, M. (2000) *Solid State Ionics*, **136–137**, 371–374.
174 Chandrashekhar, G.V., Bednovitz, A. and La Placa, S.J. (1979) *Fast Ion Transport in Solids* (eds P. Vashishta, J.N. Mundy and G.K. Shenoy), Elsevier, Amsterdam, pp. 447–450.
175 Nalbandyan, V.B., Belyaev, I.N. and Mezhzhorina, N.V. (1979) *Zh. Neorgan. Khimii*, **24**, 3207–3212. (in Russian).
176 Golubev, A.M., Molchanov, V.N., Antipin, M.Yu. and Simonov, V.I. (1981) *Kristallografiya*, **26**, 1254–1258.
177 Nalbandyan, V.B. (1991) *Neorgan. Mater.*, **27**, 1006–1010. (in Russian).
178 Yoshikado, S., Ohachi, T., Taniguchi, I., Watanabe, M., Fujiki, Y. and Onoda, Y. (1989) *Solid State Ionics*, **35**, 377–385.
179 Yoshikado, S., Funatomi, H., Watanabe, M., Onoda, Y. and Fujiki, Y. (1999) *Solid State Ionics*, **121**, 127–132.
180 Shilov, G.V., Atovmyan, L.O., Volochaev, V.A. and Nalbandyan, V.B. (1999) *Crystallogr. Reports*, **44**, 960–964.
181 Avdeev, M.Yu., Nalbandyan, V.B., Beskrovnyi, A.I., Balagurov, A.M. and Volochaev, V.A. (1999) *Russ. J. Inorg. Chem.*, **44**, 480–484.
182 Nalbandyan, V.B. (1999) *Russ. J. Inorg. Chem.*, **44**, 511–519.
183 Reid, A.F., Wadsley, A.D. and Sienko, M. (1968) *Inorg. Chem.*, **7**, 112–118.
184 Groult, D., Mercey, C. and Raveau, B. (1980) *J. Solid State Chem.*, **32**, 289–296.
185 Michiue, Y. and Sato, A. (2004) *Acta Crystallogr. B*, **60**, 692–697.
186 Mumme, W.G. (1968) *Acta Crystallogr. B*, **24**, 1114–1120.
187 Yang, H. and Konzett, J. (2004) *J. Solid State Chem.*, **177**, 4576–4581.
188 Angerer, P., Fischer, R.X., Schmücker, M. and Schneider, H. (2008) *J. Eur. Ceram. Soc.*, **28**, 493–497.
189 Thangadurai, V., Kaack, H. and Weppner, W.J.F. (2003) *J. Am. Ceram. Soc.*, **86**, 437–440.
190 Thangadurai, V. and Weppner, W. (2005) *J. Am. Ceram. Soc.*, **88**, 411–418.
191 Thangadurai, V. and Weppner, W. (2005) *Adv. Function. Mater.*, **15**, 107–112.
192 Cussen, E.J. (2006) *Chem. Commun.*, 412–413.
193 Thangadurai, V. and Weppner, W. (2006) *J. Solid State Chem.*, **179**, 974–984.
194 Murugan, R., Weppner, W., Schmid-Beurmann, P. and Thangadurai, V. (2007) *Mater. Sci. Eng. B*, **143**, 14–20.
195 O'Callaghan, M.P. and Cussen, E.J. (2007) *Chem. Commun.*, 2048–2050.
196 Cussen, E.J. and Yipa, T.W.S. (2007) *J. Solid State Chem.*, **180**, 1832–1839.
197 Murugan, R., Thangadurai, V. and Weppner, W. (2007) *Angew. Chem. Int. Ed.*, **46**, 7778–7781.
198 Percival, J., Kendrick, E. and Slater, P.R. (2008) *Solid State Ionics*, **179**, 1666–1669.
199 O'Callaghan, M.P. and Cussen, E.J. (2008) *Solid State Sciences*, **10**, 390–395.
200 O'Callaghan, M.P., Powell, A.S., Titman, J.J., Chen, G.Z. and Cussen, E.J. (2008) *Chem. Mater.*, **20**, 2360–2369.
201 Percival, J., Kendrick, E. and Slater, P.R. (2008) *Mater. Res. Bull.*, **43**, 765–770.
202 Murugan, R., Thangadurai, V. and Weppner, W. (2008) *J. Electrochem. Soc.*, **155**, A90–101.
203 Briant, J.L. and Farrington, G.C. (1981) *Solid State Ionics*, **5**, 207–210.

204 Sigarev, S.E. (1992) *Kristallografiya*, 7, 1055–1086.

205 Sigaryov, S.E. (1992) *Mater. Sci. Eng. B*, **13**, 121–123.

206 Sudreau, F., Petit, D. and Boilot, J.P. (1989) *J. Solid State Chem.*, **83**, 78.

207 Aono, H., Sugimoto, E., Sadaoka, Y., Imanaka, N. and Adachi, G. (1989) *J. Electrochem. Soc.*, **136**, 590–591.

208 Xu, X., Wen, Z., Yang, X. and Chen, L. (2008) *Mater. Res. Bull.*, **43**, 2334–2341.

209 Suzuki, T., Yoshida, K., Uematsu, K., Kodama, T., Toda, K., Ye, Z.-G., Ohashi, M. and Sato, M. (1998) *Solid State Ionics*, **113–115**, 89–96.

210 Thangadurai, V., Shukla, A.K. and Gopalakrishnan, J. (1999) *J. Mater. Chem.*, **9**, 739–741.

211 Catti, M., Morgante, N. and Ibberson, R.M. (2000) *J. Solid State Chem.*, **152**, 340–347.

212 Rambabu, G., Anantharamulu, N., Vithal, M., Raghavender, M. and Prasad, G. (2006) *J. Appl. Phys.*, **100**, 083707.

213 Catti, M. (2001) *J. Solid State Chem.*, **156**, 305–312.

214 Belous, A.G., Novitskaya, G.N., Polyanetskaya, S.V. and Gornikov, Y.I. (1987) *Russian J. Inorg. Chem.*, **32**, 156–158.

215 Belous, A.G., Novitskaya, G.N. and Polyanetskaya, S.V. (1987) *Neorgan. Mater.*, **23**, 1330–1332. (in Russian).

216 Stramare, S., Thangadurai, V. and Weppner, W. (2003) *Chem. Mater.*, **15**, 3974–3990.

217 Varez, A., Fernández-Díaz, M.T., Alonso, J.A. and Sanz, J. (2005) *Chem. Mater.*, **17**, 2404–2412.

218 Catti, M. (2007) *Chem. Mater.*, **19**, 3963–3972.

219 Inaguma, Y., Yu, J., Katsumata, T. and Itoh, M. (1997) *J. Ceram. Soc. Jpn*, **105**, 548–550.

220 Paris, M.A., Sanz, J., Leon, C., Santamaria, J., Ibarra, J. and Vares, A. (2000) *Chem. Mater.*, **12**, 1694–1701.

221 Arakawa, S., Shiotsu, T. and Hayashi, S. (2005) *J. Ceram. Soc. Jpn*, **113**, 317–319.

222 Inaguma, Y., Matsui, Y., Yu, J., Shan, Y.J., Nakamura, T. and Itoh, M. (1997) *J. Phys. Chem. Solids*, **58**, 843–852.

223 He, L.X. and Yoo, H.I. (2003) *Electrochim. Acta*, **48**, 1357–1366.

224 Morata-Orrantia, A., García-Martín, S. and Alario-Franco, M.Á. (2003) *Chem. Mater.*, **15**, 3991–3995.

225 Bohnke, O. (2008) *Solid State Ionics*, **179**, 9–15.

226 Mizumoto, K. and Hayashi, S. (1999) *Solid State Ionics*, **116**, 263–269.

227 Belous, A.G., Gavrilenko, O.N., Pashkova, E.V. and Mirnyi, V.N. (2002) *Russ. J. Electrochem.*, **38**, 425–430.

228 Chen, C.H., Xie, S., Sperling, E., Yang, A.S., Henriksen, G. and Amine, K. (2004) *Solid State Ionics*, **167**, 263–272.

229 Phama, Q.N., Bohnke, C., Emery, J., Bohnke, O., Le Berre, F., Crosnier-Lopez, M.-P., Fourquet, J.-L. and Florian, P. (2005) *Solid State Ionics*, **176**, 495–504.

230 Hong, H.Y.-P. (1978) *Mater. Res. Bull.*, **13**, 117–124.

231 Burmakin, E.I., Shekhtman, G.Sh., Alikin, V.N. and Stepanov, G.K. (1981) *Elektrokhimiya*, **17**, 1734–1739. (in Russian).

232 Li-quan, C., Lian-zhong, W., Guang-can, C., Gang, W. and Zi-rong, L. (1983) *Solid State Ionics*, **9/10**, 149–152.

233 Bose, D.N. and Majumdar, D. (1984) *Bull. Mater. Sci.*, **6**, 223–230.

234 Abrahams, I., Bruce, P.G., David, W.I.F. and West, A.R. (1989) *Acta Crystallogr.*, **45**, 457–462.

235 Burmakin, E.I., Voronin, V.I. and Shekhtman, G.Sh. (2003) *Russ. J. Electrochem.*, **39**, 1124–1129.

236 Du, Y.A. and Holzwarth, N.A.W. (2007) *Phys. Rev. B*, **76**, 174302.

237 Ivanov-Shitz, A.K. and Kireev, V.V. (2003) *Crystallogr. Reports*, **48**, 112–123.

238 Kanno, R. and Murayama, M. (2001) *J. Electrochem. Soc.*, **148**, A742–A746.

239 Kanno, R., Hata, T., Kawamoto, Y. and Irie, M. (2000) *Solid State Ionics*, **130**, 97–104.

240 Murayama, M., Kanno, R., Irie, M., Ito, S., Hata, T., Sonoyama, N. and

Kawamoto, Y. (2002) *J. Solid State Chem.*, **168**, 140–148.

241 Murayama, M., Sonoyama, N., Yamada, A. and Kanno, R. (2004) *Solid State Ionics*, **170**, 173–180.

242 Schulz, H. and Thiemann, K.H. (1979) *Acta Crystallogr. A*, **35**, 309–314.

243 Partik, M., Schneider, M. and Lutz, H.D. (1994) *Z. Anorg. Allg. Chem.*, **620**, 791–795.

244 Schneider, M., Kuske, P. and Lutz, H.D. (1993) *Z. Naturforschung B*, **48**, 1–6.

245 Lutz, H.D., Kuske, P. and Wussow, K. (1987) *Z. Anorg. Allg. Chem.*, **553**, 172–178.

246 Kanno, R., Takeda, Y., Mori, M. and Yamamoto, O. (1987) *Chem. Lett.*, 1465–1468.

247 Kanno, R., Takeda, Y. and Yamamoto, O. (1981) *Mater. Res. Bull.*, **16**, 999–1005.

248 Yamada, K., Kumano, K. and Okuda, T. (2006) *Solid State Ionics*, **177**, 1691–1695.

249 Tomita, Y., Matsushita, H., Kobayashi, K., Maeda, Y. and Yamada, K. (2008) *Solid State Ionics*, **179**, 867–870.

250 Deiseroth, H.-J., Kong, S.-T., Eckert, H., Vannahme, J., Reiner, C., Zaiß, T. and Schlosser, M. (2008) *Angew. Chem. Int. Ed.*, **47**, 755–758.

251 Jeitschko, W., Bither, T.A. and Bierstedt, P.E. (1977) *Acta Crystallogr. B*, **33**, 2767–2775.

252 Reau, J.-M., Portier, J., Levasseur, A., Villeneuve, G. and Pouchard, M. (1978) *Mater. Res. Bull.*, **13**, 1415–1423.

253 Shannon, R.D., Taylor, B.E., English, A.D. and Berzins, T. (1977) *Electrochim. Acta*, 22, 783–796.

254 Idemoto, Y., Ricardson, J.W., Koura, N., Kohara, S. and Loong, C.-K. (1998) *J. Phys. Chem. Solids*, **59**, 363–376.

255 Kreuer, K.-D., Rabenau, A. and Weppner, W. (1982) *Angew. Chem. Int. Ed. Engl.*, **21**, 208–209.

256 Nakamura, O., Kodama, T., Ogino, I. and Miyake, Y. (1979) *Chem. Lett.*, **8**, 17–18.

257 Korosteleva, A.P., Leonova, L.S. and Ukshe, E.A. (1987) *Elektrokhimiya*, **23**, 1349–1353. (in Russian).

258 Kreuer, K.D., Hampele, M., Dolde, K. and Rabenau, A. (1988) *Solid State lonics*, **28–30**, 589–593.

259 Bradley, J.A. and Illingworth, J.W. (1936) *Proc. R. Soc. London, Ser. A*, **157**, 113–131.

260 Nakamura, O., Ogino, I. and Kodama, T. (1981) *Solid State Ionics*, **3/4**, 347–351.

261 Denisova, T.A., Leonidov, O.N., Maksimova, L.G. and Zhuravlev, N.A. (2001) *Russ. J. Inorg. Chem.*, **46**, 1553–1558.

262 Howe, A.T. and Shilton, M.G. (1979) *J. Solid State Chem.*, **28**, 345–361.

263 Kreuer, K.D., Stoll, I. and Rabenau, A. (1983) *Solid State Ionics*, **9–10**, 1061–1064.

264 Weppner, W. (1981) *Solid State Ionics*, **5**, 3–8.

265 Sinitsyn, V.V., Privalov, A.I., Lips, O., Baranov, A.I., Kruk, D. and Fujara, F. (2008) *Ionics*, **14**, 223–226.

266 Belushkin, A.V., David, W.I.F., Ibberson, R.M. and Shuvalov, L.A. (1991) *Acta Crystallogr. B*, **47**, 161–166.

267 Baranov, A.I., Merinov, B.V., Tregubchenko, A.V., Khiznichenko, V.P., Shuvalov, L.A. and Schagina, N.M. (1989) *Solid State lonics*, **36**, 279–282.

268 Haile, S.M., Chisholm, C.R.I., Sasaki, K., Boysen, D.A. and Uda, T. (2007) *Faraday Discuss.*, **134**, 17–39.

269 Yamane, Y., Yamada, K. and Inoue, K. (2008) *Solid State Ionics*, **179**, 483–488.

270 Zhou, W., Bondarenko, A.S., Boukamp, B.A. and Bouwmeester, H.J.M. (2008) *Solid State Ionics*, **179**, 380–384.

271 Iwahara, H., Esaka, T., Uchida, H. and Maeda, N. (1981) *Solid State Ionics*, **3/4**, 359–363.

272 Iwahara, H. (1996) *Solid State Ionics*, **86–88**, 9–15.

273 Bonanos, N. (2001) *Solid State Ionics*, **145**, 265–274.

274 Sammes, N., Phillips, R. and Smirnova, A. (2004) *J. Power Sources*, **134**, 153–159.

275 Fabbri, E., D'Epifanio, A., Di Bartolomeo, E., Licoccia, S. and Traversa, E. (2008) *Solid State Ionics*, **179**, 558–564.

276 Schober, T. and Bohn, H.G. (2000) *Solid State Ionics*, **127**, 351–360.

277 Nomura, K., Takeuchi, T., Kamo, S., Kageyama, H. and Miyazaki, Y. (2004) *Solid State Ionics*, **175**, 553–555.

278 Shirsat, A.N., Kaimal, K.N.G., Bharadwaj, S.R. and Das, D. (2004) *J. Solid State Chem.*, **177**, 2007–2013.

279 Tanner, C.W. and Virkar, A.V. (1996) *J. Electrochem. Soc.*, **143**, 1386–1389.

280 Shimada, T., Wen, C., Taniguchi, N., Otomo, J. and Takahashi, H. (2004) *J. Power Sources*, **131**, 289–292.

281 Reichel, U., Arons, R.R. and Schilling, W. (1996) *Solid State lonics*, **86–88**, 639–645.

282 Kosacki, I. and Tuller, H.L. (1995) *Solid State Ionics*, **80**, 223–229.

283 Kumar, R.V., Cobb, L.J. and Fray, D.J. (1996) *Ionics*, **2**, 162–168.

284 Schober, T., Bohn, H.G., Mono, T. and Schilling, W. (1999) *Solid State Ionics*, **118**, 173–178.

285 Sharova, N.V. and Gorelov, V.P. (2003) *Russian J. Electrochem.*, **39**, 461–466.

286 Qi, X. and Lin, Y.S. (1999) *Solid State Ionics*, **120**, 85–93.

287 Bannykh, A.V. and Kuzin, B.L. (2003) *Ionics*, **9**, 134–139.

288 Wells, A.F. (1986) *Structural Inorganic Chemistry*, Clarendon Press, Oxford, Chapter 8.

289 Haugsrud, R. and Norby, T. (2006) *Nat. Mater.*, **5**, 193–196.

290 Shimura, T., Fujimoto, S. and Iwahara, H. (2001) *Solid State Ionics*, **143**, 117–123.

291 Li, S., Schönberger, F. and Slater, P. (2003) *Chem. Commun.*, 2694–2695.

292 Kitamura, N., Amezawa, K., Tomii, Y. and Yamamoto, N. (2003) *Solid State Ionics*, **162–163**, 161–165.

293 Amezawa, K., Takahashi, N., Kitamura, N., Tomii, Y. and Yamamoto, N. (2004) *Solid State Ionics*, **175**, 575–579.

294 Labrincha, J.A., Frade, J.R. and Marques, F.M.B. (1997) *Solid State Ionics*, **99**, 33–40.

295 Wang, J.-D., Xie, Y.-H., Zhang, Z.-F., Liu, R.-Q. and Li, Z.-J. (2005) *Mater. Res. Bull.*, **40**, 1294–1302.

296 Baffier, S., Badot, J.C. and Colomban, Ph. (1980) *Solid State Ionics*, **2**, 107–113.

297 Baffier, S., Badot, J.C. and Colomban, Ph. (1984) *Solid State Ionics*, **13**, 233–236.

298 DeNuzzio, J.D. and Farrington, G.C. (1989) *J. Solid State Chem.*, **79**, 65–74.

299 Nicholson, P.S., Munshi, M.Z.A., Singh, G., Sayer, M. and Bell, M.F. (1986) *Solid State Ionics*, **18/19**, 699–703.

300 Kuo, C.K., Tan, A., Sarkar, P. and Nicholson, P.S. (1990) *Solid State Ionics*, **38**, 251–254.

301 Brunner, D.G. and Tomandl, G. (1987) *Adv. Ceram. Mater.*, **2**, 794–797.

302 Kuntz, M., Tomandl, G. and Hoser, A. (1988) *Solid State Ionics*, **27**, 211–220.

303 Hinrichs, R. and Tomandl, G. (1998) *Solid State Ionics*, **107**, 117–122.

304 Nalbandyan, V.B. and Trubnikov, I.L. S. (2002) 11th International Conference on Solid State Protonic Conductors, University of Surrey, Guildford, UK, p. C19.

305 Clearfield, A. (1991) *Solid State Ionics*, **46**, 35–43.

306 Stenina, I.A., Kislitsyn, M.N., Ghuravlev, N.A. and Yaroslavtsev, A.B. (2008) *Mater. Res. Bull.*, **43**, 377–383.

307 Alberti, G., Casciola, M., Costantino, U. and Leonardi, M. (1984) *Solid State Ionics*, **14**, 289–295.

308 Nagao, M., Kamiya, T., Heo, P., Tomita, A., Hibino, T. and Sano, M. (2006) *J. Electrochem. Soc.*, **153**, A1604–A1609.

309 Tomita, A., Kajiyama, N., Kamiya, T., Nagao, M. and Hibino, T. (2007) *J. Electrochem. Soc.*, **154**, B1265–B1269.

8
Conducting Solids: In the Search for Multivalent Cation Transport

Nobuhito Imanaka and Shinji Tamura

Editorial Preface

Vladislav V. Kharton

The interfacial and transport phenomena related to multivalent cation migration in solids are of major interest in our understanding of many fundamental mechanisms and for numerous potential applications. The exact identification and quantitative analysis of these phenomena are, however, among the most complex tasks in solid-state electrochemistry. In those systems which comprise several types of ion that may be mobile, a precise separation of the contributions provided by different ionic species is only possible by combining several complementary electrochemical, spectroscopic, and microscopic techniques. These experiments require also careful examination of their reproducibility, an assessment of the relative roles of all factors that influence ionic transport, a thorough characterization of the materials, the elimination of various parasitic effects, and the correct accounting of possible parallel and alternative mechanisms. For multivalent cation conductors, typical experimental problems include the presence of minor amounts of impurities (e.g., protons or alkaline metal cations), interfacial transport (via extended defects, grain boundaries or surfaces), electronic conduction and/or phase decomposition processes induced by the applied electrical field, cation demixing under chemical potential gradients, possible volatilization of components, phase separation at the interfaces, and other complex electrode phenomena. In particular, due to the electrostatic and stereological factors discussed in previous chapters, the mobility of multivalent cations is always lower compared to that of monovalent species. Consequently, even microscopic amounts of ionic impurities, barely distinguishable by standard analytical methods, may have non-negligible effects on the transport properties. In the case of electrochemical measurements involving direct currents, high electrical and polarization resistances may lead to excessively high voltages, the generation of electronic charge carriers, and/or decomposition. These factors are in part responsible for the contradictory conclusions, various different opinions, doubts and questions reported with regards to migration- and intercalation-related phenomena when the cation

Solid State Electrochemistry I: Fundamentals, Materials and their Applications. Edited by Vladislav V. Kharton

ISBN: 978-3-527-32318-0

oxidation state exceeds 1+. For further information on this topic, the reader is directed to Refs. [1–16] and references therein.

There exist also numerous theoretical limitations. For example, the effects expected on macroscopic transfer and discharge of M^{3+} cations and BO_4^{2-} anions in $M_2(BO_4)_3$ (M = Al, In, Y; B = W, Mo) under applied DC voltage in simple electrochemical cells are similar. The identification of mobile species in such systems requires the use of sophisticated methods, including systematic investigations on the diffusion/reaction fronts in various diffusion couples, or on the transport of isotope tracers in single crystals, combined with the careful elimination of all secondary effects. Moreover, the diffusion of relatively large ionic species with a substantially high charge may only occur via various cooperative mechanisms, which are difficult for modeling. As a result, the type of migrating species and the diffusion mechanisms remain the subject of much debate, even for $M_2(BO_4)_3$ [10, 11, 15, 16].

Irrespective of these controversies, research in this intriguing area is crucially important from both fundamental and applied points of view. For instance, solid electrolytes and mixed conductors with mobile Al^{3+} cations may find applications in new types of rechargeable battery, and also in the aluminum industry; alkaline-earth cation conductors may be used, in particular, for precise humidity control in gaseous media and for CO_2 sequestration. This chapter includes a brief introduction to the field, with emphasis placed primarily on the authors' own studies. Although many aspects of materials electrochemical behavior require detailed investigation and further validation, the chapter provides an excellent overview of the phases where multivalent cation conduction may occur.

8.1 Introduction

The relationship between a current passing through a solid and the resultant chemical changes, and their compliance with Faraday's law, was first noted during the nineteenth century [17]. Notably, one of the most famous solid electrolytes – stabilized zirconia – was first developed in 1897 as a light Source, through the use of resistive-heating, and became known as the "Nernst glower". An extended knowledge of electrical conduction in these ionic solids was acquired by Wagner in 1943 [18], while between 1897 and 1943 many other important discoveries relating to ionic conducting behavior in solids were made, mainly in halides. Especially important were the very high ion-conducting characteristics demonstrated for silver iodide (AgI) [19], which is known to undergo a structural phase transformation at 149 °C, from a low-temperature γ phase to a high-temperature α phase (see Chapters 2 and 3). Coincident with this transition, the Ag^+ ion conductivity jumps more than five orders of magnitude, to achieve a value almost equivalent to that of the liquid AgI phase. It was while attempts were being made to stabilize the highly Ag^+ ion-conductive α-AgI phase at room temperature that compounds containing both Ag^+ and I^- ions (e.g., $RbAg_4I_5$) were first synthesized, in 1965 [20]. Unfortunately, $RbAg_4I_5$ was found to be highly unstable and, as yet, has not found any commercial application. At about the same time, the details of a La compound, LaF_3, which exhibits F^- ion-conducting characteristics, were first reported [21]. Another very important discovery, made by Kummer *et al.*, was that of a two-dimensional, high Na^+ ion-conducting β-alumina [22]; this was not identified until shortly after 1943, despite knowledge of the material dating back to the 1930s. As these mobile ionic species have a mass and also occupy volume, it was realized that such characteristics should be taken into account in order to improve their transport through a three-dimensional (3-D) tunnel structure. One of the optimum candidate structures to be generally accepted as suitable for ion migration was developed by Goodenough *et al.* in 1976, in order to achieve a high ionic conduction [23]. This "tailored solid electrolyte", with the chemical formula $Na_{1+x}Zr_2P_{3-x}Si_xO_{12}$, was subsequently named NASICON (*Na*$^+$ *S*uper *I*onic *CON*ductor).

As ion mobility in a solid lattice depends on the valence of the conducting ionic species, those cations with valence states higher than divalent were considered to be poor migrating species in a solid crystal lattice, due to their extraordinarily strong electrostatic interactions with the surrounding anion species. Consequently, until 1995 the migrating ion species in solids were believed to be limited to certain mono- and divalent ions. Although various materials, including Ln^{3+}-β/β''-Al_2O_3 (Ln: lanthanoids) [24–39], β-alumina-related materials [40–44] and β-$LaNb_3O_9$ [45, 46], were reported to be possible trivalent cation-conducting solids, the only demonstration of cation transport was based on total conductivity measurements. The first direct evidence of trivalent cation conduction in a solid was reported for the $Sc_2(WO_4)_3$ single crystal in 1995 [47]. Subsequently, various types of trivalent cation-conducting solids with $Sc_2(WO_4)_3$-type [48–86] or NASICON-type structures [87–103] were developed. In the case of tetravalent cation conduction in solids, $Zr_2O(PO_4)_2$ [104] and the NASICON-type Zr^{4+} [105–108] and Hf^{4+} [106, 109–111] were first identified experimentally shortly after 2000. In this chapter, where the details of novel multivalent cation-conducting solids are described, the symbols Ln and R are used to denote the lanthanoids (La-Lu) and the rare earths of Sc, Y and the lanthanoids, respectively.

8.2 Analysis of Trivalent Cation Transport

Until 1995, it had been accepted that mobile ionic species in solids were limited to monovalent and divalent ions, because ionic transport depended heavily on the valence of the migrating ions. For trivalent cations, the electrostatic interaction between the conducting trivalent cation and the surrounding anions was considered to be so strong that a trivalent ion could not migrate within a solid lattice. Therefore, trivalent cations were commonly used in solid electrolytes in order to adjust defect concentrations and lattice size. In order to realize trivalent cation conduction, it is essential to reduce the strong interactions between the mobile trivalent cation and the surrounding anion framework. Although it has been reported that Ln^{3+}-β/β''-alumina [24–39], β-alumina-related materials [40–44] and Ln^{3+}-β-$LaNb_3O_9$ [45, 46] are trivalent ion conductors, such reports have neither directly nor quantitatively demonstrated trivalent ion migration in solids. Among these solids, Ln^{3+}-β″-alumina solid electrolytes in particular were prepared by substituting Ln^{3+} onto a Na^+ site in Na^+-β″-alumina. As a consequence, a small amount of Na^+ ions would remain within the solid and perhaps contribute to ion conduction (see Ref. [32]). Solid electrolytes where trivalent cation conduction has been directly demonstrated include $M_2(M'O_4)_3$ (M = Al, In, Sc, Y, Er–Lu; M′ = W, Mo) with $Sc_2(WO_4)_3$-type structures [47–86], and NASICON-type $R_{1/3}Zr_2(PO_4)_3$ [87–93] and $(M_xZr_{1-x})_{4/(4-x)}Nb(PO_4)_3$ (M: rare earths or Al) [94–102].

8.2.1 β/β″-Alumina

β-Alumina (general composition: $Na_{1+x}Al_{11}O_{17+x/2}$) has a layered structure with a conducting NaO plane sandwiched between $Al_{11}O_{16}$ blocks; the lattices of β″-alumina (general composition: $Na_{1+x}M_xAl_{11-x}O_{17}$; M: divalent cations such as Mg) also comprise a Na_2O plane and MAl_2O_4 blocks. Trivalent cation exchange into the conduction planes of Na^+-β″-alumina crystals is possible, and has been performed for various cations, including Bi [24, 25, 30], Cr [25], La [26], Pr [24, 27–30], Nd [24, 25, 30], Sm [24, 25, 30], Eu [24, 25, 30], Gd [24, 25, 30], Tb [24, 25, 30], Dy [24, 30], Ho [31], Er [25, 30], and Yb [24, 25]. Gd^{3+}-β″-Al_2O_3 was prepared by the usual exchange reaction method (the degree of ion exchange is listed in Table 8.1), and claimed to be a Gd^{3+} ion-conducting solid electrolyte. Although Gd^{3+}-β″-alumina is an insulator at room temperature ($\sigma_{rt} < 10^{-11}$ S cm^{-1}), because Gd^{3+} ions within the conduction plane are bonded to oxygen between the spinel blocks, the conductivity at elevated temperatures of 600–700 °C increases to a value of 10^{-3} S cm^{-1}, which is assumed to result from Gd^{3+} ion transport [30]. For other lanthanoid ions (Ln = La, Pr, or Nd) exchanged Ln^{3+}-β″-alumina crystals, the conductivities of 10^{-4}–10^{-6} S cm^{-1} have been reported at temperatures of approximately 400–500 °C. However, the charge carrier type in these solids was not conclusively identified. As these Ln^{3+}-β″-aluminas

Table 8.1 Degree of ion exchange for M^{3+}-β″-Al_2O_3 crystals.

Trivalent ion species	Degree of ion exchange (%)	Reference
La^{3+}	98	[26]
Pr^{3+}	43	[24]
Nd^{3+}	95	[32]
Sm^{3+}	90	[24]
Eu^{3+}	95	[24]
Gd^{3+}	100	[24]
Tb^{3+}	90	[24]
Ho^{3+}	98	[31]
Er^{3+}	96	[25]
Yb^{3+}	90	[24]
Bi^{3+}	70	[24]
Cr^{3+}	75	[25]

are prepared using an ion-exchange method, a small amount of Na^+ ion may still remain, despite the authors' claim that the ion-exchange ratio is approximately 100% (e.g., in the case of the Gd^{3+}-β″-alumina crystal). For other highly ion-exchanged β″-aluminas with La^{3+} (98%), Ho^{3+} (98%), Er^{3+} (96%) (see Table 8.1), there remains at least 2% or 4% Na^+, and these remaining Na^+ ions contribute to the ionic transport. In addition, the presence of divalent Mg^{2+} cations in the spinel blocks may also be speculated as having a non-negligible contribution. Therefore, the candidate species for conduction are Ln^{3+}, Mg^{2+}, remaining Na^+ cations, protons, and oxide anions. Unfortunately, the ion-exchanged Ln^{3+}-β″-aluminas have not been directly identified as Ln^{3+} ion conductors by experimental procedures, as only electrical conductivity measurements and X-ray diffraction (XRD) studies were performed. The activation energies for all samples in the low-temperature range were derived from Arrhenius plots, and were modest in comparison with those in the higher temperature region. As an example, temperature dependence of conductivity in the ion-exchanged Na^+/Ho^{3+}-β″-aluminas [31] are depicted in Figure 8.1. The activation energies for 98% exchanged Na^+/Ho^{3+}-β″-alumina are 5.6 kJ mol^{-1} at 150 °C, and 44 kJ mol^{-1} at 500 °C. For the purpose of investigating this unique behavior in ion-exchanged Ln^{3+}-β″-alumina, the Ln^{3+} cation occupation in the conduction plane of the β″-alumina structure was studied using XRD for Pr^{3+}-β″-alumina at temperatures between 25 and 500 °C [29]. Based on these measurements, it became clear that only 10% Pr^{3+} occupied the Beevers–Ross (BR) sites at room temperature, while the BR position occupancy was increased on heating up to 250 °C, where both the mid-oxygen (mO) and BR sites were occupied to near-equal extent. At higher temperatures, a redistribution of Pr^{3+} back into

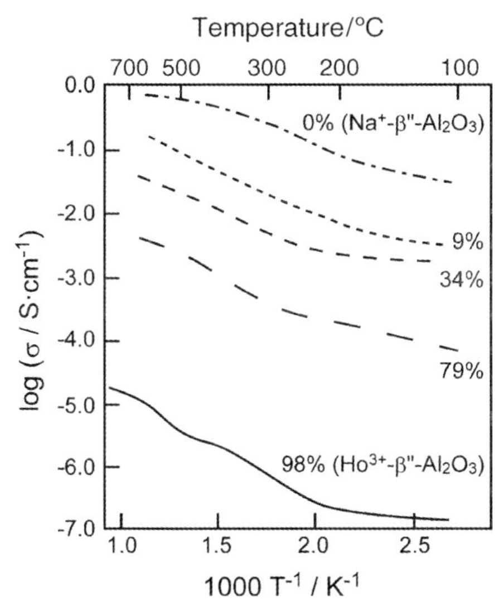

Figure 8.1 Temperature dependence of the conductivity for ion-exchanged Na^+/Ho^{3+}-β″-alumina in an air atmosphere. (Reproduced with permission from Ref. [31].)

the mO sites was again observed. Thus, it would appear difficult for the Pr^{3+} ions to migrate at low temperature because they are trapped within the BR sites and unable to exhibit any long-range migration due to the considerably high potential barriers between the nonequivalent BR and mO sites. The result was that residual Na^+ ions may still have migrated. However, at elevated temperatures, the Pr^{3+} ion could obtain sufficient energy so as to overcome the barriers and migrate out from the BR sites into adjacent mO sites.

In 2000, the DC electrolysis of Nd^{3+}-β″-alumina (the degree of ion exchange was ca. 98%) was performed at both 250 and 650 °C in order to directly identify the migrating species in ionically exchanged Ln^{3+}-β″-alumina [32]. A schematic illustration of the DC electrolysis set-up for Nd^{3+}-β″-alumina is shown in Figure 8.2. The elemental distribution inside Nd^{3+}-β″-alumina electrolyzed at 250 °C (Figure 8.2a) and 650 °C (Figure 8.2b) was investigated using line electron probe microanalysis (EPMA). No significant segregation of the Nd atoms was observed at the cathodic surface for the sample electrolyzed at 250 °C, whereas the Na atoms segregated to a significant degree. In contrast, when a sample was electrolyzed at 650 °C a strong Nd peak was also recognized, together with the Na peak. These results clearly indicated that the migrating species in Nd^{3+}-β″-alumina changed from only Na^+ to both

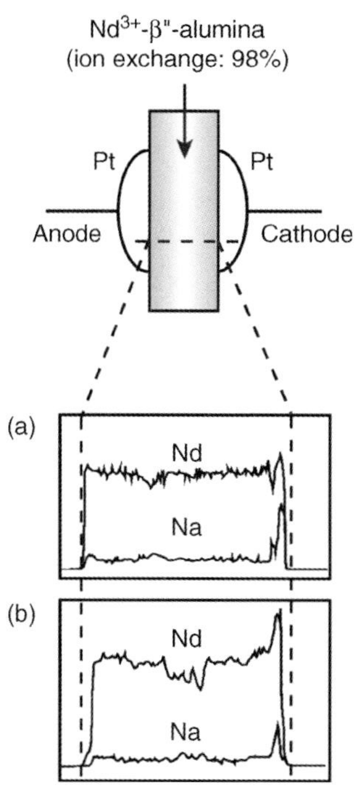

Figure 8.2 Schematic illustration of the experimental set-up for DC electrolysis of Nd^{3+}-β″-alumina, and the EPMA line analysis results for Nd^{3+}-β″-alumina electrolyzed at 250 °C (a) and 650 °C (b) in an air atmosphere.

Na^+ and Nd^{3+} with increasing temperature, and that the ion-exchanged Nd^{3+}-β″-alumina was not a pure trivalent Nd^{3+} cationic conductor, but rather a Na^+ and Nd^{3+} mixed ionic conductor at elevated temperatures. The existence of other ion-exchanged Ln^{3+}-β″-aluminas has also been in doubt, notably with regards to whether Ln^{3+} is the only migrating cationic species, or not, and for the same reason as considered for Nd^{3+}-β″-alumina.

Trivalent ion-exchange reactions for polycrystalline Na^+-β″-alumina have rarely been reported (e.g., for Na^+/Nd^{3+}-β″-alumina [38]), mainly because ion exchange into Na^+-β″-alumina polycrystals is much more difficult than for a single crystalline material, especially in the case of trivalent cations. However, no electrical conductivity measurements were conducted on these samples.

As in the case of β-alumina, the Ln^{3+}-β-aluminas (Ln = Pr, Nd, Ho) were synthesized using an ion-exchange method [33–35]. However, the ion-exchange of Ln^{3+} for a Na^+ site in β-alumina again proved to be difficult, due not only to the structural characteristics but also to the mechanisms of charge compensation; for example, the maximum degree of the exchange ratio was only 95%. Such a low ion exchange appears to be due to an excess of Na^+ ions which is in turn compensated by interstitial O^{2-} ions in the conduction planes. Notably, the ion-conducting properties of these Ln^{3+}-β-aluminas were not investigated.

8.2.2
β-Alumina-Related Materials

The β- or β″-gallates [112, 113] are recognized as isomorphous materials of β/β″-aluminas, and often exhibit a faster ionic conductivity than the corresponding β/β″-aluminas, due mainly to the larger ionic radius of Ga^{3+} (0.076 nm [114]) compared to Al^{3+} (0.0675 nm) in the spinel blocks, and which causes an enlargement of the space between two adjacent spinel blocks. Consequently, the steric hindrance of the migrating cations is reduced, and this results in a faster ion exchange. Unfortunately, as the gallates have a much inferior thermal stability compared to β-alumina, and they are also very expensive to buy, their applications have been limited.

The magnetoplumbite-type ($A^{2+}B^{3+}{}_{12}O_{19}$) compound is also related to β-alumina. In the magnetoplumbite structure (see Figure 8.3), a mirror plane is sandwiched between two spinel blocks, similar to β/β″-Al_2O_3. However, the mirror plane has no oxygen vacancies, and the ions are more densely packed compared to β/β″-Al_2O_3; consequently, the magnetoplumbite-type compounds exhibit a considerably lower conductivity and higher activation energies than β/β″-Al_2O_3. Among magnetoplumbite-type solid electrolytes containing lanthanoid cations, both $LnAl_{11}O_{18}$ (Ln = La, Ce, Pr, Nd, Sm, Eu) and $LnAl_{12}O_{18}N$ (Ln = La, Ce, Pr, Nd, Sm, Eu, Gd) have been reported [40–44]. The diffusion of trivalent lanthanoid cations was monitored using impedance spectroscopy and electromotive force (EMF) measurements, but no direct demonstration of trivalent cation conduction has yet been reported, similar to the situation for Ln^{3+}-β″-alumina.

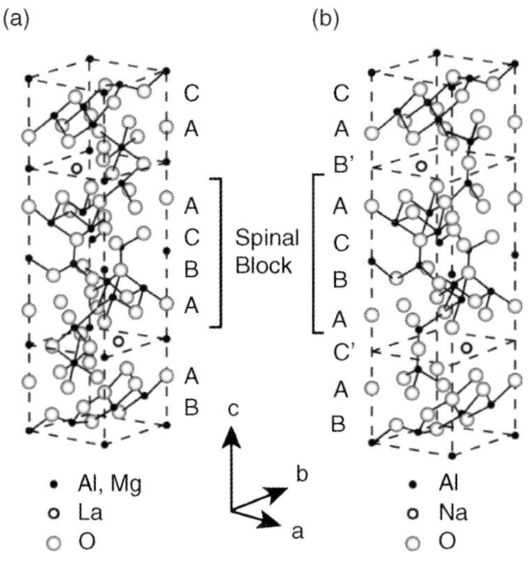

Figure 8.3 Comparison of the hexagonal crystal structures of the magnetoplumbite phase $LaMgAl_{11}O_{19}$ (a) and Na^+-β-Al_2O_3 (b).

8.2.3
Perovskite-Type Structures

Trivalent cationic conduction in perovskite structures has been reported for β-$LaNb_3O_9$ [46]. The compound is A-site deficient, which is a key requirement for high-valency cation conduction. The La^{3+} ions are surrounded by 12 oxygen atoms, and adopt an ordered distribution. One of the A-sites located in the plane parallel to the ab plane is entirely empty, while one is statistically occupied by 2/3 lanthanum ions, as illustrated in Figure 8.4. The vacant A-sites are energetically equivalent and double the number of total La^{3+} ions in the lattice, therefore providing an interconnected 3-D channel for the assumed ionic motion. β-$LaNb_3O_9$ is a mixed ionic–electronic conductor, with electronic conduction appearing at temperatures in excess of 577 °C. It has also been reported that La^{3+} ions are the predominant conduction species below 577 °C. The direct current (DC) electrolysis of β-$LaNb_3O_9$ was conducted to demonstrate La^{3+} migration. Following electrolysis, color changes of both the cathodic and anodic surfaces were observed, and a high concentration of La atoms was reported, but only at the cathodic surface. However, as the electrolysis was performed at approximately 727 °C, when significant electronic conduction occurs, the question remains as to whether β-$LaNb_3O_9$ can be referred to as a "pure" trivalent cationic conductor.

In 2000, the details of a new trivalent Y^{3+} ion-conducting solid electrolyte with an A-site-deficient perovskite structure, $Y_x(Ta_{3x}W_{1-3x})O_3$ $(0 \leq x \leq 0.33)$ [115], were reported. By substituting the pentavalent Ta^{5+} site for hexavalent W^{6+} in $Y_{1/3}TaO_3$, A-site cations (such as Y^{3+}) could be completely moved into alternate layers, and Y^{3+} vacancies introduced. In the $Y_x(Ta_{3x}W_{1-3x})O_3$ series, $Y_{0.06}(Ta_{0.18}W_{0.82})O_3$ $(x = 0.06)$ exhibited the highest conductivity (ca. 2.6×10^{-5} S cm^{-1} at 362 °C), this

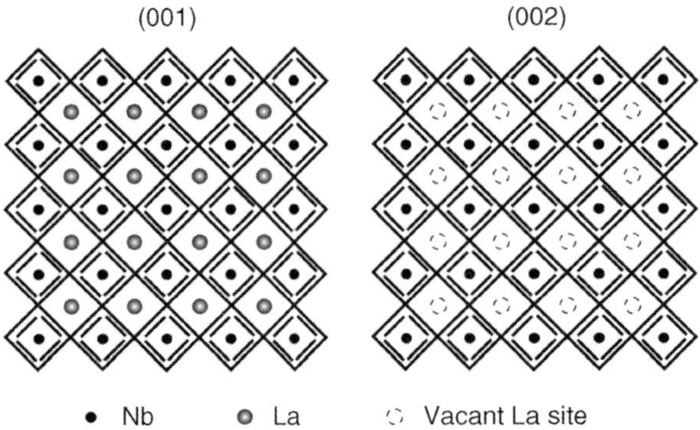

Figure 8.4 Crystal structure of $La_{1/3}NbO_3$ parallel to the ab plane. One site is completely empty, and one is statistically occupied by 2/3 Ln.

being approximately 400- and 800-fold higher than that of $La_{1/3}NbO_3$ and $La_{1/3}TaO_3$, respectively. Y^{3+} ion conduction in $Y_x(Ta_{3x}W_{1-3x})O_3$ was indicated by the polarization behavior in an oxygen atmosphere, which suggested the absence of any O^{2-} anion conduction; high-valence Ta^{5+} and W^{6+} ions were shown to barely migrate within the solid. However, polarization measurements in an O_2 atmosphere may have been indicative of only a negligible role of oxide anion conduction, while the experiments applied for $Y_x(Ta_{3x}W_{1-3x})O_3$ were insufficient to demonstrate trivalent cation migration.

8.2.4
$Sc_2(WO_4)_3$-Type Structures

The tungstate with trivalent cation, $M_2(WO_4)_3$, has two crystal structures depending on the trivalent cation size, as depicted in Figure 8.5 [116]. One structure is the $Sc_2(WO_4)_3$-type variant with orthorhombic symmetry, and the other is the monoclinic $Eu_2(WO_4)_3$-type lattice. The $Sc_2(WO_4)_3$-type structure is quasi-layered with a large space between the layers where R^{3+} ions can migrate smoothly, as shown in Figure 8.6a. In this structure, hexavalent tungsten (W^{6+}) bonds to four surrounding oxide anions to form a rigid $WO_4{}^{2-}$ tetrahedron unit, which results in a considerable reduction of the electrostatic interaction between M^{3+} and O^{2-}, and thereby allows M^{3+} ion diffusion. The $Eu_2(WO_4)_3$-type structure also contains W^{6+}. However, the trivalent cation in the $Eu_2(WO_4)_3$-type structure is 8-coordinated (M^{3+} in the $Sc_2(WO_4)_3$-type lattice has sixfold coordination); M^{3+} cation migration is hampered due to the stronger electrostatic interaction with oxide anions (Figure 8.6b).

Figure 8.7 shows the electrical conductivities at 600 °C and activation energies for $M_2(WO_4)_3$ with the $Sc_2(WO_4)_3$-type structure. The conductivity of $M_2(WO_4)_3$ changes with the variation of ionic radius, while $Sc_2(WO_4)_3$ exhibits the highest conductivity among the $M_2(WO_4)_3$ series. In solids with trivalent M^{3+} ions larger than Sc^{3+}, the conductivity decreases. On the other hand, for Al^{3+} – which has a much smaller ionic radius (0.0675 nm [114]) in comparison with Sc^{3+} (0.0885 nm) – the conductivity is

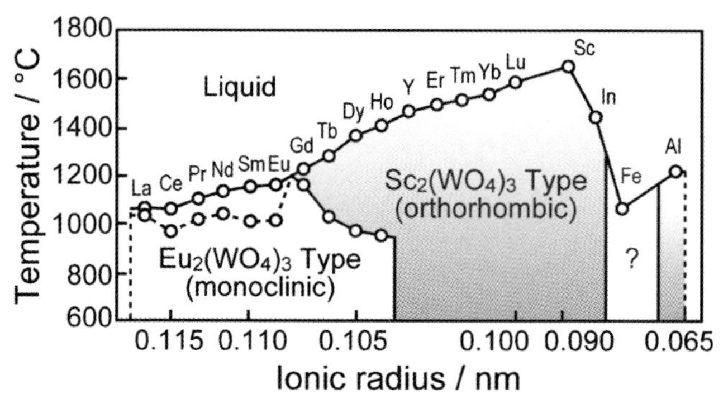

Figure 8.5 Phase relationships in tungstates with trivalent cations, $M_2(WO_4)_3$. (Reproduced with permission from Ref. [116].)

also lower, presumably due to a larger decrease in the volume of the migrating trivalent ion (from Sc^{3+} to Al^{3+}) compared to that of the crystal lattice (from $Sc_2(WO_4)_3$ to $Al_2(WO_4)_3$).

Trivalent cation conduction in the $Sc_2(WO_4)_3$-type solid was clearly demonstrated by measurements of the oxygen partial pressure dependencies of the electrical conductivity, polarization measurement, and DC electrolysis. All $M_2(WO_4)_3$ compounds with the $Sc_2(WO_4)_3$-type structure exhibited a constant conductivity over a wide range of oxygen partial pressures, indicating an absence of any essential electronic contribution. The DC conductivity was measured in atmospheres with different oxygen partial pressures, and the DC to AC conductivity ratio in any P_{O_2} atmosphere was drastically decreased over time. This phenomenon suggests that the possibility of oxide anion conduction is rather eliminated, because if O^{2-} anions are the dominant migrating species, then the $(\sigma_{dc}/\sigma_{ac})$ ratio should be rather constant; for the four-probe DC conductivity measurements, this ratio should be equal to unity. The ion transference number (t_{ion}) calculated using the 10 min electrolysis test was 0.99. Furthermore, for $Sc_2(WO_4)_3$, EMF measurements of the O_2 concentration cell

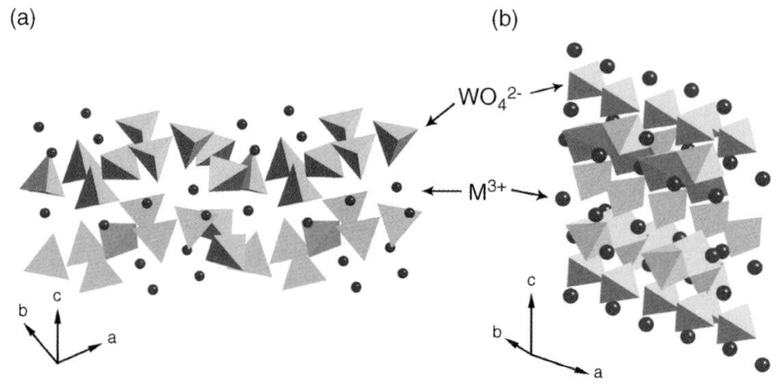

Figure 8.6 The crystal structures of the $M_2(WO_4)_3$-type solid. (a) $Sc_2(WO_4)_3$-type; (b) $Eu_2(WO_4)_3$-types.

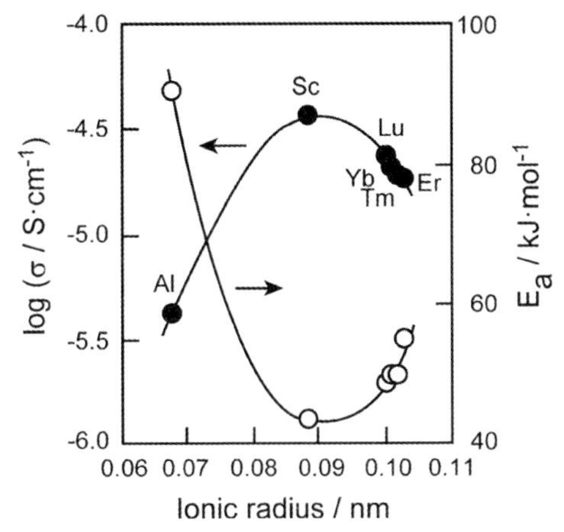

Figure 8.7 Electrical conductivity at 600 °C in an air atmosphere, and activation energies for $M_2(WO_4)_3$ with $Sc_2(WO_4)_3$-type structure. (Reproduced with permission from Ref. [52].)

using $Sc_2(WO_4)_3$ were also conducted at various temperatures, with the aim of assessing electronic conduction; the measured EMF was very close to the theoretical value calculated from the Nernst equation; t_{ion} was estimated to be unity.

For the purpose of identifying the mobile cationic species in $Sc_2(WO_4)_3$, electrolysis was carried out by applying a DC voltage of 10 V, which is higher than the decomposition voltage of $Sc_2(WO_4)_3$ (ca. 1.2 V) with Pt as an ion-blocking electrode. Figure 8.8b shows a scanning electron microscopy (SEM) image of the cathodic surface of the electrolyzed pellet sample, and the EPMA line analysis for the $Sc_2(WO_4)_3$ solid electrolyte (Figure 8.8a). After the electrolysis, needle-shaped crystallites where the Sc amount was ninefold higher than in the bulk $Sc_2(WO_4)_3$ were observed at the cathodic surface; these were identified as Sc_6WO_{12}. On the other hand, the color of the electrolyzed anodic surface was changed to yellow, and WO_3 was detected. These results indicated that the following reaction had occurred during the DC electrolysis, resulting in the macroscopic migration of Sc^{3+}:

$$\text{(Anodic surface)}\quad Sc_2(WO_4)_3 \rightarrow 2Sc^{3+} + 3WO_3 + (3/2)O_2 + 6e^- \tag{8.1}$$

$$\text{(Cathodic surface)}\quad 2Sc^{3+} + 6e^- \rightarrow 2Sc \tag{8.2}$$

$$2Sc + (3/2)O_2 \rightarrow Sc_2O_3 \tag{8.3}$$

$$(8/3)Sc_2O_3 + (1/3)Sc_2(WO_4)_3 \rightarrow Sc_6WO_{12} \tag{8.4}$$

The trivalent Al^{3+} ion transport in $Al_2(WO_4)_3$ with $Sc_2(WO_4)_3$-type structure was investigated using a large single crystal (15 mm in diameter and up to 35 mm in length) grown by the modified Czochralski (CZ) method [54], because the

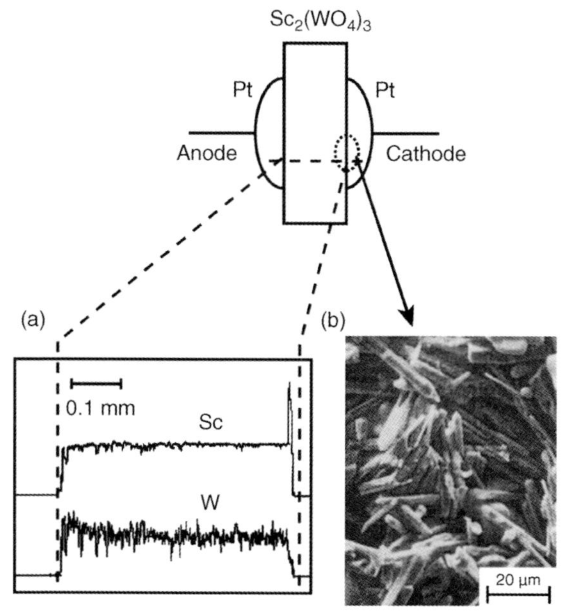

Figure 8.8 (a) EPMA line analysis for electrolyzed $Sc_2(WO_4)_3$; (b) SEM image of the cathodic surface of the electrolyzed pellet. (Reproduced with permission from Ref. [54].)

$Sc_2(WO_4)_3$-type structure is orthorhombic and the ion conductivity behavior should be anisotropic. Ion conduction along the *b*-axis was found to be highest, while conductivity parallel to the *b*-axis at temperatures <500 °C was higher than that of the polycrystalline sample, and conductivity along the *c*-axis was approximately two orders of magnitude lower.

Various solid solutions with $Sc_2(WO_4)_3$-type structures have also been extensively investigated for the purpose of enhancing trivalent ion conduction in tungstates. Figure 8.9 shows the electrical conductivities of $(1-x)Sc_2(WO_4)_3$-$xLn_2(WO_4)_3$ (Ln = Gd [50], Lu [58]) at 600 °C. For the Lu-containing system, discontinuity of the conductivity behavior is recognized between $x = 0.5$ and 0.6. Following DC electrolysis, the predominant charge carriers were found to change from Sc^{3+} at $x < 0.5$ to Lu^{3+} in the higher x region. On the other hand, the conductivity of the Gd-containing system decreased drastically at $x \approx 0.7$. This decrease occurs owing to the structural change from conductive $Sc_2(WO_4)_3$-type to the insulating $Eu_2(WO_4)_3$-type structure (see Figure 8.5).

In order to improve the conductivity of aluminum tungstate, $Al_2(WO_4)_3$-$Sc_2(WO_4)_3$-$Lu_2(WO_4)_3$ solid solutions were prepared [51]. Figure 8.10 depicts their electrical conductivity at 600 °C. Among the materials investigated, $0.1(Al_2(WO_4)_3)$–$0.9((Sc_{0.5}Lu_{0.5})_2(WO_4)_3)$ showed the highest conductivity (8.7×10^{-5} S cm^{-1}), which was approximately 25-fold higher than that of pure $Al_2(WO_4)_3$ (3.4×10^{-6} S cm^{-1}). The migrating trivalent species in $0.1(Al_2(WO_4)_3)$–$0.9((Sc_{0.5}Lu_{0.5})_2(WO_4)_3)$ were identified as only Al^{3+} by the DC electrolysis method. This indicates that Al^{3+} ion

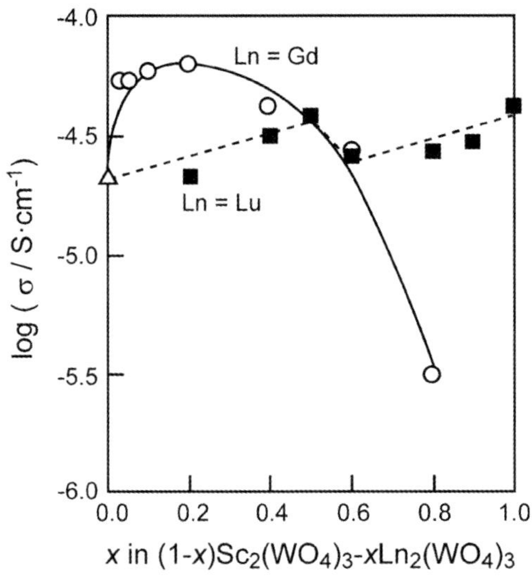

Figure 8.9 Electrical conductivity of $(1-x)Sc_2(WO_4)_3$-$xLn_2(WO_4)_3$ (Ln = Lu, Gd) at 600 °C in an air atmosphere.

migration is much easier than Sc^{3+} and Lu^{3+} migration in the $Sc_2(WO_4)_3$-type phases, if an appropriate structural arrangement for Al^{3+} ion conduction is achieved.

Trivalent cation conduction in molybdates with the $Sc_2(WO_4)_3$-type structure, $M_2(MoO_4)_3$, was also investigated. The transport of trivalent M^{3+} in molybdates was higher than that in tungstates (Figure 8.11), due to the smaller radius of Mo^{6+}

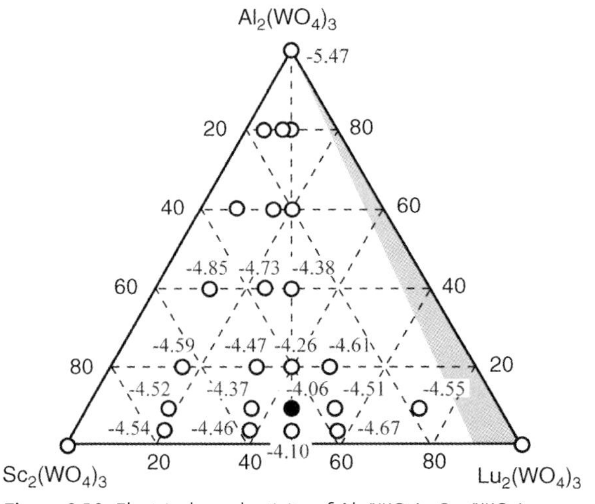

Figure 8.10 Electrical conductivity of $Al_2(WO_4)_3$-$Sc_2(WO_4)_3$-$Lu_2(WO_4)_3$ solid solutions at 600 °C in an air atmosphere. (Reproduced with permission from Ref. [51].)

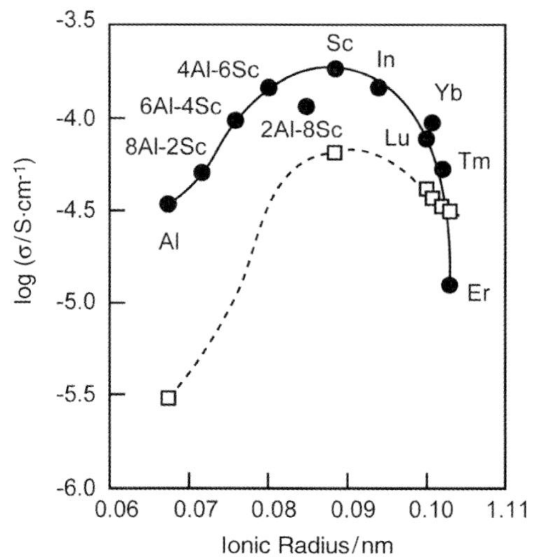

Figure 8.11 M^{3+} radii dependence of the electrical conductivity at 600 °C in an air atmosphere for tungstates (□) and molybdates (•) with the $Sc_2(WO_4)_3$-type structure.

(0.055 nm [114]) compared to W^{6+} (0.056 nm), which resulted in a smaller volume of the corresponding lattice elements suitable for trivalent ion conduction in the $Sc_2(WO_4)_3$-type structure. In $Sc_2(MoO_4)_3$, Sc^{3+} conduction was also demonstrated by DC electrolysis, similar to that for $Sc_2(WO_4)_3$. Based on the results of the electrolysis, it was clear that molybdates with $Sc_2(WO_4)_3$-type structures were trivalent cation conductors, whereas molybdates tended to be easily reduced compared with tungstates; $Sc_2(WO_4)_3$ was not reduced even at a Po_2 of 10^{-17} Pa, while $Sc_2(MoO_4)_3$ was already reduced at $Po_2 = 10^{-13}$ Pa and 700 °C (Figure 8.12). It should be noted that an increase in conductivity at low Po_2 indicated the appearance of electronic conduction, caused by the reduction of Mo^{6+}.

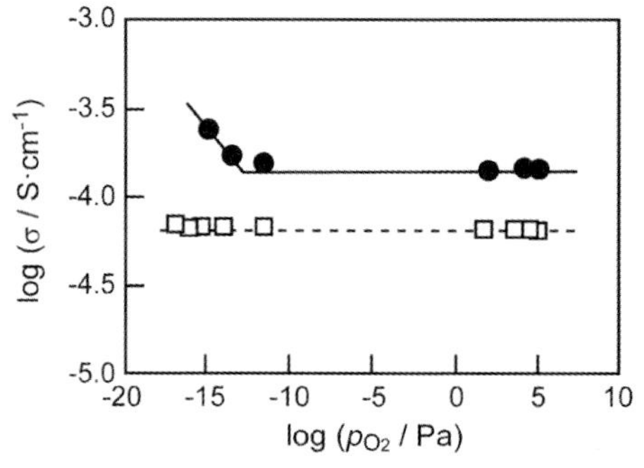

Figure 8.12 Oxygen partial pressure dependence of the electrical conductivity for $Sc_2(WO_4)_3$ (□) and $Sc_2(MoO_4)_3$ (•) at 700 °C.

8.2.5
NASICON-Type Structures

NASICON represents another family of high-conductivity Na^+-conducting solid electrolytes, as described in previous chapters. Its structure has a 3-D network, where PO_4 tetrahedra and ZrO_6 octahedra are linked by shared oxygens. In 1994, Talbi *et al.* [117] reported that $Ln_{1/3}Zr_2(PO_4)_3$ (Ln = lanthanoids) with the NASICON-type structure could be prepared using a sol–gel method. However, the Ln^{3+} cations in these materials are limited to several lanthanoids, and their ionic conductivity was not investigated. In 1999, $Sc_{1/3}Zr_2(PO_4)_3$ with a NASICON-type structure was prepared and the Sc^{3+} transport studied [87, 88]. In order to realize a high cationic conduction, the concept of the mobile ion size/lattice size ratio was introduced so as to optimize the diffusion pathway size [89]. This relationship (the *A* ratio) in $R_{1/3}Zr_2(PO_4)_3$ was seen to decrease monotonically with decreasing R^{3+} radius, and only $Sc_{1/3}Zr_2(PO_4)_3$ showed a smaller *A* ratio compared to $LiTi_2(PO_4)_3$, with the fastest conduction among the Li^+-conducting NASICON-type electrolytes. As the electrostatic interaction between the mobile trivalent ions and surrounding oxide anions becomes stronger, compared to monovalent Li^+, the optimum *A* ratio in the former case should be smaller. Therefore, a high conductivity would be expected for $Sc_{1/3}Zr_2(PO_4)_3$. Figure 8.13 shows the *A* ratio dependence of the electrical conductivity in $R_{1/3}Zr_2(PO_4)_3$ prepared by a sol–gel method at 600 °C. As expected, $Sc_{1/3}Zr_2(PO_4)_3$ possessed the highest conductivity; subsequent DC electrolysis demonstrated Sc^{3+} ion transport in $Sc_{1/3}Zr_2(PO_4)_3$, similar to the case of $Sc_2(WO_4)_3$.

However, the conductivity (1.07×10^{-5} S cm^{-1} at 600 °C) of $Sc_{1/3}Zr_2(PO_4)_3$ synthesized with the sol–gel technique was low in comparison with that of $Sc_2(WO_4)_3$ tungstate (3.74×10^{-5} S cm^{-1} at 600 °C), largely because $Sc_{1/3}Zr_2(PO_4)_3$ obtained via the sol–gel method has a low crystallinity. In addition, the crystallization (at ca. 800 °C) and decomposition temperatures (ca. 950 °C) of $Sc_{1/3}Zr_2(PO_4)_3$ were close, and any improvement in the crystallinity seemed difficult to achieve when using a

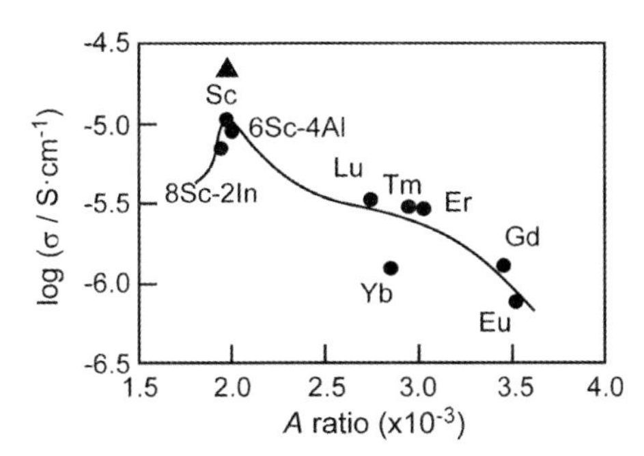

Figure 8.13 Relationship between the *A* ratio and electrical conductivity in an air atmosphere for $R_{1/3}Zr_2(PO_4)_3$ series prepared by the sol–gel method (•) and $Sc_{1/3}Zr_2(PO_4)_3$ obtained via ball milling (▲), at 600 °C. (Reproduced with permission from Ref. [93].)

conventional sol–gel method. It was also very difficult to prepare $Sc_{1/3}Zr_2(PO_4)_3$ by a conventional solid-state reaction at temperatures below the decomposition occurring at 950 °C. In order to enhance the crystallinity of $Sc_{1/3}Zr_2(PO_4)_3$, a mechanical activation method was applied to the synthesis, with ball milling being used to mix the starting powders. This application of mechanical energy led to the formation of an amorphous phase that was necessary to achieve a high crystallinity. The resultant crystallinity of the $Sc_{1/3}Zr_2(PO_4)_3$ was greatly enhanced, and the ionic conductivity increased to 2.9×10^{-5} S cm^{-1} at 600 °C (see Figure 8.13; closed triangle). The conductivity of $Sc_{1/3}Zr_2(PO_4)_3$ prepared using mechanical activation was comparable to the series with the $Sc_2(WO_4)_3$-type structure, especially at 600 °C. The activation energy in $Sc_{1/3}Zr_2(PO_4)_3$ prepared via ball milling was 72.7 kJ mol^{-1}, slightly higher than the activation energies of $Sc_2(WO_4)_3$ (56.4 kJ mol^{-1}) and $Al_2(WO_4)_3$ (65.8 kJ mol^{-1}). However, this also indicated that similar suitable pathways had been formed in both the NASICON and $Sc_2(WO_4)_3$-type structures.

The Al^{3+} ion is considered to be a more suitable trivalent cation species in the NASICON-type phases, because the ionic radius (0.0675 nm [114]) is smaller than that of Sc^{3+} (0.0885 nm), and a smaller *A* ratio would be expected for $Al_{1/3}Zr_2(PO_4)_3$, if such a solid could be synthesized. Unfortunately, however, the latter NASICON-type compound cannot be obtained due to stereological limitations. In order to realize the formation of the NASICON-type structure with mobile Al^{3+}, $(Al_xZr_{1-x})_{4/(4-x)}Nb(PO_4)_3$ series was prepared [94]. In this series, which was conducted in 2002, shrinkage of the NASICON-type lattice was achieved by partially substituting the Zr^{4+} (0.086 nm [114]) sites with the smaller, pentavalent Nb^{5+} (0.078 nm). A reduction in the electrostatic interaction between Al^{3+} ions and O^{2-} anions was also realized, due to the high valence state of Nb^{5+} compared to Zr^{4+}. Among the $(Al_xZr_{1-x})_{4/(4-x)}Nb(PO_4)_3$ series, only those samples with an aluminum content (x) ≤ 0.2 could form a single phase with the NASICON structure; the highest Al^{3+} ion conductivity was obtained for the solubility-limited composition, $(Al_{0.2}Zr_{0.8})_{20/19}Nb(PO_4)_3$. Figure 8.14 shows the conductivity of $(Al_{0.2}Zr_{0.8})_{20/19}Nb(PO_4)_3$ and the corresponding data for $Al_2(WO_4)_3$. The conductivity of $Sc_{1/3}Zr_2(PO_4)_3$, with maximum trivalent ion conductivity among the NASICON-type $R_{1/3}Zr_2(PO_4)_3$ series, is also shown. The ionic conductivity of $(Al_{0.2}Zr_{0.8})_{20/19}Nb(PO_4)_3$, at 4.5×10^{-4} S cm^{-1} (at 600 °C), was seen to be two orders of magnitude higher than that of $Al_2(WO_4)_3$, its value reaching the practical application range ($\sigma > 10^{-4}$ S cm^{-1}).

These results suggest that a suitable crystal lattice size for each trivalent cation species can be obtained by the partial replacement of constituent ions in NASICON-type compounds. The suitable sizes for rare earth ion conduction were also extensively investigated for $(R_{0.05}Zr_{0.95})_{80/79}Nb(PO_4)_3$ [102] by introducing an *A* ratio similar to that for $R_{1/3}Zr_2(PO_4)_3$. Figure 8.15 shows the relationship between the trivalent R^{3+} conductivity at 600 °C and the *A* ratio for $(R_{0.05}Zr_{0.95})_{80/79}Nb(PO_4)_3$ and $R_{1/3}Zr_2(PO_4)_3$. Although the conductivity of $R_{1/3}Zr_2(PO_4)_3$ was seen to decrease monotonically with the increasing *A* ratio, the highest conduction in the $(R_{0.05}Zr_{0.95})_{80/79}Nb(PO_4)_3$ series was obtained for R = Dy (*A* ratio: 3.29×10^{-3}), indicating that the most suitable lattice size for R^{3+} transport in NASICON-type lattices depends on the constituent species.

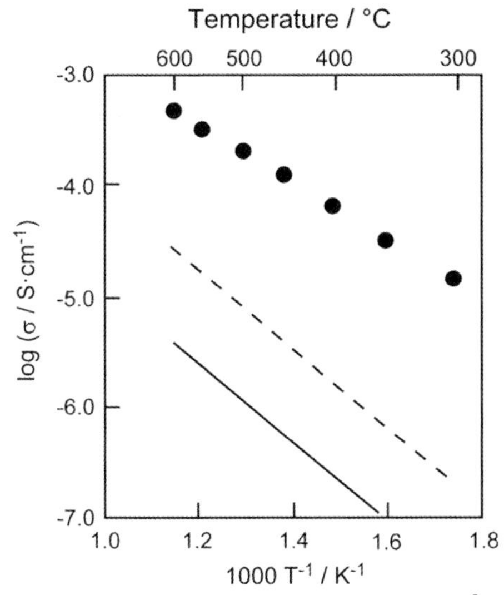

Figure 8.14 Temperature dependence of the Al^{3+} ion conductivity of $(Al_{0.2}Zr_{0.8})_{20/19}Nb(PO_4)_3$ (•) and the corresponding data of $Al_2(WO_4)_3$ with the $Sc_2(WO_4)_3$ type structure (—). The Sc^{3+} ion conductivity of $Sc_{1/3}Zr_2(PO_4)_3$ (- - -) is also shown. (Reproduced with permission from Ref. [94].)

8.3 Search for Tetravalent Cation Conductors

Although Zr^{4+} ion conduction in $Zr_2O(PO_4)_2$ was first reported in 2000 [104], the conductivity was still low compared to that of lower-valent ions. For the $Zr_2O(PO_4)_2$

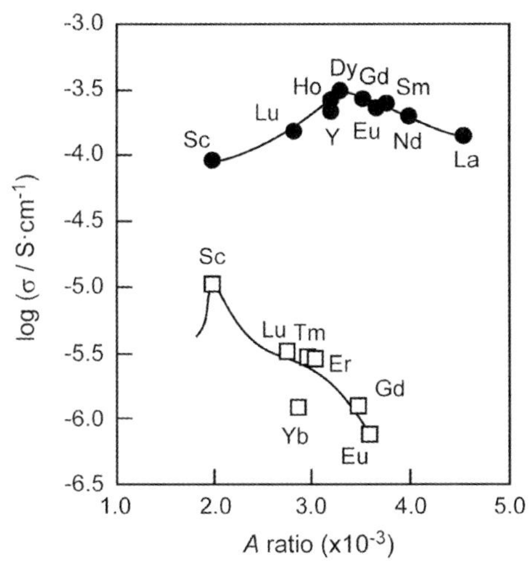

Figure 8.15 Relationship between the A ratio and electrical conductivity in an air atmosphere for $(R_{0.05}Zr_{0.95})_{80/79}Nb(PO_4)_3$ (•) and $R_{1/3}Zr_2(PO_4)_3$ solids (□), at 600 °C.

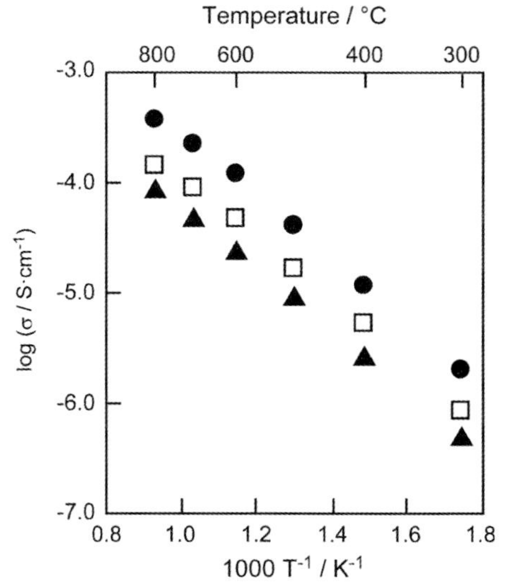

Figure 8.16 Temperature dependence of the tetravalent M^{4+} cation conductivity of $MM'(PO_4)_3$ [(M, M′) = (Zr, Ta) (▲), (Zr, Nb) (□), (Hf, Nb) (•)] in an air atmosphere.

solid, Zr^{4+} ion conduction was demonstrated with DC electrolysis, although since 2001 a high tetravalent cation conduction has been reported for solids with the NASICON-type structure. The first report to be made related to $ZrNb(PO_4)_3$ [105], and although the crystal structure was initially reported as the β-$Fe_2(SO_4)_3$-type [105], it was later identified as being of the NASICON-type. In this phase, Zr^{4+} vacancies were introduced by partially substituting Zr^{4+} in $Zr_2O(PO_4)_2$ with Nb^{5+}, accompanying the transformation into the NASICON-type polymorph. Analogously, both NASICON-type $ZrTa(PO_4)_3$ [107] and $HfNb(PO_4)_3$ [106] were found to be Zr^{4+} and Hf^{4+} conductors, respectively. The corresponding order of the conductivity variations, $HfNb(PO_4)_3 > ZrNb(PO_4)_3 > ZrTa(PO_4)_3$ (Figure 8.16), may have originated from the electronegativity of the constituent cations in $MM'(PO_4)_3$. Those cations having a high electronegativity could form stronger covalent bonds with the surrounding anions. Therefore, compared to Zr^{4+} (electronegativity 1.33) and Hf^{4+} (1.3), the bonding of Hf^{4+} with O^{2-} anions should be weaker than that of Zr^{4+}; the smaller size of Hf^{4+} (0.085 nm [114]) compared to Zr^{4+} (0.086 nm) might also promote migration. On the other hand, for M′ = Nb or Ta, both ions have a similar size (0.078 nm), whereas the electronegativity of Nb^{5+} (1.6) is higher than that of Ta^{5+} (1.5). These constituent cations could effectively reduce the electrostatic interaction between Zr^{4+} and surrounding O^{2-} by tightly bonding to the O^{2-} anions.

Recently, an even higher Zr^{4+} conductivity was reported for a $ZrTa(PO_4)_3$ derivative, namely $Zr_{39/40}TaP_{2.9}W_{0.1}O_{12}$ [108], where the P^{5+} sites were partially substituted with W^{6+}. By doping with hexavalent tungsten, a further reduction in the electrostatic interaction between Zr^{4+} and O^{2-} would be expected. Furthermore, the W^{6+} ion (ionic radius 0.056 nm [114]) was larger than P^{5+} (ionic radius 0.031 nm),

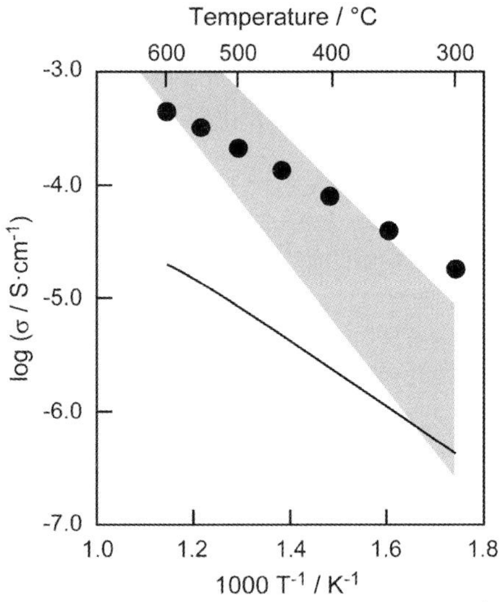

Figure 8.17 Temperature dependence of the Zr^{4+} ion conductivity in an air atmosphere for $Zr_{39/40}TaP_{2.9}W_{0.1}O_{12}$ (•) and $ZrTa(PO_4)_3$ (—). The shaded area shows the conductivities of commercially applied O^{2-} ion-conducting stabilized zirconias. (Reproduced with permission from Ref. [108].)

and therefore this type of substitution would also be expected to expand the NASICON-type crystal lattice. The highest ionic conductivity was obtained for $Zr_{39/40}TaP_{2.9}W_{0.1}O_{12}$, the solid solution end member. Figure 8.17 compares the conductivity of this composition with the corresponding data for $ZrTa(PO_4)_3$ and data on O^{2-}-conducting stabilized zirconias, which are used for as commercial oxygen sensors in the automobile and steel industries. The activation energy (48.7 kJ mol^{-1}) for cation conduction in $Zr_{39/40}TaP_{2.9}W_{0.1}O_{12}$ was lower than that in $ZrTa(PO_4)_3$ (55.9 kJ mol^{-1}), which indicated that a smooth cationic transport in the NASICON-type structure would be realized due to its expansion and reduced electrostatic interactions between Zr^{4+} and O^{2-} anions.

References

1 Dunn, B. and Farrington, G.C. (1986) *Solid State Ionics*, **18–19**, 31–9.

2 (a) Angell, C.A. (1986) *Solid State Ionics*, **18–19**, 72–78; (b) Lunden, A. (1988) *Solid State Ionics*, **28–30**, 163–7.

3 Bruce, P.G. (1991) *Phylos. Mag. A*, **64**, 1101–12.

4 Andersen, N.H., Bandaranayake, P.W.S.K., Careem, M.A., Dissanayake, M.A.K.L., Wijayasekera, C.N., Kaber, R., Lunden, A., Mellander, B.E., Nilsson, L. and Thomas, J.O. (1992) *Solid State Ionics*, **57**, 203–9.

5 (a) Armstrong, R.D. and Wang, H. (1994) *Electrochim. Acta*, **39**, 1–5; (b) Neiman, A. (1996) *Solid State Ionics*, **83**, 263–73.

6 Lesunova, R.P., Palguev, S.F. and Burmakin, E.I. (1998) *Inorg. Mater.*, **34**, 38–41.

7 Leonidov, I.A., Dontsov, G.I., Knyazhev, A.S., Slinkina, M.V., Leonidova, O.N. and Fotiev, A.A. (1999) *Russ. J. Electrochem.*, **35**, 29–33.
8 Sata, T., Sata, T. and Yang, W. (2002) *J. Membrane Sci.*, **206**, 31–60.
9 Leonidov, I.A., Leonidova, O.N., Surat, L.L. and Samigullina, R.F. (2003) *Inorg. Mater.*, **39**, 616–20.
10 Imanaka, N., Hasegawa, Y. and Hasegawa, I. (2004) *Ionics*, **10**, 385–90.
11 Driscoll, D.J., Islam, M.S. and Slater, P.R. (2005) *Solid State Ionics*, **176**, 539–46.
12 (a) Adams, S. (2005) *Mater. Res. Soc. Symp. Proc.*, **835**, 3–8; (b) Mitelman, A., Levi, M.D., Lancry, E., Levi, E., and Aurbach, D. (2007) *Chem. Commun.*, **41**, 4212–14.
13 Imanaka, N., Ando, K. and Tamura, S. (2007) *J. Mater. Chem.*, **17**, 4230–2.
14 Heuer, A.H. (2008) *J. Eur. Ceram. Soc.*, **28**, 1495–507.
15 Zhou, Y., Adams, S., Rao, R.P., Edwards, D.D., Neiman, A. and Pestereva, N. (2008) *Chem. Mater.*, **20**, 6335–45.
16 Kulikova, T., Neiman, A., Kartavtseva, A., Edwards, D. and Adams, D. (2008) *Solid State Ionics*, **178**, 1714–18.
17 Warburg, E. and Tegetmeier, F. (1988) *Wiedemann Ann. Phys.*, **32**, 455.
18 Wagner, C. (1943) *Naturwissenschaften*, **31**, 265.
19 Tubandt, C.Z. (1921) *Anorg. Allgem. Chem.*, **115**, 105.
20 Bradley, J.N. and Greene, P.D. (1965) *Trans. Faraday Soc.*, **62**, 2069.
21 Frant, M.S. and Ross, J.W. Jr (1966) *Science*, **154**, 1553.
22 Weber, N. and Kummer, J.T. (1967) *Proc. Ann. Power Sources Conf.*, **21**, 37.
23 Goodenough, J.B., Hong, H.Y.-P. and Kafalas, J.A. (1976) *Mater. Res. Bull.*, **11**, 203.
24 Dunn, B. and Farrington, G.C. (1983) *Solid State Ionics*, **9/10**, 223.
25 Sattar, S., Ghosal, B., Underwood, M.L., Mertwoy, H., Salzberg, M.A., Frydrych, W.S. and Farrington, G.C. (1986) *J. Solid State Chem.*, **65**, 231.
26 Köhler, J. and Urland, W. (1997) *Z. Anorg. Allg. Chem.*, **623**, 231.
27 Köhler, J. and Urland, W. (1996) *Z. Anorg. Allg. Chem.*, **622**, 191.
28 Köhler, J. and Urland, W. (1996) *J. Solid State Chem.*, **127**, 161.
29 Köhler, J. and Urland, W. (1996) *Solid State Ionics*, **86–88**, 93.
30 Farrington, G.C., Dunn, B. and Thomas, J.O. (1983) *Appl. Phys. A*, **32**, 159.
31 Tietz, F. and Urland, W. (1995) *Solid State Ionics*, **78**, 35.
32 Köhler, J., Imanaka, N., Urland, W. and Adachi, G. (2000) *Angew. Chem. Int. Ed.*, **39**, 904.
33 Tietz, F. and Urland, W. (1993) *J. Alloys Compd.*, **192**, 78.
34 Tietz, F. and Urland, W. (1991) *Solid State Ionics*, **46**, 331.
35 Tietz, F. and Urland, W. (1992) *J. Solid State Chem.*, **100**, 255.
36 Kumar, R.V. and Kay, D.A.R. (1985) *Metall. Trans. B*, **16**, 295.
37 Kumar, R.V. (1997) *J. Alloys Compd.*, **250**, 501.
38 Yang, D.L., Dunn, B. and Morgan, P.E.D. (1991) *J. Mater. Sci. Lett.*, **10**, 485.
39 Wen, Z.-Y., Lin, Z.-X. and Tian, S.-B. (1990) *Solid State Ionics*, **40/41**, 91.
40 Wang, X.E., Lejus, A.M., Vivien, D. and Collongues, R. (1988) *Mater. Res. Bull.*, **23**, 43.
41 Sun, W.Y., Yen, T.S. and Tien, T.Y. (1991) *J. Solid State Chem.*, **95**, 424.
42 Kahn, A., Lejus, A.M., Madsac, M., Théry, J., Vivien, D. and Bernier, J.C. (1981) *J. Appl. Phys.*, **52**, 6864.
43 Warner, T.E., Fray, D.J. and Davies, A. (1996) *Solid State Ionics*, **92**, 99.
44 Warner, T.E., Fray, D.J. and Davies, A. (1997) *J. Mater. Sci.*, **32**, 279.
45 Dyer, A.J. and White, E.A.D. (1964) *Trans. Br. Ceram. Soc.*, **63**, 301.
46 George, A.M. and Virkar, A.N. (1988) *J. Phys. Chem. Solids*, **49**, 743.
47 Imanaka, N., Kobayashi, Y. and Adachi, G. (1995) *Chem. Lett.*, **24**, 433.
48 Kobayashi, Y., Egawa, T., Tamura, S., Imanaka, N. and Adachi, G. (1997) *Chem. Mater.*, **9**, 1649.

49 Imanaka, N., Kobayashi, Y., Tamura, S. and Adachi, G. (1998) *Electrochem. Solid-State Lett.*, **1**, 271.

50 Kobayashi, Y., Egawa, T., Okazaki, Y., Tamura, S., Imanaka, N. and Adachi, G. (1998) *Solid State Ionics*, **111**, 59.

51 Tamura, S., Egawa, T., Okazaki, Y., Kobayashi, Y., Imanaka, N. and Adachi, G. (1998) *Chem. Mater.*, **10**, 1958.

52 Imanaka, N., Kobayashi, Y., Fujiwara, K., Asano, T., Okazaki, Y. and Adachi, G. (1998) *Chem. Mater.*, **10**, 2006.

53 Köhler, J., Imanaka, N. and Adachi, G. (1998) *Chem. Mater.*, **10**, 3790.

54 Imanaka, N., Hiraiwa, M., Tamura, S., Adachi, G., Dabkowska, H., Dabkowski, A. and Greedan, J.E. (1998) *Chem. Mater.*, **10**, 2542.

55 Imanaka, N. and Adachi, G. (1998) *Molten Salts*, **41**, 177.

56 Kobayashi, Y., Tamura, S., Imanaka, N. and Adachi, G. (1998) *Solid State Ionics*, **113–115**, 545.

57 Köhler, J., Kobayashi, Y., Imanaka, N. and Adachi, G. (1998) *Solid State Ionics*, **113–115**, 553.

58 Kobayashi, Y., Egawa, T., Tamura, S., Imanaka, N. and Adachi, G. (1999) *Solid State Ionics*, **118**, 325.

59 Tamura, S., Imanaka, N. and Adachi, G. (1999) *Adv. Mater.*, **11**, 64.

60 Dabkowski, A., Dabkowska, H., Greedan, J.E., Imanaka, N., Hiraiwa, M., Tamura, S. and Adachi, G. (1999) *J. Cryst. Growth*, **197**, 879.

61 Imanaka, N., Hiraiwa, M., Tamura, S., Adachi, G., Dabkowska, H. and Dabkowski, A. (1999) *J. Cryst. Growth*, **200**, 169.

62 Köhler, J., Imanaka, N. and Adachi, G. (1999) *J. Mater. Chem.*, **9**, 1357.

63 Köhler, J., Imanaka, N. and Adachi, G. (1999) *Solid State Ionics*, **122**, 173.

64 Imanaka, N., Asano, T., Tamura, S., Kobayashi, Y. and Adachi, G. (1999) *Electrochem. Solid-State Lett.*, **2**, 330.

65 Adachi, G., Köhler, J. and Imanaka, N. (1999) *Electrochemistry*, **67**, 744.

66 Imanaka, N., Okazaki, Y., Kobayashi, Y., Tamura, S., Asano, T., Egawa, T. and Adachi, G. (1999) *Solid State Ionics*, **126**, 41.

67 Köhler, J., Imanaka, N. and Adachi, G. (1999) *Z. Anorg. Allg. Chem.*, **625**, 1890.

68 Imanaka, N., Hiraiwa, M., Tamura, S. and Adachi, G. (2000) *J. Cryst. Growth*, **209**, 217.

69 Imanaka, N., Hiraiwa, M., Tamura, S., Adachi, G., Dabkowska, H. and Dabkowski, A. (2000) *J. Cryst. Growth*, **208**, 466.

70 Imanaka, N., Köhler, J., Masui, T., Taguchi, A., Mori, H. and Adachi, G. (2000) *J. Am. Ceram. Soc.*, **83**, 427.

71 Köhler, J., Imanaka, N., Urland, W. and Adachi, G. (2000) *Angew. Chem. Int. Ed.*, **39**, 904.

72 Imanaka, N., Kobayashi, Y., Tamura, S. and Adachi, G. (2000) *Solid State Ionics*, **136–137**, 319.

73 Hiraiwa, M., Tamura, S., Imanaka, N., Adachi, G., Dabkowska, H. and Dabkowski, A. (2000) *Solid State Ionics*, **136–137**, 427.

74 Imanaka, N., Tamura, S., Adachi, G. and Kowada, Y. (2000) *Solid State Ionics*, **130**, 179.

75 Imanaka, N., Tamura, S., Kobayashi, Y., Okazaki, Y., Hiraiwa, M., Ueda, T. and Adachi, G. (2000) *J. Alloys Compd.*, **303–304**, 303.

76 Okazaki, Y., Ueda, T., Tamura, S., Imanaka, N. and Adachi, G. (2000) *Solid State Ionics*, **136–137**, 437.

77 Köhler, J., Imanaka, N. and Adachi, G. (2000) *Solid State Ionics*, **136–137**, 431.

78 Imanaka, N., Ueda, T., Okazaki, Y., Tamura, S. and Adachi, G. (2000) *Chem. Mater.*, **12**, 1910.

79 Imanaka, N., Hiraiwa, M. and Adachi, G. (2000) *J. Cryst. Growth*, **220**, 176.

80 Secco, R.A., Liu, H., Imanaka, N. and Adachi, G. (2001) *J. Mater. Sci. Lett.*, **20**, 1339.

81 Secco, R.A., Liu, H., Imanaka, N., Adachi, G. and Rutter, M.D. (2002) *J. Phys. Chem. Solids*, **63**, 425.

82 Imanaka, N., Köhler, J., Tamura, S. and Adachi, G. (2002) *Eur. J. Inorg. Chem.*, **2002**, 105.

83 Imanaka, N., Yoshikawa, S., Hiraiwa, M., Tamura, S. and Adachi, G. (2002) *Mater. Lett.*, **56**, 856.

84 Imanaka, N., Hiraiwa, M., Tamura, S., Adachi, G., Dabkowska, H. and Dabkowski, A. (2002) *J. Mater. Sci.*, **37**, 3483.

85 Secco, R.A., Liu, H., Imanaka, N. and Adachi, G. (2002) *J. Phys.: Condens. Matter*, **14**, 11285.

86 Liu, H., Secco, R.A., Imanaka, N., Rutter, M.D., Adachi, G. and Uchida, T. (2003) *J. Phys. Chem. Solids*, **64**, 287.

87 Tamura, S., Imanaka, N. and Adachi, G. (1999) *Adv. Mater.*, **11**, 1521.

88 Tamura, S., Imanaka, N. and Adachi, G. (2000) *Solid State Ionics*, **136–137**, 423.

89 Tamura, S., Imanaka, N. and Adachi, G. (2001) *J. Alloys Compd.*, **323–324**, 540.

90 Tamura, S., Imanaka, N. and Adachi, G. (2001) *Chem. Lett.*, **30**, 672.

91 Tamura, S., Imanaka, N. and Adachi, G. (2001) *J. Mater. Sci. Lett.*, **20**, 2123.

92 Tamura, S., Imanaka, N. and Adachi, G. (2002) *Solid State Ionics*, **154–155**, 767.

93 Imanaka, N. and Adachi, G. (2002) *J. Alloys Compd.*, **344**, 137.

94 Imanaka, N., Hasegawa, Y., Yamaguchi, M., Itaya, M., Tamura, S. and Adachi, G. (2002) *Chem. Mater.*, **14**, 4481.

95 Hasegawa, Y., Imanaka, N. and Adachi, G. (2003) *J. Solid State Chem.*, **171**, 387.

96 Hasegawa, Y., Tamura, S., Imanaka, N., Adachi, G., Takano, Y., Tsubaki, T. and Sekizawa, K. (2004) *J. Alloys Compd.*, **375**, 212.

97 Hasegawa, Y., Tamura, S., Imanaka, N. and Adachi, G. (2004) *J. Alloys Compd.*, **379**, 262.

98 Hasegawa, Y. and Imanaka, N. (2005) *Solid State Ionics*, **176**, 2499.

99 Hasegawa, Y., Tamura, S. and Imanaka, N. (2005) *J. New Mater. Electrochem. Syst.*, 8, 203.

100 Hasegawa, Y. and Imanaka, N. (2006) *J. Alloys Compd.*, **408–412**, 661.

101 Hasegawa, Y., Hoshiyama, T., Tamura, S. and Imanaka, N. (2007) *J. New Mater. Electrochem. Syst.*, **10**, 177.

102 Hasegawa, Y., Tamura, S., Sato, M. and Imanaka, N. (2008) *Bull. Chem. Soc. Jpn*, **81**, 521.

103 Tamura, S., Yamamoto, S. and Imanaka, N. (2008) *J. New Mater. Electrochem. Syst.*, **11**, 1.

104 Imanaka, N., Ueda, T., Okazaki, Y., Tamura, S., Hiraiwa, M. and Adachi, G. (2000) *Chem. Lett.*, **29**, 452.

105 Imanaka, N., Ueda, T. and Adachi, G. (2001) *Chem. Lett.*, **30**, 446.

106 Itaya, M., Imanaka, N. and Adachi, G. (2002) *Solid State Ionics*, **154–155**, 319.

107 Imanaka, N., Ueda, T. and Adachi, G. (2003) *J. Solid State Electrochem.*, **7**, 239.

108 Imanaka, N., Tamura, S. and Itano, T. (2007) *J. Am. Chem. Soc.*, **129**, 5338.

109 Imanaka, N., Itaya, M. and Adachi, G. (2002) *Mater. Lett.*, **53**, 1.

110 Imanaka, N., Itaya, M. and Adachi, G. (2002) *Mater. Lett.*, **57**, 209.

111 Liu, H., Secco, R.A., Imanaka, N. and Adachi, G. (2002) *Solid State Commun.*, **123**, 411.

112 Foster, L.M. and Arbach, G.V. (1977) *J. Electrochem. Soc.*, **124**, 164.

113 Foster, L.M. and Stumpf, H.C. (1973) *J. Am. Chem. Soc.*, **73**, 1590.

114 Shannon, R.D. (1976) *Acta Crystallogr. A*, **32**, 751.

115 Sakai, N., Toda, K. and Sato, M. (2000) *Electrochemistry*, **68**, 504.

116 Nassau, K., Shiever, J.W. and Keve, E.T. (1971) *J. Solid State Chem.*, **3**, 411.

117 Talbi, M.A., Brochu, R., Parent, C., Rabardel, L. and Flem, G.L. (1994) *J. Solid State Chem.*, **110**, 350.

9
Oxygen Ion-Conducting Materials

Vladislav V. Kharton, Fernando M.B. Marques, John A. Kilner, and Alan Atkinson

Abstract

Oxygen ionic and mixed ionic-electronic conductors find important applications in solid-state electrochemical devices, including sensors, solid oxide fuel cells, high-temperature electrolyzers, and oxygen separation membranes. This chapter presents a brief overview of oxide phases with high diffusivity of O^{2-} anions, providing an introduction to this fascinating topic. Particular emphasis is centered on the comparative analysis of ionic and electronic conductivity variations in the major groups of solid oxide electrolytes and mixed conductors, such as perovskite- and fluorite-related compounds, apatite-type silicates, and derivatives of γ-$Bi_4V_2O_{11}$ and β-$La_2Mo_2O_9$. The defect chemistry mechanisms relevant to the oxygen ion migration processes are briefly discussed.

9.1
Introduction

Technologies based on the use of high-temperature electrochemical cells with oxygen anion- or mixed-conducting ceramics provide important advantages with respect to the conventional industrial processes [1–8]. In particular, solid oxide fuel cells (SOFCs) are considered as alternative electric power generation systems due to high energy-conversion efficiency, fuel flexibility, and environmental safety [1–3]. Dense ceramic membranes with mixed oxygen-ionic and electronic conductivity have an infinite theoretical separation factor with respect to oxygen, and can be used for gas separation and the partial oxidation of light hydrocarbons [4, 5, 7, 8] (these and other applications are reviewed in Chapters 12 and 13). For all types of electrochemical cells, the key properties which determine the use of a material are the partial ionic and electronic conductivities. As an example, solid electrolytes for SOFCs, oxygen pumps and electrochemical sensors should exhibit a maximum oxygen ionic conductivity, while the electronic transport should be minimal. In contrast, for SOFC electrodes

Solid State Electrochemistry I: Fundamentals, Materials and their Applications. Edited by Vladislav V. Kharton

ISBN: 978-3-527-32318-0

and oxygen separation membranes, both ionic and electronic conductivities should be as high as possible (see Chapter 3). In the case of SOFC interconnectors, both minimum ionic and maximum electronic conduction are necessary.

In this chapter we present a brief overview of the major groups of solid oxide electrolytes and mixed conductors, placing special emphasis on their ion transport properties. The main features of the crystal structures enabling fast anion diffusion and the basic defect-chemistry mechanisms relevant for these materials are addressed in Chapters 2–4. Since it has been impossible to cover all promising compositions and migration-affecting factors within the chapter, priority has been given to those single-phase oxide ceramics which exhibit a high ionic conductivity sufficient for practical applications. Further information regarding conventional and new materials for high-temperature electrochemical devices is available in a variety of reviews and monographs [4, 6–17].

9.2 Oxygen Ionic Transport in Acceptor-Doped Oxide Phases: Relevant Trends

Perhaps one of the most widely studied topics in high-temperature electrochemistry is the oxygen ion conductivity of the fluorite-structured binary oxide systems. Except for the δ-phase of Bi_2O_3, these are formed by preparing solid solutions of tetravalent metal oxides (e.g., zirconia and ceria) with lower-valent metal oxides, most notably the trivalent rare earths (Ln^{3+}). In these solid solutions, addition of the lower-valent cation is charge-compensated by the formation of oxygen vacancies, which are highly mobile at elevated temperatures. The combination of this ability of the fluorite lattice to accept high concentrations of vacancies with the fact that they are mobile at high temperatures is the origin of this high oxygen ion conductivity. A prominent feature seen in all investigations is the maximum in the conductivity isotherms with respect to the content of the aliovalent additive. A simple analysis [12] shows that a maximum is to be expected when the oxygen sublattice is half occupied, although experimentally this is found to occur at much lower concentrations. As an example, this concentration in $Zr_{1-x}Y_xO_{2-x/2}$ corresponds to $x \approx 0.07$–0.11, depending on temperature, preparation route, pre-history, purity, and other factors. For zirconia, this effect is complicated by the fact that it only adopts the fluorite structure (at moderate temperatures) when the dopant concentration is already high, because the small Zr^{4+} cation (0.84 Å in eightfold coordination) is not large enough to stabilize the fluorite structure to low temperatures. In contrast, in pure ceria the Ce^{4+} cation (0.97 Å in eightfold coordination) is large enough to be stable in the fluorite lattice, and thus provides an ideal opportunity for examining any effects from dilute to concentrated solid solutions. The example of ceria-based electrolytes is used in this section to illustrate the key effects of acceptor-type doping, which are observed in most fluorite- and perovskite-related phases.

For most ionic conductors, the conductivity activation energy (E_a) may be considered to consist of two parts, namely the enthalpy of ion migration (ΔH_m) and any terms caused by the interaction of the point defects and/or by the charge carrier formation. For low concentrations of the trivalent additive in ceria lattice, the interaction term is thought to be due to the formation of defect associates of the oxygen vacancy with the substitutional cation(s). In a simplest case, this type of defect interaction in yttria-doped ceria can be presented as

$$Y'_{Ce} + V_O^{\bullet\bullet} \leftrightarrow \{Y'_{Ce}V_O^{\bullet\bullet}\}^{\bullet} \quad (9.1)$$

with the association enthalpy ΔH_a. Shown here is the simple dimer, or associated pair, but it should be noted that the concentration of substitutional cations is twice that of the oxygen vacancies, and so there is a high probability of forming trimers and higher clusters as the concentration increases (a point to which we will return). If simple dimers were to form, then the activation energy for conductivity would be given by the sum $(\Delta H_a + \Delta H_m)$. Irrespective of the association type, two vacancy concentrations can be defined: the *stoichiometric concentration*, which is determined by the electroneutrality condition; and the *mobile* or *"free" vacancy* concentration given by the association equilibria, which can be very different. Only the "free" vacancies are mobile and can contribute significantly to oxygen ionic transport.

What is found by experiment is that, as a general rule, at substitutional concentrations close to the maximum in the conductivity isotherms, there is a minimum in the activation energy. In an early (but very comprehensive) study of ceria solid solutions with the trivalent rare earths, Faber *et al.* [18] showed that the depth of the minimum, and the concentration at which it occurs, depends upon the identity of the rare earth cation (Figure 9.1). The minima have been ascribed [19] to competitive defect interactions. Initially, the effect is a weakening of the association energy of the dimers caused by an electrostatic interaction between the cluster and the unassociated substitutionals having an opposite effective charge in the lattice; note, however, that

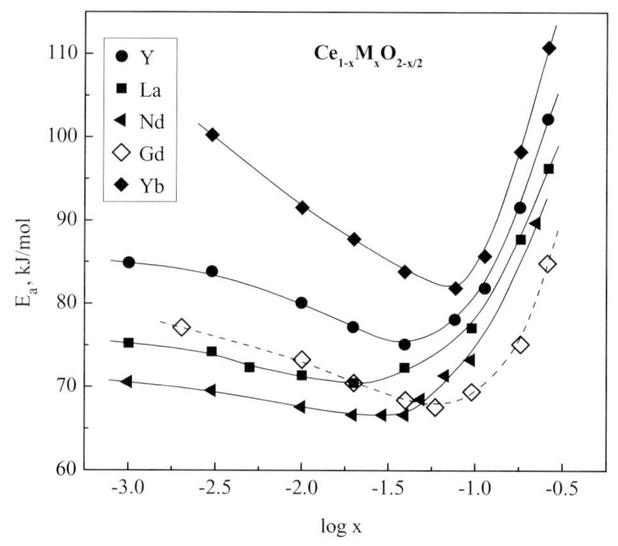

Figure 9.1 Activation energy versus dopant concentration in ceria solid solutions. Data from Ref. [18].

the intrinsic vacancy-formation process cannot be neglected at low dopant concentrations. On further doping, higher-order clusters form, which act as much deeper traps for the oxygen vacancies and thus increase the association term.

There are distinct differences between the compositional dependence of E_a for each of the substitutional species, both in terms of the magnitude and shape of the dependence, even though they have the same effective charge and give rise to identical stoichiometric vacancy concentrations (see Figure 9.1). In particular, there is a further minimum here as a function of the size of the substitutional cation. This is shown much better in Figure 9.2, where the minima with concentration are plotted as a function of the dopant radius. The global minimum corresponding to the lowest total activation energy is an important feature for the technological application of ceria solid solutions. It is usually assumed that, to a first approximation, the migration enthalpy in these materials changes very little from system to system, and that the observed changes in E_a reflect changes in the association term. If this is the case, then Figures 9.1 and 9.2 demonstrate that the association energy must contain terms that reflect both an electrostatic and an elastic interaction between the components of any defect cluster, as the dopant size is so important.

The explanation of the minimum in activation energy with dopant radius was initially made in terms of the elastic component of the association energy of the simple pairs – that is, in terms of the size mismatch between the host and the substitutional ion (see early references in Ref. [12]). Later computer simulations of

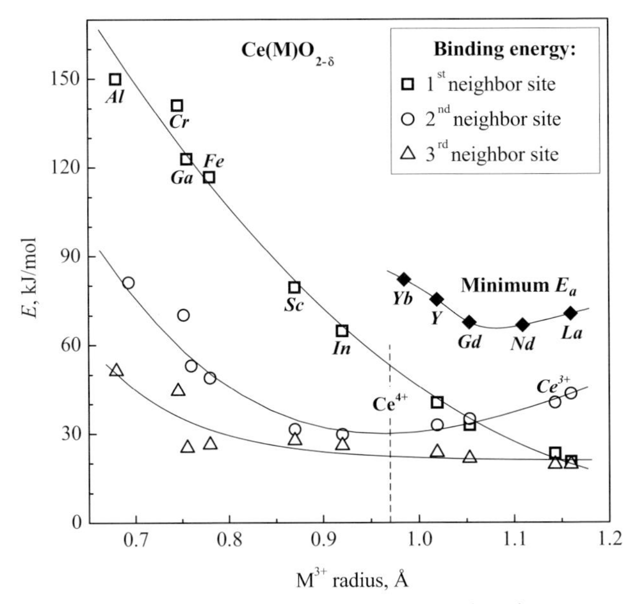

Figure 9.2 The minimum activation energy in ceria-based systems (closed symbols) and calculated binding energies of vacancy substitutional pairs in ceria (open symbols) plotted against the M^{3+} dopant cation radius in eightfold coordination. Data from Refs. [18, 20].

ceria-based solid solutions [20] showed that the situation is more complicated and that, in fact, the minimum is caused by the switchover of the vacancy from a first ($r_{ion} \leq Gd^{3+}$) to a second nearest-neighbor site ($r_{ion} \geq Gd^{3+}$) to the substitutional ion in the fluorite lattice, as the dopant size exceeds that of the host (see Figure 9.2). A similar dependence is operative in zirconia electrolytes [21]; moreover, the same type of size dependence could be found for larger clusters. More recent density functional calculations for ceria-based materials have confirmed this trend [22], and have shown that at the minimum in the association (binding) energy where the crossover takes place, there is a balance between the electrostatic and elastic interactions that removes the energy difference between the first- and second-neighbor configurations. It would seem, therefore, that the above-mentioned theoretical studies present a consistent account of the size dependence. However, recent calorimetric studies performed by Navrotsky *et al.* [23] have shown that this explanation might not yet be complete. In fact, their measurements would indicate that the vacancy is, as expected, in the first nearest-neighbor position in CeO_2-Y_2O_3, and CeO_2-Gd_2O_3, but the same seems to be true for CeO_2-La_2O_3 solid solutions, in contrast to the calculated predictions.

A word of caution is needed at this point on the above observations and following comments. It is clear from Figure 9.1 that the defect structure has a strong concentration dependence, and thus for meaningful comparisons between different materials it is necessary to compare like with like. However, the concentration dependence of the activation energy curve is different for each substitutional cation; indeed, Andersson *et al.* [22] showed that the interplay between the electronic and elastic component depends heavily on the dopant type. It is probably the case that the type of defect clusters observed is very dependent upon concentration, and that typically around the concentration for the minimum activation energy a change in the defect structure takes place.

In the case of mixed ionic-electronic conductors, an analysis of the lattice strain-promoted clustering phenomena becomes even more complicated as the oxygen vacancy concentration – and, consequently, the ionic conductivity – depend on both the oxygen partial pressure and cation size mismatch. In this situation, any quantitative comparison is often difficult, even when the experimental data were obtained for essentially similar materials and under identical external conditions. Nonetheless, the data available on ferrite-, cobaltite-, and nickelate-based systems with perovskite-related structures (e.g., [24–28]) seem to confirm that, when long-range ordering in the oxygen sublattices can be neglected, the ionic transport tends to increase with decreasing difference between the host and dopant cation sizes. As an example, Figure 9.3 compares the variations of partial ionic and p-type electronic conductivities in perovskite-type $Ln_{0.5}A_{0.5}FeO_{3-\delta}$ (A = Sr,Ba) [24]. Under oxidizing conditions, increasing the mismatch between Ln^{3+} and A^{2+} radii weakens the metal–oxygen bonding, increases the oxygen nonstoichiometry, and decreases the hole concentration due the electroneutrality condition; the ionic transport then exhibits a trend opposite to the vacancy concentration variations. A similar tendency is observed for the ionic conductivity in reducing atmospheres, when the oxygen deficiency becomes similar for all $Ln_{0.5}A_{0.5}FeO_{3-\delta}$ perovskites; the activation energy

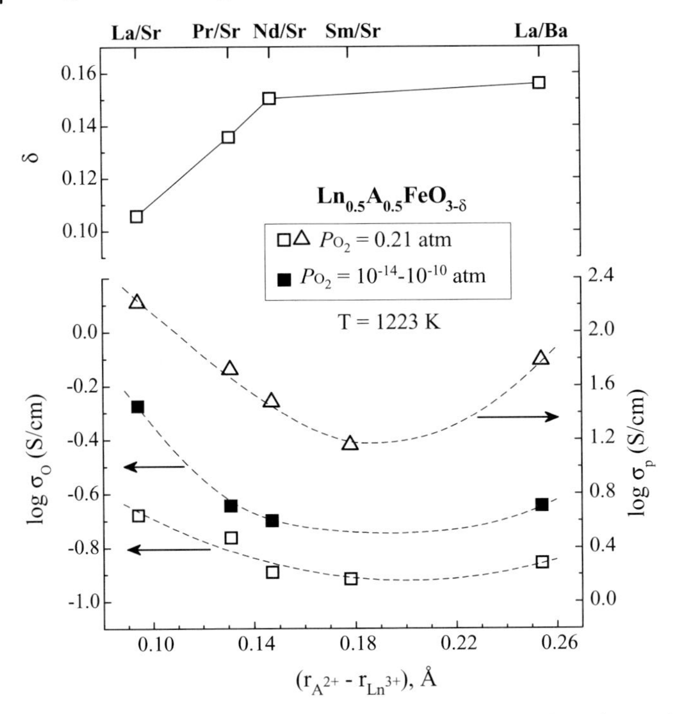

Figure 9.3 Oxygen nonstoichiometry at atmospheric oxygen pressure (top), and values of partial p-type electronic and ionic conductivities in air, and ionic conductivity in reducing atmospheres when the oxygen content in the perovskite lattice $(3-\delta)$ is close to 2.75 atoms per formula unit (bottom), versus the difference between A^{2+} and Ln^{3+} cation radii in $Ln_{0.5}A_{0.5}FeO_{3-\delta}$ (Ln = La-Sm, A = Sr, Ba) at 1223 K. Data from Ref. [24]. All A-site cation radii correspond to ninefold coordination.

increases with lattice strains and displays no direct correlations with the stereological parameters which might affect ion diffusivity [24].

One way to confirm the defect cluster configurations is to use localized probes such as the synchrotron-based extended X-ray absorption fine structure (EXAFS) and X-ray absorption near edge structure (XANES) spectroscopic analyses to examine the local cation coordination [29, 30]. Alternatively, electron beam-based techniques such as high-resolution electron energy loss spectroscopy (EELS) and selected-area electron diffraction (SAED) can be used [31]. Most of these techniques have been applied to materials with the acceptor-type dopant concentration >5%, where trimers and higher-order clusters are to be expected. Interestingly, even in concentrated solid solutions based on ceria, the different trivalent substitutionals show differing behaviors. The study of yttria- and gadolinia-doped $CeO_{2-\delta}$ using EXAFS and XANES [29] revealed a greater tendency towards defect association in the former system, with the most likely configurations comprising either two or four trivalent cations clustered with one or two neighboring oxygen vacancies. Similar conclusions were drawn when the study was extended to the CeO_2-La_2O_3 system [30]. Deguchi *et al.* [30] also suggested a similarity between the radial distribution functions for the dopant cations (e.g., R_{Ln-Ln}) and those found in the corresponding rare earth sesquioxides, concluding that "*...it is interesting to note that the original structure of each dopant oxide is kept even in the ceria lattice*". This theme was echoed by Ou *et al.* [31], who showed that the defect structure in ceria electrolytes does indeed change with the dopant content, albeit for rather concentrated materials containing 15–25% additives; these authors proposed the existence of ordered nanodomains – essentially large vacancy-dopant clusters that resemble the parent C-type structure of Ln_2O_3.

Some final comments are needed to qualify the previous analysis. For the experimental observations, it has already been stated that care must be taken to ensure that like materials are compared (although the term "like" materials is not quite clear). It is also instructive to remember that the different techniques probe the materials over different length scales, and are sensitive to the different properties of the material. The theoretical techniques also differ in their strengths and weaknesses to simulate these complex lattices, and will be more successful in their description of one aspect of the problem. However, the most problematic issue is the comparability of even the most "identical" samples.

In most oxygen ion-conducting materials used for practical applications, the lattice diffusivity of the cations is particularly sluggish. This brings its own problems for understanding the microscopic mechanisms and materials optimization. Usually, the distribution of dopant cations is assumed to be random among the available lattice sites – indeed, this is the aim of the (mainly solution based) fabrication processes designed to maximize cation mixing. Yet, this is far from the case in the ceramic samples that are subsequently investigated. The computer simulations show that there is an energetic advantage for the formation of clusters. However, these need to form from the initial random distribution imposed by fabrication and thermal prehistory; the redistribution kinetics may be very slow, and it is unlikely that a true "equilibrium" structure will result. Whichever metastable cation distribution results from the prehistory, the oxygen sublattice will rapidly adjust to it. It is probable, therefore, that subtle differences in the observed defect cluster structure can occur between ostensibly similar materials, and this should be taken into account when evaluating the experimental data relating to this fascinating topic.

9.3 Stabilized Zirconia Electrolytes

The maximum ionic conductivity in ZrO_2-based systems is observed when the content of acceptor-type dopant cations with the smallest radii (Sc, Yb, Y) is close to the minimum necessary to completely stabilize the cubic fluorite-type phase in the operating temperature range [9, 11, 16, 32–35]. This concentration (often referred to as the low stabilization limit) and the conductivity of the ceramic electrolytes are dependent, to a finite extent, on the pre-history and various microstructural features. In addition to the metastable states discussed above, critical microstructural factors include the dopant segregation at grain boundaries, impurities, porosity, and the formation of ordered microdomains. Nevertheless, despite minor contradictions still existing in the literature, for most important systems the dopant concentration ranges providing maximum ionic transport are well established. For example, the highest conductivity in $Zr_{1-x}Y_xO_{2-x/2}$ and $Zr_{1-x}Sc_xO_{2-x/2}$ is observed at $x=$ 0.07–0.11 and 0.09–0.11, respectively. Although further additions decrease the ionic conduction due to progressive defect clustering, typical dopant concentrations in the solid electrolyte ceramics used for most practical applications are moderately higher, for example, $x \approx 0.15$ in $Zr_{1-x}Y_xO_{2-x/2}$ (8 mol.% Y_2O_3). This doping strategy aims, in particular, to partly suppress ageing at 900–1300 K associated with kinetically limited phase changes and time degradation of the conductivity. In the case of scandia-stabilized zirconia (SSZ), where a relatively fast decomposition of metastable cubic and rhombohedral solid solutions and/or partial ordering occur at moderate temperatures, this strategy is less effective compared to Y-containing analogues. Taking into account the precursor costs, yttria-stabilized zirconia (YSZ) is among the most commonly used solid electrolytes to date. Numerous attempts have been made to identify new electrolyte compositions in ternary systems, particularly to increase the conductivity by optimizing the average size of dopant cations, to increase the stability of Sc-containing materials by co-doping, or to decrease the cost of Ln^{3+}-stabilized phases by mixing rare- and alkaline-earth dopants [33]. However, no worthwhile improvement has been observed. Figure 9.4a displays typical values of the total conductivity (predominantly ionic) for selected ZrO_2-based phases and one YSZ-Al_2O_3 composite. Small additions of highly dispersed alumina make it possible to improve the mechanical strength and to reduce the grain-boundary resistance due to the "scavenging" of silica-rich phases [45–47]; this approach has been used successfully to prepare a variety of solid electrolyte ceramics (e.g., [48–50]).

Another necessary comment is that, if compared to other oxide electrolytes, stabilized zirconia ceramics without variable-valence additives exhibit, as an average, a minimum electronic contribution to total conductivity in the oxygen partial pressure range most important for practical applications [1, 6, 10, 11, 13, 14, 16, 33]. This approximate P_{O_2} range is from 100–200 atm (oxygen compressors, impurity sensors in oxygen, high-pressure SOFCs) down to 10^{-25}–10^{-20} atm (standard SOFC systems, water vapor electrolyzers, exhaust gas sensors). Whilst in reducing environments, the n-type electronic transport in ThO_2- and $LaGaO_3$-based electrolytes is lower than that of stabilized ZrO_2, shows lower p-type electronic conduction and, thus, a higher performance under oxidizing conditions. It should be noted also that the performance of lanthanum gallate at low P_{O_2} is limited by Ga^{3+} reduction and gallium oxide volatilization, rather than the n-type electronic conductivity. For applications in oxygen-separation membranes and SOFC electrodes, the electronic transport in stabilized zirconias can be moderately enhanced by incorporating variable-valence cations (e.g., Ti, Nb, Cr, Tb, Pr) into the fluorite-type lattice [4, 9, 17, 33, 51, 52]. However, such doping leads usually to decreasing ionic conductivity; the solubility of transition metal cations is relatively low and temperature-dependent, whereas the addition of praseodymium oxide causes thermomechanical instability.

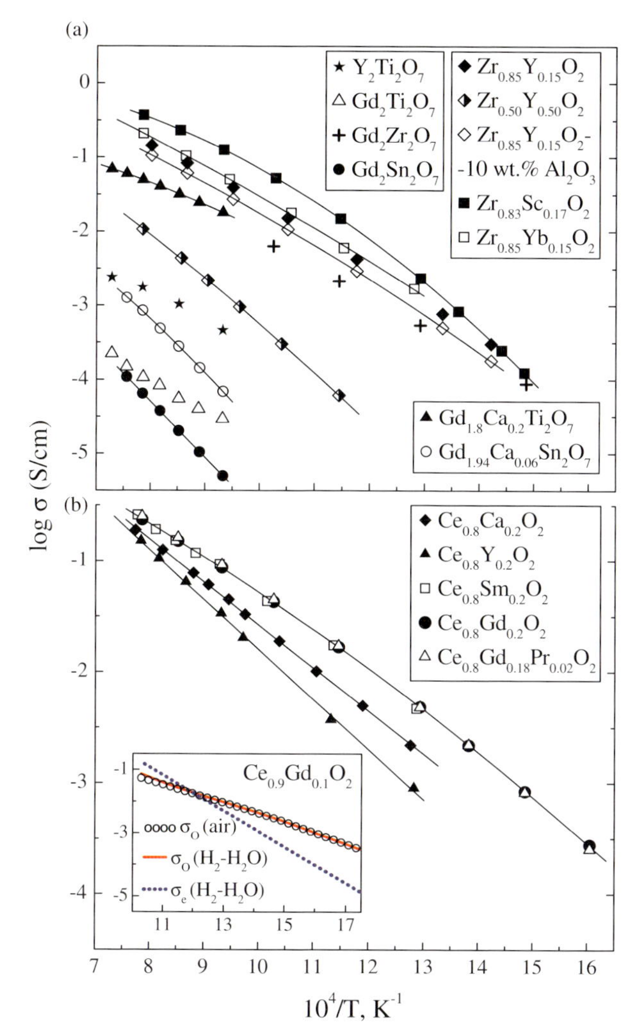

Figure 9.4 Total conductivity of stabilized zirconia (a) and doped ceria (b) solid electrolytes [34–40], compared to the oxygen ionic conductivity of pyrochlore-type $Gd_2Zr_2O_{7-\delta}$ [41], $(Gd,Ca)_2Ti_2O_{7-\delta}$ [42], $(Gd,Ca)_2Sn_2O_{7-\delta}$ [43], and $Y_2Ti_2O_{7-\delta}$ [44]. All data correspond to the atmospheric oxygen pressure. The inset shows the ionic conductivity of $Ce_{0.9}Gd_{0.1}O_{2-\delta}$ in air, and the partial ionic and electronic conductivities in a reducing atmosphere containing 10% H_2 and 2.3% H_2O [14].

9.4 Doped Ceria

The main advantages of fluorite-type $Ce(Ln)O_{2-\delta}$ [38, 53–61] include a faster oxygen-ionic transport with respect to stabilized ZrO_2, a lower cost than that of La(Sr)Ga(Mg) $O_{3-\delta}$ (LSGM) ceramics, and modest reactivity and volatility of the components in comparison with $LaGaO_3$-, $La_{10}Si_6O_{27}$-, and Bi_2O_3-based materials. In addition, the superior catalytic properties of ceria are advantageous for increasing the exchange currents of the SOFC anodes. Among CeO_2-based phases, the highest level of ionic conduction is characteristic of the solid solutions $Ce_{1-x}M_xO_{2-\delta}$ (M = Gd or Sm; $x = 0.10–0.20$) and their analogues obtained by co-doping, where the concentrations of substitutional cations are adjusted to optimize the average radius and hence, to minimize the tendency to vacancy clustering. Selected data on the total conductivity, which is predominantly ionic under oxidizing conditions, are shown in Figure 9.4b. As with all ceramic electrolytes, the grain boundaries are partially blocking to ionic transport across them; this is an extra contribution to the total resistance that is dependent on impurities and dopants that segregate to the boundaries, and is therefore highly variable from one source to another. This has been an origin of numerous contradictions in the reported data, particularly that relating to the optimum ceria-based compositions. Thus, while $Ce_{0.9}Gd_{0.1}O_{1.95}$ has the highest lattice conductivity, $Ce_{0.8}Gd_{0.2}O_{1.90}$ often has higher total conductivity because its grain boundary contribution seems to be more tolerant to impurities [57]. Due to the dopant segregation and interaction with impurities, analogous discrepancies can also be found on the effects of co-doping (e.g., Refs [60–62] and references cited therein).

The p-type electronic conductivity of gadolinia-doped ceria (CGO) in air is 0.5 to 3.0-fold lower than that of LSGM, and this difference increases with decreasing temperature (Figure 9.5). The main problems in using doped ceria as an SOFC electrolyte arise, however, from the partial reduction of Ce^{4+} to Ce^{3+} under the reducing conditions of the anode [11, 13, 16, 55, 57, 69]. This has two main effects: (i) it produces n-type electronic conductivity, which causes a partial internal electronic short circuit in a cell; and (ii) it generates oxygen deficiency and an expansion of the lattice, which can lead to mechanical failure. For $Ce_{0.9}Gd_{0.1}O_{2-\delta}$, the electronic conductivity at the anode side is greater than the ionic conductivity for temperatures higher than about 823 K (inset in Figure 9.4). Such properties mean that ceria electrolytes are viable only for low-temperature operation; at the same time, the appearance of significant mixed conductivity in reducing environments is advantageous for ceria-containing anodes. The behavior of doped $CeO_{2-\delta}$ with respect to variable-valence additives which have been incorporated to further enhance electronic transport in electrodes and ceramic membranes, is generally similar to other fluorite-type oxides, such as zirconia [4, 7, 8, 17, 40, 70–73]. In particular, maximum oxygen permeability is achieved in those systems with the highest solubility of variable-valence cations, such as $Pr^{4+/3+}$ [73]; however, the thermomechanical stability of such ceramic materials remains an open issue.

9.5 Anion Conductors Based on Bi_2O_3

Oxide phases derived from Bi_2O_3 are particularly interesting due to their high ionic conductivity in comparison to other solid electrolytes [4, 74–78]. The maximum ionic transport is known for stabilized δ-Bi_2O_3 with a highly oxygen-deficient fluorite

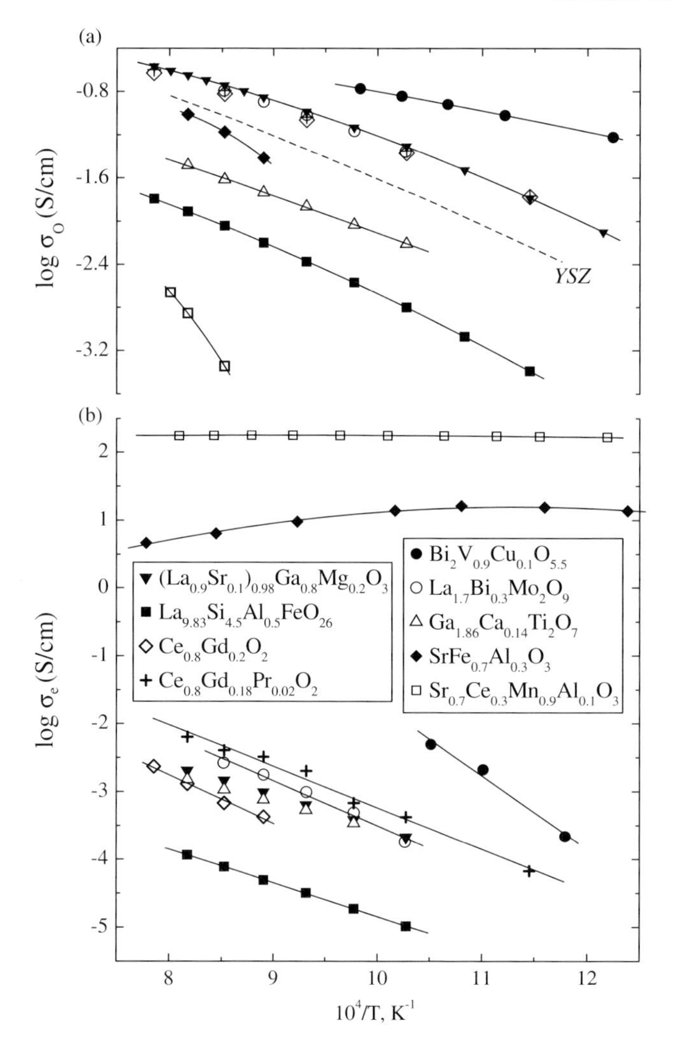

Figure 9.5 Comparison of the oxygen ionic (a) and electronic (p- and n-type) (b) conductivity of selected solid electrolytes and mixed conductors under oxidizing conditions. The partial ionic conductivity of perovskite-type $Sr_{0.7}Ce_{0.3}Mn_{0.9}Al_{0.1}O_{3-\delta}$ [63] correspond to the oxygen partial pressure gradient of 0.21/0.021 atm; data on electronic conduction in $Ce_{0.8}Gd_{0.18}Pr_{0.02}O_{2-\delta}$ [40], $Bi_2V_{0.9}Cu_{0.1}O_{5.5-\delta}$ [64] and $La_{1.7}Bi_{0.3}Mo_2O_9$ [65] correspond to the P_{O_2} range of 1.0/0.21 atm. All other data [36, 40, 64–68] correspond to $P_{O_2} = 0.21$ atm. The ionic conductivity of 8 mol% yttria-stabilized zirconia (*YSZ*) [36] is shown for comparison.

structure, and for one member of the Aurivillius series, γ-$Bi_4V_2O_{11}$, which is the parent compound of the so-called BIMEVOX family. Unfortunately, Bi_2O_3-based materials exhibit a number of disadvantages that hamper their practical application, notably a very high thermal expansion, instability in reducing atmospheres, volatilization of bismuth oxide, a high corrosion activity, and low mechanical strength.

Stabilization of the high-diffusivity δ-Bi_2O_3 phase down to intermediate and low temperatures can be achieved by doping with rare-earth and/or higher-valency cations, such as Y, Dy, Er, W, or Nb [11, 13, 74–76]. Figure 9.6 displays representative data on the total conductivity (mainly ionic) for two δ-Bi_2O_3-based compositions. As with zirconia and hafnia electrolytes, the highest ionic transport is observed for materials containing the minimum addition necessary to stabilize the δ-polymorph; further doping decreases oxygen ion mobility due to a decreasing unit cell volume and increasing the average energy of the metal–oxygen bonding. As Bi^{3+} cations are relatively large, at a given doping level the ionic transport increases with increasing Ln^{3+} radius. However, the minimum stabilization limit also increases with Ln^{3+} size, and this leads to a decrease in the maximum ionic conductivity with increasing radius of the stabilizing cation. As a result, the highest conductivity in the Bi_2O_3-Ln_2O_3 systems is observed for Er- and Y-containing solid solutions, namely $Bi_{1-x}Er_xO_{1.5}$ ($x \approx 0.20$) and $Bi_{1-x}Y_xO_{1.5}$ ($x = 0.23$–0.25). It should be noted, however, that all known δ-Bi_2O_3-based phases with the disordered fluorite structure are metastable at temperatures below 770–870 K, and exhibit a slow degradation in conductivity with time due to sluggish polymorphic transformations. The solid

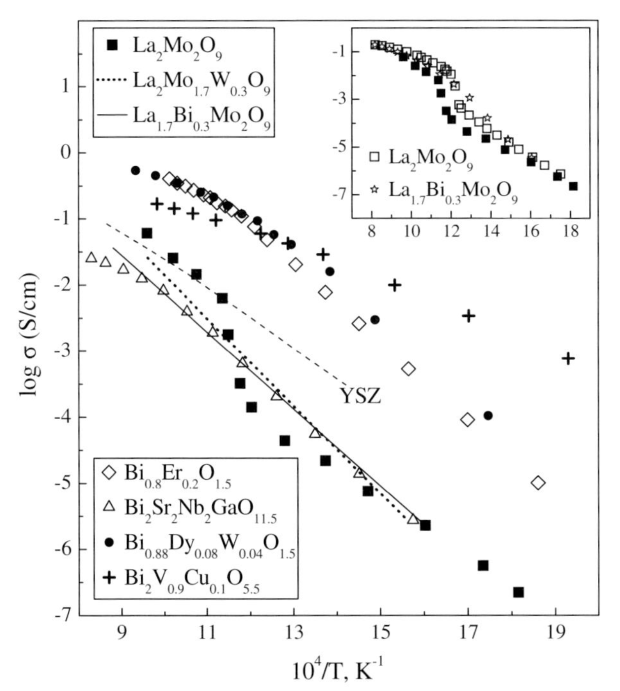

Figure 9.6 Total conductivity of Bi_2O_3-based [64, 75, 79, 80] and $La_2Mo_2O_9$-based [81] materials in air, compared to the ionic transport in 8 mol% yttria-stabilized zirconia (YSZ) [36]. The inset presents the conductivity data on undoped $La_2Mo_2O_9$ and $La_{1.7}Bi_{0.3}Mo_2O_9$ [65], for comparison.

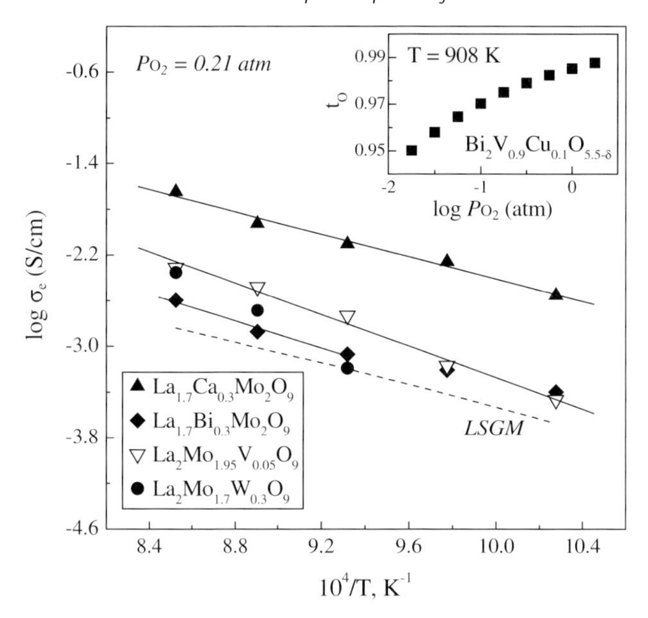

Figure 9.7 Temperature dependence of the electronic conductivity of $La_2Mo_2O_9$-based ceramics, determined by the faradaic efficiency and total conductivity measurements at atmospheric oxygen pressure [65]. The dashed line corresponds to the p-type electronic conductivity of $(La_{0.9}Sr_{0.1})_{0.98}Ga_{0.8}Mg_{0.2}O_{3-\delta}$ (LSGM) electrolyte in air [66]. The inset shows the oxygen partial pressure (P_{O_2}) dependence of the ion transference numbers of $Bi_2V_{0.9}Cu_{0.1}O_{5.5-\delta}$ at 908 K [64].

solutions based on γ-$Bi_4V_2O_{11}$ [49, 77, 78, 82, 83] stabilized by the substitution of 7–15% vanadium with transition metal cations such as Cu, Ni, or Co, possess a better phase stability at moderate temperatures, superior ionic conductivity and oxygen ion transference numbers (t_0) close to unity below 900 K under oxidizing conditions (Figures 9.5 and 9.6). The electrolytic domain of BIMEVOX ceramics is, however, very narrow. For example, decreasing the P_{O_2} to 10^{-2} atm results in the electron transference numbers of $Bi_2V_{0.9}Cu_{0.1}O_{5.5-\delta}$ increasing to 0.05 at 908 K (inset in Figure 9.7); any further reduction will cause phase decomposition.

9.6
Transport Properties of Other Fluorite-Related Phases: Selected Examples

The pyrochlore-type compounds, where the crystal structure is usually considered as a cation-ordered fluorite derivative with $^1/_2$ vacant oxygen site per fluorite formula unit, constitute another large family of oxygen anion conductors [9, 33, 41–43, 84–88]. The unoccupied sites provide pathways for oxygen migration; furthermore, the $A_2B_2O_{7\pm\delta}$ pyrochlore structure may tolerate formation of cation and anion vacancies, doping in both cation sublattices, and antistructural cation disorder. Regardless of these factors, the oxide pyrochlores possess, in general, worse transport properties compared to the fluorite-type compounds (Figures 9.4 and 9.5). At elevated temperatures (typically up to 1650–2500 K), most pyrochlores disorder into fluorite polymorphs. A decreasing A-site cation radius favors this transition. As a rule, partially cation-disordered pyrochlores exhibit relatively high activation energies for anion transport in the intermediate-temperature range; the maximum ionic conductivity (which can be further enhanced by acceptor-type doping within the solid solution formation limits) occurs for cation stoichiometry, for example, $Gd_2Ti_2O_7$ and $Gd_2Zr_2O_7$. Until now, the highest values of oxygen ion diffusivity in pyrochlore-type compounds have been achieved for $Gd_{2-x}Ca_xTi_2O_{7-\delta}$ with $x \approx 0.2$ [42, 84]. These materials also display a significant electronic contribution to the total conductivity under oxidizing and strongly reducing conditions, close to the upper limit acceptable for solid electrolytes (Figure 9.5). However, due to the limited solubility of variable-valence cations in most pyrochlore phases, the electronic transport cannot be further enhanced by doping to a substantial extent, as for the fluorite-type solid electrolytes.

Among other ion-conducting phases with fluorite-like structures, note should be taken of $Y_4NbO_{8.5}$, $(Y,Nb,Zr)O_{2-\delta}$ solid solutions, and their derivatives (see Refs. [89–91] and references cited therein). The total conductivity of $Y_4NbO_{8.5-\delta}$ is essentially independent of the oxygen partial pressure, which may suggest a dominant ionic transport [90]. However, the conductivity level in this system is rather low, and similar to that in pyrochlore-type titanates and zirconates, although some improvements can be achieved by the addition of zirconia.

9.7 Perovskite-Type $LnBO_3$ (B = Ga, Al, In, Sc, Y) and their Derivatives

Perovskite-type phases derived from $LaGaO_3$ via acceptor-type doping into both cation sublattices of the ABO_3 perovskite structure exhibit a higher ionic conductivity than that of stabilized zirconia, and are thus promising materials for electrochemical cells operating in the intermediate-temperature range (Figures 9.5, 9.8 and 9.9). In oxidizing atmospheres, the p-type electronic transport in the gallate solid electrolytes is moderately higher with respect to CeO_2-based solid solutions (Figure 9.5), but the electrolytic domain of doped $LaGaO_3$ extends to substantially lower oxygen chemical potentials. Following the principle of minimum strain giving maximum oxygen ion mobility, doping with Sr^{2+} leads to a higher ionic conductivity in comparison with either Ca^{2+} or Ba^{2+} [104–106]. As for the fluorite-type electrolytes, there is an optimum concentration of acceptor cations, depending on their size; further additions result in progressive vacancy association processes. In the case of the $La_{1-x}Sr_xGa_{1-y}Mg_yO_{3-\delta}$ (LSGM) series, the maximum ionic transport is achieved at $x = 0.10$–0.20 and $y = 0.15$–0.20, whilst the introduction of smaller Ln^{3+} cations and the creation of an A-site deficiency decrease the ionic conduction [103–108]. It should be noted that qualitatively similar trends are known for numerous perovskite-type systems with either predominant ionic transport or mixed conductivity, such as rare-earth aluminates [109, 110], ferrites, and cobaltites [24–28, 98, 102, 111],

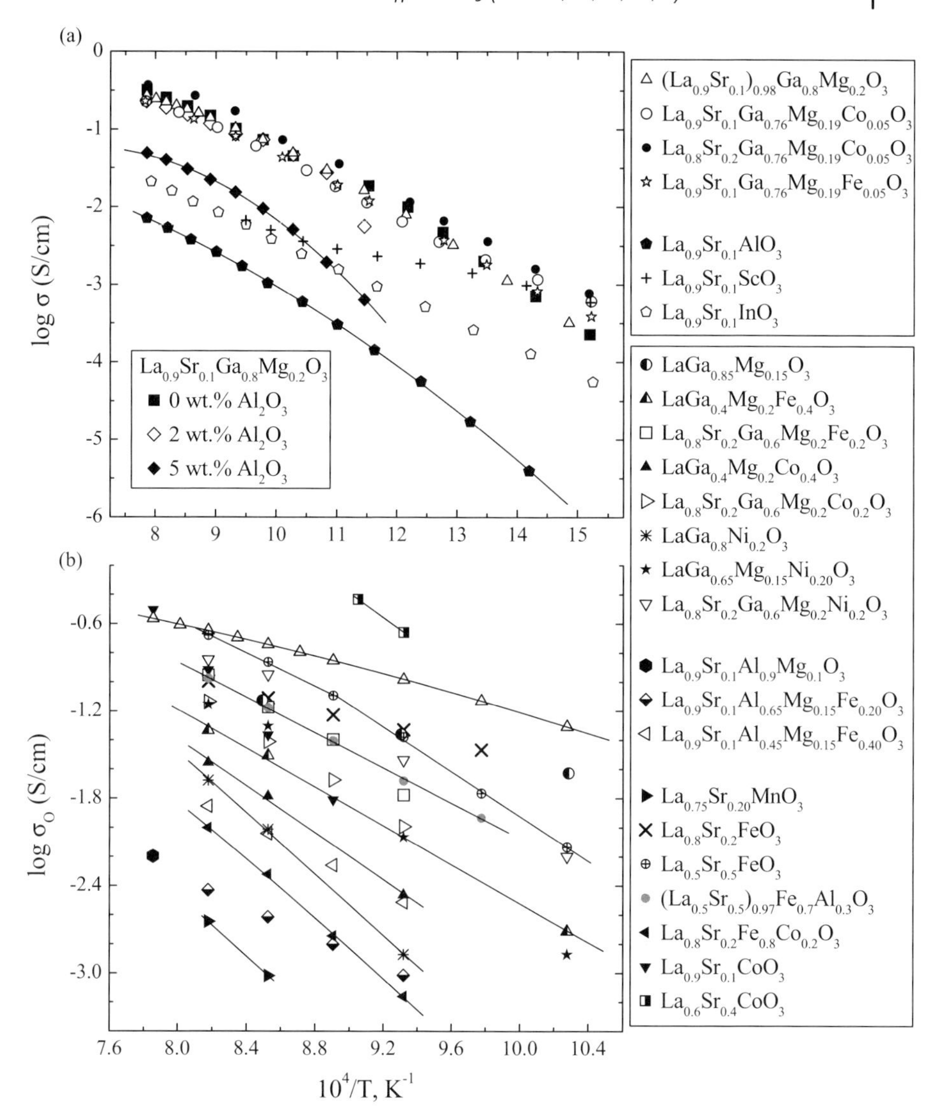

Figure 9.8 (a) Comparison of the total conductivity of $LaGaO_3$-based solid electrolytes in air [48, 66, 92] and $La_{0.9}Sr_{0.1}MO_{3-\delta}$ (M = Al, Sc, In) in a nitrogen atmosphere, where the conduction is mainly ionic [93]; (b) Oxygen ionic conductivity of various perovskite-type materials [24, 27, 94–102]. The data on $La_{0.9}Sr_{0.1}CoO_{3-\delta}$ [100] and $La_{0.8}Sr_{0.2}FeO_{3-\delta}$ [98] correspond to the oxygen partial pressure of 0.044 atm and to the P_{O_2} range 10^{-9}–10^{-19} atm, respectively. All other data shown in panel (b) correspond to atmospheric oxygen pressure.

although the optimum concentrations of acceptor-type dopants are always unique for each system.

The introduction of small amounts of variable-valence cations, such as iron or cobalt, into the B sites increases ionic conduction in the gallate-based electrolytes and

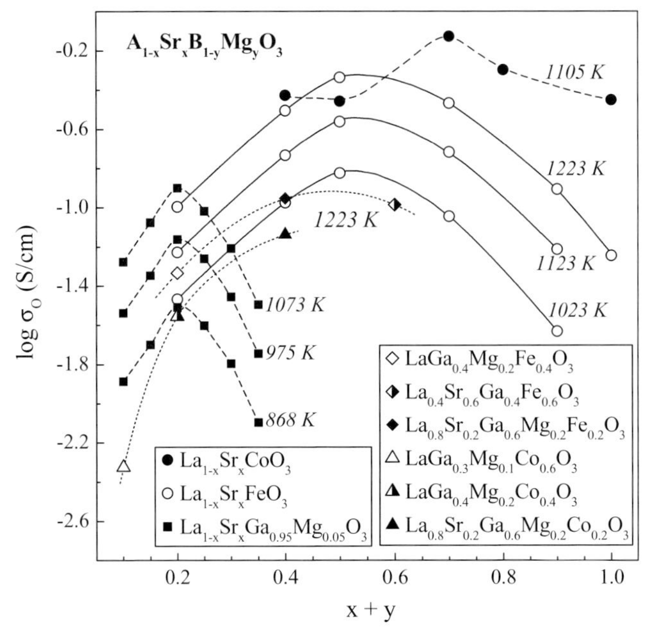

Figure 9.9 Partial oxygen-ionic conductivity of $La_{1-x}Sr_xCoO_{3-\delta}$ [27], $La_{1-x}Sr_xFeO_{3-\delta}$ [98], $La_{1-x}Sr_xGa_{1-y}Mg_yO_{3-\delta}$ [103], and $La_{1-x}Sr_x(Ga,M)_{1-y}Mg_yO_{3-\delta}$ (M = Fe,Co) [97] ceramics as a function of acceptor-type dopant concentration. All lines are for visual guidance only.

their analogues, and produces only a small increase in the electronic conductivity [95, 112, 113]. However, the concentration of transition metal dopants in the solid electrolyte ceramics should be limited to below 3–7% due to rising electronic transport. Also, in many cases the ionic conductivity of heavily doped gallates is lower than that of LSGM, particularly in the intermediate-temperature range and under oxidizing conditions, when the average oxidation state of transition metal cations is higher than 3+ (Figure 9.8). Nevertheless, optimization of the acceptor-type and variable-valence dopants content makes it possible to develop mixed-conducting membranes with a high oxygen permeability and improved dimensional stability [114–117]. The disadvantages of $LaGaO_3$-based electrolytes and membrane materials include the possible reduction and volatilization of gallium oxide, the relatively high cost of gallium, and significant reactivity with electrodes and catalysts.

Perovskite-type aluminates based on acceptor-substituted $LnAlO_3$ [93, 94, 106, 109, 110, 118, 119] possess a better stability with respect to reduction and volatilization as compared to doped $CeO_{2-\delta}$ and $LaGaO_{3-\delta}$, and also moderate thermal expansion and relatively low cost. Their major disadvantages relate to a rather low level of oxygen ionic transport, a significant p-type electronic conductivity under oxidizing conditions, and, in many cases, poor sinterability. Although the published data are somewhat contradictory due to microstructural differences and high grain-boundary resistivity of aluminate ceramics, the highest ionic conduction is known for compositions close to $La_{0.9}Sr_{0.1}AlO_{3-\delta}$, possibly substituted with less than 10% Mg^{2+} on the B sublattice. Other materials of this large family comprising perovskite-like $LnBO_3$ (B = In, Sc, Y) and their derivatives [93, 94, 106, 120–124], represent mainly academic interest, although a considerable level of protonic conduction in several $LaYO_3$- and $LaScO_3$-based phases might also be of interest for some applications. The oxygen ionic conductivity of In-, Sc-, and Y-containing perovskites is comparable or even lower than that of aluminates, while the costs are substantially higher. In addition, these perovskites exhibit a high electron-hole contribution to total conductivity under oxidizing conditions, as for aluminates.

Similar properties are also known for brownmillerite- and perovskite-type phases derived from $Ba_2In_2O_5$ and having a substantially high electrical conductivity [125–132]. The conductivity is typically oxygen-ionic in dry atmospheres with moderate Po_2, mixed ionic and p-type electronic under oxidizing conditions, and protonic in H_2O-containing gas mixtures. The parent compound, brownmillerite-like $Ba_2In_2O_5$, exhibits a mixed conductivity with a dominant oxygen ionic transport in dry air; the ion transference number at 773 K is approximately 0.93. Heating up to 1140–1230 K causes a transition into the disordered perovskite phase, which leads to a drastic increase in the anion conduction. Doping with higher-valence cations (e.g., Zr^{4+}, Ce^{3+}, Sn^{4+} or Hf^{4+} in the indium sites, or La^{3+} in the barium sublattice) makes it possible to stabilize the disordered cubic perovskite structure, and thus to increase the ionic conductivity in the intermediate temperature range (Figure 9.10);

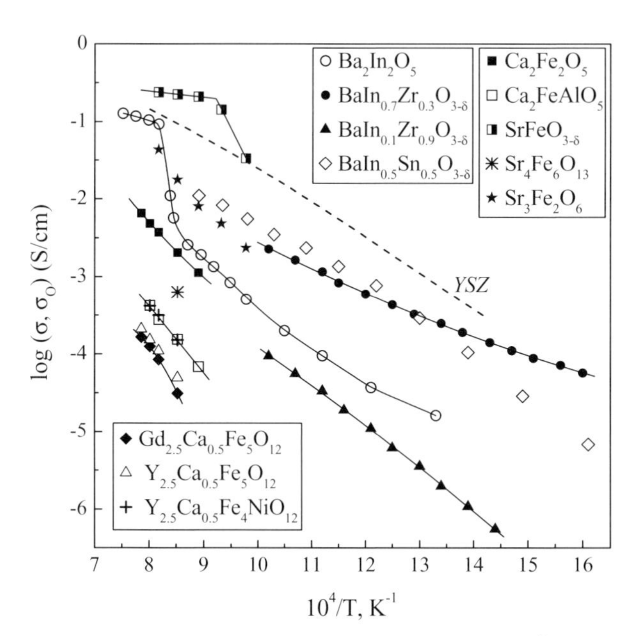

Figure 9.10 Comparison of the oxygen ionic conductivity of several ferrite-based materials with layered, perovskite and garnet structures [68, 133, 134], and total conductivity of $Ba_2In_2O_5$-based ceramics [125, 127, 129]. Data for $Ba_2In_2O_5$ [125] correspond to $Po_2 = 10^{-6}$ atm, where the conductivity is predominantly oxygen ionic. The ionic conductivity of $SrFeO_{3-\delta}$ and $Sr_3Fe_2O_{6\pm\delta}$ [134] correspond to moderately reducing conditions, where the average oxidation state of iron cations is 3+. All other data correspond to dry air.

the substitution with transition metal cations (V^{5+}, W^{6+}, Mo^{6+}) promotes hole transport. Another important advantage is related to the high protonic conductivity under SOFC operation conditions [128, 129]. However, due to the instability of $Ba_2In_2O_5$-based ceramics in humid atmospheres, high reactivity with CO_2 and easy reducibility [127, 129, 130, 132], it is difficult to imagine practical applications, taking into account that both the stability and ionic conductivity of LSGM are higher [126]. Again, the easy hydration and phase instability are also typical for analogous compounds, such as $Ba_2Sc_2O_5$ [135].

9.8
Perovskite-Related Mixed Conductors: A Short Overview

The complex phase diagrams and rich crystal chemistry of the transition metal-containing oxide systems, and great diversity in the defect chemistry and transport properties of mixed-conducting materials known in these systems, make it impossible to systematize all promising compositions in a brief survey. The primary attention here is therefore centered on the comparison of major families of the oxide mixed conductors used for dense ceramic membranes and porous electrodes of SOFCs and other high-temperature electrochemical devices.

Perovskite-type manganites $(Ln,A)MnO_{3\pm\delta}$ (Ln = La-Yb or Y; A = Ca, Sr, Ba, Pb) and their derivatives possess a high electronic conductivity, substantial electrocatalytic activity towards oxygen reduction at temperatures above 1000–1100 K, and moderate thermal expansion coefficients (TECs) that are compatible with commonly used solid electrolytes such as YSZ [1–4, 11, 136–138]. Although the total conductivity of manganites is lower compared to their Co- and Ni-containing analogues (Figure 9.11), the latter perovskite families exhibit other important disadvantages, including excessively high TECs and/or limited thermodynamic stability, even under oxidizing conditions. In fact, $(La_{1-x}Sr_x)_{1-y}MnO_{3\pm\delta}$ (LSM) and LSM-based composites are still considered as state-of-the-art cathode materials for SOFCs operating at 1070–1270 K. All perovskite-related manganites exhibit a predominant electronic conduction in combination with low oxygen ion diffusivity; their transport properties and electrochemical activity are heavily dependent on the oxygen nonstoichiometry. In particular, the electrocatalytic behavior under high cathodic polarization and the specific oxygen permeability, which are limited by both surface exchange kinetics and bulk ionic conduction, are usually correlated with oxygen vacancy generation at the manganite surface [136, 137, 148–152]. At atmospheric Po_2 and temperatures below 1270 K, $La_{1-x}Sr_xMnO_{3\pm\delta}$ perovskites are oxygen-hyperstoichiometric at $x \leq 0.2$, and become deficient on further doping [152, 153]. The p-type electronic conductivity of $Ln_{1-x}A_xMnO_{3\pm\delta}$ at moderate A^{2+} concentrations increases with x as the Mn^{4+} fraction increases; the maximum lies in the range $x = 0.2$–0.5 and shifts towards a lower dopant content on heating. This feature is qualitatively similar to most other perovskite systems with predominant hole transport, whilst the acceptor-type dopant concentrations corresponding to maximum conductivity are unique for each solid solution and depend on

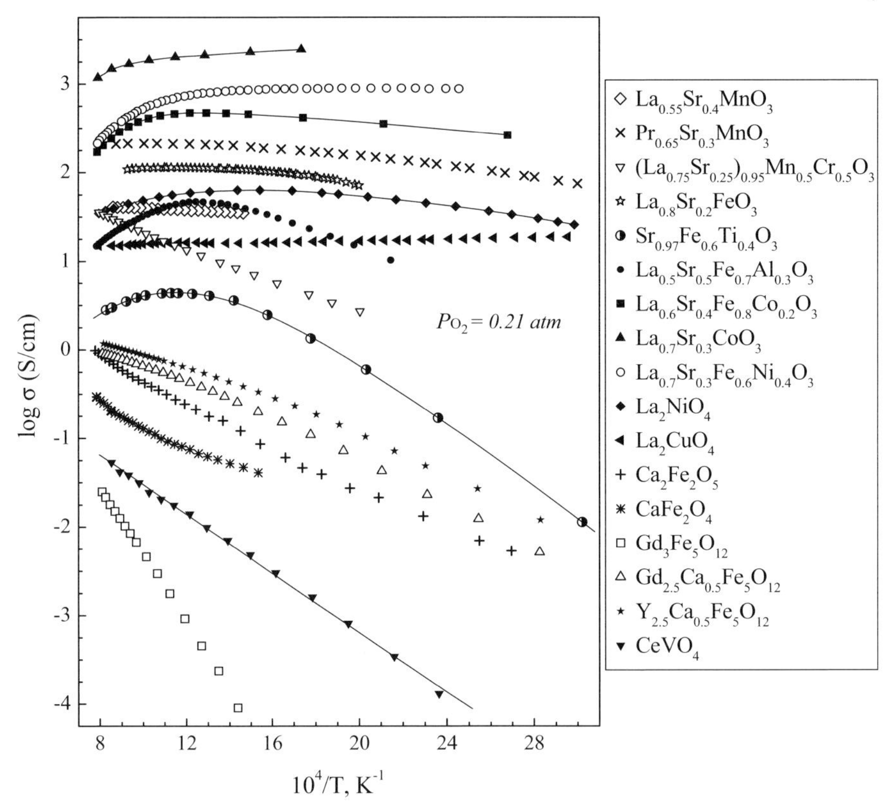

Figure 9.11 Total conductivity of various mixed-conducting materials in air [68, 101, 133, 139–147].

temperature and oxygen chemical potential. In the case of $Ln_{0.7}Sr_{0.3}MnO_{3\pm\delta}$ (Ln = La-Gd) series, the highest conductivity was reported for $Nd_{0.7}Sr_{0.3}MnO_{3\pm\delta}$ above 760 K and for $Pr_{0.7}Sr_{0.3}MnO_{3\pm\delta}$ at lower temperatures [154]. The minimum polarization resistance of manganite cathodes at 1070–1270 K, observed in the range $x = 0.3$–0.7, also tends to shift towards lower dopant concentrations when the temperature increases, and is close enough to the conductivity maximum. The introduction of an A-site cation deficiency increases the electrode performance due to a suppressed reactivity with zirconia electrolytes, a higher oxygen-vacancy content and, often, a faster ionic conduction [139, 151, 152, 155, 156].

The number of iron-containing oxide phases with a significant mixed conductivity, which are stable under the operating conditions of SOFC cathodes and ceramic membranes, is larger than that in the manganite systems. These primarily include perovskite-like $(Ln,A)FeO_{3-\delta}$ and their derivatives existing in all Ln-A-Fe-O systems, $A_2Fe_2O_{5\pm\delta}$ brownmillerites, $(Ln,A)_3Fe_5O_{12\pm\delta}$ garnets in the systems with relatively small Ln^{3+} cations, Ruddlesden–Popper series $(Ln,A)_{n+1}Fe_nO_z$, and a variety of other intergrowth compounds such as $Sr_4Fe_6O_{13\pm\delta}$ (see Refs [4, 8, 17, 98, 134, 152, 157–168] and references cited therein). However, due to structural constrains and defect chemistry features limiting both ionic and electronic transport, in most cases an extensive iron substitution is necessary to achieve the total conductivity higher than 10–30 S cm^{-1} and partial ionic conductivity higher than 0.1 S cm^{-1} at temperatures above 700 K. For B-site-undoped ferrites, the maximum conductivity is characteristic of perovskite-related solid solutions, such as $Ln_{1-x}Sr_xFeO_{3-\delta}$, where the highest level of electronic and ionic transport is known for Ln = La and $x \approx 0.5$. Selected data on the ionic conductivity of these perovskites are presented in Figures 9.3, 9.8 and 9.9. Figure 9.12 compares the steady-state oxygen permeation fluxes through dense ceramic membranes of various perovskite-related compounds with predominant electronic conductivity. For most materials the oxygen fluxes are governed by both surface exchange and bulk ionic conduction, but can still be used to evaluate variations of the ionic transport properties in different systems due to the well-known correlations between the exchange kinetics and ion diffusivity [173–175]. It should also be noted that, in the

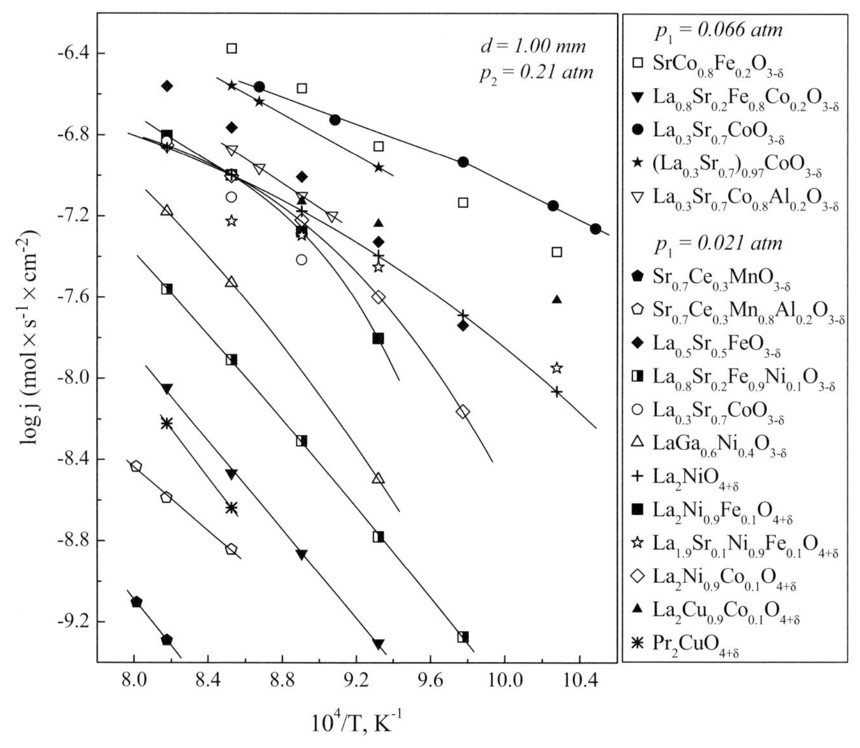

Figure 9.12 Temperature dependence of the oxygen permeation fluxes through various mixed-conducting membranes, made of perovskite-related compounds with predominant electronic conductivity, under fixed oxygen chemical potential gradients [24, 63, 111, 162, 169–172]. p_1 and p_2 are the oxygen partial pressures at the membrane permeate and feed sides, respectively. All data correspond to the ceramic membranes without surface modification.

Ruddlesden–Popper series, the partial electronic and ionic conductivities tend to decrease when the concentration of rock salt-type $(Ln,A)_2O_2$ layers increases; the general trends observed on acceptor doping are similar to those in the perovskite systems (e.g., Refs [134, 165, 167, 168]). If compared to manganite electrode materials, one important disadvantage of perovskite-related ferrites relates to a high chemical expansion, which provides a critical contribution to the apparent TECs due to oxygen losses at elevated temperatures, and may lead to a thermomechanical incompatibility with common solid electrolytes [117, 162, 176, 177]. The layered ferrite phases – particularly the Ruddlesden–Popper series and brownmillerites – display modest stoichiometry changes and better thermomechanical properties compared to the perovskite compounds, although their compatibility with electrolyte ceramics is still limited [133, 134, 165, 167, 168, 178]. For the membrane reactor materials, the thermomechanical and thermodynamic instabilities typical for undoped ferrite ceramics require extensive compositional modifications, such as the substitution of more than 20–30% iron with other cations (e.g., Ti, Cr, Ga, Al) and/or the fabrication of ferrite-based composites containing dimensionally stable components [4, 7, 8, 17, 102, 114–117, 152, 157, 162–166].

Perovskite-related cobaltites exhibit substantially better transport properties and electrochemical activity in comparison with their ferrite and manganite analogues, but possess also a lower thermodynamic stability and higher thermal and chemical expansion [4, 8, 27, 101, 111, 136, 138, 152, 156, 157, 159, 168, 179]. Due to relatively high mixed conductivity and fast exchange kinetics, significant attention is drawn to the layered cobaltites where the state of Co cations is often more stable with respect to disordered perovskite phases and the TECs are lower; important compositional families are $LnBaCo_2O_{5+\delta}$ (Ln = Pr, Gd-Ho, Y), $LnBaCo_4O_{7+\delta}$ (Ln = Dy-Yb,Y), and also Ruddlesden–Popper type $(Ln,A)_4Co_3O_{10-\delta}$ and $(Ln,A)_2CoO_{4\pm\delta}$ (Ln = La-Nd) existing at moderately reduced oxygen pressures (see Refs [179–184] and references cited therein). Information on the long-term performance of such materials under SOFC cathodic conditions is, however, scarce. In the well-studied $Ln_{1-x}A_xCoO_{3-\delta}$ systems, the maximum total conductivity at 800–1300 K is observed for Ln = La-Sm, A = Sr and $x = 0.25$–0.50, shifting towards a lower x on heating; the highest level of the oxygen ionic transport corresponds to $x = 0.65$–0.70 and Ln = La. The level of ionic conduction in cubic $(Sr,La)CoO_{3-\delta}$ is not exceptional, but lies close to the maximum known for the oxide mixed conductors where a higher oxygen permeability was only reported for Bi_2O_3-containing composites and for perovskite-like phases derived from $A(Co,Fe)O_{3-\delta}$ (A = Sr, Ba). The roles of A^{2+} cation size and its matching to Ln^{3+} radius in these complex systems, where long- and short-range vacancy ordering in the highly oxygen-deficient lattices is often observed even at elevated temperatures, are still disputed. In addition to experimental limitations associated with the partial conductivity measurements when the ion transference numbers are as low as 10^{-7}–10^{-5}, the discrepancies in the available published data may partly originate from local ordering, cation demixing under nonequilibrium conditions, and phase separation in the intermediate-temperature range [4, 185–188]. Analogously to ferrite-based systems, these phenomena are all affected by the cobaltites thermodynamic properties correlating with the tolerance factors and A-site cation radii, and also by the material's microstructure. As a consequence, whilst the positive impact of Ba^{2+} doping on oxygen reduction kinetics was established by numerous research groups [24, 184, 189–192], data on the ionic transport in (Ba,Sr)(Co,Fe)$O_{3-\delta}$ perovskites are quite contradictory (see Refs. [193–196] and references therein). On the other hand, an extensive Ba^{2+} substitution for Sr^{2+} in the electrode and membrane materials may lead to a fast degradation due to reactions with CO_2 and water vapor present in the air [197, 198].

At oxygen pressures close to atmospheric, perovskite-type $LaNiO_{3-\delta}$ is only stable below 1130–1250 K; further heating leads to the decomposition into K_2NiF_4-type $La_2NiO_{4+\delta}$ and NiO via the separation of Ruddlesden–Popper-type $La_4Ni_3O_{10-\delta}$ and $La_3Ni_2O_{7-\delta}$ phases at intermediate stages [152, 199, 200]. As for $LaNiO_{3-\delta}$, these layered compounds display attractive electrochemical and transport properties, but suffer from an insufficient phase stability in the range of temperatures and oxygen chemical potentials necessary for practical applications [199–203]. Decreasing the Ln^{3+} cation radius and acceptor-type doping both have a negative influence on the stability of the nickelate. The highest stability in ternary Ln-M-O (Ln = La, Pr, Nd; M = Ni, Cu) systems is observed for the K_2NiF_4-type compounds, which attract significant interest due to their moderate thermal and chemical expansion, in combination with a relatively high ionic conductivity. The total conductivity of the K_2NiF_4-type nickelates remains predominantly p-type electronic over the entire P_{O_2} range where these phases exist; the hole transport is lower than that in the perovskite analogues, but still sufficient for practical applications (see Figures 9.11 and 9.13). A moderate A^{2+} doping leads usually to a higher electronic conductivity and a lower oxygen content in $Ln_2NiO_{4+\delta}$, with a negative impact on the oxygen diffusivity which is substantially determined by the interstitial anion migration [28, 206–210]. On the other hand, although the substitution of nickel with higher-valence cations raises the interstitials' concentration, an increase in the ionic transport is only observed at temperatures above 1100 K. Within the operating temperature range of SOFCs and ceramic membranes, maximum ionic conduction in the Ln_2NiO_4-based systems is thus characteristic of the compositions with a modest dopant content, particularly for the parent nickelate phases. The corresponding values of the ionic conductivity and oxygen permeability are 4 to 12-fold lower if compared to highly oxygen-deficient SrCo(Fe)$O_{3-\delta}$ and Sr(La)$CoO_{3-\delta}$ perovskites, but are quite similar to those in (La,Sr)$FeO_{3-\delta}$.

Again, the number of high-conductivity cuprates that are stable under SOFC cathodic conditions is not very large, including primarily the K_2NiF_4-type $(Ln,A)_2CuO_{4\pm\delta}$ (Ln = La–Sm; Ln = Sr, Ba) and various derivatives of layered $LnA_2Cu_3O_{7-\delta}$ formed in the systems with Y^{3+} and smallest lanthanide cations. The total conductivity of $La_2CuO_{4+\delta}$, which exhibits a small oxygen excess in air, has a pseudometallic character and is considerably lower than that of lanthanum nickelate (see Figure 9.11); moderate acceptor-type doping enhances the p-type electronic transport, whereas decreasing the Ln^{3+} size has an opposite effect. The oxygen diffusion and surface exchange in all cuprates are rather slow, and tend to decrease when the Ln^{3+} radius decreases [36, 204, 220–225]. In the case of $La_{2-x}Sr_xCuO_{4\pm\delta}$, the ion diffusivity

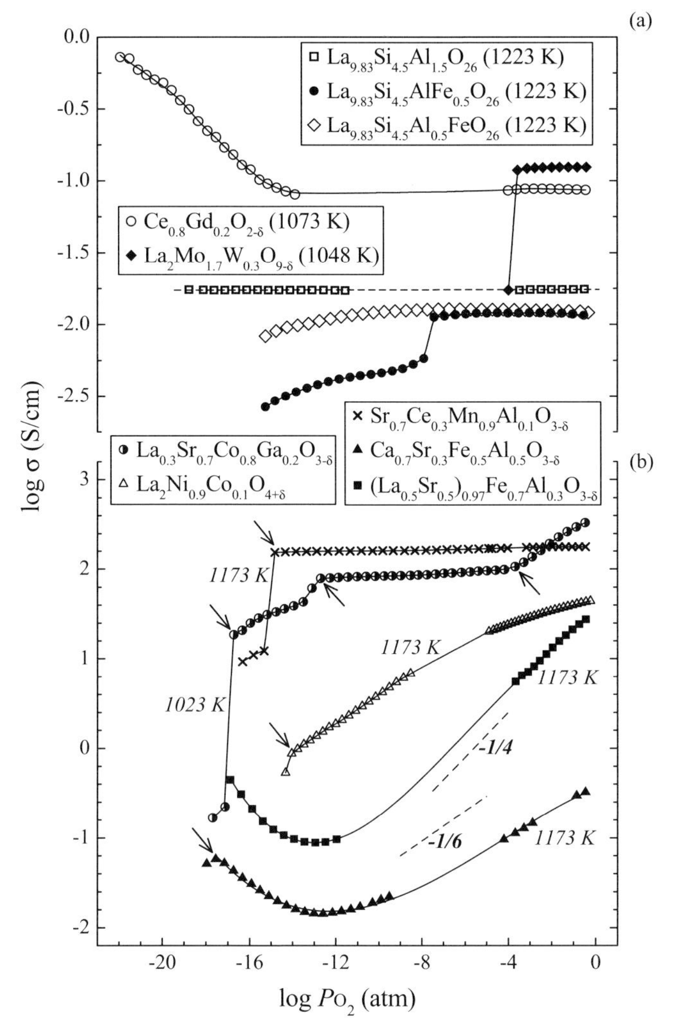

Figure 9.13 Examples of the oxygen partial pressure (Po_2) dependencies of total conductivity of various solid electrolytes and mixed conductors, including apatite-type $La_{9.83}Si_{4.5}Al_{1-x}Fe_xO_{26-\delta}$ [204, 205], $La_2Mo_{1.7}W_{0.3}O_{9-\delta}$ [65], fluorite-type $Ce_{0.8}Gd_{0.2}O_{2-\delta}$, perovskite-like $Sr_{0.7}Ce_{0.3}Mn_{0.9}Al_{0.1}O_{3-\delta}$ [63], $Ca_{0.7}Sr_{0.3}Fe_{0.5}Al_{0.5}O_{3-\delta}$ [67], $(La_{0.5}Sr_{0.5})_{0.97}Fe_{0.7}Al_{0.3}O_{3-\delta}$ [102] and $La_{0.3}Sr_{0.7}Co_{0.8}Ga_{0.2}O_{3-\delta}$ [111], and K_2NiF_4-type $La_2Ni_{0.9}Co_{0.1}O_{4+\delta}$ [206]. The arrows indicate phase decomposition on reduction, except for the right-hand arrow in the σ versus Po_2 curve of $La_{0.3}Sr_{0.7}Co_{0.8}Ga_{0.2}O_{3-\delta}$ which shows the perovskite → brownmillerite transition. All lines are for visual guidance only.

becomes slightly higher on modest Sr doping ($x < 0.1$) and drops with further additions. The substitution of copper with higher-valence cations often has positive effects on the ionic transport, thus indicating the relevance of both vacancy and interstitial migration mechanisms. It should be noted that increasing the oxygen content in another well-known cuprate, $YBa_2Cu_3O_{7-\delta}$, also results in a higher oxygen permeability and ionic conductivity [213].

Finally, it should be noted that numerous perovskite-related materials with relatively low oxygen ionic conductivity at 700–1200 K have been excluded from consideration in this brief survey, but may have potential electrochemical applications in fuel cell anodes, current collectors, sensors, and catalytic reactors. Further information on these applications is available elsewhere [1–4, 11, 159, 217–219].

9.9
$La_2Mo_2O_9$-Based Electrolytes

Another group of oxygen ionic conductors (the so-called LAMOX series) is based on the high-temperature polymorph of lanthanum molybdate, β-$La_2Mo_2O_9$, having a cubic lattice which is isostructural with β-$SnWO_4$ [13, 16, 65, 81, 220–225]. The fast anion migration in β-$La_2Mo_2O_9$ was explained by the lone-pair substitution concept [81, 220, 221] by which lone electron pairs of cations could stabilize oxygen vacancies. In particular, the outer electrons of Sn^{2+} in β-$SnWO_4$ project into a vacant oxygen site, whereas the La^{3+} cations have no lone-pair electrons and half of the sites occupied by lone pairs in β-$SnWO_4$ are vacant in lanthanum molybdate. Whatever the microscopic mechanisms of anion diffusion, the partial substitution of La^{3+} with Bi^{3+} and smaller rare-earth cations (Gd, Dy, Y) and/or the incorporation of V^{5+} or W^{6+} in the molybdenum sublattice can suppress the $\beta \leftrightarrow \alpha$ transition, which normally occurs at approximately 853 K and results from long-range oxygen vacancy ordering. Examples of the cubic solid solutions with highest conductivity and stability in these systems are $La_{2-x}Ln_xMo_{2-y}W_yO_{9-\delta}$ (Ln = Y, Gd; $x = 0.1$–0.3; $y = 0.3$–0.6), $La_{1.7}Bi_{0.3}Mo_2O_{9-\delta}$, and $La_2Mo_{1.95}V_{0.05}O_{9-\delta}$. Figures 9.5–9.7 and 9.13 compare the transport properties of $La_2Mo_2O_9$-based materials and other solid electrolytes. The disadvantages of LAMOX include a relatively high electronic contribution to the total conductivity under ambient conditions, degradation in moderately reducing atmospheres, high TECs, and reactivity with electrode materials [65, 222, 225–228]. These problems are less prominent if compared to the Bi_2O_3-based compositions, but the ionic conductivity of LAMOX is also lower. The electronic conduction in LAMOX, mainly n-type, is comparable to the p-type transport in LSGM; the electron transference numbers are, however, considerably higher for the former family and further increase with temperature and with reducing P_{O_2}.

9.10
Solid Electrolytes with Apatite Structure

Due to non-negligible levels of the oxygen ionic and/or protonic conduction in numerous crystalline and amorphous silicates and germanates free of alkaline metal cations [229–244], the electrochemical and transport phenomena in these materials has became a focus of research interest since the 1930s. However, extensive studies of

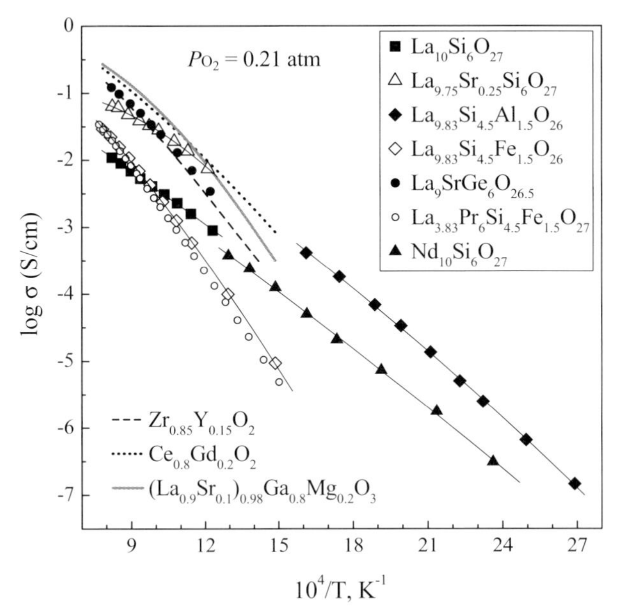

Figure 9.14 Total conductivity of several apatite-type phases in air [204, 205, 247–249]. Data for 8 mol% yttria-stabilized zirconia [36], $Ce_{0.8}Gd_{0.2}O_{2-\delta}$ [40] and $(La_{0.9}Sr_{0.1})_{0.98}Ga_{0.8}Mg_{0.2}O_{3-\delta}$ [66] electrolytes are shown for comparison.

the apatite-type phases $(Ln,A)_{10-x}M_6O_{26\pm\delta}$, where M = Si or Ge, and their derivatives began only during the 1990s [204, 205, 245–257]. The germanate apatites exhibit better ion transport properties compared to their Si-containing analogues, but seem unlikely to find practical applications due to high volatilization, a tendency to glass formation, and the high costs of GeO_2. These disadvantages are less pronounced for the apatite-type silicates, which may be of practical interest owing to low cost and substantial ionic conductivity comparable to that of stabilized zirconia (Figures 9.5, 9.13 and 9.14). The oxygen ionic transport in $Ln_{10}Si_6O_{27}$ (Ln = La, Pr, Nd, Sm, Gd, Dy) increases with increasing radius of Ln^{3+} cations, being maximum for the La-containing phase [245, 246]. Irrespective of the wide scatter among experimental data associated with relatively poor sintering and different processing routes, the highest ionic conductivity is observed for the apatite phases containing more than 26 oxygen ions per unit formula. This tendency suggests a significant role for the interstitial migration mechanism, in agreement with computer simulations [258, 259]. Other important factors which influence oxygen diffusion include the deficiency of Ln-sublattice and Si-site doping, which affect the unit cell volume and may promote the formation of ionic charge carriers via the Frenkel disorder mechanism. Decreasing the oxygen concentration below a stoichiometric value leads to the vacancy mechanism becoming dominant, and there is also a considerable drop in ionic conductivity. Similar effects are typical for transition metal-containing apatites at low oxygen pressures (see

Figure 9.13), although degradation in reducing atmospheres is also contributed by the silicon oxide volatilization and local surface decomposition [253, 254]. The latter factor, and the fast surface diffusion of silica blocking interfacial reactions, result in a poor electrochemical performance of porous electrodes in contact with the silicate solid electrolytes [256, 260].

References

1 Minh, N.Q. and Takahashi, T. (1995) *Science and Technology of Ceramic Fuel Cells*, Elsevier, Amsterdam.

2 Kinoshita, K. (1992) *Electrochemical Oxygen Technology*, Wiley-Interscience, New York, Chichester, Brisbane, Toronto, Singapore.

3 Yamamoto, O. (2000) *Electrochim. Acta*, **45**, 2423–35.

4 Bouwmeester, H.J.M. and Burgraaf, A.J. (1996) in *Fundamentals of Inorganic Membrane Science and Technology* (eds A.J. Burggraaf and L. Cot), Elsevier, Amsterdam-Lausanne-NY-Oxford-Shannon-Tokyo, pp. 435–528.

5 Dyer, P.N., Richards, R.E., Russek, S.L. and Taylor, D.M. (2000) *Solid State Ionics*, **134**, 21–33.

6 Badwal, S.P.S. and Ciacchi, F.T. (2001) *Adv. Mater.*, **13**, 993–6.

7 Fontaine, M.-L., Larring, Y., Norby, T., Grande, T. and Bredesen, R. (2007) *Ann. Chim. Sci. Mat.*, **32**, 197–212.

8 Liu, Y., Tan, X. and Li, K. (2006) *Catal. Rev.*, **48**, 145–98.

9 Etsell, T.H. and Flengas, S.N. (1970) *Chem. Rev.*, **70**, 339–76.

10 Rickert, H. (1982) *Electrochemistry of Solids. An Introduction*, Springer-Verlag, Berlin, Heidelberg, New York.

11 Perfilyev, M.V., Demin, A.K., Kuzin, B.L. and Lipilin, A.S. (1988) *High-Temperature Electrolysis of Gases*, Nauka, Moscow.

12 Kilner, J.A. (2000) *Solid State Ionics*, **129**, 13–23.

13 Goodenough, J.B. (2003) *Annu. Rev. Mater. Res.*, **33**, 91–128.

14 Kharton, V.V., Marques, F.M.B. and Atkinson, A. (2004) *Solid State Ionics*, **174**, 135–49.

15 Mogensen, M., Lybye, D., Bonanos, N., Hendriksen, P.W. and Poulsen, F.W. (2004) *Solid State Ionics*, **174**, 279–86.

16 Fergus, J.W. (2006) *J. Power Sources*, **162**, 30–40.

17 Sunarso, J., Baumann, S., Serra, J.M., Meulenberg, W.A., Liu, S., Lin, Y.S. and Diniz da Costa, J.C. (2008) *J. Membrane Sci.*, **320**, 13–41.

18 Faber, J., Geoffroy, C., Roux, A., Sylvestre, A. and Abelard, P. (1989) *Appl. Phys. A*, **49**, 225–32.

19 Wang, D.Y., Park, D.S., Griffith, J. and Nowick, A.S. (1981) *Solid State Ionics*, **2**, 95–105.

20 Minervini, L., Zacate, M.O. and Grimes, R.W. (1999) *Solid State Ionics*, **116**, 339–49.

21 Zacate, M.O., Minervini, L., Bradfield, D.J., Grimes, R.W. and Sickafus, K.E. (2000) *Solid State Ionics*, **128**, 243–54.

22 Andersson, D.A., Simak, S.I., Skorodumova, N.V., Abrikosov, I.A. and Johansson, B. (2006) *Proc. Natl Acad. Sci. USA*, **103**, 3518–21.

23 Navrotsky, A., Simoncic, P., Yokokawa, H., Chen, W. and Lee, T. (2007) *Faraday Discuss.*, **134**, 171–80.

24 Kharton, V.V., Kovalevsky, A.V., Patrakeev, M.V., Tsipis, E.V., Viskup, A.P., Kolotygin, V.A., Yaremchenko, A.A., Shaula, A.L., Kiselev, E.A. and Waerenborgh, J.C. (2008) *Chem. Mater.*, **20**, 6457–67.

25 Bidrawn, F., Lee, S., Vohs, J.M. and Gorte, R.J. (2008) *J. Electrochem. Soc.*, **155**, B660–5.

26 Stevenson, J.W., Armstrong, T.R., Carneim, R.D., Pederson, L.R. and Weber, W.J. (1996) *J. Electrochem. Soc.*, **143**, 2722–9.

27 Kharton, V.V., Naumovich, E.N., Vecher, A.A. and Nikolaev, A.V. (1995) *J. Solid State Chem.*, **120**, 128–6.

28 Kharton, V.V., Viskup, A.P., Kovalevsky, A.V., Naumovich, E.N. and Marques, F.M.B. (2001) *Solid State Ionics*, **143**, 337–53.

29 Yamazaki, S., Matsui, T., Ohashi, T. and Arita, Y. (2000) *Solid State Ionics*, **136–137**, 913–20.

30 Deguchi, H., Yoshida, H., Inagaki, T. and Horiuchi, M. (2005) *Solid State Ionics*, **176**, 1817–25.

31 Ou, D.R., Mori, T., Ye, F., Zou, J., Auchterlonie, G. and Drennan, J., (2008) *Phys. Rev. B*, **77**, 024108-8.

32 Badwal, S.P.S. (1992) *Solid State Ionics*, **52**, 23–32.

33 Kharton, V.V., Naumovich, E.N. and Vecher, A.A. (1999) *J. Solid State Electrochem.*, **3**, 61–81.

34 Yamamoto, O., Arati, Y., Takeda, Y., Imanishi, N., Mizutani, Y., Kawai, M. and Nakamura, Y. (1995) *Solid State Ionics*, **79**, 137–42.

35 Badwal, S.P.S., Ciacchi, F.T. and Milosevic, D. (2000) *Solid State Ionics*, **136–137**, 91–9.

36 Mori, M., Abe, T., Itoh, H., Yamomoto, O., Takeda, Y. and Kawahara, T. (1994) *Solid State Ionics*, **74**, 157–64.

37 Lee, J.-H. and Yoshimura, M. (1999) *Solid State Ionics*, **124**, 185–91.

38 Steele, B.C.H. (2001) *J. Mater. Sci.*, **36**, 1053–68.

39 Yahiro, H., Eguchi, K. and Arai, H. (1989) *Solid State Ionics*, **36**, 71–5.

40 Kharton, V.V., Viskup, A.P., Figueiredo, F.M., Naumovich, E.N., Yaremchenko, A.A. and Marques, F.M.B. (2001) *Electrochim. Acta*, **46**, 2879–89.

41 van Dijk, M.P., de Vries, K.J. and Burggraaf, A.J. (1985) *Solid State Ionics*, **16**, 211–24.

42 Kramer, S.A. and Tuller, H.L. (1995) *Solid State Ionics*, **82**, 15–23.

43 Yu, T.-H. and Tuller, H.L. (1998) *J. Electroceram.*, **2**, 49–55.

44 Yamaguchi, S., Kobayashi, K., Abe, K., Yamazaki, S. and Iguchi, Y. (1998) *Solid State Ionics*, **113–115**, 393–402.

45 Drennan, J. and Auchterlonie, G. (2000) *Solid State Ionics*, **134**, 75–87.

46 Lee, J.-H., Mori, T., Li, J.-G., Ikegami, T., Komatsu, M. and Haneda, H. (2000) *J. Electrochem. Soc.*, **147**, 2822–9.

47 Yuzaki, A. and Kishimoto, A. (1999) *Solid State Ionics*, **116**, 47–51.

48 Yasuda, I., Matsuzaki, Y., Yamakawa, T. and Koyama, T. (2000) *Solid State Ionics*, **135**, 381–8.

49 Steil, M.C., Fouletier, J., Kleitz, M. and Labrune, P. (1999) *J. Eur. Ceram. Soc.*, **19**, 815–18.

50 Hu, H., Yan, H. and Chen, Z. (2007) *Mater. Sci. Eng. B.*, **145**, 85–90.

51 Han, P. and Worrell, W.L. (1995) *J. Electrochem. Soc.*, **142**, 4235–46.

52 Fagg, D.P., Feighery, A.S.J. and Irvine, J.T.S. (2003) *J. Solid State Chem.*, **172**, 277–87.

53 Kudo, T. and Obayashi, H. (1976) *J. Electrochem. Soc.*, **123**, 415–19.

54 Inaba, H. and Tagawa, H. (1996) *Solid State Ionics*, **83**, 1–16.

55 Goedickemeier, M. and Gauckler, L.J. (1998) *J. Electrochem. Soc.*, **145**, 414–21.

56 Mogensen, M., Sammes, N.M. and Tompsett, G.A. (2000) *Solid State Ionics*, **129**, 63–94.

57 Steele, B.C.H. (2000) *Solid State Ionics*, **129**, 95–110.

58 Kharton, V.V., Figueiredo, F.M., Navarro, L., Naumovich, E.N., Kovalevsky, A.V., Yaremchenko, A.A., Viskup, A.P., Carneiro, A., Marques, F.M.B. and Frade, J.R. (2001) *J. Mater. Sci.*, **36**, 1105–17.

59 Kilner, J.A. (2007) *Faraday Discuss.*, **134**, 9–15.

60 Omar, S., Wachsman, E.D. and Nino, J.C. (2008) *Solid State Ionics*, **178**, 1890–7.

61 Dudek, M., Bogusz, W., Zych, L. and Trybalska, B. (2008) *Solid State Ionics*, **179**, 164–7.

62 Van Herle, J., Horita, T., Kawada, T., Sakai, N., Yokokawa, H. and Dokiya, M. (1996) *Solid State Ionics*, **86–88**, 1255–8.

63 Marozau, I.P., Kharton, V.V., Viskup, A.P., Frade, J.R. and Samakhval, V.V. (2006) *J. Eur. Ceram. Soc.*, **26**, 1371–8.

64 Yaremchenko, A.A., Avdeev, M., Kharton, V.V., Kovalevsky, A.V., Naumovich, E.N. and Marques, F.M.B. (2002) *Mater. Chem. Phys.*, **77**, 552–8.

65 Marozau, I.P., Marrero-Lopez, D., Shaula, A.L., Kharton, V.V., Tsipis, E.V., Nunez, P. and Frade, J.R. (2004) *Electrochim. Acta*, **49**, 3517–24.

66 Kharton, V.V., Shaula, A.L., Vyshatko, N.P. and Marques, F.M.B. (2003) *Electrochim. Acta*, **48**, 1817–28.

67 Shaula, A.L., Kharton, V.V., Vyshatko, N.P., Tsipis, E.V., Patrakeev, M.V., Marques, F.M.B. and Frade, J.R. (2005) *J. Eur. Ceram. Soc.*, **25**, 489–99.

68 Kharton, V.V., Tsipis, E.V., Yaremchenko, A.A., Vyshatko, N.P., Shaula, A.L., Naumovich, E.N. and Frade, J.R. (2003) *J. Solid State Electrochem.*, **7**, 468–76.

69 Atkinson, A. (1997) *Solid State Ionics*, **95**, 249–58.

70 Nauer, M., Ftikos, C. and Steele, B.C.H. (1994) *J. Eur. Ceram. Soc.*, **14**, 493–9.

71 Kang, C.Y., Kusaba, H., Yahoro, H., Sasaki, K. and Teraoka, Y. (2006) *Solid State Ionics*, **177**, 1799–802.

72 Fagg, D.P., Kharton, V.V. and Frade, J.R. (2002) *J. Electroceram.*, **9**, 199–207.

73 Fagg, D.P., Shaula, A.L., Kharton, V.V. and Frade, J.R. (2007) *J. Membrane Sci.*, **299**, 1–7.

74 Takahashi, T., Iwahara, H. and Arao, T. (1975) *J. Appl. Electrochem.*, **5**, 187–195.

75 Sammes, N.M., Tompsett, G.A., Nafe, H. and Aldinger, F. (1999) *J. Eur. Ceram. Soc.*, **19**, 1801–26.

76 Kharton, V.V., Naumovich, E.N., Yaremchenko, A.A. and Marques, F.M.B. (2001) *J. Solid State Electrochem.*, **5**, 160–87.

77 Abraham, F., Boivin, J.C., Mairesse, G. and Nowogrocki, G. (1990) *Solid State Ionics*, **40–41**, 934–7.

78 Vanier, R.N., Mairesse, G., Abraham, F. and Nowogrocki, G. (1994) *Solid State Ionics*, **70–71**, 248–52.

79 Kendall, K.R., Thomas, J.K. and zur Loye, H.-C. (1994) *Solid State Ionics*, **70–71**, 221–4.

80 Jiang, N., Wachsman, E.D. and Jung, S.-H. (2002) *Solid State Ionics*, **150**, 347–53.

81 Goutenoire, F., Isnard, O., Suard, E., Bohnke, O., Laligant, Y., Retoux, R. and Lacorre, Ph. (2000) *J. Mater. Chem.*, **10**, 1–7.

82 Francklin, A.J., Chadwick, A.V. and Couves, J.W. (1994) *Solid State Ionics*, **70–71**, 215–20.

83 Watanabe, A. and Das, K. (2002) *J. Solid State Chem.*, **163**, 224–30.

84 Kramer, S., Spears, M. and Tuller, H.L. (1994) *Solid State Ionics*, **72**, 59–66.

85 Yamaguchi, S., Kobayashi, K., Abe, K., Yamazaki, S. and Iguchi, Y. (1998) *Solid State Ionics*, **113–115**, 393–402.

86 Wuensch, B.J., Eberman, K.W., Heremans, C., Ku, E.M., Onnerud, P., Yeo, E.M.E., Haile, S.M., Stalick, J.K. and Jorgensen, J.D. (2000) *Solid State Ionics*, **129**, 111–33.

87 Pirzada, M., Grimes, R.W., Minervini, L., Maguire, J.F. and Sickafus, K.E. (2001) *Solid State Ionics*, **140**, 201–8.

88 Norby, T. (2001) *J. Mater. Chem.*, **11**, 11–18.

89 Lee, J.-H. and Yoshimura, M. (1999) *Solid State Ionics*, **124**, 185–91.

90 Lee, J.-H., Yashima, M. and Yoshimura, M. (1998) *Solid State Ionics*, **107**, 47–51.

91 Yamamura, H., Matsui, K., Kakinuma, K. and Mori, T. (1999) *Solid State Ionics*, **123**, 279–85.

92 Stevenson, J.W., Hasinska, K., Canfield, N.L. and Armstrong, T.R. (2000) *J. Electrochem. Soc.*, **147**, 3213–18.

93 Nomura, K. and Tanase, S. (1997) *Solid State Ionics*, **98**, 229–36.

94 Lybye, D., Poulsen, F.W. and Mogensen, M. (2000) *Solid State Ionics*, **128**, 91–103.

95 Kharton, V.V., Viskup, A.P., Yaremchenko, A.A., Baker, R.T., Gharbage, B., Mather, G.C., Figueiredo, F.M., Naumovich, E.N. and Marques, F.M.B. (2000) *Solid State Ionics*, **132**, 119–30.

96 Kharton, V.V., Yaremchenko, A.A., Viskup, A.P., Mather, G.C., Naumovich, E.N. and Marques, F.M.B. (2001) *J. Electroceram.*, **7**, 57–66.

97 Yaremchenko, A.A., Shaula, A.L., Logvinovich, D.I., Kharton, V.V., Kovalevsky, A.V., Naumovich, E.N., Frade, J.R. and Marques, F.M.B. (2003) *Mater. Chem. Phys.*, **82**, 684–690.

98 Patrakeev, M.V., Bahteeva, J.A., Mitberg, E.B., Leonidov, I.A., Kozhevnikov, V.L. and Poeppelmeier, K.R. (2003) *J. Solid State Chem.*, **172**, 219–31.

99 Kharton, V.V., Viskup, A.P., Marozau, I.P. and Naumovich, E.N. (2003) *Mater. Lett.*, **57**, 3017–21.

100 Ishigaki, T., Yamauchi, S., Kishio, K., Mizusaki, J. and Fueki, K. (1988) *J. Solid State Chem.*, **73**, 179–87.

101 Ullmann, H., Trofimenko, N., Tietz, F., Stöver, D. and Ahmad-Khanlou, A. (2000) *Solid State Ionics*, **138**, 79–90.

102 Kharton, V.V., Waerenborgh, J.C., Viskup, A.P., Yakovlev, S.O., Patrakeev, M.V., Gaczyński, P., Marozau, I.P., Yaremchenko, A.A., Shaula, A.L. and Samakhval, V.V. (2006) *J. Solid State Chem.*, **179**, 1273–84.

103 Huang, K., Tichy, R.S. and Goodenough, J.B. (1998) *J. Am. Ceram. Soc.*, **81**, 2565–75.

104 Ishihara, T., Matsuda, H. and Takita, Y. (1994) *J. Am. Chem. Soc.*, **116**, 3801–3.

105 Stevenson, J.W., Armstrong, T.R., McGready, D.E., Pederson, L.R. and Weber, W.J. (1997) *J. Electrochem. Soc.*, **144**, 3613–20.

106 Hayashi, H., Inaba, H., Matsuyama, M., Lan, N.G., Dokiya, M. and Tagawa, H. (1999) *Solid State Ionics*, **122**, 1–15.

107 Huang, P. and Petric, A. (1996) *J. Electrochem. Soc.*, **143**, 1644–48.

108 Drennan, J., Zelizko, V., Hay, D., Ciacci, F.T., Rajendran, S. and Badwal, S.P. (1997) *J. Mater. Chem.*, **7**, 79–83.

109 Ranlov, J., Mogensen, M. and Poulsen, F.W. (1993) in *Proceedings, 14th Riso International Symposium on Materials Science* (eds F.W. Poulsen, J.J. Bentzen, T. Jacobsen, E. Skou and M.J.L. Ostergard), Riso National Laboratory, Roskilde, pp. 389–98.

110 Nguen, T.L., Dokiya, M., Wang, S., Tagawa, H. and Hashimoto, T. (2000) *Solid State Ionics*, **130**, 229–41.

111 Kharton, V.V., Tsipis, E.V., Yaremchenko, A.A., Marozau, I.P., Viskup, A.P., Frade, J.R. and Naumovich, E.N. (2006) *Mater. Sci. Eng. B*, **134**, 80–8.

112 Ishihara, T., Akbay, T., Furutani, H. and Takita, Y. (1998) *Solid State Ionics*, **113–115**, 585–91.

113 Trofimenko, N. and Ullmann, H. (1999) *Solid State Ionics*, **118**, 215–27.

114 Lee, S., Lee, K.S., Woo, S.K., Kim, J.W., Ishihara, T. and Kim, D.K. (2003) *Solid State Ionics*, **158**, 287–96.

115 Politova, E.D., Aleksandrovskii, V.V., Zaitsev, S.V., Kaleva, G.M., Mosunov, A.V., Stefanovich, S.Y., Avetisov, A.K., Suvorkin, S.V., Kosarev, G.V., Sukhareva, I.P., Sung, J.S., Choo, K.Y. and Kim, T.H. (2006) *Mater. Sci. Forum*, **514–516**, 412–16.

116 Juste, E., Julian, A., Etchegoyen, G., Geffroy, P.M., Chartier, T., Richet, N. and Del Gallo, P. (2008) *J. Membrane Sci.*, **319**, 185–91.

117 Kharton, V.V., Yaremchenko, A.A., Patrakeev, M.V., Naumovich, E.N. and Marques, F.M.B. (2003) *J. Eur. Ceram. Soc.*, **23**, 1417–26.

118 Mizusaki, J., Yasuda, I., Shimoyama, J., Yamauchi, S. and Fueki, K. (1993) *J. Electrochem. Soc.*, **140**, 467–71.

119 Anderson, P.S., Marques, F.M.B., Sinclair, D.C. and West, A.R. (1999) *Solid State Ionics*, **118**, 229–39.

120 He, H., Huang, X. and Chen, L. (2000) *Solid State Ionics*, **130**, 183–93.

121 Yamamura, H., Yamazaki, K., Kakinuma, K. and Nomura, K. (2002) *Solid State Ionics*, **150**, 255–61.

122 Thangadurai, V. and Weppner, W. (2001) *J. Electrochem. Soc.*, **148**, A1294–301.

123 Fujii, H., Katayama, Y., Shimura, T. and Iwahara, H. (1998) *J. Electroceram.*, **2**, 119–25.

124 Kato, H. and Yugami, H. (2008) *Electrochemistry*, **76**, 334–7.

125 Goodenough, J.B., Ruiz-Dias, J.E. and Zhen, Y.S. (1990) *Solid State Ionics*, **44**, 21–31.

126 Goodenough, J.B. (1997) *Solid State Ionics*, **94**, 17–25.

127 Manthiram, A., Kuo, J.F. and Goodenough, J.B. (1993) *Solid State Ionics*, **62**, 225–34.

128 Zhang, G.B. and Smyth, D.M. (1995) *Solid State Ionics*, **82**, 153–60.

129 Schober, T. (1998) *Solid State Ionics*, **109**, 1–11.

130 Hashimoto, T., Inagaki, Y., Kishi, A. and Dokiya, M. (2000) *Solid State Ionics*, **128**, 227–31.

131 Rolle, A., Vannier, R.N., Giridharan, N.V. and Abraham, F. (2005) *Solid State Ionics*, **176**, 2095–103.

132 Rolle, A., Fafilek, G. and Vannier, R.N. (2008) *Solid State Ionics*, **179**, 113–119.

133 Shaula, A.L., Pivak, Y.V., Waerenborgh, J.C., Gaczyński, P., Yaremchenko, A.A. and Kharton, V.V. (2006) *Solid State Ionics*, **177**, 2923–30.

134 Patrakeev, M.V., Leonidov, I.A., Kozhevnikov, V.L. and Kharton, V.V. (2004) *Solid State Sci.*, **6**, 907–13.

135 Omata, T., Fuke, T. and Otsuka-Yao-Matsuo, S. (2006) *Solid State Ionics*, **177**, 2447–51.

136 Adler, S.B. (2004) *Chem. Rev.*, **104**, 4791–843.

137 Fleig, J. (2003) *Annu. Rev. Mater. Res.*, **33**, 361–82.

138 Tsipis, E.V. and Kharton, V.V. (2008) *J. Solid State Electrochem.*, **12**, 1367–91.

139 Huang, X., Liu, J., Lu, Z., Liu, W., Pei, L., He, T., Liu, Z. and Su, W. (2000) *Solid State Ionics*, **130**, 195–201.

140 Simner, S.P., Shelton, J.P., Anderson, M.D. and Stevenson, J.W. (2003) *Solid State Ionics*, **161**, 11–18.

141 Petric, A., Huang, P. and Tietz, F. (2000) *Solid State Ionics*, **135**, 719–25.

142 Kostogloudis, G.C. and Ftikos, C. (1999) *Solid State Ionics*, **126**, 143–51.

143 Chiba, R., Yoshimura, F. and Sakurai, Y. (2002) *Solid State Ionics*, **152–153**, 575–82.

144 Boehm, E., Bassat, J.-M., Steil, M.C., Dordor, P., Mauvy, F. and Grenier, J.-C. (2003) *Solid State Sci.*, **5**, 973–81.

145 Kharton, V.V., Tsipis, E.V., Marozau, I.P., Viskup, A.P., Frade, J.R. and Irvine, J.T.S. (2007) *Solid State Ionics*, **178**, 101–13.

146 Kharton, V.V., Kovalevsky, A.V., Tsipis, E.V., Viskup, A.P., Naumovich, E.N., Jurado, J.R. and Frade, J.R. (2002) *J. Solid State Electrochem.*, **7**, 30–6.

147 Tsipis, E.V., Patrakeev, M.V., Kharton, V.V., Vyshatko, N.P. and Frade, J.R. (2002) *J. Mater. Chem.*, **12**, 3738–45.

148 Jiang, S.P. (2006) *J. Solid State Electrochem.*, **11**, 93–102.

149 Hammouche, A., Siebert, E., Hammou, A., Kleitz, M. and Caneiro, A. (1991) *J. Electrochem. Soc.*, **138**, 1212–16.

150 Lee, H.Y., Cho, W.S., Oh, S.M., Wiemhöfer, H.D. and Göpel, W. (1995) *J. Electrochem. Soc.*, **142**, 2659–64.

151 Kharton, V.V., Nikolaev, A.V., Naumovich, E.N. and Vecher, A.A. (1995) *Solid State Ionics*, **81**, 201–9.

152 Kharton, V.V., Yaremchenko, A.A. and Naumovich, E.N. (1999) *J. Solid State Electrochem.*, **3**, 303–26.

153 Yasumoto, K., Mori, N., Mizusaki, J., Tagawa, H. and Dokiya, M. (2001) *J. Electrochem. Soc.*, **148**, A105–11.

154 Sakaki, Y., Takeda, Y., Kato, A., Imanishi, N., Yamamoto, O., Hattori, M., Iio, M. and Esaki, Y. (1999) *Solid State Ionics*, **118**, 187–94.

155 Schachtner, R., Ivers-Tiffée, E., Weppner, W., Männer, R. and Wersing, W. (1995) *Ionics*, **1**, 63–9.

156 Huebner, W., Reed, D.M. and Anderson, H.U. (1997) *SOFC V*, in (eds U. Stimming, S.C. Singhal, H. Tagawa and W. Lehnert), The Electrochemical Society, Pennington, NJ, pp. 411–20.

157 Oleynikov, N.N. and Ketsko, V.A. (2004) *Russ. J. Inorg. Chem.*, **49**, 1–21.

158 McCammon, C. (1996) *Phase Transitions*, **58**, 1–26.

159 Gauckler, L.J., Beckel, D., Buergler, B.E., Jud, E., Muecke, U.P., Prestat, M., Rupp, L.J.M. and Richter, J. (2004) *Chimia*, **58**, 837–50.

160 Mai, A., Haanappel, V.A.C., Uhlenbruck, S., Tietz, F. and Stöver, D. (2005) *Solid State Ionics*, **176**, 1341–50.

161 Lashtabeg, A. and Skinner, S.J. (2006) *J. Mater. Chem.*, **16**, 3161–70.

162 Kharton, V.V., Yaremchenko, A.A., Shaula, A.L., Viskup, A.P., Marques, F.M.B., Frade, J.R., Naumovich, E.N., Casanova, J.R. and Marozau, I.P. (2004) *Defect Diffus. Forum*, **226–228**, 141–60.

163 Ramos, T. and Atkinson, A. (2004) *Solid State Ionics*, **170**, 275–86.

164 Fisher, C.A.J. and Islam, M.S. (2005) *J. Mater. Chem.*, **15**, 3200–7.

165 Markov, A.A., Patrakeev, M.V., Kharton, V.V., Pivak, Y.V., Leonidov, I.A. and Kozhevnikov, V.L. (2007) *Chem. Mater.*, **19**, 3980–7.

166 Tsipis, E.V., Patrakeev, M.V., Kharton, V.V., Yaremchenko, A.A., Mather, G.C., Shaula, A.L., Leonidov, I.A., Kozhevnikov, V.L. and Frade, J.R. (2005) *Solid State Sci.*, **7**, 355–65.

167 Jennings, A.J., Skinner, S.J. and Helgason, O. (2003) *J. Solid State Chem.*, **175**, 207–17.

168 Al Daroukh, M., Vashook, V.V., Ullmann, H., Tietz, F. and Arual Raj, I. (2003) *Solid State Ionics*, **158**, 141–50.

169 Kharton, V.V., Sobyanin, V.A., Belyaev, V.D., Semin, G.L., Veniaminov, S.A., Tsipis, E.V., Yaremchenko, A.A., Valente, A.A., Marozau, I.P., Frade, J.R. and Rocha, J. (2004) *Catal. Commun.*, **5**, 311–16.

170 Kharton, V.V., Tsipis, E.V., Naumovich, E.N., Thursfield, A., Patrakeev, M.V., Kolotygin, V.A., Waerenborgh, J.C. and Metcalfe, I.S. (2008) *J. Solid State Chem.*, **181**, 1425–33.

171 Kharton, V.V., Viskup, A.P., Kovalevsky, A.V., Naumovich, E.N. and Marques, F.M.B. (2001) *Solid State Ionics*, **143**, 337–53.

172 Tsipis, E.V., Kiselev, E.A., Kolotygin, V.A., Waerenborgh, J.C., Cherepanov, V.A. and Kharton, V.V. (2008) *Solid State Ionics*, **179**, 2170–80.

173 Steele, B.C.H. (1995) *Solid State Ionics*, **75**, 157–165.

174 Merkle, R., Maier, J. and Bouwmeester, H.J.M. (2004) *Angew. Chem. Int. Ed.*, **43**, 5069–73.

175 De Souza, R.A. (2006) *Phys. Chem. Chem. Phys.*, **8**, 890–7.

176 Lein, H.L., Wiik, K. and Grande, T. (2006) *Solid State Ionics*, **177**, 1795–8.

177 Søgaard, M., Hendriksen, P.V. and Mogensen, M. (2007) *J. Solid State Chem.*, **180**, 1489–503.

178 Prado, F., Gurunathan, K. and Manthiram, A. (2002) *J. Mater. Chem.*, **12**, 2390–5.

179 Petrov, A.N., Cherepanov, V.A. and Zuev, A.Yu. (2006) *J. Solid State Electrochem.*, **10**, 517–37.

180 Taskin, A.A., Lavrov, A.N. and Ando, Y. (2005) *Appl. Phys. Lett.*, **86**, 1–3.

181 Tarascon, A., Skinner, S.J., Chater, R.J., Hernandez-Ramirez, F. and Kilner, J.A. (2007) *J. Mater. Chem.*, **17**, 3175–81.

182 Lee, K.T. and Manthiram, A. (2006) *Chem. Mater.*, **18**, 1621–6.

183 Kawada, T., Sase, M., Kudo, M., Yashiro, K., Sato, K., Mizusaki, J., Sakai, N., Horita, T., Yamaji, K. and Yokokawa, H. (2006) *Solid State Ionics*, **177**, 3081–6.

184 Tsipis, E.V., Kharton, V.V. and Frade, J.R. (2006) *Solid State Ionics*, **177**, 1823–6.

185 Bucher, E., Sitte, W., Rom, I., Rapst, I., Grogger, W. and Hofer, F. (2002) *Solid State Ionics*, **152–153**, 417–21.

186 van Doorn, R.H.E. and Burggraaf, A.J. (2000) *Solid State Ionics*, **128**, 65–78.
187 Svarcova, S., Wiik, K., Tolchard, J., Bouwmeester, H.J.M. and Grande, T. (2008) *Solid State Ionics*, **178**, 1787–91.
188 Vente, J.F., McIntosh, S., Haije, W.G. and Bouwmeester, H.J.M. (2006) *J. Solid State Electrochem.*, **10**, 581–8.
189 Ishihara, T., Fukui, S., Nishiguchi, H. and Takita, Y. (2002) *Solid State Ionics*, **152–153**, 609–13.
190 Chang, A., Skinner, S.J. and Kilner, J.A. (2006) *Solid State Ionics*, **177**, 2009–11.
191 Baumann, F.S., Fleig, J., Cristiani, G., Stuhlhofer, B., Habermeier, H.U. and Maier, J. (2007) *J. Electrochem. Soc.*, **154**, B931–41.
192 Peña-Martinez, J., Marrero-Lopez, D., Perez-Coll, D., Ruiz-Morales, J.C. and Nuñez, P. (2007) *Electrochim. Acta*, **52**, 2950–8.
193 Shao, Z., Xiong, G., Tong, J., Dong, H. and Yang, W. (2001) *Sep. Purif. Technol.*, **25**, 419–29.
194 Vente, J.F., Haije, W.G. and Rak, Z.S. (2006) *J. Membrane Sci.*, **276**, 178–84.
195 Zeng, P., Chen, Z., Zhou, W., Gu, H., Shao, Z. and Liu, S. (2007) *J. Membrane Sci.*, **291**, 148–56.
196 Fisher, C.A.J., Yoshiya, M., Iwamoto, Y., Ishii, J., Asanuma, M. and Yabuta, K. (2007) *Solid State Ionics*, **177**, 3425–31.
197 Yan, A., Cheng, M., Dong, Y., Yang, W., Maragou, V., Song, S. and Tsiarakas, P. (2006) *Appl. Catal. B*, **66**, 64–71.
198 Arnold, M., Wang, H. and Feldhoff, A. (2007) *J. Membrane Sci.*, **293**, 44–52.
199 Zinkevich, M. and Aldinger, F. (2004) *J. Alloys Compd.*, **375**, 147–61.
200 Bannikov, D.O. and Cherepanov, V.A. (2006) *J. Solid State Chem.*, **179**, 2721–7.
201 Amow, G. and Skinner, S.J. (2006) *J. Solid State Electrochem.*, **10**, 538–46.
202 Amow, G., Davidson, I.J. and Skinner, S.J. (2006) *Solid State Ionics*, **177**, 1205–10.
203 Tsipis, E.V., Patrakeev, M.V., Waerenborgh, J.C., Pivak, Y.V., Markov, A.A., Gaczyński, P., Naumovich, E.N. and Kharton, V.V. (2007) *J. Solid State Chem.*, **180**, 1902–10.
204 Yaremchenko, A.A., Shaula, A.L., Kharton, V.V., Waerenborgh, J.C., Rojas, D.P., Patrakeev, M.V. and Marques, F.M.B. (2004) *Solid State Ionics*, **171**, 51–9.
205 Kharton, V.V., Shaula, A.L., Patrakeev, M.V., Waerenborgh, J.C., Rojas, D.P., Vyshatko, N.P., Tsipis, E.V., Yaremchenko, A.A. and Marques, F.M.B. (2004) *J. Electrochem. Soc.*, **151**, A1236–46.
206 Kharton, V.V., Yaremchenko, A.A., Shaula, A.L., Patrakeev, M.V., Naumovich, E.N., Logvinovich, D.I., Frade, J.R. and Marques, F.M.B. (2004) *J. Solid State Chem.*, **177**, 26–37.
207 Vashook, V., Yushkevich, I., Kokhanovsky, L., Makhnach, L., Tolochko, S., Kononyuk, I., Ullmann, H. and Altenburg, H. (1999) *Solid State Ionics*, **119**, 23–30.
208 Skinner, S.J. and Kilner, J.A. (2000) *Solid State Ionics*, **135**, 709–12.
209 Cleave, A.R., Kilner, J.A., Skinner, S.J., Murphy, S.T. and Grimes, R.W. (2008) *Solid State Ionics*, **179**, 823–6.
210 Tsipis, E.V., Naumovich, E.N., Shaula, A.L., Patrakeev, M.V., Waerenborgh, J.C. and Kharton, V.V. (2008) *Solid State Ionics*, **179**, 57–60.
211 Routbort, J.L. and Rothman, S.J. (1994) *J. Appl. Phys.*, **76**, 5616–28.
212 Opila, E.J., Tuller, H.L., Wuensch, B.J. and Maier, J. (1993) *J. Am. Ceram. Soc.*, **76**, 2363–9.
213 Patrakeev, M.V., Leonidov, I.A., Kozhevnikov, V.L., Tsidilkovskii, V.I., Demin, A.K. and Nikolaev, A.V. (1993) *Solid State Ionics*, **66**, 61–67.
214 Mozhaev, A.P., Mazo, G.N., Galkin, A.A. and Khramova, N.V. (1996) *Russ. J. Inorg. Chem.*, **41**, 881–5.
215 Bochkov, D.M., Kharton, V.V., Kovalevsky, A.V., Viskup, A.P. and Naumovich, E.N. (1999) *Solid State Ionics*, **120**, 281–8.
216 Mazo, G.N., Savvin, S.N., Abakumov, A.M., Hadermann, J., Dobrovol'skii,

Yu.A. and Leonova, L.S. (2007) *Russ. J. Electrochem.*, **43**, 436–42.

217 Singhal, S.C. (2000) *Solid State Ionics*, **135**, 305–13.

218 Fergus, J.W. (2006) *Solid State Ionics*, **177**, 1529–41.

219 Fergus, J.W. (2007) *Sens. Actuators B*, **123**, 1169–79.

220 Lacorre, P., Goutenoire, F., Bohnke, O., Retoux, R. and Laligant, Y. (2000) *Nature*, **404**, 856–8.

221 Goutenoire, F., Isnard, O. and Lacorre, P. (2000) *Chem. Mater.*, **12**, 2575.

222 Georges, S., Goutenoire, F., Laligant, Y. and Lacorre, P. (2003) *J. Mater. Chem.*, **13**, 2317–21.

223 Basu, S., Devi, P.S. and Maiti, H.S. (2004) *Appl. Phys. Lett.*, **85**, 3486–8.

224 Georges, S., Skinner, S.J., Lacorre, P. and Steil, M.C. (2004) *Dalton Trans.*, **19**, 3101–5.

225 Jin, T.-Y., Madhava Rao, M.V., Cheng, C.L., Tsai, D.S. and Hung, M.H. (2007) *Solid State Ionics*, **178**, 367–74.

226 Subasri, R., Näfe, H. and Aldinger, F. (2003) *Mater. Res. Bull.*, **38**, 1965–77.

227 Marrero-Lopez, D., Peña-Martinez, J., Ruiz-Morales, J.C., Perez-Coll, D., Martin-Sedeno, M.C. and Nuñez, P. (2007) *Solid State Ionics*, **178**, 1366–78.

228 Georges, S. and Salaun, M. (2008) *Solid State Ionics*, **178**, 1898–906.

229 Bockris, J.O'M., Kitchener, J.A., Ignatowicz, S. and Tomlison, J.W. (1948) *Discuss. Faraday Soc.*, **4**, 265–81.

230 Schwartz, M. and Mackenzie, J.D. (1966) *J. Am. Ceram. Soc.*, **49**, 582–5.

231 Evstropiev, K.K. and Petrovskii, G.T. (1978) *Dokl. Akad. Nauk SSSR*, **241**, 1334–6.

232 Doi, A. (1983) *Jpn. J. Appl. Phys.*, **22**, 228–32.

233 Raveau, B. (1986) *J. Chem. Sci.*, **96**, 419–48.

234 Litovchenko, A.S., Ishutina, O.D. and Kalinichenko, A.M. (1991) *Phys. Stat. Sol. A*, **123**, k57–9.

235 Lee, W.K., Lim, B.S., Liu, J.F. and Nowick, A.S. (1992) *Solid State Ionics*, **53–56**, 831–6.

236 Nogami, M., Miyamura, K. and Abe, Y. (1997) *J. Electrochem. Soc.*, **144**, 2175–8.

237 Liebau, F. (1999) *Angew. Chem. Int. Ed.*, **38**, 1733–7.

238 Desaglois, F., Follet-Houttemane, C. and Boivin, J.C. (2000) *J. Mater. Chem.*, **10**, 1673–7.

239 Jacob, S., Javornizky, J., Wolf, G.H. and Angel, C.A. (2001) *Int. J. Inorg. Mater.*, **3**, 241–51.

240 Martin-Sedeno, M.C., Marrero-Lopez, D., Losilla, E.R., Bruque, S., Nuñez, P. and Aranda, M.A.G. (2006) *J. Solid State Chem.*, **179**, 3445–55.

241 Aoki, Y., Muto, E., Onoue, S., Nakao, A. and Kunitake, T. (2007) *Chem. Commun.*, **23**, 2396–98.

242 Porras-Vazquez, J.M., De la Torre, A.G., Losilla, E.R. and Aranda, M.A.G. (2007) *Solid State Ionics*, **178**, 1073–80.

243 Kuang, X., Green, M.A., Niu, H., Zajdel, P., Dickinson, C., Claridge, J.B., Jantsky, L. and Rosseinsky, M.J. (2008) *Nature Mater.*, **7**, 498–504.

244 Leon-Reina, L., Porras-Vazquez, J.M., Losilla, E.R., Moreno-Real, L. and Aranda, M.A.G. (2008) *J. Solid State Chem.*, **181**, 2501–6.

245 Nakayama, S., Kageyama, T., Aono, H. and Sadaoka, Y. (1995) *J. Mater. Chem.*, **5**, 1801–5.

246 Nakayama, S. and Sakamoto, M. (1998) *J. Eur. Ceram. Soc.*, **18**, 1413–8.

247 Nakayama, S., Sakamoto, M., Higuchi, M., Kodaira, K., Sato, M., Kakita, S., Suzuki, T. and Itoh, K. (1999) *J. Eur. Ceram. Soc.*, **19**, 507–510.

248 Abram, E.J., Sinclair, D.C. and West, A.R. (2001) *J. Mater. Chem.*, **11**, 1978–9.

249 Arikawa, H., Nishiguchi, H., Ishihara, T. and Takita, Y. (2000) *Solid State Ionics*, **136–137**, 31–7.

250 Tolchard, J.R., Islam, M.S. and Slater, P.R. (2003) *J. Mater. Chem.*, **13**, 1956–61.

251 Sansom, J.E.H., Najib, A. and Slater, P.R. (2004) *Solid State Ionics*, **175**, 353–5.

252 Yoshioka, H. and Tanase, S. (2005) *Solid State Ionics*, **176**, 2395–8.

253 Shaula, A.L., Kharton, V.V. and Marques, F.M.B. (2005) *J. Solid State Chem.*, **178**, 2050–61.

254 Brisse, A., Sauvet, A.L., Barthet, C., Georges, S. and Fouletier, J. (2007) *Solid State Ionics*, **178**, 1337–43.

255 Vincent, A., Savignat, S.B. and Gernais, F. (2007) *J. Eur. Ceram. Soc.*, **27**, 1187–92.

256 Tsipis, E.V., Kharton, V.V. and Frade, J.R. (2007) *Electrochim. Acta*, **52**, 4428–35.

257 Bechade, E., Julien, I., Iwata, T., Masson, O., Thomas, P., Champion, E. and Fukuda, K. (2008) *J. Eur. Ceram. Soc.*, **28**, 2717–24.

258 Tolchard, R., Slater, P.R. and Islam, M.S. (2007) *Adv. Funct. Mater.*, **17**, 2564–71.

259 Kendrick, E., Islam, M.S. and Slater, P.R. (2008) *Chem. Commun.*, **6**, 715–7.

260 Brisse, A., Sauvet, A.L., Barthet, C., Beaudet-Savignat, S. and Fouletier, J. (2006) *Fuel Cells*, **6**, 59–63.

10
Polymer and Hybrid Materials: Electrochemistry and Applications

Danmin Xing and Baolian Yi

Abstract

Due to substantial practical interest in proton-exchange membrane fuel cells (PEMFCs) during the past few decades, significant improvements have been achieved in polymer and hybrid materials. Much of the research effort has centered on the development of composite membranes and new proton-conducting materials, aimed to improving their performance, durability, and cost. Attention has also been focused on new membranes operating under high-temperature/low-humidity conditions. This chapter presents a brief review on the motivation for novel proton-exchange membrane developments, together with details of recent progress and technology gaps.

10.1
Introduction

The fuel cell is an electrochemical device that directly converts the chemical energy of a fuel and an oxidant to electrical energy. Compared with the internal combustion engine (ICE), fuel cells have many advantages, including a high energy conversion efficiency (40–50%), a high power density, no pollution, and low noise. Hence, fuel cells are considered to be the new form of "green" power sources during the twenty-first century, and undoubtedly will achieve many promising applications in areas such as stationary and mobile power sources, electric vehicles, submarines and spaceflight [1–4].

The wide variety of fuel cells that have been developed during the past few decades can be classified according to their operating temperature. For example, fuel cells can be classified as low-temperature (<100 °C), medium-temperature (100–300 °C), or high-temperature (600–1000 °C). When considering the electrolyte types, fuel cells can be grouped into the following classes: alkaline fuel cell (AFC); phosphoric acid fuel cell (PAFC); molten carbonate fuel cell (MCFC); solid oxide fuel cell (SOFC); and proton-exchange membrane fuel cell (PEMFC). Detailed descriptions of these fuel cells are available in Ref. [5].

Solid State Electrochemistry I: Fundamentals, Materials and their Applications. Edited by Vladislav V. Kharton

ISBN: 978-3-527-32318-0

The PEMFC has a solid ionomer membrane as the electrolyte, and a platinum, carbon-supported Pt or Pt-based alloy as the electrocatalyst. Within the cell, the fuel is oxidized at the anode and the oxidant reduced at the cathode. As the solid proton-exchange membrane (PEM) functions as both the cell electrolyte and separator, and the cell operates at a relatively low temperature, issues such as sealing, assembly, and handling are less complex than with other fuel cells. The PEMFC has also a number of other advantages, such as a high power density, a rapid low-temperature start-up, and zero emission. With highly promising prospects in both civil and military applications, PEMFCs represent an ideal future alternative power source for electric vehicles and submarines [6].

The PEM, which is a key component of the PEMFC, has dual functions of: (i) conducting the protons; and (ii) separating the anode and cathode chambers. The PEM is also critically important with regards to the performance and cost of a PEMFC. In this chapter, attention is focused on the main issues and progress in polymer and hybrid materials for PEMFC applications.

10.2 Fundamentals

10.2.1 The Proton-Exchange Membrane Fuel Cell (PEMFC)

The PEMFC is also referred to as a polymer electrolyte fuel cell or solid polymer electrolyte fuel cell. A schematic drawing of a single PEMFC is shown in Figure 10.1. In this system, hydrogen fuel supplied to the anode reacts electrochemically at the electrode

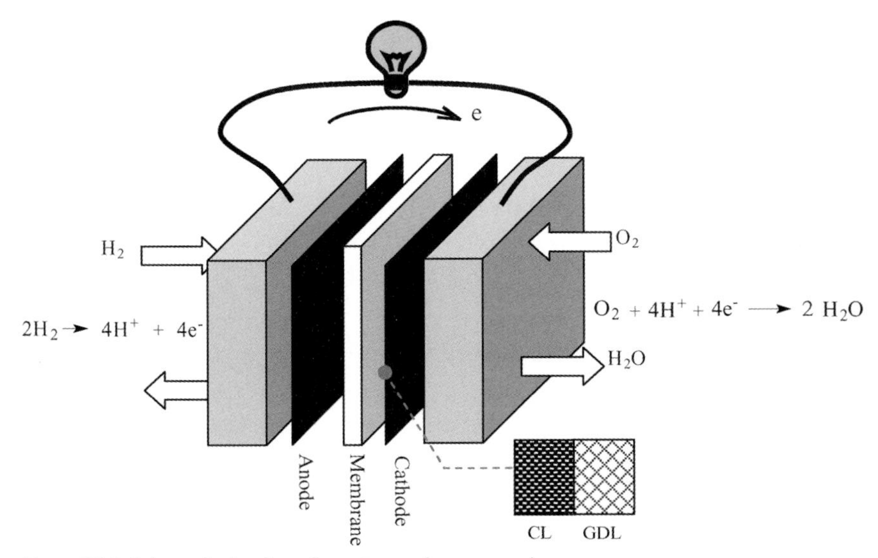

Figure 10.1 Schematic drawing of a proton-exchange membrane fuel cell. CL = catalyst layer; GDL = gas diffusion layer.

to form hydrogen ions (protons) and electrons. The protons are then transferred through the polymer electrolyte membrane to the cathode, while electron transport occurs through an external load – such as the motor of an electric vehicle – to the cathode. At the cathode of the PEMFC, the oxygen combines with hydrogen ions and electrons to form water. Thus, the reactions that occur in a PEMFC can be represented as:

- At the anode:

$$2H_2 \rightarrow 4H^+ + 4e^- \tag{10.1}$$

- At the cathode:

$$O_2 + 4H^+ + 4e^- \rightarrow 2H_2O \tag{10.2}$$

- Overall:

$$2H_2 + O_2 \rightarrow 2H_2O \tag{10.3}$$

Typically, carbon electrodes with a platinum electrocatalyst are used for both the anode and cathode. The PEM used in the fuel cell serves to keep the fuel separate from the oxidant, and also as an electrolyte.

The PEMFCs not only generate power with a high power density but also have a low operating temperature. This is especially advantageous as the cell can be started quickly under ambient conditions, especially when pure hydrogen fuel is available. As the only liquid present in the PEMFC is the water formed at the cathode side, issues such as sealing, assembly and handling are less complex than with other fuel cells. One disadvantage, however, is that a platinum catalyst is required to promote the electrochemical reaction. When reformed hydrocarbon fuels which typically contain carbon monoxide (CO) as a contaminant are used, the ppm levels of CO formed will be tolerated by the platinum catalyst at 80 °C.

Another issue for PEMFCs is that of water management. Whereas, dehydration of the membrane leads to a reduction in proton conductivity, an excess of water may lead to the electrode being flooded.

An additional, but critical, disadvantage of PEMFCs is their high cost, this being associated primarily with the platinum catalyst and fluorinated polymer electrolyte membrane.

10.2.2 Proton-Exchange Membranes for Fuel Cells

The PEM, which serves as both the electrolyte and reactant gas separator, represents the core component of the PEMFC. When serving as the electrolyte, the PEM transfers hydrogen ions from the anode to the cathode, whereas when serving as the reactant gas separator it prevents the fuel and the oxidizer from mixing. Consequently, the key requirements for the PEM will include:

- A high proton conductivity for maximum cell performance.
- A low permeability to the reactant gases, so as to maximize the coulombic efficiency.

- An excellent thermal and chemical stability, for PEMFC reliability and durability.
- An adequate mechanical strength and dimensional stability to ensure reliability in both assembly and operation.
- A low cost, to aid commercialization.

10.2.3 Membrane Characterization

The performance of PEMs can be characterized by either their *electrochemical parameters*, which include their equivalent weight (EW) and proton conductivity, or by their *physical properties*, which include thickness, gas permeability, mechanical strength, water uptake, and swelling, and so on.

10.2.3.1 Electrochemical Parameters

The EW (expressed as $g\,mol^{-1}$) describes the weight of dry ion-exchange resin per mole unit of ion-exchange groups, and can be determined via titration [7]. It is the reciprocal parameter with respect to the ion-exchange capacity (IEC; expressed as $mEq.\,g^{-1}$), and reflects the acid concentration in the PEM. In general, for PEMs with identical chemical structures, those with a small EW value possess a high proton conductivity.

The proton conductivity (σ; expressed as $S\,cm^{-1}$) is the most important property of the PEMFC membranes for minimizing cell resistance and maximizing cell efficiency. In general, the proton conductivities of sulfonated PEMs tend to increase with the acid group concentration and hydration level.

10.2.3.2 Physical Properties

Because the ionic resistance of the PEM is proportional to its thickness, this parameter is extremely important for cell performance [8]. It is necessary to minimize membrane thickness, while maintaining an acceptable mechanical strength.

Another important parameter of the PEM is that of *gas permeability*. The active oxygen species are produced in the membrane when oxygen transfers from cathode to anode, which in turn leads to membrane degradation. On the other hand, if the fuel gas is able to diffuse to the cathode and react chemically with oxygen, this will cause efficiency losses. Therefore, reactant gas leakage will results in a reduced cell performance in all cases.

During the course of membrane electrode assembly (MEA), manufacture and PEMFC operation, the membranes are exposed to the impacts of temperature, humidity, and pressure. Consequently, it is important that the membranes possess a good mechanical stability, and in particular a high mechanical strength and minimal swelling.

10.2.3.3 Evaluation of Durability

High durability is a major requirement for PEMFCs when used commercially. For example, a typical lifetime requirement for a stationary application would be 40 000 h, but for transportation purposes this would be 5000 h for cars and 10 000 h for buses. The main mechanism of membrane degradation is radical attack via H_2O_2 formation. Within the fuel cell environment, H_2O_2 is formed via a two-electron oxygen reduction reaction at the Pt catalyst, although due to the high potential in the cathode the H_2O_2 will be unstable and peroxide radicals will form as intermediate species [9]. Not surprisingly, many reports have confirmed that electrolyte degradation does indeed occur mainly in the cathode.

During the past few years, the durability characteristics of PEMs have attracted much attention, with many forms of accelerated evaluation techniques for membranes having been developed to avoid the typical long-term approach to durability testing. As a result, many such studies have been conducted under open-circuit conditions [10–13]. In order to minimize the testing time and resources involved, simulated conduction tests may offer an alternative approach, and are typically based on the use of H_2O_2 solution [14], Fenton's reagent [15], and the simulation of cell environments [16–18]. As an example, Liu *et al.* [19] reported the details of an electrochemical technique that used a thin Pt wire as the working electrode, for determining H_2O_2 concentrations within the PEM of operating fuel cells. An alternative parameter, the fluoride ion emission rate (FER), can also be monitored in the drain of exhaust gases or liquids, using ion chromatography and ion-selective electrodes. The degradation patterns of PEMs have also been investigated using thermogravimetry, as well as a variety of spectroscopic methods and electron spin resonance (ESR). Gas leakage, membrane weight and mechanical strength have also been monitored, both before and after durability testing, in order to evaluate the degree of membrane degradation.

10.3 Fluorinated Ionomer Membranes

10.3.1 Perfluorosulfonate Membranes

The current state-of-the-art PEM is Nafion®, which was developed during the late 1960s by DuPont, primarily as a perm-selective separator in chlor-alkali electrolyzers [20, 21]. Nafion is based on the copolymerization of tetrafluoroethylene with a perfluorinated vinyl ether monomer containing a sulfonyl fluoride in the side chain. Subsequent conversion of the sulfonyl fluoride to the sulfonic acid group is required. The perfluorinated sulfonic acid (PFSA) structure combines an excellent proton conductivity and exceptional chemical stability, both of which are especially important in fuel cell applications. Similar polymers are produced by the Asahi Chemical Company and Asahi Glass Company. The Dow Chemical Company has also produced a similar polymer with shorter side chains and a higher ratio of SO_3H to CF_2 groups [22]. The chemical structures of Nafion and the Dow membranes are compared in Figure 10.2.

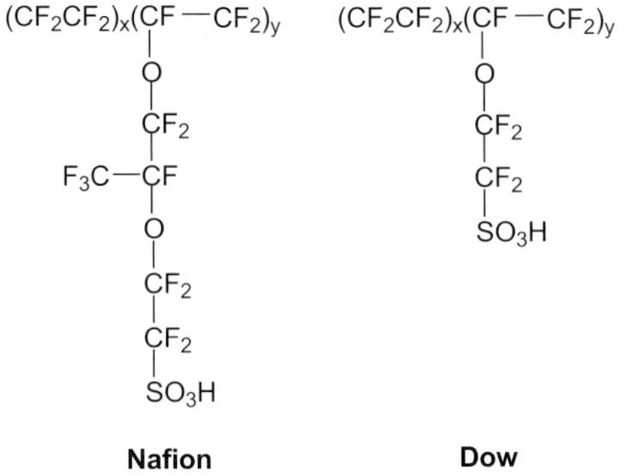

$(CF_2CF_2)_x(CF—CF_2)_y$ | O | CF_2 | $F_3C—CF$ | O | CF_2 | CF_2 | SO_3H

Nafion

$(CF_2CF_2)_x(CF—CF_2)_y$ | O | CF_2 | CF_2 | SO_3H

Dow

Figure 10.2 Chemical structures of Nafion and Dow membranes.

The EW (or IEC) can be varied by changing the ratio of the two components (x and y in Figure 10.2). Although, previously, Nafion was produced commercially with EW-values of 900, 1100, and 1200, currently Nafion 1100 EW membranes with thicknesses of 2, 5, 7, and 10 mil (1 mil = 25.4 μm), known as Nafion 112, 115, 117, and 1110, appear to be the only grades used widely in fuel cells.

The above-described Nafion membranes were produced using the extrusion-cast process which was developed for the creation of "thick" films. In recent years, however, technical progress in fuel cell design and performance has increased the demand for mass-produced, thin membranes with lower overall costs, for fuel cell applications. There has also been a growing demand for larger lot sizes, increased roll lengths, and improved physical properties, and in order to meet these requirements DuPont has subsequently developed a solution-casting process. This has several key advantages over the extrusion-cast process, namely that:

- Large dispersion batches can be pre-qualified in terms of their quality; that is, they will be free from contamination and have a predictable performance (e.g., acid capacity).
- The production rates for H^+ membranes are increased overall, as compared to polymer extrusion followed by chemical treatment.
- There is improved thickness control and uniformity of the membrane, and most notably an ability to produce very thin membranes (e.g., 12.7 μm) [23].

These solution-cast membranes are registered as NRE membranes. The effects of the manufacturing processes on membrane properties are compared in Table 10.1 [24, 32].

Kreuer *et al.* [25] investigated the membrane properties, including water sorption, transport (proton conductivity, electro-osmotic water drag and water diffusion), microstructure and viscoelasticity of the short-side-chain (SSC) perfluorosulfonic acid ionomers (PFSA, Dow 840 and Dow 1150) with different IEC-values. The data were compared to those for Nafion 117, and the implications for using such ionomers as separator materials in direct methanol and hydrogen fuel cells discussed. The major advantages of PFSA membranes were seen to be: (i) a high proton conductivity,

Table 10.1 Properties of Nafion membranes prepared using different processes.

Membrane type		Nafion 112	NRE-212
Manufacturing process		Extrusion-cast	Solution-cast
Thickness (μm)		51	51
Tensile strength (Max./MPa)	Transverse	32	32
	Lengthways	43	32
Swelling in water at 100 °C (%)	Transverse	6	15
	Lengthways	17	15
Water uptake in water at 100 °C (%)		38	36
Conductivity ($S\,cm^{-1}$)		0.083	—

which could achieve $0.2\ S\,cm^{-1}$; and (ii) an excellent chemical stability. However, a limited operation temperature (due to a requirement for humidification), a low glass transition temperature (T_g), high methanol permeability and cost, and insufficient mechanical strength and dimension stability proved to be severe disadvantages for these membranes.

10.3.2
Partially Fluorosulfonated Membranes

During recent years, much effort has been focused on the development of new membranes, with the Ballard Advanced Materials Corporation having introduced a styrene membrane based on a novel family of sulfonated copolymers incorporating α, β,β- trifluorostyrene and a series of substituted α,β,β-trifluorostyrene comonomers [26]. These have been registered as BAM membranes; their general formula is shown in Figure 10.3. The water content of a series of membranes ranging from 375 to 920 EW was found to be much higher than that of Nafion and the Dow membranes [27]. It has also been reported that this generation of BAM membranes exhibits a superior performance with respect to PFSA at current densities in excess of $0.6\ A\,cm^{-2}$ and low cost (US\$ 0.5–1.5 m^{-2}; US\$ 5–15 ft^{-2}).

Büchi *et al.* [28] reported a series of differently crosslinked FEP-g-polystyrene PEMs [FEP = poly(tetrafluoroethylene-*co*-hexafluoropropylene)], synthesized by preradiation grafting of the monomers onto a base film and subsequent sulfonation of the grafted component. Both, divinylbenzene (DVB) and/or triallyl cyamirate (TAC) were used as crosslinkers. The physical properties of these membranes, such as water

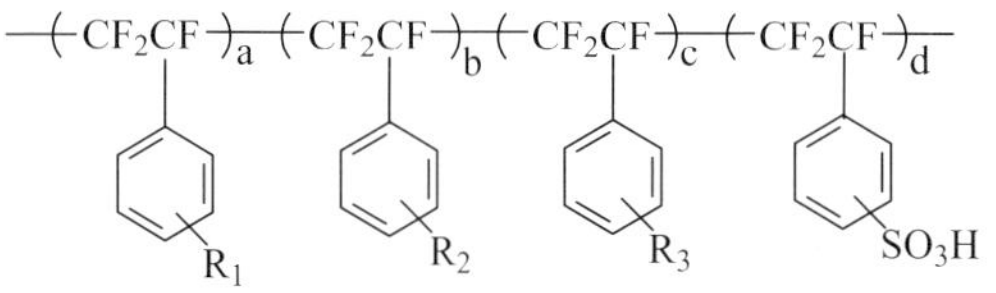

$-(CF_2CF)_a-(CF_2CF)_b-(CF_2CF)_c-(CF_2CF)_d-$ with pendant phenyl rings bearing R_1, R_2, R_3, SO_3H

R_1, R_2, R_3 = alkyls, halogens, or $CF{=}CF_2$, CN, NO_2, OH

Figure 10.3 Chemical structures of BAM membranes.

uptake and specific resistance, were found to be heavily influenced by the nature of the crosslinker. When both the stability and performance of the membranes were tested in H_2/O_2 fuel cells, the thick (170 μm) DVB crosslinked membranes showed a stable operation of 1400 h at temperatures of up to 80 °C. The highest power density in the model PEMFC was found for the DVB and TAC double-crosslinked membrane, with corresponding values more than 60% higher compared to those for the Nafion 117 membrane.

Soresi *et al.* [29] performed a structural modification of poly(vinylidene fluoride) (PVDF), which is hydrophobic in nature, by radiation grafting with styrene comonomers, thus enabling a linking of the sulfonic groups. The critical parameters of this process proved to be the energy of the radiation field (electrons, γ-rays), the absorbed dose, and the physico-chemical properties of the starting polymer (crystallization degree, porosity, etc.). The membranes obtained from PVDF with 5 mol% HFP copolymer, and from commercial porous and dense PVDF homo-polymer films, were compared. Both, high grafting degrees and substantial water uptake were achieved, depending on the nature of the polymer matrix. A room-temperature conductivity in excess of 0.060 S cm^{-1} at 90% relative humidity (RH) was observed in the case of sulfonated membranes obtained from the PVDF-based copolymer.

Yamakia *et al.* [30] prepared PEMs by the radiation-induced grafting of styrene into crosslinked polytetrafluoroethylene (PTFE) films and subsequent sulfonation. The degree of grafting was controlled over a range of 5 to 75% by the crosslinking density of the PTFE matrix, as well as by the grafting conditions. The resultant membranes showed a large IEC (up to 2.6 mEq /g^{-1}), which exceeded the performance of commercially available PFSA films, such as Nafion.

10.3.3 Reinforced Composite Membranes

The main driving force for the development of reinforced composite membranes has been the possibility of increasing both mechanical strength and dimensional stability, leading to the feasible production of very thin membranes. As mentioned in Section 10.2.2, as the cell resistance increases proportionally to the thickness, an improvement in PEMFC performance could easily be achieved by using thin, reinforced composite membranes. From a cost perspective, such thin reinforced composites would contain much less of the expensive ionomer resin than conventional membranes.

10.3.3.1 PFSA/PTFE Composite Membranes

More recently, W. L. Gore & Associates, Inc. [31] have used expanded polytetrafluoroethylene (ePTFE) porous films and PFSA resin solutions to produce a composite membrane. In this process, the PFSA solution was brushed onto both sides of the ePTFE film so as to impregnate and substantially occlude the interior volume of the film; a nonionic surfactant (5%, w/v) was then added into the PFSA solution as a penetrant. The surfactant was removed by soaking in isopropanol after drying at 140 °C (this procedure was repeated several times so as to fully occlude the interior

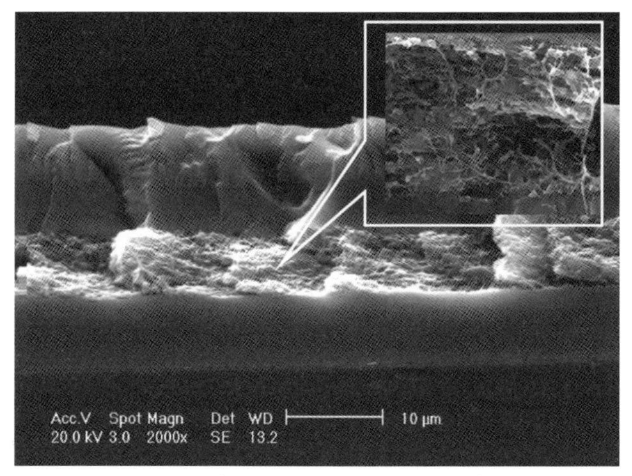

Figure 10.4 Scanning electron microscopy images of cross-sections of the PFSA/PTFE composite membrane.

volume). When Liu *et al.* [32] used porous PTFE as the support for Nafion/PTFE composite membranes, the composites were synthesized by impregnating porous PTFE with a Nafion solution on a hot plate. Tang *et al.* [33] fabricated Nafion/PTFE composite membranes by using a hydrophilically modified expanded PTFE matrix to increase the PFSA resin loading; these composites exhibited an improved mechanical strength in the dry state, and dimensional stability in the wet state. Recently, Sunrise Power Co., Ltd developed a PFSA/PTFE composite membranes by using a spray-paint process. In these composite membranes (SPCM), the PFSA resin is distributed uniformly in the microporous PTFE layer, there being two homogeneous thin PFSA films on the PTFE surface (as shown in Figure 10.4). When compared to the NRE-212, this composite membrane possessed a greater mechanical strength and demonstrated less swelling. Moreover, the performance of a small PEMFC stack with SPCM membranes (Figure 10.5) proved to be comparable to that of NRE-212.

10.3.3.2 **PFSA/CNT Composite Membranes**

Liu *et al.* [34] developed a carbon nanotube (CNT) -reinforced PFSA composite membrane for PEMFCs, where the PFSA/CNT composite was prepared using a solution-casting method. The reinforced composite contained a small amount of CNT (1 wt%), and also showed an excellent mechanical strength and lower dimensional changes compared to Nafion. Yet, the performance of the CNT-reinforced membranes (50 µm) was virtually identical to that of commercial NRE-212. Wang *et al.* [35] subsequently used an improved solution-cast method to prepare multiwall CNT (MWCNT)/PFSA-reinforced membranes. For this, the MWCNTs (1–4 wt%) were oxidized with H_2O_2 and then added to a MWCNT/PFSA/*N,N*-dimethylacetamide (DMAC) solution with sodium hydroxide (NaOH); the long-term stability of the resultant dispersions was substantially improved, with scanning electron microscopy (SEM) inspections of the as-cast membrane demonstrating a uniform distribution of

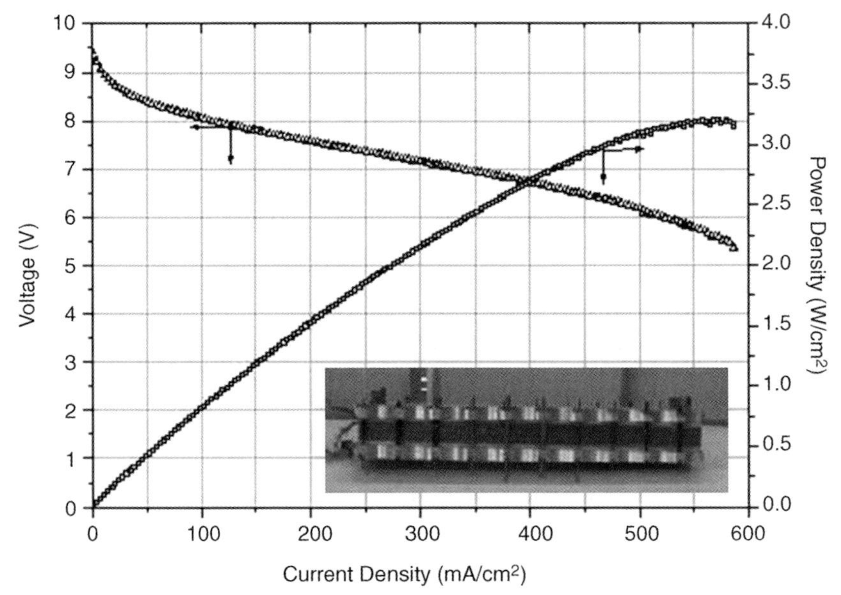

Figure 10.5 Performance of short stack with SPCM membranes (40 μm thickness). The stack (inset) comprises 10 cells × 290 cm^2; H_2/air at atmospheric pressure; stack temperature at 45 °C. Δ = voltage; □ = power density.

MWCNTs in the PFSA resin. Compared to the recast Nafion membranes, both stretching and tensile strength values were greatly improved (by 27% and 54%, respectively). A comparison of the mechanical properties and electronic conductivity showed the optimum MWCNT content to be approximately 3 wt%.

10.3.4
Hybrid Organic–Inorganic Membranes

The performance degradation of PFSA membranes at elevated temperatures represents another serious drawback, as membrane dehydration, a reduction in ionic conductivity and a decreased affinity for water are all observed above 80 °C. In order to overcome these disadvantages of PFSA, many research groups have focused their attention on a variety of on promising alternatives [36, 37]. One solution to this problem would be to develop hybrid organic–inorganic, self-humidifying membranes [38–40]. Subsequently, Yang *et al.* [41] prepared a type of homogeneous proton-conducting WO_3/PFSA composite based on a dynamic conductivity variations concept, where the membrane resistance could be reduced during fuel cell operation due to the formation of highly conducting hydrogen–tungsten bronzes. Indeed, the resistivity of the WO_3-containing membranes, as measured with *in-situ* AC impedance spectroscopy during the single cell operation, was seen to be significantly lower compared to Nafion-112, and led to a superior performance of those fuel cells with WO_3/PFSA membranes.

10.3.4.1 Hygroscopic Material/PFSA Composite Membranes

In this case, the hygroscopic components are normally oxides such as SiO_2, TiO_2, and ZrO_2. All such compounds display a high water-retention ability which is crucial for proton conductivity at elevated temperatures. In these studies, Kwak *et al.* [42] hot-pressed a mixture of melt-fabricated perfluorosulfonylfluroride copolymer resin (Nafion resin) with a fine mordenite powder to prepare composite membranes for high-temperature PEMFCs. The proton conductivity was found to increase in line with increasing mordenite content at temperatures >90 °C, due to the presence of water in the mordenite. The current–voltage relationships for single cells at various operation temperatures also showed that the composite membrane containing 10 wt% mordenite had maximum performance. Since the use of oxide particles to produce nanocomposite membranes is often difficult due to conglomeration, an *in-situ* sol–gel reaction may be an effective approach. The SiO_2/Nafion composite membranes can be produced via the *in-situ*, acid-catalyzed sol–gel reaction of tetraethylorthosilicate (TEOS) in PFSA films [43–45]. TEOS, when reacted with water in an acidic medium, undergoes polymerization to form a mixture of SiO_2 and a siloxane polymer with hydroxide and ethoxide groups; the TEOS uptake content in a composite membrane is varied according to the sol–gel reaction time. The addition of SiO_2 ($<$10 wt%) improved the water retention of the composite membranes, and also increased the proton conductivity at elevated temperatures. Nanosized TiO_2 powder was synthesized via a sol–gel procedure by a rapid hydrolysis of $Ti(OiPr)_4$; the membranes were then prepared by mixing the Nafion–dimethylacetamide (DMAc) dispersion with 3 wt% TiO_2, and casting. The results for water uptake, IEC and single cell performance were compared with the commercial Nafion-115 and home-made, recast Nafion membranes. Power density values of 0.51 and 0.26 $W\,cm^{-2}$ at 0.56 V were obtained at 110 and 130 °C, respectively, for the composite Nafion–titania membrane [46].

The incorporation of nanosized inorganic particles into PEMs, whilst producing an effective humidifying effect, would certainly lower the IEC of the membrane. Some groups have used organic substances containing sulfonic acid ($-SO_3H$) to modify the hygroscopic material, to enhance the proton conductivity and improve the performance of Nafion. Rhee *et al.* [47] used a variety of thiol and sulfone groups to graft onto the surface of titanate nanosheets in order to render organic sulfonic acid (HSO_3^-) functionality. The nanocomposite membranes, cast together with Nafion using surface-sulfonated titanates as inorganic fillers, showed a high proton conductivity and a better mechanical and thermal stability when compared to Nafion. The methanol permeability of the nanocomposite membranes was shown to decrease on addition of the sulfonated titanate. Functionalized montmorillonite (MMT) was also employed to improve PFSA [58, 59]; these composite membranes provide a low methanol crossover, without sacrificing proton conductivity due to the introduction of sulfonic acid groups at the MMT surface, followed by blending with the Nafion ionomer.

10.3.4.2 Catalyst Material/PFSA Composite Membranes

Watanabe *et al.* [38–40] developed a form of self-humidifying composite membrane with nanometer-size Pt and/or hygroscopic materials dispersed into the PFSA resin.

The Pt particles catalyze hydrogen oxidation in order to generate water and to humidify the membrane. However, when hygroscopic materials (e.g., SiO_2 or TiO_2) were added into the membranes synchronously, the water produced at the Pt particles was adsorbed by these metal oxides [48–52]. When Kwak *et al.* [53] investigated the effect of Pt loading in the self-humidifying PEMs, the optimum level of Pt particles embedded in the membrane was found to be approximately 0.15 mg cm^{-2}, based on measurements of single-cell performance characteristics and resistance. Lee *et al.* [54] used Pt–zirconium phosphate–PFSA composite membranes for self-humidifying PEMFCs; here, the composite was prepared by adding ZrP to the Pt-containing membrane with 1 *M* H_3PO_4 solution after soaking in 0.5 *M* $ZrOCl_2$. The zirconium phosphate particles provided stability at elevated temperatures due to the water- retaining and attractive electrochemical properties of this component.

10.3.4.3 Heteropolyacid/PFSA Composite Membranes

$Cs_{2.5}H_{0.5}PW_{12}O_{40}$ is insoluble in water and organic solvents, and also has a high surface area. In addition, the acidity per unit acid site of $Cs_{2.5}H_{0.5}PW_{12}O_{40}$ was superior to Nafion-H as well as to homogeneous acids, such as H_2SO_4, and $H_3PW_{12}O_{40}$. These facts demonstrate that $Cs_{2.5}H_{0.5}PW_{12}O_{40}$ is a promising solid acid for use in PEMs. Li *et al.* [55] subsequently prepared a $Cs_{2.5}H_{0.5}PW_{12}O_{40}$/Nafion/PTFE self-humidifying composite by recasting the $Cs_{2.5}H_{0.5}PW_{12}O_{40}$/Nafion layer onto both sides of a Nafion/PTFE membrane, using a solution-recast method. Due to the strong acidic, hydrophilic, and redox properties of $Cs_{2.5}H_{0.5}PW_{12}O_{40}$, the $Cs_{2.5}H_{0.5}PW_{12}O_{40}$/Nafion layer not only contributed to the membrane humidification but also improved the open-circuit voltage of the fuel cell. When the self-humidifying composite membrane was employed as an electrolyte in a H_2/O_2 PEMFC, a higher maximum power density value (1.1 W cm^{-2}) was obtained than for the Nafion/PTFE composite (0.6 W cm^{-2}), at operating and humidifying temperatures of 70 °C and 20 °C, respectively. Wang *et al.* [56] used a $Cs_{2.5}H_{0.5}PWO_{40}/SiO_2$ composite to achieve the best performance under dry conditions at 60 °C; the stable characteristics were provided due to the substitution of H^+ with Cs^+ and interaction between the components.

10.3.4.4 Self-Humidifying Reinforced Composite Membranes

Liu *et al.* [57] developed a preparation method for self-humidifying composite membranes. When using the solution-cast method, porous PTFE substrates in these membranes may increase their strength and distribute self-humidifying layers adjacent to the anode side. The resultant performance of PEMFCs is dramatically improved in terms of both the cell voltage and current density under dry conditions; the cell with Pt/C-PEM showed the best, and most stable, performance. Wang *et al.* [58] subsequently developed a Pt/SiO_2/PFSA/PTFE self-humidifying reinforced composite that minimized membrane conductivity loss under dry conditions due to the presence of a catalyst and hydrophilic Pt/SiO_2.

A novel, thin, three-layer reinforced and self-humidifying composite membrane, comprising two outer layers of plain Nafion and a middle layer of Pt/CNT-dispersed Nafion, has been developed [59]. The Pt/CNTs present in the membrane provide sites for the catalytic recombination of H_2 and O_2 permeating through the membrane to produce water, simultaneously improving the mechanical properties. The water directly humidifies the membrane, thus enabling the operation of PEMFCs with dry reactants.

Matos *et al.* [60] developed Nafion–titanate nanotube composites as the PEMFC electrolyte operating at elevated temperatures. The addition of 5–15 wt% nanotubes to the ionomer allowed the PEMFC performance essentially to be sustained up to 130 °C. The polarization curves of PEMFCs using composite electrolytes reflected a competing effect between an increase in water uptake due to the extremely large surface area of the nanotubes, and a decrease in the proton conductivity of the composites.

10.4 Non-Fluorinated Ionomer Membranes

Despite perfluorinated polymer electrolytes (Nafion, Flemion, Aciplex) having been used extensively in PEMFCs, their poor thermal mechanical stability (low T_g), high methanol permeability, and extremely high production costs have led to limitations of their large-scale application. Therefore, a variety of hydrocarbon polymer electrolytes have been developed during the past decade, with the greatest emphasis placed on the costs, conductivity, and durability of these materials. At the same time, the poor mechanical stability and inadequate durability of the hydrocarbon polymer membranes were identified as the main barriers to their practical application.

10.4.1 Materials, Membranes, and Characterization

Poly(arylene sulfone)s and poly(arylene ketone)s are important engineering thermoplastics, and display high T_g-values, high thermal stabilities, good mechanical properties, and an exceptional resistance to both oxidation and acid-catalyzed hydrolysis. It is only during the past decade that the sulfonated aromatic polymers have been considered to be well-suited as PEM candidates for fuel cells [61–64].

10.4.1.1 Post-Sulfonated Polymers

Several types of industrially produced polymer, including poly(ether ether ketone) (PEEK), poly(ether sulfone) (PES), polyimide (PI), and poly(benzimidazole) (PBI) (see Figure 10.6), have attracted much attention owing to their excellent physical, chemical, and other properties. In order to prepare proton-conductive membrane materials by introducing sulfonic acid groups into the polymer backbone, the initial route is related to post-sulfonation. The pioneering studies of Noshay and Robeson [65] resulted in a mild sulfonation procedure for the commercially available bisphenol-A-based poly(ether sulfone). Different sulfonation agents have been employed for post-modification reactions, including concentrated sulfuric acid, fuming sulfuric acid, chlorosulfonic acid, and a sulfur trioxide–triethyl phosphate

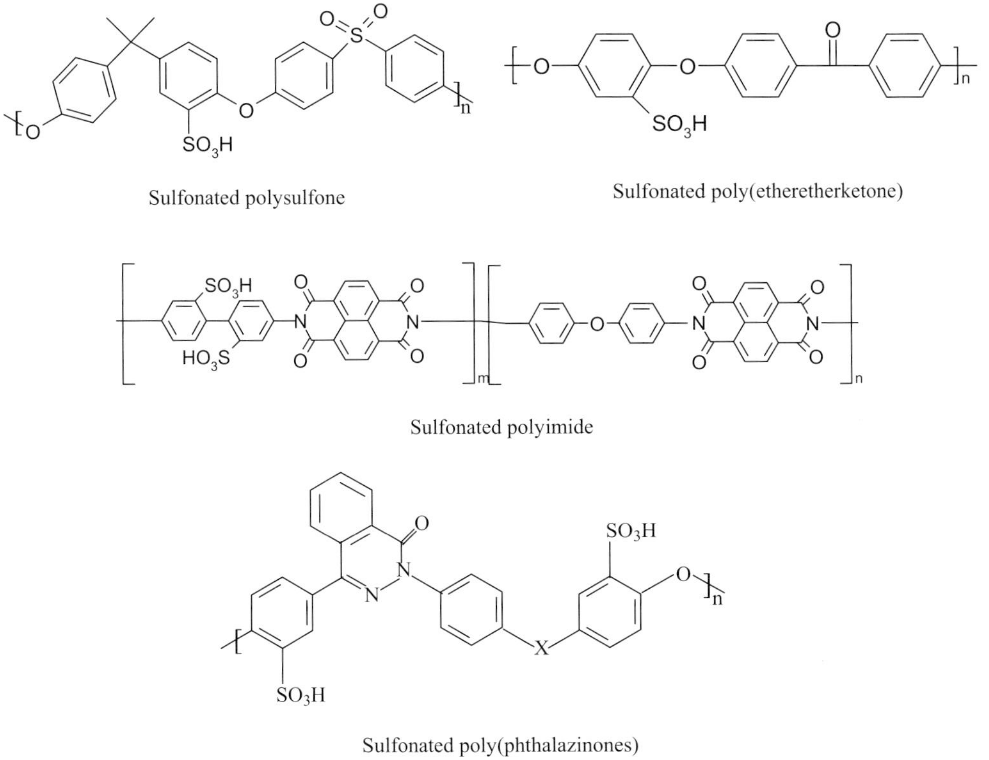

Sulfonated polysulfone

Sulfonated poly(etheretherketone)

Sulfonated polyimide

Sulfonated poly(phthalazinones)

Figure 10.6 Structures of nonfluorinated polymer electrolyte materials.

complex. Genova-Dimitrova *et al.* [66] reported the comparative analysis of sulfonating agents, while Gao *et al.* [67] prepared a class of sulfonated poly(phthalazinones), including sulfonated poly(phthalazinone ether sulfones), sulfonated poly(phthalazinone ether ketones) and sulfonated poly(phthalazinone ether sulfone ketones) by modification of the corresponding poly(phthalazinones). The sulfonation reactions were conducted at room temperature using mixtures of 95–98% concentrated sulfuric acid and 27–33% fuming sulfuric acid, with different acid ratios in order to achieve a degree of sulfonation (DS) of 1.00–1.37. The conductivity of all sulfonated polymers is known to increase with DS, reaching $>10^{-2}\ S\ cm^{-1}$ at DS values close to 1.0.

In these post-sulfonation reactions, the sulfonic acid group is usually restricted to the activated position, or to the aromatic ether bond; the precise control over its location and the degree of sulfonation is often difficult. Moreover, secondary reactions and degradation of the polymer backbone may occur during sulfonation [68]. Kerres *et al.* [69] reported an alternative sulfonation process for commercial polysulfone based on a series of steps including metalation–sulfination–oxidation reactions. The advantages of the partial oxidation process using NaOCl are as follows: (i) the oxidation degree can be adjusted finely to a necessary level; (ii) no side reactions take place during oxidation; and (iii) the partially oxidized polymers are stable at ambient temperature. However, this method is still restricted by the molecular structures of existing commercial polymers, which cannot control the ionic monomer units and distribution of the ionic groups. Another novel post-sulfonation method was reported by Karlsson *et al.* [70], where the polymers were prepared via lithiation and sulfination of polysulfone (PSU), followed by the grafting of sulfoethyl, sulfopropyl, or sulfobutyl chains onto the sulfonate units.

10.4.1.2 Direct Polymerization from the Sulfonated Monomers

Generally, in the post-sulfonation reaction the sulfonic acid group is restricted to the activated position *ortho* to the aromatic ether bond (structure I in Figure 10.7). For the bisphenol-A based systems, no more than one sulfonic acid group per repeat unit could be introduced into the polymer. It would be of great interest to investigate sulfonation on the deactivated portion of the repeat unit (structure II), as one might expect an enhanced stability and, perhaps, a modestly higher acidity due to the strong electron-withdrawing properties of the sulfone group [71, 72]. The deactivated rings are more stable against desulfonation, as the anticipated intermediate carbonation required for desulfonation is more difficult for stabilization on a deactivated ring. Another shortcoming of the post-sulfonating approach is a difficult control of the sulfonation process, which may occasionally lead to side reactions. Typically, the degree and position of sulfonation will depend on the process conditions, such as temperature, time, and concentration.

One alternative approach to obtain sulfonated aromatic polymers is the direct synthesis via aromatic nucleophilic substitution reaction from the sulfonated-functionalized monomers. This direct synthesis process has been proven to be more advantageous than post-sulfonation, namely:

- The concentration and positions of the sulfonate groups within the directly-synthesized sulfonated aromatic polymers can be easily controlled.

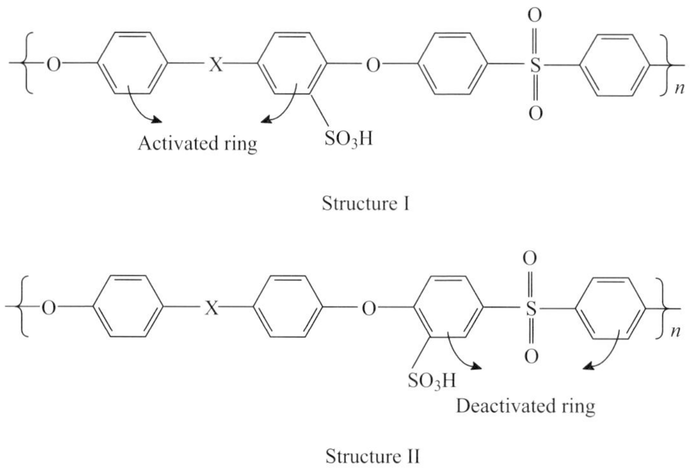

Activated ring SO_3H

Structure I

SO_3H Deactivated ring

Structure II

Figure 10.7 Structures of sulfonated poly(arylene ether sulfone)s possess different sulfonic acid positions.

- The *meta*-positioned sulfonate group may enhance proton conductivity due to increased acidity.
- The direct synthesis method avoids the crosslinking and other side reactions, which could result in better thermal stability and mechanical properties.

Most sulfonated aromatic polymers can be directly synthesized via aromatic nucleophilic substitution reaction. The direct synthesis has been widely investigated by McGrath *et al.* [71–79], who showed that the carbonate base process includes two steps: the phenols first react with carbonate to produce phenoxides, after which the phenoxides undergo polycondensation with halide. For the first reaction, there are two factors that cause incomplete formation of the phenoxide, namely poor solubility of the K_2CO_3 in the reaction solvent (e.g., DMAc), and water formation. The water formed may act as a nucleophilic agent and, after reacting with the activated halide, produce unreactive bisphenol. Hence, a typical polymerization process was performed at a high temperature (150–190 °C) for a long duration (16–30 h). In this procedure, toluene or chlorobenzene must be used as the azeotropic agent to remove any water formed during the polymerization. Xing *et al.* [80] synthesized a series of poly(arylene ether sulfone)s and sulfonated copolymers via the improved procedure, when the polymerizations were performed with 3.0 molar equivalents of potassium carbonate per OH group and the phenoxide could be completely formed due to a high excess of potassium carbonate. The latter system produced no water; thus, no azeotropic distillation was required and the polymerization could be performed at a lower temperature (~130 °C). It should be noted also that sulfonated poly(arylene ether sulfone)s show excellent oxidative and thermal stability, good mechanical properties, a high proton conductivity, and thus represent a very promising materials for PEMFC applications [79, 80]. The proton conductivity of the copolymers containing 40–50% sulfonic acid was 0.12–0.18 $S\,cm^{-1}$.

Sulfonated polyimides are synthesized by polycondensation using a commercial naphthalenic dianhydride monomer and different sulfonated and nonsulfonated diamines, the proportions of which are adjusted so as to control the IEC value. However, common five-membered ring polyimides are generally unstable towards acid due to the easy hydrolysis of imido rings, and sulfonated polyimides are generally less stable than their nonsulfonated counterparts. Six-membered ring polyimides derived from 1,4,5,8-naphthalenetetracarboxylic dianhydride (NTDA) have been found to be fairly stable in both acid and pure water [81]. Faure and coworkers [82, 83] were the first to synthesize a series of sulfonated five-membered and six-membered ring polyimides from oxydiphthalic dianhydride (ODPA) and NTDA, respectively, and subsequently to test these in fuel cell systems. Their results indicated that ODPA-based polyimides were not sufficiently stable under the PEMFC conditions, whereas the NTDA-based polyimides were stable while the sulfonation degree was controlled to an appropriate level.

The main developments in the field of sulfonated polyimide membranes are currently related to the use of chemically modified diamines in order to enhance stability against hydrolytic and oxidative attack [84, 85]. The relationships between membrane stability (especially water stability) and polyimide structure (diamine moiety) have been studied [86–88]. Those factors having significant impact on the water stability include: (i) the structure of the sulfonated diamine moieties including the flexibility, basicity and configuration; (ii) the ratio of the sulfonated diamine moiety; and (iii) the type of backbone of the polymers (aliphatic/aromatic, main chain/side chain, branched/crosslinked).

10.4.1.3 **Microstructures and Proton Transportation**

Kreuer *et al.* [89, 90] investigated the proton-transport mechanisms in different materials. The transport properties and swelling behavior of Nafion and different sulfonated polyetherketones were explained in terms of distinct differences in their microstructures, and in the acid–base ionization/dissociation constants (pK_a) of the acidic functional groups. The less-pronounced hydrophobic/hydrophilic separation of sulfonated polyetherketones compared to Nafion corresponded to narrower, less-connected hydrophilic channels and to larger separations between less-acidic sulfonic acid functional groups. At high water contents, this is shown to significantly reduce electro-osmotic drag and water permeation while maintaining a high proton conductivity. The blending of sulfonated polyetherketones with other polyaryls further reduced the solvent permeation (by a factor of 20 compared to Nafion), increased the membrane flexibility in the dry state, and also led to an improved swelling behavior. Therefore, polymers based on sulfonated polyetherketones appear not only to be promising low-cost alternative membrane materials for hydrogen PEMFCs, but they may also help to reduce the problems associated with high water drag and high methanol cross-over in the direct methanol fuel cell (DMFC). These studies were in progress just as McGrath and colleagues [91] were reaching many of the same conclusions about sulfonated poly(arylene ether sulfones). PFSA membranes such as Nafion have a pK_a of about –6 as compared to an aromatic sulfonic acid moiety in sulfonated PEEK, of about –1. The implication of this difference is that there may be a lower ionic charge-carrier concentration in the Nafion membrane as compared to PEEK to produce the same conductivity under identical conditions; the microstructural differences between the two systems are somewhat more subtle.

10.4.1.4 **Durability Issues**

Unfortunately, one general issue of most nonfluorinated PEMs is their insufficient stability for fuel cell applications, the chemical degradation of PEM being a major factor responsible for PEMFC degradation. The hydroxyl and hydroperoxyl radicals are generated in either the anode or cathode after a two-electron reduction reaction between H_2 and O_2, and cross over the PEM to the opposite electrode [92, 93]. In the PFSA backbone, the C–F bond energy is high (485 kJ mol^{-1}), and the fluorine atoms wrap the C–C main chain closely in order to avoid attack from oxidative free radicals, which in turn leads to a good oxidation resistance. However, in the F-free PEMs, the C–H bond dissociation enthalpy is lower and oxidative degradation occurs more easily.

Both, Yu and Wang and coworkers [94, 95] have developed multilayer composite membranes to improve the stability of PEMs. The PSSA–Nafion composite membrane was designed to prevent the oxidative degradation of PSSA; the performance of fuel cells with this composite membrane, where the Nafion membrane was bonded with the PSSA and located at the cathode side, was shown to be stable. The chemical stability of sulfonated poly(biphenyl ether sulfone) (SPSU-BP)/PBI crosslinked blend membranes in H_2O_2 solution have been investigated [96]. The weight losses and IEC changes of SPSU-BP/PBI were lower than those of SPSU-BP, indicating a higher chemical stability of the blend membranes.

Since an alternative approach would be to develop the inorganic–organic composite membranes, Haugen [97] prepared PFSA membranes doped with heteropolyacids (HPAs) such as 12-HSW, 21-HPW, and 18-HPW, and tested their stability performance under fuel cell operating conditions. The results of the F^- release test showed that HPAs reduced the degradation of the PFSA membrane. As a consequence, Ballard Power Systems, Inc. [98] has recently published the details of a method based on additives to the anode, cathode or PEM so as to reduce membrane degradation in fuel cells. The additive may be a radical scavenger, a crosslinker, a H_2O_2 decomposition catalyst, and/or a H_2O_2 stabilizer. In order to provide an enhanced lifetime of the MEA, UTC Fuel Cells LLC [99] reported the details of a H_2O_2 decomposition catalyst which could be placed in at least one position chosen from the anode, cathode, intermediate layers between the membrane and electrode, and/or within the membrane. Xing *et al.* [100] prepared a Ag-SiO_2/SPSU-BP composite to increase the chemical stability, whereby the Ag/SiO_2 catalyzes the decomposition of H_2O_2. Accelerated fuel cell life tests conducted under open-circuit voltage (OCV) conditions showed that the durability of PEMFC with the Ag-SiO_2/SPSU-BP membrane was indeed improved. It should be noted that the configurational stability of the above composite membranes, as well as the mechanisms providing improved durability, all require further investigation.

10.4.2
Reinforced Composite Membranes

Although hydrocarbon polymer membranes exhibit certain advantages, they also demonstrate barriers against practical applications, including a higher water uptake, a brittle nature at elevated temperatures, and poor dimensional and chemical/mechanical stabilities in the wet state [101]. The proton conductivity of sulfonated hydrocarbon membranes depends on the degree of sulfonation. although the mechanical properties tend to deteriorate progressively with an increasing DS. Highly sulfonated polymers may swell extensively at high temperature and humidity [102–104]. In order to overcome these problems and to improve the PEM's properties, a method of reinforcement by forming a composite membrane, either by introducing reinforcing materials or with chemical crosslinking, is generally adopted. The reinforcement materials used have included PTFE porous films, woven PEEKs, and fiber-glasses. The composite membranes have demonstrated improved properties, such as mechanical strength and dimensional stability [105]. Despite the lower EW of these composites, due to the addition of insulating reinforcement materials, a low resistance can be achieved by using thin membranes [106].

Kerres's group synthesized several different types of basic, sulfonated or sulfinated polymer, and developed blend concepts for the fuel cell ionomer membranes [107–122]. Different types of polyaryl blend membranes – that is, ionic, ionic-covalent, and covalent crosslinking – have been studied, and the results have shown that, in the case of ionic crosslinking, the ionic bonds have an unsatisfactory thermal stability which makes the use of such membranes questionable. In contrast, the ionic–covalent blend membrane had a good mechanical stability, a reduced fuel permeability, and suitable proton conductivity. Ultra-violet irradiation-induced crosslinking is thought to be an effective method for generating three-dimensional (3-D) polymer networks, owing to its ease of use, relative safety, low cost, and high initiation rate under intense illumination [123–125].

10.4.3 Hybrid Organic–Inorganic Membranes

In recent years, organic–inorganic hybrids have received great attention due to their advantage in improving proton transport and mechanical properties. Many research groups have investigated organic–inorganic hybrid materials, where functional groups were covalently/physically attached to the organic part of the hybrid, including Nafion and hydrocarbon membranes. The functional inorganic additives make it possible to: (i) prevent membrane (anode side) dry out; that is, so-called *self-humidifying* [126–128]; (ii) improve the mechanical strength and dimensional stability [129, 130]; (iii) enhance the pathways of proton conduction [131–134]; and (iv) suppress methanol permeation [135–139].

Zhang *et al.* [128] synthesized a self-humidifying membrane based on a sulfonated poly(ether ether ketone) (SPEEK) hybrid with a sulfated zirconia (SO_4^{2-}/ ZrO_2, SZ) -supported platinum catalyst (Pt-SZ catalyst). This type of composite membrane has a higher proton conductivity than plain SPEEK, due to the effect of the Pt-SZ catalyst; the membrane also provided excellent single cell performance at low humidity.

Liang *et al.* [131] have developed an effective approach to improve protonic transport in sulfonated poly(phthalazinone ether sulfone ketone) (SPPESK) membranes; the SPPESK/PWA/SiO_2 composite membrane was prepared by supporting PWA on silica gel via the sol–gel process. The composite showed not only a higher proton conductivity (0.11 S cm^{-1} at 95 °C) than that of the original membrane, but also excellent mechanical properties (46.8 MPa tensile strength and 9.6% swelling at 80 °C). Another significant improvement was the reduced loss of PWA after immersion of the membrane in water.

Su *et al.* [138] prepared composite PEMs from sulfonated poly(phthalazinone ether ketone) (sPPEK) and various amounts of sulfonated silica nanoparticles (silica-SO_3H). The use of silica-SO_3H was seen to compensate for the reduction in the IEC of the membrane, while the strong –SO_3H/–SO_3H interaction between sPPEK chains and silica-SO_3H particles led to ionic crosslinking in the membrane structure, which improved not only the thermal stability but also the methanol resistance.

10.5 High-Temperature PEMs

In an aim to increase PEMFC efficiency and reduce complexity, an increasing amount of research effort has been focused on the development of high-temperature PEMs [140–143]. Operating at elevated temperatures allows improvements to be made in the kinetics of fuel oxidation and oxygen reduction, thus significantly enhancing the fuel cell efficiency. Another advantage relates to simplifying the water management system. As no liquid water is present in the fuel cell above 100 °C, the two-phase flow need not be considered, and this (at least potentially) will lead to a reduction in both the size and complexity of the system. However, several challenges emerge when operating above 100 °C. First, the proton conductivity of the membrane may degrade due to dehydration. The above-mentioned membranes can only function within the temperature range 25–100 °C because their conductivity is dependent on their water content. For example, whereas the conductivity of Nafion at 100% relative humidity (RH) was 0.1–0.2 S cm^{-1} at 30–85 °C, it fell to only 0.00014 S cm^{-1} when the RH was decreased to 34% at 30 °C [144, 145].

The currently available PEMs for high-temperature fuel cells may be classified as two distinct categories, in relation to the temperature range. The inorganic–organic hybrid membranes referred to in Sections 10.2.3 and 10.3.3, which potentially offer operation at temperatures above 80 °C, still depend on humidification to maintain an acceptable conductivity. Accordingly, these membranes might provide a maximum operational temperature of only 120 °C. A brief summary of possible candidate membranes for PEMFCs operating at temperatures of up to 200 °C is provided in the following sections.

10.5.1 Acid-Doped Polybenzimidazole

Polybenzimidazole (PBI) has been suggested as a useful polymer candidate for PEMFCs on the basis of its excellent thermochemical stability and mechanical properties. Commercially available PBI has a T_g of 425–435 °C, and shows a long-term thermal stability above 350 °C. As shown in Figure 10.8, PBI is a basic polymer

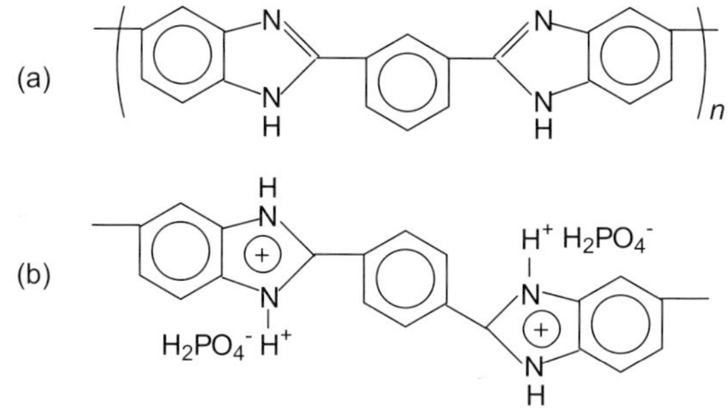

(a)

(b)

Figure 10.8 Chemical structure of (a) polybenzimidazole (PBI) and (b) phosphoric acid-doped PBI.

(pK_a = 5.5) which is able to form a complex with a strong acid or a very strong base. Aharomi *et al.* [146] were the first to recognize that PBI, when doped with phosphoric or sulfuric acid, was a potential electrolyte. Thereafter, phosphoric acid-doped PBI membranes have been intensively investigated for use in high-temperature PEMFCs.

Phosphoric acid-doped PBI membranes may be prepared by either of two methods [147–153]. The first method involves casting from a polymer solution in NaOH/ethanol or LiCl/DMAC, followed by water washing and doping by immersion in phosphoric acid solution. The acid loading is calculated from the weight difference of PBI before and after immersion, with the loading being determined by the molarity of the acid. For example, immersion in a 5 *M* acid solution provides a doping level of 3 mol acid per PBI repeat unit (designated as a 300% doping level), while immersion in an 11 *M* acid solution provides a 520% doping level. The second method involves direct casting from a PBI and phosphoric acid solution in a suitable solvent, usually trifluoroacetic acid (TFA). The latter method enables direct control of the acid loading.

The membranes formed via these two processes demonstrate different performance levels. Typically, the post-doped PEMs are stronger and tougher than directly cast membranes, which require a polymer of a higher inherent viscosity in order to prepare membranes with reasonable strength. In general, the conductivity of the directly cast membrane will be significantly higher than that of post-doped membranes [154].

Conduction in such a PBI electrolyte is heavily dependent on the doping level [155]. Over the acid doping range of 200% to 560%, the activation energies range from 18 to 25 $kJ\,mol^{-1}$. At a doping level of 200%, membrane conductivity is approximately $2.5 \times 10^{-2}\,S\,cm^{-1}$ at 200 °C, and the presence of free or unbound acid is necessary to improve the transport. At a doping level of 570% phosphoric acid, the conductivity is $4.6 \times 10^{-3}\,S\,cm^{-1}$ at room temperature, $4.8 \times 10^{-2}\,S\,cm^{-1}$ at 170 °C, and $7.9 \times 10^{-2}\,S\,cm^{-1}$ at 200 °C.

An electro-osmotic drag due to proton migration is defined as the number of water molecules moved with each proton in the absence of a concentration gradient. For comparison, the electro-osmotic drag coefficient for vapor or liquid-equilibrated Nafion membranes ranges from 0.9 to 3.2 at room temperature [146]. For phosphoric acid-doped PBI membranes, however, the water drag coefficient is close to zero [149, 151], indicating that the conductivity of the doped PBI is less demanding on fuel humidification during fuel cell operation; a hopping mechanism has been suggested in order to explain the proton transport in H_3PO_4/PBI.

In H_2/O_2 fuel cell tests with phosphoric acid-doped PBI membranes [156], power densities of 0.28 and 0.55 $W\,cm^{-1}$ were reported for a H_3PO_4/PBI membrane operating at atmospheric pressure using dry H_2 and O_2 at temperatures of 125 °C and 200 °C, respectively. The CO tolerance was assessed by supplying different amounts of CO in the hydrogen feed gas [157]; a 3% CO admixture was found to be permissible when the operating temperature was 200 °C.

The oxidative decay and mechanical degradation, which resulted in failure of the MEA, proved to be the main reasons for poor long-term performance of the single cell. When Zhai *et al.* [158] studied the lifetimes and performance degradation mechanism of H_3PO_4/PBI high-temperature PEMFCs, their results showed that the dissolution and re-deposition of platinum caused agglomeration of the Pt particles and Pt deposition onto H_3PO_4/PBI. This in turn resulted in a smaller electrochemically active surface area of the cathode, and degradation of the single cells. The loss of H_3PO_4 from the H_3PO_4/PBI membranes had only a minor influence on cell performance.

10.5.2 Nitrogen-Containing Heterocycles

The nitrogen-containing heterocycles, such as imidazole, pyrazole, or benzimidazole (Figure 10.9), exhibit a moderate intrinsic proton conductivity [159]. The basic nitrogen site acts as a strong proton acceptor for Brønsted acids, such as sulfonic acid groups, to form protonic charge carriers $(C_3H_3N_2H_2)^+$. The compact molecules are advantageous for extended local dynamics, while the protonated and unprotonated nitrogen functionals may act as donors and acceptors in the proton-transfer mechanisms. The aromatic ring within imidazole is itself rather nonpolar, and this tends to promote proton-transfer reactions. As the melting point is higher than that of water and phosphoric acid, imidazole becomes an interesting candidate for supporting proton conductivity at elevated temperatures (120–200 °C).

The Grotthus-type diffusion mechanism was used to explain the proton diffusion process in imidazole chains [160]. The protonic defects cause local rather than long-range disorder by forming (... HIm - $(HImH)^+$ - ImH ...) and ^{t}(ImH) configurations. At 117 °C, the proton-transfer step is fast, with a time scale of 0.3 ps; the reorientation step is rate-determining and the corresponding time scale is approximately 30 ps.

The ionic conductivities of sulfonated polyetherketone/imidazole (pyrazole) systems were first studied by Kreuer *et al.* [161]. The intercalation of imidazole (pyrazole) into the polymer with Brønsted acid functions was shown to produce a high protonic conductivity (ca. 10^{-2} S cm^{-1}). However, the volatility of these heterocycles hampers their application for high-temperature PEMFCs. In order to improve the thermal stability, the immobilization of these heterocycle systems was evaluated [162–164]. When Schuster *et al.* characterized the conductivity of imidazole-terminated ethylene

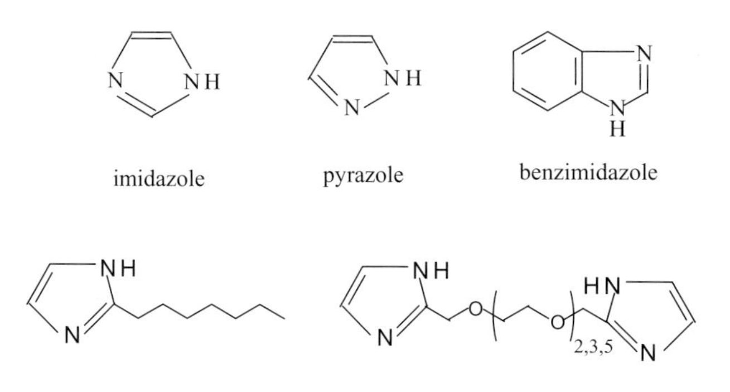

Figure 10.9 Structures of nitrogen-containing heterocycles.

oxide oligomers, the conductivity of the completely water-free materials was 10^{-5} S cm^{-1}, and further enhanced (up to 10^{-3} S cm^{-1}) after triflic acid-doping at 120 °C. Based on recent experimental results, the 2-heptyl-imidazole showed a large electrochemical stability window and proton conductivity of 10^{-3} S cm^{-1} at 150 °C. However, the insufficient thermo-oxidative stability and high overpotential for oxygen reduction on platinum are severe disadvantages for high-temperature PEMFC applications.

10.5.3
Room-Temperature Ionic Liquids

Room-temperature ionic liquids (RTILs), which are liquid at ambient temperature, contain essentially only organic *cations*, such as 1-alkyl-3-methylimidazolium, 1-alkylpyridinium, *N*-methyl-*N*-alkylpyrrolidinium and ammonium ions, and *anions* such as halides, inorganic anions, and large organic anions. RTILs can be combined with polymers to form ionic liquid–polymer composite membranes. Because of their unique properties, which include high thermal stability, good mechanical properties, negligible volatility and, especially, relatively high ionic conductivity, much effort is currently being centered on the use of RTIL–polymer composite membranes as new candidate proton-conducting materials. An example would be a fuel cell electrolyte capable of operating under anhydrous conditions and at elevated temperatures.

Many approaches have been developed for the production of ionic liquid–polymer composite membranes. For example, Doyle *et al.* [165] prepared RTILs/PFSA composite membranes by swelling the Nafion with ionic liquids. When 1-butyl, 3-methyl imidazolium trifluoromethane sulfonate was used as the ionic liquid, the ionic conductivity of the composite membrane exceeded 0.1 S cm^{-1} at 180 °C. A comparison between the ionic liquid-swollen membrane and the liquid itself indicated substantial proton mobility in these composites. Fuller *et al.* [166] prepared ionic liquid–polymer gel electrolytes by blending hydrophilic RTILs into a poly(vinylidene fluoride)–hexafluoropropylene copolymer [PVdF(HFP)] matrix. The gel electrolytes prepared with an ionic liquid : PVdF(HFP) mass ratio of 2 : 1 exhibited ionic conductivities $>10^{-3}$ S cm^{-1} at room temperature, and $>10^{-2}$ S cm^{-1} at 100 °C. When Noda and Watanabe [167] investigated the *in situ* polymerization of vinyl monomers in the RTILs, they produced suitable vinyl monomers that provided transparent, mechanically strong and highly conductive polymer electrolyte films. As an example, a 2-hydroxyethyl methacrylate network polymer in which $BPBF_4$ was dissolved exhibited an ionic conductivity of 10^{-3} S cm^{-1} at 30 °C.

Noda *et al.* [168] reported the details of Brønsted acid-based ionic liquids consisting of a monoprotonic acid and an organic base, in particular solid bis(trifluoromethanesulfonyl)amide (HTFSI) and solid imidazole (Im) mixed at various molar ratios to form liquid fractions. Studies of the conductivity, ^{1}H NMR chemical shift, self-diffusion coefficient, and electrochemical polarization results indicated that, for the Im excess compositions, the proton conductivity increased with an increasing Im molar fraction, with rapid proton-exchange reactions taking place between the protonated Im cation and Im. Proton conduction was found to occur via a combination of Grotthuss- and vehicle-type mechanisms. Recently, Nakamoto [169] reported the creation of a simple protic ionic liquid obtained by combining diethylmethylamine (dema) and trifluoromethanesulfonic acid (T_fOH). This liquid ([dema]/[T_fOH] = 1/1) exhibited a high thermal stability (onset weight-loss temperature of 360 °C, liquefying at −13.1 °C), and substantial ionic conductivity (4.3×10^{-2} S cm^{-1} at 120 °C). In addition, a facile oxygen reduction – which led to a high, stable OCV (1.03 V) – was reported for the H_2/O_2 fuel cell. It should be noted that the critical issue for the use of protic ionic liquids in fuel cells is how to process thin membranes with a satisfactory H_2/O_2 leakage.

10.5.4
Inorganic Membranes: A Brief Comparison

Many solid acids, such as $CsHSO_4$ and CsH_2PO_4, have attracted much attention for applications in intermediate-temperature fuel cells. As these materials are considered in greater detail in Chapter 7, only a few relevant examples are listed here which provide experimental data that can be compared with PEMFC performance. In particular, Haile *et al.* [170] demonstrated a continuous, stable power generation for H_2/O_2 fuel cells operated at 160 °C with a humidity-stabilized $CsHSO_4$ electrolyte. The stable conductivity of $CsHSO_4$ (1.3×10^{-4} S cm^{-1}) was determined at 146 °C and under humidified air. Of note, two critical factors – namely operation above 100 °C and harmless impact of steam in the case of $CsHSO_4$ electrolyte – were apparent. Analogously, CsH_2PO_4 was tested as an electrolyte in a solid acid fuel cell [171]. The thermal analysis of CsH_2PO_4 in H_2 and O_2 atmospheres confirmed that the phosphate-based compound cannot be reduced to form solid phosphorus or gaseous H_xP species. The H_2/O_2 fuel cell with a 260 μm-thick CsH_2PO_4 membrane was operated at 235 °C, in which both the anode and the cathode gases were humidified to prevent dehydration; during the subsequent 100 h period, both the polarization performance and OCV were remarkably stable. Despite the much higher power and current densities than for the $CsHSO_4$ electrolyte, the electrolyte thickness was the primary performance-limiting factor in the case of CsH_2PO_4. Although these solid acids demonstrate excellent proton transport, they have not been used on a practical basis owing to their unstable conductivity and poor mechanical properties, and the introduction of an inorganic porous support matrix might, therefore, represent an effective approach. For instance, Bocchetta *et al.* [172] fabricated thin nanoporous alumina membranes filled with $CsHSO_4$, while Wang *et al.* [173] prepared a $CsHSO_4$–SiO_2 nanocomposite. Although the results of these studies showed significant improvements in conductivity, the stability and tightness of these inorganic pore-filling membranes remain a major challenge in fuel cell development.

10.6
Conclusions

Performance, durability, and cost are the three basic requirements for the commercial application of PEMFCs. Whilst state-of-the-art PEMFC technology is based on the use of PFSA membranes, extensive investigations have been undertaken in the field of alternative PEM systems. These developments have been focused primarily on reducing the costs and increasing the temperature of PEMFC operation. In particular, composite membranes have attracted much interest and have shown great promise as electrolytes for PEMs. Aspects related to the structures and characterization of the composite membranes are also very important for further developments. Although a number of proton-conducting electrolyte systems have demonstrated good performance, their use in PEMFCs remains very limited. The more serious issues are related to the unsatisfactory durability of novel, non-fluorinated PEMs and the various instabilities of the membranes. The long-term durability of composite membranes has not yet been sufficiently investigated, and this has in turn led to the need for further optimization, modeling and PEMFC performance analysis.

Acknowledgments

The authors thank Dr Liang Wang and Dr Dan Zhao, Dalian Institute of Chemical Physics (CAS), for collating the references used in this review, and Dr Hongmei Yu for her critical reading of the manuscript.

References

1 Prater, K.B. (1994) *J. Power Sources*, **51**, 129–44.
2 Joon, K. (1998) *J. Power Sources*, **71**, 12–18.
3 Kumm, W.H. (1990) *J. Power Sources*, **29**, 169–79.
4 Adams, V.W. (1990) *J. Power Sources*, **29**, 181–92.
5 U.S. Department of Energy, Office of Fossil Energy, National Energy Technology Laboratory (2004) *Fuel Cell Handbook*, 7th edn, EG&G Technical Services, Inc., Morgantown, West Virginia, Contract No. DE-AM26-99FT40575.
6 Appleby, A.J. (1996) *Philos. Trans. R. Soc. London, Ser. A*, **354**, 1681–93.
7 Kerres, J., Zhang, W., Jörissen, L. *et al.* (2002) *J. New Mater. Electrochem. Systems*, **5**, 97–107.
8 Du, X., Yu, J., Yi, B. *et al.* (2001) *Phys. Chem. Chem. Phys.*, **3**, 3175–9.
9 Mittal, V.O., Kunz, H.R. and Fenton, J.M. (2007) *J. Electrochem. Soc.*, **154**, B652–56.
10 Paik, C., Skiba, T., Mittal, V. *et al.* (2005) Proceedings of The Electrochemical Society Meeting, May 15–20, Vol. 2005-1, Quebec City, Quebec, Canada, Abstract 771.
11 Endoh, E., Terazono, S., Widjaja, H. *et al.* (2004) *Electrochem. Solid-State Lett.*, **7**, A209–11.
12 Yu, J., Matsuura, T., Yoshikawa, Y. *et al.* (2005) *Electrochem. Solid-State Lett.*, **8**, A156–58.
13 Teranishi, K., Kawata, K., Tsushima, S. *et al.* (2006) *Electrochem. Solid-State Lett.*, **9**, A475–77.
14 Kinumoto, T., Inaba, M., Nakayama, Y. *et al.* (2006) *J. Power Sources*, **158**, 1222–28.
15 Kundu, S., Simon, L.C. and Fowler, M.W. (2008) *Polym. Degrad. Stab.*, **93**, 214–24.
16 Aoki, M., Uchida, H. and Watanabe, M. (2005) *Electrochem. Commun.*, **7**, 1434–8.
17 Ramaswamy, N., Hakim, N., Mukerjee, S. *et al.* (2008) *Electrochim. Acta*, **53**, 3279–95.

18 Mittal, V.O., Kunz, H.R. and Fenton, J.M. (2006) *J. Electrochem. Soc.*, **153**, A1755–9.

19 Liu, W. and Zuckerbrod, D. (2005) *J. Electrochem. Soc.*, **152**, A1165–70.

20 Ibrahim, S.M., Price, E.H., Smith, R.A. (1983) *Proc. Electrochem. Soc.*, 83–6.

21 Resnick, P.R. and Grot, W.G. (1978) US Patent 4,113, 585.

22 Steck, A. (1995) in Proceedings of the First International Symposium on New Materials for Fuel Cell Systems (eds O. Savadogo, P.R. Roberge and T.N. Veziroglu), July 9–13, Montreal, Canada.

23 Curtin, D.E., Lousenberg, R.D. and Henry, T.J. (2004) *J. Power Sources*, **131**, 41–8.

24 http://www.dupont.com/fuelcells/products/nafion.html.

25 Kreuer, K.D., Schuster, M., Obliers, B. *et al.* (2008) *J. Power Sources*, **178**, 499–509.

26 Wei, J., Stone, C. and Steck, A.E. (1995) US Patent 5,422,411.

27 Savadogo, O. (1998) *J. New Mater. Electrochem. Systems*, **1**, 47–66.

28 Büchi, F.N., Gupta, B., Haas, O. *et al.* (1995) *J. Electrochem. Soc.*, **142** (9), 3044–8.

29 Soresi, B., Quartarone, E., Mustarelli, P. *et al.* (2004) *Solid State Ionics*, **166**, 383–9.

30 Yamakia, T., Asano, M., Maekawa, Y. *et al.* (2003) *Radiat. Phys. Chem.*, **67**, 403–7.

31 Bahar, B., Hobson, A.R. and Kolde, J.A. (1997) US Patent 5,599,614.

32 Liu, F., Yi, B., Xing, D. *et al.* (2003) *J. Membr. Sci.*, **212**, 213–23.

33 Tang, H., Pan, M., Jiang, S. *et al.* (2007) *Electrochim. Acta*, **52**, 5304–11.

34 Liu, Y., Yi, B., Shao, Z. *et al.* (2006) *Electrochem. Solid-State Lett.*, **9** (7), A356–9.

35 Wang, L., Xing, D., Zhang, H. *et al.* (2008) *J. Power Sources*, **176**, 270–5.

36 Smitha, B., Sridhar, S. and Khan, A.A. (2003) *J. Membr. Sci.*, **225**, 63–76.

37 Smitha, B., Sridhar, S. and Khan, A.A. (2004) *Macromolecules*, **37**, 2233–9.

38 Watanabe, M., Uchida, H., Seki, Y. *et al.* (1996) *J. Electrochem. Soc.*, **143**, 3847–52.

39 Watanabe, M., Uchida, H. and Emori, M. (1998) *J. Phys. Chem. B*, **192**, 3129–37.

40 Uchida, H., Mizuno, Y. and Watanabe, M. (2002) *J. Electrochem. Soc.*, **149**, A682–9.

41 Yang, J., Li, Y., Huang, Y. *et al.* (2008) *J. Power Sources*, **177**, 56–60.

42 Kwak, S., Yang, T., Kim, C. *et al.* (2004) *Electrochim. Acta*, **50**, 653–7.

43 Adjemian, K.T., Lee, S.J., Srinivasan, S. *et al.* (2002) *J. Electrochem. Soc.*, **149** (3), A256–61.

44 Jung, D.H., Cho, S.Y., Peck, D.H. *et al.* (2002) *J. Power Sources*, **106**, 173–7.

45 Liu, Y., Yi, B. and Zhang, H. (2005) *Chin. J. Power Sources*, **29**, 92–4.

46 Saccà, A., Carbone, A., Passalacqua, E. *et al.* (2005) *J. Power Sources*, **152**, 16–21.

47 Rhee, C., Kim, Y., Lee, J. *et al.* (2006) *J. Power Sources*, **159**, 1015–24.

48 Ramani, V., Kunz, H.R. and Fenton, J.M. (2004) *J. Membr. Sci.*, **232**, 31–44.

49 Watanabe, M., Sakairi, M. and Inoue, K. (1994) *J. Electroanal. Chem.*, **375**, 415–18.

50 Lee, H.K., Kim, J.I., Park, J.H. *et al.* (2004) *Electrochim. Acta*, **50**, 761–8.

51 Watanabe, M., Uchida, H. and Emori, M. (1998) *J. Electrochem. Soc*, **145**, 1137–41.

52 Yang, T., Yoon, Y., Kim, C. *et al.* (2002) *J. Power Sources*, **106**, 328–32.

53 Kwak, S., Yang, T., Kim, C. *et al.* (2003) *J. Power Sources*, **118**, 200–4.

54 Lee, H., Kim, J., Park, J. *et al.* (2004) *Electrochim. Acta*, **50**, 761–8.

55 Li, M., Shao, Z., Zhang, H. *et al.* (2006) *Electrochem. Solid-State Lett.*, **9** (2), A92–5.

56 Wang, L., Yi, B., Zhang, H. *et al.* (2007) *Electrochim. Acta*, **52**, 5479–83.

57 Liu, F., Yi, B., Xing, D. *et al.* (2003) *J. Power Sources*, **124**, 81–9.

58 Wang, L., Xing, D., Liu, Y. *et al.* (2006) *J. Power Sources*, **161**, 61–7.

59 Liu, Y., Yi, B., Shao, Z. *et al.* (2007) *J. Power Sources*, **163**, 807–13.

60 Matos, B.R., Santiago, E., Fonseca, F.C. *et al.* (2007) *J. Electrochem. Soc.*, **154** (12), B1358–61.

61 Lufrano, F., Squadrito, G., Patti, A. *et al.* (2000) *J. Appl. Polym. Sci.*, **77**, 1250–7.

62 Zaidi, S.M.J., Mikhailenko, S.D., Robertson, G.P. *et al.* (2000) *J. Membr. Sci.*, **173**, 17–34.

63 Kerres, J. (2001) *J. Membr. Sci.*, **185**, 3–27.

64 Alberti, G., Casciola, M., Massinelli, L. *et al.* (2001) *J. Membr. Sci.*, **185**, 73–81.

65 Noshay, A. and Robeson, L.M. (1976) *J. Appl. Polym. Sci.*, **20** (7), 1885–903.

66 Genova-Dimitrova, P. (2001) *J. Membr. Sci.*, **185**, 59–71.

67 Gao, Y., Robertson, G.P., Guiver, M.D. *et al.* (2003) *J. Membr. Sci.*, **227**, 39–50.

68 Hickner, M.A., Ghassemi, H., Kim, Y.S. *et al.* (2004) *Chem. Rev.*, **104**, 4587–612.

69 Kerres, J., Zhang, W. and Cui, W. (1998) *J. Polym. Sci.: Part A: Polym. Chem.*, **36**, 1441–8.

70 Karlsson, L.E. and Jannasch, P. (2004) *J. Membr. Sci.*, **230**, 61–70.

71 Wang, F., Hickner, M., Kim, Y.S. *et al.* (2002) *J. Membr. Sci.*, **197**, 231–42.

72 Wang, F., Hickner, M., Ji, Q. *et al.* (2001) *Macromol. Symp.*, **175**, 387–95.

73 Mecham, J.B. (2001) Direct polymerization of sulfonated poly (arylene ether) random copolymers and poly(imide) sulfonated poly(arylene ether) segmented copolymers: new candidates for proton exchange membrane fuel cell material systems. Ph.D. Thesis, Virginia Polytechnic Institute and State University.

74 McGrath, J.E., Hickner, M., Kim, Y.S. *et al.* (2003) Advances in Materials for PEMFC systems, Asilomar Conference Grounds, February 23–26, Pacific Grove, California, p. 22.

75 Gil, M., Ji, X. and Li, X. (2004) *J. Membr. Sci.*, **234**, 75–81.

76 Kim, Y.S., Wang, F., Hickner, M. *et al.* (2003) *J. Membr. Sci.*, **212**, 263–82.

77 Harrison, W.L., Wang, F., Mecham, J.B. *et al.* (2003) *J. Polym. Sci., Part A: Polym. Chem.*, **41**, 2264–76.

78 Kim, Y.S., Hickner, M.A., Dong, L. *et al.* (2004) *J. Membr. Sci.*, **243**, 317–26.

79 Wiles, K.B., Wang, F. and McGrath, J.E. (2005) *J. Polym. Sci. Part A: Polym. Chem.*, **43**, 2964–76.

80 Xing, D. and Kerres, J. (2006) *J. New Mater. Electrochem. Syst.*, **9**, 51–60.

81 Genies, C., Mercier, R., Sillion, B. *et al.* (2001) *Polymer*, **42**, 5097–105.

82 Faure, S., Mercier, R., Aldebert, P. *et al.* (1996) French Patent 9605707.

83 Faure, S., Cornet, N., Gebel, G. *et al.* (1997) in Proceedings of the Second International Symposium on New Materials for Fuel Cell and Modern Battery Systems (eds O. Savadogo and P.R. Roberge), Montreal, Canada, p. 818.

84 Guo, X., Tanaka, K., Kita, H. *et al.* (2004) *J. Polym. Sci., Part A: Polym. Chem.*, **42**, 1432–40.

85 Miyatake, K., Zhou, H. and Watanabe, M. (2004) *Macromolecules*, **37**, 4956–60.

86 Yin, Y., Suto, Y., Sakabe, T. *et al.* (2006) *Macromolecules*, **39**, 1189–98.

87 Meyer, G., Perrot, C., Gebel, G. *et al.* (2006) *Polymer*, **47**, 5003–11.

88 Fang, J., Guo, X., Xua, H. *et al.* (2006) *J. Power Sources*, **159**, 4–11.

89 Kreuer, K.D. (2001) *J. Membr. Sci.*, **185**, 29–39.

90 Kreuer, K.D., Paddison, S.J., Spohr, E. *et al.* (2004) *Chem. Rev.*, **104**, 4637–78.

91 Hickner, M., Wang, F., Kim, Y. *et al.* (2001) Chemistry–morphology–property relationships of novel proton exchange membranes for direct methanol fuel cells. ACS Fuel (Part 1), Vol. 222, August 26–30, Chicago, p. 51.

92 Yeager, E. (1984) *Electrochim. Acta.*, **29**, 1527–37.

93 LaConti, A.B., Hamdan, M. and McDonald, R.C. (2003) in *Handbook of Fuel Cells: Fundamentals, Technology, and Applications*, Vol. 3 (eds W. Vielstich, A. Lamm and H.A. Gasteiger), John Wiley & Sons, New York, pp. 647–62.

94 Yu, J., Yi, B., Xing, D. *et al.* (2003) *Phys. Chem. Chem. Phys.*, **5**, 611–51.

95 Wang, L., Yi, B., Zhang, H. *et al.* (2007) *J. Power Sources*, **164**, 80–5.

96 Xing, D. and Kerres, J. (2006) *Polym. Adv. Technol.*, **17**, 591–7.

97 Haugen, G.M., Meng, F.Q., Aieta, N. *et al.* (2006) The 210th Electrochemical Society

Meeting, Cancun, Mexico, October 29–November 3, Electrochemical Society.

98 Andrews, N.R., Knights, S.D., Murray, K.A. *et al.* (2005) US Patent 0,136,308.

99 Cipollini Ned, E., Condit, D.A., Hertzberg, J.B. *et al.* (2006) US Patent 7,112,386.

100 Xing, D., Zhang, H., Wang, L. *et al.* (2007) *J. Membr. Sci.*, **296**, 9–14.

101 Jung, H. and Park, J. (2007) *Electrochim. Acta*, **52**, 7464–8.

102 Huang, R.Y.M., Shao, P., Burne, C.M. *et al.* (2001) *J. Appl. Polym. Sci.*, **82**, 2651–60.

103 Ulrich, H. and Rafler, G. (1998) *Die Angew. Makromol. Chemie*, **263**, 71–8.

104 Bauer, B., Jones, D.J., Roziere, J. *et al.* (2000) *J. New Mater. Electrochem. Systems*, **3**, 93–8.

105 Xing, D., Yi, B., Liu, F. *et al.* (2005) *Fuel Cells*, **5** (3), 406–411.

106 Zhu, X., Zhang, H., Liang, Y. *et al.* (2007) *J. Mater. Chem.*, **17**, 386–97.

107 Kerres, J., Cui, W. and Reichle, S. (1996) *J. Polym. Sci.: Part A: Polym. Chem.*, **34**, 2421–38.

108 Zhang, W., Tang, C.-M. and Kerres, J. (2001) *Sep. Purif. Technol.*, **22**, 209–21.

109 Kerres, J., Tang, C.-M. and Graf, C. (2004) *Ind. Eng. Chem. Res.*, **43** (16), 4571–9.

110 Kerres, J.A. and van Zyl, A.J. (1999) *J. Appl. Polym. Sci.*, **74**, 428–38.

111 Kerres, J., Ullrich, A. and Hein, M. (2001) *J. Polym. Sci.: Part A: Polym. Chem.*, **39**, 2874–88.

112 Kerres, J., Hein, M., Zhang, W. *et al.* (2003) *J. New Mater. Electrochem. Systems*, **6** (4), 223–9.

113 Tang, C.M., Zhang, W. and Kerres, J. (2004) *J. New Mater. Electrochem. Systems*, **7** (4), 287–98.

114 Kerres, J. (2005) *Fuel Cells*, **5** (2), 230–47.

115 Kerres, J., Tang, C.M. and Graf, C. (2004) *Ind. Eng. Chem. Res.*, **43**, 4571–9.

116 Kerres, J. (2001) *J. Membr. Sci.*, **185** (1), 3–27.

117 Kerres, J., Ullrich, A., Meier, F. *et al.* (1999) *Solid State Ionics*, **125**, 243–9.

118 Cui, W., Kerres, J. and Eigenberger, G. (1998) *Sep. Purif. Technol.*, **14**, 145–54.

119 Zhang, W., Tang, C.M. and Kerres, J. (2001) *Sep. Purif. Technol.*, **22–23**, 209–21.

120 Schonberger, F., Hein, M. and Kerres, J. (2007) *Solid State Ionics*, **178** (7–10), 547–54.

121 Kerres, J., Ullrich, A., Harin, T. *et al.* (2000) *J. New Mater. Electrochem. Systems*, **3** (3), 229–39.

122 Kerres, J., Zhang, W. and Jörissen, L. (2002) *J. New. Mater. Electrochem. Systems*, **5**, 97–107.

123 Doytcheva, M., Stamenova, R., Zvetkov, V. *et al.* (1998) *Polymer*, **39**, 6715–21.

124 Decker, C. and Viet, T.N.T. (1999) *Macromol. Chem. Phys.*, **200**, 358–67.

125 Liu, Y., Lee, J.Y. and Hong, L. (2004) *J. Power Sources*, **129**, 303–11.

126 Xing, D., Yi, B., Fu, Y. *et al.* (2004) *Electrochem. Solid-State Lett.*, **7** (10), A315–17.

127 Wang, L., Yi, B., Zhang, H. *et al.* (2007) *Electrochim. Acta*, **52**, 5479–83.

128 Zhang, Y., Zhang, H., Zhai, Y. *et al.* (2007) *J. Power Sources*, **168**, 323–9.

129 Zhang, Y., Zhang, H., Zhua, X. *et al.* (2007) *J. Power Sources*, **165**, 786–92.

130 Wang, L., Yi, B., Zhang, H. *et al.* (2007) *J. Power Sources*, **167**, 47–52.

131 Zhang, Y. (2007) Studies on sulfonated poly(ether ether ketone) and its composite membrane for proton exchange membrane fuel cells. Ph.D. Thesis, Dalian University of Technology, Dalian, P.R. China.

132 Shahi, V. (2007) *Solid State Ionics*, **177**, 3395–404.

133 Di Vona, M.L., Ahmed, Z. and Bellitto, S. (2007) *J. Membr. Sci.*, **296**, 156–61.

134 Sambandam, S. and Ramani, V. (2007) *J. Power Sources*, **170**, 259–67.

135 Kjar, J., Yde-Andersen, S., Knudsen, N.A. *et al.* (1991) *Solid State Ionics*, **46**, 169–73.

136 Plotarzewski, Z., Wieczorek, W., Przyluski, J. *et al.* (1999) *Solid State Ionics*, **119**, 301–4.

137 Miyake, N., Wainright, J.S. and Savinell, R.F. (2001) *J. Electrochem. Soc.*, **148**, A898–904.

138 Su, Y., Liu, Y. and Sun, Y. (2007) *J. Membr. Sci.*, **296**, 21–8.

139 Zhang, G. and Zhou, Z. (2005) *J. Membr. Sci.*, **261**, 107–13.

140 Yang, C., Costamagna, P., Srinivasan, S. *et al.* (2001) *J. Power Sources*, **103**, 1–9.

141 Li, Q., He, R., Jensen, J.O. *et al.* (2003) *Chem. Mater.*, **15**, 4896–915.

142 Zhang, J., Xie, Z., Zhang, J. *et al.* (2006) *J. Power Sources*, **160**, 872–91.

143 Savadogo, O. (2004) *J. Power Sources*, **127**, 135–61.

144 Anantaraman, A.V. and Gardner, C.L. (1996) *J. Electroanal. Chem.*, **414**, 115–20.

145 Ma, C., Zhang, L., Mukerjee, S. *et al.* (2003) *J. Membr. Sci.*, **219**, 123–36.

146 Aharomi, S.M. and Litt, M.H. (1974) *J. Polym. Sci., Part A: Polym. Chem.*, **12**, 639–50.

147 Wainright, J.S., Wang, J.T., Savinell, R.F. *et al.* (1995) *J. Electrochem. Soc.*, **142**, L121–3.

148 Fontanella, J.J., Wintersgill, M.C., Wainright, J.S. *et al.* (1998) *Electrochim. Acta*, **43**, 1289–94.

149 Weng, D., Wainright, J.S., Landau, U. *et al.* (1996) *J. Electrochem. Soc.*, **143**, 1260–3.

150 Samms, S.R., Wasmus, S. and Savinell, R.F. (1996) *J. Electrochem. Soc.*, **143**, 1225–32.

151 Glipa, X., Bonnet, B., Jones, D.J. *et al.* (1999) *J. Mater. Chem.*, **9**, 3045–9.

152 Li, Q., Hjuler, H.A. and Bjerrum, N.J. (2001) *J. Appl. Electrochem.*, **31**, 773–9.

153 Bouchet, R. and Siebert, E. (1999) *Solid State Ionics*, **118**, 287–99.

154 Wainright, J.S., Litt, M. and Savinell, R.F. (2003) *Handbook of Fuel Cell, Fundamentals, Technology, and Application*, Vol. 3 (eds W. Vielstich, A. Lamm and H.A. Gasteiger), John Wiley & Sons, New York.

155 He, R., Li, Q., Xiao, G. *et al.* (2003) *J. Membr. Sci.*, **226**, 169–84.

156 Jensen, J.O., Li, Q., He, R. *et al.* (2003) 1st European Hydrogen Energy Conference, September 2–5, Grenoble, France.

157 Li, Q., He, R., Gao, J. *et al.* (2003) *J. Electrochem. Soc.*, **150**, A1599–605.

158 Zhai, Y., Zhang, H., Liu, G. *et al.* (2007) *J. Electrochem. Soc.*, **154**, B72–6.

159 Kawada, A., McGhie, A.R. and Labes, M.M. (1970) *J. Chem. Phys.*, **52**, 3121.

160 Münch, W., Kreuer, K. and Silvestri, W. (2001) *Solid State Ionics*, **145**, 437–43.

161 Kreuer, K., Fuchs, A., Ise, M. *et al.* (1998) *Electrochim. Acta*, **43**, 1281–8.

162 Schuster, M., Meyer, W., Wegner, G. *et al.* (2001) *Solid State Ionics*, **145**, 85–92.

163 Schuster, M., Rager, T., Noda, A. *et al.* (2005) *Fuel Cell*, **5**, 355–65.

164 Persson, J.C. and Jannasch, P. (2003) *Chem. Mater.*, **15**, 3044–5.

165 Doyle, M., Choi, S.K. and Proulx, G. (2000) *J. Electrochem. Soc.*, **147** (1), 34–7.

166 Fuller, J., Breda, A.C. and Carlin, R.T. (1998) *J. Electroanal. Chem.*, **459**, 29–34.

167 Noda, A. and Watanabe, M. (2000) *Electrochim. Acta*, **45**, 1265–70.

168 Noda, A., Susan, M.A.B.H., Kudo, K. *et al.* (2003) *J. Phys. Chem. B*, **107**, 4024–33.

169 Nakamoto, H. and Watanabe, M. (2007) *Chem. Commun.*, 2539–41.

170 Haile, S.M., Boysen, D.A., Chisholm, C.R.I. *et al.* (2001) *Nature*, **410**, 910–13.

171 Boysen, D.A., Uda, T., Chisholm, C.R.I. *et al.* (2004) *Science*, **303**, 68–70.

172 Bocchetta, P., Chiavarotti, G., Masi, E. *et al.* (2004) *Electrochem. Commun.*, **6**, 923–8.

173 Wang, S., Otomo, J., Ogura, M. *et al.* (2005) *Solid State Ionics*, **176**, 755–60.

11
Electrochemistry of Electronically Conducting Polymers

Mikhael Levi and Doron Aurbach

Abstract

This chapter is devoted to the preparation, structural, and electrochemical characterization of π-conjugated organic polymers as a typical example of mixed electron–ionic conductors among a vast class of solid, electrochemically active materials. A short review of general features of doping-induced changes in π-conjugated polymers, as well as of their common and specific mechanisms, is provided. The chapter also provides an introduction into the thermodynamics and kinetics of electrochemical doping processes in organic polymers, and emphasizes some common features as well as distinctions with that of ion-insertion into inorganic intercalation cathodes and anodes.

11.1
Introduction

Early studies on π-conjugated molecular organic semiconductors date back to the period between the 1940s and 1970s [1]. However, due to a lack of any effective means of controlling the structure–property relationships, the organic materials studied at the time displayed poorer characteristics as compared to those of conventional, inorganic semiconductors. Two truly outstanding discoveries which were made between the 1970s and 1980s led to a major breakthrough in the field, with the investigators – H. Shirakawa, A. McDiarmid, and A. Heeger – being awarded the Nobel Prize for chemistry in 2000.

By carrying out experiments on the polymerization of acetylene by using a concentrated Ziegler–Natta catalyst solution, Shirakawa and Ikeda were the first to report on the preparation of partially crystalline polyacetylene (PA) films under well-defined conditions [2]. Some six years later, Shirakawa, McDiarmid, and Heeger and colleagues [3] presented unequivocal evidence that, after doping with iodine vapor, the room-temperature electronic conductivity of PA film was

Solid State Electrochemistry I: Fundamentals, Materials and their Applications. Edited by Vladislav V. Kharton

ISBN: 978-3-527-32318-0

increased by seven orders of magnitude, approaching that of typical metals. In 1981, Diaz *et al.* [4] showed, for the first time, that electronically conducting polypyrrole (PPy) films of well-defined structure could be obtained by the direct anodic oxidation (polymerization) of pyrrole in suitable electrolyte solutions. Both, PA and PPy are typical representatives of the rapidly expanding class of π-conjugated organic polymers, usually referred to as electronically conducting polymers (ECPs).

As intensive studies on the ECPs have been carried out for almost 30 years, a vast knowledge of the methods of preparation and the physico-chemical properties of these materials has accumulated [5–17]. The electrochemistry of the ECPs has been systematically and repeatedly reviewed, covering many different and important topics such as electrosynthesis, the elucidation of mechanisms and kinetics of the doping processes in ECPs, the establishment and utilization of structure–property relationships, as well as a great variety of their applications as novel electrochemical systems, and so forth [18–23]. In this chapter, a classification is proposed for electroactive polymers and ion-insertion inorganic hosts, emphasizing the unique feature of ECPs as mixed electronic–ionic conductors. The analysis of thermodynamic and kinetic properties of ECP electrodes presented here is based on a combined consideration of the potential-dependent differential capacitance of the electrode, chemical diffusion coefficients, and the partial conductivities of related electronic and ionic charge carriers.

11.2 Solid Organic and Inorganic Electrochemically Active Materials for Galvanic Cells Operating at Moderate Temperatures

Solid organic and inorganic redox-active materials can be divided into two major classes

- Low-dimensional charge-transfer (CT) complexes, formed by the coupling of π-conjugated electron-donor and electron-acceptor molecules (see Scheme 11.1a).
- Electroactive polymers and inorganic solids (Scheme 11.1b).

This classification takes into account the character of the chemical bonds in the related materials, the prevailing type of conductivity, and the ability of the compounds to release, reversibly, the electronic and ionic charges stored in the bulk of the electrodes, accompanied by pronounced changes in their electronic and ionic conductivities.

11.2.1 Molecular, Low-Dimensional CT Complexes and π-Conjugated Organic Oligomers

A general feature of the complexes under consideration is that they exhibit a metallic or semiconducting type of conductivity as a result of charge transfer between the parent electron donor and electron acceptor organic, π-conjugated molecules [24].

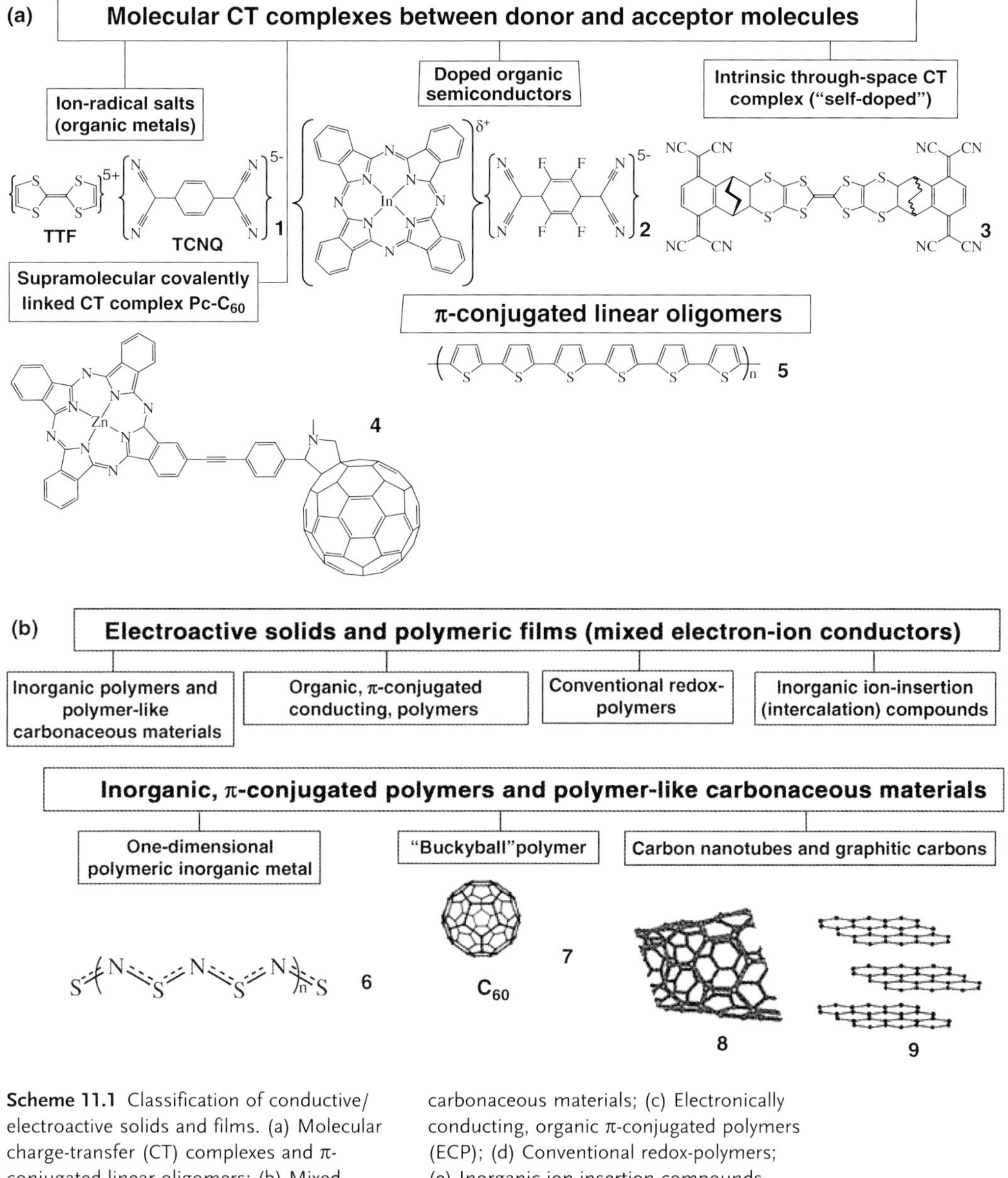

Scheme 11.1 Classification of conductive/electroactive solids and films. (a) Molecular charge-transfer (CT) complexes and π-conjugated linear oligomers; (b) Mixed electronic-ionic conductors: inorganic π-conjugated polymers and polymer-like carbonaceous materials; (c) Electronically conducting, organic π-conjugated polymers (ECP); (d) Conventional redox-polymers; (e) Inorganic ion-insertion compounds. TTF = tetrathiafulvalene; TCNQ = tetracyanoquinodimethane.

The extent of charge transfer is invariable with the applied potential, and thus the charge cannot be reversibly released from their bulk. This subclass (Scheme 11.1a) includes: (i) a highly crystalline, and conducting low-dimensional ion-radical salt (**1**), and covalently linked phthalocyanine (Pc)-fullerene (Pc-C_{60}) conjugate (**4**); and

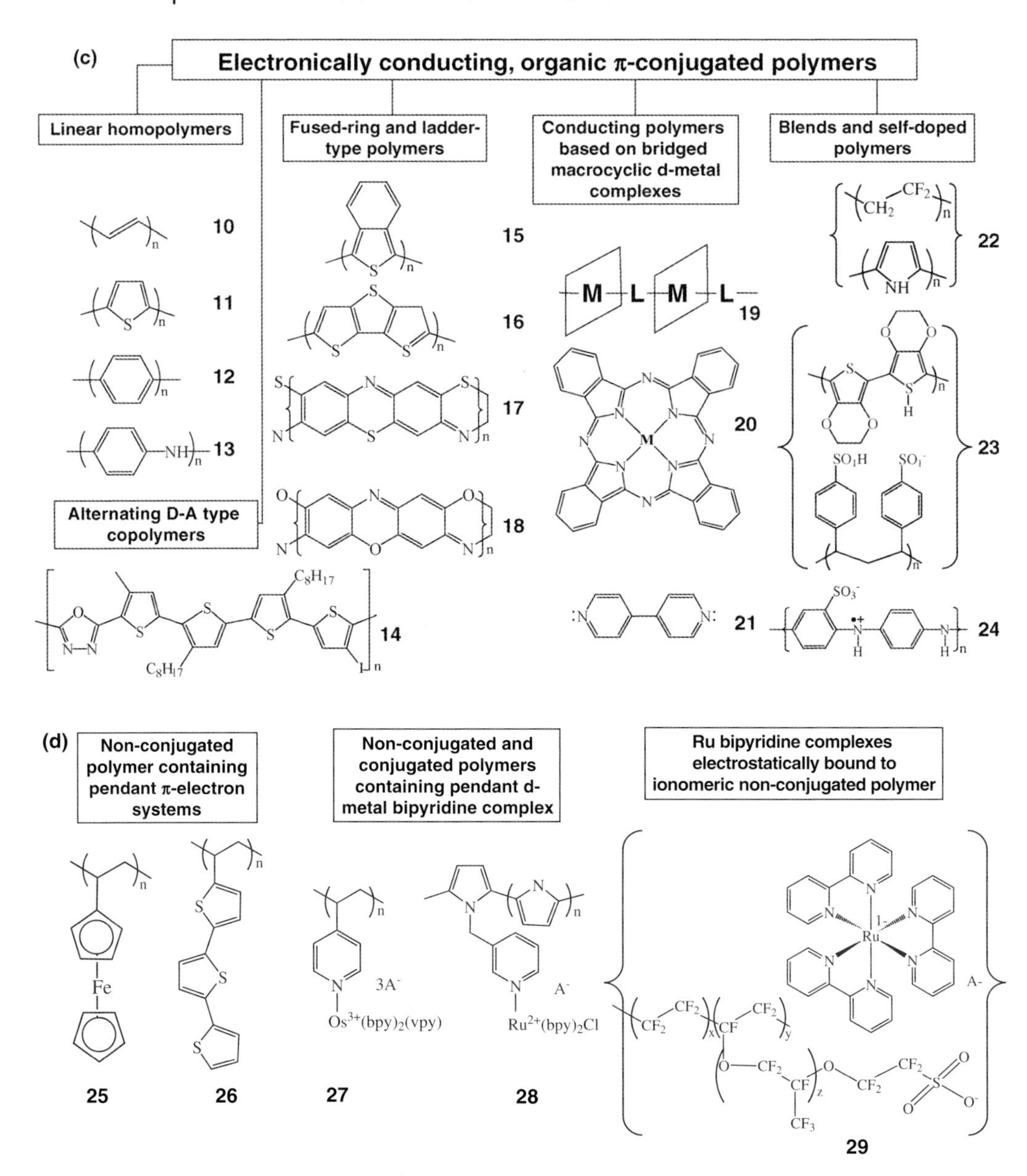

Scheme 11.1 (*Continued*)

(ii) moderately conducting, doped organic semiconductors (**2**), and novel *intramolecular* donor–acceptor systems, linked *covalently* through a rigid, saturated space (**3**).

Linear, π-conjugated organic oligomers (**5**) with well-defined crystalline structures should be separated in a special subclass, bridging the typical molecular charge

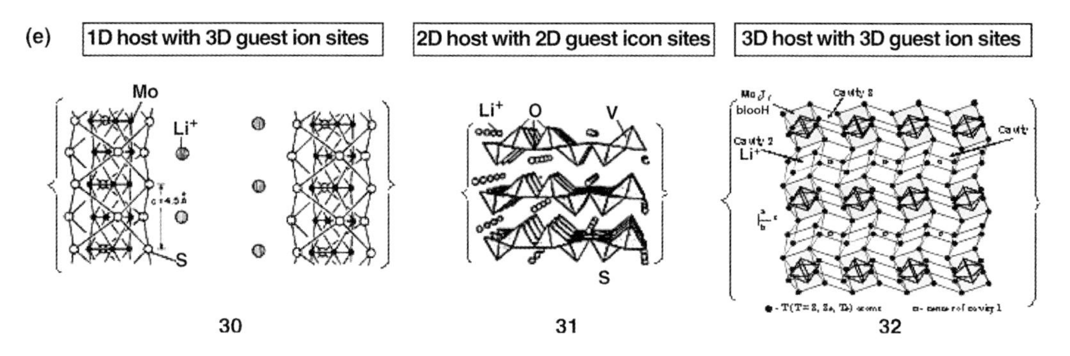

Scheme 11.1 (*Continued*)

transfer complexes and ECPs. The dual nature of the oligomers is revealed by the fact that they form charge-transfer complexes with strong enough acceptor molecules; on the other hand, they can be doped, similar to the ECPs [25].

11.2.2
Electroactive Solids and Polymeric Films with Mixed Electronic–Ionic Conductivity

These are presented by two subclasses of electroactive polymer: (i) π-conjugated polymers of both organic and inorganic nature [5–15]; and (ii) conventional redox polymers [26], and by inorganic ion-insertion (intercalation) compounds [27, 28] (see the top of Scheme 11.1b). Despite the different nature of their chemical bonds, all of these compounds are mixed, electronic–ionic conductors [29], and hence, their electronic and/or ionic conductivity is expected to change with the applied potential in a predictable, characteristic manner (see Section c11.4).

11.2.2.1 Inorganic π-Conjugated Polymers and Polymer-Like Carbonaceous Materials

Poly(sulfur nitride) (compound **6** in Scheme 11.1b) is a crystalline, one-dimensional conductor which was known for its metallic conductivity long before the highly conductive, π-conjugated polymers attracted so much attention. However, in contrast to the organic ECPs, its high conductivity does not change practically with the doping level.

Fullerene C_{60} and its derivatives, with their characteristic "buckyball" structure (**7**), can be formally considered as carbonaceous polymers of a specific three-dimensional (3-D) structure.

In turn, a single-wall carbon nanotube (SWCNT; see **8**) presents an intermediate structure between C_{60} and a well-known crystalline graphitic carbon (**9**). It should be noted that the latter comprises a vast family of graphite intercalation compounds.

11.2.2.2 Organic π-Conjugated Polymers

In Scheme 11.1c, five subclasses of organic, π-conjugated polymers are defined: (i) compounds **10–13** relate to linear homopolymers; (ii) **14** presents a conducting copolymer comprising two different monomeric units (with electron-donor and electron-acceptor properties); (iii) compounds **15–18** comprise the group of fused-ring and ladder-type polymers; (iv) phthalocyanine (Pc) complex (**20**) bridged by a conjugate spacer, bipyridil (**21**), forms a special Pc-based ECP (**19**); and (v) the group of polymeric blends: these include a blend of conducting polymer with a conventional, nonconducting polymer (**22**); a blend of conducting polymer with a polymeric ionomer (**23**); and a "self-doped" ECP (**24**).

11.2.2.3 Conventional Redox-Polymers

As shown in Scheme 11.1d, these polymers consist of the main backbone of: (i) a nonconductive polymer (**25–27**), or a polymeric ionomer (**29**); or (ii) a backbone of an ECP (**28**) to which pendant, localized redox-centers, such as ferrocene (Fc), bipyridine-complexes of Ru, Os, and so forth, or even low-molecular-weight thiophene oligomers, are covalently attached (**25**, **27**, and **26**, respectively). Covalent attachment is characteristic of the structure **28**, whereas **29** contains a typical electrostatic bond between the electroactive bipyridine-complex of Ru and the polymeric ionomer's backbone.

11.2.2.4 Inorganic Ion-Insertion (Intercalation) Compounds

These compounds, which are shown in Scheme 11.1e as a separate subclass, are represented by a vast group of transition metal-containing oxides, chalcogenides (S, Se, Te), oxochalcogenides, clustered chalcogenides (Chevrel phases, etc.) of different dimensionality [27–29]: compound **30** has a typical one-dimensional (polymer-like) host with 3-D ordered guest atom sites; an example is $Li_xMo_6S_6$; compound **31** comprises a two-dimensional (2-D) host lattice with a 2-D set of intercalation sites (e.g., $Li_xV_2O_5$ in polyhedral representation); **32** is a 3-D host with a set of 3-D intercalation sites (e.g., Chevrel phase, $Li_xMo_6S_8$). Additional information on such compounds can be found in Chapters 2, 5 and 7.

Scheme 11.2 summarizes the typical values of room-temperature electronic conductivity of various electroactive materials (classified as presented in this section) in comparison with the conductivity of a typical metallic conductor (Cu) and insulator (diamond). As already discussed, the conductivity ratio for the doped and undoped π-conjugated oligomers is less than that for the ECP (as well as for the n-doped C_{60}). For the last two groups of materials a drastic increase in conductivity, by more than 10 orders of magnitude, takes place upon doping. This change exceeds by orders of magnitude the changes typical for the SWCNT (a good electronic conductor) and for Li-insertion inorganic conductors. Conventional redox-polymers do not reveal high conductivity, and the changes in this quantity upon their redox-switching are far less than that for ECPs. The electronic band diagrams of the ECP considered in Section 11.3.1 explains their conductivity behavior very well.

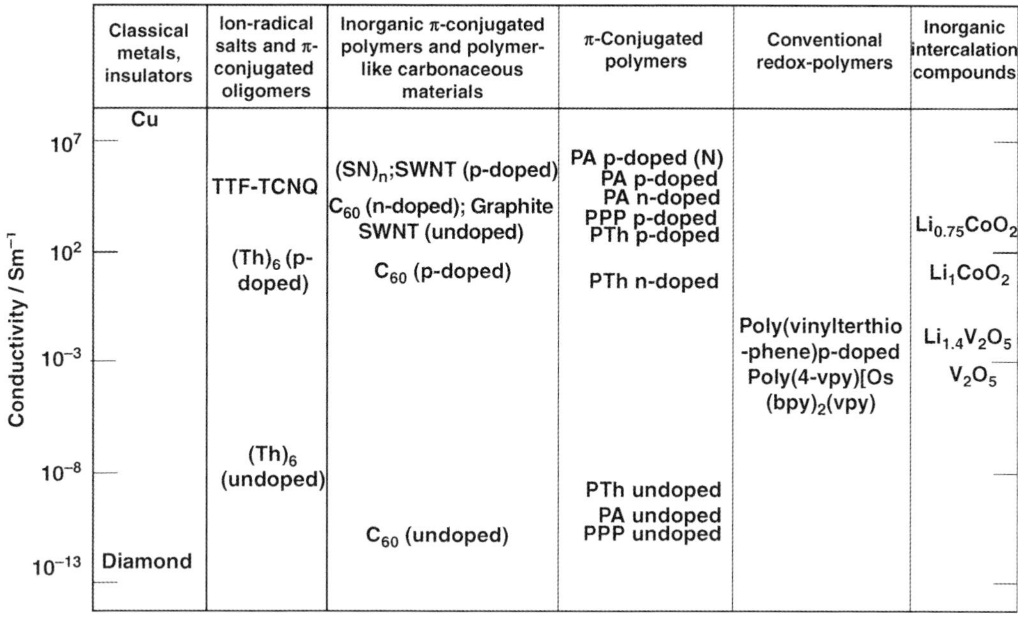

Scheme 11.2 Typical values of the electronic conductivity of the various classes of conducting solids, including ECPs in their doped and undoped form, in comparison with metal (Cu) and insulator (diamond) under ambient conditions.

11.3
General Features of Doping-Induced Changes in π-Conjugated Polymers

11.3.1
The Electronic Band Diagram of ECP as a Function of Doping Level

The term "polymer doping", which is used to denote the transition of a pristine (neutral) form of a π-conjugated polymer to its highly conducting form, differs fundamentally from the doping of inorganic materials. During the course of doping of π-conjugated polymers, oxidation/reduction processes on the polymeric backbone result in the appearance of electronic and counterbalancing ionic charge carriers in their bulk. This transforms the polymer from an insulating to a highly electronically and ionically conducting state. Such oxidation/reduction can be performed chemically in the absence of solvents (e.g., in vapors of Cl_2, Br_2, I_2, $FeCl_3$, etc.), or electrochemically, using either two- or three-electrode cells (with a metallic current collector to which polymeric film is attached). Clearly, a variety of aqueous and nonaqueous solutions can be used for electrochemical doping.

When considering one-dimensional (1-D) structures of π-conjugated polymers within the framework of Paierls' approach to electronic conductivity [30], one should expect the ground (neutral) state of these polymers to be a perfect insulator.

The electrons located at neighboring C atoms are coupled, causing a local deformation of the initial lattice, which results in alternation of the C=C and C−C bond lengths. A simplified electronic band diagram of the neutral state of the polymer is shown in Figure 11.1a, where the doping-induced changes are presented in panels a to d, following the results of Ref. [31]. The conducting and valence bands (CB and VB) of the neutral polymer are separated by a gap shown by the two-sided arrow in Figure 11.1a. The changes in the electronic structure of the π-conjugated polymer when an electron is taken from its valence band during

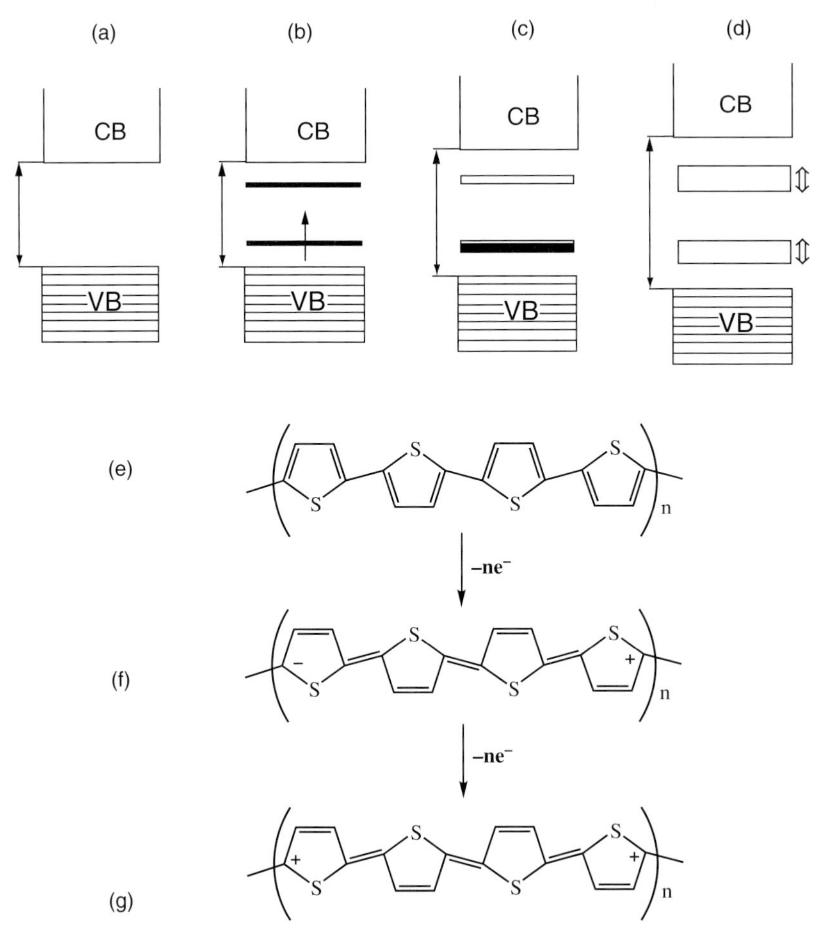

Figure 11.1 Electronic band diagrams of a non-degenerate π-conjugated polymer related to different doping levels. (a) Undoped (neutral state); (b) Slightly doped polymer with localized polaronic levels; (c) Moderately doped polymer with polaronic bands; (d) Heavily doped polymer with bipolaronic bands. The benzenoid (e) and quinoid (f and g) chemical structures correspond to neutral, slightly, and heavily doped material, respectively. Note the formation of a cation-radical (i.e., of a spin-bearing, polaronic-type charge carrier, (f)), and dication species (spinless bipolaronic charge carrier, (g)) upon polymer doping.

oxidation (thus leaving an unpaired electron on the macromolecule) are shown in Figure 11.1b and d.

As a result of oxidation and the accompanying deformation of the chemical bonds, the polymer transforms from its benzenoid- to quinoid-type structure [19] (see the reactions in Figure 11.1e and f, respectively), creating an electronic level for the unpaired electron inside the gap, closer to the top of the valence band. From a chemical point of view, a cation-radical in the polymer backbone is delocalized within several repeat units. The location of the electronic level of unpaired electron in the gap, correlated with appearance of positive charge and the related transformation of the chemical bonds around this level, is equivalent to formation of a positively charged polaron. Since it localizes in the gap, the doping signifies a *local* suppression of the Paierls transition [30]. With further oxidation, more electrons are withdrawn from the valence band; the polarons interact with each other, merging the localized polaronic levels into a continuous polaronic band (see Figure 11.1c).

Reduction (cathodic polarization) causes the appearance of negatively charged polarons in the gap near the bottom of the conduction bands, similar to the formation of positively charged polarons considered above. At high levels of doping, polaronic charge carriers may recombine their spins forming spinless, double-charged bipolaronic charges carriers (dications from a chemical point of view; see their formation in Figure 11.1f and g). These expand their own bands in the gap at the expense of the polaronic and valence bands, as shown in Figure 11.1d (the width of the former bands is indicated by double-ended arrows). The evolution of the electronic band structure with doping level suggests the appearance of a metal-like conductivity, which, however, must be confirmed by the relevant experimental techniques [30].

11.3.2
The Effect of Morphology on the Conductivity of the Polymeric Films

The conductivity of a polymeric film cannot be simply reduced to 1-D conductivity of the constituting macromolecules; rather, it should be noticeably dependent on the film's morphology. Taking as an example PA with its typical fibrillar, "spaghetti"-like morphology containing semi-crystalline domains (shown schematically by a set of parallel lines in Figure 11.2), it is easy to see that the conductivity implies both intrachain and interchain pathways [30]. Stretching of the PA film results in an increase of both the film's crystallinity and the number of interfibrillar crosslinks per unit cross-section, thus enhancing the conductivity.

The effect of the inserted counterbalancing ions was not taken into account in the electronic band diagram of Figure 11.1. Due to the electroneutrality of the film's bulk at any moment of the doping process, the counter-ions are distributed between the positively charged macromolecular chains of the polymer, forming their sublattice, as shown schematically in the upper part of Figure 11.2. When a film is chemically doped, the counterbalancing anions are formed by electron transfer from the polymer chain to an oxidizing species (see, e.g., the chemical reaction of doping

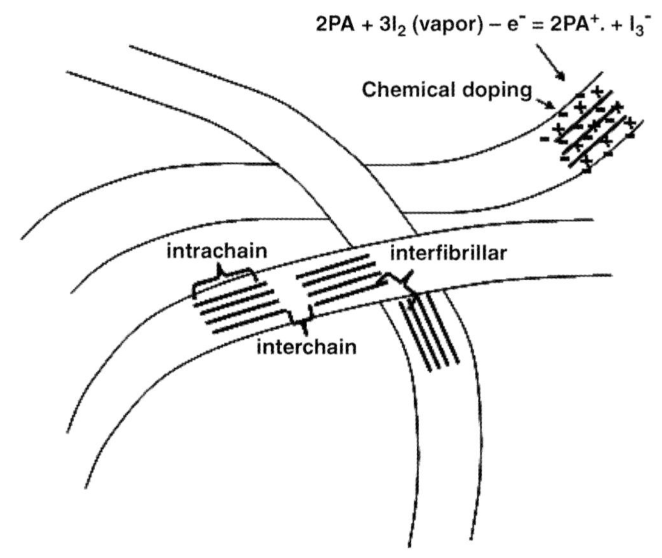

Figure 11.2 Schematic representation of the fibrillar structure of PA film with intrachain, interchain, and interfibrillar conduction pathways. The semicrystalline character of the separate parts of the PA, consisting of sublattices of polaronic and counter-ionic charge carriers, is shown in the upper-right corner of the figure. The reaction of chemical doping of PA with I_2 vapor is indicated as an example.

of PA with iodine vapor in Figure 11.2). During the course of electrochemical doping, the counter-ions are inserted from the equilibrating solution.

11.3.3 Electrochemical Synthesis and Doping

11.3.3.1 Selection of Suitable Electrolyte Solutions

In a sense, electrochemical synthesis provides means for the effective oxidation of a suitable monomeric species, the formation of oligomeric species, and their further grafting onto the electrode surface to form continuous polymeric films. Electrochemical doping can be expressed as coupling of oxidation/reduction in the neutral polymeric films to the simultaneous counter-ions (and solvent molecules) insertion from the equilibrating electrolyte solutions. This can be induced by polarization of the electrodes in either potentiostatic or galvanostatic mode. The electrochemical methods were shown to be particularly flexible for the preparation of conducting polymer films in a controlled manner (in terms of their thickness, morphology, doping level, mechanical properties, etc.), as well as for the effective doping when structure–property relationship is required to be under continuous control.

Many polar organic aprotic solvents, such as acetonitrile, propylene carbonate, and sulfolane, and less polar solvents such as tetrahydrofuran, dimethylchloride, and nitrobenzene, with soluble salts containing small (perchlorates, perfluoroborates, etc.) or large (tetraphenylborate, *p*-toluenesulfonate, etc.) anions, small (Li^+, Na^+) or large (tetra-alkylammonium) cations, may be suitable. The requirements for their purification are usually those recommended in classical handbooks of electrochemistry, in those sections related to non-aqueous systems [32]. Some polymeric materials can be electrochemically doped in both nonaqueous and aqueous solutions (e.g., polypyrrole, polyaniline, etc.), whereas others should preferably be doped in very dry solutions of nonaqueous solvents, with all manipulations carried out under an inert atmosphere in a glove box (PA is a typical example of this).

The correct choice of solvent and salt depends on the sensitivity of the electronic charge carriers (i.e., of polycation- and polyanion-radicals) towards electrophilic or nucleophilic reactions with solution species. A rough rule is the same as that known to determine the stability of low-molecular-weight organic cation- and anion-radicals in solutions [33]. Namely, cation-radicals are mostly stable in solvents of low nucleophilicity. For example, the cation-radical of polyparaphenylene (PPP) is formed at relatively high anodic potential. Applications of aqueous solutions are excluded because of their high nucleophilicity and thermodynamic instability in the required potential range. However, an exotic solvent such as concentrated H_2SO_4, which forms an emulsion with pristine benzene monomer, appears to be suitable for both electropolymerization and p-doping of the PPP films [34]. The next example refers to the heavily doped poly(3-methoxythiophene), which remains stable in a high charge density state if a low-nucleophilicity solvent such as CH_2Cl_2 at $0\,^{\circ}C$ is used [35].

The opposite case is the n-doping of π-conjugated organic polymers. The anion-radicals, which are readily formed on the macromolecular chains (negatively charged polarons and bipolarons) in highly nucleophilic solvents, are destroyed in solvents of low nucleophilicity. For example, it was possible to reach a relatively high degree of n-doping of PTh in liquid ammonia at $-55\,^{\circ}C$ [36]. Hexamethylphosphorustriamide (HMPA), an organic solvent with the highest ever known electron donor number, was the only choice to obtain n-doping of PTh in the presence of highly electrophilic cations such as Li^+ [37]. This last example clearly shows that a careful consideration of nucleophilic and electrophilic properties relates not only to solvent molecules but also to all components of the solution, including the cations and anions of the salt used.

11.3.3.2 A Short Survey on In Situ Techniques used for Studies of Mechanisms of Electrochemical Doping of π-Conjugated Polymers

The relevant *in situ* techniques can be divided into two groups (for a recent review, see Ref. [23]).

The first group consists of conventional electroanalytical techniques such as cyclic voltammetry (CV), chronoamperometry, chronopotentiometry, coulometry, and electrochemical impedance spectroscopy (EIS), all of which provide general information about the doping process (see also Chapters 4 and 5). Below are listed some typical questions that can be answered using the above group of techniques:

- A distinction between p- and n-doping processes, and evaluation of the electronic band gap of the polymer under consideration.

- An estimation of the maximum doping level of a polymeric film; measurement of charging–discharging characteristics of the polymeric electrode at slow rates for their further fitting with doping isotherms.
- A general view of the mechanisms and kinetics of the doping processes (using EIS), and a quantitative evaluation of the chemical diffusion coefficients of electronic and ionic charge carriers with the use of small-amplitude techniques (often called potentiostatic and galvanostatic intermittent titration techniques; PITT and GITT, respectively [29]).
- Characterization of the film's porosity.

The second group of electrochemical methods is aimed at measuring any special characteristics of the films, which cannot be directly obtained from standard electrochemical measurements in the potential–current–time (E–i–t) domain. In this case, in addition to a potentiostat/galvanostat, some specialized equipment and suitable electrochemical cells are required. This group of techniques includes:

- *In situ* electronic conductivity measurements of polymeric film electrodes as a function of the potential, solvent and/or electrolyte nature.
- *In situ* atomic force microscopy (AFM) and scanning tunneling microscopy (STM), which can be used to follow the morphology of the ECP films.
- Scanning electrochemical microscopy (SECM), which has an intermediate resolution between that of conventional lithography and AFM/STM, is useful for mapping the electrochemically active areas of electronically conductive patterns of the films [38]; it also can be used for the special mode of coating of the surface of metallic electrodes with ECP films by electroless deposition [39].
- *In situ* optical absorption characterization (UV-Vis-NIR) provides a direct clue as to the optical band gap of neutral forms of the polymers, and to the development of intergap electronic states as a function of the doping level.
- *In situ* Fourier transform infra-red spectroscopy (FTIR) can be used to trace the appearance of surface-active groups on the polymer surface, and to characterize the insertion of counter-ions as a function of the doping level.
- *In situ* Raman spectroscopy can be used when the characterization of doping-induced changes in symmetry and polymeric chains conformation is required.
- Electrochemical quartz-crystal microbalance (EQCM) methods (which allow for high-precision measurements of mass changes during the course of polymerization and doping) are very effective for discriminating between counter-ion and co-ion fluxes and fluxes of solvent molecules, when combined with simultaneous CV measurements.
- *In situ* electron-spin resonance measurements enable the qualitative and quantitative detection of ion-radicals, and tracking of their relaxation dynamics.

11.3.3.3 Mechanisms of Electrochemical Synthesis of Conducting Polymer Films

Electrosynthesis is a complicated process which begins at the electrode surface with an electron-transfer reaction and results in the formation of a monomeric cation-radical (step 1 in Figure 11.3). This is followed by two subsequent chemical steps (dimerization of cation-radicals with the subsequent deprotonation to form neutral dimers, steps 2 and 3) [40]. In the following, the intermediate dimers are oxidized at a lower potential compared to the monomer (step 4), reacting with monomeric cation-radicals (step 5), so that the polymeric chain grows by the repetition of ECC cycles (E = electrochemical step, C = chemical step; see the right-hand column in Figure 11.3). Electropolymerization continues further until insoluble thiophene oligomers are formed. Adhering to the electrode surface, they give rise to the formation of growing films on the electrodes surfaces [41, 42]. The growth of the polymeric film can be studied using chronoamperometry. An example [43] is shown in Figure 11.4, where a characteristic rising current transient is due to a gradual increase of the internal surface of the film through which the current passes (similar to the classical mechanism of electrodeposition of metals [44]). The higher is the potential, the faster is the nucleation step and the following film's growth. The current reaches approximately a plateau as the electrode surface is completely covered with the polymeric film, so that it continues to grow in a steady-state condition.

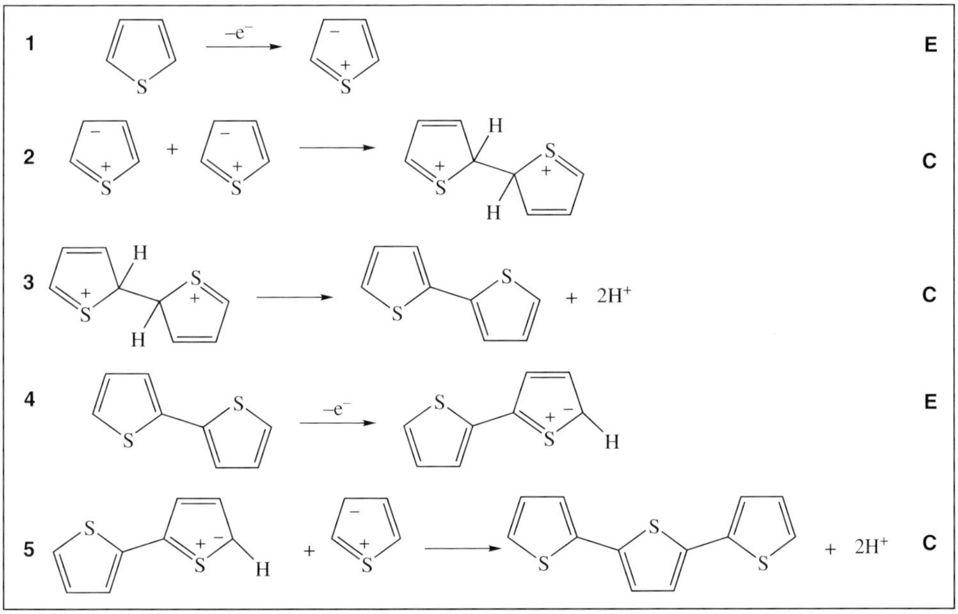

Figure 11.3 Typical $(ECC)_n$ mechanism of the anodic oxidation of thiophene with the formation of a PTh film. E and C in the right-hand part of the figure denote electrochemical and chemical steps of the entire oxidation mechanism, respectively.

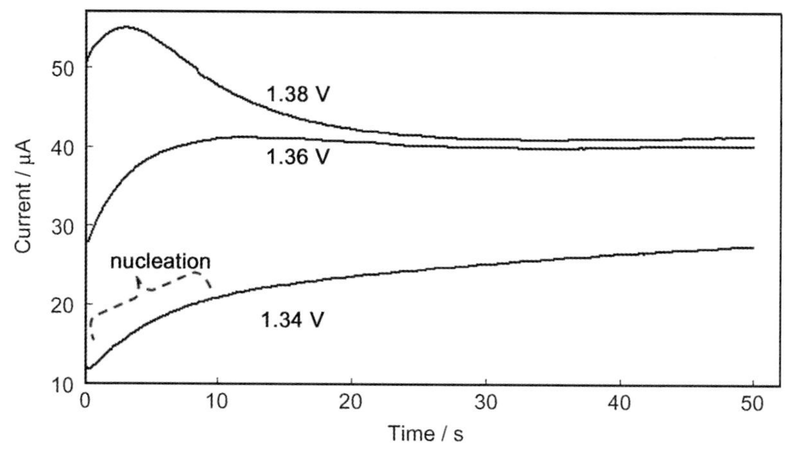

Figure 11.4 Potentiostatic deposition of polyfluorene (octylthiophene) film (PFDOBT-HH) onto a Pt wire by electro-oxidation of a 0.002 *M* solution of 2,7-bis(4-octylthien-2-yl)-fluoren-9-one in 0.25 *M* $TEABF_4/CH_2Cl_2$.

It is important to note that several parasitic processes may compete with electrodeposition reactions, namely: (i) a rapid diffusion of the intermediate oligomeric species to the solution bulk (this occurs when cation-radicals are especially stable in the chosen solution); and (ii) when, in contrast to the previous case, the cation-radicals formed by oxidation of the monomers are very reactive towards solution components. This may substantially decrease the Faradaic efficiency of the electrodeposition process [40]. The reactivity of π-conjugated oligomers towards solution species depends on the monomers' electron donor and/or electron acceptor strength of possible aliphatic or aromatic substituents in the 3-position with respect to the heteroatom in the 5-membered heterocyclic molecule. Thus, the careful molecular design of ECP films showing stable and highly reversible behavior upon their doping/undoping will always imply that the electron donor/acceptor abilities of: (i) the related heteroatomic repeat unit; (ii) the alkyl-substituent in the ring; and (iii) the components of the electrolyte solutions, are well tuned [19–22].

It should be remembered that the polymeric films are obtained electrochemically in already doped forms; that is, in a general case the anions (as the principal counter-ions), co-ions, and solvent molecules may participate in film deposition. The complicated mechanism of doping is best studied by the combined application of CV and EQCM. A typical example is the galvanostatic polymerization of PPy films [43] (see Figure 11.5a). The potential first increases, and then decreases, due to the development of an internal surface of the deposited oligomeric nuclei. During the deposition process, the mass of the coating increases almost linearly with the charge involved, thereby demonstrating mass per electron (mpe) values of 52–56 g mol^{-1} [43]. The stoichiometry of the electropolymerization of pyrrole is the same as that for thiophene (see Figure 11.4), and estimated as $2.3e^-$ per mol monomer units. This should result in an mpe of 43 g mol^{-1} in EQCM experiments with

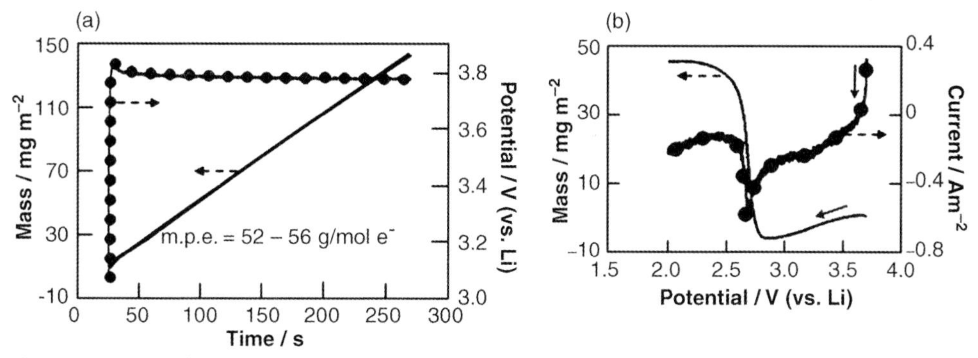

Figure 11.5 (a) Simultaneous potential and mass changes during the deposition of a 0.1 μm-thick PPy film ($1\,Am^{-2}$, 250 s) in a 1 *M* pyrrole + 0.25 *M* $TBAClO_4$/PC solution, measured by EQCM; (b) Current and mass changes during a first cathodic potentiodynamic polarization of the film immediately after its electrodeposition. Scan rate $v = 5\,mV\,s^{-1}$.

electrodeposited PPy films. It might be suggested that the higher experimentally determined mpe values are due to the insertion of co-ions and solvent molecules. Moreover, the competition between these three processes depends on the potential and the direction of the potential scan. For instance, let us consider the previous case of the PPy film deposited galvanostatically, and then polarized cathodically by a linear potential scanning [43] (see Figure 11.5b). The current response was peak-shaped, as would be expected for undoping under these conditions (the curve in Figure 11.5b marked by solid circles). The film's mass will first decrease, signifying undoping accompanied by deinsertion of the perchlorate counter-anions. However, if undoping is continued the mass begins to increase considerably increase, indicating the insertion of solvent molecules and co-ions (in this case, TBA cations).

11.3.3.4 Dynamics of the Micromorphological Changes in ECP Films as a Function of their Doping Level

It is logical to expect that the complicated character of ions and solvent molecules insertion/deinsertion upon doping/undoping of the polymeric films should also be reflected by changes in their micromorphological structure. Demonstrated here is a typical case of a freshly prepared PPy film in a $TBAClO_4$/PC solution [43] (see Figure 11.6), studied by the combined application of CV and EQCM. This combination allows the charging characteristics of the polymeric film to be connected to a variety of ion and solvent molecule insertion phenomena. The anodic–cathodic redox peak of approximately 2.8 V (versus Li) is clearly expressed in the CV curve (marked by the thick curve in Figure 11.6), whereas the second peak is less distinctive because of the relatively low anodic vertex potential. Nevertheless, the change in film mass (thin curve in Figure 11.6) on doping/undoping can be clearly rationalized in terms of the following two processes. At the start of the anodic scan, the mass will decrease due to the deinsertion of cations and solvent molecules, whereas at more positive potentials the charge of the film will be compensated by

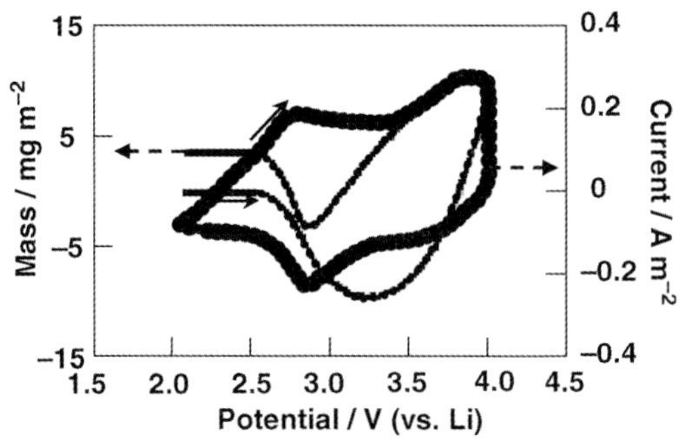

Figure 11.6 Current (thick line) and mass changes (thin line) during the first potentiodynamic cycle experiments with a PPy film electrode in a fresh, monomer-free 0.25 *M* $TBAClO_4$/PC solution in an EQCM experiment. Scan rate $v = 5\ mV\ s^{-1}$.

inserted anions. In the negative potential scan the processes occur in a reversed order.

In addition to mass changes when the film is doped/undoped, drastic morphological changes were directly tracked by *in situ* AFM imaging [43], taken both in the doped and undoped states (i.e., at 4.0 V and 2.0 V, respectively; see Figure 11.7). In its neutral (reduced) form, the film had a granular morphology; however, after the removal of co-ions and solvent molecules, as the anions are inserted into an oxidized polymeric film, some short-length oriented fibrils are formed (see Figure 11.7b). Obviously, the anions screen very well the positively charged (polaronic-type) carriers on the macromolecular chains and, in the absence of an excess of solvent molecules, are located closer to the polymer backbones (as shown schematically in Figure 11.2), favoring a more ordered, and apparently partially crystalline, structure.

An additional factor has also been identified as being responsible for the micromorphological changes described above. It is known that π-conjugation of the doped ECP results in a planar configuration of the repeat units (which favors ordering of the neighboring macromolecules). Electrically neutral macromolecules will normally have a non-planar conformation (helical, globular, etc.), which is additionally stabilized by the presence of an excess of solvent molecules and neutral salt [45]. The doping-induced conformational changes of the macromolecules can be kinetically sluggish, and contribute together with the ionic and solvent fluxes to a complicated dynamics of the electrochemical doping processes.

11.3.3.5 The Maximum Attainable Doping Levels and the Conductivity Windows

Both, stable and reversible charge storage at high concentrations of charge carriers in ECP, depends not only on the internal stability of the polaronic and bipolaronic charge carriers, but also on their reactions with the solution components. At ambient conditions (i.e., for typical organic solvents with rather low nucleophilicity, room temperature measurements, limiting potential up to 1.2 V versus Ag/Ag^+), many polyheterocyclic ECPs exhibit a doping level of $1e^-$ per three to four rings [19]. The maximum doping

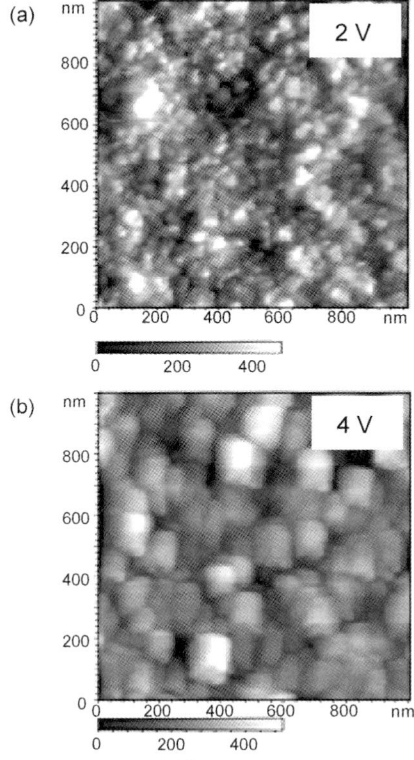

Figure 11.7 Atomic force microscopy images measured *in situ* from fresh PPy film formed in a 0.25 M $TBAClO_4$/PC solution (the conditions of galvanostatic polymerization are indicated in the caption to Figure 11.5a). The images were taken during two consecutive potentiodynamic cycles, at (a) 2 V and (b) 4 V, as indicated. Area scanned: 1 × 1 μm.

level of PPy and PTh can be increased by a factor of 2 in liquid SO_2 at −40 °C, with the limiting potential up to 2.0 V (versus Ag/Ag^+) – that is, $1e^-$ per two rings [46].

The maximum doping level depends, to a great extent, on the film's morphology. A slow electropolymerization at low anodic potentials will increase the maximum doping level [47–49]. As an example, Figure 11.8 shows the CV curves for a PPy film with a gradual shift of the anodic vertex potential up to 4.8 V (versus Ag/Ag^+), until the maximum doping level of $1e^-$ per ring was achieved [49]. The stoichiometry of the two redox processes observed, and also a possible mechanism of a parasitic anodic reaction at very high potentials (exceeding 5 V versus Li), are shown in this figure. An alternative strategy based on a careful selection of the substituent in the thiophene ring (e.g., in poly-3-methoxydithiophene) also resulted in the maximum doping level of $1e^-$ per ring [35].

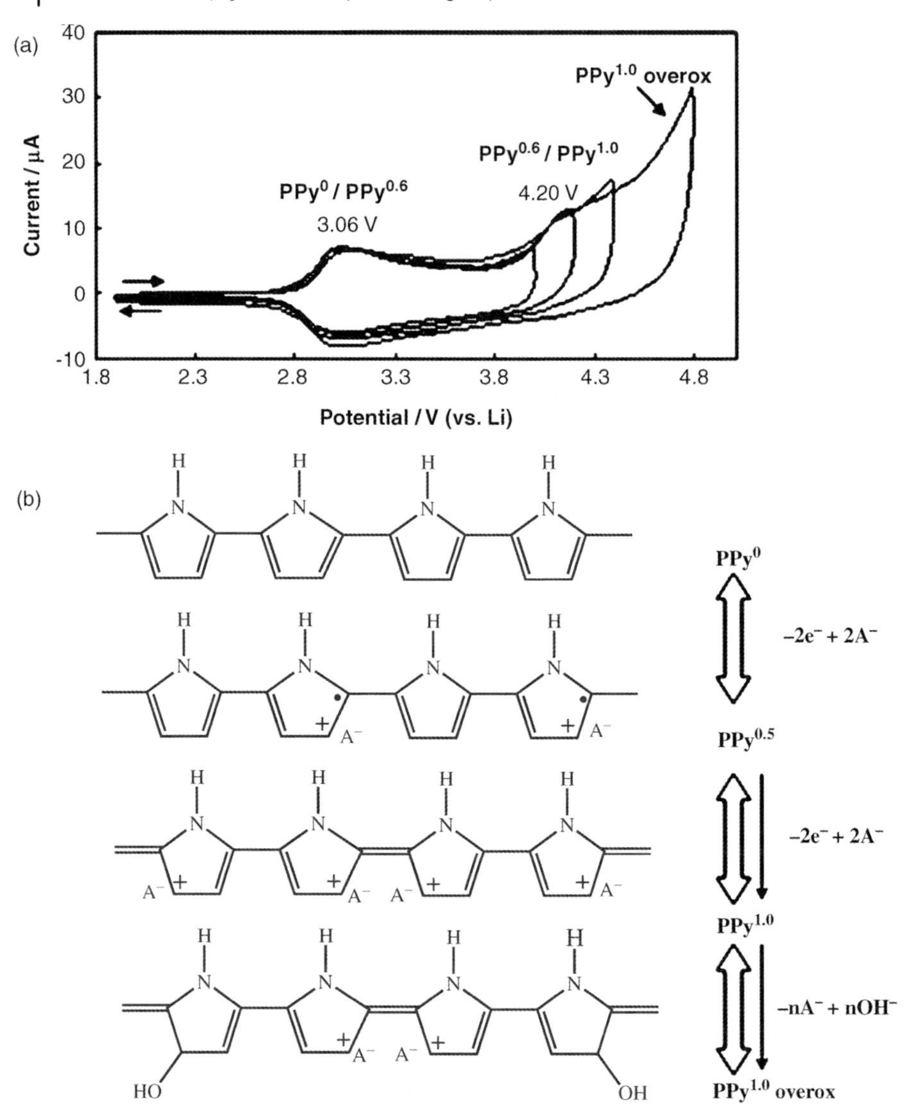

Figure 11.8 (a) Anodic behavior of a 0.2 μm-thick PPy film prepared with the use of a current density of approximately $0.8\,A\,m^{-2}$ during 610 s, and afterward cycled in 0.25 *M* $LiClO_4$/PC at a scan rate of $5\,mV\,s^{-1}$ to different vertex potentials, as indicated; (b) Electrochemical reactions occurring in the vicinity of the first and second anodic peaks at 3.06 and 4.20 V (versus Li), respectively.

In contrast to p-doping, the reported intercalation levels for n-doped ECP have never exceeded the level of $1e^-$ per three to four rings [50–53]. A special strategy was developed to obtain electronically balanced, donor–acceptor-type copolymers [54], for example, comprising octylthiophene and oxadiazole units (see **14** in Scheme 11.1c) [55, 56].

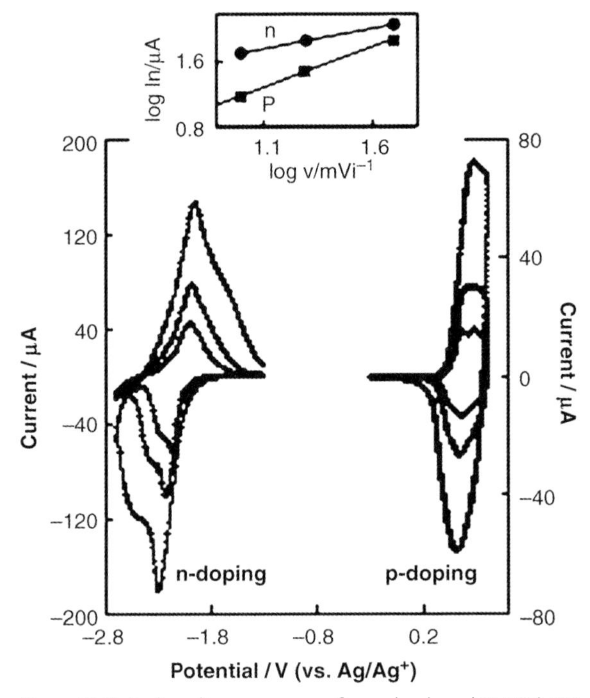

Figure 11.9 Cyclic voltammograms of p- and n-doped PMOThOD (structure **14** in Scheme 11.1c) film electrode 0.5 μm thick on a Pt current collector measured at 10, 20, and 50 mV s^{-1} scan rates. The inset at the top of the figure shows a double logarithmic plot of the n- and p-doping peak heights as a function of the scan rate (the slopes were found to be 0.5 and 1.0, respectively).

Figure 11.9 shows the families of CV curves measured for p- and n-doped films of this polymer at different scan rates, ν [56]. The dependence of the peaks currents, I_p, on the scan rate reveals linear and square-root plots for the p- and n-doped films, respectively. A linear plot of I_p on ν implies the absence of a diffusion control of the kinetics of the p-doping reaction. In contrast, the square-root dependence of I_p on ν suggests a diffusion-like movement of either electronic (i.e., negatively charged polarons) or counter-ionic (i.e., cationic) charge carriers on the film doping/undoping.

The p-doped PMOThOD film reveals a sigmoid-like dependence of the conductivity versus potential (this is in good agreement with the essentially flat peak on the related CV curve; the upper curve in Figure 11.10) [56]. The conductivity near the plateau is relatively high, as can be seen from the lower curve in Figure 11.10. In contrast, the n-doped PMOThOD film reveals a narrow (finite) conductivity window (lower curves in Figure 11.11) which is linked to a higher localization of the negatively charged polaron on the macromolecular chain compared to that for the positively charged carriers [56]. This is also reflected by the different shapes of the related CV curves (compare the upper curves in Figures 11.10 and 11.11). In addition, consecutive cycling of the n-doped film reveals typical features of the charge-trapping

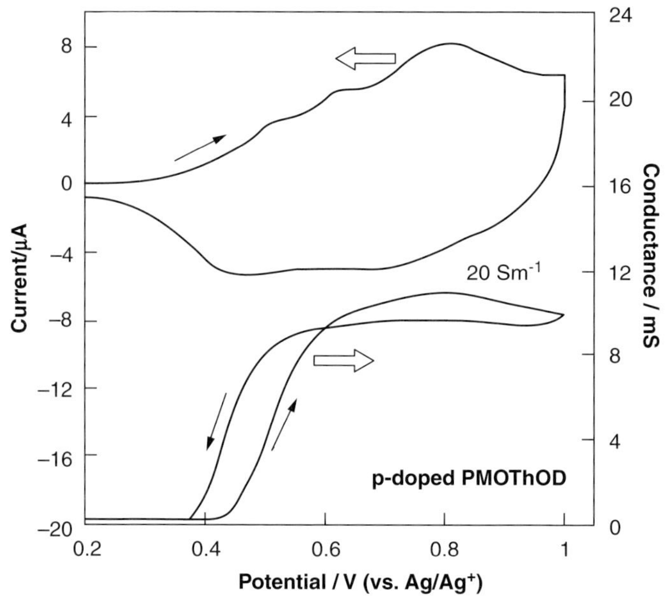

Figure 11.10 Cyclic voltammograms and related electrical *in situ* conductivity curves of a p-doped PMOThOD film electrode polarized to 1 V (versus Ag/Ag^+). The potential scan rate was $20\,mV\,s^{-1}$.

phenomenon, which is characteristic of many conducting polymers in their n-doped form (this is further discussed in the following section).

The differences in the mechanisms of conductivity of p- and n-doped film can be summarized as follows. The n-doping process is characterized by a substantial localization of the negatively charged carriers [57], and for this reason the absolute value of electronic conductivity and characteristic width of the conductivity window is less than that for the p-doped film. At intermediate doping levels, the p-type conductivity of some ECPs may reach values typical of metals, however, as the valence band of the p-doped polymeric film becomes depleted in electrons (i.e., as they are withdrawn from the CB at high enough electrode potentials; see the electronic band diagram in Figure 11.1d). The localization of the remaining charge carriers then again increases, so that the metallic state of the polymer degrades, and the polymer again becomes an insulator (as is the case for the derivatized PTh, PPy, etc., in liquid SO_2 at $-40\,°C$ [46]).

It should be noted that *in situ* conductivity measurements of ECP films allow for an estimation of the absolute values of conductivity, although high values of conductivity are not the only feature of a metallic state. *Ex situ* direct current electronic conductivity measurements of the films should be carried out in order to examine any temperature dependence of the conductivity and thermoelectric power over a wide range of temperatures, starting with very low values (of a few K) [30]. Typical metals have negative temperature coefficients for their electronic conductivity, and positive temperature coefficients for their thermopower, which contrasts with

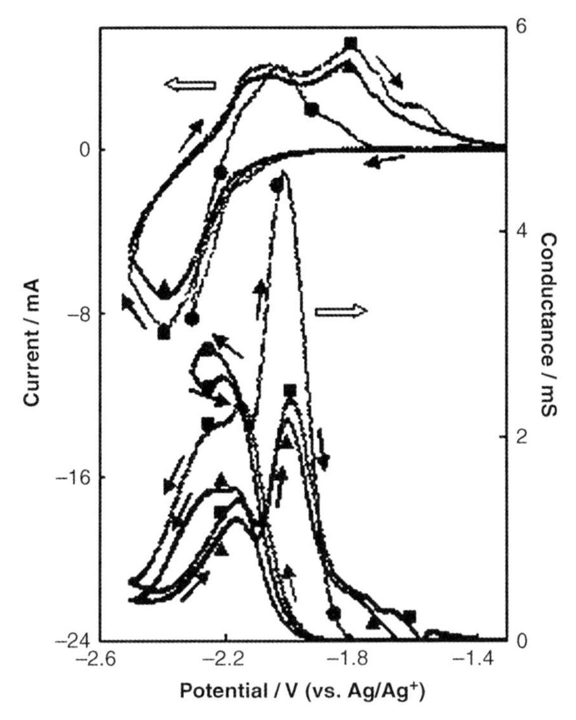

Figure 11.11 Cyclic voltammetry curves (top) measured simultaneously with *in situ* electrical conductivity (bottom) for the n-doped PMOThOD film electrode. The curve marked by filled circles relates to the 3rd cycle measured in a narrow potential window, down to −2.3 V. The following curves (marked by filled squares and triangles) relate to the 3rd and the 10th cycles measured down to the vertex potential −2.5 V. The potential scan rate was 20 mV s^{-1}.

conventional semiconductors [30, 58]. In reality, the wide variety of ECPs studied to date have revealed a thermoelectric power which, indeed, is indicative of the metallic behavior [58]. In contrast, the electronic conductivity of p-doped conjugated polymers is characterized by a positive temperature coefficient [58], similar to that for inorganic semiconductors. Variations in the electronic conductivity of doped, π-conjugated polymers with temperature are similar to those of amorphous metals (e.g., Mg-Zn, Ca-Al), and are ascribed to a hopping conductivity mechanism (Mott-type variable range hopping [59]). The metallic domains in 1-D organic conductors (crystalline and semicrystalline domains) are believed to be linked together through disordered regions. As the resistance of these films depends on the various domains, the disordered regions predetermine the Mott-type conductivity mechanism [58]. In contrast, the thermoelectric power appears to be basically insensitive to the electrical barriers between the metallic and disordered domains.

11.3.3.6 **Charge Trapping in n-Doped Conducting Polymers**

Charge-trapping during consecutive n-doping of ECP was abundantly reported for the conducting PTh and its derivatives [50, 51, 56, 60]. From Figure 11.11, it can be seen that the consecutive cycling of n-doped PMOThOD films causes an abrupt decrease in the mobility of the electronic charge carriers, whereas the total concentration of charge carriers is decreased only to a small extent [56]. Moreover, when opening the negative vertex potential from −2.3 to −2.5 V (versus Ag/Ag^+) in that case, the total injected charge increases, as can be seen from the CV curve in the upper part of Figure 11.11, whereas the conductivity continuously decreases. The appearance of parallel transient peaks of conductivity versus potential, and the anodic peak in the CV curves at potentials around −1.8 V (which are caused by opening the potential window down to −2.5 V), relates to a partial oxidation of the charge carriers trapped at very negative electrode potentials. However, the other part of the charge carriers remains trapped, and accumulate on the consecutive n-doping. These trapped charges can be only released during a first p-redoping cycling – that is, at much higher potentials within the characteristic potential window of the p-doping process, as shown in Figure 11.12. Similar voltammetric evidence for the charge trapping of negatively charged carriers was found for the various derivatives of PTh [50, 51]. Based on the EIS features of the polymeric films, additional evidence for the existence of charge carriers trapping was furnished in Refs [56, 60, 61]. Indeed it was found that, upon charge trapping, the electrochemical impedance is increased enormously, due to losses of electronic conductivity; however, charge detrapping that occurs on a first p-redoping potential scan will result in a dramatic decrease in impedance which, basically, will return to its initial, low value. A simple relaxation model describing the charge trapping effect in n-doped conducting polymers, which is well reflected by their EIS behavior, has been proposed [60]. This model takes into account the different mobilities of the "shallow" (mobile) and "hollow" (trapped) charge carriers, and hence the related different relaxation times constants, which can easily be revealed by EIS measurements.

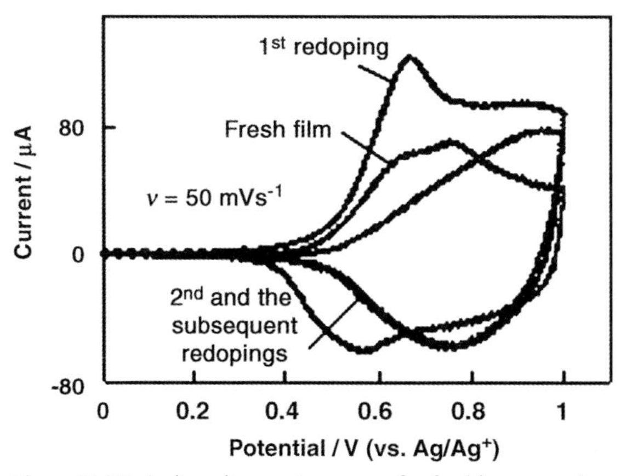

Figure 11.12 Cyclic voltammetry curves for freshly prepared p-doped PMOThOD film, and for the 1st and the 2nd p-redoping cycles after accumulation of negatively charged carriers during consecutive n-doping. The scan rate is indicated.

11.4 The Thermodynamics and Kinetics of Electrochemical Doping of Organic Polymers and Ion-Insertion into Inorganic Host Materials

At this point, a brief discussion is held of the approach [62, 63] which permits an understanding of the thermodynamics of charge storage mechanisms, and the kinetics of doping (insertion) reactions for different classes of redox-active solids and polymeric films (see Chapter 1). In Figure 11.2, a crystalline fragment of a PA fibril indicates (schematically) the location of the positively charged electronic carriers (polarons) on the macromolecular chains and the counterbalancing anions between the polymeric chains of the fibril. Suppose now that such a fibril or, equivalently, a thin polymer film, is in electronic equilibrium with a metallic current collector on one side, and in ionic equilibrium with the electrolyte solution containing the counterbalancing anion on the other side (see Figure 11.13a). It is assumed that the polymeric film has a

(a) M — Polymer film — Solution

e_m^- / e_p^- $\quad PT^o \rightleftharpoons PT^+ + e_p^- \quad$ A_p^- / A_s^-

$[PT^+]=[A_p^-]$

${}_m\phi_p$ $\qquad$ ${}_p\phi_s$

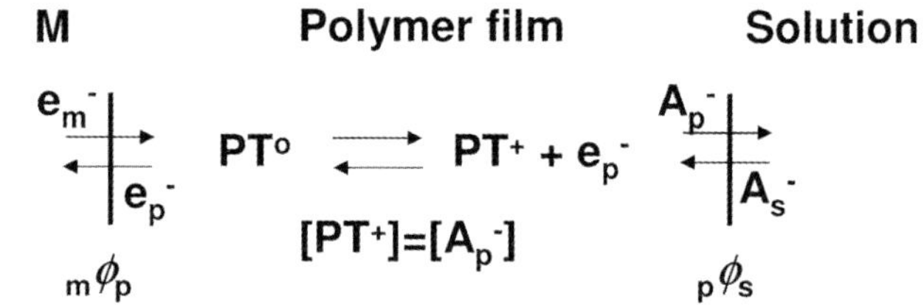

(b)

$$\mu_{PT^+} = \mu^{(o)}_{PT^+} + kT\ln\frac{X}{1-X} + kTgX \quad (11.8)$$

$$\mu_{A_p^-} = \mu^{(o)}_{A_p^-} + kT\ln X \quad (11.9)$$

$$ {}_m\phi_p = \frac{\mu^e_m + \mu_{PT^+} - \mu_{PT^o}}{e} \quad (11.10)$$

$$ {}_p\phi_s = \frac{\mu_{A_p^-} - \mu_{A_s^-}}{e} \quad (11.11)$$

$$(E - E'_o) = {}_m\phi_p + {}_p\phi_s \quad (11.12)$$

Perm-selective behaviour:

$$f(E - E'_o) = \ln\frac{X^2}{1-X} + gX \quad (11.13)$$

Breakdown of perm-selectivity:

$$f(E - E'_o) = \ln\frac{X}{1-X} + gX \quad (11.14)$$

Figure 11.13 (a) A schematic picture of a current collector/polymer film/electrolyte solution system with electron and ion equilibria across the interfaces; (b) Equations for the chemical potentials of polarons and anions, and the related Galvani potentials. The effect of perm-selectivity is taken into account; (c) Derivation of variations of the Galvani potentials as a function of the applied voltage; (d–f) Potential profiles for the system at different doping levels and types of limitation of the redox-capacity under the various conditions indicated.

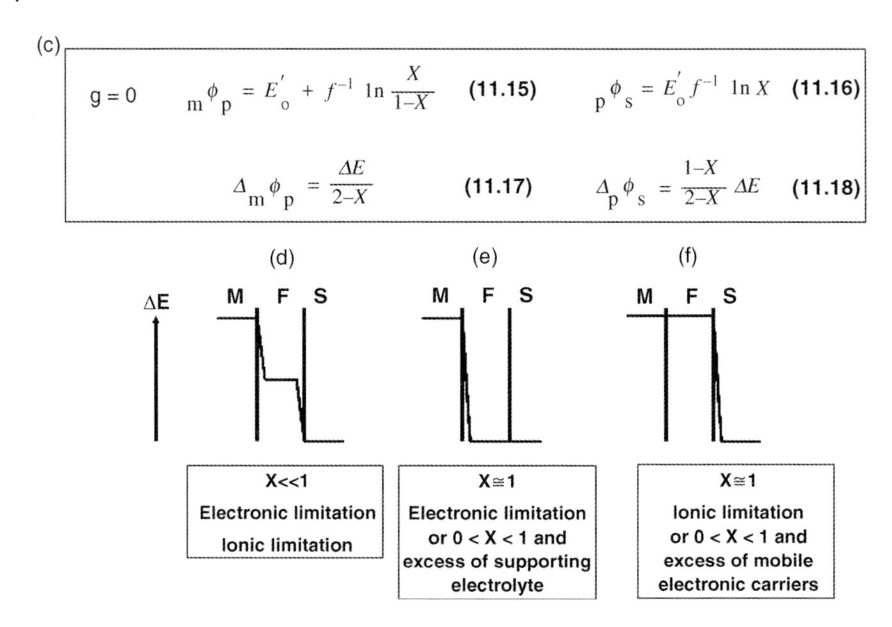

Figure 11.13 (*Continued*)

rather dense morphology, permitting the anions A_s^-, to enter the film from the solution in order to preserve its electroneutrality. However, the dense morphology of the film prevents it from a considerable partitioning of the electrolyte as a whole (i.e., the film is considered as perm-selective). The model which presented here is valid, first of all, if localization of the electronic and (or) ionic charge carrier states takes place. It is thus expected to be relevant not only for the redox-polymers such as polyvinylferrocene (structure **25** in Scheme 11.1d) and inorganic ion-insertion compounds (structures **30–32** in Scheme 11.1e) but also for n-doping (full range), and for the start and finish of the p-doping of π-conjugated organic polymers, as discussed previously. The starting points for brief discussion are the following assumptions.

In Figure 11.13a, the simplest case of the injection of singly-charged polarons in polythiophene, PT^+, is considered. This injected charge carrier can be formerly presented by a chemical equilibrium between a neutral PT^o species of the film, a charge carrier, PT^+, and electrons in the film, e_p^-, as shown in the center of Figure 11.13a. By applying a positive external potential to the electrochemical cell, electrons are withdrawn from the film to the current collector, thus driving the PT^o dissociation into PT^+ and e_p^- until an equilibrium state at this potential is reached. In order to preserve the electroneutrality of the film, single-charged anions from solution phase, A_s^-, with a concentration equal to that of the PT^+, are inserted into the film's bulk at equilibrium conditions, thus becoming A_p^-.

We first consider a situation where the number of accessible sites for PT^+ is limited, while the occupation of an equal number of sites by counter-ions is far from saturation of the latter type of site. In other words, we assume two sublattices – one for the electronic species, and the other for the ionic moieties, for which we need to write expressions for their chemical potentials in the framework of the Bragg–Williams approximation to regular solutions [64], or, equivalently, by using lattice gas models with interactions between the lattice sites [27]. The chemical potentials of PT^+ polarons and the counterbalancing anions, A_p^-, in the film are given by Equations 11.8 and 11.9 in Figure 11.13b. Here, the symbol $^{(o)}$ denotes the standard potential, X is the dimensionless doping level, and g is the interaction constant for the polaronic lattice (the condition $g = 0$ indicates an absence of interactions, signifying a Nernstian (Langmurian) behavior. In contrast, the cases of $g < 0$ and $g > 0$ describe attractive and repulsive interactions, respectively; the isotherm which takes into account g is called a Frumkin-type isotherm [65]); k and T are the Boltzmann constant and absolute temperature, respectively.

Note that the chemical potential of the counter-ions, μ_{Ap}^-, is written in a dilute solution approximation as saturation of concentration at high doping level is postulated for the polarons only.

The electronic and ionic equilibria on the left- and right-hand sides of the polymeric film create two Galvani potentials, ${}_m\phi_p$ and ${}_p\phi_s$ (see Figure 11.13a), each of which is determined by the difference in the chemical potentials of the electrons and anions in the film, and in the current collector and solution, respectively (see Equations 11.10 and 11.11, respectively, in Figure 11.13b). The applied voltage across the galvanic cell, $(E - E_o')$ is, of course, equal to the sum of these two Galvani potentials (the potential of the reference electrode, which is reversible with respect to the cations in solution, M^+, is constant and is included in standard potential of the cell, E_o') (see Equation 11.12). A combination of Equations 11.9–11.12 (follow the directions of the arrows in Figure 11.13b) results in a doping (insertion) isotherm of a non-Frumkin-type because of the perm-selective properties of the film (i.e., the so-called Donnan equilibrium was taken into account) [66] (see Equation 11.13).

Let us assume now that the perm-selectivity is broken down; that is, an excess of MA electrolyte is present in the film, so that the change in the concentration of A_p^- induced by doping, is small compared to the excessive electrolyte concentration. This leads to an invariance of the chemical potential of A_p^- instead of the condition of perm-selectivity (Equation 11.9), so that a Frumkin-type isotherm (Equation 11.14) appears instead of the non-Nernstian isotherm (Equation 11.13).

As seen from Equations 11.9–11.14, the different types of isotherm are internally linked to the different distribution of Galvani potentials across the cell. Let us find this distribution, assuming for simplicity, $g = 0$. By combining Equations 11.8–11.11, and collecting all terms which are independent of X in E_o', we obtain for the Galvani-potentials, $\Delta_m\phi_p$ and $\Delta_p\phi_s$ (Equations 11.15 and 11.16, respectively; see Figure 11.13c).

The variations of these two Galvani-potentials with the applied potential $(E - E_o')$ can easily be found by differentiation of both parts of the isotherm (Equation 11.13) with respect to E, and substituting the result into the differentiated (with respect to E) Equations 11.15 and 11.16 (see Equations 11.17 and 11.18 in Figure 11.13c).

On the basis of these equations, important conclusions can be drawn with regards to the potential profile across the galvanic cell comprising the polymeric film, which is described herein. At a low doping level, as $X \ll 1$, the variation of the applied potential ΔE results in equal variations of both Galvani-potentials, $\Delta_m\phi_p = \Delta_p\phi_s = \Delta E/2$, as shown in Figure 11.13d. In contrast, at high doping levels, the concentration of PT^+ saturates, and the potential across the polymer/solution interface ceases to change because the chemical potential of the anions no longer changes (a consequence of the bulk electroneutrality). This means that the applied potential will polarize a single interface (i.e., the metal/polymer interface), as shown in Figure 11.13e. Thus, if the film is characterized by a breaking down of the perm-selectivity, there will be always a single polarizable interface (i.e., the metal/polymer contact) at *all doping levels,* as μ_{Ap}^- is supposed to be constant. In other words, the classical Frumkin isotherm (Equation 11.14) implies a single polarizable interface.

In view of the symmetry of the systems, which consist of charged polarons and single-charged anions, the alternative assumption, that there are fixed numbers of sites for the accommodation of anions (and a dilute solution approximation for the polarons), leads simply to an interchange of Equations 11.17 and 11.18. Hence, at a low doping level the potential profile remains as shown in Figure 11.13d, whereas at high doping levels – as the chemical potential of electrons in the polymer film becomes independent of the applied potential – the latter drops across the single polymer/metal interface (see Figure 11.13f). Note that, quite symmetrically to the previously considered case of breaking down the perm-selectivity effect, the assumption that the film contains "excessive" and mobile electrons in the film (or host) – that is, the film (host) is electronically conducting – results immediately in the single potential drop across the polymer/solution interface at all the doping levels, from 0 to 1 (see Figure 11.13e).

A question arises concerning the analysis of the shapes of the isotherms (Equations 11.13 and 11.14), in order to determine which species – electronic or ionic – will limit the maximum doping level. The derivative of the fQ_mX (where $f = e/kT$; Q_m is the maximum charge that the film can store) with respect to the applied potential, E, is called the differential capacitance, C_{dif} (the dimension $C/V = F$), which can easily be obtained by differentiation of the isotherms (see Equations 11.19 and 11.20) in Figure 11.14a and b, respectively. The calculated potential dependences of C_{dif} for both isotherms are shown in Figure 11.15.

At first glance, the asymmetry of the C_{dif} versus E curve corresponding to the perm-selective isotherm is large enough to identify the type of carrier limitation. However, on a practical basis these curves are significantly distorted by slow interfacial kinetics and Ohmic potential drops, in addition to differences in the carriers' interactions in the host.

As an example, Figure 11.16a–c shows the potential dependencies of C_{dif} for Li^+ insertion into V_2O_5, the electrochemical doping of fresh films of PPy I (prepared at low potentials), and of the aged PPy films (PPy II, cycled continuously to the anodic vertex potential of 4.0 V, versus Li), respectively. The potential scale is the same for all three curves; hence, it can be seen that the half-peak widths of the differential capacitance curves for the three materials increase in the sequence: $V_2O_5 \ll$ PPy I < PPy II. The extreme narrowness of the curves for the V_2O_5 electrode is due to an intercalation process (Li-ions insertion into V_2O_5) which occurs via a first-order phase transition (a highly negative attraction constant in Equation 11.19, $g < -4$) [28, 67].

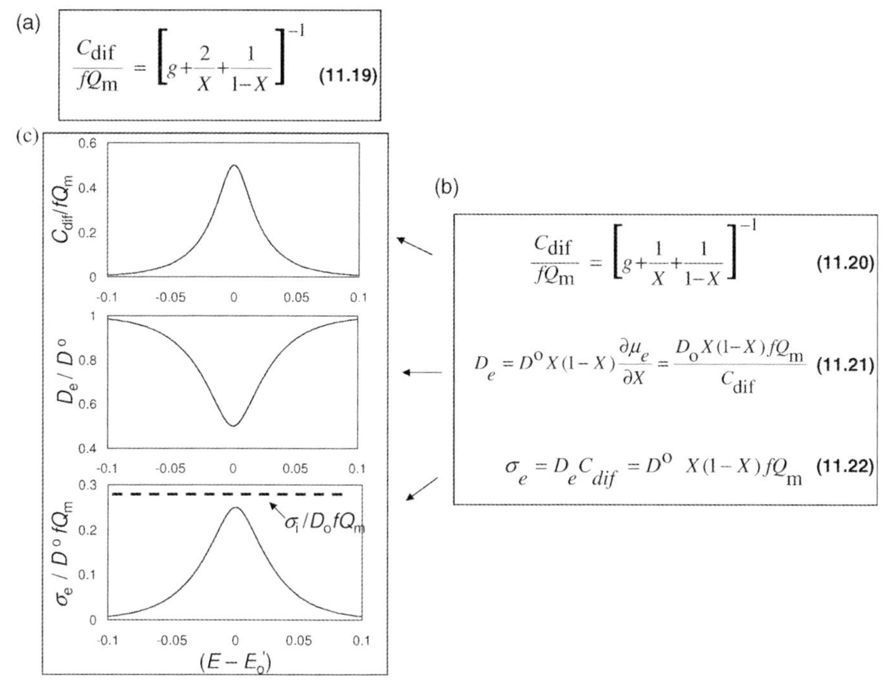

Figure 11.14 Equations for the differential capacitance for a perm-selective polymeric film (a) and for a film with the excess of background electrolyte in its bulk (b). In the latter case, the chemical diffusion coefficient and electronic conductivity are explicitly written down. D_e is the chemical diffusion coefficient of electronic species; D^O is the self-diffusion coefficient. The potential dependences of C_{dif}, D, and σ obtained with the use of Equations 11.20–11.22, assuming $g = -2$) are shown in panel c.

The behavior of PPy I [49] is very similar to that of the short-length oligomers with predominantly localized electronic carriers [25], whereas its continuous cycling results in a further solid-state polymerization of PPy I with the final structure (PPy II), which is less prone to electronic charge localization [47, 49]. This example shows that the intercalation-induced phase transition in V_2O_5 dominates in determining the shape of the differential capacitance curve.

Thus, an approach is required which can discriminate unambiguously between the different types of limitation of the differential redox-capacitance in electroactive materials [62, 63]. We refer to a simple particular case of the limitation of the redox-capacity by single-charged electronic carriers with an excess of electrolyte solution inside the film. Then, the Frumkin-type isotherm (Equation 11.14) is valid, with C_{dif} determined by Equation 11.20. At this point, we also need to define two important kinetic characteristics – the chemical diffusion coefficient, D, and the conductivity σ (parameterized as a product of D and C_{dif} [63]) which, in the particular case considered, should be assigned entirely to the electronic species (see Equations 11.21 and 11.22, respectively, in Figure 11.14b). Figure 11.14c shows the calculated potential dependences of C_{dif}, D, and σ. The criterion for distinguishing the type of

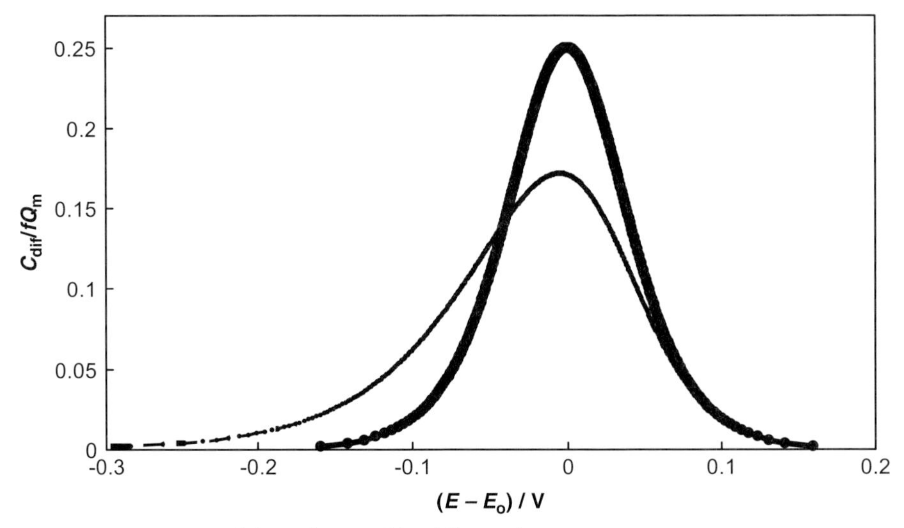

Figure 11.15 Potential dependences of the differential capacitance of a perm-selective film (thin line) and in the case of the presence of an excess of electrolyte solution in the film (thick line) calculated with the use of Equations 11.19 and 11.20, respectively, assuming $g = 0$.

limitation of the redox-capacity is related to comparing the shapes of the potential dependences of C_{dif} and σ. It is seen from Figure 11.14c that a limitation of the capacitance by the availability of the electronic species requires a peak-shaped potential dependence of σ, characteristic of a redox-type conductivity. The ionic conductivity in this particular case is constant (due to an excess of electrolyte in the film). In the opposite limiting case, when the insertion of ions limits the capacitance, and the host has a metallic conductivity, it is the potential dependence of the ionic conductivity that has a peak corresponding to the peak of C_{dif} versus E (the case related to many Li intercalation host materials).

Note also that we took as an example particular, simplified cases, when the total mixed electron-ionic conductivity, σ_{ei}, is equal either to σ_e or σ_i. In general, both σ_{ei} and D_{ei} consist of partial contributions of the electronic and ionic charge carriers, which do not necessarily have the same values. EIS represents one of the most powerful techniques available for discriminating between partial electronic and ionic contributions to D_{ei} and σ_{ei}. Using this technique, it was found [68] that the modulation of interfacial potentials across the metal/film/solution interface with a small amplitude applied potential, may effectively probe the relative mobilities of the both types of charge carrier. In particular, the classical Warburg region appears in the related Nyquist plots from high to moderate frequencies, when both charge carriers have the same mobilities (the equivalent electrical circuit analogue includes a frequency-independent resistance and the distributed capacitance). In the case when the mobilities are different, a non-homogeneous electrical field arises in the film [68]. This results in a certain "delay" of the concerted electronic and ionic transport in the

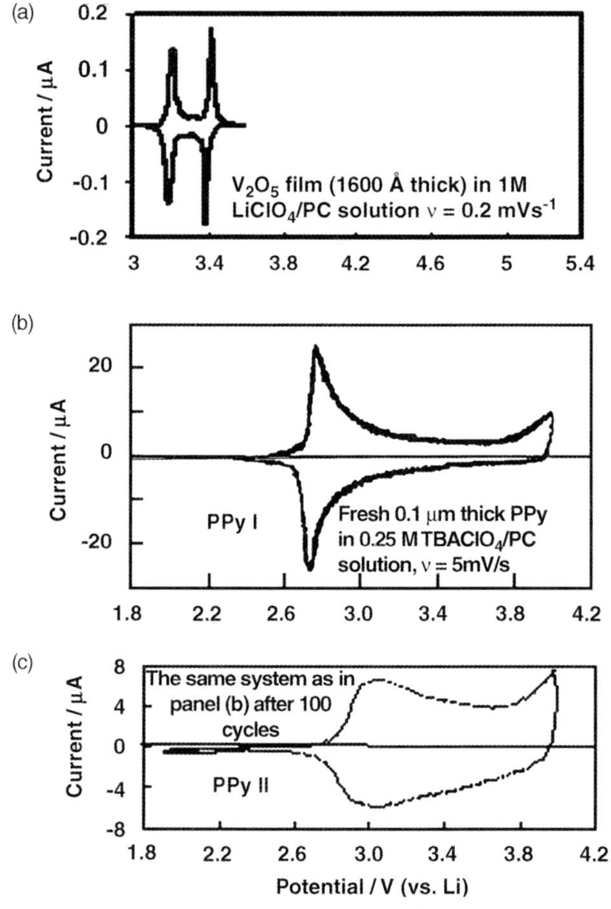

Figure 11.16 Potential dependences of the differential capacitance for three different electronic–ionic conductors. (a) V_2O_5 film (1600 Å thick) in 1 *M* $LiClO_4$/PC solution. Scan rates of 0.2 mV s^{-1}; (b) Freshly prepared 0.1 μm-thick PPy cycled in 0.25 *M* $TBAClO_4$/PC solution up to 4 V (versus Li), $\nu = 5$ mV s^{-1}; (c) The same PPy film as in panel b, after 100 consecutive cycles.

film, so that the related equivalent circuit includes both distributed capacitance and resistance.

11.5 Concluding Remarks

Progress in the synthesis of high-quality thin films and powders of π-conjugated polymers, together with a better understanding of the mechanisms of their doping, have been made, along with permanent attempts to introduce them into practice. On the basis of their unique electronic, ionic, optical, conformational, and crystalline properties, coupled with an extreme flexibility to control their characteristics via the organic synthesis, ECPs have become one of the most important classes of electrochemically active, solid-state materials. The commercialization of novel materials and devices always occurs amid bitter competition with pre-existing materials and devices, and in this respect two major groups relating to the application of ECPs can be distinguished (note – we have omitted here some very interesting, albeit stand-alone, specific applications such as ECP-based chemical sensors [69] or ECP actuators such as artificial muscles [70]. The first group relates to the use of ECPs for energy storage and conversion, and the second group to their use in electronics, including the manufacture of organic light-emitting diodes (OLEDs), solid-state memory, photovoltaic cells, and organic thin-film transistors (OTFTs) [71]. The history of Li batteries provides a clear demonstration of the difficulties encountered when introducing new materials, such as ECPs, to the field of energy storage and conversion. During the early 1990s, polyaniline was commercialized as the cathode for non-aqueous batteries with Li anodes; However, production was very soon discontinued when batteries with Li anodes were replaced by Li-ion batteries that comprised typical inorganic intercalation electrodes (Li-graphite anodes and lithiated transition metal oxide cathode, Li_xMO_2 cathodes) and also displayed a much better performance than Li-batteries with a polyaniline cathode. Yet, this did not discourage many research groups from attempting to utilize ECPs as supercapacitors, or as hybrid supercapacitors with inorganic materials [72].

Among the many applications of ECPs, perhaps the most successful has been the commercialization of OLEDs in automobile radios, electric shavers, cell phones, and cameras displays (in limited venues) [73]. In the case of OTFTs, those produced from ECPs show certain advantages: their softness is beneficial to their low-temperature processability and fine-tuning of their structure–property relationships, and they are inexpensive to produce. However, their morphology and microcrystallinity, and hence other important physico-chemical characteristics, may be difficult to control, as might also be their reproducibility of performance. Clearly, the many ongoing investigations with ECPs will contribute much to the future success of these materials.

References

1 Bruting, W. (ed.) (2005) *Physics of Organic Semiconductors*, Wiley-VCH, Weinheim.

2 Shirakawa, H. and Ikeda, S. (1971) *Polymer J.*, **2**, 231–44.

3 Shirakawa, H., Louis, E.J., Mac-Diarmid, A.G., Chiang, C.K. and Heeger, A.J. (1977) *Chem. Commun.*, 578–80.

4 Diaz, A.F., Castillo, J.I., Logan, J.A. and Lee, W.-Y. (1981) *J. Electroanal. Chem.*, **129**, 115–32.

5 Skotheim, T.A. and Reynolds, J.R. (eds) (1986) *Handbook of Conducting Polymers*, 1st edn, Marcel Dekker, New York.

6 Skotheim, T.A., Elsenbaumer, R. and Reynolds, J.R. (eds) (1998) *Handbook of Conducting Polymers*, 2nd edn, Marcel Dekker, New York.

7 Skotheim, T.A. and Reynolds, J.R. (eds) (2007) *Handbook of Conducting Polymers*, 3rd edn, CRC Press, New York.

8 Salaneck, W.R., Seki, K., Kahn, A. and Pireaux, J. (eds) (2002) *Conjugated Polymer and Molecular Interfaces*, Marcel Dekker, New York.

9 Hanziioannou, G. and van Hutten, P. (eds) (2000) *Semiconducting Polymers*, Wiley-VCH, Weinheim.

10 Takemoto, K., Offenbrite, R. and Kamachi, M. (eds) (1997) *Functional Monomers and Polymers*, Marcel Dekker, New York.

11 Rupprecht, L. (ed.) (1999) *Conductive Polymers and Plastics in Industrial Applications*, Plastics Design Library, New York.

12 Reynolds, J.R. and Epstein, A.J. (2000) *Adv. Mater.*, **12**, 1565–70.

13 Reddinger, J.L. and Reynolds, J.R. (1999) *Adv. Polym. Sci.*, **12**, 57–123.

14 Chandrasekhar, P. (1999) *Conducting Polymers, Fundamentals and Applications*, Kluwer Academic Publishers, New York.

15 Wallace, G.G., Spinks, G.M., Kane-Maquire, L.A.P. and Teasdale, P.R. (eds) (2003) *Conductive Electroactive Polymers: Intelligent Materials Systems*, 2nd edn, CRS Press, New York.

16 Leclerc, M. (2007) *Macromol. Rapid Commun. (Special Issue)*, **28**, 1665–824.

17 Freund, M.S. and Deore, B.A. (2007) *Self-Doped Conducting Polymers*, John Wiley & Sons, New York.

18 Hillman, A.R. (1987) *Electrochemical Science and Technology of Polymers*, Vol. 1 (ed. R.G. Linford), Elsevier, London, pp. 103–239 and 241–91.

19 Heinze, J. (1990) *Topics in Current Chemistry*, Vol. 2, Springer, Berlin, pp. 1–47.

20 (a) Roncali, J. (1992) *Chem. Rev.*, **92**, 711–38. (b) Roncali, J. (1997) *Chem Rev.*, **97**, 173–205.

21 Novak, P., Muller, K., Santhanam, K.S.V. and Haas, O. (1997) *Chem. Rev.*, **97**, 207–82.

22 Pickup, P.G. (1999) *Modern Aspects of Electrochemistry*, Vol. 33 (eds R.E. White, J.O'M. Bockris and B.E. Conway), Kluwer Academic/Plenum Publishers, New York, pp. 549–97.

23 Inzelt, G. (2008) *Conducting Polymers. A New Era in Electrochemistry* (ed. F. Scholz), Springer-Verlag, Berlin.

24 Farges, J.-P. (ed.) (1994) *Organic Conductors. Fundamentals and Applications*, Marcel Dekker, New York.

25 Fichou, D. (ed.) (1999) *Handbook of Oligo- and Polythiophenes*, Wiley VCH, Weinheim.

26 Murray, R.W. (ed.) (1992) *Molecular Design of Electrode Surfaces. Techniques of Chemistry*, Vol. 22, John Wiley & Sons, New York.

27 McKinnon, W.R., and Haering, R.R. (1983) *Modern Aspects of Electrochemistry*, Vol. 15 (ed. J.O'M. Bockris), Plenum Press, New York, pp. 235–304.

28 Bruce, P.G. (ed.) (1995) *Solid State Electrochemistry*, Cambridge University Press, Cambridge.

29 Gelling, P.J. and Bouwmeester, H.J.M. (eds) (1997) *The CRC Handbook of Solid State Electrochemistry*, CRC Press, New York.

30 Roth, S. and Carrol, D. (2004) *One-Dimensional Metals*, 2nd edn, Wiley-VCH, Weinheim.

31 van Haare, J.A.E.H., Havinga, E.E., van Dongen, J.L.J., Janssen, R.A.J., Cornil, J. and Bredas, J.-L. (1998) *Chem. Eur. J.*, **4**, 1509–22.

32 Kissinger, P.T. and Heineman, W.R. (eds) (1996) *Laboratory Techniques in Electroanalytical Chemistry*, 2nd edn, Marcel Dekker, New York.

33 Lund, H. and Hammerich, O. (eds) (2001) *Organic Electrochemistry*, 4th edn, Marcel Dekker, New York.

34 Levi, M.D. and Pisarevskaya, E.Yu. (1991) *Synth. Met.*, **45**, 309–22.

35 Heinze, J. (1991) *Synth. Met.*, **41–43**, 2805–23.

36 Crooks, R.M., Chyan, O.M.R. and Wrighton, M.S. (1989) *Chem. Mater.*, **1**, 2–4.

37 Mastragostino, M. and Soddu, L. (1990) *Electrochim. Acta*, **35**, 463–6.

38 Bard, A.J. and Mirkin, M.V. (eds) (2001) *Scanning Electrochemical Microscopy*, Marcel Dekker, New York.

39 Borgwarth, K., Rohde, N., Ricken, C., Hallensleben, M.L., Mandler, D. and Heinze, J. (1999) *Adv. Mater.*, **11**, 1221–5.

40 Waltman, R.J. and Bargon, J. (1986) *Can. J. Chem.*, **64**, 76–95.

41 Chao, F., Costa, M. and Tian, C. (1993) *Synth. Met.*, **53**, 127–47.

42 Hwang, B.J., Santhanam, R. and Lin, Y.L. (2001) *Electrochim. Acta*, **46**, 2843–53.

43 Cohen, Y.S., Levi, M.D. and Aurbach, D. (2003) *Langmuir*, **19**, 9804–11.

44 Greef, R., Peat, R., Peter, L.M., Pletcher, D. and Robinson, J. (1985) *Instrumental Methods in Electrochemistry*, Ellis Horwood, Chichester, UK.

45 Otero, T.F. (1999) *Modern Aspects of Electrochemistry*, Vol. 33 (eds J.O'M. Bockris, R.E. White and B.E. Conway), Plenum Press, New York, pp. 307–434.

46 Ofer, D., Crooks, R.M. and Wrighton, M.S. (1990) *J. Am. Chem. Soc.*, **112**, 7869–79.

47 Zhou, M., Pagels, M., Geschke, B. and Heinze, J. (2002) *J. Phys. Chem. B*, **106**, 10065–73.

48 Novak, P. and Vielstich, W. (1990) *J. Electrochem. Soc.*, **137**, 1681–9.

49 Levi, M.D., Lankri, E., Gofer, Y., Aurbach, D. and Otero, T. (2002) *J. Electrochem. Soc.*, **149**, E204–14.

50 Levi, M.D., Gofer, Y., Aurbach, D., Lapkowski, M., Vieil, E. and Serose, J. (2000) *J. Electrochem. Soc.*, **147**, 1096–104.

51 Rudge, A., Raistrick, I., Gottesfeld, S. and Ferraris, J.P. (1994) *Electrochim. Acta*, **39**, 273–87.

52 Zotti, G. and Schiavon, G. (1994) *Synth. Met.*, **63**, 53–6.

53 Sarker, H., Gofer, Y., Killian, J.G., Poehler, T.O. and Searson, P.C. (1997) *Synth. Met.*, **88**, 179–85.

54 Zotti, G., Zecchin, S., Schiavon, G., Berlin, A., Pagani, G., Borgonovo, M. and Lazzaroni, R. (1997) *Chem. Mater.*, **9**, 2876–86.

55 Fisyuk, A.S., Demadrille, R., Querner, C., Zagorska, M., Bleuse, J. and Pron, A. (2005) *New J. Chem.*, **29**, 707–13.

56 Pomerantz, Z., Levi, M.D., Salitra, G., Demadrille, R., Fisyuk, A., Zaban, A., Aurbach, D. and Pron, A. (2008) *Phys. Chem. Chem. Phys.*, **10**, 1032–42.

57 Zotti, G., Zecchin, S., Schiavon, G., Vercelli, B. and Zanelli, A. (2004) *Chem. Mater.*, **16**, 3667–76.

58 Kaiser, A.B. (2001) *Rep. Prog. Phys.*, **64**, 1–49.

59 Mott, N.F. and Davis, E.A. (1979) *Electronic Processes in Non-Crystalline Materials*, 2nd edn, Oxford, Clarendon.

60 Levi, M.D., Gofer, Y., Aurbach, D. and Berlin, A. (2004) *Electrochim. Acta*, **49**, 433–44.

61 Levi, M.D. and Aurbach, D. (2007) *Electrochem. Soc. Trans.*, **3** (27), 259–63.

62 Vorotyntsev, M.A., Daikhin, L.I. and Levi, M.D. (1992) *J. Electroanal. Chem.*, **332**, 213–35.

63 Chidsey, C.E.D. and Murray, R.W. (1986) *J. Phys. Chem.*, **90**, 1479–84.

64 Moore, W.J. (1972) *Physical Chemistry*, Prentice-Hall, Englewood Cliffs, New Jersey, p. 273.

65 Levi, M.D. and Aurbach, D. (1999) *Electrochim. Acta*, **45**, 167–85.

66 Vorotyntsev, M.A. and Badiali, J.P. (1994) *Electrochim. Acta*, **39**, 289–306.

67 Lu, Z., Levi, M.D., Salitra, G., Gofer, Y., Levi, E. and Aurbach, D. (2000) *J. Electroanal. Chem.*, **491**, 211–21.

68 Vorotyntsev, M.A., Daikhin, L.I. and Levi, M.D. (1994) *J. Electroanal. Chem.*, **364**, 37–49.

69 McQuade, D.T., Pullen, A.E. and Swage, T.M. (2000) *Chem. Rev.*, **100**, 2537–74.

70 Kim, K.J. and Tadokoro, S. (eds) (2007) *Electroactive Polymers for Robotics Applications. Artificial Muscles and Sensors*, Springer-Verlag, London.

71 Horowitz, G. (2004) *J. Mater. Res.*, **19**, 1946–62.

72 Mastragostino, M., Arbizzani, C. and Soavi, F. (2002) *Solid State Ionics*, **148**, 493–8.

73 Sheats, J.R. (2004) *J. Mater. Res.*, **19**, 1974–94.

12
High-Temperature Applications of Solid Electrolytes: Fuel Cells, Pumping, and Conversion

Jacques Fouletier and Véronique Ghetta

Abstract

High-temperature, solid-state electrochemical reactors have numerous current and potential applications. In this chapter, we briefly review the various modes of operation of cells involving a solid electrolyte conducting by oxide ions or protons, either for energy generation or fine chemicals production. The specific requirements of the materials constituting the cell core are outlined, after which examples of current applications are briefly described. These include oxygen and hydrogen pumping, atmosphere control, intermediate- and high-temperature solid oxide fuel cells, and catalytic membrane reactors.

12.1
Introduction

Ceramic electrochemical reactors are currently undergoing intense investigation, the aim being not only to generate electricity but also to produce chemicals. Typically, ceramic dense membranes are either pure ionic (solid electrolyte; SE) conductors or mixed ionic-electronic conductors (MIECs). In this chapter we review the developments of cells that involve a dense solid electrolyte (oxide-ion or proton conductor), where the electrical transfer of matter requires an external circuitry. When a dense ceramic membrane exhibits a mixed ionic-electronic conduction, the driving force for mass transport is a differential partial pressure applied across the membrane (this point is not considered in this chapter, although relevant information is available in specific reviews).

Solid-state electrochemical cells based on pure ionic conductors – which often are referred to as solid-electrolyte membrane reactors (SEMRs) – have numerous current or potential applications, including the production of high-purity oxygen by separating nitrogen and oxygen from the air; the control of oxygen in a gas or in a molten metal (oxygen pump or an oxygen "getter", a form of oxygen-trapping device);

Solid State Electrochemistry I: Fundamentals, Materials and their Applications. Edited by Vladislav V. Kharton

ISBN: 978-3-527-32318-0

high-temperature water electrolysis; high-temperature fuel cells for energy production (e.g., solid oxide fuel cells; SOFCs) or for the production of synthetic gases; and catalytic membrane reactors (CMRs) [1–5]. The majority of these industrially important catalytic reactions are oxidations, hydrogenations, and dehydrogenations.

The fields of applications range from metallurgy to the semiconductor industry, to medicine, the petrochemical industry, electricity production, chemical cogeneration (i.e., the simultaneous production of electricity, heat, and chemicals), atmospheric control within the laboratory, and so on. However, virtually no large-scale industrial applications have yet been reported, and for a variety of reasons, including:

- the relatively high price of membrane units;
- the low electrode surface area compared to the large surface area of a porous catalyst;
- an insufficient demonstration of the durability and sustained performances of the reactors (in the range of 25 000–30 000 h); and
- the high production costs (notably the high temperature requirements and consequent consumption of electrical energy) [5].

Current and future advancements in materials engineering might, however, lead to a significant reversal of this trend, and in this context the lowering of operating temperatures represents the main target. By comparison, compared to conventional catalytic reactors, SEMRs could be used to produce expensive fine chemicals, with attractive yields.

The single cells consist of a dense solid electrolyte membrane and two porous electrodes. In most cases, at least one of the electrodes is exposed to an oxygen-containing gas (often, ambient air), while the other electrode is exposed to an inert gas, a liquid metal, a partial vacuum, or a reacting mixture (hydrogen, water vapor, hydrocarbons, CO, CO_2, etc.). The single-chamber reactor (SCR) has been also proposed either as a membrane reactor or as a fuel cell. In this case, the solid-electrolyte disk, with two different electrodes that are coated either on opposite sides or on the same side of the pellet, is suspended in a flow of the reacting mixture (see Section 12.6.3).

Basically, four types of set-up can be considered:

- Open-circuit cells (chemical sensors) (for further details, see Chapter 13).
- Electrochemical pumps and compressors; that is, devices for dosage, separation, compression or removal of oxygen (or hydrogen), according to Faraday's law. The set-up devoted to separate oxygen from air has been also referred to as either the ion transport membrane (ITM) or solid electrolyte oxygen separation (SEOS) [6].
- Catalytic reactors, where the incorporation of an appropriate catalyst can increase the conversion and lead to an increase in selectivity. These reactors can operate according to either a "pumping mode" or a "fuel-cell mode".
- Catalytic membrane reactors which exhibit a nonfaradaic electrochemical modification of catalytic activity (NEMCA) effect.

The development of novel oxygen solid electrolytes conducting at temperatures below 800 °C is of major importance in the large-scale development of applications, such as intermediate-temperature (IT) SOFCs, and electrocatalytic reactors for natural gas conversion. Decreasing the membrane thickness also represents a means of reducing the ohmic loss, albeit to a limited extent (i.e., sufficient mechanical strength and no defects) until the surface exchange processes become rate limiting.

During the past ten years, a great deal of research has been devoted to the improvement of electrode microstructure [7, 8], notably of the thickness, porosity and dual phase nature, as well as the use of functionally graded materials and interlayers.

Although protonic conductors have not yet achieved the same degree of development, intensive effort is currently being applied to the development of hydrogen sensors, low-temperature SOFCs, the removal of hydrogen from coal gas, the production of pure hydrogen from hydrocarbons or via steam electrolysis, and the hydrogenation and dehydrogenation of organic compounds. These materials possess a number of theoretical advantages compared to oxide ions conductors, notably in terms of their efficiency and fuel utilization. However, further investigations will be required in order to improve their chemical stability (under a hydrogen atmosphere and air) and electrical conductivity. In addition, the majority of the currently proposed proton-conducting ceramics depend heavily on both temperature and atmosphere – that is, on protons, oxygen vacancies, and electronic defects.

In this chapter, the aim is to provide a brief review of the main applications of cells based on oxide-ion and protonic conductors.

12.2 Characteristics of a Current-Carrying Electrode on an Oxide Electrolyte

In the case of an electrolyte purely conducting by oxide ions, the role of a direct current (DC) through an electrode can be described within the "microsystem" concept (Figure 12.1) [9]. The electrode formed by the contact between the metal,

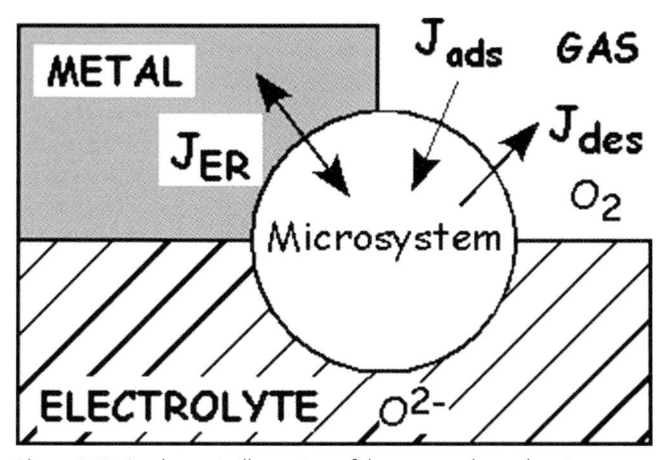

Figure 12.1 A schematic illustration of the oxygen electrode microsystem concept.

the solid electrolyte and the gas is viewed as measuring the oxygen activity in the microsystem that exchanges oxygen from the surrounding gas, relatively easily.

The local balance of oxygen in the microsystem takes into account the fluxes of adsorption (J_{ads}) and desorption (J_{des}) of oxygen, and the flux linked to the electrode reaction (J_{ER}), corresponding to either an intake or an uptake of oxygen, according to:

$$\frac{1}{2}O_2 + 2e^- \Leftrightarrow O^{2-} \tag{12.1}$$

Under a stationary state, the balance of the fluxes can be written as:

$$J_{ads} \pm J_{ER} = J_{des}. \tag{12.2}$$

On the cathode side of the cell, the gas fixes the maximum adsorption flux; this flux corresponds to a limiting cathodic current (I_{lim}), for which the oxygen activity in the microsystem is nil. For a current intensity higher than I_{lim}, a direct injection of electrons within the solid electrolyte will occur simultaneously with the oxygen reaction (Equation 12.1). Here, two situations are encountered:

- Case 1: the DC current is lower than $|I_{lim}|$. The mass balance within a layer of the solid electrolyte near the cathode indicates that no stoichiometry change is obtained; rather, only an oxide ions flux from the cathode to the anode is observed. This flux obeys the Faraday law (i.e., $J = I/2F$, and the cell functions as an oxygen pump (Figure 12.2).
- Case 2 (Figure 12.3): here, the applied cathodic current I is higher than $|I_{lim}|$, which corresponds to the maximum oxide ions flux. The current obeys the following equation:

$$I = I_{lim} + I_e. \tag{12.3}$$

The current I_e corresponds to the injection of electrons that are trapped on the point defects, thus inducing a coloration (the formation of "color centers" such as $V_O^{\bullet}$ or $V_O^{\times}$;

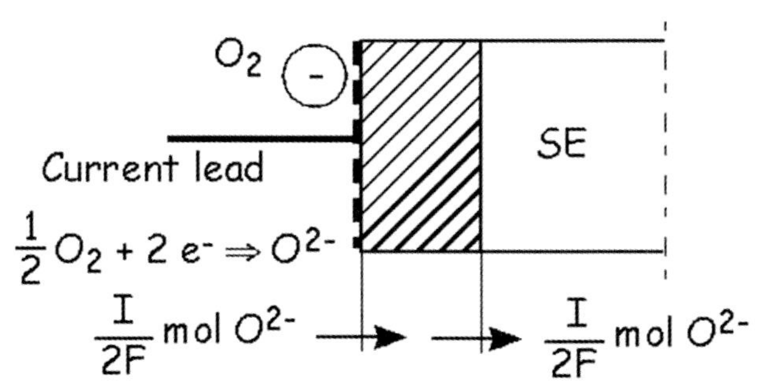

Figure 12.2 Schematic representations of the cathode processes on an oxide electrolyte: pumping mode.

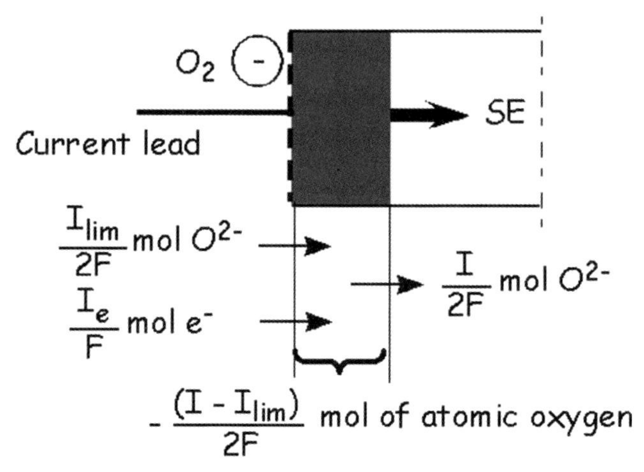

Figure 12.3 Schematic representations of the cathode processes on an oxide electrolyte: electrochemical reduction of the oxide electrolyte.

see previous chapters). It could easily be demonstrated that this nonstoichiometric zone progresses within the oxide, from the cathode to the anode [10, 11].

The shape of the current–voltage curve at the cathode of the cell illustrates the two cases, as shown in Figure 12.4. Two domains can be observed: (i) the A–B interval corresponds to the case 1 (the reduction of oxygen takes place within the "electrolytic

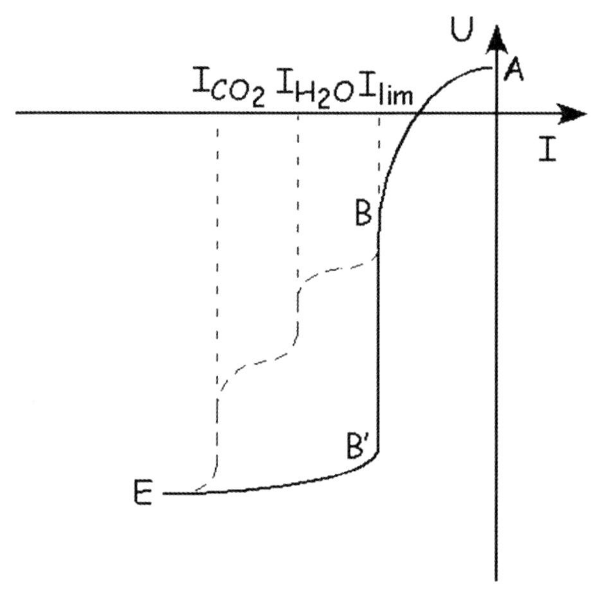

Figure 12.4 Typical cathode characteristics. The solid line indicates oxygen reduction; the dashed line indicates the $O_2 + H_2O + CO_2$ mixture.

domain" of the electrolyte), (ii) in the B′–E interval, the solid electrolyte is electrochemically reduced. It should be pointed out that, when other oxygenated species than oxygen are present in the gas (H_2O, CO_2, etc.), limiting currents for each species can be defined (corresponding to their maximum adsorption fluxes). As a consequence, the shape of the current–voltage curve is modified, as shown in Figure 12.4 (dashed line).

12.3 Operating Modes

Let us consider a solid oxide electrolyte reactor in which one side of the membrane is in contact with ambient air, and the other side (referred to as the working electrode) is fed either by an inert or a reactive gas (a mixture of hydrogen, natural gas, hydrocarbons, CO, etc. and H_2O and/or CO_2). Figure 12.5 shows, in graphical terms, the voltage variation of the working electrode referred to air as a function of the current.

Four situations can be considered here:

- The *fuel cell mode*: the driving force is the Gibbs energy of fuel oxidation reactions. The cell is used for the generation of electricity and/or for the production of chemicals.
- The *pumping mode*, by applying an external potential. As shown in Figure 12.5, the oxygen feed can be increased (according to Faraday's law or with the NEMCA

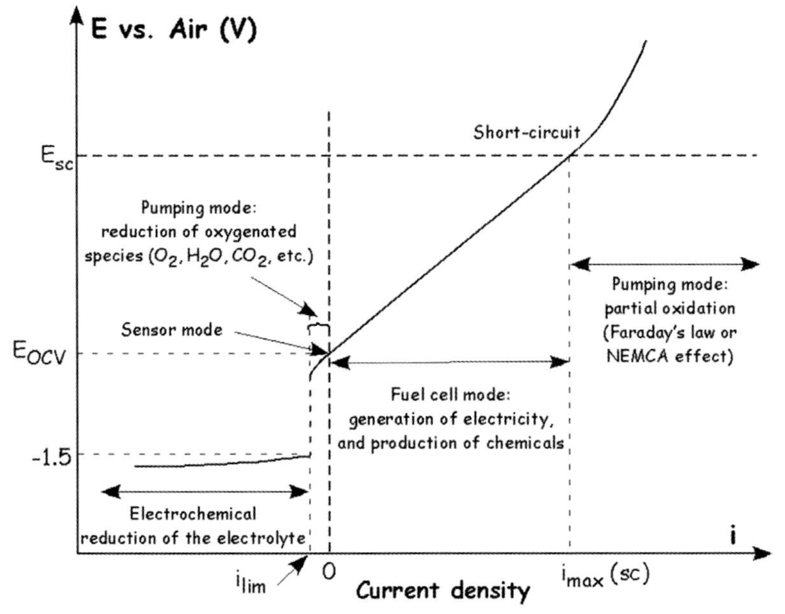

Figure 12.5 Current–voltage curve of a membrane reactor (air is provided at one side of the membrane), with an indication of the operating modes.

effect). When the reducing applied voltage is less than the open-circuit voltage (E_{OCV}), the oxygenated species such as O_2, CO_2 and H_2O are reduced.

- *Electrochemical reduction of the electrolyte* (the applied current density is higher than the limiting current, i_{lim}).
- The *sensor mode* at OCV (see Chapter 13).

12.3.1 Electrochemical Pumping

The oxygen flux through the membrane is controlled by imposing an external current, in both directions, between the electrodes. A current corresponds to $I/2\,F$mol of O^{2-} ions per second transported through the electrolyte, according to Faraday's law. Most technological applications require current densities approaching $1\,A\,cm^{-2}$, which corresponds to oxygen fluxes of $2.6\,\mu mol\ O_2\,cm^{-2}\,s^{-1}$ or $3.5\,cm^3\,cm^{-2}\,min^{-1}$. As described in Section 12.6.1, the same device can be used for oxygen pumping, the reduction of carbon dioxide or water vapor, or the oxidation of hydrogen. The device can also be used for controlling the oxygen pressure in a reactor or a flowing gas, or as a catalytic reactor.

Investigations carried out on membrane reactors based on pure oxide ion-conducting membranes have been reviewed [12]. Compared to the MIEC-membrane reactor, SEMRs allow for direct control of the oxygen permeation rate due to the faradaic coupling of oxygen flux and cell current. The SEMR has advantages over conventional catalytic reactors [3, 5], including:

- The ionic flux can easily be controlled by the current passing through the cell.
- The oxygen (or hydrogen) produced at the working electrode are of the highest purity (there is no poisoning or deactivation of the catalyst by impurities contained in the gas feed, such as ppm-level H_2S in hydrogen).
- The reactivity and selectivity of the ionic species supplied at the electrode can be very different from that of the corresponding gaseous component (O_2, H_2, etc.).
- A high oxygen pressure can be achieved without need for a compressor.
- Air instead of pure oxygen can be used for catalytic oxidation in a single reactor.
- By controlling the oxygen flux, the local reactants composition can be kept outside the flammability region, so that the risk of explosion is eliminated, and hot spots can be avoided.
- The simultaneous production of a useful chemical and generation of electricity can be very attractive from an economics point of view [5].

12.3.2 Fuel Cell Mode

The SEMR combines the membrane reactor concept with the working principle of a SOFC, with oxygen being reduced to O^{2-} anions at the cathode side and transported via the solid electrolyte membrane to the anode. At the anodic catalyst, hydrocarbons react with the supplied oxygen species to form oxidized products.

Hydrogen is the preferred feedstock for use in the present generation of fuel cells for energy production, due to its high electrochemical reactivity compared to the more common fossil fuels, such as hydrocarbons, alcohols, or coal. In the latter cases, a gas treatment is necessary before entering the stack (*external reforming*) or within the stack itself (*internal reforming*) for producing hydrogen or synthesis gas.

In this process, air and fuel are fed to the reactor at opposite sides of a dense oxygen ion-conducting solid electrolyte membrane. The theoretical electromotive force (EMF), E_{th} or E_{OCV}, is calculated from the Nernst equation (1.01 V at 800 °C with pure hydrogen at the anode and air at the cathode). The voltage output (U) under load conditions obeys the following equation:

$$U = E_{\mathrm{th}} - R.I + \eta_c - \eta_a \tag{12.4}$$

where R is the cell resistance, and η_c ($\eta_c < 0$) and η_a ($\eta_a > 0$) are the polarization losses associated with the cathode and anode, respectively. As indicated in Figure 12.6, various phenomena contribute to the irreversible losses – that is, activation polarization, ohmic drop, and concentration polarization. The SOFC performance depends mainly on the ohmic loss of the electrolyte and on the cathodic polarization. These polarization phenomena are minimized by an appropriate choice of the materials (conductivity, exchange kinetics, etc.) and by optimization of the electrode microstructure.

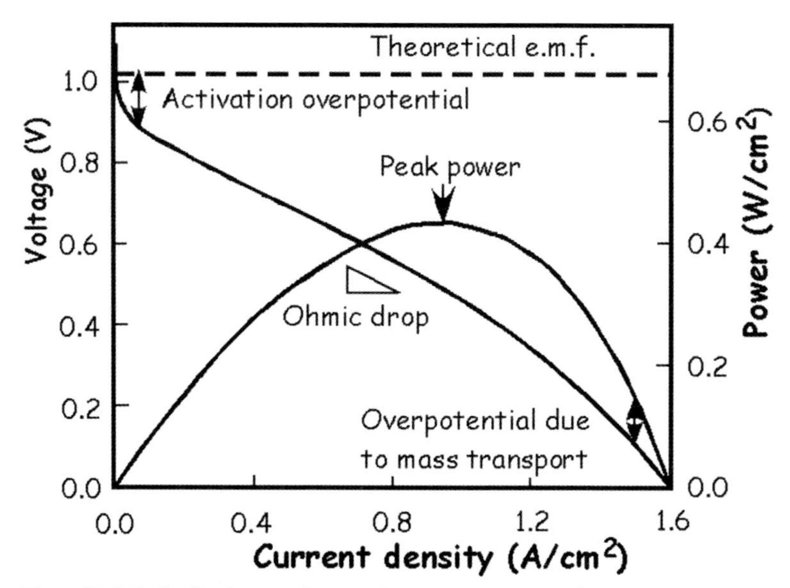

Figure 12.6 Fuel cell voltage and power density versus current density.

Because of the high temperature of operation, one major advantage of the SOFC, in comparison with other fuel cell types, is its flexibility in the choice of fuel: natural gas and other hydrocarbons can be reformed within the cell stack [13]. Internal reforming involves the conversion of hydrocarbons to hydrogen and carbon monoxide:

- *steam reforming*:

$$CH_4 + H_2O = 3H_2 + CO \tag{12.5}$$

- *dry reforming*:

$$CH_4 + CO_2 = 2CO + 2H_2 \tag{12.6}$$

followed by electrochemical reactions:

$$H_2 + O^{2-} = H_2O + 2e^- \tag{12.7}$$

$$CO + O^{2-} = CO_2 + 2e^- \tag{12.8}$$

Generally, a pre-reforming of the fuel is carried out, whereby part of hydrocarbon is reformed in an external reactor and the remainder is internally reformed. Methane and other hydrocarbons can also be converted to hydrogen by partial oxidation (POX); this is an exothermic process which is often combined with endothermic steam reforming, and leads to an autothermic conversion of methane.

Internal reforming has a number of practical issues, notably that a high excess of steam is necessary in order to avoid carbon deposition, and the cooling effect due to the endothermicity of the reforming reaction may induce thermal stress within the cell. Vernoux *et al.* [14, 15] have proposed a new concept – *gradual internal reforming* – which is based on a local coupling of the steam reforming on the catalyst dispersed on the anode material and the electrochemical oxidation of the *in situ*-produced hydrogen (see Figure 12.7). This concept has been demonstrated both experimentally [16] and, more recently, in model form [17].

Two recent concepts have also generated significant excitement:

- The *direct oxidation* of hydrocarbon without reforming [18–21]; the key issues here are the lower electrochemical oxidation rate with respect to hydrogen, and carbon deposition on the anode surface.

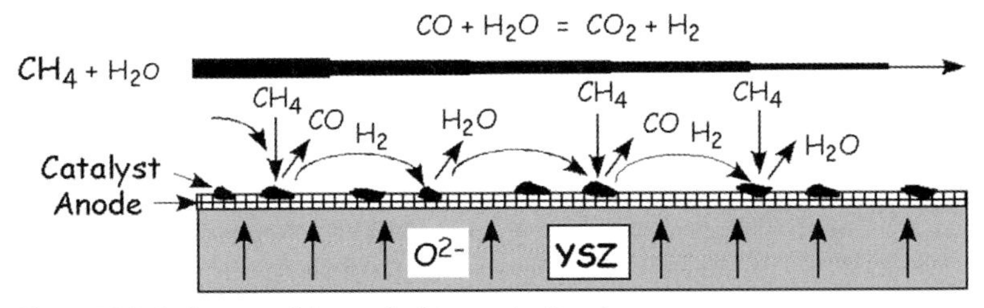

Figure 12.7 A diagram of the gradual internal reforming process.

- A *single-chamber SOFC*, based on the highly specific catalytic activity of the two electrodes [22–25]. This concept offers a variety of advantages: the gas-sealing problem is overcome; there is less carbon deposition risk due to the presence of a large amount of oxygen; the system can withstand higher mechanical and thermal forces; and the solid electrolyte can be porous. Although an impressive performance of this system has recently been reported ($0.6\,W\,cm^{-2}$), future studies are necessary, notably to improve electrode selectivity, fuel utilization, and durability.

12.3.3 The NEMCA Effect

The concept of the NEMCA effect or electrochemical promotion catalysis (EPOC) was proposed first by Vayenas *et al.* [26–30]. It has been shown that the catalytic activity and selectivity of a catalyst, when deposited on a solid electrolyte, can be greatly affected by the application of small currents or potentials, supplying or removing ions at the catalyst surface. The induced steady-state change in the reaction rate may be up to 90-fold higher than the open-circuit catalytic rate, and up to 3×10^5-fold higher than the steady-state of ions supply predicted by the Faraday's law. The NEMCA effect has been observed for tens of catalytic reactions on porous catalysts, including metals such as Pt, Pd, Rh, Au, and Ag, or oxides such as IrO_2 and RuO_2. The solid electrolytes used were O^{2-} ionic conductors (stabilized zirconia and doped ceria), sodium ion conductors (such as β''-alumina or NASICON), various protonic conductors, fluoride electrolytes (CaF_2), or mixed ionic-electronic conductors, such as TiO_2 or CeO_2.

Following Vayenas *et al.*, the effect of current or potential on the catalysis activity is usually described by two parameters (we consider here the use of an O^{2-} conductor):

- the rate enhancement ratio (ρ), defined as: $\rho = r/r_o$, where r_o is the catalytic rate at open circuit and r the catalytic rate under polarization;
- the apparent faradaic efficiency or enhancement factor, Λ, defined as:

$$\Lambda = \Delta r/(I/2F) = (r - r_o)/(I/2F), \tag{12.9}$$

where I is the applied current, F is the Faraday constant, and $I/2F$ equals the rate of O^{2-} ions supply to the catalyst.

In the case of a Faradaic effect, all oxygen which is transported electrochemically through the electrolyte reacts at the anode ($\Lambda = 1$). A reaction exhibits the NEMCA effect when $|\Lambda| > 1$. When $\Lambda > 1$, the reaction is termed electrophobic (the catalytic reaction is promoted by a positive current or overpotential), while a reaction accelerated by a negative current or potential is termed electrophilic.

The NEMCA effect was explained by taking into account the increase in the catalyst work function during oxygen pumping, and the consequent weakening of the binding strength of the adsorbed species. A typical example is given in Figure 12.8 [31–33].

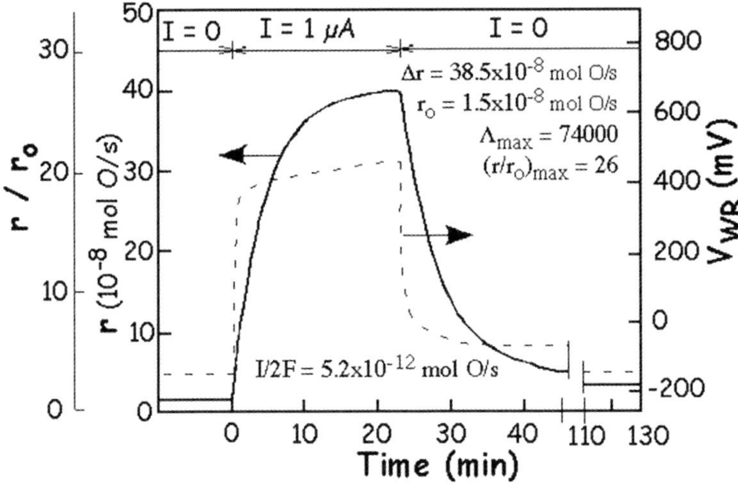

Figure 12.8 Example of the electrochemical promotion catalysis (ethylene oxidation on platinum catalyst) using a YSZ membrane: oxidation rate (r) and catalyst potential (V_{WR}) response to step changes in applied current.

An important characteristic of NEMCA is that, for any reaction, the magnitude of $|\Lambda|$ can be estimated by the equation:

$$|\Lambda| = \frac{2Fr_o}{i_o} \tag{12.10}$$

where i_o is the exchange current density of the catalyst/solid electrolyte interface.

This means that, in contrast to conventional fuel cell applications, highly polarizable interfaces (low i_o values) are required in order to obtain high Λ values – that is, in order to observe the NEMCA effect.

12.3.4 Electrolyte Reduction

It has been demonstrated that, in the case of a potentiostatic electrolysis and for a cathodic voltage less than -1.5 V (versus air reference electrode; see Figure 12.5) a stationary current is obtained [34]. In fact, for a more cathodic voltage an electrochemical reduction of the electrolyte may occur, inducing a continuous current increase. In the case of a galvanostatic mode, when the corresponding current is smaller than the limiting currents associated with the possible gas electrode reactions (involving species such as O_2, CO_2, H_2O), a stationary state is rapidly obtained. By forcing a current higher than the sum of the limiting currents, the solid electrolyte is reduced. It has been demonstrated that the reduced oxide can be reversibly reoxidized, provided that the reoxidation rate is moderate [35, 36]. An electrically renewable and controllable oxygen getter based on reversible reduction–oxidation cycles of stabilized zirconia has been developed [37].

12.4
Cell Materials

12.4.1
Electrolytes

As the details of solid electrolytes are considered elsewhere in this book, we will in this chapter list only the major groups of electrolyte membranes and critical issues related to their applications. The two main difficulties that arise during the utilization of SOFCs or catalytic membranes concern the requirement for a high ionic conductivity and a high oxygen partial pressure gradient between the fuel compartment and the air chamber. This tends to induce a number of problems in the membrane materials, such as structural modification and phase segregation [38]. Typically, the membrane must withstand an oxidizing atmosphere (air) on one side, and a highly reducing atmosphere (methane, hydrocarbons), with oxygen pressures down to $\sim 10^{-19}$ bar on the other side, and at high temperature.

Consequently, considerable attention has been paid to lowering the operating temperature, in either of two ways:

- By decreasing the membrane thickness, although only to a limited extent, until the surface reactions become rate limiting [40].
- The development of electrolytes that exhibit a higher conductivity at low temperature. Three candidates have emerged, namely doped ceria, doped lanthanum gallate, and doped barium zirconate. The first two of these are oxygen ion electrolytes, and the latter is a proton conductor.

It is considered that the bulk area specific resistance R_0 must be lower than $R_0 = L/\sigma = 0.15\,\Omega\,\text{cm}^2$, where L is the electrolyte thickness and σ is its total conductivity, predominantly ionic [39]. At present, fabrication technology allows the preparation of reliable supported structures with film thicknesses in the range 10–15 μm; consequently, the electrolyte ionic conductivity must be higher than $10^{-2}\,\text{S}\,\text{cm}^{-1}$. As shown in Figure 12.9, a few electrolytes (ceria-based oxides, stabilized zirconias, and doped gallates) exceed this minimum ionic conductivity above 500 °C.

In all ceramic high-temperature (HT) SOFC systems, $LaCr(Mg)O_3$ and $La(Ca)CrO_3$ are the interconnect materials used. The disadvantages of this include an expensive manufacturing route and a poor ability to withstand rapid temperature changes. Ferritic stainless steel can be used in IT SOFCs (see Figure 12.9).

12.4.1.1 Oxide Electrolytes

Fully stabilized zirconias doped with Y_2O_3 (8–9 mol%) or CaO (15 mol%), or partially stabilized zirconia doped with MgO (3 mol%), are the most widely used electrolytes in devices operating at temperatures higher than 800–850 °C.

For temperatures less than 800 °C, a variety of ceramic materials have been proposed, including doped ceria, lanthanum gallates ($LaGaO_3$), or bismuth-based oxides (e.g., Bi_2O_3-Er_2O_3 or doped $Bi_2VO_{5.5}$ family). None of these materials fulfils all

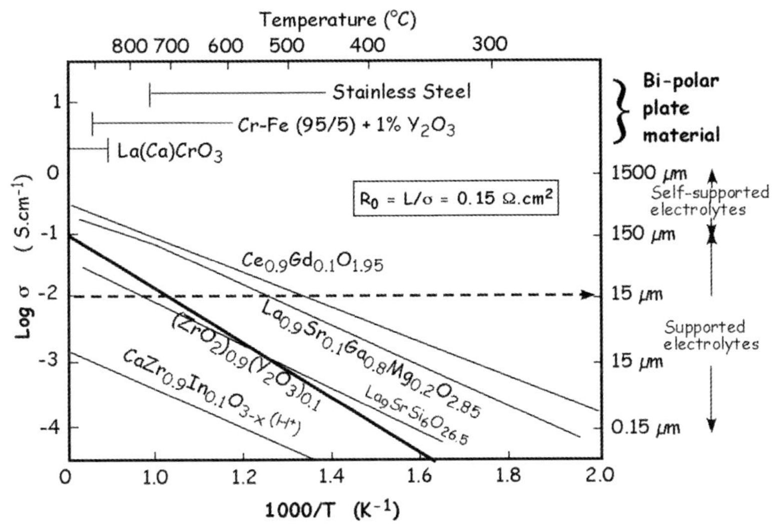

Figure 12.9 Arrhenius plots of the ionic conductivity of selected electrolytes. The temperature ranges of utilization of interconnects materials are also indicated. For electrolyte thicknesses >150 μm, the cell can be supported by the ionic membrane.

of the requirements in terms of chemical stability, stability under reducing conditions, and mechanical strength.

One promising group of solid electrolytes are the apatite-type silicates $A_{10-x}(SiO_4)_6O_{2\pm\delta}$ (where A = rare- and alkaline-earth cations). These exhibit a high ionic transport number, a moderate thermal expansion, and show no chemical reactivity with the cathode materials [41].

12.4.1.2 **Proton-Conducting Electrolytes**

Since the pioneering studies of Iwahara *et al.* [42, 43], most of the HT (>500 °C) proton-conducting ceramic materials that have been developed have been either cerates or zirconates, with the general formula $(Sr,Ba)Ce_{1-x}M_xO_{3-\delta}$, where M is a rare earth element, x (the molar site fraction) is less than its upper solubility limit (usually <0.2), and δ is the oxygen deficiency per unit formula [44]. When these ceramics are exposed to a hydrogen-containing atmosphere at an elevated temperature, the electronic conductivity is decreased and protonic conduction appears. The conductivity in either a humid or a hydrogen-containing atmosphere at elevated temperatures is of the order of 10^{-2}–10^{-3} S cm^{-1} over a temperature range of 1000 to 600 °C, due to the incorporation of protons by the reactions shown in Equations 12.11 and 12.12:

$$H_2O_{(g)} + V_O^{\bullet\bullet} + O_O^{\times} \Leftrightarrow 2OH_O^{\bullet} \quad (12.11)$$

$$H_{2(g)} + 2O_O^{\times} \Leftrightarrow 2OH_O^{\bullet} + 2e' \quad (12.12)$$

The protons migrate mainly as lone protons, and jump from lattice site to lattice site by the Grotthus mechanism. The protonic conductivity of zirconates [(Ca,Sr,Ba) ZrO_3] is approximately one order of magnitude lower than that of the cerates; however, their chemical and mechanical stabilities are superior. In contrast to cerates, zirconates barely react with carbon dioxide gas. The stability of barium cerates can be increased when cerium atoms are partially substituted by zirconium atoms.

On recent years these proton-conducting ceramics have attracted increasing interest, not only as SOFC materials but also for HT electrolysis, for sensors, and for electrochemical processes. However, their widespread application depends on a number of factors, such as material stability, electrochemical and energetic efficiency and, of course, their cost.

12.4.2 Electrodes

Three types of material are used as electrodes, namely metals, mixed ionic-electronic oxides, and composite materials [45].

In laboratory-prepared devices (such as oxygen pumps), porous platinum layers that function as both the electrode and current collector are generally deposited on both surfaces of the membrane.

12.4.2.1 Cathode

In the case of a porous electrode prepared from an inert metal or a semiconducting oxide, the electrode reaction is limited to the so-called triple phase boundaries (TPBs) between the electronic conductor, the ionic conductor, and oxygen gas (air). $La_{0.85}Sr_{0.15}MnO_3$ perovskite (LSM) is the most widely used cathode material at temperatures above 800 °C, based on its high thermal and chemical stabilities. However, its performance decreases rapidly at low temperature. The performance can be improved by preventing the formation of any highly resistive layer at the electrode–electrolyte interface (the nonstoichiometric compound $La_{1-x}MnO_3$ is much less reactive with zirconia than $LaMnO_3$), and also by increasing the TPB length (by mixing LSM with zirconia) [46, 47]. One strategy for improving cathode performance is to grade its composition so that the transition from the solid electrolyte to the electrode material occurs over several compositional steps. Cathodes of this type are usually described as *functionally graded* [48, 49].

For IT SOFCs, perovskite oxides $(La,Sr)(Co,Fe)O_{3-\delta}$ have attracted particular attention [50]. Due to chemical reactivity with yttria-stabilized zirconia (YSZ), the use of a protective interlayer between the cathode and electrolyte is required to increase the system's stability during long-term operation [51–53].

A new family of compounds, formulated $A_2MO_{4+\delta}$, with the K_2NiF_4 type-structure, has been recently proposed as cathode materials [54–56]. These compounds exhibit a relatively high oxygen conductivity, a level of electronic conductivity of $\sigma_e \approx 10^2$ S cm^{-1} at 700 °C, high oxygen exchange coefficients, thermal expansion coefficients (TECs) that are compatible with commonly used electrolytes such as YSZ, doped ceria and lanthanum gallates, and chemical compatibility with the SOFC components.

12.4.2.2 Anode

Within a SEMR, the anodic layer serves both as electrode and as a reaction zone for the desired selective oxidation [57]. The electrode design has a critical effect on reactor performance. Ideally, the electrode must possess both high ionic and electronic conductivity and good catalytic properties for the desired reaction, avoiding side reaction such as carbon deposition that will block the anode function. Despite the main requirements that an anode must satisfy having been reviewed recently [58], the ideal material is difficult to find, and a variety of electrode configurations has been proposed (Figure 12.10):

- A porous *mixed ionic-electronic layer* may be deposited on the solid electrolyte surface (Figure 12.10a). However, it is generally difficult to find a material which exhibits a high catalytic activity in combination with sufficient stability.
- A *composite electrode*, in which the electronic conductor (typically a metal) is mixed with an ionic conductor. A well-known example is the conventional Ni–YSZ composite electrode used as anode of SOFCs (Figure 12.10b). The composition of the anode, the particle sizes of the powders, and the manufacturing method are key to providing good adhesion of the electrode with the electrolyte, to bring its thermal coefficient closer to that of the electrolyte, to prevent sintering of the nickel particles, and to obtain a high electrical conductivity and a high activity for reforming and electrochemical reactions.
- A *bilayer electrode* in which an electronic conductor is placed between the electrolyte and the porous catalytic material. Such a configuration can be used in the case of a low electrical conductivity of the catalyst: as an example, the controlled partial oxidation of hydrocarbons has been recently modeled (Figure 12.10c) [59]. Another example concerns the decomposition of the global reaction in two independent steps: such an electrode design was developed for the partial oxidation of *iso*-octane [60], or for the internal reforming of methane in SOFCs without water vapor excess [61, 62] (Figure 12.10d). The methane is reformed on the porous catalyst, which is insensitive for methane cracking; the hydrogen produced diffuses through the catalyst layer and is oxidized on the Ni–YSZ anode, which also serves as the current collector.

12.5 Cell Designs

In most cases, the cells are either in the form of tubes (open or closed-ended) or flat plates. Although the cells based on flat plates are more compact, the stacks require high-temperature seals, they are more sensitive to thermal stress and thermal cycling, and they are more difficult to pressurize. The tubular designs are more robust, allow the use of room-temperature seals, and exhibit a good thermal cycling capability.

Two concepts for cell design have been developed: for an electrolyte thickness $>150\,\mu m$, the membrane is self-supporting (see Figure 12.8), whereas for lower

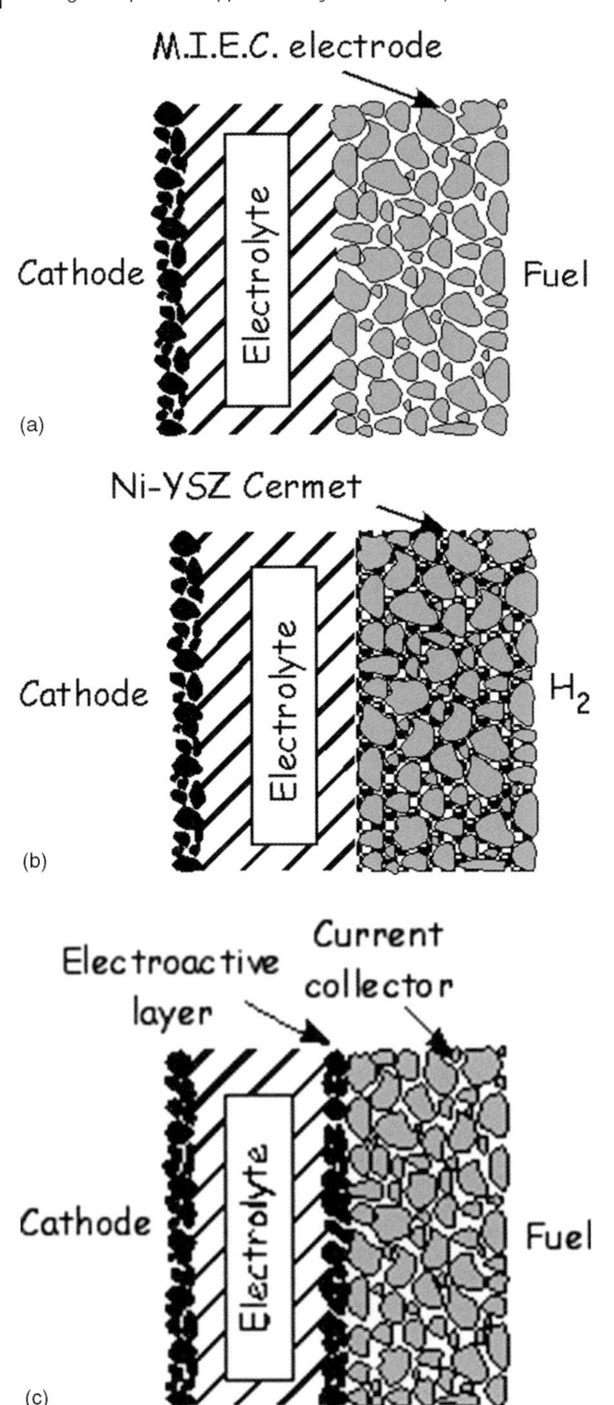

Figure 12.10 Various configurations of anodes. MIEC = mixed ionic electronic conductor; Cermet = ceramic–metal mixture.

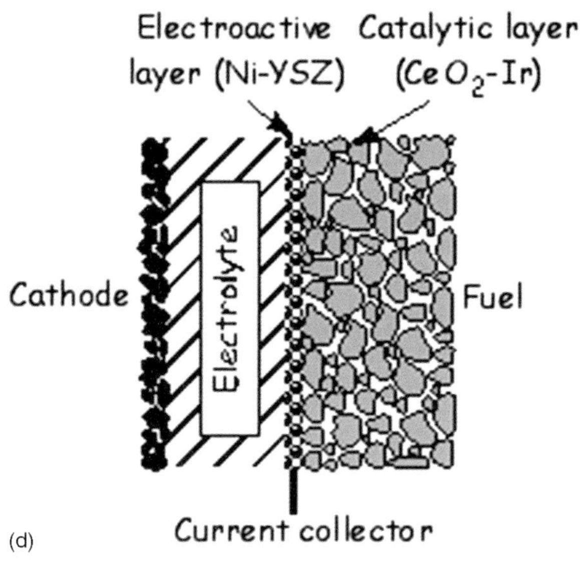

Figure 12.10 *(Continued)*

thicknesses the electrolyte membrane is deposited onto a porous electrode, which may be either the anode or cathode of the cell. Typically, the cathode is selected as a supporting layer in the tubular design, whereas the anode, the electrolyte and, more recently, the interconnect, is the more common choice for the planar design.

Large-scale manufacturing techniques are used to produce each of the layers; these include tape casting, extrusion, slip casting, calendaring, pressing, and screen-printing [82]. Thin-film deposition technologies are also used, including physical processes such as sputtering or vacuum evaporation, low-pressure metall-organic chemical vapor deposition (MO-CVD), plasma-enhanced CVD, pulsed laser deposition, magnetron sputtering and electrochemical vapor deposition (EVD) or wet chemistry, namely sol–gel, electrostatic spray deposition (ESD) or ultrasonic spray pyrolysis of sols (pyrosol methods). At present, the main challenge relates to the development of low-cost, high-deposition rate methods.

12.6 Examples of Applications

Despite the growing interest in SEMRs during the past 25 years, there is at present no industrial application of these reactors, except in the case of potentiometric sensors. Examples of the application of membrane reactors operating under closed-circuit conditions are briefly described in the following sections, with special emphasis on the laboratory set-ups used for materials studies and on membrane reactors presently in the pre-commercial phase, such as SOFCs or oxygen generators.

12.6.1 Oxygen and Hydrogen Pumping, Water Vapor Electrolysis

Recently, dense ceramic oxides have attracted great interest as potentially economical, clean, and efficient means of producing oxygen by its separation from air or other oxygen-containing gas mixtures. The envisioned applications range from small-scale oxygen pumps for medical and aerospace applications [63] to large-scale usage in combustion processes, for example, coal gasification.

In addition to SOFCs, both planar and tubular designs of cell are currently being developed for electrochemical oxygen separation and compression [6, 64–68]. Tubular designs enable a higher-pressure oxygen production compared to the planar geometry, but with a lower membrane area to device volume ratio. An advantage of this is an ability to deliver high-purity oxygen at elevated pressures, thus eliminating the need for compressors [69, 70].

The SEOS [71, 72] process employs an electrochemical stack fabricated from high-temperature solid electrolytes to produce high-purity oxygen at elevated pressure from a feed stream of ambient-pressure air (Figure 12.11). Whilst the early SEOS systems used YSZ as a solid electrolyte, at present a doped-ceria, operating at temperatures in excess of 600 °C, is preferred. The electrodes used are doped-strontium lanthanum manganite or cobaltite, and each electrochemical cell is in contact with a dense interconnect made from an electronically conductive perovskite material, with a glass ceramic being used as the seal material. When these cells were tested over periods of more than 20 000 h, relatively low degradation rates of less than 0.5% per 1000 h were encountered.

With proton-conducting electrolytes, a variety of pumping devices has been proposed, including electrochemical hydrogen separation from various gas mixtures, and the extraction of hydrogen atoms from hydrogen compounds such as water vapor and hydrogen sulfide. Figure 12.12 shows the Faraday law test of a hydrogen pump involving a cerate membrane [73]. Here, it has been shown that the pumping capacity was improved noticeably by adding a small percentage of water vapor.

Investigations have also been conducted with *steam electrolysers*. The advantage of these is that pure hydrogen gas, without water vapor, can be obtained – in contrast to those electrolysers which use an oxide-ion conductor.

12.6.2 Pump–Sensor Devices

These devices are widely used for the control and monitoring of oxygen content in gas mixtures or liquid metals. Usually, the electrochemical pump and oxygen sensor are based on YSZ tubes with platinum electrodes.

12.6.2.1 Open System: Oxygen Monitoring in a Flowing Gas

The gas circuit is shown schematically in Figure 12.13a. Nominally pure gases (Ar, N_2, He, etc.) are generally used. According to Faraday's law, the oxygen mole fraction X in the flowing gas obeys the following equation:

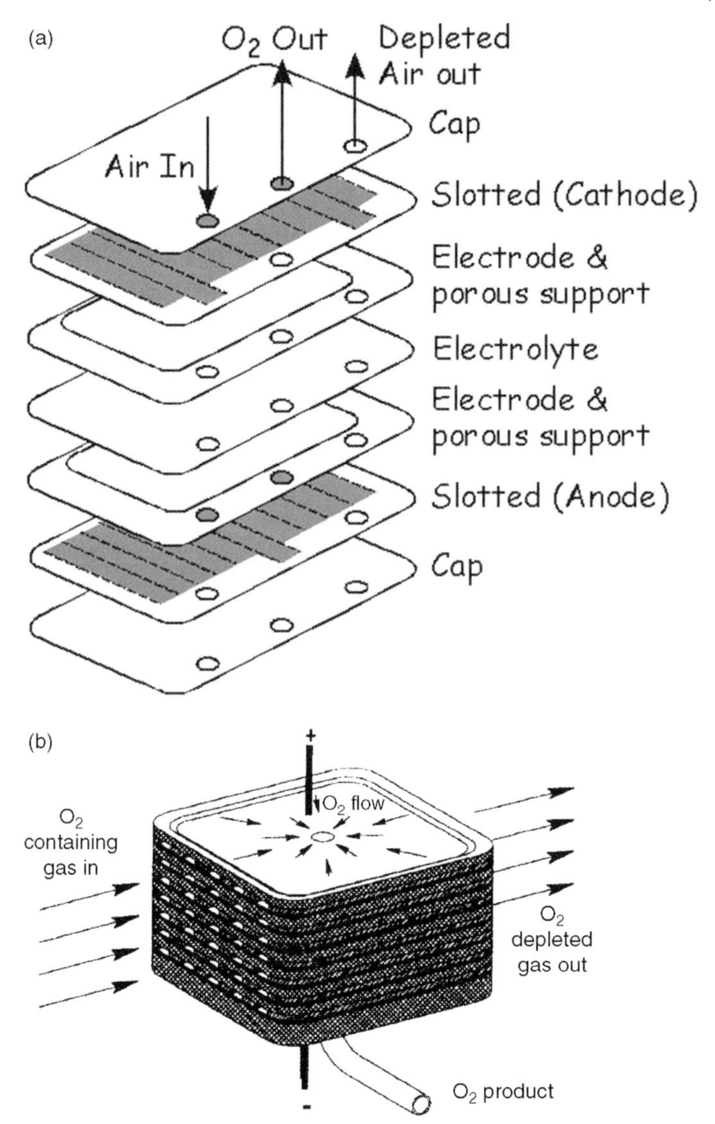

Figure 12.11 (a) Geometries and functions of the different layers in a planar electrochemical oxygen generator; (b) SEOS–oxygen generator multiple planar cell stack (for details, see the text). Further details are available in Refs. [67, 71].

$$X = X^{\circ} + 0.209\, I/D \tag{12.13}$$

where X° is the oxygen mole fraction in the gas supplied to the pump, D is the gas flow rate measured in $\mathrm{l\,h^{-1}}$ under Normal Temperature and Pressure (NTP) conditions, and I is measured in A. The *Faraday's law test* is based on a verification of the theoretical Equation 12.13. The oxygen mole fraction is determined by the oxygen

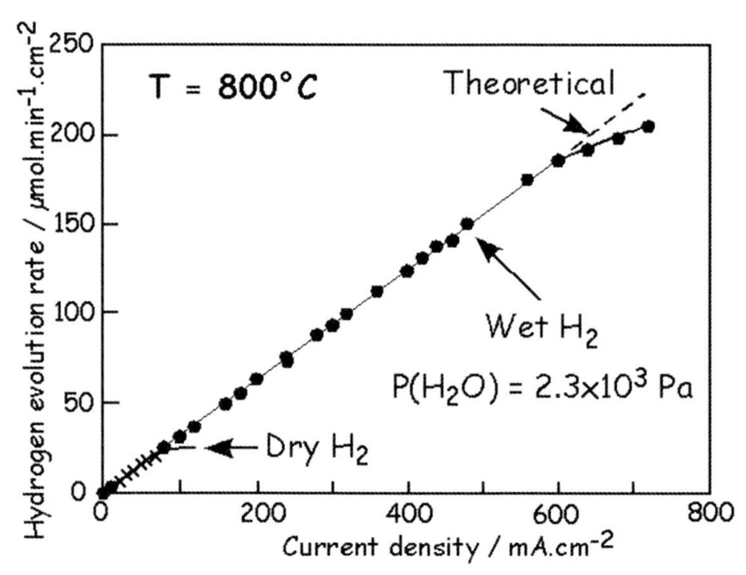

Figure 12.12 Hydrogen pumping using proton-conducting ceramics.

sensor (Nernst equation) for various current intensities I passing through the pump. As shown in Figure 12.13b, the oxygen mole fraction can be controlled in the 10^{-7} to 1 mole fraction range [74]. Such a device was also developed to monitor small variations in the oxygen molar fraction in air [75] (the oxygen fraction can be controlled within the range $\pm 300\,\mu l\, l^{-1}$ of oxygen around the normal background, with an error of less than $4\,\mu l\, l^{-1}$).

The same device can be used for monitoring the composition of CO_2–CO and Ar–H_2O–H_2 mixtures [76, 77]. In that case, pure (H_2, CO_2) or premixed gases, such as Ar–5% H_2 are used. According to Faraday's law, the equilibrium oxygen pressure versus the current intensity passing through the pump will obey the following equations:

- CO–CO_2 mixture

$$P_{O_2} = \left(2.392\frac{D}{I} - 1\right)^2 \exp\left(21.05 - \frac{68150}{T}\right) \quad \text{(in bar)} \tag{12.14}$$

Ar–H_2–H_2O mixture

$$P_{O_2} = \left(2.392\, q\frac{D}{I} - 1\right)^{-2} \exp\left(13.278 - \frac{59571}{T}\right) \quad \text{(in bar)} \tag{12.15}$$

where q is the hydrogen mole fraction in the feed gas (see Figure 12.14).

The association of an electrochemical oxygen pump and an oxygen sensor allows the monitoring of oxygen partial pressure in a flowing gas in the range from 1 bar to 10^{-25} bar, with an accuracy of 2%.

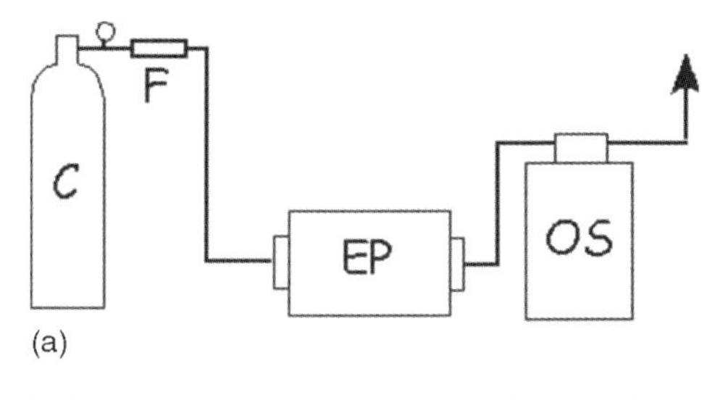

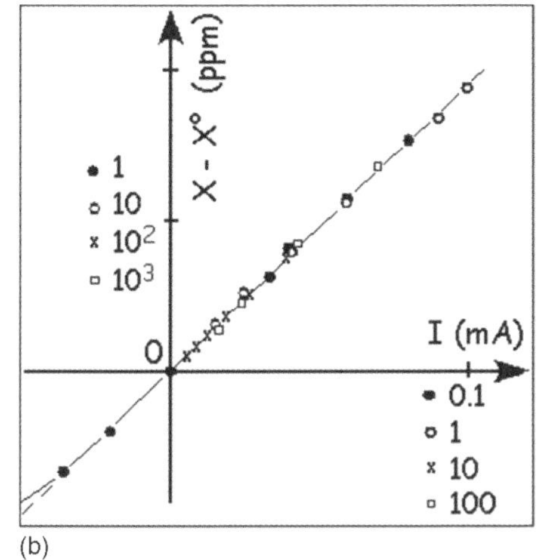

Figure 12.13 (a) The gas circuit. C = gas container, F = flowmeter, EP = electrochemical pump, OS = oxygen sensor; (b) Faraday's law test (T_{sensor} = 900 °C, $X°$: 9.1 × 10^{-7}, D = 11.1 l h^{-1}, experimental slope: 1.96 × 10^{-2} A^{-1}, theoretical slope: 1.88 × 10^{-2} A^{-1}).

12.6.2.2 Closed Systems

The zirconia-based pump–sensor device can be used for controlling the oxygen partial pressure in closed systems; typical applications include the oxygen permeation flux measurements, oxygen monitoring in molten metals, and coulometric titration.

Figure 12.15 shows a schematized set-up devoted to determining the permeation flux through mixed ionic-electronic oxides [78, 79] under various partial pressure gradients (feed oxygen pressure: 0.2 bar). For a given oxygen pressure within the chamber (controlled by the sensor EMF), the permeation flux density j_{O_2} is determined by the current passing through the oxygen pump:

$$j_{O_2} = \frac{I}{(4\,F\,S)} \tag{12.16}$$

where S is the membrane surface area.

The schematic drawing of the set-up developed for the oxygen activity control in a lead bath is shown in Figure 12.15b [80]. Here, two closed-end tubes were used, with

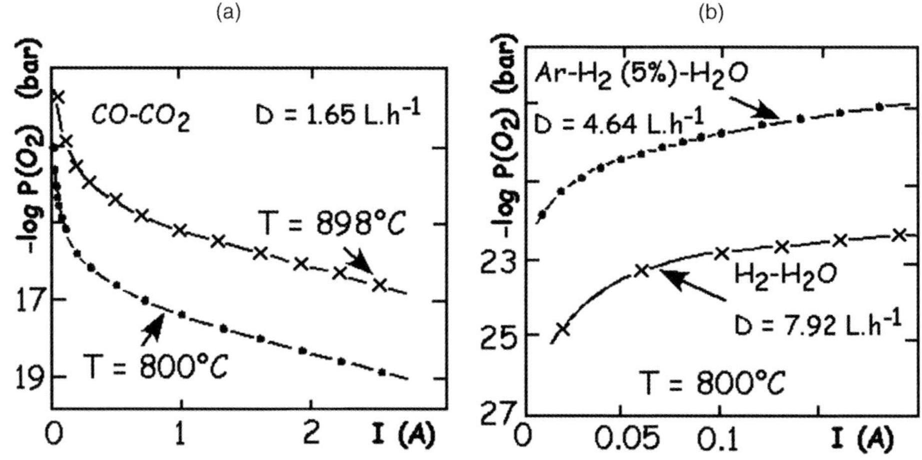

Figure 12.14 Variation of the equilibrium oxygen pressure as a function of the current passing trough the pump. (a) Reduction of flowing carbon dioxide for two temperatures of the sensor; (b) Oxidation of Ar–H_2 (5%) mixture and pure hydrogen. The full lines are the theoretical curves, according to Equations 12.13 and 12.14.

air flowing inside the tubes (reference atmospheres). With this set-up, the oxygen activity can be controlled from 10^{-10} up to the oxygen solubility limit, according to the theoretical Faraday's law (Figure 12.16):

$$\Delta n_O = \frac{Q}{2\,F} \tag{12.17}$$

where Δn_O is the oxygen mole number variation and Q is the quantity of electricity passing through the oxygen pump.

12.6.2.3 **Amperometric and Coulometric Sensors**

Amperometric sensors are based on the electrochemical reduction of oxygen, and are governed by the diffusion of the electroactive species through a barrier – that is, holes or porous layers [81, 82]. The coulometric sensors are based on a similar concept, and were developed for closed-chamber systems. Although very few amperometric and coulometric sensors have been studied to date, some relevant examples are summarized in Chapter 13.

12.6.3
HT- and IT-SOFC

In recent years, although SOFC technology has demonstrated a higher energy efficiency than conventional technologies, the costs of current SOFC systems remain prohibitive for widespread commercialization. As illustrated in Figure 12.17, the most common materials for the HT SOFCs are YSZ electrolytes, strontium-doped

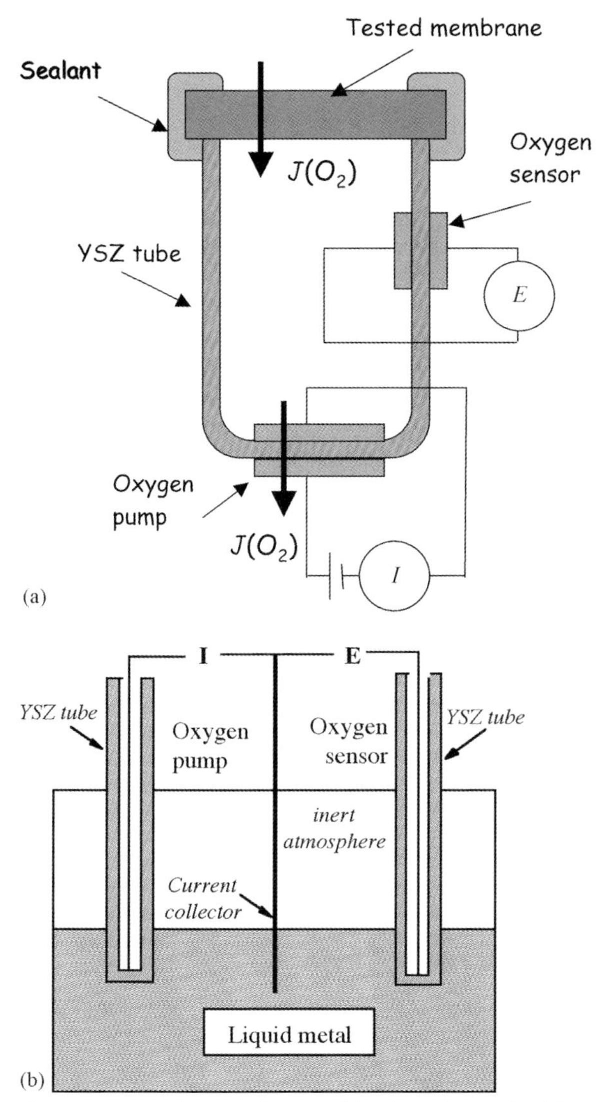

Figure 12.15 (a) Schematic drawing of the electrochemical cell for oxygen permeation measurement; (b) Schematic drawing of the set-up developed for the oxygen activity control in a lead bath.

lanthanum manganite (LSM) for the cathode, Ni–YSZ cermets for the anode, and doped lanthanum chromite (LC) for the interconnects. The cells operate at 900–1000 °C under either atmospheric or pressurized conditions. As a number of reviews are available on ceramic fuel cells [83–91], only the current status of the technology will be outlined at this point.

At present, the need to reach power densities of several hundred mW cm^{-2} to satisfy economical requirements implies high-temperature (800–1000 °C) operations

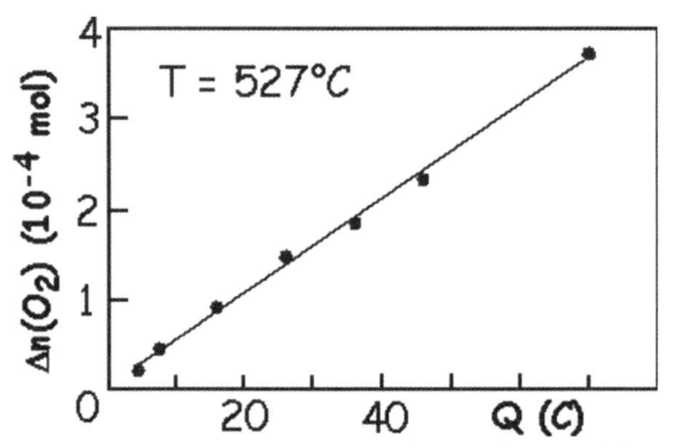

Figure 12.16 Oxygen variation in the lead bath as a function of the quantity of electricity Q passing through the pump at 527 °C.

with state-of-the-art SOFCs. In order to be commercially competitive, however, the cost of both the materials and fabrication must be dramatically reduced.

One effective method of cost reduction would be to reduce the operating temperature, and in this respect four types of stack have been developed. These include: (i) a tubular design, with the tubes closed at one end (Westinghouse prototypes) or opened at both ends; (ii) a planar technology; (iii) a segmented series arrangement of individual cells (Rolls-Royce); and (iv) a monolith concept (Argonne National Laboratory).

The main advantages of the tubular design are good thermal cycles resistance and no requirement for high-temperature seals. In contrast, the planar geometry provides a higher volumetric power density.

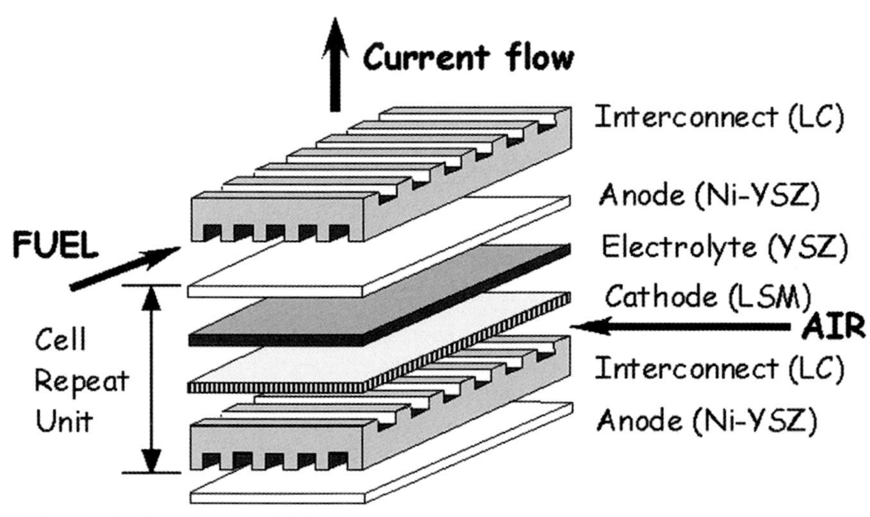

Figure 12.17 The planar SOFC design.

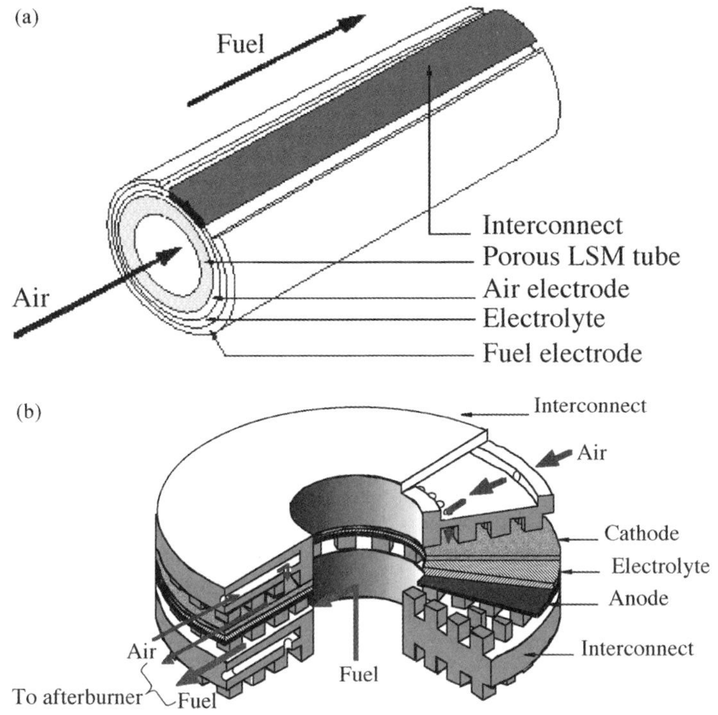

Figure 12.18 Example of SOFC geometries. (a) Tubular geometry of the Siemens-Westinghouse system; (b) Planar structure of the Sulzer-Hexis SOFC (for details, see the text).

In recent years, SOFC technology has been developed over a broad spectrum of power generation applications, with systems ranging from portable devices (e.g., 500 W battery charger) to small power systems (e.g., 5 kW residential power or automobile auxiliary power units), and to distributed generation power plants (e.g., 100–500 kW systems). SOFCs can also be integrated with a gas turbine to form large (several hundred kW to multi-MW) pressurized hybrid systems [87].

The most commonly used fuel – especially for stationary applications – is *natural gas*, although a variety of fuels have also been tested, such as waste biogas, landfill gas, or the products of coal gasification (after the removal of any sulfur-containing impurities that might poison the anode materials).

The Siemens-Westinghouse tubular cell (Figure 12.18a) is supported on a porous LSM tube, which is closed at one end; the electrolyte (YSZ), the interconnect (La(Ca) CrO_3) and the anode layers (Ni–YSZ cermet) are deposited by EVD, plasma, and liquid-spraying technologies, respectively, The tubes are 2.2 cm in diameter, and have an active length of 150 cm [92]. Air is introduced through a ceramic injector tube positioned inside the cell. Typically, 50–90% of the fuel is utilized, with part of the depleted fuel being recirculated in the fuel stream and the remainder being combusted to preheat the incoming air and/or fuel. The temperature of the exhaust gas from the fuel cell is 600–900 °C, depending on the operating conditions. The tubular SOFCs are able to withstand thermal cycles over more than 100 occasions, from room temperature to 1000 °C. An alternative geometry cell has been investigated, which utilizes flattened tubes with incorporated ribs in the air electrode (HPD-SOFC). Since 1984, a variety of prototypes have been built and tested, with outputs ranging from a few kW up to 250 kW (a 200 kW pressurized SOFC stack with a 50 kW microturbine generator). As an example, a 100 kW SOFC stack (100 kW AC + ~85 kW of hot water) which included 1152 tubes has been in operation for over 20 000 h, with an electrical efficiency of 43%.

In the Sulzer-Hexis prototype (shown schematically in Figure 12.18b), which runs on natural gas, the key component is the ceramic/metal hybrid stack with circular planar elements. The inner round aperture (2.2 cm diameter) is used as a channel for the fuel supply, while the metallic interconnect ensures an electrical contact between the individual segments of the stack, and also distributes the gases onto the surface of the electrodes. The fuel pours radially out of the channel at the anode end of the cell to the outside. Simultaneously, preheated air is fed from the outside to the interior of the stack through four channels, and then redirected so as to flow radially over the cathode end of the cell to the outside. The fuel, which is not converted on the anode, is burned off at the edge of the stack. This fuel cell has been developed to supply, simultaneously, an electrical power of 1 kW and a thermal capacity of approximately 2.5 kW.

At present, a great deal of research is being devoted to the development of intermediate-temperature protonic ceramic fuel cells (IT-PCFCs), which can simultaneously produce value-added chemicals and electrical power [93, 94]. As shown schematically in Figure 12.19, proton conduction implies that water vapor is produced at the cathode, where it is swept away by air (in contrast to the SOFC, where it dilutes the fuel). Consequently, with a purely protonic electrolyte and

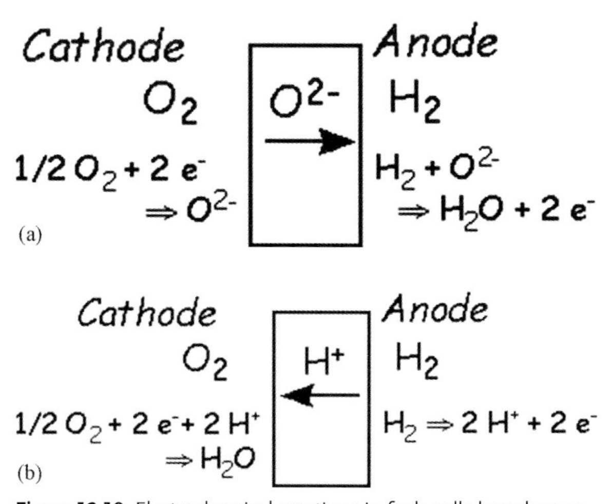

Figure 12.19 Electrochemical reactions in fuels cells based on an oxide-ion electrolyte (case 1) and on a protonic conductor (case 2).

hydrogen as fuel, the system could be operated "dead-ended" by maintaining the fuel pressure to compensate for hydrogen flux as ions through the electrolyte, allowing 100% fuel utilization. Unfortunately, one problem encountered when using oxygen ionic conductors is that the fuel cannot be fully depleted in the anode reaction due to water vapor formation.

One essential future development for these proton conductors will be to enhance their conductivity. The problem is, that most investigators in this area have used noble metal electrodes, and the performance achieved – even taking into account the lower operating temperature – cannot compete with that obtained with more conventional electrolytes.

12.6.4
Catalytic Membrane Reactors

Today, the majority of research investigations into CMRs are being conducted by many institutions, in addition to oil and chemical and utilities companies [5]. The use of mixed ionic–electronic membrane reactors for the partial oxidation of natural gas is undergoing active development by a number of consortia based around Air Products and Chemicals (USA), Praxair (USA), and/or Air Liquide (France). At present, the development of CMRs involving a pure ion-conducting electrolyte is restricted to a few reports of conceptual systems [12, 95].

A proton-conducting CMR can be used as a chemical reactor for the hydrogenation or dehydrogenation of organic compounds (e.g., the dehydrogenation of ethane to produce ethylene) [96]. The use of proton conductors brings advantages for the selective conversion of hydrocarbons because: (i) no carbon oxides are generated in the anode chamber, as the proton-conducting electrolyte permits only the transfer of protons, and no oxygen is available for further reactions of the dehydrogenation product; and (ii) as in PCFCs, proton conduction implies that water vapor is produced at the cathode, where it is swept away by air, rather that at the anode where it dilutes the hydrocarbon. The main problems to be solved are first, to improve the yield of the product desired, and second to decrease the electric loss of electrochemical reactors.

Acknowledgment

Dr Samuel Georges (LEPMI) is warmly acknowledged for his critical reading of the manuscript.

References

1 Mazanec, T.J. (1994) *Solid State Ionics*, **70/71**, 11–19.

2 Badwal, S.P.S. and Ciacchi, F. (2001) *Adv. Mater.*, **13** (12–13), 993–6.

3 Sundmacher, K., Rihko-Struckmann, L.K. and Galvita, V. (2005) *Catalysis Today*, **104**, 185–99.

4 Julbe, A., Farrusseng, D. and Guizard, C. (2005) *Catalysis Today*, **104**, 102–13.

5 Stoukides, M. (2000) *Catal. Rev. Sci. Eng.*, **42** (1 & 2), 1–70.
6 Meixner, D.L., Brengel, D.D., Henderson, B.T., Abrardo, J.M., Wilson, M.A., Taylor, D.M. and Cutler, R.A. (2002) *J. Electrochem. Soc.*, **149** (9), D132–6.
7 Kleitz, M. and Petitbon, F. (1996) *Solid State Ionics*, **92**, 65–74.
8 Weber, A. and Ivers-Tiffée, E. (2004) *J. Power Sources*, **127**, 273–83.
9 Fouletier, J., Fabry, P. and Kleitz, M. (1976) *J. Electrochem. Soc.*, **123**, 204–13.
10 Fabry, P. and Kleitz, M. (1976) in *Electrode Processes in Solid State Ionics* (eds M. Kleitz and J. Dupuy), D. Reidel Publ. Comp., Dordrecht, pp. 331–65.
11 Fouletier, J. and Kleitz, M. (1975) *Vacuum*, **25**, 307–14.
12 Munder, B., Ye, Y., Rihko-Struckmann, L. and Sundmacher, K. (2005) *Catal. Today*, **104**, 138–48.
13 Ahmed, K. and Froger, K. (2000) *Catal. Today*, **63**, 479–87.
14 Vernoux, P., Guindet, J. and Kleitz, M. (1998) *J. Electrochem. Soc.*, **145**, 3487–92.
15 Vernoux, P., Guillodo, M., Fouletier, J. and Hammou, A. (2000) *Solid State Ionics*, **135**, 425–31.
16 Georges, S., Parrour, G., Hénault, M. and Fouletier, J. (2006) *Solid State Ionics*, **177**, 2109–12.
17 Klein, J.M., Bultel, Y., Georges, S. and Pons, M. (2007) *Chem. Eng. Sci.*, **62**, 1636–49.
18 Park, S., Vohs, S. and Gorte, R.J. (2000) *Nature*, **404**, 265–7.
19 Kim, H., Park, S., Vohs, J.M. and Gorte, R.J., (2001) *J. Electrochem. Soc.*, **148**, A693.
20 Murray, E.P., Tsai, T. and Barnett, S.A. (1999) *Nature*, **400**, 649–51.
21 Gorte, R.J., Park, S., Vohs, J.M. and Wang, C. (2000) *Adv. Mater.*, **12**, 1465–9.
22 Hibino, Y.T. and Iwahara, H. (1993) *Chem. Lett.*, **7**, 1131.
23 Hibino, T., Hashimoto, A., Inoue, T., Suzuki, M., Tokuno, J.I., Yoshida, S.I. and Sano, M. (2000) *J. Electrochem. Soc.*, **147**, 2888–92.
24 Yano, M., Tomita, A., Sano, M. and Hibino, T. (2007) *Solid State Ionics*, **177**, 3351–9.
25 Riess, I. (2008) *J. Power Sources*, **175**, 325–37.
26 Vayenas, C.G., Bebelis, S. and Ladas, S. (1990) *Nature (London)*, **343**, 625–7.
27 Vayenas, C.G., Jaksic, M.M., Bebelis, S.I. and Neophytides, S.N. (1996) *Modern Aspects in Electrochemistry*, vol. **29** (eds J.O'M. Bockris, B.E. Conway and W.R.E. White), Plenum Press, New York, p. 57.
28 Vayenas, C.G., Bebelis, S.I., Yentekakis, I.V. and Neophytides, S.N. (1997) in *The CRC Handbook of Solid State Electrochemistry* (eds P.J. Gellings and H.J.M. Bouwmeester), CRC Press, London, pp. 481–553, Chap. 14.
29 Vayenas, C.G., Bebelis, S., Pliangos, C., Brosda, S. and Tsiplakides, D. (2001) *Electrochemical Activation of Catalysis: Promotion, Electrochemical Promotion and Metal-Support Interactions*, Kluwer Academic/Plenum Publishers, New York.
30 Foti, G., Bolzonella, I. and Comninellis, C. (2003) in *Electrochemical Promotion of Catalysis, Modern Aspect of Electrochemistry*, Vol. 36 (eds C.G. Vayenas, B.E. Conway and R.E. White), Kluwer Academic/ Plenum Publishers, New York, pp. 191–254.
31 Vayenas, C.G., Bebelis, S., Yentekakis, I.V. and Lintz, H.G. (1992) *Catal. Today*, **11**, 303–425.
32 Bebelis, S., Makri, M., Buekenhoudt, A., Luyten, J., Brosda, S., Petrolekas, P., Pliangos, C. and Vayenas, C.G. (2000) *Solid State Ionics*, **129**, 33–46.
33 Vayenas, C.G. (2004) *Solid State Ionics*, **168**, 321–6.
34 Fabry, P., Kleitz, M. and Déportes, C. (1973) *J. Solid State Chem.*, **6**, 230–9.
35 Levy, M., Fouletier, J. and Kleitz, M. (1980) *J. Physique*, **C6**, 335–9.
36 Levy, M., Fouletier, J. and Kleitz, M. (1988) *J. Electrochem. Soc.*, **135**, 1584–9.
37 Fouletier, J. and Kleitz, M. (1975) *Vacuum*, **25**, 307–14.

38 Bouwmeester, H.J.M. (2003) *Catal. Today*, 82, 141–50.

39 Steele, B.C.H. (2001) *J. Mater. Sci.*, **36**, 1053–68.

40 Bouwmeester, H.J.M. and Burggraaf, A.J. (1996) *Fundamentals of Inorganic Membrane Science and Technology*, Vol. 4 (eds A.J. Burggraaf and L. Cot), Membrane, Science, Technology Series, Elsevier, Amsterdam, Chap. 10.

41 Brisse, A., Sauvet, A.-L., Barthet, C., Georges, S. and Fouletier, J. (2007) *Solid State Ionics*, **178**, 1337–43.

42 Iwahara, H., Esaka, T., Uchida, H. and Maeda, N. (1981) *Solid State Ionics*, **3–4**, 359–63.

43 Iwahara, H. (1995) *Solid State Ionics*, **77**, 289–98.

44 Iwahara, H., Asakura, Y., Katahira, K. and Tanaka, M. (2004) *Solid State Ionics*, **168**, 299–310.

45 Steele, B.C.H., Hori, K.M. and Uchino, S. (2000) *Solid State Ionics*, **135**, 445–50.

46 Kenjo, T. and Nishiya, M. (1992) *Solid State Ionics*, **57**, 295–302.

47 Ostergard, M.J.L., Clausen, C., Bagger, C. and Mogensen, M. (1995) *Electrochim. Acta*, **40**, 1971–81.

48 Sasaki, K., Wurth, J.-P., Gshwend, R., Gödickemeier, R. and Gauckler, L.J.J. (1996) *J. Electrochem. Soc.*, **143**, 530–3.

49 Holtappels, P. and Bagger, C. (2002) *J. Eur. Ceram. Soc.*, **22**, 41–8.

50 Liu, Z., Han, M.F. and Miao, W.T. (2007) *J. Power Sources*, **173**, 837–41.

51 Anderson, M.D., Stevenson, J.W. and Simner, S.P. (2004) *J. Power Sources*, **129**, 188–92.

52 Mai, A., Haanappel, V.A.C., Uhlenbruck, S., Tietz, F. and Stöver, D. (2005) *Solid State Ionics*, **176**, 1341–50.

53 Kim, W.H., Song, H.S., Moon, J. and Lee, H.W. (2006) *Solid State Ionics*, **177**, 3211–16.

54 Boehm, E., Bassat, J.-M., Steil, M.C., Dordor, P., Mauvy, F. and Grenier, J.-C. (2003) *Solid State Sci.*, **5**, 973–81.

55 Velinov, N., Brashkova, N. and Kuzhukharov, V. (2004) Sixth European Solid Oxide Fuel Cell Forum, Lucerne, Switzerland, (ed. M. Mogensen.), pp. 1322–9.

56 Skinner, S.K. and Kilner, J.A. (2000) *Solid State Ionics*, **135**, 709–12.

57 Sun, C. and Stimming, U. (2007) *J. Power Sources*, **171**, 247–60.

58 Goodenough, J.B. and Huang, Y.H. (2007) *J. Power Sources*, **173**, 1–10.

59 Munder, B., Ye, Y., Rihko-Struckmann, L. and Sundmacher, K. (2005) *Catal. Today*, **104**, 138–48.

60 Zhan, Z. and Barnett, S.C. (2006) *J. Power Sources*, **157**, 422–9.

61 Klein, J.M., Hénault, M., Gélin, P., Bultel, Y. and Georges, S. (2008) *Electrochem. Solid-State Lett.*, **11**, B144–7.

62 Klein, J.M., Georges, S. and Bultel, Y. (2008) *J. Electrochem. Soc*, **155**, B333–9.

63 Waller, D., Kilner, J.A. and Steele, B.C.H. (1996) Proceedings, 1st International Symposium on Ion-Conducting Ceramic Membranes, The Electrochemical Society, Pennington, New Jersey, Vol. 95–24, pp. 309–14.

64 Suitor, J.W. (1992) *Solid State Ionics*, **52**, 277.

65 Badwal, S.P.S. and Ciacchi, F. (2001) *Adv. Mater.*, **13**, 993–6.

66 Ciacchi, F.T., Badwal, S.P.S. and Zelizko, V. (2002) *Solid State Ionics*, **152–153**, 763–8.

67 Cutler, R.A., Meixner, D.L., Henderson, B.T., Hutchings, K.N., Taylor, D.M. and Wilson, M.A. (2005) *Solid State Ionics*, **176**, 2589–98.

68 Hutchings, K.N., Bai, J., Cutler, R.A., Wilson, M.A. and Taylor, D.M. (2008) *Solid State Ionics*, **179**, 442–50.

69 Drevet, C., Hénault, M., Fouletier, J. and Feger, D. (1997) Proceedings, 1st International Symposium on Ceramic Membranes, Vol. 95–24 (eds H.U. Anderson, A.C. Khandkar and M. Liu), The Electrochemical Society, Pennington, New Jersey, pp. 309–14.

70 Drevet, C., Hénault, M. and Fouletier, J. (2000) *Solid State Ionics*, **136–137**, 807–12.

71 Joshi, A.V., Steppan, J.J., Taylor, D.M. and Elangoyan, S. (2004) *J. Electroceram.*, **13**, 619–25.

72 Dyer, P.N., Richards, R.E., Russek, S.L. and Taylor, D.L. (2000) *Solid State Ionics*, **134**, 21–33.

73 Matsumoto, H., Hamajima, S., Yajima, T. and Iwahara, H. (2001) *J. Electrochem. Soc.*, **148**, D121–4.

74 Fouletier, J., Mantel, E. and Kleitz, M. (1982) *Solid State Ionics*, **6**, 1–13.

75 Fouletier, J., Bonnat, M., Le Bot, J. and Adamowicz, S. (1997) *Sens. Actuators B*, **B45**, 155–60.

76 Caneiro, A., Bonnat, M. and Fouletier, J. (1981) *J. Appl. Electrochem.*, **11**, 83–90.

77 Fouletier, J., Siebert, E. and Caneiro, A. (1984) in *Advances in Ceramics*, Vol. 12, Science and Technology of Zirconia II, The American Ceramic Society, Columbus, Ohio, pp. 618–26.

78 Yaremchenko, A.A., Kharton, V.V., Viskup, A.P., Naumovich, E.N., Lapchuk, N.M. and Tikhonovich, V.N. (1999) *J. Solid State Chem.*, **142**, 325–35.

79 Kovalevsky, A.V., Kharton, V.V. and Naumovich, E.N. (1999) *Mater. Lett.*, **38**, 300–4.

80 Ghetta, V., Gamaoun, F., Fouletier, J., Hénault, M. and Lemoulec, A. (2001) *J. Nucl. Mater.*, **296**, 295–300.

81 Kleitz, M., Siebert, E., Fabry, P. and Fouletier, J. (1992) *Sensors - A Comprehensive Survey* (eds W. Göpel, J. Hesse and J.N. Zemel), Chemical and Biochemical Sensors, Part I, Vol. 2, VCH, Weinheim, pp. 341–428.

82 Fabry, P. and Siebert, E. (1997) *The CRC Handbook of Solid State Electrochemistry* (eds P.J. Gellings and H.J.M. Bouwmeester), CRC Press, London, pp. 329–69.

83 Minh, N.Q. (1993) *J. Am. Ceram. Soc.*, **76**, 563–88.

84 Minh, N.Q. and Takahashi, T. (1995) *Science and Technology of Ceramic Fuel Cells*, Elsevier, Amsterdam, The Netherlands.

85 Carrette, L., Friedrich, K.A. and Stimming, U. (2000) *ChemPhysChem*, **1**, 162–93.

86 Singhal, S.C. and Kendall, K. (eds) (2003) *High Temperature Solid Oxide Fuel Cells: Fundamentals, Design and Applications*, Elsevier, Oxford, UK.

87 Minh, N.Q. (2004) *Solid State Ionics*, **174**, 271–7.

88 Hammou, A. and Guindet, J. (1997) *The CRC Handbook of Solid State Electrochemistry* (eds P.J. Gellings and H.J.M. Bouwmeester), CRC Press, London, pp. 407–43.

89 Ormerod, R.M. (2003) *Chem. Soc. Rev.*, **32**, 17–28.

90 Haile, S.M. (2003) *Acta Mater.*, **51**, 5981–6000.

91 Blum, L., Buchkremer, H.P., Gross, S., Gubner, A., de Haart, L.G.J., Nabielek, H., Quadakkers, W.J., Reisgen, U., Smith, M.J., Steinberger-Wilckens, R., Steinbrech, R.W., Tietz, F. and Vinke, I.C. (2007) *Fuel Cells*, **3**, 204–10.

92 Singhal, S.C. (2000) *Solid State Ionics*, **135**, 305–13.

93 Norby, T. and Larring, Y. (1997) *Curr. Opin. Solid State Mater. Sci.*, **2**, 593–9.

94 Kreuer, K.D. (2003) *Annu. Rev. Mater. Res.*, **33**, 333–59.

95 Athanassiou, C., Pekridis, G., Kaklidis, N., Kalimeri, K., Vartzoka, S. and Marnellos, G. (2007) *Int. J. Hydrogen Energy*, **32**, 38–54.

96 Iwahara, H., Asakura, Y., Katahira, K. and Tanaka, M. (2004) *Solid State Ionics*, **168**, 299–310.

13
Electrochemical Sensors: Fundamentals, Key Materials, and Applications

Jeffrey W. Fergus

Abstract

Solid electrolytes are well suited for use in chemical sensors for high-temperature applications. The first such sensors followed from the use of solid electrolytes for thermodynamic measurements, and are based on galvanic cells, for which the open circuit voltage is related to the concentration of the mobile ionic species, or to another target species through equilibration at the electrode surface. However, sensors can also be based on nonequilibrium, but steady-state, potentials established between electrochemical reactions at the electrode. Another approach is to use the current passing through the electrolyte, rather than the voltage across the electrolyte, as the sensor signal. In this chapter we first describe the different methods for using solid electrolytes in chemical sensors, and then discuss some of the materials challenges and example applications.

13.1
Introduction

More than 50 years ago, Kiukkola and Wagner [1] demonstrated that solid electrolytes could be used in galvanic cells to measure the thermodynamic properties of oxides. Those pioneering studies subsequently led to the more practical application of solid electrolytes in chemical sensors. The ionic conduction in solid electrolytes is thermally activated and thus its rate increases with increasing temperature; consequently, solid electrolytes are well suited for high-temperature applications. On the other hand, such temperature dependence also means that sensors based on solid electrolytes can respond sluggishly or may be inoperable at low temperatures. However, one of the advantages of solid-state devices, such as those based on solid electrolytes, is that they can be miniaturized [2, 3]. Decreasing the device dimensions reduces diffusion lengths, which in turn decreases the response time and thus also the minimum operating temperature.

Ionic conducting solids can be used in a variety of different types of sensor [4–12]. Solids for which the ionic conductivity is much larger than the electronic conductivity can

Solid State Electrochemistry I: Fundamentals, Materials and their Applications. Edited by Vladislav V. Kharton

ISBN: 978-3-527-32318-0

be used as the solid electrolyte in galvanic cells, as mentioned above. However, those solids which have significant ionic and electronic conductivity (referred to as *mixed conductors*) can also be used in chemical sensors. In mixed conductors, the interaction between the ionic and electronic defects leads to an atmosphere-dependent (e.g., oxygen partial pressure-dependent for a mixed-conducting oxide) electrical resistance, which can be used as a sensor signal. The focus of this chapter, however, is on sensors that use pure ion conductors as electrolytes rather than those based on mixed-conducting solids. Following a description of the different ways in which solid electrolytes can be used in chemical sensors, some materials challenges and example applications will be discussed.

13.2 Operation Principles

In chemical sensors, solid electrolytes function as transducers by providing a relationship between chemical species and electrons, so that an electrical signal corresponding to the concentration of a particular chemical species is produced. This signal can be either a voltage or current, depending on the configuration in which the solid electrolyte is used.

13.2.1 Voltage-Based Sensors

Sensors for which the output is an open-circuit voltage are referred to as *potentiometric* sensors, and can be used for a wide variety of species [13–18]. The measured voltage can be established by a thermodynamic equilibrium or by a nonequilibrium steady state between electrochemical reactions at the electrode.

13.2.1.1 Potentiometric Sensors: Equilibrium

The most direct use of a solid electrolyte is to measure the activity or concentration of the ion that is mobile in the electrolyte. For example, for a galvanic cell with a sodium ion-conducting electrolyte, the cell voltage (E) is given by

$$E = \frac{RT}{F} \ln\left(\frac{[\mathrm{Na}^+]^S}{[\mathrm{Na}^+]^R}\right) \tag{13.1}$$

where R is the gas constant, T is absolute temperature, F is Faraday's constant, and $[\mathrm{Na}^+]^S$ and $[\mathrm{Na}^+]^R$ are the sodium ion concentrations at the sensing and reference electrodes, respectively. If sodium metal is present at the electrode, the sodium ion activity will be determined by

$$\mathrm{Na}^+ + e^- = \mathrm{Na} \tag{13.2}$$

so the cell voltage becomes

$$E = \frac{RT}{F} \ln\left(\frac{a_{\mathrm{Na}}^S}{a_{\mathrm{Na}}^R}\right) \tag{13.3}$$

and the sodium activity can be measured as shown schematically in Figure 13.1a.

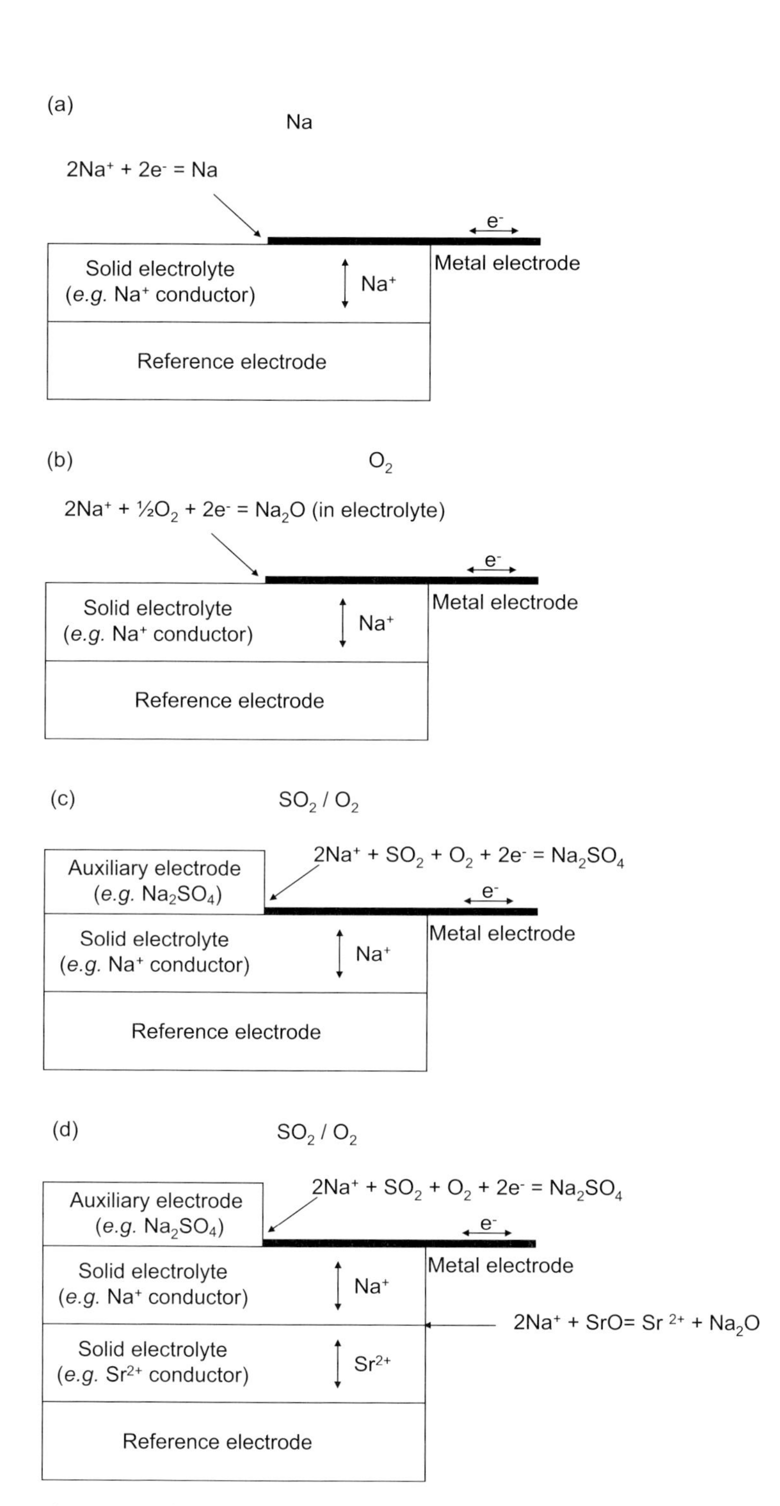

Figure 13.1 Schematic diagrams of potentiometric sensors.
(a) No auxiliary electrode; (b) Electrolyte as auxiliary electrode;
(c) Auxiliary electrode added; (d) Cation-/cation-electrolyte chain;
(e) Cation-/anion-electrolyte chain.

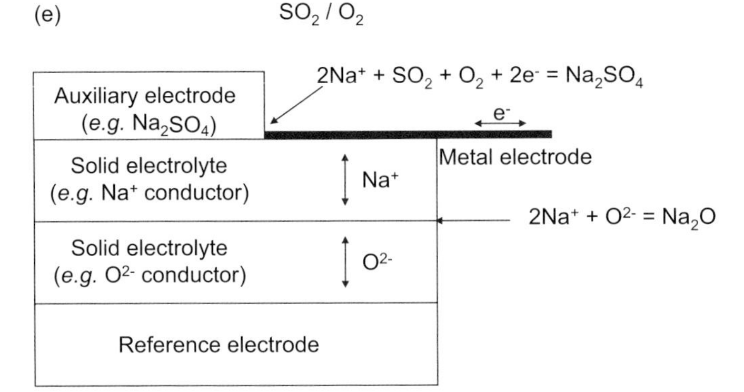

Figure 13.1 *(Continued)*

If the electrolyte is an oxide, the Na_2O in the electrolyte, typically in a compound or solid solution, can equilibrate with oxygen according to:

$$2Na^+ + \tfrac{1}{2}O_2 + 2e^- = Na_2O \tag{13.4}$$

so the sodium ion concentration – and thus the cell voltage – is related to the oxygen partial pressure. As shown in Figure 13.1b, the Na_2O need not be a separate phase, but its activity must remain constant for the relationship between sodium ion activity and oxygen partial pressure – and thus the cell voltage – to be stable.

If the species to be measured is not present in the electrolyte, an auxiliary electrode can be added and can provide a relationship between the target species and the ion that is mobile in the electrolyte. For example, Figure 13.1c shows that Na_2SO_4 can equilibrate with SO_2 according to

$$2Na^+ + SO_2 + O_2 + 2e^- = Na_2SO_4. \tag{13.5}$$

Hence, the sodium activity – and thus the cell potential – is related to the SO_2 partial pressure. The auxiliary electrode must be in contact with both the electrolyte (with the mobile sodium ions) and the metal electrode (to measure the electrical signal), as well as the gas; consequently, porous electrodes are typically used to provide a large three-phase-boundary area. It is also possible to mix the auxiliary electrode with the electrode either only near the surface (where it is needed) or throughout the electrolyte (which is sometimes easier to fabricate); this is referred to as a *composite electrolyte.*

Sensors may also use multiple electrolytes (referred to as *electrolyte chains*), and two examples of these are shown in Figure 13.1d and e. In both examples, a Na_2SO_4 auxiliary electrode is used with a sodium ion conductor so that the sensing electrode reaction is given by Equation (13.5). The difference between the two cases is that, in Figure 13.1d, the sodium ion conductor is used with another cation (strontium) conductor, whereas in Figure 13.1e the sodium ion conductor is used with an anion (oxygen) conductor. In both cases, an equilibrium reaction is required to relate the two species in the two different electrolytes. In Figure 13.1d, the exchange reaction between the two oxides:

$$2Na^{+} + SrO = Sr^{2+} + Na_2O \tag{13.6}$$

can relate the activities of the two mobile cation species. In Figure 13.1e, the oxygen ion activity is related to the sodium ion activity by a reaction similar to that in Equation (13.4), except that oxygen ions, rather than oxygen gas, equilibrate according to the following reaction:

$$2Na^{+} + O^{2-} = Na_2O. \tag{13.7}$$

The equilibration between the two electrolytes can sometimes require an additional phase, some specific examples of which will be discussed later in the chapter. One advantage of using an electrolyte chain is that the reference electrode reaction need not be the same as the sensing electrode reaction, which in turn expands the choice of potential reference electrodes. In addition, the electrolyte that provides the best performance may not be commercially available or easy to fabricate. In such cases, a commonly available, easily fabricated or low-cost electrolyte can be used for the primary structure, and a coating of the better-performing electrolyte applied to the surface.

Various combinations of electrolytes and auxiliary electrodes enable the measurement of a wide variety of species. However, as the auxiliary electrodes must be stable in the target atmosphere, there are thermodynamic limits to their use [19]. In cases where stable auxiliary electrodes are not available, electrolytes can be used to measure either steady-state or nonequilibrium potentials.

13.2.1.2 Potentiometric: Nonequilibrium

The nonequilibrium potentials measured in solid-electrolyte cells are established by electrochemical reactions, just as for equilibrium-based sensors. For example, in an environment containing CO and O_2, the CO could be oxidized chemically according to the following reaction:

$$CO + \tfrac{1}{2}O_2 = CO_2. \tag{13.8}$$

However, on an oxygen ion conduction electrolyte surface, the oxidation of CO can occur electrochemically by

$$CO + O^{2-} = CO_2 + 2e^{-}, \tag{13.9}$$

but only if the electrons produced by this oxidation reaction are used in a reduction reaction such as

$$\tfrac{1}{2}O_2 + 2e^{-} = O^{2-}. \tag{13.10}$$

Here, the sum of these two electrochemical reactions is the chemical reaction in Equation (13.8). Before reaching equilibrium, these two reactions will reach a steady state in which all electrons produced by Equation (13.9) are consumed by Equation (13.10). This process is shown schematically in Figure 13.2a, in which the lines represent the electrical currents associated with an anodic overpotential for the

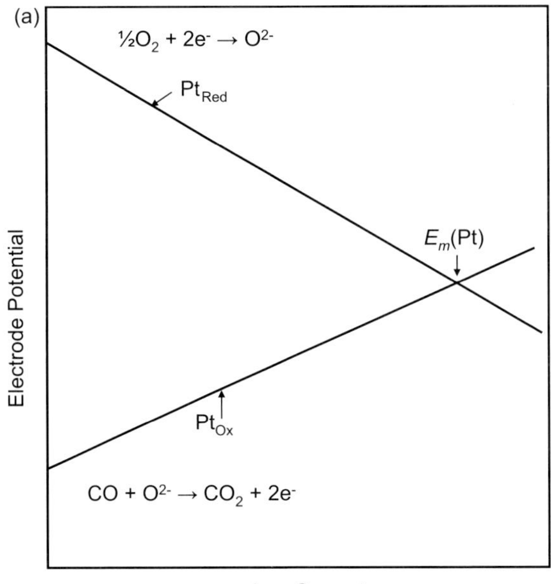

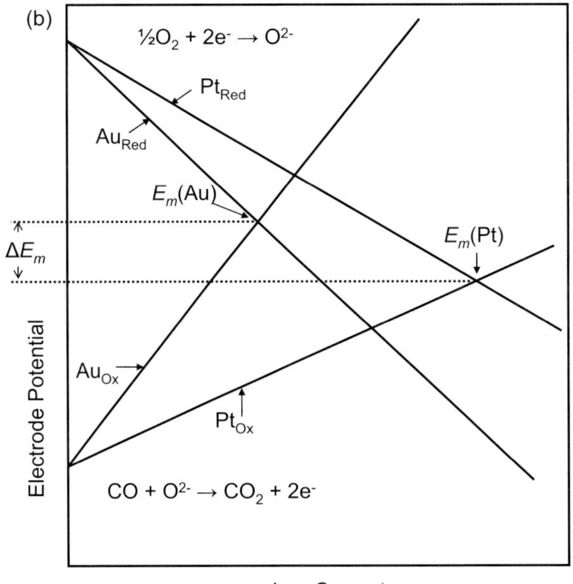

Figure 13.2 Mixed potential sensor mechanism. (a) Mixed potential at Pt electrode; (b) Mixed potentials at Pt and Au electrodes; (c) Mixed potentials at Pt and Au electrodes at lower CO partial pressure; (d) Mixed potentials at Pt and Au electrodes under concentration polarization.

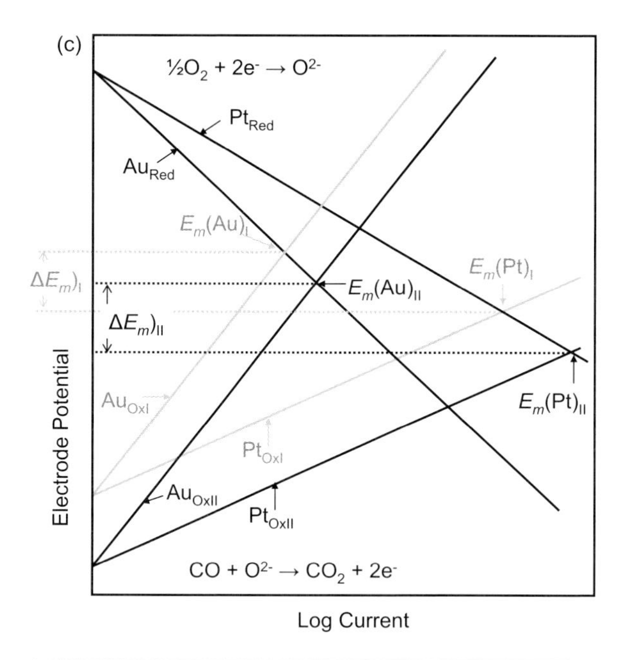

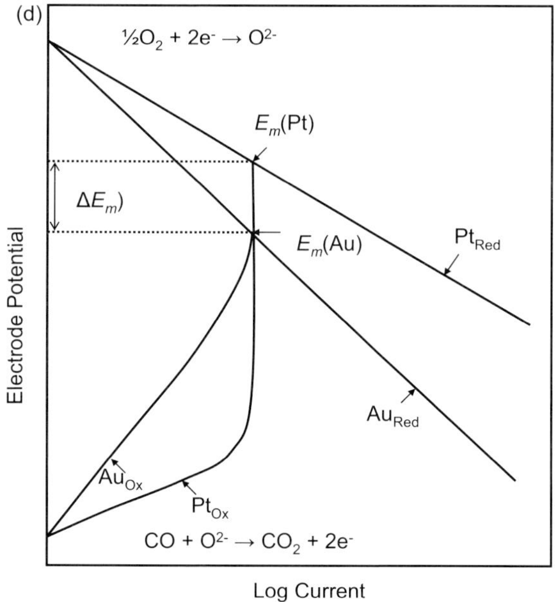

Figure 13.2 *(Continued)*

oxidation of CO (Equation 13.9) and a cathodic overpotential for the reduction of oxygen (Equation 13.10). The intersection of these two lines represents the mixed potential, E_m (Pt), established by the two reactions. At this voltage the anodic current from the oxidation reaction (Equation 13.9) is exactly equal to the cathodic current from the reduction reaction (Equation 13.10), so the net current is zero and, and thus the sensor output is an open-circuit voltage.

This mixed potential can be applied to a chemical sensor by using electrodes from two different materials with different polarization characteristics [20–23]. In Figure 13.2b, a second set of polarization curves are added (i.e., for Au) to demonstrate that the difference in polarization behavior leads to the establishment of a different mixed potential, E_m (Au). The sensor response is based on the difference between these two mixed potentials, ΔE_m (Au). The effect of CO composition on this mixed potential, as illustrated in Figure 13.2c, is due to a shift in the CO polarization curves, which leads to a change in the two mixed potentials, and thus a change in the difference between the two mixed potentials.

These examples are based on both electrodes operating in the activation polarization regime, in which the logarithm of the current is proportional to the overpotential. However, there are situations – particularly at low concentrations – in which the electrochemical reaction is limited by mass transport to the electrode surface. This is referred to as *concentration polarization*, and is illustrated in Figure 13.2d. In this case, above a critical overpotential the current becomes constant, which appears as a vertical line in the plot. A new mixed potential is established at the intersection of this vertical line and the cathode polarization for the oxygen reduction. This potential depends on the gas concentration, and thus can be used for the chemical sensor signal.

One of the advantages of mixed potential sensors is that it is possible for both electrodes to be exposed to the same gas. The elimination of a need to separate the two electrodes simplifies the sensor design, which in turn reduces fabrication costs. Although this simpler planar design is often used, the electrodes are sometimes separated to provide a more stable reference potential. As with equilibrium potentiometric sensors, the minimum operating temperature is often limited by electrolyte conductivity. However, the maximum operation temperatures for nonequilibrium sensors are typically lower than those of equilibrium sensors, because the electrode reactions tend towards equilibrium as the temperature increases. This operating temperature window depends on the electrode materials, as will be discussed later in the chapter.

13.2.2
Current-Based Sensors

The flow of electrical current in a circuit with an electrolyte causes ionic transport between the electrodes, which affects the electrode reaction. The application of a small perturbation in a potentiometric sensor has been shown to improve the sensor performance in what is referred to as the current reversal method [24–26]. An analysis of the response to such a perturbation can also be used as a diagnostic tool to evaluate sensor performance during operation [27, 28]. Similarly, the application of a bias voltage to a potentiometric sensor can be used to affect the performance of the sensor, such as the relative responses to NO and NO_2 gases [29–31]. The response to voltage or current perturbations can also be used as the sensor signal.

13.2.2.1 Sensors Based on Impedance Measurements

One current-based approach is referred to as *impedancemetric sensing* [32]. This is based on impedance spectroscopy, in which a cyclic voltage is applied to the electrode and an analysis of the resultant electrical current is used to determine the electrode impedance. As different processes have different characteristic frequencies, impedance spectroscopy can be used to identify and separate contributions from different processes, such as electron transfer at the interface from solid-state electronic conduction. The frequency range of the applied voltage in impedancemetric sensors is selected so that the measured impedance is related to the electrode reaction, rather than to transport in the electrode or electrolyte material. Thus, the response is different from that in resistance-based sensors, which are related to changes in the electrical conductivity of a semiconducting material in response to changes in the gas composition.

The equipment required to analyze the frequency response in impedance spectroscopy is sophisticated, and is thus too expensive to be practical for an operating sensor. In some systems it may be possible to identify a critical frequency (or a limited number of frequencies) that provides the desired response, so that simpler circuitry can be used. However, because of the complicated electronics, impedancemetric sensors are less common than potentiometric sensors.

13.2.2.2 Amperometric Sensors

A more common current-based sensor is an amperometric sensor [33], where a solid electrolyte is separated from the sample gas by a barrier with a small orifice or a porous barrier. A voltage is applied across the electrolyte, and this induces an ionic current. As the applied voltage is increased, the current increases until it becomes limited by the supply of gas through the small orifice or porous barrier. This limiting current is proportional to the concentration of the gas, and thus can be used as the sensor signal.

Oxygen ion conductors are used in amperometric sensors for a variety of gas species. The selectivity is controlled by selecting electrode materials that catalyze particular reactions; hence, multiple electrodes are required for some multicomponent gas mixtures. In such cases, the system is designed so that each electrode removes a particular gas from the gas stream. The particular gas removed by a particular electrode can also be controlled by the voltage at the electrode, just as in cyclic voltammetry, which has also been used in solid-electrolyte based sensors [34]. In addition to providing selectivity, control of the applied voltage can be used to improve the magnitude of the response [35].

The configuration of an amperometric sensor is compared with those of potentiometric and impedancemetric sensors in Figure 13.3. The output of a potentiometric sensor is an open-circuit voltage (E), so current does not flow through the external circuit (Figure 13.3a and b). Current (I) flows in both impedancemetric and amperometric sensors, and is used for the sensor signal. In impedancemetric sensors (Figure 13.3c), a cyclic voltage is applied so that the impedance of the electrode reaction can be discerned, whereas in amperometric sensors (Figure 13.3d), the current though the electrolyte is limited by the mass transport of the reactants or

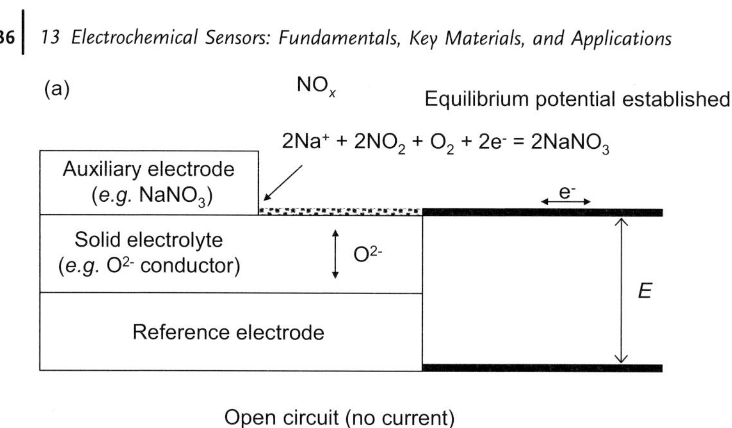

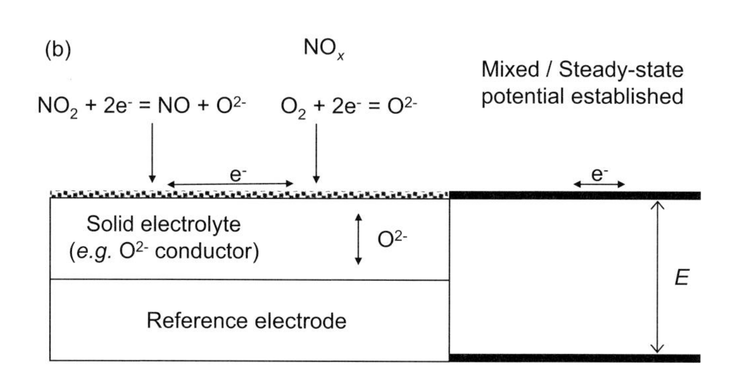

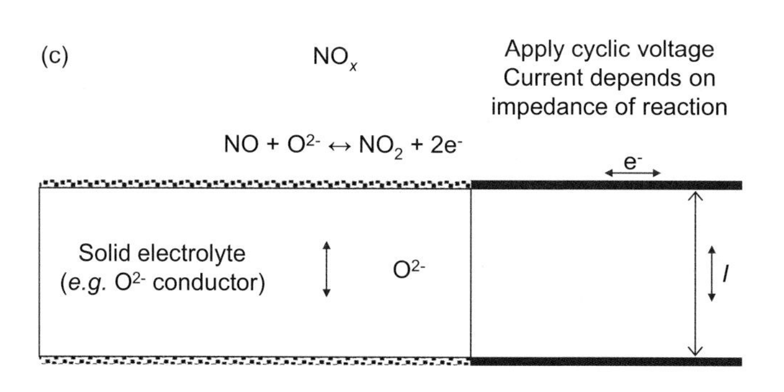

Figure 13.3 Schematic of chemical sensor operations. (a) Potentiometric: equilibrium; (b) Potentiometric: nonequilibrium; (c) Impedancemetric; (d) Amperometric.

products through the small orifice. Although amperometric sensors are more complicated than potentiometric sensors, the sensor output (limiting current) increases linearly with gas concentration, so the complication is justified in some applications.

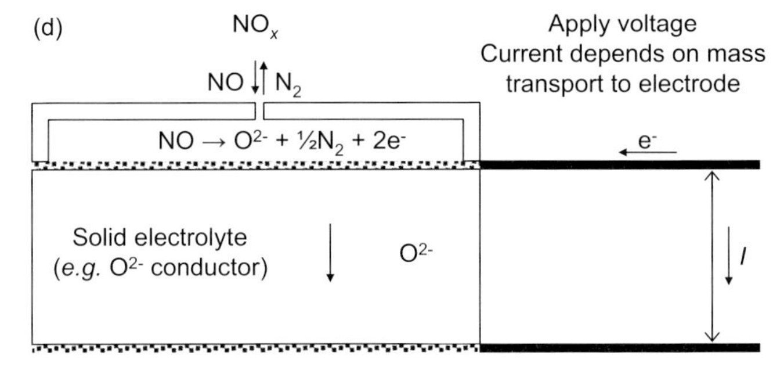

Figure 13.3 *(Continued)*

13.3 Materials Challenges

The performance of each type of electrochemical sensor described above depends on the selection of appropriate materials, for both the electrolyte and electrode. The most appropriate materials used depend not only on the sensor type, but also on the species being detected.

13.3.1 Electrolytes

The most widely used solid electrolytes are anion, specifically oxygen ion, conductors. Most oxygen ion-conducting electrolytes are oxides that form the cubic fluorite structure, although there are some promising electrolytes that form the perovskite structure [36–38] (see also Chapters 2 and 9). The most commonly used oxygen ion-conducting electrolyte is *zirconia*, which is typically stabilized with dopants, such as yttrium, scandium, and calcium. In addition to stabilizing the cubic crystal structure, the dopants have lower valences than zirconia, and so increase the concentration of oxygen vacancies, which in turn increases the ionic conductivity. The conductivity typically reaches a maximum and then decreases at high dopant levels due to association among ionic point defects, which decreases their mobility [39]. Although yttria-stabilized zirconia (YSZ) is the most common oxygen ion-conducting electrolyte, certain new, highly conducting electrolytes – such as lanthanide aluminosilicate compounds [40] and bismuth-vanadium-based oxides [41–43] – have recently been reported.

A solid electrolyte material used in a chemical sensor must have very low electronic conductivity, because electronic conduction through the electrolyte, depending on the reversibility of the electrode, can affect the electrode potential. Generally, the ionic transport number (t_{ion}) – which is the ionic conductivity divided by the total conductivity – should be at least 0.99. Oxygen ion-conducting electrolytes have a limited range of oxygen partial pressures within which this criterion is satisfied. This in turn creates challenges for oxygen sensors used at very low oxygen partial

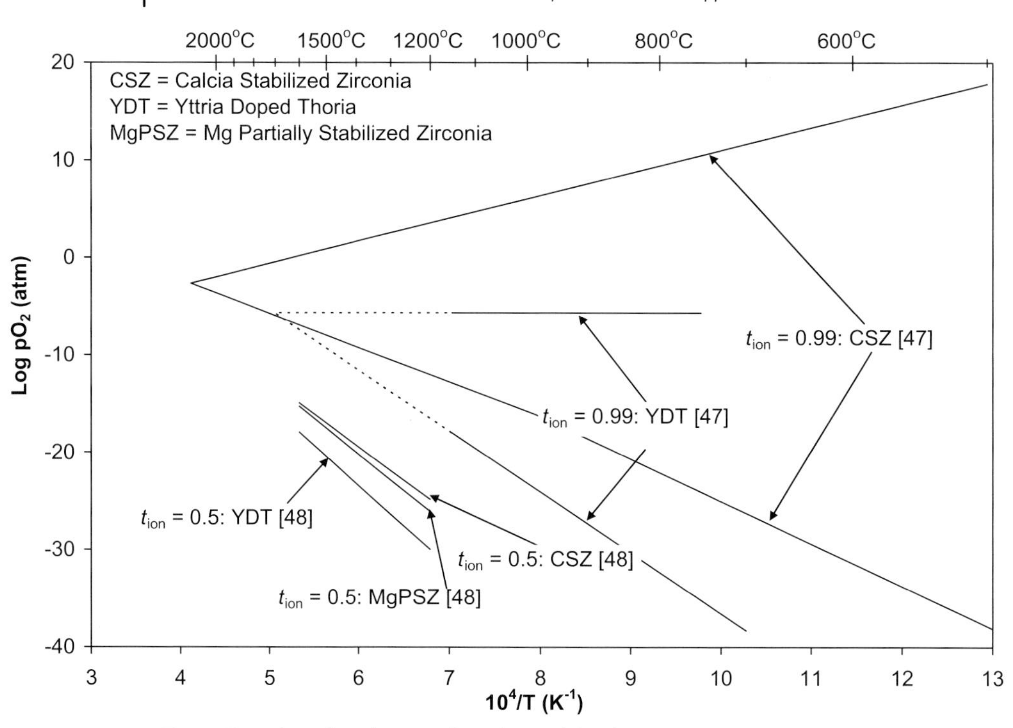

Figure 13.4 Electrolytic domains for calcia-stabilized zirconia (CSZ), yttria-doped thoria (YDT), and magnesia partially stabilized zirconia (MgPSZ) [47, 48].

pressures, where significant n-type electronic conduction can occur, such that the electrolyte becomes mixed conducting [44, 45]. In addition to increasing the electronic conduction, the reduction of zirconia can lead to a blackening of the electrolyte surface in low oxygen partial pressures [46]. As shown in Figure 13.4, the oxygen partial pressure at which electronic conduction occurs increases with increasing temperature; consequently, n-type conduction is typically an issue in high-temperature applications [47, 48]. As mentioned above, although electrochemical sensors generally require a transport number of at least 0.99, the transition is also sometimes specified by the oxygen partial pressure at which the ionic an electronic conduction are equal ($t_{ion} = 0.5$). One other cubic fluorite oxide with a high oxygen ion conductivity is doped cerium oxide, although ceria is less widely used for oxygen sensors because Ce^{4+} is easily reduced to Ce^{3+}, so that electronic conduction occurs at relatively high oxygen partial pressures [49].

Several fluorides form the cubic fluorite structure (including CaF_2 for which it is named), and are excellent fluoride ion conductors [50] (see Chapter 2). With the introduction of an oxide phase, either during fabrication or *in situ* during operation, fluoride ion conductors can be used as electrolytes in oxygen sensors [51]. One of the advantages of using fluoride electrolytes is that they tend to remain pure ionic conductors at lower oxygen partial pressures than do oxide electrolytes, which can exhibit significant electronic conduction under such conditions. Some fluoride electrolytes also have high conductivity, and can be used for sensors that operate at relatively low temperatures [52–54].

Cation-conducting electrolytes (see Chapter 7) are also used in electrochemical sensors, one of the most widely used being β alumina [55, 56]. The β alumina structure consists of relatively densely packed spinel blocks that are separated by less densely packed planes through which ionic conduction occurs. The most common example is sodium β alumina, which is a Na^+ ion conductor, although the sodium can be exchanged with other ions to create electrolytes that conduct other cations [57], as well as other species, such as NO^+ [58] (see also Chapter 8).

Another common sodium ion-conducting electrolyte is NASICON, which is a solid solution between $NaZr_2(PO_4)_3$ and $Na_4Zr_2(SiO_4)_3$ (i.e., $Na_{1+x}Zr_2P_{3-x}Si_xO_{12}$) [59]. The NASICON structure contains tunnels through which ionic conduction occurs. As in β alumina, the sodium in NASICON can be replaced with other ions [60]. The conductivity increases with increasing silicon content, and the maximum conductivity occurs at a composition of about $Na_3Zr_2P_1Si_2O_{12}$ ($x \sim 2$) [61, 62]. The conductivities of some NASICON materials are shown with the ranges of conductivities for YSZ and β alumina in Figure 13.5 [63–68].

NASICON can also be modified so that it conducts protons (see Ref. [69] and Chapter 7). However, alkaline-earth cerates and zirconates forming the perovskite structure are more commonly used as high-temperature proton conductors [70–78]. In general, cerates have a higher proton conductivity, but zirconates are more stable. As with oxygen ion-conducting electrolytes, proton conduction in these electrolytes occurs only within a limited range of hydrogen partial pressures. In addition, as they are oxides, oxygen defects can occur. Figure 13.6 shows the predominant defects in indium-doped calcium zirconate, which were calculated based on an extrapolation of conductivity measurements [79]. Hydrogen conduction occurs by interstitials $H_i^{\bullet}$ (also expressed as a OH^- on a O^{2-} site: $OH_O^{\bullet}$), which are stable at high hydrogen and oxygen partial pressures (upper right-hand corner of Figure 13.6). The hydrogen and oxygen must also be in equilibrium with water vapor, so the regions at the upper right-hand corner of the diagram can occur only at high total pressures. At lower hydrogen partial pressures, p-type electronic conduction occurs by holes ($h^{\bullet}$) and, as with the oxygen ion-conducting electrolytes described above, the tendency for electronic conduction increases with increasing temperature. In addition, at low oxygen partial pressures, oxygen vacancies ($V_O^{\bullet\bullet}$) are predominant, so calcium zirconate can also serve as an oxygen ion conductor.

Salts of the target gas species can be used as electrolytes in chemical sensors. For example, carbonates (e.g., K_2CO_3 [80] or Li_2CO_3 [81]), or mixtures of carbonates (e.g., Na_2CO_3-$BaCO_3$ [82, 83]), have been used as electrolytes for CO_2 sensors. Similarly, mixtures of sulfates (e.g., Li_2SO_4-Ag_2SO_4 [84–87] or Na_2SO_4-$BaSO_4$-Ag_2SO_4 [83]) have been used as electrolytes for SO_2 sensors. While these salts provide a direct measurement of the desired species, their stability is generally inferior to that of other electrolytes. However, as will be discussed later, these materials are used as auxiliary electrodes with other electrolytes.

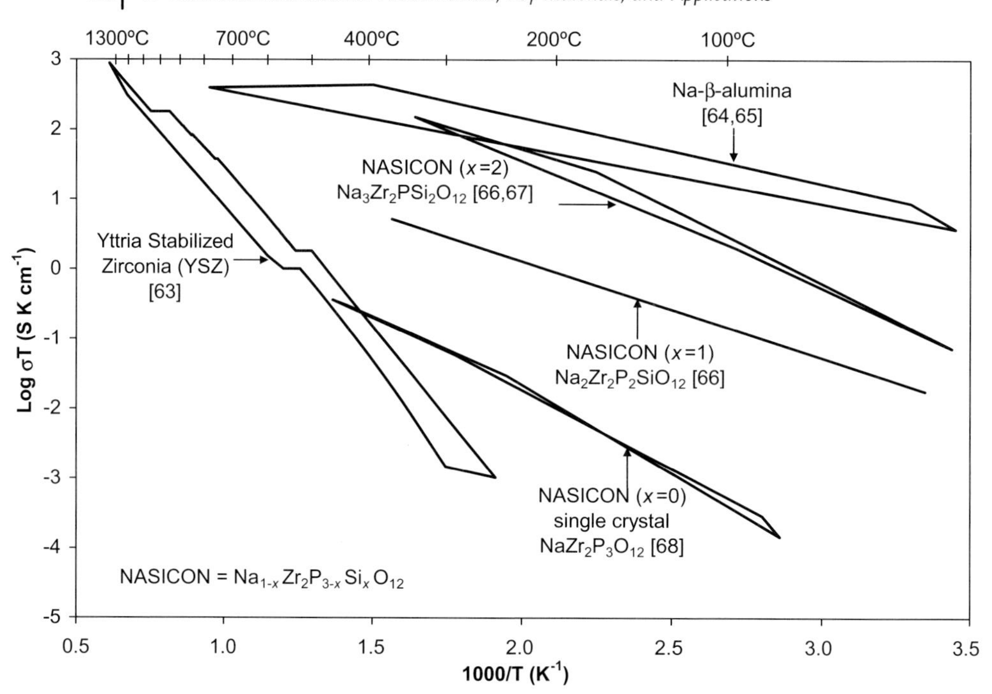

Figure 13.5 Conductivity of solid electrolyte materials [63–68].

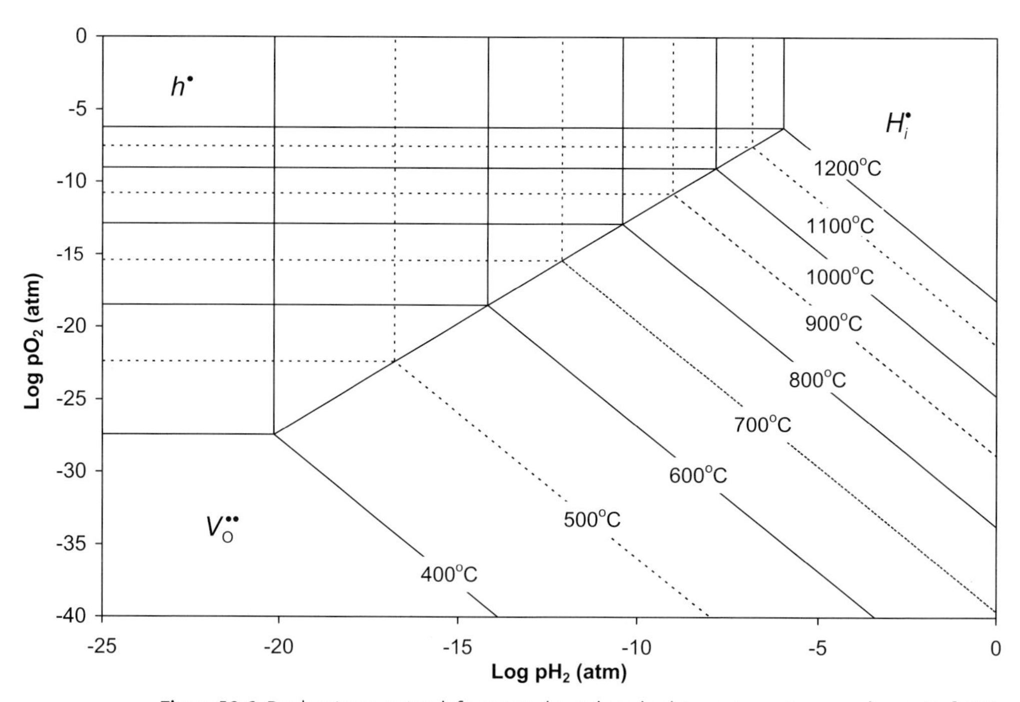

Figure 13.6 Predominant point defects in indium-doped calcium zirconate, according to Ref. [79].

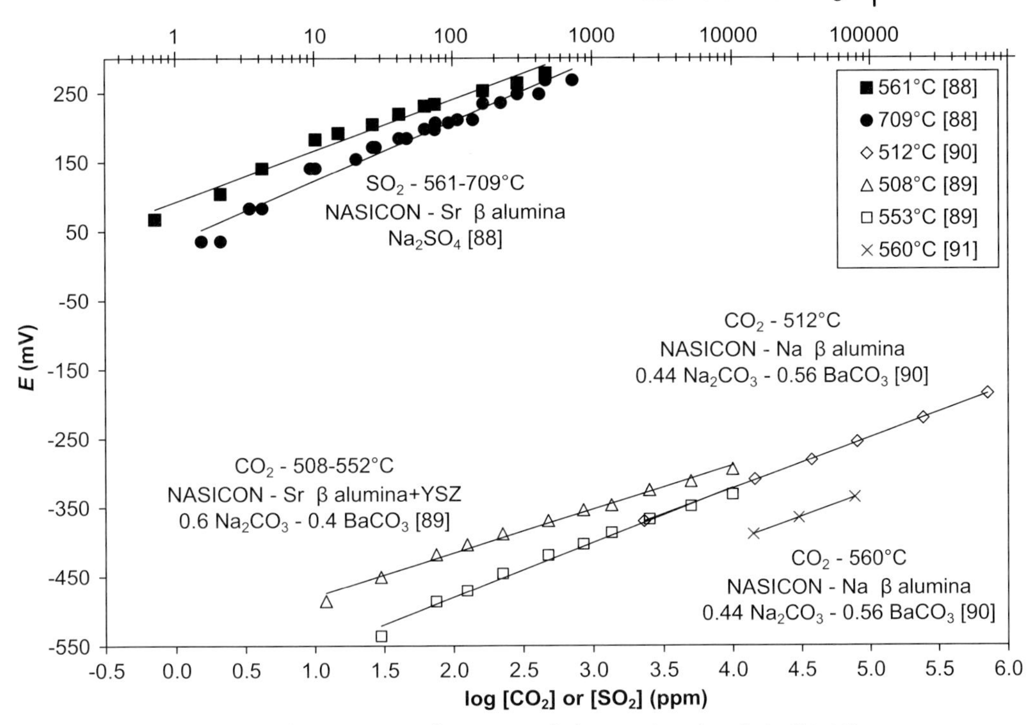

Figure 13.7 Outputs of CO_2 and SO_2 sensors with NASICON/β alumina electrolyte chains [88–91].

As discussed above, the use of multiple electrolytes in electrolyte chains can improve sensor performance. The outputs of some CO_2 and SO_2 sensors using both NASICON and β alumina are shown in Figure 13.7 [88–91]. The auxiliary electrodes used are given in the figure, but these will be discussed in more detail later. YSZ is also used with cation-conducting electrolytes, including NASICON (SO_2 [92, 93], NO_2 [94], CO_2 [95, 96]), sodium aluminosilicate glasses (NO_2 [97], CO_2 [98–100]), and the aluminum conductor ($(Al_{0.2}Zr_{0.8})_{10/19}Nb(PO_4)_3$) ($SO_2$ [101, 102], NO_2 [103–109]). The outputs of some of these sensors are shown in Figure 13.8 [95, 96, 101].

13.3.2 Electrodes

Although the transduction from chemical information occurs through the electrolyte, the electrodes are important for transferring the desired information (i.e., target species) to measurable information (i.e., the ion that is mobile in the electrolyte).

13.3.2.1 Reference Electrodes

One of the challenges in the development of reliable electrochemical sensors is the identification of stable reference electrodes, since any drift in the reference potential will lead to a drift in the sensor output. The reference electrode for an electrochemical

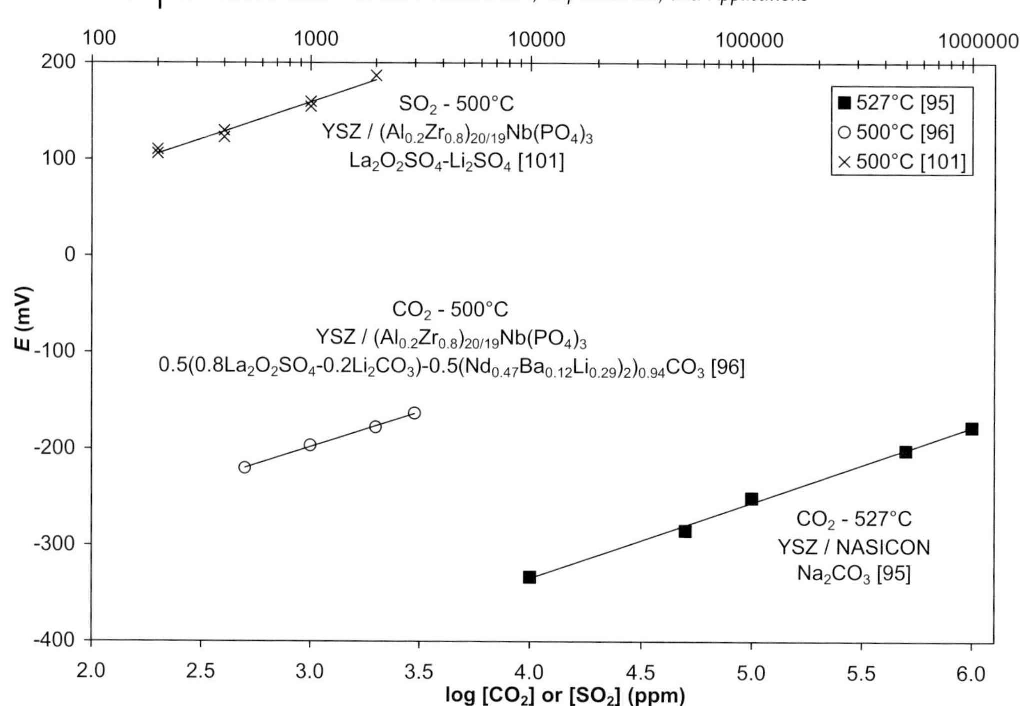

Figure 13.8 Outputs of CO_2 and SO_2 sensors with YSZ/cation-conductor electrolyte chains [95, 96, 101].

gas sensor can be the same as the working electrode, except that a gas of known composition is passed over the reference electrode. Such a gas-phase electrode requires two gas supplies that must be separated (Figure 13.9a); this complicates the sensor design and creates challenges in the development of satisfactory, gas-tight seals. The sensor design can be simplified by using a solid-state reference electrode, as shown in Figure 13.9b. For an oxygen sensor, a reference oxygen partial pressure can be established by a mixture of a metal and metal oxide, such as chromium and chromium oxide, which determines an oxygen partial pressure according to the following reaction:

$$2\mathrm{Cr} + \tfrac{3}{2}\mathrm{O}_2 = \mathrm{Cr}_2\mathrm{O}_3. \tag{13.11}$$

As mentioned above (and shown in Figure 13.9c), when electrolyte chains are used the electrolyte in contact with the reference electrolyte can be different from that used for the sensing electrode, which provides flexibility in the design of the reference electrode material.

A major factor in the selection of a reference electrode is the reference potential established. If the reference potential is close to the potential at the sensing electrode, the cell voltage will be small and thus tend to be more stable than a large voltage. For example, the oxygen partial pressure established by a Pd + PdO reference electrode

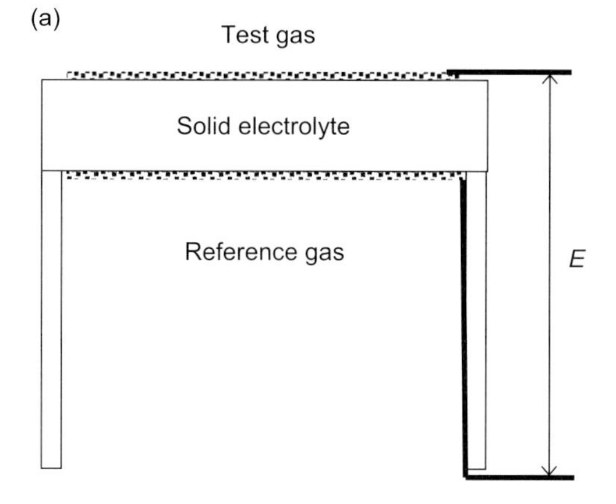

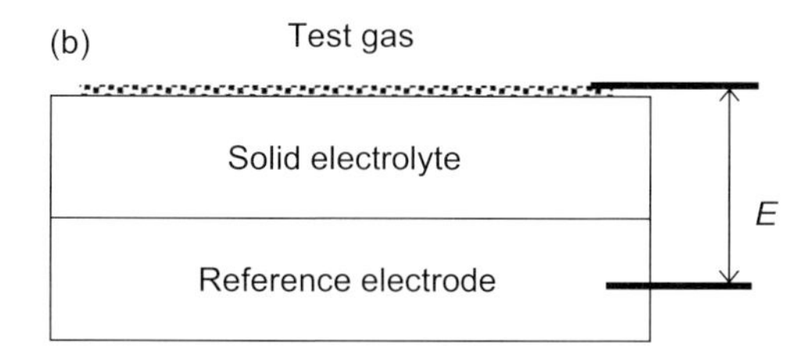

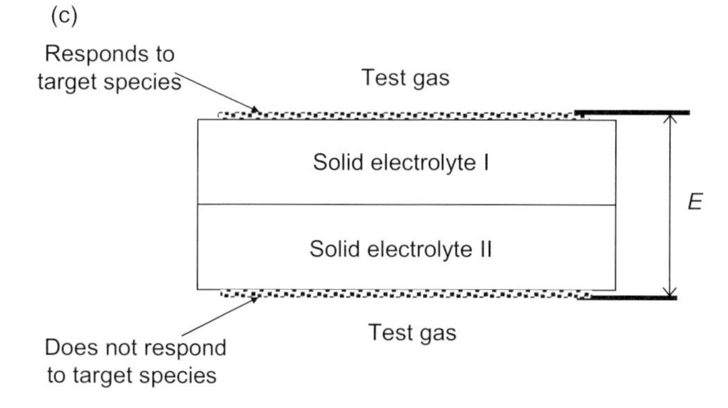

Figure 13.9 Schematic of reference electrode. (a) Gas-phase reference; (b) Solid-state reference; (c) Reference with electrolyte chain.

works well for moderate oxygen partial pressures of approximately 0.1 Pa [110, 111], while Mo + MoO_2 [112, 113] or other transition metals with oxides (simple or mixed) [114, 115] are suitable for use at the high temperatures and low oxygen partial pressures in molten steel.

Reference electrode performance also includes the *response kinetics*. For example, the addition of NiO to a Cr + Cr_2O_3 reference electrode reduces instabilities during heating of oxygen sensors used in molten steel [116]. The addition of a mixed-conducting oxide, such as $(La,Sr)(Co,Fe)O_3$ [117], can also reduce the electrode impedance and improve the response time. At lower temperatures, the reaction kinetics will decrease, which leads to a sluggish response; consequently, low-melting-temperature metals, such as indium [118, 119], bismuth [119–121], and tin [119] are used in the reference electrodes.

Reference electrode stability is important to avoid drift in the sensor signal. For example, while sodium metal would, in theory, provide a stable reference potential for a sodium ion-conducting electrolyte, sodium is very reactive; hence, more stable sodium-containing compounds, such as $NaCl\text{-}Na_2CO_3$ [122], $Na_{0.75}CoO_2$ [123], or $Na_{0.85}CoO_2$ [124] are used. Stability is also a challenge in the development of solid-state reference electrodes for use with proton-conducting electrolytes. For example, $Ti + TiH_2$ is used as a reference electrode in hydrogen sensors for use in molten aluminum [125]. Humidity may cause degradation of reference electrodes, but stability can be improved by using mixed oxides, such as $La_{0.6}Sr_{0.4}Co_{0.78}Ni_{0.02}Fe_{0.2}O_3$ [126]. The addition of mixed oxides, such as $(La,Sr)MnO_3$ [127], can be used to inhibit the sintering of porous electrodes, which also improves stability; the properties of such perovskites are briefly described in Chapter 9. Similarly, the addition of alumina can reduce the interaction between nickel and platinum after long operation times [128]. With sensor miniaturization, smaller amounts of interaction are significant relative to the component dimensions, and thus can affect the distribution of electric fields within the sensor, while reference electrode placement can affect performance [129].

13.3.2.2 **Auxiliary Electrodes**

An *auxiliary electrode* provides a link between the target species and the species to which the electrolyte responds, by equilibrating with both. In some cases, the electrolyte material can provide this function, so a separate auxiliary phase need not be added (see Figure 13.1b). One example is the use of Na β alumina as an oxygen sensor [130–132], where the Na_2O in the electrolyte equilibrates between oxygen gas and sodium ions according to Equation (13.4). One advantage of this approach is that β alumina remains a pure ionic conductor to very low oxygen partial pressures. A similar approach has been used with the fluoride ion-conducting electrolytes MgF_2 [133] and LaF_3 [134] in sensors for measuring the concentrations of magnesium and lanthanum, respectively, in molten aluminum.

An additional electrolyte can also provide the auxiliary electrode function. For example, Na_2CO_3 can be used with Na β alumina [135–137], NASICON [138–146], or a sodium aluminosilicate glass [147] as the electrolyte for a CO_2 sensor, in which case the Na_2CO_3 equilibrates with CO_2 in the gas and Na^+ in the electrolyte according to the following reaction:

$$2Na^+ + CO_2 + \tfrac{1}{2}O_2 + 2e^- = Na_2CO_3. \tag{13.12}$$

The electrode performance can be improved by using a mixed carbonate as the auxiliary electrode. As long as the auxiliary electrode contains Na_2CO_3, the electrode reaction will be the same, even though the activity of Na_2CO_3 may be less than one.

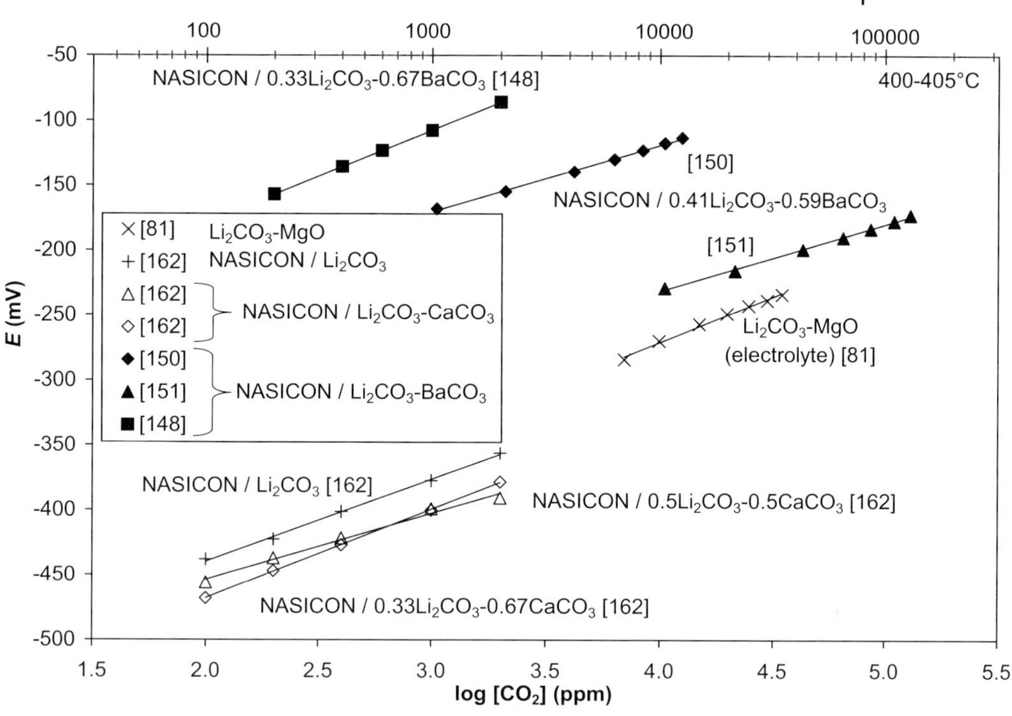

Figure 13.10 Outputs of CO_2 sensors based on NASICON or carbonate electrolytes at 400–405 °C [81, 148, 150, 151, 162].

Na_2CO_3 is most commonly mixed with $BaCO_3$ [90, 91, 140, 141, 148–158], but Na_2CO_3-$SrCO_3$ [159] and Na_2CO_3-$CaCO_3$ [160] have also been used as auxiliary electrodes. When used with a sodium ion-conducting electrolyte, the alkaline-earth carbonate is not involved in the electrode reaction. However, $SrCO_3$ has been used as the functional auxiliary electrode material with a Sr β alumina electrolyte [161].

The outputs of some sensors with NASICON electrolytes and Li_2CO_3 or Li_2CO_3-containing auxiliary [148, 150, 151, 162] electrodes are shown in Figure 13.10. As the auxiliary electrode does not contain sodium, these Li-based electrodes rely on the *in situ* formation of Na_2CO_3 during operation. The slopes of the outputs of all these sensors are similar to that of a sensor using a Li_2CO_3-MgO electrolyte [81]. The differences in the absolute magnitudes of the sensor responses are due to differences in the reference electrode used, which adds (or subtracts) a constant voltage, depending on the particular electrode reaction.

Lithium carbonate can be used more directly as the auxiliary electrode with a lithium ion conductor, since the *in situ* formation of another carbonate phase is not required. Lithium ion conductors used with Li_2CO_3 include LISICON [163] and other Li_3PO_4-based electrolytes [164–173]. As with sodium ion conductors, Li_2CO_3-containing carbonates are used with lithium ion conductors [174, 175]. The outputs of some examples of these lithium ion-conducting electrolyte-based sensors are shown in Figure 13.11 [163, 172–174].

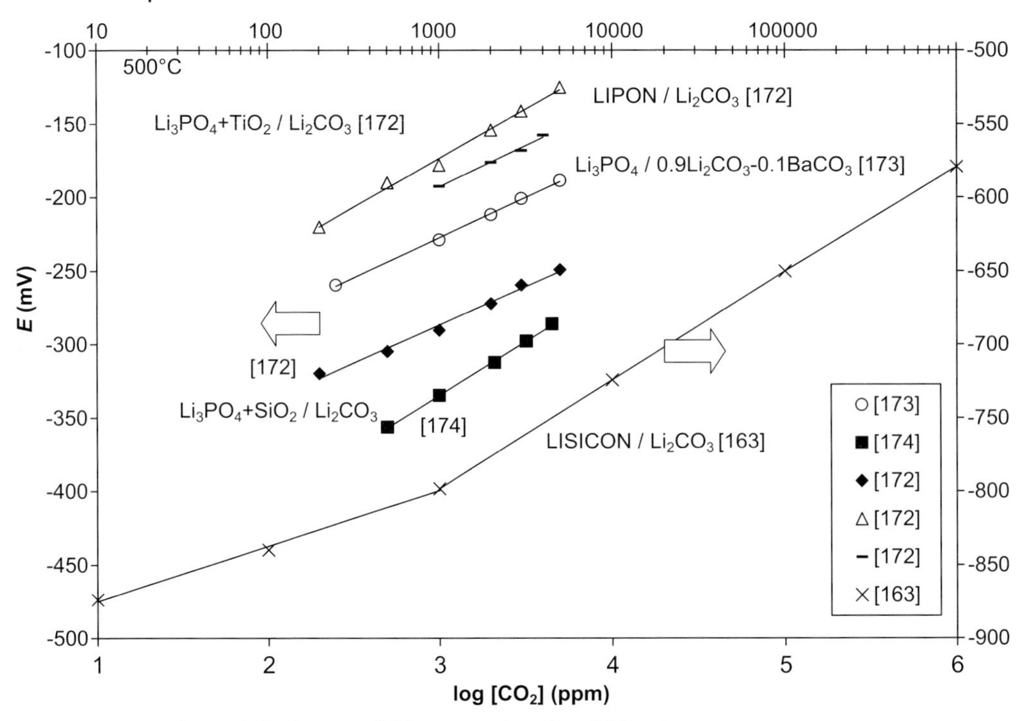

Figure 13.11 Outputs of CO_2 sensors based on lithium ion conducting electrolytes at 500 °C [163, 172–174].

The sensor output depends on the composition of the mixed carbonate auxiliary electrode. The outputs of CO_2 sensors with NASICON electrolytes and Li_2CO_3-$BaCO_3$ [151] or Li_2CO_3-$CaCO_3$ [162] auxiliary electrolytes are shown in Figure 13.12. There are shifts in the curves with changing electrode composition but, more importantly, the slopes – which represent the sensitivities – also change. Figure 13.13 shows that the addition of $BaCO_3$ to Li_2CO_3 [151, 176] or $CaCO_3$ to Na_2CO_3 [160] increases the sensitivity, with maximum sensitivity occurring when the amount of the alkaline earth carbonate is about twice that of the alkali carbonate.

Sensors for SO_2 have been developed following an approach similar to that used for CO_2 sensors. The equilibrium between a sulfate (e.g., Ag_2SO_4) and SO_3 gas establishes a cation activity (e.g., Ag^+) according to the following reaction:

$$2Ag^+ + SO_3 + \tfrac{1}{2}O_2 + 2e^- = Ag_2SO_4. \tag{13.13}$$

If excess oxygen is present, the SO_2 will be oxidized according to

$$SO_2 + \tfrac{1}{2}O_2 = SO_3 \tag{13.14}$$

and the SO_3 partial pressure (P_{SO_3}) is related to the SO_2 partial pressure (P_{SO_2}) according to [85]:

$$P_{SO_3} = \frac{P_{SO_2}}{\left(1 + \dfrac{1}{K \cdot (pO_2)^{\frac{1}{2}}}\right)}. \tag{13.15}$$

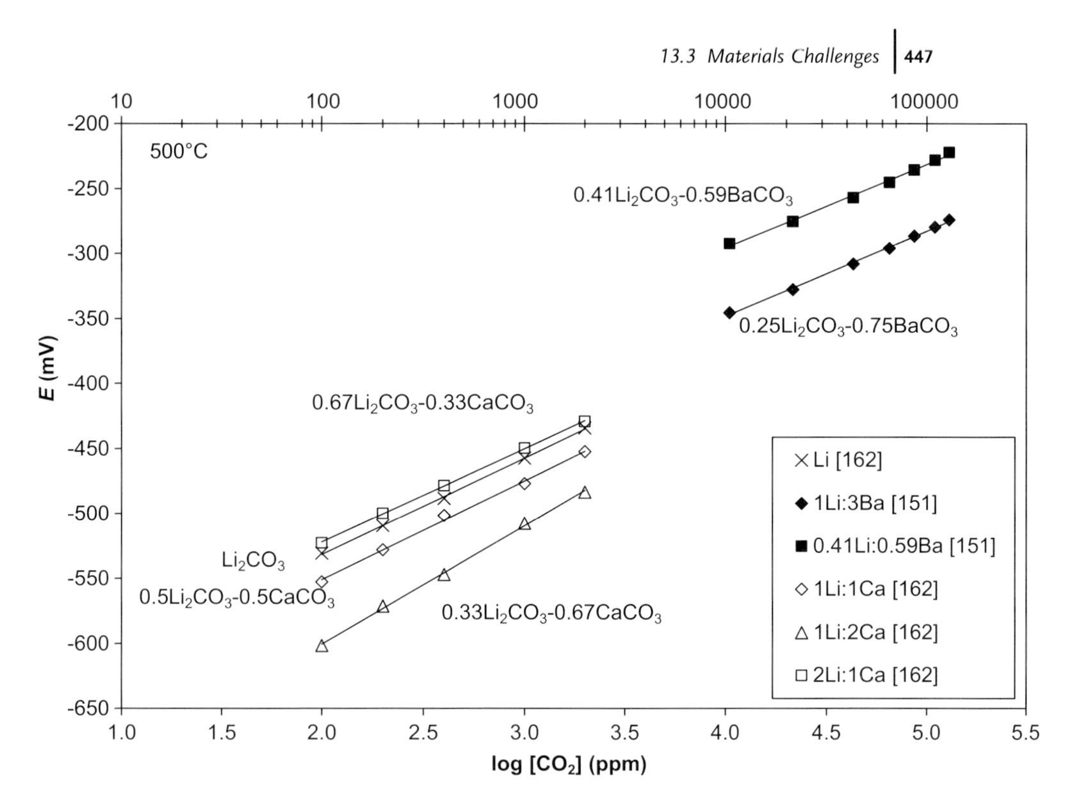

Figure 13.12 Outputs of CO_2 sensors based on NASICON electrolytes and Li_2CO_3-containing auxiliary electrodes at 500 °C [151, 162].

Hence, a cation-conducting solid electrolyte can be used for a SO_2 sensor. Figure 13.14 shows that the output of a sensor with an Ag β″ alumina electrolyte [177, 178] and an Ag_2SO_4 electrode generates a response that is similar to those of sensors using Li_2SO_4-Ag_2SO_4 electrolytes [86, 87].

In a similar way, Na_2SO_4 has been used as an auxiliary electrode with sodium ion conductors (Na β alumina [179] or NASICON [180]). As with carbonates for CO_2 sensors, mixed sulfates (e.g., Na_2SO_4-$BaSO_4$ [181]) can be used for the auxiliary electrode. At low oxygen partial pressures, sulfur may be present as H_2S, rather than as SO_x, in which case sulfides can be used as auxiliary electrodes. For example, CaS has been used with a CaF_2 electrolyte for measuring H_2S gas concentrations [182].

Similar sensors have also been developed for measuring NO_x gases. While NO sensors have been developed using NO^+-conducting electrolytes, including NO^+-exchanged β alumina [183, 184] and $Ga_{11}O_{17}$ [185] (which forms the same crystal structure as β alumina), NO_2 sensors generally require a nitrate (typically $NaNO_3$ [186–188], $Ba(NO_3)_2$ [189], or a mixture of the two [83, 94]) as the auxiliary phase. The outputs of some NO_2 sensors are shown in Figure 13.15 [83, 94, 186–189].

As mentioned above, the auxiliary electrode can form during sensor operation. The auxiliary electrode sometimes forms by an exchange reaction, such as when Na_2CO_3 forms from reaction of Na^+ with Li_2CO_3 [140, 190–192] or Li_2CO_3-containing

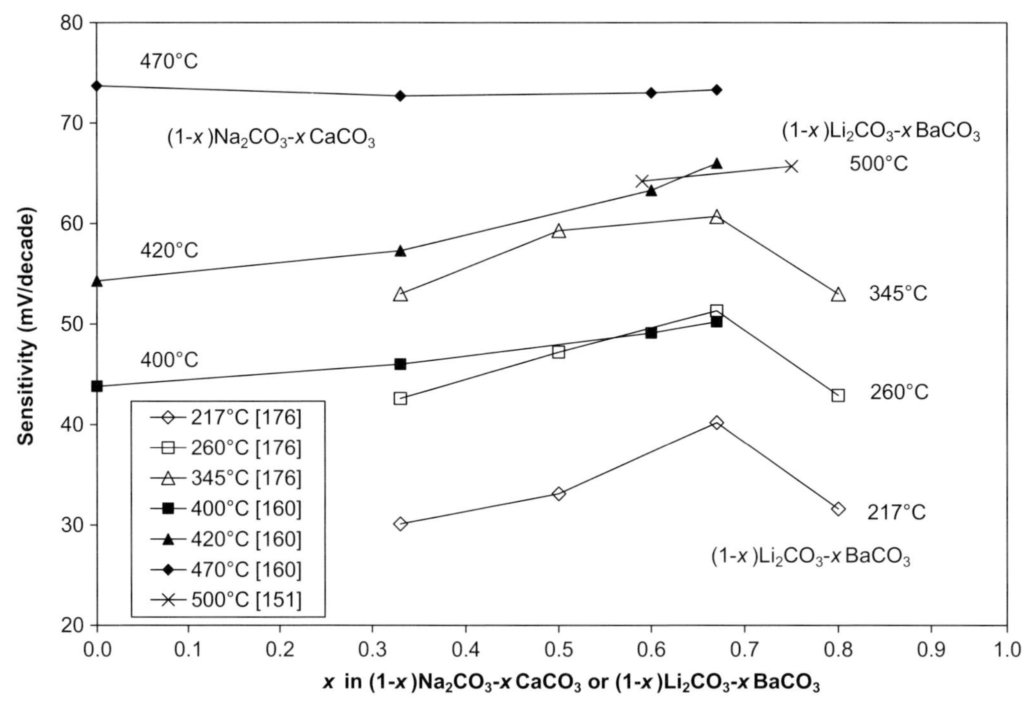

Figure 13.13 Sensitivities of CO_2 sensors based on NASICON electrolytes and mixed carbonate auxiliary electrodes [151, 160, 176].

carbonates (e.g., Li_2CO_3-$BaCO_3$ [140, 148, 150, 151, 162, 176, 192–200] or Li_2CO_3-$CaCO_3$ [150, 162, 192, 200]) when NASICON is used as the electrolyte for a CO_2 sensor. Similarly, alkaline-earth (Ca^{2+} [201, 202] and Mg^{2+} [203, 204]) conducting electrolytes have been used with a Na_2SO_4 electrode, which relies on the formation of the corresponding sulfates ($CaSO_4$ and $MgSO_4$) to provide the sensitivity to SO_2. The outputs of some of these SO_2 sensors are shown in Figure 13.16 [201–203].

The carbonate phase can also form from reaction between the cation in the electrolyte and CO_2 from the atmosphere, in which case a carbonate phase need not be added. For example, a CO_2 sensor with a NASICON electrolyte and an indium tin oxide (ITO) electrode relies on the *in situ* formation of Na_2CO_3 during sensor operation [205]. Similarly, the formation of Li_2CO_3 allows $LiLaGeO_4$ to be used as the electrolyte for a CO_2 sensor [206].

Carbon dioxide sensors using Li_2CO_3 [207–210] or Li_2CO_3-$BaCO_3$ [211] as the auxiliary electrode with zirconia electrolytes, both magnesia- [207, 208] and yttria- [209–211] stabilized zirconia, have also been reported. As zirconia does not contain lithium, an additional phase is required to relate the lithium activity in the carbonate (which is related to CO_2 according to a reaction analogous to Equation 13.12) to the oxygen ion activity in the zirconia. Although an additional phase is not added during sensor fabrication, a mixed-oxide phase – specifically Li_2ZrO_3 – forms, so that the

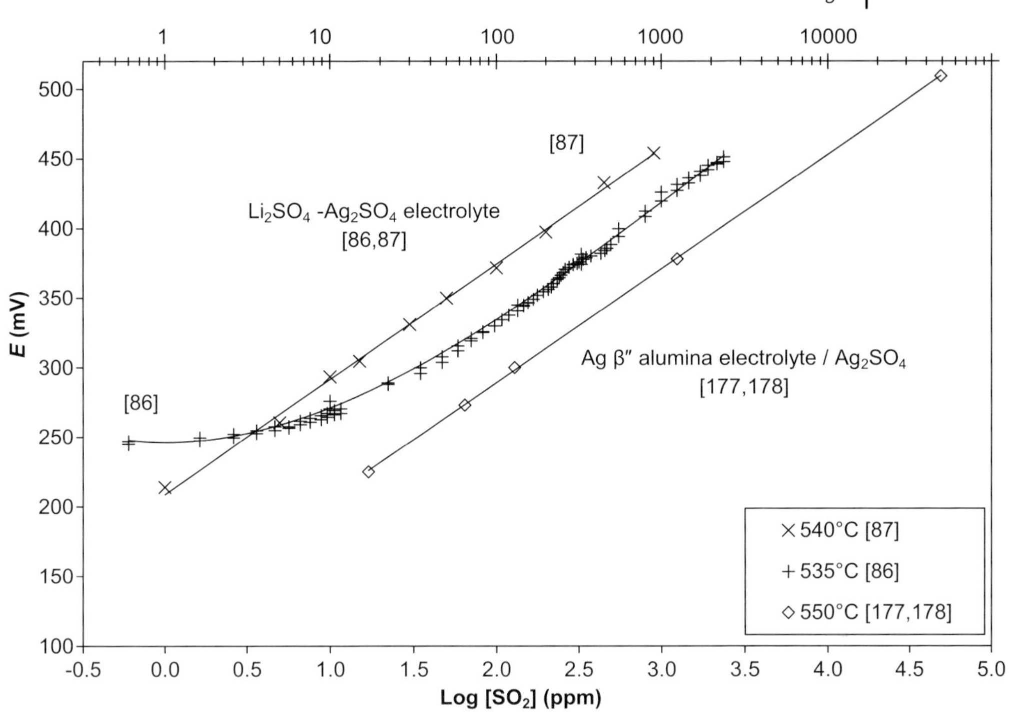

Figure 13.14 Outputs of SO_2 sensors based on using Ag_2SO_4 in the electrolyte [177, 178] or auxiliary electrodes [86, 87].

lithium ion activity in Li_2CO_3 establishes an oxygen ion activity in the electrolyte according to the following reaction:

$$2Li^+ + O^{2-} + ZrO_2 = Li_2ZrO_3. \qquad (13.16)$$

During operation, Li_2O dissolves in the carbonate electrode and the equilibrium between lithium and oxygen ions depends on the activity of the dissolved Li_2O. Any consequent changes in the concentration of dissolved Li_2O can lead to drifts in the sensor signal over time [209, 210].

A similar approach has been used in the development of zirconia-based SO_2 sensors, in which Li_2SO_4 or Li_2SO_4-containing auxiliary electrodes are used with magnesia-stabilized zirconia (MSZ) electrolytes [212–214]. As with CO_2 sensors, these sensors require the formation of Li_2ZrO_3 at the interface between the MSZ electrolyte and the carbonate electrode in order for the sensor to operate. A SO_2 sensor with a K_2SO_4-$BaSO_4$-SiO_2 auxiliary electrode and YSZ electrolyte has also been reported [215]. In this case, the sensor response is attributed to the formation of a K-Zr-Y-O phase. The outputs of some zirconia-based SO_2 sensors are shown in Figure 13.17 [213–215]. Most of the responses indicate a value of n (number of electrons from the Nernst equation) of approximately 2, although the response of the sensor with the K_2SO_4-$BaSO_4$-SiO_2 auxiliary electrode has a lower sensitivity (i.e., a higher n).

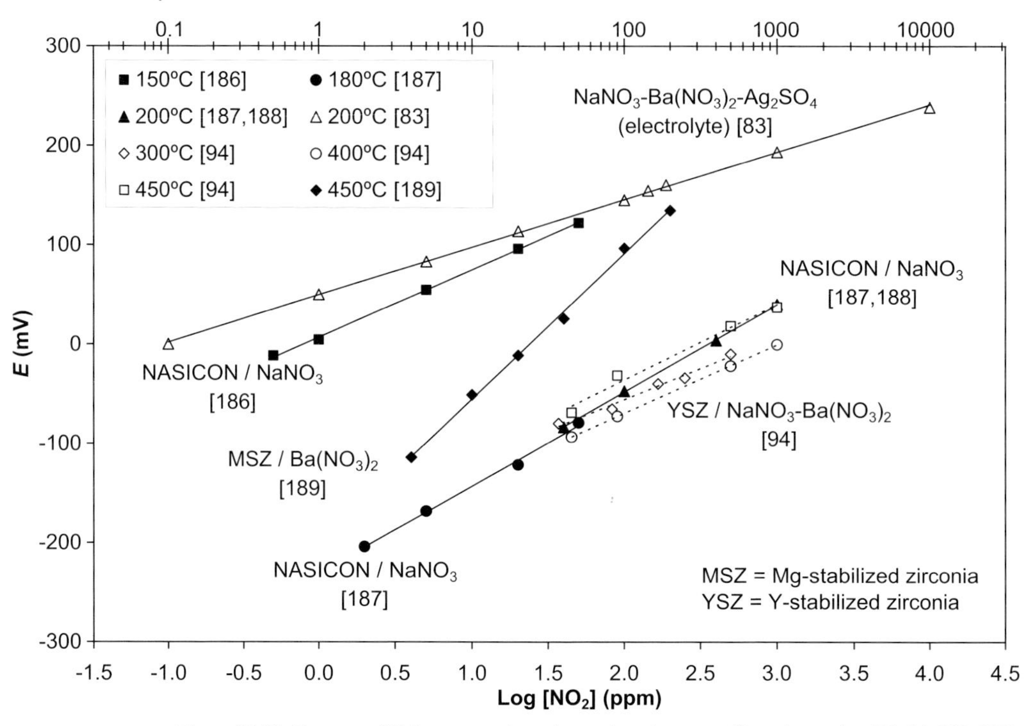

Figure 13.15 Outputs of NO_2 sensors based on using nitrate auxiliary electrodes [83, 94, 186–189].

Zirconia electrolytes have also been used for NO_x sensors by using nitrates ($NaNO_3$ [94] or $Ba(NO_3)_2$ [189]) as auxiliary electrodes. As nitrates are typically less stable than carbonates and sulfates, nonequilibrium-based NO_x sensor are more commonly used (these will be discussed later).

The equilibration between cations and oxygen ions in zirconia-based sensors suggests that zirconia can be used in the measurement of metal concentrations. Zirconia sensors for the measurement of aluminum [216, 217], iron [218], or silicon [219] in molten metals have been developed by using oxides containing the target metal to relate the metal activity to the oxygen partial pressure (e.g., Al_2O_3 for Al, Mg_2FeO_4 for Fe or $ZrSiO_4$ for Si). A similar approach can be used with fluoride ion-conducting electrolytes, where a fluoride of the target metal is used as the auxiliary electrode. For example, AlF_3 has been used as the auxiliary electrode with CaF_2 for the measurement of the aluminum activity in molten zinc [220].

Auxiliary electrodes are also required if a cation-conducting electrolyte is used for measuring a metal which is different from the ion that is mobile in the electrolyte. For example, the use of Na β alumina for the measurement of antimony in zinc requires the use of a $NaSbO_3$ auxiliary electrode [221]. Sodium β alumina has also been used for measuring magnesium and strontium concentrations in molten aluminum [222]. In this case, the auxiliary electrode forms *in situ* by an exchange reaction between

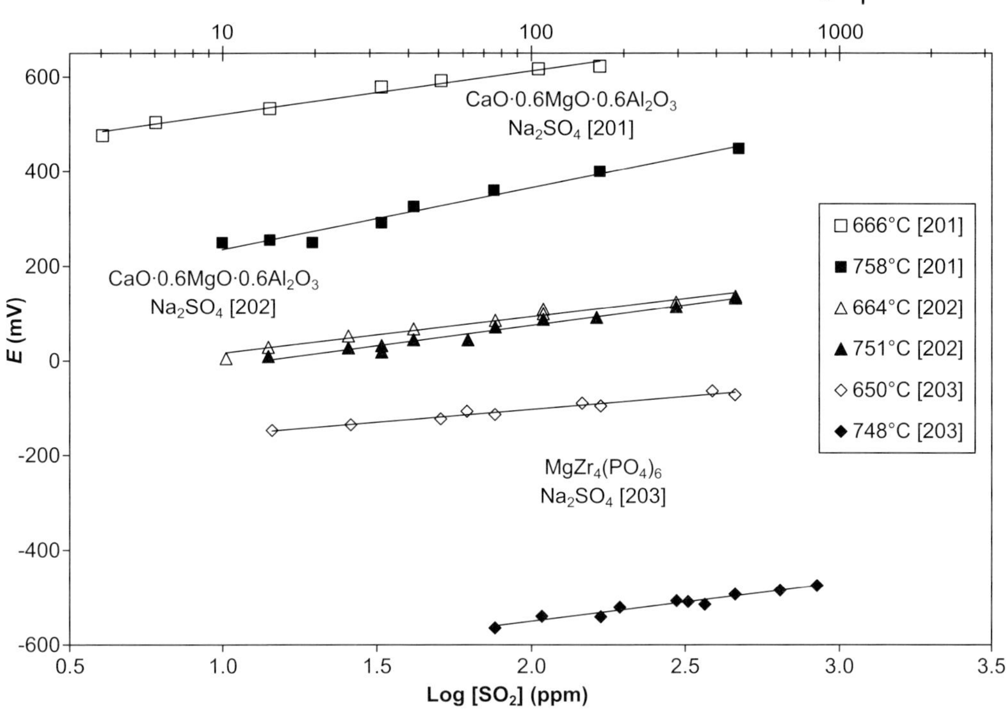

Figure 13.16 Outputs of SO_2 sensors based on using alkaline-earth ion-conducting electrolytes [201–203].

sodium in the electrolyte and magnesium or strontium in the alloy, such that the dissolved MgO or SrO acts as the auxiliary electrode.

A similar approach is used in CO_2 gas sensors based on electrolyte chains of YSZ and cation conductors. Specifically, YSZ has been used with magnesium- [223], aluminum- [96, 105, 224, 225], or scandium- [225–227] conducting electrolytes and Li_2CO_3-containing electrodes. The sensitivity to CO_2 is attributed to the dissolution of lithium in the electrolyte rather than to the formation of a new carbonate phase.

The Nernst equation indicates that the sensitivity of the sensor response (i.e., voltage/log concentration) should increase with increasing temperatures. Figure 13.18 shows that this is indeed the case for several CO_2 and SO_2 sensors [154, 157, 178, 179, 213, 215]. The expected value of n for both these sensors is 2, and most of the sensors have values of n which are close to theoretical. In addition, increasing the operating temperature will also leads to a higher electrolyte conductivity and more rapid electrode kinetics; hence, equilibrium-based electrochemical sensors are well suited to high operating temperatures. However, as stable electrode materials are not always available for equilibrium-based sensors, nonequilibrium sensors may be used. Nonequilibrium-based sensors generally have a lower maximum operating temperature because systems tend to approach equilibrium faster as the temperature increases.

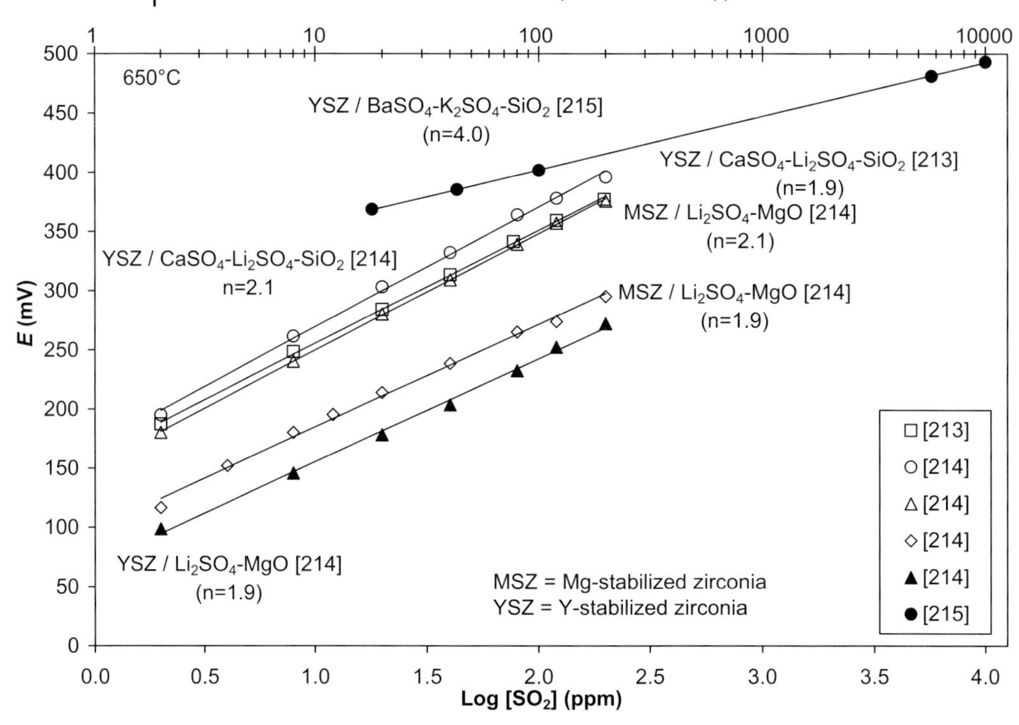

Figure 13.17 Outputs of zirconia-based SO_2 sensors based at 650 °C [213–215].

13.3.2.3 **Electrocatalytic Electrodes**

As discussed above, nonequilibrium sensors are based on differences in the reaction kinetics on different electrode materials. The ease of a reaction on the electrode can be described by the *overpotential*. For example, the overpotential for the oxidation of CO on gold is higher than that on platinum, which leads to a difference in the mixed potential established on the two electrodes [228, 229]. The outputs of mixed potential CO sensors with gold and platinum electrodes on $(Ce,Gd)O_2$ (CGO) [228–231], YSZ [232, 233] and β alumina [234, 235] are shown in Figure 13.19. The responses of sensors in a planar configuration, in which both electrodes are exposed to the same gas, and those in a tubular configuration, in which the two gases are separated, are similar.

The mixed potential mechanism was described above, using CO as an example. However, the mechanism can be applied to any pair of oxidation and reduction reactions. Thus, mixed potential sensors have been reported for other reducing gases, such as hydrocarbons. Figure 13.20 shows that gold and platinum electrodes can be used to measure the amount of propylene (C_3H_6) [228, 229, 231, 233, 236–238]. Mixed potential hydrocarbon sensors have also been reported using proton-conducting electrolytes [239–242].

As mentioned above, the poor stability of nitrate compounds leads to the use of nonequilibrium approaches for the measurement of NO_x gas [243, 244]. During the

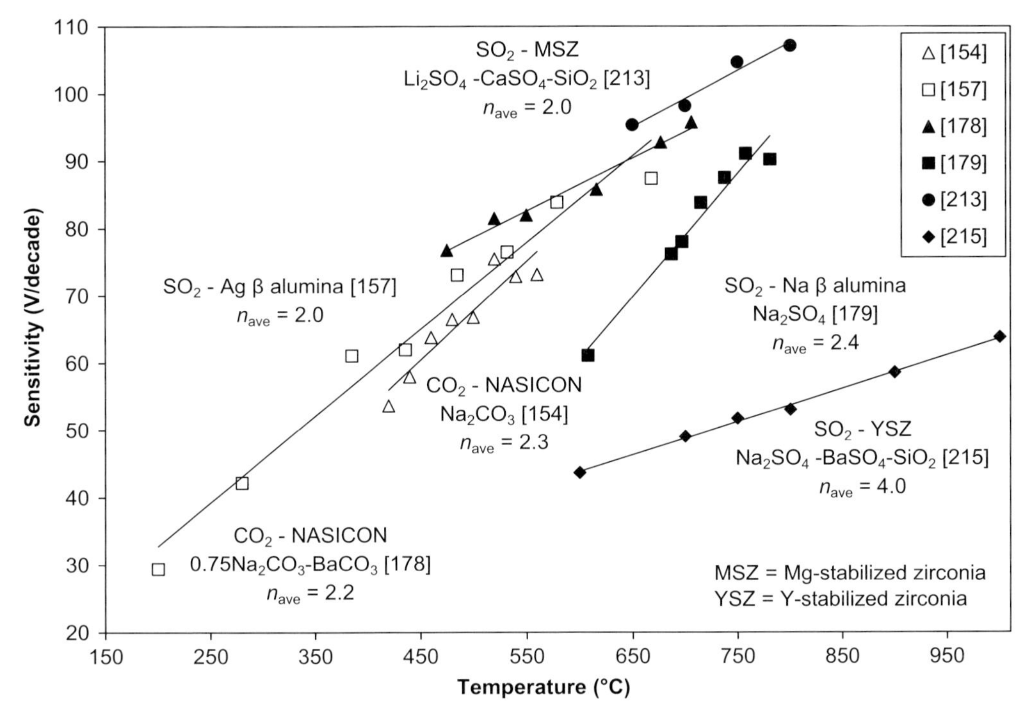

Figure 13.18 Temperature dependences of sensitivities of CO_2 and SO_2 sensors [154, 157, 178, 179, 213, 215].

operation of a mixed potential sensor in NO_x [245, 246], NO can be oxidized at the electrode by

$$NO + O^{2-} = NO_2 + 2e^-, \tag{13.17}$$

in which case the sensor would operate as described above for CO. However, at high NO_2 concentrations, the reverse reaction can occur:

$$NO_2 + 2e^- = NO + O^{2-} \tag{13.18}$$

which would need to be balanced by an oxidation reaction, such as

$$O^{2-} = \tfrac{1}{2}O_2 + 2e^-, \tag{13.19}$$

so that the responses to NO and NO_2 are opposite in sign.

The rate of dissociation of NO_x is different on different metals [247–249], so metal electrodes have been used for NO_x sensors using a variety of electrolytes, including oxygen ion conductors (e.g., YSZ [250] and scandia-stabilized zirconia (ScSZ) [251]), sodium ion conductors (e.g., β alumina [252] and NASICON [94]) and proton conductors (e.g., indium-doped $Sn_2P_2O_7$ [253]).

Although, metal electrodes provide a mixed potential response to reducing gases, the response can be improved by using semiconducting oxide electrodes. For example,

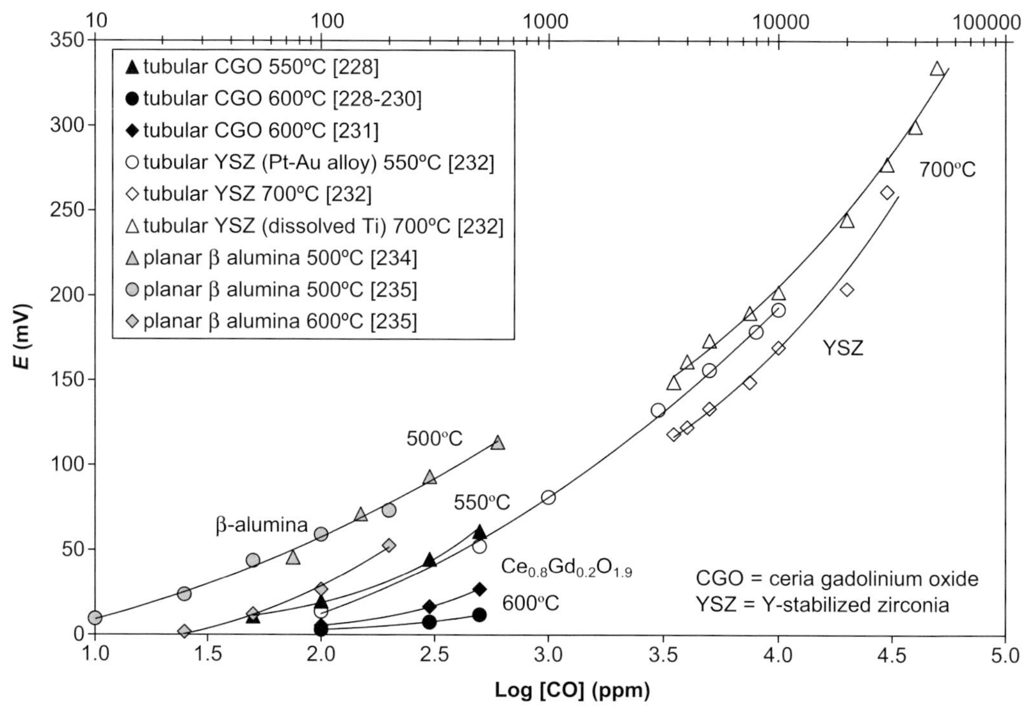

Figure 13.19 Outputs of mixed potential CO sensors with gold and platinum electrodes and yttria-stabilized zirconia (YSZ), ceria gadolinium oxide, and β alumina electrolytes [228–235].

WO_3 has been used as an electrode for CO sensors [254–260]. Figure 13.21 [254–258] shows that the outputs of CO sensors with YSZ electrolytes and WO_3 electrodes decrease with increasing temperature, which, as discussed earlier, is a general tendency for nonequilibrium sensors. Tungsten oxide has also been used as an electrode for NO_x sensors [246, 254–259, 261–263], some examples of which are shown in Figure 13.22 [254–259, 262, 263].

Nickel oxide has also been used as the electrode material for NO_x sensors [264–274]. One of the advantages of NiO electrodes is the improved sensor response at high temperatures. Figure 13.23 shows the sensitivities of NO_x sensors with WO_3 [254–259, 262] or NiO [264–269] electrodes. As shown above for CO sensors, the responses of sensors with WO_3 electrodes decrease as the temperature is increased above 600 °C. The responses of sensors with NiO electrodes, however, remain high up to operating temperatures of 900 °C. The response of NiO electrodes can be increased even further with the addition of ruthenium [268].

Another transition metal oxide used as the electrode material in NO_x sensors is Cr_2O_3 [275–281]. Other binary oxides used for NO_x sensors include ZnO [264], V_2O_5 [282], and SnO_2 [278, 279], which has also been used for CO sensors [283]. Indium oxide (In_2O_3) has been used in a CO sensor and tested in engine exhaust gas [284]. Oxides used as electrodes in sensors for hydrocarbons include Ga_2O_3 [285],

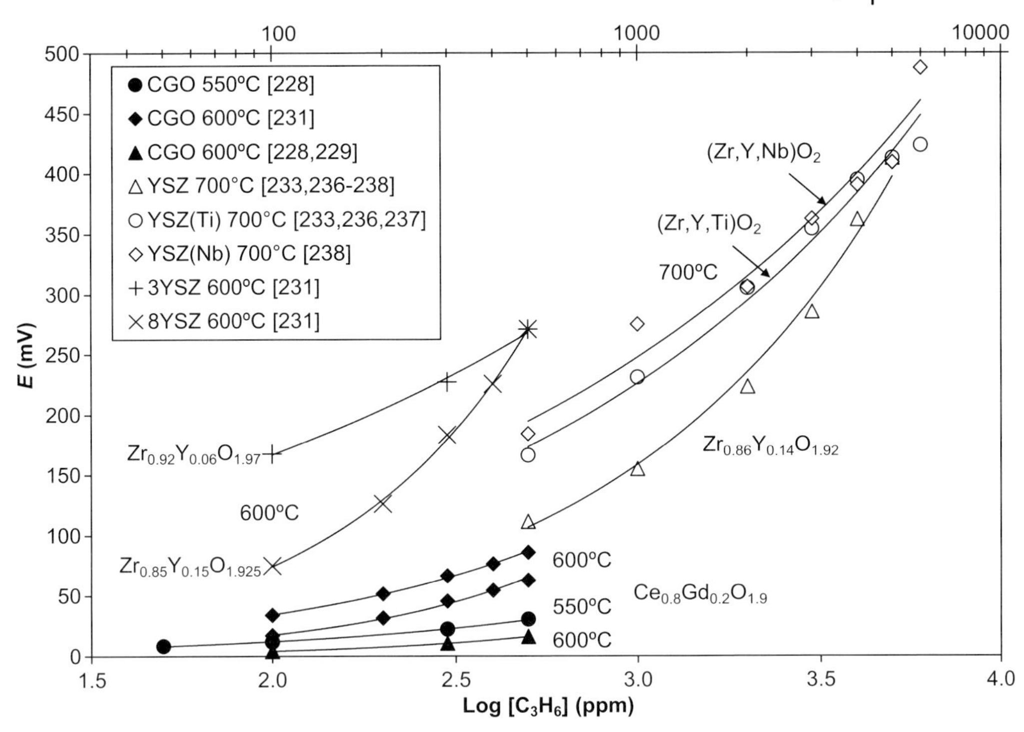

Figure 13.20 Outputs of mixed potential C_3H_6 sensors with gold electrodes and various ceria- and zirconia-based electrolytes [228, 229, 231, 233, 236–238].

Nb_2O_5 [286–289], Ta_2O_5 [289], and Pr_6O_{11} [290]. Although a number of simple oxides can be used as sensor electrodes, the use of multiple oxides often improves response.

Among mixed oxides employed in mixed potential sensors is ITO, this having been used for both NO_x [291] and CO [292–294] sensors. A further example of a doped oxide being used as an electrode is TiO_2, which has been doped with tantalum for hydrocarbon sensors [295] or vanadium for SO_2 sensors [296].

Two-phase mixtures of oxides have also been used in mixed potential sensors. Such examples include Cr_2O_3 + NiO [297] for NO_x sensors, CuO + ZnO [298, 299] or SnO_2 + CdO [300] for CO sensors, and In_2O_3 + MnO_2 [301, 302] for hydrocarbon sensors. Some examples of the outputs of NO_x and CO sensors with two-phase oxide mixtures as electrodes are shown in Figure 13.24 [270, 297, 298, 300].

The *perovskite structure* contains two differently sized cations, which not only expands dopant flexibility but also provides a means for tailoring the properties for sensor applications. For example, (La,Sr)MnO_3 (LSM) [229, 303–305] and $LaCoO_3$ (LCo) [306] electrodes have been used for CO sensors, while LSM [307] and (La,Sr)CrO_3 (LSCr) [308] have each been used for propylene sensors. Some example results of these sensors are shown in Figure 13.25 [229, 303–308]. (La,Sr)CrO_3 [264, 309–311] has also been used as the electrode for NO_x sensors. Several other chromites,

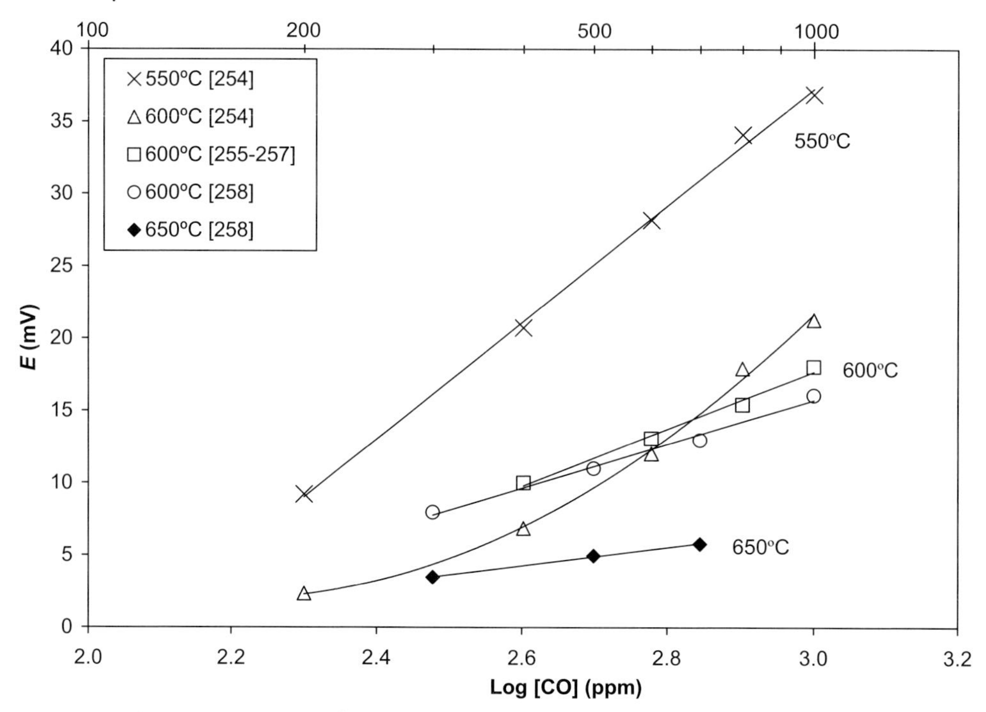

Figure 13.21 Outputs of CO sensors with YSZ electrolytes and WO_3 electrodes [254–258].

including (La,Ca)CrO_3 [275], (La,Ba)CrO_3 [264], La(Cr,Mg)O_3 [312], and (Sm,Sr)CrO_3 [313], have shown sensitivity towards NO_x.

The outputs of some mixed potential-type chemical sensors correlate with the type of electronic defect (i.e., n-type versus p-type), so the response has been attributed to the semiconducting behavior of the electrode material [314]. $LaFeO_3$, which has been used as a semiconductor-type gas sensor [315, 316], has also been used as an electrode with YSZ [255, 263, 317] or NASICON [317, 318] electrolytes for potentiometric NO_x sensors. Strontium (i.e., (La,Sr)FeO_3 [255, 256, 284]) or strontium and cobalt (i.e., (La,Sr)(Co,Fe)O_3 [275, 280, 309]) have been added to $LaFeO_3$ to improve electrode performance. (La,Ca)MnO_3 doped with either cobalt or nickel on the manganese site has been used as the electrode for NO_x sensors [319]. The outputs of some NO_x sensors with perovskite electrodes are shown in Figure 13.26 [255, 256, 264, 275, 309, 312].

Spinel oxides have also been used as electrodes for NO_x sensors. As with perovskites, several chromites forming the spinel structure, including $NiCr_2O_4$ [29, 272, 275, 309, 320], $CuCr_2O_4$ [321–323], $CoCr_2O_4$ [275], $ZnCr_2O_4$ [313, 324–326], and $CdCr_2O_4$ [320, 327], are sensitive to NO_x gas. Other spinel electrode materials include $ZnFe_2O_4$ [327–331] and $CdMn_2O_4$ [325]. The outputs of some NO_x sensors with spinel electrodes are shown in Figure 13.27 [29, 275, 309, 313, 320, 324, 325, 327–330].

Recently, La_2CuO_4 [332–334] and $(La,Sr)_2CuO_4$ [335] have shown promise as electrodes for NO_x sensors. The response of La_2CuO_4 is attributed to the effect of the adsorbed intermediate N–O ionic species on the Fermi level in the electrode. This

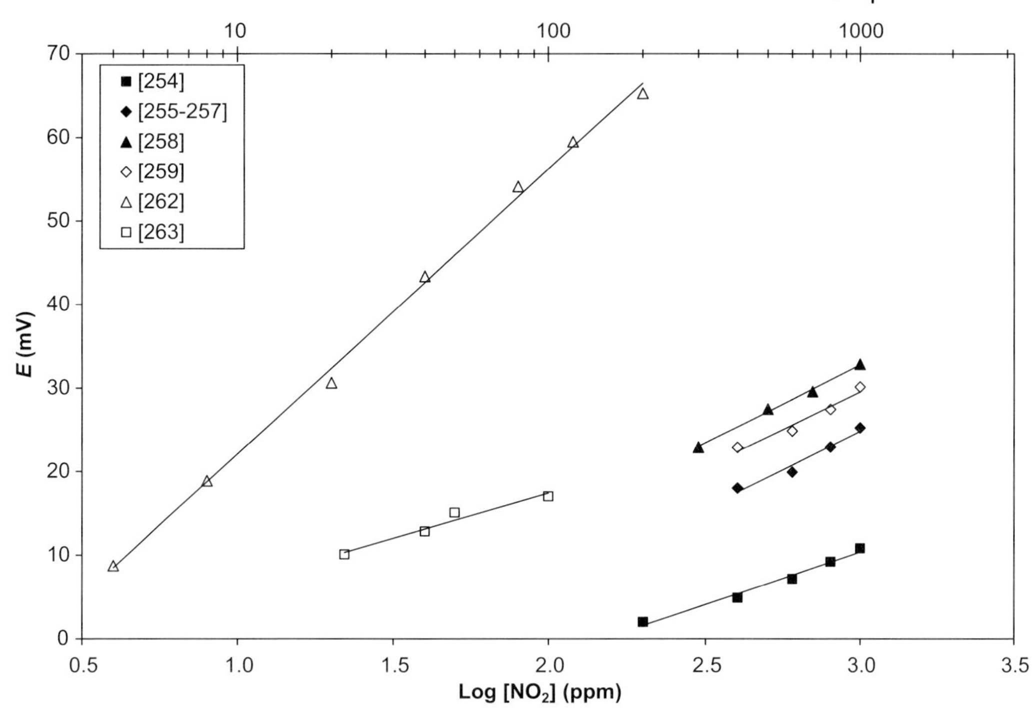

Figure 13.22 Outputs of NO_2 sensors with YSZ electrolytes and WO_3 electrodes at 600 °C [254–259, 262, 263].

mechanism has been applied to other electrodes [336], and used to explain the effect of the semiconducting nature of the electrode material on the sensor response.

Nonequilibrium SO_2 sensors have been reported using a sodium ion-conducting electrolyte ($Na_5DySi_4O_{12}$) with sulfide electrodes. Among a large number of binary and mixed sulfides, CdS provides the best response [337, 338].

A major challenge in the development of nonequilibrium sensors is that of *selectivity*. Many of the electrodes used in these sensors respond to multiple gases, and consequently interpretation of the sensor signal in mixed gases may be difficult. Evaluations of sensors in mixed gases have shown that, while the sensors respond to the desired gas species, the response is often different from that in laboratory testing. For example, a shift in the response of NO_x sensor in engine exhaust, as compared to that in a laboratory testing, has been attributed to interference from CO or hydrocarbon gases [339]. Similarly, the calibration curve for a NO_x sensor in diesel exhaust is different from that determined in the laboratory [168]. In addition, the presence of a relatively small amount of oxygen (3%) can affect the response of a NO_x sensor [340]. The effect of oxygen gas content on the outputs of NO_x [341] and CO [23, 303, 341] sensors is shown in Figure 13.28. This effect can be corrected for if the oxygen content is known, but the presence of oxygen decreases the magnitude of the sensor response, which may reduce the signal-to-noise ratio of the sensor output.

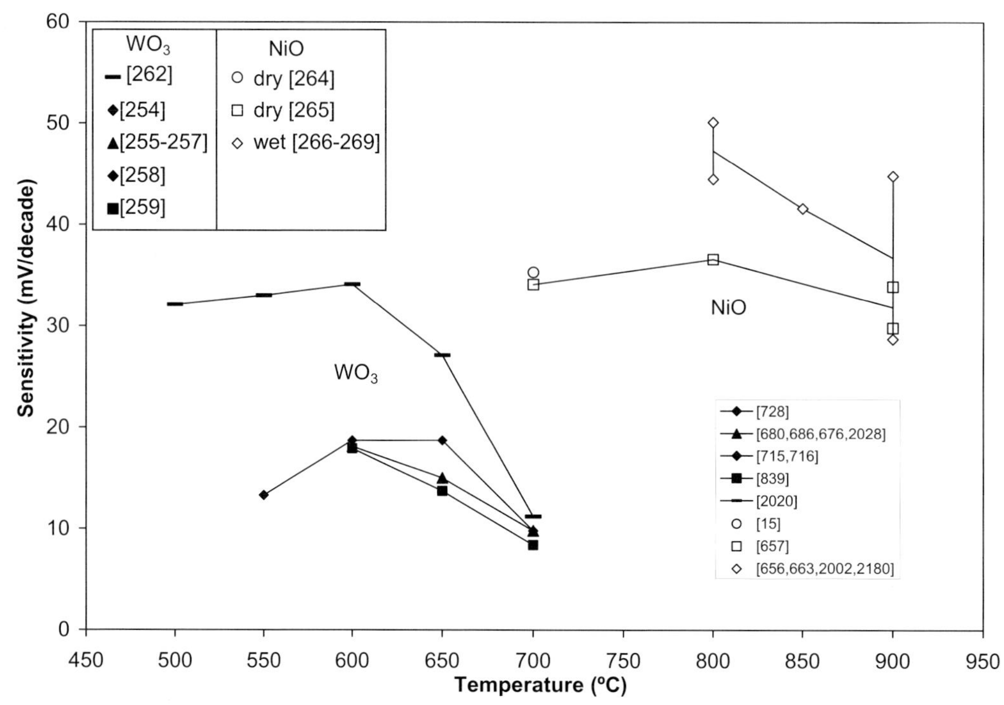

Figure 13.23 Sensitivities of NO_2 sensors with YSZ electrolytes and WO_3 [254–259, 262] or NiO [264–269].

One approach to improving selectivity is to use a *porous zeolite* on the electrode [168, 281, 341, 342]. In addition to using the controlled porosity to preferentially control the relative diffusion rates of different species, the zeolite can be loaded with catalysts to promote the desired chemical reaction.

Selectivity can also be controlled by the strategic selection of the electrode material. Selectivity among CO and different hydrocarbons represents a challenge because electrodes that respond to one reducing gas will typically respond to others. However, the relative magnitudes of the responses often differ, and these differences can be used to improve selectivity [343–345]. For example, $LaCoO_3$ [306], $LaMnO_3$ [307], and In_2O_3-MnO_2 [301, 302] respond more strongly to propylene than to CO, while the reverse is the case for Nb_2O_5 [343]. Similarly, $LaCoO_3$ [306], In_2O_3-MnO_2 [301, 302], and Nb_2O_5 [343] respond more strongly to propylene than to propane, while the reverse is the case for $LaCrO_3$ [345]. As each electrode responds to multiple gases, a single electrode would not provide a clear selective response. However, as the relative responses to different gases differ among electrodes, the patterns of responses in a multiple-electrode sensor array can be used to recognize individual gas concentrations within a mixed gas environment. The relative selectivities of the different electrodes can change with temperature, which provides an additional parameter to improve selectivity [228, 312].

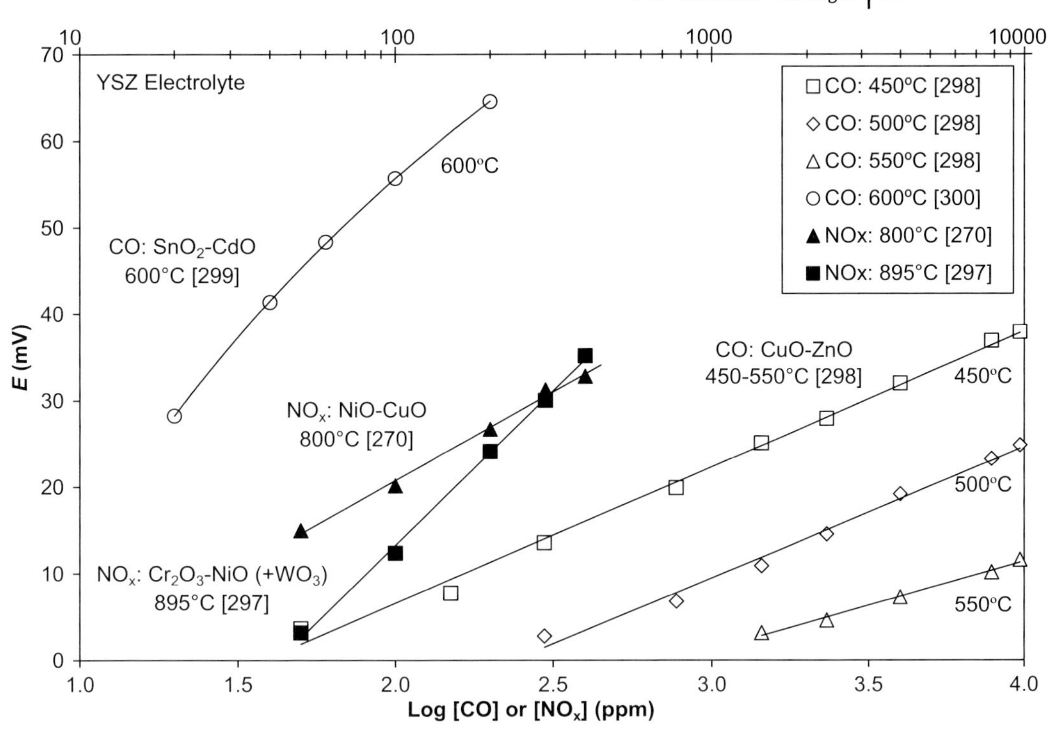

Figure 13.24 Outputs of CO [298, 300] and NO_x [270, 297] sensors with YSZ electrolytes and two-phase oxide mixtures electrodes.

13.3.2.4 Electrodes for Current-Based Sensors

The properties of electrode materials are also important in *amperometric sensors*. If the sensor is designed to measure the neutral species that correspond to mobile ions in the electrolyte (i.e., an oxygen ion conductor for oxygen [110, 346–348] or a proton conductor for hydrogen [349]), then the electrode selection is relatively straightforward. However, amperometric sensors are also used to measure species other than those that are mobile in the electrolyte. For example, oxygen ion-conducting electrolytes, including YSZ [350–356] and LSGM [357–361], have been used in amperometric sensors for NO detection, in which case the source of oxygen for the electrolyte is NO rather than O_2. For the amperometric measurement of a reducing species, on the other hand, oxygen can be pumped through the electrolyte to oxidize the reducing gas; consequently, YSZ has been used in amperometric sensors for measuring CO [362] and hydrocarbon gases [363, 364]. Although oxygen ion conductors are most commonly used for amperometric sensors, sodium ion-conducting electrolytes have also been used [365]. The outputs of some example amperometric gas sensors are shown in Figure 13.29 [350, 351, 354, 358, 359, 362, 363, 365].

Selectivity in amperometric sensors can be achieved by catalyzing specific reactions. The addition of oxide electrodes, such as $CdCr_2O_4$ [354] or $MgAl_2O_4$ [355, 356], can be used to catalyze the NO decomposition reaction. The decomposition can

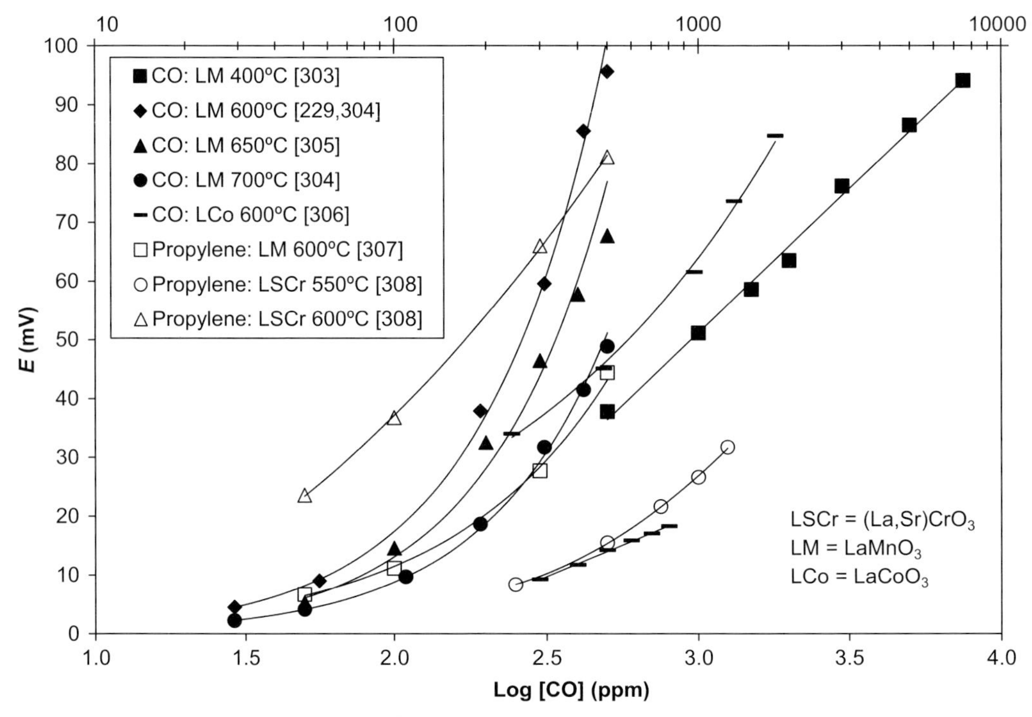

Figure 13.25 Outputs of CO [229, 303–306] and propylene [307, 308] sensors with YSZ electrolytes and perovskite electrodes.

also be promoted by doping the electrolyte. In particular, the addition of nickel to (La,Sr)(Ga,Mg)O_3 has been shown to improve selectivity of the sensor to NO [358–361]. An alternative approach is to add a catalyst that eliminates the interfering species. For example, a zeolite layer has been used in a NO_x sensor to promote the oxidation of CO to CO_2, which does not interfere with the sensor signal.

Oxygen can interfere with amperometric sensors. Figure 13.30 shows that the sensor outputs for NO sensors increase with increasing oxygen content, since the reduction of O_2 provides oxygen ions that contribute to the cell current [350, 363]. On the other hand, the sensor output for an amperometric CO sensor decreases slightly with increasing oxygen content, because a portion of the CO may be chemically oxidized to CO_2. This means that less oxygen – and hence less ionic current – will be needed to oxidize the remaining CO molecules [366].

Selectivity can be improved through the use of multiple electrodes in series, where one species is removed by the first electrode, so that the current at the second electrode is due only to the remaining species [236, 350, 366–368]. For example, in a gas containing NO and O_2, with the addition of an oxygen-permeable membrane that inhibits NO decomposition (e.g., $Ce_{0.8}Gd_{0.2}O_{1.9}$ + $Gd_{0.7}Ca_{0.3}CoO_3$ [369–371]), the first electrode will remove oxygen and the current at this electrode will be related to

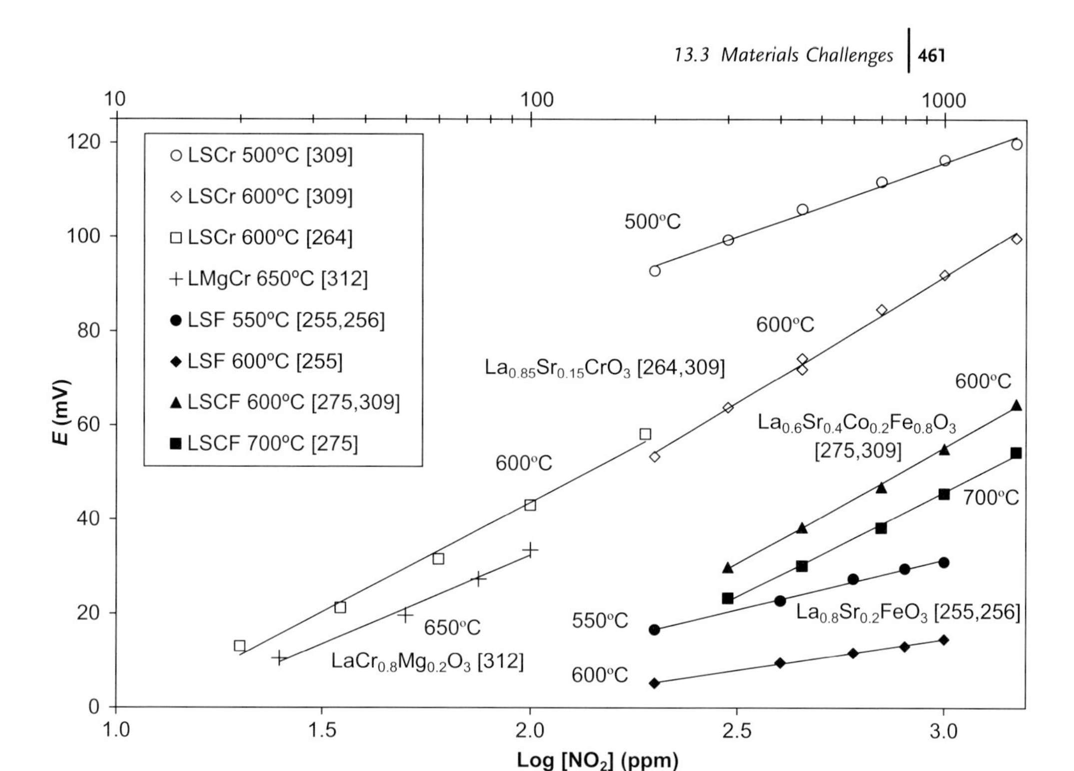

Figure 13.26 Outputs of NO_x sensors with perovskite electrodes [255, 256, 264, 275, 309, 312].

the amount of oxygen. This eliminates the oxygen interference and allows the second electrode to be used for measuring NO concentrations. As discussed earlier, the reactions that occur at the individual electrodes can also be controlled by applying a voltage [263, 363, 366]. This is essentially the same principle as that used in cyclic voltammetry, where the reaction is identified by the voltage, and the amount of reaction by the current. Solid electrolytes, including YSZ [372] and LISICON [373], have been used in this mode.

The catalytic activity of electrode materials is also important in impedancemetric sensors, which use many of the same electrode materials as mixed potential sensors. For example, $LaFeO_3$ [374], Cr_2O_3 [375], $ZnCr_2O_4$ [245, 320, 328, 376, 377] and $NiCr_2O_4$ [272, 378] have been used in both potentiometric and impedancemetric NO_x sensors, although the performance of each electrode material differs between the two types of sensor. For example, $ZnCr_2O_4$ does not perform particularly well in the potentiometric mode, but is one of the best electrodes in the impedancemetric mode. Some example outputs of impedancemetric NO_x sensors are shown in Figure 13.31 [320, 328, 376, 378]. Similarly, many of the electrode materials used for CO and hydrocarbon impedancemetric sensors are the same as are used for potentiometric sensors. Electrode materials used include Ga_2O_3 [379], In_2O_3 [380], and ZnO [381], and some example sensor outputs are shown in Figure 13.32 [379, 381].

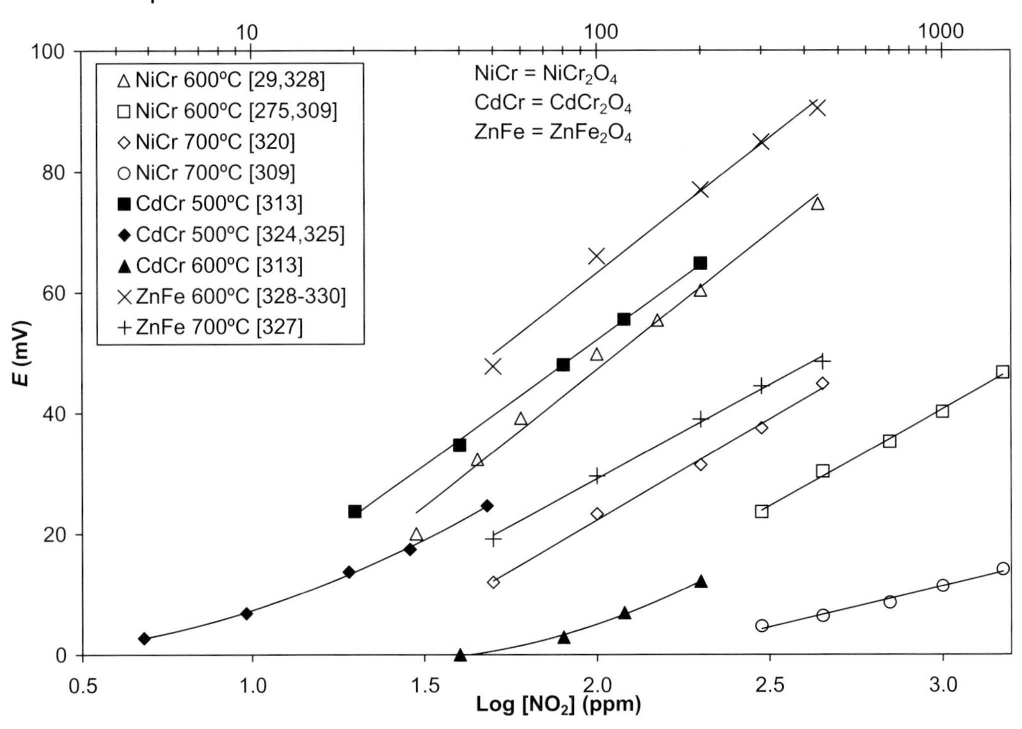

Figure 13.27 Outputs of NO_x sensors with spinel electrodes [29, 275, 309, 313, 320, 324, 325, 327–330].

13.4 Applications

Solid electrolyte sensors are particularly well suited to high-temperature aggressive environments, and have been used to meet chemical sensing demands in both gaseous and molten metal environments.

13.4.1 Gaseous Medium

Information on the gas composition in combustion processes is important for improving efficiency and reducing emissions [382]. The closer the measurement is made to the combustion, the more accurately the result reflects the combustion conditions. Thus, the good high-temperature performance of solid electrolyte-based chemical sensors is valuable when developing sensors for the *in situ* monitoring of combustion processes.

Combustion is important in many industrial processes. While sensors used in automotive combustion processes [383–386] have in the past received the most attention, combustion gas sensors are also important for improving efficiency and

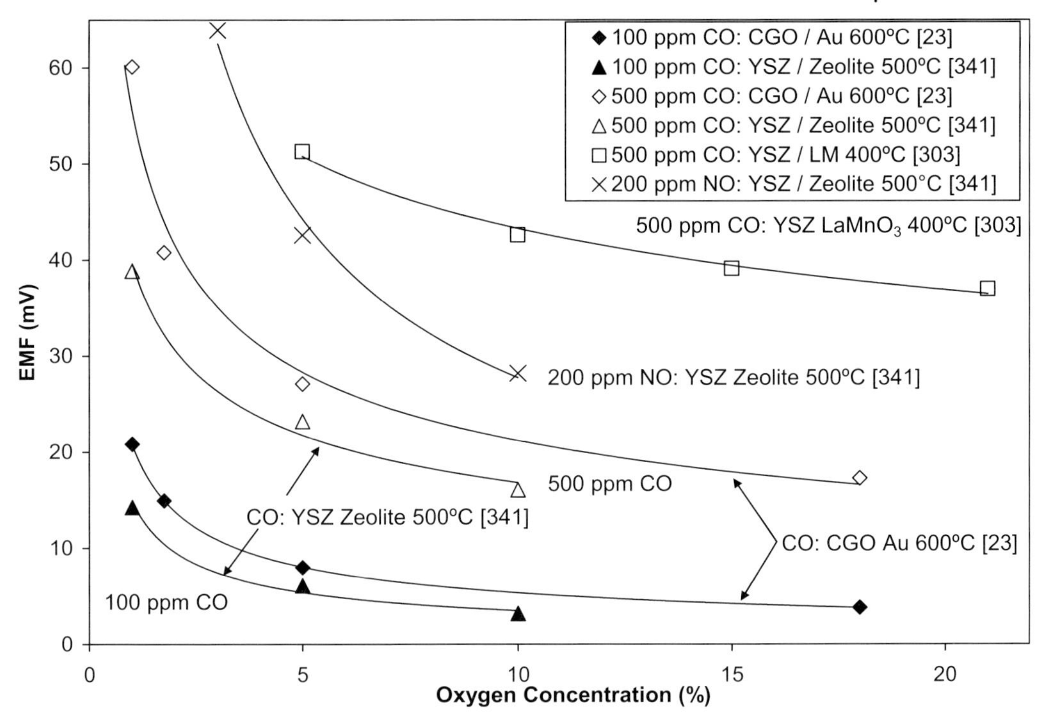

Figure 13.28 Effect of oxygen on output or CO and NO sensors [23, 303, 341].

reducing emissions in other processes, such as glass processing [387, 388], liquid petroleum gas (LPG) boilers [389], and even wood burning [390]. The sensors developed for these applications can also be used in other areas, where information on gas compositions is important, such as controlling the atmosphere during sintering [391] or characterizing the gas flow and distribution in a fluidized-bed reactor [392].

Oxygen sensors are the most widely used solid electrolyte-based sensors [393–395], because the control of oxygen concentrations is critical to controlling the combustion process. For automotive applications, exhaust gas oxygen (EGO) sensors provide critical information for controlling the air-to-fuel ratio for internal combustion engines [396, 397]. The use of an optimal air-to-fuel ratio leads to increased efficiency and reduced emissions.

Electrochemical sensors are also used to monitor the emissions from combustion processes. Because the oxidizer for most combustion processes is air, which contains 78% nitrogen, NO_x gases are common components of the exhaust and their concentrations must be minimized [398, 399]. In addition, incomplete combustion can result in CO or hydrocarbon gaseous compounds in the exhaust gas, which both represent unconverted chemical energy and are hazardous to the environment. The use of sulfur-containing fuels can lead to the formation of SO_x gases [400], which have a detrimental impact on the environment, such as promoting acid rain.

The need for chemical sensors is continually increasing as new energy conversion technologies are being developed and implemented. Today, the measurement of CO_2

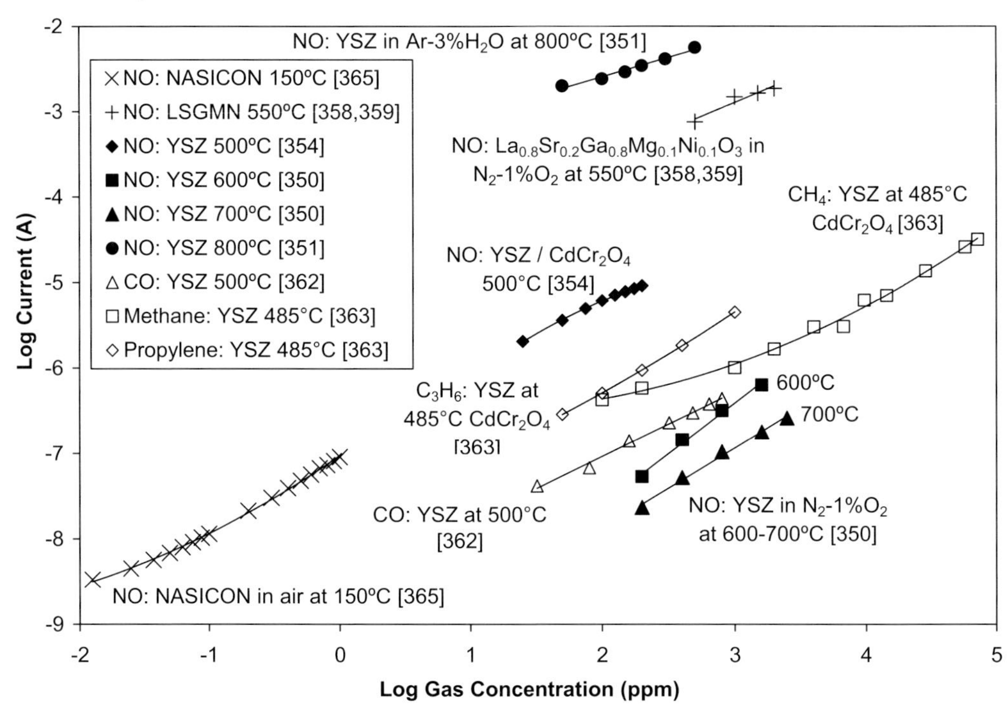

Figure 13.29 Outputs of amperometric gas sensors [350, 351, 354, 358, 359, 362, 363, 365].

is becoming increasingly important for determining the "carbon footprints" of industrial processes, and also for monitoring/controlling carbon sequestration processes. The increased use of hydrogen as a fuel will create additional needs for sensors to measure both hydrogen and water vapor. In addition to the increased demand for new types of sensor, improved performances will be required as emission requirements become increasingly stringent, requiring in turn sensors with lower detection limits.

13.4.2
Molten Metals

Solid electrolyte-based sensors have been used in the characterization of molten metals [401–403]. The most widely used electrochemical sensor for molten metals is the oxygen sensor for *molten steel* [404–406]. In addition to being used to monitor oxygen content, the zirconia electrolyte can be used to electrochemically remove oxygen from the steel during processing, and thus improve process control [407–410]. Two challenges in this application are the very low oxygen partial pressure in molten steel [411, 412] and the rapid heating during insertion into the molten steel, which can cause thermal shock [413]. Satisfying both these criteria would require designs consisting of multiple electrolytes with complementary properties [414–416]. The high operating temperature can also lead to the degradation of seals; hence, nonisothermal designs, in

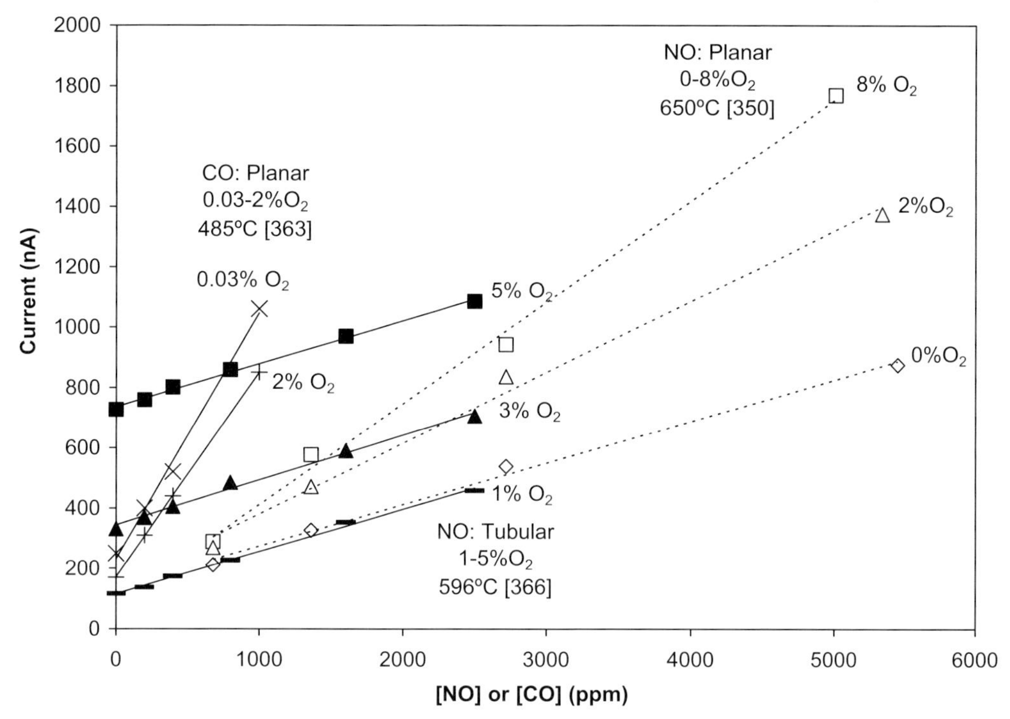

Figure 13.30 Effect of oxygen content on output of amperometric CO [363] and NO [350, 366] sensors.

which the reference electrode temperature is reduced, have been used [417, 418]. In addition to oxygen sensors, sensors for measuring the amounts of sulfur [419–421] and nitrogen [421, 422] dissolved in steel have also been developed.

In *molten aluminum*, the most important dissolved gas is hydrogen, which may be introduced due to reaction with water vapor in the atmosphere, and lead to metal porosity during the casting process. This has led to the development of hydrogen sensors for molten aluminum. These sensors are based on proton-conducting oxides [423–427] although, as discussed above, at low oxygen partial pressures (as occur in molten aluminum) oxygen ion conduction can occur in these oxides [428, 429]. This may lead to the sensor output being affected by the water vapor partial pressure.

Hydrogen and oxygen contents are also important in the processing of *copper* and *copper alloys*; hence, the hydrogen sensors developed for molten aluminum, and the oxygen sensors for molten steel, have both been adapted for use in molten copper [430–432]. As with the sensors for steel, nonisothermal oxygen sensors with a reference electrode at a lower temperature have been used in molten copper [433]. Oxygen sensors have also been used for a variety of other metals, including Ni [434], Ag-Pb [435], Pb-Bi [436], and also molten glasses [437].

Solid electrolyte-based sensors can also be used to measure the concentration of metallic components in *molten alloys*. For example, as sodium and strontium are used to control the cast microstructure of aluminum alloys, sensors have been developed

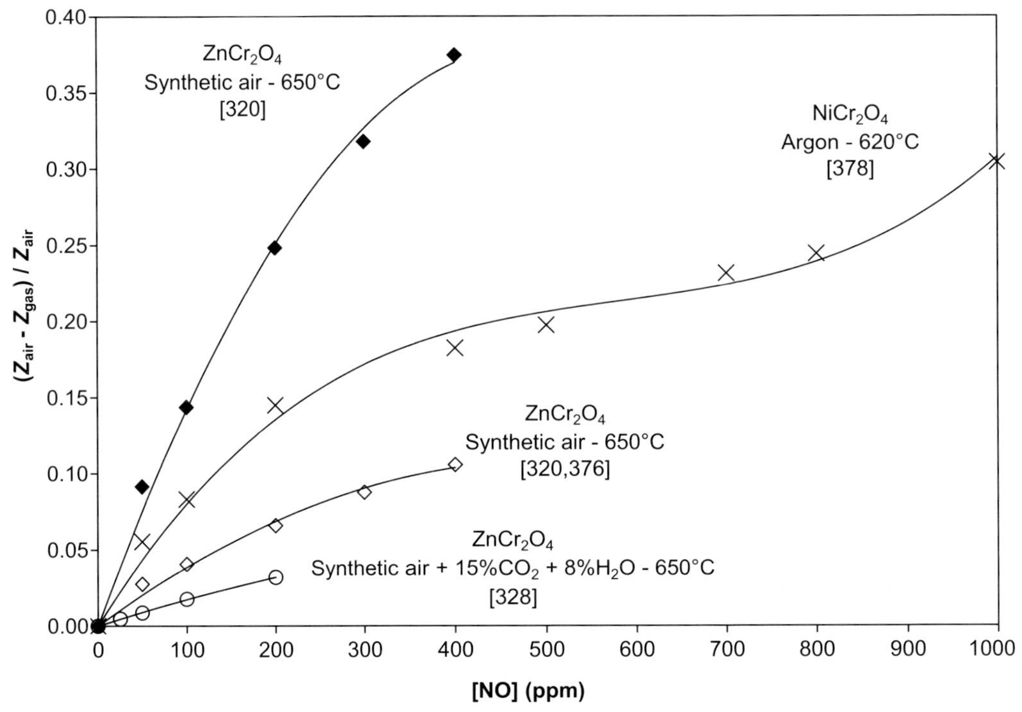

Figure 13.31 Outputs of impedancemetric NO_x sensors [320, 328, 376, 378].

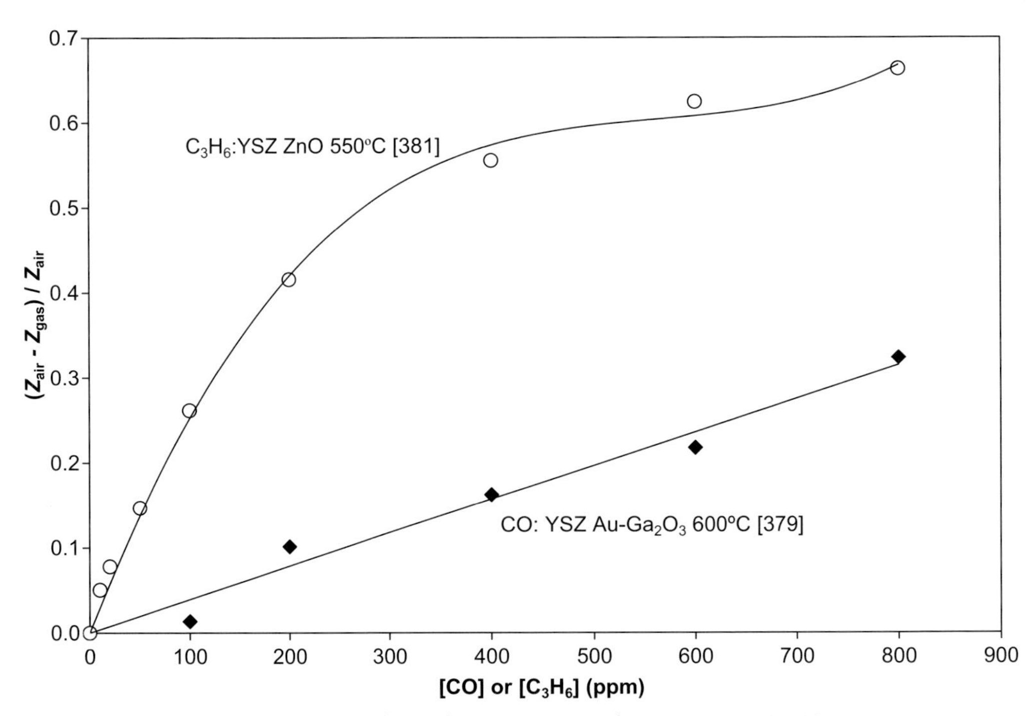

Figure 13.32 Outputs of impedancemetric CO and C_3H_6 sensors [379, 381].

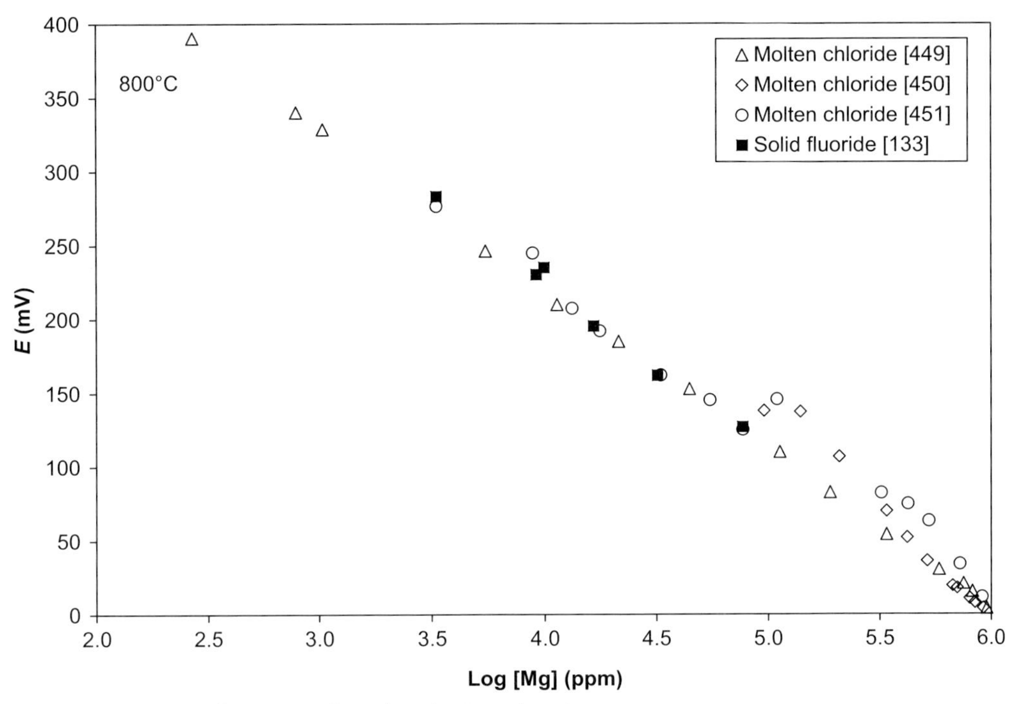

Figure 13.33 Outputs of Mg sensors for molten aluminum, based on solid fluoride [133] and molten chloride [448–450] electrolytes.

to measure the levels of sodium [222, 438–440] and strontium [441–444] in molten aluminum. Another example is lithium, which is used to increase the specific strength of alloys; this has led to the development of sensors to measure the lithium content of molten aluminum [445–447].

One of the main attractions of solid electrolyte-based sensors is that the electrolyte can be used as a structural component, and this in turn provides a clear advantage over liquid electrolytes. As an example, molten chloride-based sensors have been used to measure magnesium concentrations in molten aluminum during its recycling. In the process the molten chlorides must be contained within a porous crucible, but this complicates the sensor design. This has led to solid electrolyte-based sensors, which have a simple design, being used to monitor magnesium levels in molten aluminum [133, 444]. Figure 13.33 shows the output of one such sensor to be in excellent agreement with the outputs of molten chloride-based sensors [133, 448–450].

13.5
Summary and Conclusions

Solid electrolytes are well suited to high-temperature aggressive environments, and can be used in several different ways for measuring a wide variety of chemical species.

Potentiometric sensors generate a voltage, which can be related to an equilibrium electrochemical potential, or to a potential established by a steady-state – but nonequilibrium – electrode process. Solid electrolytes can also be used in dynamic modes, where a current is passed through the electrolyte and the magnitude of the current or impedance of the electrode reaction is related to the concentration of a target species in the surrounding atmosphere. The application of these techniques to the measurement of different species depends critically on the available electrolyte and electrode materials. While the details of many successful sensors have been reported, the increasing demand for more accurate process information will continue to create challenges and opportunities for the future development of chemical sensors.

References

1 Kiukkola, K. and Wagner, C. (1957) Measurements of galvanic cells involving solid electrolytes. *J. Electrochem. Soc.*, **104** (6), 379–87.

2 Radhakrishnan, R., Virkar, A.V., Singhal, S.C., Dunham, G.C. and Marina, O.A. (2005) Design, fabrication and characterization of a miniaturized series-connected potentiometric oxygen sensor. *Sens. Actuators B*, **105** (2), 312–21.

3 Dubbe, A. (2003) Fundamentals of solid state ionic micro gas sensors. *Sens. Actuators B*, **88**, 138–48.

4 Azad, A.-M., Akbar, S., Mhaisalkar, S.G., Birkefeld, L.D. and Goto, K.S. (1992) Solid-state gas sensors: A review. *J. Electrochem. Soc.*, **139** (12), 3690–704.

5 Yamazoe, N. and Miura, N. (1994) Environmental gas sensing. *Sens. Actuators B*, **20**, 95–102.

6 Mukundan, R. and Garzon, F. (2004) Electrochemical sensors for energy and transportation. *Electrochem. Soc. Interface*, **13** (2), 30–5.

7 Moos, R. (2005) A brief overview on automotive exhaust gas sensors based on electroceramics. *Int. J. Appl. Ceram. Tech.*, **2** (5), 401–13.

8 Akbar, S., Dutta, P. and Lee, C. (2006) High-temperature ceramic gas sensors: A review. *Int. J. Appl. Ceram. Technol.*, **3** (4), 302–11.

9 Ivers-Tiffee, E., Hardtl, K.H., Menesklou, W. and Riegel, J. (2001) Principles of solid state oxygen sensors for lean combustion gas control. *Electrochim. Acta*, **47** (5), 807–14.

10 Lundström, I. (1996) Approaches and mechanisms to solid state based sensing. *Sens. Actuators B*, **35** (1–3), 11–9.

11 Lee, D.-D. and Lee, D.-S. (2001) Environmental gas sensors. *IEEE Sensors J.*, **1** (3), 214–24.

12 Takeuchi, T. (1988) Oxygen sensors. *Sens. Actuators B*, **14** (2), 109–24.

13 Weppner, W. (1987) Solid-state electrochemical gas sensors. *Sens. Actuators* **12**, 107–19.

14 Weppner, W. (1992) Advanced principles of sensors based on solid state ionics. *Mater. Sci. Eng.*, **B15**, 48–55.

15 Park, C.O., Akbar, S.A. and Weppner, W. (2003) Ceramic electrolytes and electrochemical sensors. *J. Mater. Sci.*, **38**, 4639–60.

16 Zhuiykov, S. and Miura, N. (2005) Solid-state electrochemical gas sensors for emission control, in *Materials for Energy Conversion Devices* (eds C.C. Sorrell, S. Sugihara and J. Nowotny), Woodhead Publ. Ltd, Cambridge, UK, pp. 303–35.

17 Park, C.O. and Akbar, S.A. (2003) Ceramics for chemical sensors. *J. Mater. Sci.*, **38**, 4611–37.

18 Yamazoe, N. and Miura, N. (1998) Potentiometric gas sensors for oxidic gases. *J. Electroceram.*, **2** (4), 243–55.

19 Jacob, K.T., Swaminathan, K. and Sreedharan, O.M. (1989) Stability constraints in the design of galvanic cells using composite electrolytes and auxiliary electrodes. *Solid State. Ionics*, **34**, 167–73.

20 Miura, N., Lu, G. and Yamazoe, N. (2000) Progress in mixed-potential type devices based on solid electrolyte for sensing redox gases. *Solid State Ionics*, **136–7**, 533–42.

21 Kotzeva, V.P. and Kumar, R.V. (1999) The response of yttria stabilized zirconia oxygen sensors to carbon monoxide gas. *Ionics*, **5** (3–4), 220–6.

22 Fergus, J.W. (2007) Solid electrolyte based sensors for the measurement of CO and hydrocarbon gases. *Sens. Actuators B*, **122**, 683–93.

23 Garzon, R., Mukundan, R. and Brosha, E.L. (2001) Modeling the response of mixed potential electrochemical sensors. Proceedings of the Electrochemical Society, 2000–32; Solid-State Ionic Devices II Ceramic Sensors, The Electrochemical Society, Pennington, New Jersey, pp. 305–13.

24 Varamban, S.V., Ganesan, R. and Periaswami, G. (2005) Simultaneous measurement of emf and short circuit current for a potentiometric sensor using perturbation technique. *Sens. Actuators B*, **104** (1), 94–102.

25 Gibson, R.W., Kumar, R.V. and Fray, D.J. (1999) Novel sensors for monitoring high oxygen concentrations. *Solid State Ionics*, **121** (1–4), 43–50.

26 Hills, M.P., Schwandt, C. and Kumar, R.V. (2006) An investigation of current reversal mode for gas sensing with a solid electrolyte. *Int. J. Appl. Ceram. Technol.*, **3** (3), 200–9.

27 Zhuiykov, S. (2006) "*In situ*" diagnostics of solid electrolyte sensors measuring oxygen activity in melts by a developed impedance method. *Meas. Sci. Technol.*, **17**, 1570–8.

28 Zhuiykov, S. (2005) Sensors measuring oxygen activity in melts: Development of impedance method for "in-situ" diagnostics and control electrolyte/liquid-metal electrode interface. *Ionics*, **11**, 352–61.

29 Zhuiykov, S., Nakano, T., Kunimoto, A., Yamazoe, N. and Miura, N. (2001) Potentiometric NO_x sensor based on stabilized zirconia and $NiCr_2O_4$ sensing electrode operating at high temperatures. *Electrochem. Commun.*, **3** (2), 97–101.

30 West, D.L., Montgomery, F.C. and Armstrong, T.R. (2006) 'Total NO_x' sensing elements with compositionally identical oxide electrodes. *J. Electrochem. Soc.*, **153** (2), H23–8.

31 West, D.L., Montgomery, F.C. and Armstrong, T.R. (2005) Electrically biased NO_x sensing elements with coplanar electrodes. *J. Electrochem. Soc.*, **152** (6), H74–9.

32 Woo, L.Y., Martin, L.P., Glass, R.S. and Gorte, R.J. (2007) Impedance characterization of a model Au/yttria-stabilized zirconia/Au electrochemical cell in varying oxygen and NO_x concentration. *J. Electrochem. Soc.*, **154** (4), J129–35.

33 Reinhardt, G., Mayer, R. and Rösch, M. (2002) Sensing small molecules with amperometric sensors. *Solid State Ionics*, **150** (1–2), 79–92.

34 Jasinski, G., Jasinski, P., Nowakowski, A. and Chachulski, B. (2006) Properties of a lithium solid electrolyte gas sensor based on reaction kinetics. *Meas. Sci. Technol.*, **17**, 17–21.

35 Ueda, T., Plashnitsa, V.V., Elumalai, P. and Miura, N. (2007) Novel measuring method for detection of propene using zirconia-based amperometric sensor with oxide-based sensing electrode. *Sens. Mater.*, **19** (6), 333–45.

36 Etsell, T.H. and Flengas, S.N. (1970) The electrical properties of solid oxide electrolyte. *Chem. Rev.*, **70** (3), 339–76.

37 Mogensen, M., Lybye, D., Bonanos, N., Hendriksen, P.V. and Poulsen, F.W.

(2004) Factors controlling the oxide ion conductivity of fluorite and perovskite structured oxides. *Solid State Ionics*, **174** (1–4), 279–86.

38 Steele, B.C.H. (1992) Oxygen ion conductors and their technological applications. *Mater. Sci. Eng.*, **B13**, 79–87.

39 Fergus, J.W. (2003) Doping and defect association in oxides for use in oxygen sensors. *J. Mater. Sci.*, **38** (21), 4259–70.

40 Takeda, N., Itagaki, Y., Aono, H. and Sadaoka, Y. (2006) Preparation and characterization of $Ln_{9.33+x/3}Si_{6-x}Al_xO_{26}$ (Ln = La, Nd and Sm) with apatite-type structure and its application to a potentiometric O_2 gas sensor. *Sens. Actuators B*, **115** (1), 455–9.

41 Cho, H.S., Sakai, G., Shimanoe, K. and Yamazoe, N. (2005) Preparation of $BiMeVO_x$ (Me = Cu, Ti, Zr, Nb, Ta) compounds as solid electrolyte and behavior of their oxygen concentration cells. *Sens. Actuators B*, **109** (2), 307–14.

42 Cho, H., Sakai, G., Shimanoe, K. and Yamazoe, N. (2005) Behavior of oxygen concentration cells using $BiCuVO_x$ oxide-ion conductor. *Sens. Actuators B*, **108** (1–2), 335–40.

43 Pasciak, G., Prociow, K., Mielcarek, W., Bornicka, B. and Mazurek, B. (2001) Solid electrolytes for gas sensors and fuel cells application. *J. Eur. Ceram. Soc.*, **21**, 1867–70.

44 Ramanarayanan, T.A. and Worrell, W.L. (1974) Limitations in the use of solid state electrochemical cells for high-temperature equilibrium measurements. *Can. Metall. Q.*, **13** (2), 325–9.

45 Sridhar, K.R. and Blanchard, J.A. (1999) Electronic conduction in low oxygen partial pressure measurements using an amperometric zirconia oxygen sensor. *Sens. Actuators B*, **59**, 60–7.

46 Guo, X., Sun, Y.-Q. and Cui, K. (1996) Darkening of zirconia: A problem arising from oxygen sensors in practice. *Sens. Actuators B*, **31**, 139–45.

47 Patterson, J.W. (1971) Conduction domains for solid electrolytes. *J. Electrochem. Soc.*, **118** (7), 1033–9.

48 Seetharaman, S. and Sichen, D. (1996) Development and application of electrochemical sensors for molten metals processing, in *Emerging Separation Technologies for Metals II* (ed. R.G. Bautista), The Minerals, Metals and Materials Society, Warrendale, PA, pp. 317–40.

49 Chueh, W.C., Lai, W. and Haile, S.M. (2008) Electrochemical behavior of ceria with selected metal electrodes. *Solid State Ionics*, **179**, 1036–41.

50 Réau, J.-M. and Portier, J. (1978) Fluorine ion conductors, in *Solid Electrolytes: General Principles, Characterization, Materials, Applications* (eds P. Hagenmuller and W. Van Gool), Academic Press, New York, pp. 313–33.

51 Fergus, J.W. (1997) The application of solid fluoride electrolytes in chemical sensors. *Sens. Actuators B*, **42** (2), 119–30.

52 Lee, J.-H. and Lee, D.-D. (1998) A SrF_2-based oxygen sensor operative at low temperature. *Sens. Actuators B*, **46**, 169–73.

53 Eguchi, T. and Kuwano, J. (1995) Towards a room temperature, solid state, oxygen gas sensor. *Mater. Res. Bull.*, **30** (11), 1351–7.

54 Siebert, E., Fouletier, J. and Kleitz, M. (1987) Oxygen sensing with solid electrolyte cells from room temperature up to 250 °C. *J. Electrochem. Soc.*, **134** (6), 1573–8.

55 Farrington, G.C. and Briant, J.L. (1979) Fast ionic transport in solids. *Science*, **204**, 1371–9.

56 Bates, J.B., Wang, J.-C. and Dudney, N.J. (1982) Solid electrolytes – the beta aluminas. *Phys. Today*, **35** (7), 46–53.

57 Louey, R., Langberg, D.E., Swinbourne, D.R. and van Deventer, J.S.J. (2000) *Solid Electrolyte – New Techniques in Metal Refining*, Australasian Institute of Mining Metallurgy, Carlton, Australia, pp. 559–70.

58 Radzilowski, R.H. and Kummer, J.T. (1969) The preparation of nitrosonium β-alumina. *Inorg. Chem.*, **8** (11), 2531–3.

59 Hong, H.Y.-P. (1976) Crystal structures and crystal chemistry in the system $Na_{1+x}Zr_2Si_xP_{3-x}O_{12}$. *Mater. Res. Bull.*, **11** (2), 173–82.

60 Novoselov, A., Zhuravleva, M., Zakalyukin, R., Fomichev, V. and Zimina, G. (2008) Phase stability and ionic conductivity of solid solutions with NASICON structure. *J. Am. Ceram. Soc.*, **91** (4), 1377–9.

61 Kumar, P.P. and Yashonath, S. (2002) Structure, conductivity, and ionic motion in $Na_{1+x}Zr_2Si_xP_{3-x}O_{12}$: A simulation study. *J. Phys. Chem. B*, **106**, 7081–9.

62 Mazza, D. (2001) Modeling ionic conductivity in Nasicon structures. *J. Solid State Chem.*, **156**, 154–60.

63 Fergus, J.W. (2006) Electrolytes for solid oxide fuel cells. *J. Power Sources*, **162** (1–2), 30–40.

64 Bates, J.B., Brown, G.M., Kaneda, T., Brundage, W.E., Wang, J.C. and Engstrom, H. (1979) Properties of single crystal beta″ aluminas, in *Fast Ion Transport in Solids* (eds P. Vashishta, J.N. Mundy and G.K. Shenoy), Elsevier North-Holland, New York, pp. 261–6.

65 Whittingham, M.S. and Huggins, R.A. (1971) Measurement of sodium ion transport in beta alumina using reversible solid electrodes. *J. Chem. Phys.*, **54** (1), 414–16.

66 Bohnke, O., Ronchetti, S. and Mazza, D. (1999) Conductivity measurements on nasicon and nasicon-modified materials. *Solid State Ionics*, **122**, 127–6.

67 Kudo, T. (1997) Survey of types of solid electrolytes, in *The CRC Handbook of Solid State Electrochemistry* (eds P.J. Gellings and H.J.M., Bouwmeester), CRC Press, Boca Raton, FL, pp. 195–221.

68 Ivanov-Schitz, A.K. and Bykov, A.B. (1997) Ionic conductivity of the $NaZr_2(PO_4)_3$ single crystals. *Solid State Ionics*, **100**, 153–5.

69 Maffei, N. and Kuriakose, A.K. (1999) A hydrogen sensor based on a hydrogen ion conducting solid electrolyte. *Sens. Actuators B*, **56**, 243–6.

70 Iwahara, H., Asakura, Y., Katahira, K. and Tanaka, M. (2003) Prospect of hydrogen technology using proton-conducting ceramics. *Solid State Ionics*, **168** (3–4), 299–310.

71 Bonanos, N. (2001) Oxide-based protonic conductors: Point defects and transport properties. *Solid State Ionics*, **145**, 265–74.

72 Poulsen, F.W. (1989) Proton conduction in solids, in *High Conductivity Solid Ionic Conductors: Recent Trends and Applications* (ed. Y. Takahashi), World Scientific, Singapore, pp. 166–200.

73 Norby, T. (1989) Hydrogen defects in inorganic solids, in *Selected Topics in High Temperature Chemistry: Defect Chemistry of Solids*, Vol. 9 (eds Ø. Johannesen and A.G. Anderson), Elsevier, Amsterdam, pp. 101–42.

74 Norby, T. and Larring, Y. (1997) Concentration and transport of protons in oxides. *Curr. Opin. Solid State Mater. Sci.*, **2**, 593–9.

75 Kreuer, K.-D. (1996) Proton conductivity: materials and applications. *Chem. Mater.*, **8**, 610–41.

76 Kreuer, K.D. (1997) On the development of proton conducting materials for technological applications. *Solid State Ionics*, **97**, 1–15.

77 Alberti, G. and Casciola, M. (2001) Solid state protonic conductors, present main applications and future prospects. *Solid State Ionics*, **145**, 3–16.

78 Iwahara, H., Yajima, T., Hibino, T., Ozaki, K. and Suzuki, H. (1993) Protonic conduction in calcium, strontium and barium zirconates. *Solid State Ionics*, **61**, 65–9.

79 Kurita, N., Fukatsu, N., Ito, K. and Ohashi, T. (1995) Protonic conduction domain of indium-doped calcium zirconate. *J. Electrochem. Soc.*, **142** (5), 1552–9.

80 Côté, R., Bale, C.W. and Gauthier, M. (1984) K_2CO_3 solid electrolyte as a CO_2 probe: Decomposition measurements of $CaCO_3$. *J. Electrochem. Soc.*, **131** (1), 63–7.

81 Salam, F., Birke, P. and Weppner, W. (1999) Solid-state CO_2 sensor with

Li_2CO_3-MgO electrolyte and $LiMn_2O_4$ as a solid reference electrode. *Electrochem. Solid-State Lett.*, **2** (4), 201–4.

82 Dubbe, A., Wiemhöfer, H.-D., Sadaoka, Y. and Göpel, W. (1995) Microstructure and response behaviour of electrodes for CO_2 gas sensors based on solid electrolytes. *Sens. Actuators B*, **24–25**, 600–2.

83 Currie, J.F., Essalik, A. and Marusic, J.-C. (1999) Micromachined thin film solid state electrochemical CO_2, NO_2 and SO_2 gas sensors. *Sens. Actuators B*, **59**, 235–41.

84 Worrell, W.L. and Liu, Q.G. (1984) A new sulfur dioxide sensor using a novel two-phase solid-sulfate electrolyte. *J. Electroanal. Chem.*, **168**, 355–62.

85 Eastman, C.D. and Etsell, T.H. (2006) Performance of a platinum thin film working electrode in a chemical sensor. *Thin Solid Films*, **515**, 2669–72.

86 Eastman, C.D. and Etsell, T.H. (2000) Thick/thin film sulfur oxide chemical sensor. *Solid State Ionics*, **136–137**, 639–45.

87 Mari, C.M., Beghi, M., Pizzini, S. and Faltemier, J. (1990) Electrochemical solid-state sensor for SO_2 determination in air. *Sens. Actuators B*, **2**, 51–5.

88 Wang, L. and Kumar, R.V. (2005) A SO_2 gas sensor based upon composite Nasicon/Sr-β-Al_2O_3 bielectrolyte. *Mater. Res. Bull.*, **40**, 1802–15.

89 Wang, L., Zhou, H., Liu, K., Wu, Y., Dai, L. and Kumar, R.V. (2008) A CO_2 gas sensor based upon composite Nasicon/ Sr–β–Al_2O_3 bielectrolyte. *Solid State Ionics*, **179**, 1662–5.

90 Wang, L. and Kumar, R.V. (2004) Cross-sensitivity effects on a new carbon dioxide gas sensor based on solid bielectrolyte. *Meas. Sci. Technol.*, **15**, 1005–10.

91 Leonhard, V., Fischer, D., Erdmann, H., Ilgenstein, M. and Köppen, H. (1993) Comparison of thin- and thick-film CO_2 sensors. *Sens. Actuators B*, **13–14**, 530–1.

92 Slater, D.J., Kumar, R.V. and Fray, D.J. (1996) A bielectrolyte solid-state sensor which detects SO_3 independently of O_2. *Solid State Ionics*, **86–88**, 1063–7.

93 Kale, G.M., Wang, L. and Hong, Y.R. (2003) Planar SO_x sensor incorporating a bi-electrolyte couple. *Solid State Ionics*, **161**, 155–63.

94 Kale, G.M., Wang, L., Hayes, J.E., Congjin, J. and Hong, Y.R. (2003) Solid-state sensors for in-line monitoring of NO_2 in automobile exhaust emission. *J. Mater. Sci.*, **38** (21), 4293–300.

95 Wang, L., Pan, L., Sun, J., Hong, Y.R. and Kale, G.M. (2005) Fabrication, characterization and application of (8 mol% Y_2O_3) ZrO_2 thin film on $Na_3Zr_2Si_2PO_{12}$ substrate for sensing CO_2 gas. *J. Mater. Sci.*, **40** (7), 1717–23.

96 Tamura, S., Hasegawa, I., Imanaka, N., Maekawa, T., Tsumiishi, T., Suzuki, K., Ishikawa, H., Ikeshima, A., Kawabata, Y., Sakita, N. and Adachi, G.-y. (2005) Carbon dioxide gas sensor based on trivalent cation and divalent oxide anion conducting solids with rare earth oxycarbonate based auxiliary electrode. *Sens. Actuators B*, **108** (1–2), 359–63.

97 Grilli, M.L., Di Bartolomeo, E., Aono, H., Sadaoka, Y., Billi, E., Montanaro, L. and Traversa, E. (2001) Solid electrolyte sensors for NO_2 detection without using a reference atmosphere, Proceedings of the Electrochemical Society, 2000–32; Solid-State Ionic Devices II Ceramic Sensors, The Electrochemical Society, Pennington, New Jersey, pp. 361–7.

98 Shimamoto, Y., Okamoto, T., Aono, H., Montanaro, L. and Sadaoka, Y. (2004) Deterioration phenomena of electrochemical CO_2 sensor with Pt, Na_2CO_3/Na_2O-Al_2O_3-$4SiO_2$//YSZ/Pt structure. *Sens. Actuators B*, **99** (1–2), 141–8.

99 Okamoto, T., Shimamoto, Y., Tsumura, N., Itagaki, Y., Aono, H. and Sadaoka, Y. (2005) Drift phenomena of electrochemical CO_2 sensor with Pt, Na_2CO_3/Na^+-electrolyte//YSZ/Pt structure. *Sens. Actuators B*, **108** (1–2), 346–51.

100 Shimamoto, Y., Okamoto, T., Itagaki, Y., Aono, H. and Sadaoka, Y. (2004)

Performance and stability of potentiometric CO_2 gas sensor based on the Pt, Li_2CO_3/Na_2O-Al_2O_3-$4SiO_2$//YSZ/Pt electrochemical cell. *Sens. Actuators B*, **99** (1–2), 113–17.

101 Hasegawa, I., Hasegawa, Y. and Imanaka, N. (2005) SO_2 gas sensor based on Al^{3+} and O^{2-} ion-conducting solids with $La_2O_2SO_4$ auxiliary electrode. *Sensors Lett.*, **3** (1), 27–30.

102 Inaba, Y., Tamura, S. and Imanaka, N. (2007) New type of sulfur dioxide gas sensor based on trivalent Al^{3+} ion-conducting solid electrolyte. *Solid State Ionics*, **179**, 1625–7.

103 Hasegawa, I., Tamura, S. and Imanaka, N. (2005) Solid electrolyte nitrogen monoxide gas sensor operating at intermediate temperature range. *Sens. Actuators B*, **108**, 314–18.

104 Imanaka, N., Hasegawa, I. and Tamura, S. (2006) Nitrogen monoxide gas sensor operable as low as 250 °C. *Sensors Lett.*, **4** (3), 340–3.

105 Imanaka, N. (2005) Novel multivalent cation conducting ceramics and their application. *J. Ceram. Soc. Jpn.*, **113** (6), 387–93.

106 Imanaka, N., Oda, A., Tamura, S. and Adachi, G.-Y. (2004) Total nitrogen oxides gas sensor based on solid electrolytes with refractory oxide-based auxiliary electrode. *J. Electrochem. Soc.*, **151** (5), H113–16.

107 Oda, A., Imanaka, N. and Adachi, G.-Y. (2003) New type of nitrogen oxide sensor with multivalent cation- and anion-conducting solid electrolytes. *Sens. Actuators B*, **93**, 229–32.

108 Tamura, S., Hasegawa, I. and Imanaka, N. (2008) Nitrogen oxides gas sensor based on Al^{3+} ion conducting solid electrolyte. *Sens. Actuators B*, **130** (1), 46–51.

109 Tamura, S. and Imanaka, N. (2007) Nitrogen oxide gas sensor based on multivalent ion-conducting solids. *Sens. Mater.*, **19** (6), 347–63.

110 Kaneko, H., Okamura, T., Taimatsu, H., Matsuki, Y. and Nishida, H. (2005) Performance of a miniature zirconia oxygen sensor with a Pd-PdO internal reference. *Sens. Actuators B*, **108** (1–2), 331–4.

111 Spirig, J.V., Ramamoorthy, R., Akbar, S.A., Routbort, J.L., Singh, D. and Dutta, P.K. (2007) High temperature zirconia oxygen sensor with sealed metal/metal oxide internal reference. *Sens. Actuators B*, **124** (1), 192–201.

112 Li, F., Tang, Y. and Li, L. (1996) Distribution of oxygen potential in ZrO_2-based solid electrolyte and selection of reference electrode of oxygen sensor. *Solid State Ionics*, **86–88**, 1027–31.

113 Li, F., Zhu, Z. and Li, L. (1994) A new way extending working-life of oxygen sensors in melt. *Solid State Ionics*, **70/71**, 555–8.

114 Dimitrov, S., Ranganathan, S., Weyl, A. and Janke, D. (1993) New reference materials for electrochemical on-line sensors to measure steel bath composition. *Steel Res.*, **64** (1), 63–70.

115 Weyl, A., Tu, S.W. and Janke, D. (1994) Sensors based on new oxide electrolyte and oxygen reference materials for online measurement in steel melts. *Steel Res.*, **65** (5), 167–72.

116 Lou, T.-J., Kong, X.-H., Huang, K.-Q. and Liu, Q.-G. (2006) Solid reference electrode of metallurgical oxygen sensor. *J. Iron Steel Res. Int.*, **13** (5), 18–20, 46.

117 Ramamoorthy, R., Akbar, S.A. and Dutta, P.K. (2006) Dependence of potentiometric oxygen sensing characteristics on the nature of electrodes. *Sens. Actuators B*, **113** (1), 162–8.

118 Colominas, S., Abella, J. and Victori, L. (2004) Characterisation of an oxygen sensor based on In/In_2O_3 reference electrode. *J. Nucl. Mater.*, **335** (2), 260–3.

119 Kaneko, H., Okamura, T. and Taimatsu, H. (2003) Characterization of zirconia oxygen sensors with a molten internal reference for low-temperature operation. *Sens. Actuators* **93** (1–3), 205–8.

120 Courouau, J.-L. (2004) Electrochemical oxygen sensors for on-line monitoring in lead-bismuth alloys: status of development. *J. Nucl. Mater.*, **335** (2), 254–9.

121 Konys, J., Muscher, H., Voß, Z. and Wedemeyer, O. (2004) Oxygen measurements in stagnant lead-bismuth eutectic using electrochemical sensors. *J. Nucl. Mater.*, **335** (2), 249–53.

122 Zhang, L., Fray, D.J., Dekeyser, J.C. and Schutter, F.D. (1996) Reference electrode of simple galvanic cells for developing sodium sensors for use in molten aluminum. *Metall. Mater. Trans. B*, **27**, 794–800.

123 Yao, P.C. and Fray, D.J. (1985) Sodium activity determinations in molten 99.5% aluminium using solid electrolytes. *J. Appl. Electrochem.*, **15**, 379–86.

124 Dubbe, A. (2008) Influence of the sensitive zeolite material on the characteristics of a potentiometric hydrocarbon gas sensor. *Solid State Ionics*, **179**, 1645–7.

125 Schwandt, C. and Fray, D.J. (2006) The titanium/hydrogen system as the solid-state reference in high-temperature proton conductor-based hydrogen sensors. *J. Appl. Electrochem.*, **36**, 557–65.

126 Kida, T., Kishi, S., Yuasa, M., Shimanoe, K. and Yamazoe, N. (2008) Planar NASICON-based CO_2 sensor using $BiCuVO_x$/perovskite-type oxide as a solid-reference electrode. *J. Electrochem. Soc.*, **155** (5), J117–21.

127 Garzon, F.H., Mukundan, R., Lujan, R. and Brosha, E.L. (2004) Solid state ionic devices for combustion gas sensing. *Solid State Ionics*, **175** (1–4), 487–90.

128 Chowdhury, A.K.M.S., Akbar, S.A., Kapileshwar, S. and Schorr, J.R. (2001) A rugged oxygen gas sensor with solid reference for high temperature applications. *J. Electrochem. Soc.*, **148** (2), G91–4.

129 Rutman, J. and Riess, I. (2008) Reference electrodes for thin-film solid-state ionic devices. *Solid State Ionics*, **179** (1–6), 108–12.

130 Davies, A., Fray, D.J. and Witek, S.R. (1995) Measurement of oxygen in steel using β alumina electrolytes. *Ironmaking Steelmaking*, **22** (4), 310–15.

131 Kumar, R.V. and Fray, D.J. (1994) Application of novel sensors in the measurement of very low oxygen potentials. *Solid State Ionics*, **70/71**, 588–94.

132 Sun, J., Jin, C., Li, L. and Hong, Y. (2001) A novel electrochemical oxygen sensor for determination of ultra-low oxygen contents in molten metal. *J. Univ. Sci. Technol. Beijing*, **8** (2), 137–40.

133 Fergus, J.W. and Hui, S. (1995) Solid electrolyte sensor for measuring magnesium in molten aluminum. *Metall. Mater. Trans. B*, **26**, 1289–91.

134 Lisheng, X., Zhitong, S. and Changzhen, W. (1995) Activity of dissolved La in liquid Al. *Scand. J. Metall.*, **24**, 86–90.

135 Hong, H.S., Kim, J.W., Jung, S.J. and Park, C.O. (2005) Thick film planar CO_2 sensors based on Na β-alumina solid electrolyte. *J. Electroceram.*, **15** (2), 151–7.

136 Möbius, H.H. (2004) Galvanic solid electrolyte cells for the measurement of CO_2 concentrations. *J. Solid State Electrochem.*, **8** (2), 94–109.

137 Holzinger, M., Maier, J. and Sitte, W. (1996) Fast CO_2-selective potentiometric sensor with open reference electrode. *Solid State Ionics*, **86–88**, 1055–62.

138 Hyodo, T., Tadashi, T., Kumazawa, S., Shimizu, Y. and Egashira, M. (2007) Effects of electrode materials on CO_2 sensing properties of solid-electrolyte gas sensors. *Sens. Mater.*, **19** (6), 365–76.

139 Obata, K., Kumazawa, S., Shimanoe, K., Miura, N. and Yamazoe, N. (2001) Potentiometric sensor based on NASICON and In_2O_3 for detection of CO_2 at room temperature – modification with foreign substances. *Sens. Actuators B*, **76**, 639–43.

140 Obata, K., Shimanoe, K., Miura, N. and Yamazoe, N. (2003) NASICON devices attached with Li_2CO_3-$BaCO_3$ auxiliary phase for CO_2 sensing under ambient conditions. *J. Mater. Sci.*, **38** (21), 4283–8.

141 Alonso-Porta, M. and Kumar, R.V. (2000) Use of NASICON/Na_2CO_3 system for measuring CO_2. *Sens. Actuators B*, **71**, 173–8.

142 Kale, G.M., Davidson, A.J. and Fray, D.J. (1996) Investigation into an improved design of CO_2 sensor. *Solid State Ionics*, **86–88**, 1107–10.

143 Sadaoka, Y., Sakai, Y. and Manabe, T. (1993) Detection of CO_2 using solid-state electrochemical sensor on sodium ionic conductors. *Sens. Actuators B*, **15–16**, 166–70.

144 Chu, W.F., Leonhard, V., Erdmann, H. and Ilgenstein, M. (1991) Thick-film chemical sensors. *Sens. Actuators B*, **4**, 321–4.

145 Feller, C., Kretzschmar, C. and Westphal, D. (2003) CO_2 solid electrolyte sensor with screen printing, 48th Internationales Wissenschaftliches Kolloquium, Technische Universität, Ilmenau, pp. 13–14.

146 Näfe, H. (1994) The feasibility of a potentiometric CO_2 sensor based on Na-beta-alumina in the light of the electronic conductivity of the electrolyte. *Sens. Actuators B*, **21**, 79–82.

147 Aono, H., Miyanaga, S. and Sadaoka, Y. (2005) Potentiometric responses of CO_2 and Cl_2 gas sensors using Ni reference electrode closed in Na^+ conducting Na_2O-Al_2O_3-$4SiO_2$ glass. *Electrochemistry (Tokyo, Japan)*, **73** (9), 791–7.

148 Miyachi, Y., Sakai, G., Shimanoe, K. and Yamazoe, N. (2005) Improvement of warming-up characteristics of potentiometric CO_2 sensor by using solid reference counter electrode. *Sens. Actuators B*, **108** (1–2), 364–7.

149 Miura, N., Yao, S., Shimizu, Y. and Yamazoe, N. (1992) Carbon dioxide sensor using sodium ion conductor and binary carbonate auxiliary phase. *J. Electrochem. Soc.*, **139** (5), 1384–8.

150 Pasierb, P., Komornicki, S., Koziński, S., Gajerski, R. and Rękas, M. (2004) Long-term stability of potentiometric CO_2 sensors based on Nasicon as a solid electrolyte. *Sens. Actuators B*, **101** (1–2), 47–56.

151 Pasierb, P., Komornicki, S., Gajerski, R., Koziński, S. and Rękas, M. (2003) The performance and long-time stability of potentiometric CO_2 gas sensors based on the (Li-Ba)CO_3|NASICON|(Na-Ti-O) electrochemical cells. *Solid State Ionics*, **157** (1–4), 357–63.

152 Lee, J.-S., Lee, J.-H. and Hong, S.-H. (2003) NASICON-based amperometric CO_2 sensor using Na_2CO_3-$BaCO_3$ auxiliary phase. *Sens. Actuators B*, **96** (3), 663–8.

153 Ward, B.J., Liu, C.C. and Hunter, G.W. (2003) Novel processing of NASICON and sodium carbonate/barium carbonate thin and thick films for a CO_2 microsensor. *J. Mater. Sci.*, **38** (21), 4289–92.

154 Lang, T., Wiemhöfer, H.-D. and Göpel, W. (1996) Carbonate based CO_2 sensors with high performance. *Sens. Actuators B*, **34**, 383–7.

155 Lang, Th., Caron, M., Izquierdo, R., Ivanov, D., Currie, J.F. and Yelon, A. (1996) Material characterization of sputtered sodium-ion conductive ceramics for a prototype CO_2 micro-sensor. *Sens. Actuators B*, **31**, 9–12.

156 Bang, Y.-I., Song, K.-D., Joo, B.-S., Huh, J.-S., Choi, S.-D. and Lee, D.-D. (2004) Thin film micro carbon dioxide sensor using MEMS process. *Sens. Actuators B*, **102**, 1–6.

157 Baliteau, S., Sauvet, A.-L., Lopez, C. and Fabry, P. (2005) Characterization of a NASICON based potentiometric CO_2 sensor. *J. Eur. Ceram. Soc.*, **25** (12), 2965–8.

158 Hong, H.S., Kim, J.W., Jung, S.J. and Park, C.O. (2006) Suppression of NO and SO_2 cross-sensitivity in electrochemical CO_2 sensors with filter layers. *Sens. Actuators B*, **113** (1), 71–9.

159 Ramírez-Salgado, J. and Fabry, P. (2003) Study of CO_2 electrodes in open devices of potentiometric sensors. *Solid State Ionics*, **158** (3–4), 297–308.

160 Shim, H.B., Kang, J.H., Choi, J.W. and Yoo, K.S. (2006) Characteristics of thick-film CO_2 sensors based on NASICON with Na_2CO_3-$CaCO_3$ auxiliary phases. *J. Electroceram.*, **17**, 971–4.

161 Goto, T., He, G., Harushima, T. and Iguchi, Y. (2003) Application of Sr β-alumina to a CO_2 gas sensor. *Solid State Ionics*, **156**, 329–36.

162 Kida, T., Kawate, H., Shimanoe, K., Miura, N. and Yamazoe, N. (2000) Interfacial structure of NASICON-based sensor attached with Li_2CO_3-$CaCO_3$ auxiliary phase for detection of CO_2. *Solid State Ionics*, **136–137**, 647–53.

163 Ménil, F., Daddah, B.O., Tardy, P., Debéda, H. and Lucat, C. (2005) Planar LiSICON-based potentiometric CO_2 sensors: influence of the working and reference electrodes relative size on the sensing properties. *Sens. Actuators B*, **107** (2), 695–7.

164 Zhang, P., Lee, C., Verweij, H., Akbar, S.A., Hunter, G. and Dutta, P.K. (2007) High temperature sensor array for simultaneous determination of O_2, CO, and CO_2 with kernel ridge regression data analysis. *Sens. Actuators B*, **123** (2), 950–63.

165 Kim, D.-H., Yoon, J.-Y., Park, H.-C. and Kim, K.-H. (2001) Fabrication and characteristics of CO_2-gas sensor using Li_2CO_3-Li_3PO_4-Al_2O_3 electrolyte and $LiMn_2O_4$ reference electrode. *Sens. Actuators B*, **76**, 591–9.

166 Narita, H., Can, Z.Y., Mizusaki, J. and Tagawa, H. (1995) Solid state CO_2 sensor using an electrolyte in the system Li_2CO_3-Li_3PO_4-Al_2O_3. *Solid State Ionics*, **79**, 349–3.

167 Zhang, Y.C., Kaneko, M., Uchida, K., Mizusaki, J. and Tagawa, H. (2001) Solid electrolyte CO_2 sensors with lithium-ion conductor and Li transition metal complex oxide as solid reference electrode. *J. Electrochem. Soc.*, **148** (8), H81–4.

168 Figueroa, O.L., Lee, C., Akbar, S.A., Szabo, N.F., Trimboli, J.A., Dutta, P.K., Sawaki, N., Soliman, A.A. and Verweij, H. (2005) Temperature-controlled CO, CO_2 and NO_x sensing in a diesel engine exhaust stream. *Sens. Actuators B*, **107** (2), 839–48.

169 Näfe, H. (2005) Potentiometric solid-state CO_2 sensor and the role of electronic conductivity of the electrolyte. *Sens. Actuators B*, **105** (2), 119–23.

170 Imanaka, N., Murata, T., Kawasato, T. and Adachi, G.-Y. (1993) CO_2 detection with lithium solid electrolyte sensors. *Sens. Actuators B*, **13–14**, 476–9.

171 Lee, C., Akbar, S.A. and Park, C.O. (2001) Potentiometric CO_2 gas sensor with lithium phosphorous oxynitride electrolyte. *Sens. Actuators B*, **76**, 591–9.

172 Park, C.O., Lee, C., Akbar, S.A. and Hwang, J. (2003) The origin of oxygen dependence in a potentiometric CO_2 sensor with Li-ion conducting electrolytes. *Sens. Actuators B*, **88**, 53–9.

173 Noh, W.S., Satyanarayana, L. and Park, J.S. (2005) Potentiometric CO_2 sensor using Li^+ ion conducting Li_3PO_4 thin film electrolyte. *Sensors*, **5** (11), 465–72.

174 Satyanarayana, L., Choi, G.P., Noh, W.S., Lee, W.Y. and Park, J.S. (2007) Characteristics and performance of binary carbonate auxiliary phase CO_2 sensor based on Li_3PO_4 solid electrolyte. *Solid State Ionics*, **177** (39–40), 3485–90.

175 Ambekar, P., Randhawa, J., Bhoga, S.S. and Singh, K. (2004) Galvanic CO_2 sensor with Li_2O: B_2O_3 glass ceramics based composite. *Ionics*, **10** (1–2), 45–9.

176 Zhu, Q., Qiu, F., Quan, Y., Sun, Y., Liu, S. and Zou, Z. (2005) Solid-electrolyte NASICON thick film CO_2 sensor prepared on small-volume ceramic tube substrate. *Mater. Chem. Phys.*, **91** (2–3), 338–42.

177 Jianhua, Y., Pinghua, Y. and Guangyao, M. (1996) A fully solid-state SO_x ($x=2,3$) gas sensor utilizing Ag-β-alumina as solid electrolyte. *Sens. Actuators B*, **31**, 209–12.

178 Yang, P.H., Yang, J.H., Chen, C.S., Peng, D.K. and Meng, G.Y. (1996) Performance evaluation of SO_x ($x=2,3$) gas sensors using Ag-β-alumina solid electrolyte. *Solid State Ionics*, **86–88**, 1095–9.

179 Wang, L., Bu, J., Kumar, R.V. and Hong, Y. (2005) A potentiometric SO_2 gas sensor with solid reference electrode. *Key Eng. Mater.*, **280–283** (Pt 1, High-Performance Ceramics III), 323–6.

180 Akila, R. and Jacob, K.T. (1988) Use of the Nasicon/Na_2SO_4 couple in a solid state sensor for SO_x ($x = 2, 3$). *J. Appl. Electrochem.*, **18**, 245–51.

181 Min, B.-K. and Choi, S.-D. (1993) SO_2-sensing characteristics of Nasicon sensors with Na_2SO_4-$BaSO_4$ auxiliary electrolytes. *Sens. Actuators B*, **93**, 209–13.

182 Jacob, K.T., Rao, D.B. and Nelson, H.G. (1978) Some studies on a solid-state sulfur probe for coal gasification studies. *J. Electrochem. Soc.*, **125** (5), 758–62.

183 Moos, R., Reetmeyer, B., Hürland, A. and Plog, C. (2006) Sensor for directly determining the exhaust gas recirculation rate – EGR sensor. *Sens. Actuators B*, **119** (1), 57–63.

184 Huerland, A., Moos, R., Mueller, R., Plog, C. and Simon, U. (2001) A new potentiometric sensors based on a NO^+ cation conducting, ceramic membrane. *Sens. Actuators B*, **77**, 287–92.

185 Imanaka, N., Yamamoto, T. and Adachi, G. (1999) Nitrogen monoxide sensing with nitrosonium ion conducting solid electrolytes. *Electrochem. Solid-State Lett.*, **2** (8), 409–11.

186 Yao, S., Shimizu, Y., Miura, N. and Yamazoe, N. (1992) Use of sodium nitrate auxiliary electrode for solid electrolyte sensor to detect nitrogen oxides. *Chem. Lett.*, 1587–90.

187 Yao, S. and Stetter, J.R. (2004) Modification of NASICON solid electrolyte for NO_x measurements. *J. Electrochem. Soc.*, **151** (4), H75–80.

188 Yao, S. and Stetter, J.R. (2003) Solid-state NO_x sensor based on surface modifications of the solid electrolyte, Proceedings of the Electrochemical Society, 2000–32; Solid-State Ionic Devices II Ceramic Sensors, The Electrochemical Society, Pennington, New Jersey, pp. 252–60.

189 Kurosawa, H., Yan, Y., Miura, N. and Yamazoe, N. (1995) Stabilized zirconia-based NO_x sensor operative at high temperature. *Solid State Ionics*, **79**, 338–43.

190 Sadaoka, Y. (2007) NASICON based CO_2 gas sensor with an auxiliary electrode composed of $LiCO_3$-metal oxide mixtures. *Sens. Actuators B*, **121**, 194–9.

191 Aono, H., Itagaki, Y. and Sadaoka, Y. (2007) $Na_3Zr_2Si_2PO_{12}$-based CO_2 gas sensor with heat-treated mixture of Li_2CO_3 and Nd_2O_3 as an auxiliary electrode. *Sens. Actuators* **126** (2), 406–14.

192 Chen, S., Jeng, D.-Y., Hadano, H., Ishiguro, Y., Nakayama, M. and Watanabe, K. (2003) A Nasicon CO_2 gas sensor with drift-detection electrode. *IEEE Trans. Instrumentation and Measurement*, **52** (3), 1494–500.

193 Miyachi, Y., Sakai, G., Shimanoe, K. and Yamazoe, N. (2004) Stabilization of counter electrode for NASICON based potentiometric CO_2 sensor. *Ceram. Eng. Sci. Proc.*, **25** (3), 471–6.

194 Kida, T., Miyachi, Y., Shimanoe, K. and Yamazoe, N. (2001) NASICON thick film-based CO_2 sensor prepared by a sol-gel method. *Sens. Actuators B*, **80**, 28–32.

195 Yao, S., Shimizu, Y., Miura, N. and Yamazoe, N. (1993) Solid electrolyte carbon dioxide sensor using sodium ionic conductor and lithium carbonate-based auxiliary phase. *Appl. Phys. A.*, **57**, 25–9.

196 Obata, K., Kumazawa, S., Matsushima, S., Shimanoe, K. and Yamazoe, N. (2005) NASICON-based potentiometric CO_2 sensor combined with new materials operative at room temperature. *Sens. Actuators B*, **108** (1–2), 352–8.

197 Lee, D.-D., Choi, S.-D. and Lee, K.-W. (1995) Carbon dioxide sensor using NASICON prepared by the sol-gel method. *Sens. Actuators B*, **24–25**, 607–9.

198 Bhoga, S.S. and Singh, K. (2005) Performance of electrochemical CO_2 gas sensor with NASICON dispersed in a binary solid electrolyte system. *Indian J. Phys.*, **79** (7), 725–6.

199 Zhou, M. and Ahmad, A. (2007) Synthesis, processing and characterization of nasicon solid electrolytes for CO_2 sensing applications. *Sens. Actuators B*, **122** (2), 419–26.

200 Yao, S., Hosohara, S., Shimizu, Y., Miura, N., Futata, H. and Yamazoe, N. (1991) Solid electrolyte CO_2 sensor using NASICON and Li-based binary carbonate electrode. *Chem. Lett.*, 2069–72.

201 Wang, L. and Kumar, R.V. (2003) Potentiometric SO_2 gas sensor based on a Ca^{2+} conducting solid electrolyte. *Mater. Sci. Technol.*, **19**, 1478–82.

202 Wang, L. and Kumar, R.V. (2006) Potentiometric SO_2 gas sensor based on a thick film of Ca^{2+} ion conducting solid electrolyte. *J. Appl. Electrochem.*, **36**, 173–8.

203 Wang, L. and Kumar, R.V. (2003) A new SO_2 gas sensor based on a Mg^{2+} conducting solid electrolyte. *J. Electroanal. Chem.*, **543**, 109–14.

204 Ikeda, S., Kondo, T., Kato, S., Ito, K., Nomura, K. and Fujita, Y. (1995) Carbon dioxide sensor using solid electrolytes with zirconium phosphate framework (2). Properties of the CO_2 gas sensor using $Mg_{1.15}Zr_4P_{5.7}Si_{0.3}O_{24}$ as electrolyte. *Solid State Ionics*, **79**, 354–7.

205 Obata, K., Shimanoe, K. and Yamazoe, N. (2003) Influences of water vapor on NASICON-based CO_2 sensor operative at room temperature. *Sens. Actuators B*, **93** (1–3), 243–9.

206 Aung, Y.L., Nakayama, S. and Sakamoto, M. (2005) Electrical properties of $MREGeO_4$ (M = Li, Na, K; RE = rare earth) ceramics. *J. Mater. Sci.*, **40** (1), 129–33.

207 Miura, N., Yan, Y., Sato, M., Yao, S., Nonaka, S., Shimizu, Y. and Yamazoe, N. (1995) Solid state potentiometric CO_2 sensors using anion conductor and metal carbonate. *Sens. Actuators B*, **24–25**, 260–5.

208 Miura, N., Yan, Y., Nonaka, S. and Yamazoe, N. (1995) Sensing properties and mechanism of a planar carbon dioxide sensor using magnesia-stabilized zirconia and lithium carbonate auxiliary phase. *J. Mater. Chem.*, **5** (9), 1391–4.

209 Näfe, H. (1997) On the electrode reaction of the Au|CO_2, O_2, Me_2CO_3 (Me = Li, Na, K)| yttria-stabilized zirconia electrode. *J. Electrochem. Soc.*, **144** (3), 915–22.

210 Wierzbicka, M., Pasierb, P. and Rekas, M. (2007) CO_2 sensor studies by impedance spectroscopy. *Physica B*, **387**, 302–12.

211 Bak, T., Nowotny, J., Rekas, M. and Sorrell, C.C. (2002) Dynamics of solid-state cell for CO_2 monitoring. *Solid State Ionics*, **152–153**, 823–6.

212 Yan, Y., Shimizu, Y., Miura, N. and Yamazoe, N. (1992) Solid-state sensor for sulfur oxides based on stabilized zirconia and metal sulfate. *Chem. Lett.*, 635–8.

213 Yan, Y., Shimizu, Y., Miura, N. and Yamazoe, N. (1994) High-performance solid-electrolyte SO_x sensor using MgO-stabilized zirconia tube and Li_2SO_4-$CaSO_4$-SiO_2 auxiliary phase. *Sens. Actuators B*, **20**, 81–7.

214 Yan, Y., Miura, N. and Yamazoe, N. (1996) Construction and working mechanism of sulfur dioxide sensor utilizing stabilized zirconia and metal sulfate. *J. Electrochem. Soc.*, **143** (2), 609–13.

215 Zhuiykov, S. (2000) Development of dual sulfur oxides and oxygen solid state sensor for 'in situ' measurements. *Fuel*, **79** (10), 1255–65.

216 Matsubara, S., Tsutae, T., Nakamoto, K., Hirose, Y., Katayama, I. and Iida, T. (1995) Determination of aluminum concentration in molten zinc by the E.M.F. method using zirconia solid electrolyte. *ISIJ Int.*, **35** (5), 512–18.

217 Li, F., Hongpeng, H., Liansheng, L., Lifen, L. and Daotze, L. (2000) A study of aluminum sensor for steel melt. *Sens. Actuators B*, **62**, 31–4.

218 Hong, Y.R., Li, F.S., Mao, Y.W., Li, L.S., Zhou, Y.Z. and Wang, P. (1990) A solid electrochemical ferro sensor for molten matter, in *Solid State Ionic Materials* (eds. B.V.R. Chowdary *et al.*), World Scientific Publishing, Singapore, pp. 375–9.

219 Okimura, T., Fukui, K. and Maruhashi, S. (1990) Development of zirconia electrolyte sensor with auxiliary electrode for the *in situ* measurement of dissolved silicon in molten iron. *Sens. Actuators B*, **1**, 203–9.

220 Matsubara, S., Tsutae, T., Nakamoto, K., Katayama, I. and Iida, T. (1995) Determination of aluminum concentration in molten zinc by the E.M.F. method using CaF_2 solid electrolyte. *Mater. Trans. JIM*, **36** (10), 1255–62.

221 Kale, G.M., Davidson, A.J. and Fray, D.J. (1996) Solid-state sensor for measuring antimony in non-ferrous metals. *Solid State Ionics*, **86–88**, 1101–5.

222 Larose, S., Dubreuil, A. and Pelton, A.D. (1991) Solid electrolyte probes for magnesium, calcium and strontium in molten aluminum. *Solid State Ionics*, **47**, 287–95.

223 Imanaka, N., Kamikawa, M. and Adachi, G.-y. (2002) A carbon dioxide gas sensor by combination of multivalent cation and anion conductors with a water-insoluble oxycarbonate-based auxiliary electrode. *Anal. Chem.*, **74** (18), 4800–4.

224 Imanaka, N., Oda, A., Tamura, S., Adachi, G.-y., Maekawa, T., Tsumeishi, T. and Ishikawa, H. (2003) Carbon dioxide gas sensor suitable for in situ monitoring. *Electrochem. Solid-State Lett.*, **7** (2), H12–14.

225 Imanaka, N., Kamikawa, M., Tamura, S. and Adachi, G. (2001) Carbon dioxide gas sensor with multivalent cation conducting solid electrolytes. *Sens. Actuators B*, **77** (1–2), 301–6.

226 Imanaka, N., Kamikawa, M., Tamura, S. and Adachi, G. (1999) CO_2 sensor based on the combination of trivalent Sc^{3+} ion-conducting $Sc_2(WO_4)_3$ and O^{2-} ion-conducting stabilized zirconia solid electrolytes. *Electrochem. Solid-State Lett.*, **2** (11), 602–4.

227 Tamura, S., Imanaka, N., Kamikawa, M. and Adachi, G.-Y. (2001) A CO_2 sensor based on a Sc^{3+} conducting $Sc_{1/3}Zr_2(PO_4)_3$ solid electrolyte. *Sens. Actuators B*, **73**, 205–10.

228 Mukundan, R., Brosha, E.L., Brown, D.R. and Garzon, F.H. (2000) A mixed-potential sensor based on a $Ce_{0.8}Gd_{0.2}O_{1.9}$ electrolyte and platinum and gold electrodes. *J. Electrochem. Soc.*, **147** (4), 1583–8.

229 Garzon, F.H., Munkundan, R. and Brosha, E.L. (2000) Solid-state mixed potential gas sensors: Theory, experiments and challenges. *Solid State Ionics*, **136–137**, 633–8.

230 Mukundan, R., Brosha, E.L. and Garzon, F.H. (2002) Solid-state electrochemical sensors for automotive applications. *Ceram. Trans.*, **130** (Chemical Sensors for Hostile Environments), 1–10.

231 Mukundan, R., Brosha, E.L., Brown, D.R. and Garzon, F.H. (1999) Ceria-electrolyte-based mixed potential sensors for the detection of hydrocarbons and carbon monoxide. *Electrochem. Solid-State Lett.*, **2** (12), 412–14.

232 Vogel, A., Baier, G. and Schüle, V. (1993) Non-Nernstian potentiometric zirconia sensors: Screening of potential working electrode materials. *Sens. Actuators B*, **15–16**, 147–50.

233 Thiemann, S., Hartung, R., Wulff, H., Klimke, J., Möbius, H.-H., Guth, U. and Schönauer, U. (1996) Modified Au-YSZ electrodes – preparation, characterization and electrode behaviour at higher temperatures. *Solid State Ionics*, **86–88**, 873–6.

234 Lalauze, R., Visconte, E., Montanaro, L. and Pijolat, C. (1993) A new type of mixed potential sensor using a thick film of beta alumina. *Sens. Actuators B*, **13–14**, 241–3.

235 Guillet, N., Lalauze, R. and Pijolat, C. (2004) Oxygen and carbon monoxide role on the electrical response of a non-Nernstian potentiometric gas sensor; proposition of a model. *Sens. Actuators B*, **98** (2–3), 130–9.

236 Somov, S., Reinhardt, G., Guth, U. and Göpel, W. (1996) Gas analysis with arrays of solid state electrochemical sensors:

Implications to monitor HCs and NO_x in exhausts. *Sens. Actuators B*, **35–36**, 409–18.

237 Göpel, W., Reinhardt, G. and Rösch, M. (2000) Trends in the development of solid state amperometric and potentiometric high temperature sensors. *Solid State Ionics*, **136–137**, 519–31.

238 Thiemann, S., Hartung, R., Guth, U. and Schönauer, U. (1996) Chemical modifications of Au-electrodes on YSZ and their influence on the non-Nernstian behavior. *Ionics*, **2** (5–6), 463–7.

239 Hibino, T., Hashimoto, A., Mori, K.-t. and Sano, M. (2001) A mixed-potential gas sensor using a $SrCe_{0.95}Yb_{0.05}O_{3-\alpha}$ electrolyte with a platinum electrode for detection of hydrocarbons. *Electrochem. Solid-State Lett.*, **4** (5), H9–11.

240 Hibino, T., Hashimoto, A., Suzuki, M. and Sano, M. (2001) Mixed potentials at metal-electrode and proton-conducting electrolyte interfaces in hydrocarbon-oxygen mixtures. *J. Phys. Chem.*, **105**, 10648–52.

241 Narayanan, B.K., Akbar, S.A. and Dutta, P.K. (2002) A phosphate-based proton conducting solid electrolyte hydrogen gas sensor. *Sens. Actuators B*, **87**, 480–6.

242 Hashimoto, A., Hibino, T., Mori, K.-T. and Sano, M. (2001) High-temperature hydrocarbon sensors based on a stabilized zirconia electrolyte and proton conductor-containing platinum electrode. *Sens. Actuators B*, **81**, 55–63.

243 Miura, N. and Yamazoe, N. (2001) Approach to high-performance electrochemical NO_x sensors based on solid electrolytes. *Sensors Update*, **6**, 191–210.

244 Ménil, F., Coillard, V. and Lucat, C. (2000) Critical review of nitrogen monoxide sensors for exhaust gases of lean burn engines. *Sens. Actuators B*, **67**, 1–23.

245 Miura, N., Wang, J., Nakatou, M., Elumalai, P., Zhuiykov, S., Terada, D. and Ono, T. (2005) Zirconia-based gas sensors using oxide sensing electrode for monitoring NO_x in car exhaust. *Ceram. Eng. Sci. Proc.*, **26** (5), 3–13.

246 Park, C.O. and Miura, N. (2006) Absolute potential analysis of the mixed potential occurring at the oxide/YSZ electrode at high temperature in NO_x-containing air. *Sens. Actuators B*, **113**, 316–19.

247 Gessner, M.A., Nagy, S.G. and Michaels, J.N. (1988) Multiple charge-transfer reaction in zirconia electrolytic cells; NO_x reduction on platinum. *J. Electrochem. Soc.*, **135** (5), 1294–301.

248 Skelton, D.C., Tobin, R.G., Lambert, D.K., DiMaggio, C.L. and Fisher, G.B. (2003) A surface-science-based model for the selectivity of platinum-gold alloy electrodes in zirconia-based NO_x sensors. *Sens. Actuators B*, **96** (1–2), 46–52.

249 Nakamura, T., Sakamoto, Y., Saji, K. and Sakata, J. (2003) NO_x decomposition mechanism on the electrodes of a zirconia-based amperometric NO_x sensor. *Sens. Actuators B*, **93**, 214–20.

250 Kubinski, D.J., Visser, J.H., Soltis, R.E., Parsons, M.H., Nietering, K.E. and Ejakov, S.G. (2002) Zirconia based potentiometric NO_x sensor utilizing Pt and Au electrodes. *Ceram. Trans.*, **130** (Chemical Sensors for Hostile Environments), 11–18.

251 Xiong, W. and Kale, G.M. (2006) Microstructure, conductivity, and NO_2 sensing characteristics of α-Al_2O_3-doped (8 mol% Sc_2O_3)ZrO_2 composite solid electrolyte. *Int. J. Appl. Ceram. Technol.*, **3** (3), 210–17.

252 Guillet, N., Lalauze, R., Viricelle, J.-P., Pijolat, C. and Montanaro, L. (2002) Development of a gas sensor by thick film technology for automotive applications: Choice of materials – realization of a prototype. *Mater. Sci. Eng. C*, **21**, 97–103.

253 Nagao, M., Namekata, Y., Hibino, T., Sano, M. and Tomita, A. (2006) Intermediate-temperature NO_x sensor based on an In^{3+}-doped SnP_2O_7 proton conductor. *Electrochem. Solid-State Lett.*, **9** (6), H48–51.

254 Di Bartolomeo, E. and Grilli, M.L. (2005) YSZ-based electrochemical sensors: From materials preparation to testing in the exhausts of an engine bench test. *J. Eur. Ceram. Soc.*, **25** (12), 2959–64.

255 Di Bartolomeo, E., Kaabbuathong, N., Grilli, M.L. and Traversa, E. (2004) Planar electrochemical sensors based on tape-cast YSZ layers and oxide electrodes. *Solid State Ionics*, **171** (3–4), 173–81.

256 Di Bartolomeo, E., Grilli, M.L. and Traversa, E. (2004) Sensing mechanism of potentiometric gas sensors based on stabilized zirconia with oxide electrodes. *J. Electrochem. Soc.*, **151** (5), H133–9.

257 Di Bartolomeo, E., Kaabbuathong, N., D'Epifanio, A., Grilli, M.L., Traversa, E., Aono, H. and Sadaoka, Y. (2004) Nano-structures perovskite oxide electrodes for planar electrochemical sensors using tape casted YSZ layer. *J. Eur. Ceram. Soc.*, **24**, 1187–90.

258 Dutta, A., Kaabbuathong, N., Grilli, M.L., Di Bartolomeo, E. and Traversa, E. (2003) Study of YSZ-based electrochemical sensors with WO_3 electrodes in NO_2 and CO environments. *J. Electrochem. Soc.*, **150** (2), H33–7.

259 Grilli, M.L., Chevallier, L., Di Vona, M.L., Licoccia, S. and Di Bartolomeo, E. (2005) Planar electrochemical sensors based on YSZ with WO_3 electrode prepared by different chemical routes. *Sens. Actuators B*, **111–112**, 91–5.

260 Grilli, M.L., Di Bartolomeo, E., Lunardi, A., Chevallier, L., Cordiner, S. and Traversa, E. (2005) Planar non-Nernstian electrochemical sensors: Field test in the exhaust of a spark ignition engine. *Sens. Actuators B*, **108** (1–2), 319–25.

261 Folch, J., Capdevila, X.G., Segarra, M. and Morante, J.R. (2007) Solid electrolyte multisensor system for detecting O_2, CO, and NO_2. *J. Electrochem. Soc.*, **154** (7), J201–8.

262 Lu, G., Miura, N. and Yamazoe, N. (2000) Stabilized zirconia-based sensors using WO_3 electrode for detection of NO or NO_2. *Sens. Actuators B*, **65**, 125–7.

263 Di Bartolomeo, E., Grilli, M.L., Yoon, J.W. and Traversa, E. (2004) Zirconia-based electrochemical NO_x sensors with semiconducting oxide electrodes. *J. Am. Ceram. Soc.*, **87** (10), 1883–9.

264 West, D.L., Montgomery, F.C. and Armstrong, T.R. (2005) 'NO-selective' NO_x sensing elements for combustion exhausts. *Sens. Actuators B*, **111–112**, 84–90.

265 Elumalai, P., Wang, J., Zhuiykov, S., Terada, D., Hasei, M. and Miura, N. (2005) Sensing characteristics of YSZ-based mixed-potential-type planar NO_x sensors using NiO sensing electrodes sintered at different temperatures. *J. Electrochem. Soc.*, **152** (7), H95–101.

266 Miura, N., Wang, J., Nakatou, M., Elumalai, P. and Hasei, M. (2005) NO_x sensing characteristics of mixed-potential-type zirconia sensor using NiO sensing electrode at high temperatures. *Electrochem. Solid-State Lett.*, **8** (2), H9–11.

267 Miura, N., Wang, J., Nakatou, M., Elumalai, P., Zhuiykov, S. and Hasei, M. (2006) High-temperature operating characteristics of mixed-potential-type NO_2 sensor based on stabilized-zirconia tube and NiO sensing electrode. *Sens. Actuators B*, **114**, 903–9.

268 Wang, J., Elumalai, P., Terada, D., Hasei, M. and Miura, N. (2006) Mixed-potential-type zirconia-based NO_x sensor using Rh-loaded NiO sensing electrode operating at high temperatures. *Solid State Ionics*, **177** (26–32), 2305–11.

269 Elumalai, P. and Miura, N. (2005) Performances of planar NO_2 sensor using stabilized zirconia and NiO sensing electrode at high temperature. *Solid State Ionics*, **176**, 2517–22.

270 Plachnitsa, V.V. and Miura, N. (2005) Effects of different additives on the sensing properties of NiO electrode used for mixed-potential-type YSZ-based gas sensors. *Chemical Sensors*, **21** (Suppl. B), 94–6.

271 Plashnitsa, V.V., Ueda, T. and Miura, N. (2006) Improvement of NO_2 sensing performances by an additional second component to the nano-structured NiO sensing electrode of a YSZ-based mixed-potential-type sensor. *Int. J. Appl. Ceram. Technol.*, **3** (2), 127–33.

272 Saruhan, B., Stranzenbach, M. and Mondragón Rodríguez, G.C. (2007) An integrated solution for NO_x-reduction and -control under lean-burn conditions. *Materialwiss. Werkstofftech.*, **38** (9), 725–33.

273 Hamamoto, K., Fujishiro, Y. and Awano, M. (2008) Gas sensing property of the electrochemical cell with a multilayer catalytic electrode. *Solid State Ionics*, **179**, 1648–51.

274 Elumalai, P., Terada, D., Hasei, M. and Miura, N. (2005) YSZ-based mixed-potential-type planar NO_2 sensors using NiO sensing electrodes sintered at different temperatures. *Chemical Sensors*, **21** (Suppl. A), 40–2.

275 West, D.L., Montgomery, F.C. and Armstrong, T.R. (2004) Electrode materials for mixed-potential NO_x sensors. *Ceram. Eng. Sci. Proc.*, **25** (3), 493–8.

276 Martin, L.P., Pham, A.Q. and Glass, R.S. (2003) Effect of Cr_2O_3 electrode morphology on the nitric oxide response of a stabilized zirconia sensor. *Sens. Actuators B*, **96**, 53–60.

277 Ono, T., Hasei, M., Kunimoto, A. and Miura, N. (2004) Improvement of sensing performances of zirconia-based total NO_x sensor by attachment of oxidation-catalyst electrode. *Solid State Ionics*, **75** (1–4), 503–6.

278 Yoo, J., Yoon, H. and Wachsman, E.D. (2006) Sensing properties of MO_x/YSZ/Pt ($MO_x = Cr_2O_3$, SnO_2, CeO_2) potentiometric sensor for NO_2 detection. *J. Electrochem. Soc.*, **153** (11), H217–21.

279 Yoo, J. and Wachsman, E.D. (2007) NO_2/NO response of Cr_2O_3- and SnO_2-based potentiometric sensors and temperature-programmed reaction evaluation of the sensor elements. *Sens. Actuators B*, **123** (2), 915–21.

280 Szabo, N.F. and Dutta, P.K. (2004) Correlation of sensing behavior of mixed potential sensors with chemical and electrochemical properties of electrodes. *Solid State Ionics*, **171** (3–4), 183–90.

281 Szabo, N. and Dutta, P.K. (2003) Strategies for total NO_x measurement with minimal CO interference utilizing a microporous zeolitic catalytic filter. *Sens. Actuators B*, **88** (2), 168–77.

282 Käding, S., Jakobs, S. and Guth, U. (2003) YSZ-cells for potentiometric nitric oxide sensors. *Ionics*, **9** (1–2), 151–4.

283 Joo, J.H. and Choi, G.M. (2006) The effect of transition-metal addition on the non-equilibrium E.M.F.-type sensor. *J. Electroceram.*, **17**, 1019–22.

284 Di Bartolomeo, E., Grilli, M.L., Antonias, N., Cordiner, S. and Traversa, E. (2005) Testing planar gas sensors based on yttria-stabilized zirconia with oxide electrodes in the exhaust gases of a spark ignition engine. *Sensor Lett.*, **3** (1), 22–6.

285 Westphal, D., Jakobs, S. and Guth, U. (2001) Gold-composite electrodes for hydrocarbon sensors based on YSZ solid electrolyte. *Ionics*, **7** (3), 182–6.

286 Zosel, J., Müller, R., Vashook, V. and Guth, U. (2004) Response behavior of perovskites and Au/oxide composites as HC-electrodes in different combustibles. *Solid State Ionics*, **175** (1–4), 531–3.

287 Zosel, J., Westphal, D., Jakobs, S., Müller, R. and Guth, U. (2002) Au-oxide composites as HC-sensitive electrode materials for mixed potential gas sensors. *Solid State Ionics*, **152–153**, 525–9.

288 Zosel, J. and Guth, U. (2004) Electrochemical solid electrolyte gas sensors-hydrocarbon and NO_x analysis in exhaust gases. *Ionics*, **105** (5–6), 366–77.

289 Hibino, T., Kakimoto, S. and Sano, M. (1999) Non-Nernstian behavior at modified Au electrodes for hydrocarbon gas sensing. *J. Electrochem. Soc.*, **146** (9), 3361–6.

290 Inaba, T., Saji, K. and Sakata, J. (2005) Characteristics of an HC sensor using a Pr_6O_{11} electrode. *Sens. Actuators B*, **108** (1–2), 374–8.

291 Li, X., Xiong, W. and Kale, G.M. (2005) Novel nanosized ITO electrode for mixed potential gas sensor. *Electrochem. Solid-State Lett.*, **8** (3), H27–30.

292 Li, X. and Kale, G.M. (2006) Influence of thickness of ITO sensing electrode film on sensing performance of planar mixed potential CO sensor. *Sens. Actuators B*, **120** (1), 150–5.

293 Li, X. and Kale, G.M. (2007) Influence of sensing electrode and electrolyte on performance of potentiometric mixed-potential gas sensors. *Sens. Actuators B*, **123** (1), 254–61.

294 Li, X. and Kale, G.M. (2006) Planar mixed-potential CO sensor utilizing novel $(Ba_{0.4}La_{0.6})_2In_2O_{5.6}$ and ITO interface. *Electrochem. Solid-State Lett.*, **9** (2), H12–15.

295 Chevallier, L., Grilli, M.L., Di Bartolomeo, E. and Traversa, E. (2006) Non-Nernstian planar sensors based on YSZ with Ta (10 at %)-doped nanosized titania as a sensing electrode for high-temperature applications. *Int. J. Appl. Ceram. Technol.*, **3** (5), 393–400.

296 Liang, X., Zhong, T., Quan, B., Wang, B. and Guan, H. (2007) Solid-state potentiometric SO_2 sensor combining NASICON with V_2O_5-doped TiO_2 electrode. *Sens. Actuators B*, **134**, 25–30.

297 Elumalai, P., Plashnitsa, V.V., Ueda, T. and Miura, N. (2008) Sensing characteristics of mixed potential- type zirconia-based sensor using laminated-oxide sensing electrode. *Electrochem. Commun.*, **10** (5), 745–8.

298 Li, N., Tan, T.C. and Zeng, H.C. (1993) High-temperature carbon monoxide potentiometric sensor. *J. Electrochem. Soc.*, **140** (4), 1068–73.

299 Wu, N., Zhao, M., Zheng, J.-G., Jiang, C., Myers, B., Li, S. and Mao, S.X. (2005) Porous CuO-ZnO nanocomposite for sensing electrode of high-temperature CO solid-state electrochemical sensor. *Nanotechnology*, **16** (12), 2878–81.

300 Miura, N., Raisen, T., Lu, G. and Yamazoe, N. (1998) Highly selective CO sensor using stabilized zirconia and a couple of oxide electrodes. *Sens. Actuators B*, **47**, 84–91.

301 Hibino, T., Tanimoto, S., Kakimoto, S. and Sano, M. (1999) High-temperature hydrocarbon sensors based on a stabilized zirconia electrolyte and metal oxide electrodes. *Electrochem. Solid-State Lett.*, **2** (12), 651–3.

302 Hibino, T., Hashimoto, A., Kakimoto, S. and Sano, M. (2001) Zirconia-based potentiometric sensors using metal oxide electrodes for detection of hydrocarbons. *J. Electrochem. Soc.*, **148** (1), H1–5.

303 Sorita, R. and Kawano, T. (1997) A highly selective CO sensor using $LaMnO_3$ electrode-attached zirconia galvanic cell. *Sens. Actuators B*, **40**, 29–32.

304 Brosha, E.L., Mukundan, R., Brown, D.R. and Garzon, F. (2002) Mixed potential sensors using lanthanum manganate and terbium yttrium zirconium oxide electrodes. *Sens. Actuators B*, **87** (1), 47–57.

305 Brosha, E.L., Mukundan, R., Brown, D.R., Garzon, F.H., Visser, J.H., Thompson, D.J. and Schonberg, D.H., Zirconia-based mixed potential CO/HC sensors with $LaMnO_3$ and Tb-doped YSZ electrode, Proceedings of the Electrochemical Society, 2000–32; Solid-State Ionic Devices II Ceramic Sensors, The Electrochemical Society, Pennington, New Jersey, pp. 314–21.

306 Brosha, E.L., Mukundan, R., Brown, D.R., Garzon, F.H., Visser, J.H., Zanini, M., Zhou, Z. and Logothetis, E.M. (2000) CO/HC sensors based on thin films of $LaCoO_3$ and $La_{0.8}Sr_{0.2}CoO_{3-\delta}$ metal oxides. *Sens. Actuators B*, **69**, 171–82.

307 Brosha, E.L., Mukundan, R., Brown, D.R., Garzon, F. and Visser, J.H. (2002) Development of ceramic mixed potential sensors for automotive applications. *Solid State Ionics*, **148** (1–2), 61–9.

308 Mukundan, R., Brosha, E.L. and Garzon, F. (2003) Mixed potential hydrocarbon sensors based on a YSZ electrolyte and oxide electrodes. *J. Electrochem. Soc.*, **150** (1–2), H279–84.

309 West, D.L., Montgomery, F.C. and Armstrong, T.R. (2005) Use of $La_{0.85}Sr_{0.15}CrO_3$ in high-temperature NO_x

sensing elements. *Sens. Actuators B*, **106** (2), 758–65.

310 Song, S.-W., Martin, L.P., Glass, R.S., Murray, E.P., Visser, J.H., Soltis, R.E., Novak, R.F. and Kubinski, D.J. (2006) Aging studies of Sr-doped $LaCrO_3$/YSZ/Pt cells for an electrochemical NO_x sensor. *J. Electrochem. Soc.*, **153** (9), H171–80.

311 Mukundan, R., Teranishi, K., Brosha, E.L. and Garzon, F.H. (2007) Nitrogen oxide sensors based on yttria-stabilized zirconia electrolyte and oxide electrodes. *Electrochem. Solid-State Lett.*, **10** (2), J26–9.

312 Brosha, E.L., Mukundan, R., Lujan, R. and Garzon, F.H. (2006) Mixed potential NO_x sensors using thin film electrodes and electrolytes for stationary reciprocating engine type applications. *Sens. Actuators* **119** (2), 398–408.

313 Miura, N., Lu, G. and Yamazoe, N. (1998) High-temperature potentiometric/amperometric NO_x sensors combining stabilized zirconia with mixed-metal oxide electrodes. *Sens. Actuators B*, **52**, 169–78.

314 Wachsman, E.D. (2003) Selective potentiometric detection of NO_x by differential electrode equilibria, Proceedings of the Electrochemical Society, 2000–32; Solid-State Ionic Devices II Ceramic Sensors, The Electrochemical Society, Pennington, New Jersey, pp. 215–21.

315 Toan, N.N., Saukko, S. and Lantto, V. (2003) Gas sensing with semiconducting perovskite oxide $LaFeO_3$. *Physica B: Condensed Matter*, **327** (2–4), 279–82.

316 Lantto, V., Saukko, S., Toan, N.N., Reyes, L.F. and Granqvist, C.G. (2004) Gas sensing with perovskite-like oxides having ABO_3 and BO_3 structures. *J. Electroceram.*, **13** (1–3), 721–6.

317 Grilli, M.L., Di Bartolomeo, E. and Traversa, E. (2001) Electrochemical NO_x sensors based on interfacing nanosized $LaFeO_3$ perovskite-type oxide and ionic conductors. *J. Electrochem. Soc.*, **148** (9), H98–102.

318 Di Bartolomeo, E., Traversa, E., Baroncini, M., Kotzeva, V. and Kumar, R.V. (2000) Solid state ceramic sensors based on interfacing ionic conductors with semiconducting oxides. *J. Eur. Ceram. Soc.*, **20**, 2691–9.

319 Zosel, J., Franke, D., Ahlborn, K., Gerlach, F., Vashook, V. and Guth, U. (2008) Perovskite related electrode materials with enhanced no sensitivity for mixed potential sensors. *Solid State Ionics*, **179**, 1628–31.

320 Miura, N., Nakatou, M. and Zhuiykov, S. (2003) Impedancemetric gas sensor based on zirconia solid electrolyte and oxide sensing electrode for detecting total NO_x at high temperature. *Sens. Actuators B*, **93** (1–3), 221–8.

321 Xiong, W. and Kale, G.M. (2006) Novel high-selectivity NO_2 sensor incorporating mixed-oxide electrode. *Sens. Actuators B*, **114**, 101–8.

322 Xiong, W. and Kale, G.M. (2005) Novel high-selectivity NO_2 sensor for sensing low-level NO_2. *Electrochem. Solid-State Lett.*, **8** (6), H49–53.

323 Xiong, W. and Kale, G.M. (2006) High-selectivity mixed-potential NO_2 sensor incorporating Au and CuO + $CuCr_2O_4$ electrode couple. *Sens. Actuators B*, **119** (2), 409–14.

324 Miura, N., Lu, G., Yamazoe, N., Kurosawa, H. and Hasei, M. (1996) Mixed potential type NO_x sensor based on stabilized zirconia and oxide electrode. *J. Electrochem. Soc.*, **143** (2), L33–5.

325 Miura, N., Kurosawa, H., Hasei, M., Lu, G. and Yamazoe, N. (1996) Stabilized zirconia-based sensor using oxide electrode for detection of NO_x in high-temperature combustion-exhausts. *Solid State Ionics*, **86–87**, 1069–73.

326 Lu, G., Miura, N. and Yamazoe, N. (1997) High-temperature sensors for NO and NO_2 based on stabilized zirconia and spinel-type oxide electrodes. *J. Mater. Chem.*, **7** (8), 1445–9.

327 Zhuiykov, S., Ono, T., Yamazoe, N. and Miura, N. (2002) High-temperature NO_x

sensors using zirconia solid electrolyte and zinc-family oxide sensing electrode. *Solid State Ionics*, **152–153**, 801–7.

328 Miura, N., Nakatou, M. and Zhuiykov, S. (2004) Development of NO_x sensing devices based on YSZ and oxide electrode aiming for monitoring car exhausts. *Ceram. Int.*, **30** (7), 1135–9.

329 Miura, N., Zhuiykov, S., Takashi, H. and Yamazoe, N. (2002) Mixed potential type sensor using stabilized zirconia and $ZnFe_2O_4$ sensing electrode for NO_x detection at high temperature. *Sens. Actuators B*, **83** (1–3), 222–9.

330 Zhuiykov, S., Muta, M., Ono, T., Hasei, M., Yamazoe, N. and Miura, N. (2001) Stabilized zirconia-based NO_x sensor using $ZnFe_2O_4$ sensing electrode. *Electrochem. Solid-State Lett.*, **4** (9), H19–21.

331 Zhang, G., Li, C., Cheng, F. and Chen, J. (2007) $ZnFe_2O_4$ tubes: Synthesis and application to gas sensors with high sensitivity and low-energy consumption. *Sens. Actuators B*, **120** (2), 403–10.

332 Yoo, J., Chatterjee, S., Van Assche, F.M. and Wachsman, E.D. (2007) Influence of adsorption and catalytic reaction on sensing properties of a potentiometric La_2CuO_4/YSZ/Pt sensor. *J. Electrochem. Soc.*, **154** (7), J190–5.

333 White, B., Chatterjee, S., Macam, E. and Wachsman, E. (2008) Effect of electrode microstructure on the sensitivity and response time of potentiometric NO_x sensors. *J. Am. Ceram. Soc.*, **91** (6), 2024–31.

334 Van Assche, F.M., IV, Nino, J.C. and Wachsman, E.D. (2008) Infrared and x-ray photoemission spectroscopy of absorbates on La_2CuO_4 to determine potentiometric NO_x sensor response mechanism. *J. Electrochem. Soc.*, **155** (7), J198–204.

335 Simonsen, V.L.E., Nørskov, L. and Hansen, K.K. (2008) Electrochemical reduction of NO and O_2 on $La_{2-x}Sr_xCuO_4$-based electrodes. *J. Solid State Electrochem.*, **12**, 1573–7.

336 Yoo, J., Van Assche, F.M. and Wachsman, E.D. (2006) Temperature-programmed reaction and desorption of the sensor elements of a WO_3/YSZ/Pt potentiometric sensors. *J. Electrochem. Soc.*, **153** (6), H115–21.

337 Shimizu, Y., Okimoto, M. and Souda, N. (2006) Solid-state SO_2 sensor using a sodium-ionic conductor and a metal-sulfide electrode. *Int. J. Appl. Ceram. Technol.*, **3** (3), 193–9.

338 Souda, N. and Shimizu, Y. (2003) Sensing properties of solid electrolyte SO_2 sensor using metal-sulfide electrode. *J. Mater. Sci.*, **38** (21), 4301–5.

339 Ono, T., Hasei, M., Kunimoto, A., Yamamoto, T. and Noda, A. (2001) Performance of the NO_x sensor based on mixed potential for automobiles in exhaust gases. *JSAE Review*, **22** (1), 49–55.

340 Yoo, J., Chatterjee, S. and Wachsman, E.D. (2007) Sensing properties and selectivities of a WO_3/YSZ/Pt potentiometric NO_x sensor. *Sens. Actuators B*, **122** (2), 644–52.

341 Szabo, N.F., Du, H., Akbar, S.A., Soliman, A. and Dutta, P.K. (2002) Microporous zeolite modified yttria stabilized zirconia (YSZ) sensors for nitric oxide (NO) determination in harsh environments. *Sens. Actuators B*, **82** (2–3), 142–9.

342 Yang, J.-C. and Dutta, P.K. (2006) Promoting selectivity and sensitivity for a high temperature YSZ-based electrochemical total NO_x sensor by using a Pt-loaded zeolite Y filter. *Sens. Actuators B*, **125** (1), 30–9.

343 Zosel, J., Ahlborn, K., Müller, R., Westphal, D., Vashook, V. and Guth, U. (2004) Selectivity of HC-sensitive electrode materials for mixed potential gas sensors. *Solid State Ionics*, **169** (1–4), 115–19.

344 Hibino, T., Kuwahara, Y., Wang, S., Kakimoto, S. and Sano, M. (1998) Nonideal electromotive force of zirconia sensors for unsaturated hydrocarbon gases. *Electrochem. Solid-State Lett.*, **1** (4), 197–9.

345 Brosha, E.L., Mukundan, R. and Garzon, F. (2003) The role of heterogeneous catalysis in the gas-sensing selectivity of high-temperature mixed potential sensors, Proceedings of the Electrochemical Society, 2002–26; Solid-State Ionic Devices III, The Electrochemical Society, Pennington, New Jersey, pp. 261–71.

346 Yagi, H. and Hideaki, K. (1995) High-temperature humidity sensor using a limiting-current-type plane multi-oxygen sensor for direct firing system. *Sens. Actuators B*, **B25** (1–3), 701–4.

347 Yun, D.H., Kim, D.I. and Park, C.O. (1993) YSZ oxygen sensor for lean burn combustion control system. *Sens. Actuators B*, **13** (1–3), 114–16.

348 Logothetis, E.M., Visser, J.H., Soltis, R.E. and Rimai, L. (1992) Chemical and physical sensors based on oxygen pumping with solid-state electrochemical cells. *Sens. Actuators B*, **9**, 183–9.

349 Katahira, K., Matsumoto, H., Iwahara, H., Koide, K. and Iwamoto, T. (2001) A solid electrolyte hydrogen sensor with an electrochemically-supplied hydrogen standard. *Solid State Ionics*, **73**, 130–4.

350 Schmidt-Zhang, P., Sandow, K.-P., Adolf, F., Göpel, W. and Guth, U. (2000) A novel thick film sensor for simultaneous O_2 and NO monitoring in exhaust gases. *Sens. Actuators B*, **70**, 25–9.

351 Hibino, T., Kuwahara, Y., Otsuka, T., Ishia, N. and Oshima, T. (1998) NO_x detection using the electrolysis of water vapour in a YSZ cell. Part I. NO_x detection. *Solid State Ionics*, **107**, 213–16.

352 Ueda, T., Plashnitsa, V.V., Nakatou, M. and Miura, N. (2007) Zirconia-based amperometric sensor using ZnO sensing-electrode for selective detection of propene. *Electrochem. Commun.*, **9**, 197–200.

353 Yang, J.-C. and Dutta, P.K. (2007) High temperature amperometric total NO_x sensors with platinum-loaded zeolite Y electrodes. *Sens. Actuators B*, **123** (2), 929–36.

354 Miura, N., Lu, G., Ono, M. and Yamazoe, N. (1999) Selective detection of NO by using an amperometric sensor based on stabilized zirconia and oxide electrode. *Solid State Ionics*, **117** (3–4), 283–90.

355 Coillard, V., Juste, L., Lucat, C. and Ménil, F. (2000) Nitrogen-monoxide sensing with a commercial zirconia lambda gauge biased in amperometric mode. *Meas. Sci. Technol.*, **11**, 212–20.

356 Coillard, V., Debéda, H., Lucat, C. and Ménil, F. (2001) Nitrogen monoxide detection with a planar spinel coated amperometric sensor. *Sens. Actuators B*, **78**, 113–18.

357 Dutta, A. and Ishihara, T. (2006) An amperometric solid state NO sensor using a $LaGaO_3$ electrolyte for monitoring exhaust gas. *Mater. Manuf. Processes*, **21**, 225–8.

358 Dutta, A. and Ishihara, T. (2005) Amperometric NO_x sensor based on oxygen pumping current by using $LaGaO_3$-based solid electrolyte for monitoring exhaust gas. *Sens. Actuators B*, **108** (1–2), 309–13.

359 Dutta, A. and Ishihara, T. (2005) Sensitive amperometric NO sensor using $LaGaO_3$-based oxide ion conducting electrolyte. *Electrochem. Solid-State Lett.*, **8** (5), H46–8.

360 Ishihara, T., Fukuyama, M., Kabemura, K., Nishiguchi, H. and Takita, Y. (2003) Hydrocarbon sensor for monitoring exhaust gas using oxygen pumping current, Proceedings of the Electrochemical Society, 2002–26; Solid-State Ionic Devices III, The Electrochemical Society, Pennington, New Jersey, pp. 287–98.

361 Ishihara, T., Fukuyama, M., Dutta, A., Kabemura, K., Nishiguchi, H. and Takita, Y. (2003) Solid state amperometric hydrocarbon sensor for monitoring exhaust gas using oxygen pumping current. *J. Electrochem. Soc.*, **150** (10), H241–5.

362 Can, Z.Y., Narita, H., Mizusaki, H. and Tagawa, H. (1995) Detection of carbon monoxide by using zirconia oxygen sensor. *Solid State Ionics*, **79**, 344–8.

363 Somov, S.I. and Guth, U. (1998) A parallel analysis of oxygen and combustibles in solid electrolyte amperometric cells. *Sens. Actuators B*, **B47** (1–2), 131–8.

364 Dutta, A., Ishihara, T. and Nishiguchi, H. (2004) An amperometric solid-state gas sensor using a $LaGaO_3$-based perovskite oxide electrolyte for detecting hydrocarbon in exhaust gas. A bimetallic anode for improving sensitivity at low temperatures. *Chem. Mater.*, **16** (24), 5198–204.

365 Ono, M., Shimanoe, K., Miura, N. and Yamazoe, N. (1999) Solid-state amperometric sensor based on a sodium ion conductor for detection of total NO_x in an atmospheric environment. *Electrochem. Solid-State Lett.*, **2** (7), 349–51.

366 Somov, S.I., Reinhardt, G., Guth, U. and Göpel, W. (2000) Multi-electrode zirconia electrolyte amperometric sensors. *Solid State Ionics*, **136–137**, 543–7.

367 Reinhardt, G., Rohlfs, I., Mayer, R. and Göpel, W. (2000) Selectivity-optimization of planar amperometric multi-electrode sensors: Identification of O_2, NO_x and combustible gases in exhausts at high temperatures. *Sens. Actuators B*, **65**, 76–8.

368 Somov, S., Reinhardt, G., Guth, U. and Göpel, W. (2000) Tubular amperometric high-temperature sensors: simultaneous determination of oxygen, nitrogen oxides and combustible components. *Sens. Actuators B*, **65**, 68–9.

369 Nigge, U., Wiemhöfer, H.-D., Römer, E.W.J., Bouwmeester, H.J.M. and Schulte, T.R. (2002) Composites of $Ce_{0.8}Gd_{0.2}O_{1.9}$ and $Gd_{0.7}Ca_{0.3}CoO_{3-\delta}$ as oxygen permeable membranes for exhaust gas sensors. *Solid State Ionics*, **146** (1–2), 163–74.

370 Schulte, T., Waser, R., Römer, E.W.J., Bouwmeester, H.J.M., Nigge, U. and Wiemhöfer, H.-D. (2001) Development of oxygen-permeable ceramic membranes for NO_x-sensors. *J. Eur. Ceram. Soc*, **21** (10–11), 1971–5.

371 Römer, E.W.J., Nigge, U., Schulte, T., Wiemhöfer, H.-D. and Bouwmeester, H.J.M. (2001) Investigations towards the use of $Gd_{0.7}Ca_{0.3}CoO_x$ as membrane in an exhaust gas sensor for NO_x. *Solid State Ionics*, **140** (1–2), 97–103.

372 Shoemaker, E.L., Vogt, M.C., Dudek, F.J. and Turner, T. (1997) Gas microsensors using cyclic voltammetry with a cermet electrochemical cell. *Sens. Actuators B*, **42**, 1–9.

373 Jasinski, G., Jasinski, P., Chachulski, B. and Nowakowski, A. (2005) Lisicon solid electrolyte electrocatalytic gas sensor. *J. Eur. Ceram. Soc.*, **25**, 2969–72.

374 Yoon, J.W., Grilli, M.L., Di Bartolomeo, E., Polini, R. and Traversa, E. (2001) The NO_2 response of solid electrolyte sensors made using nano-sized $LaFeO_3$ electrodes. *Sens. Actuators B*, **76**, 483–8.

375 Martin, L.P., Woo, L.Y. and Glass, R.S. (2007) Impedancemetric NO_x sensing using YSZ electrolyte and YSZ/Cr_2O_3 composite electrodes. *J. Electrochem. Soc.*, **154** (3), J97–104.

376 Miura, N., Nakatou, M. and Zhuiykov, S. (2002) Impedance-based total-NO_x sensor using stabilized zirconia and $ZnCr_2O_4$ sensing electrode operating at high temperature. *Electrochem. Commun.*, **4** (4), 284–7.

377 Miura, N., Koga, T., Nakatou, M., Elumalai, P. and Hasei, M. (2006) Electrochemical NO_x sensors based on stabilized zirconia: Comparison of sensing performances of mixed-potential-type and impedancemetric NO_x. *J. Electroceram.*, **17**, 979–86.

378 Stranzenbach, M., Gramckow, E. and Saruhan, B. (2007) Planar, impedance-metric NO_x-sensor with spinel-type SE for high temperature applications. *Sens. Actuators B*, **127**, 224–30.

379 Wu, N., Chen, Z., Xu, J., Chyu, M. and Mai, S.X. (2005) Impedance-metric Pt/YSZ/Au-Ga_2O_3 sensor for CO detection at high temperature. *Sens. Actuators B*, **110** (1), 49–53.

380 Nakatou, M. and Miura, N. (2005) Detection of combustible hydrogen-containing gases by using impedancemetric zirconia-based water-vapor sensor. *Solid State Ionics*, **176**, 2511–15.

381 Nakatou, M. and Miura, N. (2006) Detection of propene by using new-type impedancemetric zirconia-based sensor attached with oxide sensing-electrode. *Sens. Actuators B*, **120** (1), 57–62.

382 Docquier, N. and Candel, S. (2002) Combustion control and sensors: A review. *Prog. Energy Combust. Sci.*, **28** (2), 107–50.

383 MacLean, H.L. and Lave, L.B. (2003) Evaluating automobile fuel/propulsion system technologies. *Prog. Energy Combust. Sci.*, **29**, 1–69.

384 Szabo, N., Lee, C., Trimboli, J., Figueroa, O., Ramamoorthy, R., Midlam-Mohler, S., Soliman, A., Verweij, H., Dutta, P. and Akbar, S. (2003) Ceramic-based chemical sensors, probes and field-tests in automobile engines. *J. Mater. Sci.*, **38** (21), 4239–45.

385 Wang, D.Y. and Detwiler, E. (2006) Exhaust oxygen sensor dynamic study. *Sens. Actuators B*, **120**, 200–6.

386 Burgard, D.A., Dalton, T.R., Bishop, G.A., Starkey, J.R. and Stedman, D.H. (2006) Nitrogen dioxide, sulfur dioxide, and ammonia detector for remote sensing of vehicle emissions. *Rev. Sci. Instrum.*, **77**, 014101-1–-014101.

387 Walsh, P.M., Gallagher, R.J. and Henry, V.I. (2001) The glass furnace combustion and melting user research facility, Ceramic Engineering and Science Proceedings, 161st Conference on Glass Problems, 2000, The American Ceramic Society, Westerville, Ohio, Vol. 22, pp. 231–46.

388 Akbar, S.A. (2003) Ceramic sensors for the glass industry. *Ceramic Engineering and Science Proceedings, 163rd Conference on Glass Problems, 2002,* The American Ceramic Society, Westerville, Ohio, Vol. 24, pp. 91–100.

389 Saji, K., Kondo, H., Takahashi, H., Futata, H., Angata, K. and Suzuki, T. (1993) Development of a thin-film oxygen sensor for combustion control of gas appliances. *Sens. Actuators B*, **14** (1–3), 695–6.

390 Kubler, H. (1992) Oxygen sensor for control of wood combustion: A review. *Wood Fiber Sci.*, **24** (2), 141–6.

391 Burgermeister, A., Benisek, A. and Sitte, W. (2004) Electrochemical device for the precise adjustment of oxygen partial pressures in a gas stream. *Solid State Ionics*, **170** (1–2), 99–104.

392 Solimene, R., Marzocchella, A., Passarelli, G. and Salatino, P. (2006) Assessment of gas-fluidized beds mixing and hydrodynamics by zirconia sensors. *AIChE J.*, **52** (1), 185–98.

393 Subbarao, E.C. and Maiti, H.S. (1988) Oxygen sensors and pumps. *Adv. Ceram.*, **24** (Sci. Technol. Zirconia 3), 731–47.

394 Ramamoorthy, R., Dutta, P.K. and Akbar, S.A. (2003) Oxygen sensors: Materials, methods, designs and applications. *J. Mater. Sci.*, **38** (21), 4271–82.

395 Goto, K.S. (1988) *Solid State Electrochemistry and Its Application to Sensors and Electronic Devices*, Elsevier, Amsterdam, pp. 266–73.

396 Lee, J.-H. (2003) Review on zirconia air-fuel ratio sensors for automotive applications. *J. Mater. Sci.*, **38** (21), 4247–57.

397 Riegel, J., Neumann, H. and Wiedenmann, H.-M. (2002) Exhaust gas sensors for automotive emission control. *Solid State Ionics*, **152–153**, 783–800.

398 Zhuiykov, S. and Miura, N. (2007) Development of zirconia-based potentiometric NO_x sensors for automotive and energy industries in the early 21st century: What are the prospects for sensors? *Sens. Actuators B*, **121** (2), 639–51.

399 Zosel, J., De Blauwe, F. and Guth, U. (2001) Chemical sensors for automotive application. *Adv. Eng. Mater.*, **3** (10), 797–801.

400 Skeaff, J.M. and Dubreuil, A.A. (1993) Electrochemical measurement of SO_3–SO_2 in process gas streams. *Sens. Actuators B*, **10**, 161–8.

401 Turkdogan, E.T. (2001) Novel concept of disposable emf sensor for in situ measurements of solute contents of liquid metals in metal-refining processes. *Scand. J. Metall.*, **30**, 193–203.

402 Kale, G.M. and Kurchania, R. (1999) On-line electrochemical sensors in molten metal processing technology: A review. *Ceram. Trans.*, **92** (Electrochemistry of Glass and Ceramics), 195–220.

403 Iwase, M. (1988) Zirconia electrochemical sensors for monitoring oxygen, silicon, sulfur, aluminum, and phosphorous in iron and steel. *Adv. Ceram.*, **24** (Sci. Technol. Zirconia 3), 871–8.

404 Janke, D. (1990) Recent development of solid ionic sensors to control iron and steel bath composition. *Solid State Ionics*, **40/41**, 764–9.

405 Liu, Q. (1996) The development of high temperature electrochemical sensors for metallurgical processor. *Solid State Ionics*, **86–88**, 1037–43.

406 McLean, A. (1990) Sensor aided process control in iron and steelmaking. *Solid State Ionics*, **40/41**, 737–42.

407 Iwase, M., Tanida, M., McLean, A. and Mori, T. (1981) Electronically driven transport of oxygen from liquid iron to $CO + CO_2$ gas mixtures through stabilized zirconia. *Metall. Trans. B*, **12**, 517–24.

408 Oberg, K.E., Friedman, L.M., Boorstein, W.M. and Rapp, R.A. (1973) Electrochemical deoxidation of induction stirred copper melts. *Metall. Trans.*, **4**, 75–82.

409 Fischer, W.A. and Janke, D. (1972) Electrolytic deoxidation of liquid metals at 1600 °C. *Scr. Metall.*, **6**, 923–8.

410 Hasham, Z., Pal, U., Chou, K.C. and Worrell, W.L. (1995) Deoxidation of molten steel using a short-circuited solid electrochemical cell. *J. Electrochem. Soc.*, **142** (2), 469–75.

411 Iwase, M., Ichise, E. and Jacob, K.T. (1984) Mixed ionic and electronic conduction in zirconia and its application in metallurgy. *Adv. Ceram.*, **12** (Sci Technol Zirconia 2), 646–59.

412 Scaife, P.H., Swinkels, D.A.J. and Richards, S.R. (1976) Characterisation of zirconia electrolytes for oxygen probes used in steelmaking. *High Temp. Sci.*, **8**, 31–47.

413 Liu, Q., An, S. and Qiu, W. (1999) Study on thermal expansion and thermal shock resistance of MgO-PSZ. *Solid State Ionics*, **121**, 61–5.

414 Huang, K., Xia, Y., Wu, W., Li, Q., Liu, W. and Clauss, H., (1994) Investigations on commercially available MgO-PSZ electrolytes with Y_2O_3-FSZ coating for low oxygen determination in liquid steel. *Solid State Ionics*, **70/71**, 563–8.

415 Huang, K., Xia, Y. and Liu, Q. (1994) Rapid determination of electronic conductivity in MgO-partially stabilized ZrO_2 electrolytes using EMF method. *Solid State Ionics*, **73**, 41–8.

416 Caproni, E., Carvalho, F.M.S. and Muccillo, R. (2008) Development of zirconia–magnesia/zirconia–yttria composite solid electrolytes. *Solid State Ionics*, **179**, 1652–4.

417 Etsell, T.H. and Alcock, C.B. (1981) Non-isothermal probe for continuous measurement of oxygen in steel. *Solid State Ionics*, **3/4**, 621–6.

418 Ramasesha, S.K. and Jacob, K.T. (1989) Studies on nonisothermal solid state galvanic cells – effect of gradients on EMF. *J. Appl. Electrochem.*, **19**, 394–400.

419 Hong, Y.R., Jin, C.J., Li, L.S. and Sun, J.L. (2002) An application of the electrochemical sulfur sensor in steelmaking. *Sens. Actuators B*, **87**, 13–17.

420 Swetnam, M.A., Kumar, R.V. and Fray, D.J. (2006) Sensing of sulfur in molten metal using strontium β-alumina. *Metall. Mater. Trans. B*, **37**, 381–8.

421 Fray, D.J. (2003) The use of solid electrolytes in the determination of activities and the development of sensors. *Metall. Mater. Trans. B*, **34**, 589–94.

422 Hong, Y.R., Li, L.S., Li, F.X., Zhang, Q.X. and Mao, Y.W. (1998) An electrochemical nitrogen sensor for iron and steel melts. *Sens. Actuators B*, **53**, 54–7.

423 Zheng, M. and Zhen, X. (1993) $SrCeO_3$-based solid electrolyte probe sensing hydrogen content in molten aluminum. *Solid State Ionics*, **59**, 167–9.

424 Yajima, T., Koide, K., Takai, H., Fukatsu, N. and Iwahara, H. (1995) Application of hydrogen sensor using proton conductive ceramics as a solid electrolyte to aluminum casting industries. *Solid State Ionics*, **79**, 333–57.

425 Fukatsu, N. and Kurita, N. (2005) Hydrogen sensor based on oxide proton conductors and its application to metallurgical engineering. *Ionics*, **11**, 54–65.

426 Schwandt, C. and Fray, D.J. (2000) Hydrogen sensing in molten aluminum using a commercial electrochemical probe. *Ionics*, **6** (3–4), 222–9.

427 Gee, R. and Fray, D.J. (1978) Instantaneous determination of hydrogen content in molten aluminum and its alloys. *Metall. Trans. B*, **9**, 427–30.

428 Virkar, A.V. (2001) Transport of H_2, O_2 and H_2O through single-phase, two-phase and multi-phase mixed proton, oxygen and electron hole conductors. *Solid State Ionics*, **140**, 275–83.

429 Róg, G., Dudek, M., Kozłowska-Róg, A. and Bućko, M. (2002) Calcium zirconate, properties and application to the solid oxide galvanic cells. *Electrochim. Acta*, **47**, 4523–9.

430 Kurita, N., Fukatsu, N., Miyamoto, S., Sato, F., Nakai, H., Irie, K. and Ohashi, T. (1996) The measurement of hydrogen activities in molten copper using oxide protonic conductor. *Metall. Mater. Trans. B*, **27**, 929–35.

431 Katahira, K., Koide, K. and Iwamoto, T. (2000) Structure and performance evaluation of hydrogen sensor for molten copper under industrial conditions, in *Second International Conference on Processing Materials for Properties* (eds B. Mishra and C. Yamauchi), The Minerals, Metals & Materials Society, Warrendale, PA, pp. 347–52.

432 Tominaga, H., Yajima, K. and Nishiyama, S. (1981) Application of oxygen probes for controlling oxygen content in molten copper. *Solid State Ionics*, **3/4**, 571–4.

433 Jacob, K.T. and Ramasesha, S.K. (1989) Design of temperature-compensated reference electrodes for non-isothermal galvanic sensors. *Solid State Ionics*, **34**, 161–6.

434 Heinz, M., Koch, K. and Janke, D. (1989) Oxygen activities in Cr-containing Fe and Ni-based melts. *Steel Res.*, **60** (6), 246–54.

435 Kurchania, R. and Kale, G.M. (2001) Measurement of oxygen potentials in Ag-Pb system employing oxygen sensor. *Metall. Mater. Trans. B*, **32** (3), 417–21.

436 Konys, J., Muscher, H., Voß, Z. and Wedemeyer, O. (2001) Development of oxygen meters for the use in lead-bismuth. *J. Nucl. Mater.*, **296**, 289–94.

437 Vargas-García, R., Romero-Serrano, A., Angeles-Hernandez, M., Chavez-Alcala, F. and Bomez-Yañez, C. (2002) Pt electrode-based sensor prepared by metal organic chemical vapor deposition for oxygen activity measurements in glass melts. *Sens. Mater.*, **14** (1), 47–56.

438 Brisley, R.J. and Fray, D.J. (1983) Determination of the sodium activity in aluminum and aluminum silicon alloys using sodium beta alumina. *Metall. Trans. B*, **14**, 435–40.

439 Dekeyser, J.C., De Schutter, F., Van der Poorten, C. and Zhang, L. (1995) An electrochemical sensor for aluminum melts. *Sens. Actuators B*, **24–25**, 273–5.

440 Doughty, G., Fray, D.J., Van der Poorten, C. and Dekeyser, J. (1996) β-Alumina for controlling the rate of sodium addition to aluminum alloys. *Solid State Ionics*, **86–88**, 193–6.

441 Kirchnerova, A.J. and Pelton, A.D. (1996) Solid electrolyte probe for strontium employing a $SrCl_2$-AgCl/Ag reference and a thermodynamic evaluation of the $SrCl_2$-AgCl system. *Solid State Ionics*, **93**, 165–70.

442 Fray, D.J. (1990) Developments in on-line sensing in molten metals using solid electrolytes. *ATB Metallurgie*, **3**, 63–8.

443 Fray, D.J. (1996) The use of solid electrolytes as sensors for applications in molten metals. *Solid State Ionics*, **86–88**, 1045–54.

444 Kale, G.M., Wang, L. and Hong, Y. (2004) High-temperature sensor for in-line monitoring of Mg and Li in molten Al employing ion-conducting ceramic electrolytes. *Int. J. Appl. Ceram. Technol.*, **1** (2), 180–7.

445 Dubreuil, A.A. and Pelton, A.D. (1999) Probes for the continuous monitoring of sodium and lithium in molten aluminum, in *Light Metals 1985* (ed. H.O. Bohner), The Minerals, Metals & Materials Society, Warrendale, PA, pp. 1197–205.

446 Dekeyser, J.D. and De Schutter, F. (1992) An electrochemical lithium sensor for aluminum-lithium melts, in *Aluminum-Lithium* (eds M. Peters and P.J. Winkler), DGM Informations GmbH, pp. 831–6.

447 Yao, P.C. and Fray, D.J. (1985) Determination of the lithium content of molten aluminum using a solid electrolyte. *Metall. Trans. B*, **16**, 41–6.

448 Tiwari, B.L. (1987) Thermodynamic properties of liquid Al-Mg alloys measured by the EMF method. *Metall. Trans. A*, **18**, 1645–51.

449 Belton, G.R. and Rao, Y.K. (1969) A galvanic cell study of activity in Mg-Al liquid alloys. *Trans. Metall. Soc. AIME*, **245**, 2189–93.

450 Tsyplakova, M.M. and Strelets, Kh.L. (1969) Study of the thermodynamic properties of the magnesium-aluminum system. *J. Appl. Chem. USSR*, **42** (11), 2354–9.

Index

Solid State Electrochemistry I: Fundamentals, Materials and their Applications. Edited by Vladislav V. Kharton

ISBN: 978-3-527-32318-0

d

e

p

t